Metal Cutting Theory and Practice
Second Edition

MANUFACTURING ENGINEERING AND MATERIALS PROCESSING
A Series of Reference Books and Textbooks

SERIES EDITOR

Geoffrey Boothroyd
Boothroyd Dewhurst, Inc.
Wakefield, Rhode Island

1. Computers in Manufacturing, *U. Rembold, M. Seth, and J. S. Weinstein*
2. Cold Rolling of Steel, *William L. Roberts*
3. Strengthening of Ceramics: Treatments, Tests, and Design Applications, *Harry P. Kirchner*
4. Metal Forming: The Application of Limit Analysis, *Betzalel Avitzur*
5. Improving Productivity by Classification, Coding, and Data Base Standardization: The Key to Maximizing CAD/CAM and Group Technology, *William F. Hyde*
6. Automatic Assembly, *Geoffrey Boothroyd, Corrado Poli, and Laurence E. Murch*
7. Manufacturing Engineering Processes, *Leo Alting*
8. Modern Ceramic Engineering: Properties, Processing, and Use in Design, *David W. Richerson*
9. Interface Technology for Computer-Controlled Manufacturing Processes, *Ulrich Rembold, Karl Armbruster, and Wolfgang Ülzmann*
10. Hot Rolling of Steel, *William L. Roberts*
11. Adhesives in Manufacturing, *edited by Gerald L. Schneberger*
12. Understanding the Manufacturing Process: Key to Successful CAD/CAM Implementation, *Joseph Harrington, Jr.*
13. Industrial Materials Science and Engineering, *edited by Lawrence E. Murr*
14. Lubricants and Lubrication in Metalworking Operations, *Elliot S. Nachtman and Serope Kalpakjian*
15. Manufacturing Engineering: An Introduction to the Basic Functions, *John P. Tanner*
16. Computer-Integrated Manufacturing Technology and Systems, *Ulrich Rembold, Christian Blume, and Ruediger Dillman*
17. Connections in Electronic Assemblies, *Anthony J. Bilotta*
18. Automation for Press Feed Operations: Applications and Economics, *Edward Walker*
19. Nontraditional Manufacturing Processes, *Gary F. Benedict*
20. Programmable Controllers for Factory Automation, *David G. Johnson*
21. Printed Circuit Assembly Manufacturing, *Fred W. Kear*
22. Manufacturing High Technology Handbook, *edited by Donatas Tijunelis and Keith E. McKee*

23. Factory Information Systems: Design and Implementation for CIM Management and Control, *John Gaylord*
24. Flat Processing of Steel, *William L. Roberts*
25. Soldering for Electronic Assemblies, *Leo P. Lambert*
26. Flexible Manufacturing Systems in Practice: Applications, Design, and Simulation, *Joseph Talavage and Roger G. Hannam*
27. Flexible Manufacturing Systems: Benefits for the Low Inventory Factory, *John E. Lenz*
28. Fundamentals of Machining and Machine Tools: Second Edition, *Geoffrey Boothroyd and Winston A. Knight*
29. Computer-Automated Process Planning for World-Class Manufacturing, *James Nolen*
30. Steel-Rolling Technology: Theory and Practice, *Vladimir B. Ginzburg*
31. Computer Integrated Electronics Manufacturing and Testing, *Jack Arabian*
32. In-Process Measurement and Control, *Stephan D. Murphy*
33. Assembly Line Design: Methodology and Applications, *We-Min Chow*
34. Robot Technology and Applications, *edited by Ulrich Rembold*
35. Mechanical Deburring and Surface Finishing Technology, *Alfred F. Scheider*
36. Manufacturing Engineering: An Introduction to the Basic Functions, Second Edition, Revised and Expanded, *John P. Tanner*
37. Assembly Automation and Product Design, *Geoffrey Boothroyd*
38. Hybrid Assemblies and Multichip Modules, *Fred W. Kear*
39. High-Quality Steel Rolling: Theory and Practice, *Vladimir B. Ginzburg*
40. Manufacturing Engineering Processes: Second Edition, Revised and Expanded, *Leo Alting*
41. Metalworking Fluids, *edited by Jerry P. Byers*
42. Coordinate Measuring Machines and Systems, *edited by John A. Bosch*
43. Arc Welding Automation, *Howard B. Cary*
44. Facilities Planning and Materials Handling: Methods and Requirements, *Vijay S. Sheth*
45. Continuous Flow Manufacturing: Quality in Design and Processes, *Pierre C. Guerindon*
46. Laser Materials Processing, *edited by Leonard Migliore*
47. Re-Engineering the Manufacturing System: Applying the Theory of Constraints, *Robert E. Stein*
48. Handbook of Manufacturing Engineering, *edited by Jack M. Walker*
49. Metal Cutting Theory and Practice, *David A. Stephenson and John S. Agapiou*
50. Manufacturing Process Design and Optimization, *Robert F. Rhyder*
51. Statistical Process Control in Manufacturing Practice, *Fred W. Kear*
52. Measurement of Geometric Tolerances in Manufacturing, *James D. Meadows*
53. Machining of Ceramics and Composites, *edited by Said Jahanmir, M. Ramulu, and Philip Koshy*
54. Introduction to Manufacturing Processes and Materials, *Robert C. Creese*

55. Computer-Aided Fixture Design, *Yiming (Kevin) Rong and Yaoxiang (Stephens) Zhu*
56. Understanding and Applying Machine Vision: Second Edition, Revised and Expanded, *Nello Zuech*
57. Flat Rolling Fundamentals, *Vladimir B. Ginzburg and Robert Ballas*
58. Product Design for Manufacture and Assembly: Second Edition, Revised and Expanded, *Geoffrey Boothroyd, Peter Dewhurst, and Winston A. Knight*
59. Process Modeling in Composites Manufacturing, *edited by Suresh G. Advani and E. Murat Sozer*
60. Integrated Product Design and Manufacturing Using Geometric Dimensioning and Tolerancing, *Robert Campbell*
61. Handbook of Induction Heating, *edited by Valery I. Rudnev, Don Loveless, Raymond Cook and Micah Black*
62. Re-Engineering the Manufacturing System: Applying the Theory of Constraints, Second Edition, *Robert Stein*
63. Manufacturing: Design, Production, Automation, and Integration, *Beno Benhabib*
64. Rod and Bar Rolling: Theory and Applications, *Youngseog Lee*
65. Metallurgical Design of Flat Rolled Steels, *Vladimir B. Ginzburg*
66. Assembly Automation and Product Design: Second Edition, *Geoffrey Boothroyd*
67. Roll Forming Handbook, *edited by George T. Halmos*
68. Metal Cutting Theory and Practice, Second Edition, *David A. Stephenson and John S. Agapiou*
69. Fundamentals of Machining and Machine Tools, Third Edition, *Geoffrey Boothroyd and Winston A. Knight*

Metal Cutting Theory and Practice

Second Edition

David A. Stephenson

John S. Agapiou

Taylor & Francis
Taylor & Francis Group
Boca Raton   London   New York

A CRC title, part of the Taylor & Francis imprint, a member of the
Taylor & Francis Group, the academic division of T&F Informa plc.

Published in 2006 by
CRC Press
Taylor & Francis Group
6000 Broken Sound Parkway NW, Suite 300
Boca Raton, FL 33487-2742

© 2006 by Taylor & Francis Group, LLC
CRC Press is an imprint of Taylor & Francis Group

No claim to original U.S. Government works
Printed in the United States of America on acid-free paper
10 9 8 7 6 5 4 3 2 1

International Standard Book Number-10: 0-8247-5888-9 (Hardcover)
International Standard Book Number-13: 978-0-8247-5888-2 (Hardcover)
Library of Congress Card Number 2005053103

This book contains information obtained from authentic and highly regarded sources. Reprinted material is quoted with permission, and sources are indicated. A wide variety of references are listed. Reasonable efforts have been made to publish reliable data and information, but the author and the publisher cannot assume responsibility for the validity of all materials or for the consequences of their use.

No part of this book may be reprinted, reproduced, transmitted, or utilized in any form by any electronic, mechanical, or other means, now known or hereafter invented, including photocopying, microfilming, and recording, or in any information storage or retrieval system, without written permission from the publishers.

For permission to photocopy or use material electronically from this work, please access www.copyright.com (http://www.copyright.com/) or contact the Copyright Clearance Center, Inc. (CCC) 222 Rosewood Drive, Danvers, MA 01923, 978-750-8400. CCC is a not-for-profit organization that provides licenses and registration for a variety of users. For organizations that have been granted a photocopy license by the CCC, a separate system of payment has been arranged.

Trademark Notice: Product or corporate names may be trademarks or registered trademarks, and are used only for identification and explanation without intent to infringe.

Library of Congress Cataloging-in-Publication Data

Stephenson, David A., 1959-
  Metal cutting theory and practice / David A. Stephenson, John S. Agapiou. -- 2nd ed.
    p. cm. -- (Manufacturing engineering and materials processing ; 68)
  Includes bibliographical references and index.
  ISBN 0-8247-5888-9 (alk. paper)
  1. Metal-cutting. I. Agapiou, John S. II. Title. III. Series.

TJ1185.S815 2005
671.5--dc22   2005053103

Visit the Taylor & Francis Web site at
http://www.taylorandfrancis.com

and the CRC Press Web site at
http://www.crcpress.com

Preface to the First Edition

Metal cutting is a subject as old as the Industrial Revolution, but one that has evolved continuously as technology has advanced. The first metal-cutting machine tools, built some 450 years ago, were powered by water and employed iron and carbon steel tools. Over time, these gave way to machines powered by steam and leather belts employing high-speed steel tools, to electrically powered machines using sintered carbide tools, and most recently to computer-controlled machines using ceramic and diamond tools. The pace of change seems to have increased over the last 20 years, with progressively rapid advances in materials science and computer technology. For example, since the beginning of our own careers, typical production rates have doubled in many operations, and numerous new tool materials, work materials, and machine architectures have been introduced — and we are still many years from retirement.

Our purpose in writing this book is twofold. First, many of the books from which we learned much of our trade were written in the 1970s or earlier, and despite recent updated editions they are showing inevitable signs of age. We hoped to write a reference book that would provide a fuller treatment of recent developments than is currently available. Second, the literature in this field is somewhat dichotomous, consisting on the one hand of scientific books and articles read largely by academics and researchers, and on the other hand of trade journals, handbooks, and sales brochures read by practicing engineers. We also hoped to write a book that would appeal to both audiences by covering both research results and the current industrial practice. To make the project manageable we had to limit the technical topics to be covered. We have chosen to consider only metallic work materials, to concentrate on the traditional chip-forming cutting processes with limited material on abrasive processes, and to largely ignore subjects such as machine tool control, which could fill entire books in their own right. Even with these limitations, we recognize that we have taken on an ambitious subject. Readers, of course, will have to decide how well we have covered it.

We have been fortunate over the years to have worked with, and learned from, many fine engineers in academia and in the automotive, machine tool, and cutting tool industries. We received valuable feedback on drafts of portions of this book from Robin Stevenson, Pulak Bandyopadhyay, Yhu-Tin Lin, David W. Yen, I. S. Jawahir, Jochen S. Zenker, and Ellen D. Kock. We are grateful to these colleagues for correcting many errors and inconsistencies in the manuscript, and are responsible for all those that still remain. We also thank Simon Yates, Dawn Wechsler, Walter Brownfield, and Vivian Jao of Marcel Dekker, Inc., for their courteous and helpful editing.

Finally and most importantly, we thank our wives, Maria Clelia Milletti and Christina Agapiou, and our children, Stylianos Ioannis Agapiou, Alexandra Maria Agapiou, Adonis Ioannis Agapiou, Daphne Elizabeth Agapiou, Francesca Laura Stephenson, and Luke Andrew Stephenson, for their patience during the many times when our preoccupation with this project inconvenienced them.

David A. Stephenson
John S. Agapiou

Preface to the Second Edition

We were pleased with the reception of the first edition of this book, and are very pleased to have an opportunity to correct some of its deficiencies in a new version.

In addition to updating material throughout, we made several structural changes. Chapter 2 in the first edition, covering both machining operations and machine tools, has been split into two separate chapters in the current edition. On the other hand, Chapter 5, on chip formation, was eliminated, with the material being distributed between the current Chapters 6 and 11. We added three chapters: Cutting Fluids, High Throughput and Agile Machining, and Design for Machining. These are areas of significant recent development, and also reflect areas of emphasis in our recent professional practice. Finally, since the first edition was unexpectedly used as a university textbook, we have added examples and problems at the end of chapters to make the new edition more suitable for this purpose.

Due to lack of space, we reluctantly decided not to include a planned chapter on gear machining. We apologize to any readers disappointed by this omission.

We are grateful to John Rutz, David Yen, and Albert Shih for useful feedback on the first edition, and to David N. Dilley and Mikhail Lundblad for valuable technical input for the new version. We are also grateful to Rita Lazazzaro and Barbara Mathieu of Marcel Dekker and Cindy Carelli, Preethi Cholmondeley, and Naomi Lynch of CRC Press for their patient editing; and especially to Kavitha Kuttikan and the staff at SPI for their outstanding production work. Finally and most importantly, we again thank our families for their forbearance during our second attempt at this project.

David A. Stephenson
John S. Agapiou

The Authors

David A. Stephenson is technical fellow and manager of Manufacturing Process Analysis at General Motors Powertrain, Pontiac, Michigan. He is a member of the American Society of Mechanical Engineers (ASME) 5 Manufacturing Engineering Division Executive Committee, and was an associate technical editor for the ASME *Journal of Engineering for Industry* from 1994 to 2000. He is also a member of the Society of Manufacturing Engineers (SME) Journals Committee, and has been an associate technical editor for the SME *Journal of Manufacturing Processes*. He received the ASME Blackall Machine Tool and Gage Award in 1994, the SME Outstanding Young Manufacturing Engineer Award in 1994, and the M. Eugene Merchant Manufacturing Medal of SME/ASME in 2004. He received his bachelor's and master's degrees in mechanical engineering at the Massachusetts Institute of Technology in 1981 and 1983, respectively, and his Ph.D. from the University of Wisconsin in 1985.

John S. Agapiou is technical fellow at the Manufacturing Systems Research Lab, at General Motors R&D Center, Warren, Michigan. He is also part-time professor in the Department of Mechanical Engineering at Wayne State University. He received the SME Outstanding Young Manufacturing Engineer Award in 1992. His research focus involves developing and implementing world-class manufacturing, quality, and process validation strategies in the production and development of the automotive powertrain. His research and teaching interests include modeling and optimization of metal cutting operations — including the cutting tool and machining system, and modeling manufacturing part quality for machining lines to improve part quality, process, and productivity. He received his bachelor's and master's degrees in mechanical engineering at the University of Louisville in 1980 and 1981, respectively, and his Ph.D. from the University of Wisconsin in 1985.

Contents

1. Introduction 1
   1.1 Scope of the Subject 1
   1.2 Historical Development 1
   1.3 Types of Production 14
   References 14

2. Metal Cutting Operations 17
   2.1 Introduction 17
   2.2 Turning 17
   2.3 Boring 20
   2.4 Drilling 21
   2.5 Reaming 27
   2.6 Milling 28
   2.7 Planing and Shaping 33
   2.8 Broaching 33
   2.9 Tapping and Threading 35
   2.10 Grinding and Related Abrasive Processes 44
   2.11 Roller Burnishing 52
   2.12 Deburring 54
   2.13 Examples 54
   2.14 Problems 66
   References 68

3. Machine Tools 71
   3.1 Introduction 71
   3.2 Production Machine Tools 72
   3.3 CNC Machine Tools and Cellular Manufacturing Systems 77
   3.4 Machine Tool Structures 91
   3.5 Slides and Guideways 102
   3.6 Axis Drives 106
   3.7 Spindles 111
   3.8 Coolant Systems 126
   3.9 Tool Changing Systems 126
   3.10 Examples 129
   References 133

4. Cutting Tools 141
   4.1 Introduction 141
   4.2 Cutting Tool Materials 141

	4.3	Tool Coatings	155
	4.4	Basic Types of Cutting Tools	162
	4.5	Turning Tools	163
	4.6	Boring Tools	173
	4.7	Milling Tools	179
	4.8	Drilling Tools	192
	4.9	Reamers	223
	4.10	Threading Tools	228
	4.11	Grinding Wheels	236
	4.12	Microsizing and Honing Tools	243
	4.13	Burnishing Tools	245
	4.14	Examples	245
	4.15	Problems	258
		References	258
5.	Toolholders and Workholders	265
	5.1	Introduction	265
	5.2	Toolholding Systems	265
	5.3	Toolholder–Spindle Connection	270
	5.4	Cutting Tool Clamping Systems	313
	5.5	Balancing Requirements for Toolholders	341
	5.6	Fixtures	347
	5.7	Examples	353
	5.8	Problems	366
		References	366
6.	Mechanics of Cutting	371
	6.1	Introduction	371
	6.2	Measurement of Cutting Forces and Chip Thickness	371
	6.3	Force Components	374
	6.4	Empirical Force Models	378
	6.5	Specific Cutting Power	380
	6.6	Chip Formation and Primary Plastic Deformation	382
	6.7	Tool–Chip Friction and Secondary Deformation	389
	6.8	Shear Plane and Slip-Line Theories for Continuous Chip Formation	394
	6.9	Shear Plane Models for Oblique Cutting	398
	6.10	Shear Zone Models	399
	6.11	Minimum Work and Uniqueness Assumptions	403
	6.12	Finite Element Models	404
	6.13	Discontinuous Chip Formation	408
	6.14	Built-up Edge Formation	411
	6.15	Examples	413
	6.16	Problems	415
		References	416
7.	Cutting Temperatures	425
	7.1	Introduction	425
	7.2	Measurement of Cutting Temperatures	425
	7.3	Factors Affecting Cutting Temperatures	432

	7.4	Analytical Models for Steady-State Temperatures 434
	7.5	Finite Element and Other Numerical Models 437
	7.6	Temperatures in Interrupted Cutting 441
	7.7	Temperatures in Drilling 444
	7.8	Thermal Expansion 446
	7.9	Examples 448
	7.10	Problems 451
	References 451

8. Machining Process Analysis 459
    8.1 Introduction 459
    8.2 Turning 460
    8.3 Boring 462
    8.4 Milling 465
    8.5 Drilling 470
    8.6 Force Equations and Baseline Data 478
    8.7 Process Simulation Application Examples 482
    8.8 Finite Element Analysis for Clamping, Fixturing, and Workpiece
        Distortion Applications 484
    8.9 Finite Element Application Examples 489
    8.10 Examples 493
    8.11 Problems 499
    References 499

9. Tool Wear and Tool Life 503
    9.1 Introduction 503
    9.2 Types of Tool Wear 504
    9.3 Measurement of Tool Wear 508
    9.4 Tool Wear Mechanisms 512
    9.5 Tool Wear — Material Considerations 514
    9.6 Tool Life Testing 521
    9.7 Tool Life Equations 522
    9.8 Prediction of Tool Wear Rates 525
    9.9 Tool Fracture and Edge Chipping 528
    9.10 Drill Wear and Breakage 530
    9.11 Thermal Cracking and Tool Fracture in Milling 534
    9.12 Tool Wear Monitoring 537
    9.13 Examples 537
    9.14 Problems 543
    References 545

10. Surface Finish and Integrity 551
    10.1 Introduction 551
    10.2 Measurement of Surface Finish 552
    10.3 Surface Finish in Turning and Boring 558
    10.4 Surface Finish in Milling 562
    10.5 Surface Finish in Drilling and Reaming 565
    10.6 Surface Finish in Grinding 566
    10.7 Residual Stresses in Machined Surfaces 568

	10.8	White Layer Formation	569
	10.9	Surface Burn in Grinding	570
	10.10	Examples	571
	10.11	Problems	573
		References	574

11. Machinability of Materials 577
    - 11.1 Introduction 577
    - 11.2 Machinability Criteria, Tests, and Indices 578
    - 11.3 Chip Control 582
    - 11.4 Burr Formation and Control 588
    - 11.5 Machinability of Engineering Materials 591
    - References 611

12. Machining Dynamics 617
    - 12.1 Introduction 617
    - 12.2 Vibration Analysis Methods 617
    - 12.3 Vibration of Discrete (Lumped Mass) Systems 618
    - 12.4 Types of Machine Tool Vibration 630
    - 12.5 Forced Vibration 632
    - 12.6 Self-Excited Vibrations (Chatter) 636
    - 12.7 Chatter Prediction 653
    - 12.8 Vibration Control 658
    - 12.9 Active Vibration Control 662
    - 12.10 Examples 669
    - 12.11 Problems 691
    - References 697

13. Machining Economics and Optimization 705
    - 13.1 Introduction 705
    - 13.2 Role of a Computerized Optimization System 707
    - 13.3 Economic Considerations 708
    - 13.4 Optimization of Machining Systems — Basic Factors 710
    - 13.5 Optimization of Machining Conditions 711
    - 13.6 Formulation of the Optimization Problem 712
    - 13.7 Optimization Techniques 719
    - 13.8 Examples 741
    - 13.9 Problems 758
    - References 760

14. Cutting Fluids 767
    - 14.1 Introduction 767
    - 14.2 Types of Cutting Fluids 768
    - 14.3 Coolant Application 771
    - 14.4 Filtering 773
    - 14.5 Condition Monitoring and Waste Treatment 778
    - 14.6 Health and Safety Concerns 779
    - 14.7 Dry and Near-Dry Machining Methods 781

	14.8	Test Procedure for Cutting Fluid Evaluation	783
		References	783
15.		High Throughput and Agile Machining	787
	15.1	Introduction	787
	15.2	High Throughput Machining	787
	15.3	Agile Machining Systems	789
	15.4	Tooling and Fixturing	791
	15.5	Materials Handling Systems	800
		References	802
16.		Design for Machining	805
	16.1	Introduction	805
	16.2	Machining Costs	805
	16.3	General Design for Machining Rules	806
	16.4	Special Considerations for Specific Types of Equipment and Operations	813
	16.5	CAPP and DFM Programs	817
	16.6	Part Quality Modeling	819
	16.7	Examples	822
		References	831
Index			833

1 Introduction

1.1 SCOPE OF THE SUBJECT

Metal cutting processes are industrial processes in which metal parts are shaped by removal of unwanted material. In this book, we will primarily consider traditional chip-forming processes such as turning, boring, drilling, and milling. In these operations, metal is removed as a plastically deformed chip of appreciable dimensions, and a fairly unified physical analysis can be carried out using basic orthogonal and oblique cutting models (Figure 1.1).

Related metal removal processes include abrasive processes, such as grinding and honing, and nontraditional machining processes, such as electrodischarge, ultrasonic, electrochemical, and laser machining. In the abrasive processes, metal is removed in the form of small chips produced by a combination of cutting, plowing, and friction mechanisms; in the nontraditional processes, metal is removed on a much smaller scale by mechanical, thermal, electrical, or chemical means. In all cases, the physical mechanisms of removal differ considerably from those of chip formation, so that different physical analyses are required. Basic information on abrasive processes, tools, and surface finish capabilities is included in Chapters 2, 4, and 10. Physical analyses of abrasive and nontraditional machining processes are not considered in this book but are available in the literature [1–6].

Metal cutting processes can also be applied to nonmetallic work materials such as polymers, wood, and ceramics. When these applications are considered, the subject is more commonly called machining. Because of differences in thermomechanical properties, the analyses of metal cutting discussed in this book provide only limited insight into the machining of nonmetals. More relevant information can be found in the literature on the machining of specific classes of materials [7, 8].

The objective in this book is to provide a physical understanding of conventional and high-speed cutting processes applied to metallic workpieces. The mechanics of chip formation, temperature generation, tribology, dynamics, and material interactions are emphasized. We also include significant descriptive information on modern machinery, tooling, and fixtures. Hopefully this information, summarized with reference to the large research and trade literature on this subject, will provide the reader with sufficient physical insight and understanding to design, operate, troubleshoot, and improve high-quality, cost-effective metal cutting operations.

1.2 HISTORICAL DEVELOPMENT

Metal cutting is a subject in which the industrial practice has always led the theory. The study of metal cutting processes necessarily postdates the development of modern machine tools. Moreover, advances in the field have generally resulted from changes in practice, particularly the introduction of new tool materials.

The early development of machine tools has been described by Smiles [9], Roe [10], Burlingame [11], Rolt [12], Woodbury [13–17], and Steeds [18]. Woodworking tools with

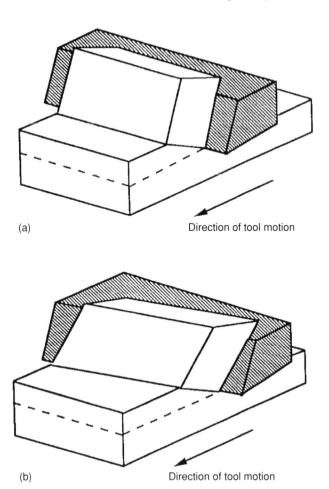

FIGURE 1.1 (a) Orthogonal cutting of a flat workpiece by a wedge-shaped tool. (b) Oblique cutting of a flat workpiece by a wedge-shaped tool.

the same kinematic motions as several modern machine tools, such as the cord lathe (Figure 1.2) and the bow drill, were developed in ancient times [16, 19, 20]. The first machines for shaping metal accurately were gear and spindle cutting machines used by European clockmakers in the early middle ages. The design of these machines, and the economic constraints on their operation, differ greatly from those of modern machine tools. Leonardo da Vinci drew designs for a number of machine tools [12, 14, 16, 18, 21], including lathes, grinders, screw-cutting machines, and boring mills (Figure 1.3). Although these machines are closer in design to modern machine tools, they seem to have served no contemporary industrial purpose and were apparently never built.

Medieval craftsmen had the basic knowledge needed to build crude machine tools, but had no economic incentive to do so. Large machine tools require substantial capital investments and concentrated power sources, and can only be operated profitably if raw materials and finished parts can be reliably transported to and from the work site in quantity. Water wheels and horse gins could provide serviceable power sources, but the lack of private capital and reliable transportation generally presented insurmountable obstacles to machine tool development.

None of these considerations, however, applied to weapons production. The first large metal cutting machine tools were water-powered cannon boring mills built in Italy in the early

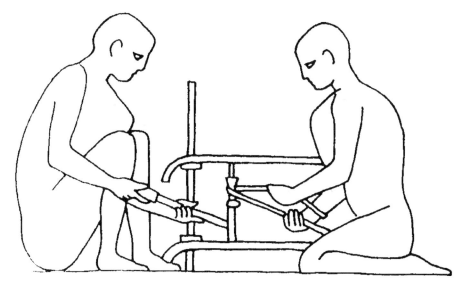

FIGURE 1.2 Low relief from the tomb of Petosiris, Egyptian Ptolemaic period (ca. 300 BCE), depicting a cord-powered lathe. The engineer on the left holds the cutting tool, while his assistant turns the work with a cord. The workpiece is mounted between centers, and the cutting tool is held against the lathe frame for support. The spindle is vertical in this stylized layout, but in practice horizontal setups were probably used, especially for larger jobs. (After H. Hodges [20].)

1500s. Over time, the early Italian designs were widely copied and improved, especially by Swiss and Dutch craftsmen. By 1750 they could be found in armories throughout Europe (Figure 1.4). As these tools were developed it was discovered that a more accurate cannon could be produced from a solid casting than from a cored one, and machines which could accommodate either type of workpiece were ultimately built. These machines are therefore the ancestors of modern drilling as well as boring machines.

The first scientific study involving metal cutting also resulted from cannon boring. Count Rumford reported his experiences in boring cannon in Bavaria in a paper presented in 1798 to the Royal Society in London [22]. He observed that boring produced a great deal of heat, and that water poured on boring tools as a coolant frequently boiled away. At the time heat was widely thought to be carried by a fluid called "caloric," and it was assumed that the heat in boring resulted from caloric being liberated from the material being cut. Rumford noted, however, that the coolant boiled away even when the tools were so dull that they burnished rather than cut the work material, and moreover that the supply of heat was inexhaustible, since the experiment could be continued indefinitely and still produce boiling. This contradicted the caloric theory of heat and led him to theorize that heat, like mechanical work, was a form of energy. Rumford modified a cannon boring mill and conducted experiments in which dulled boring bars were forced against short cylinders immersed in water (Figure 1.5) He measured both the force on the bar and the temperature rise of the water bath; based on these measurements, he estimated the conversion factor between units of heat and mechanical work, which we now call the mechanical equivalent of heat.

Beginning about 1700, machine tools began to be used in a few special commercial applications. The bores of atmospheric steam engines of the type invented by Thomas Newcomen in 1712 were machined using cannon boring mills in England. Heavy lathes formachining rolling mill rolls were used in Swedish iron mills in the early 1700s. A large

FIGURE 1.3 Leonardo Da Vinci's lathe for boring holes in logs, with complex gearing and four-jaw self-centering chuck. (a) Drawings from the *Codex Altanticus*. (After W.B. Parsons [21].) (b) Model in the Science Museum, London. (Science Museum/Science & Society Picture Library.)

lathe was also built by Jacques de Vaucanson in France in about 1765; it was apparently used to machine the cylinders of an automatic loom. In these applications, the machine tools were constructed at an existing source of water power and the machined components were used at the site, so that the additional costs associated with machining were small. Wood screws were also produced using lathes of advanced design in both England and Sweden by 1776; in this case, the machined product was of high value and easily transported.

Widespread commercial development of machine tools was made possible by James Watt's invention of the pressurized steam engine. This engine operated at higher pressures than the Newcomen engine and consequently required more accurately machined bores. In fact, Watt, an instrument maker by trade, made small model engines, but could not produce a full-sized engine for 10 years because existing boring mills could not machine large cylinders to the required accuracy [10]. This problem was solved when John Wilkinson invented a more accurate horizontal boring mill in 1776 (Figure 1.6). Wilkinson used a much heavier boring

Introduction 5

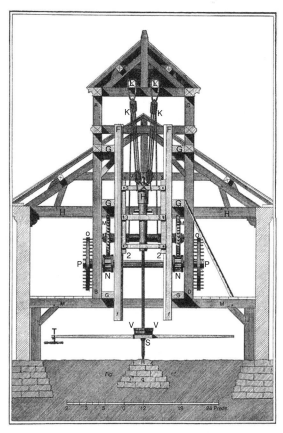

FIGURE 1.4 Horse-powered vertical cannon boring mill, ca. 1713, from Diderot's *Encylopedia* (1751–1771). Note details of foundation. (After W. Steeds [18].)

bar than was used in cannon boring mills, and supported the bar at both ends, greatly increasing rigidity and accuracy. This machine was the first recognizably modern machine tool, since it could perform heavy cuts with reasonable accuracy and employed a basic design that was replicated into the twentieth century [23]. The availability of increasingly powerful steam engines freed machine tools from dependence on water power, so that machine shops could be built at any convenient location. A number of heavy machine shops were soon built in England; two of the earliest were the Soho Foundry in Birmingham, opened in 1795, and the Round Foundry in Leeds, opened in 1797. The commercial development of machine tools was also aided by the inventions of the crucible process for producing steel, which provided more durable tools, and the steam locomotive, which increased demand for steam engines.

With these inventions in place, the English machine tool industry developed rapidly. The most influential early English tool builder was Henry Maudslay, who developed screw-cutting lathes with slide rests (Figure 1.7), the ancestors of later engine lathes, and set new standards of accuracy for the profession by inventing the screw micrometer and pioneering the use of surface plates. He also trained many of the leading machine builders of the day, most notably Joseph Whitworth, who standardized screw threads, developed many feed and drive mechanisms, and was among the first to appreciate the primary importance of structural rigidity in

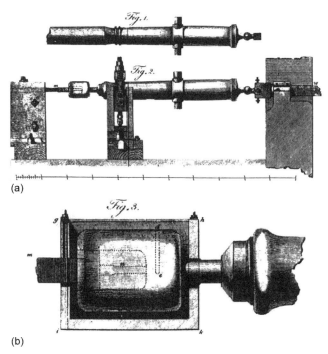

FIGURE 1.5 Count Rumford's cannon boring experiment for measuring the mechanical equivalent of heat [23]. (a) Cannon casting mounted in a boring mill. The bar marked "W" on the right is the drive shaft from a horse treadmill; the bar marked "m" on the left is a blunted boring bar going into the test chamber. (b) Detail of the test chamber. The blunted boring bar contacts a short hollow cylinder encased in a wooden box filled with water; a thermometer was inserted through hole "dc" to monitor the water temperature.

FIGURE 1.6 Model of John Wilkinson's horizontal mill for boring steam engine cylinders. This mill, built in 1775, finally permitted James Watt to build a working full-size engine after 10 years of failure resulting from inaccurately machined bores. (a) Side view. (Science Museum/Science & Society Picture Library.) (b) Detail of boring bar support and cutter. (Science Museum/Science & Society Picture Library.)

machine tool design. By 1850, the early English tool builders had perfected designs of lathes, planers, and shapers, as well as drilling, slotting, and gear cutting machines. With these tools, they manufactured steam engines for ships and steam locomotives, textile machinery, railway tires, and machine tools for sale throughout Europe.

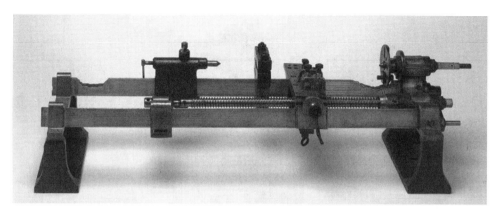

FIGURE 1.7 Henry Maudslay's original screw-cutting lathe with slide rest. The basic model with the slide rest was developed in 1794; the change gears and lead screw for cutting threads were added about 1800. (Science Museum/Science & Society Picture Library.)

Because of a government ban, English machine tools were not exported to America. In spite of this ban, or perhaps partly because of it, the machine tool industry came to be dominated by American companies after 1850. The earliest of these companies were founded to produce textile machinery to process American cotton. But the most important reason for the rise of the American tool industry was the development of manufacture using interchangeable parts. This system was initially used by French gunmakers in the 1780s [10, 24], but was perfected by American small arms manufacturers beginning about 1800, and became known as the "American System" following the British government's decision to equip the Enfield rifle factory with American machine tools in 1855. Interchangeable manufacture required accurate gages, dimensioning methods, and machine tools with automatic feeds and built-in jigs or tool guides. Several small companies were founded in New England to design and build increasingly refined tools by Amos Whitney, Francis Pratt, Elisha Root, Joseph Brown, Lucian Sharpe, Frederick Howe, and others. Engineers at these companies made important contributions to metrology and perfected advanced types of lathes, notably turret lathes and screw machines, as well as a variety of versatile milling (Figure 1.8) and grinding machines. The availability of grinding machines was important to the development of milling machines, since the ability to sharpen hardened steel cutters reduced milling costs. In turn, milling machines made a number of significant advances in tooling possible. The best example is the twist drill with milled flutes, which the universal mill was developed to produce [10]. The small arms industry remained a leading market for American tools, especially during the American civil war, but they were also used to produce textile machinery, railway locomotives, clocks and watches, sewing machines, typewriters, bicycles, and agricultural machinery.

The first scientific studies of metal cutting were conducted beginning about 1850 and have been reviewed in detail by Finnie [25] and Zorev [26]. The principal goal of the early work was to establish power requirements for various operations so that steam engines of appropriate size could be selected for tools. A number of researchers constructed crude dynamometers and conducted systematic experiments to measure cutting forces. The best known was E. Hartig, whose 1873 book [27] was a standard reference on the subject for many years. Development of more advanced dynamometers occupied researchers after Hartig's book was published. In addition, several studies of the mechanism of chip formation were carried out, most notably by Time [28], Tresca [29], and Mallock [30] (Figure 1.9). By carefully examining chips, these researchers recognized that chip formation was a shearing process, although the

FIGURE 1.8 The original Browne and Sharp Universal Milling Machine, developed in 1862. (Courtesy of Browne and Sharp.)

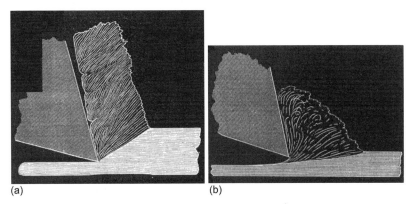

FIGURE 1.9 Drawings from Mallock's early study of chip formation [30]. Mallock used a microscope to observe chips being form in a variety of metals, as well as paraffin, soap, and clay. (a) Chip formation when cutting wrought iron (armor plate). (b) Incipient chip formation when cutting copper.

primary importance of plastic deformation in the process was not generally appreciated until much later.

Steelmaking was also advancing in Europe, and in 1868 Robert Mushet, an English steel maker, developed an improved tool steel. Mushet experimented with a number of alloying elements and eventually found a tungsten alloy which proved to be self-hardening. Mushet never patented his discovery but was secretive about it and took extraordinary measures to prevent the theft of his recipe; he worked with a few picked associates in a small workshop in the Forest of Dean (which was imposingly named the Titanic Iron and Steel Company), had materials shipped through multiple intermediaries, and habitually referred to the main ingredients by codenames [12]. He was successful in guarding his secret, for the details of his processing methods are unknown to this day. His material, however, proved to be superior to carbon steel for cutting tools and was widely used in both Europe and America.

The great historical figure in the field of metal cutting, Frederick W. Taylor, was active at the end of the nineteenth century. Taylor is best known as the founder of scientific management [31], and many popular accounts of his life do not mention his work in metal cutting [32]. The metal cutting work, however, was crucial to the implementation of his management theories. Taylor became foreman of the machine shop at the Midvale Iron Works near Philadelphia in 1880. At that time the tools and cutting speeds used to perform machining operations were selected by individual machine operators, so that practices and results varied considerably. He felt that shop productivity could be greatly increased if standard best practices were dictated by a central planning department. Putting this idea into practice, however, required the development of a quantitative understanding of the relation between speeds, feeds, tool geometries, and machining performance. Taylor embarked on a series of methodical experiments to gather the data (mainly tool life values) necessary to develop this understanding. The experiments continued over a number of years at Midvale and the nearby Bethlehem Steel Works, where he worked jointly with metallurgist Maunsel White. As a result of these experiments, Taylor was able to increase machine shop productivity at Midvale by as much as a factor of 5.

Taylor's most important practical contribution was his invention, with White, of high speed steel cutting tools. Starting with Mushet's steel, which they had chemically analyzed, Taylor and White began varying the alloy composition and heat treatment. Eventually, they developed alloys containing tungsten, chromium, and silicon, which were self-hardening and stable at much higher temperatures than Mushet's steel. Tools made of these high speed steels were vastly superior to anything available at the time and permitted a great increase in cutting speeds.

Taylor also established that the power required to feed the tool could equal the power required to drive the spindle, especially when worn tools were used. Machine tools of the day were underpowered in the feed direction (Figure 1.10), and he had to modify all the machines at the Midvale plant to eliminate this flaw. He also demonstrated the value of coolants in metal cutting and fitted his machines with recirculating fluid systems fed from a central sump (or "suds tank"). Finally, he developed a special slide rule for determining feeds and speeds for various materials.

Taylor summarized his research results in the landmark paper *On the Art of Cutting Metals* [33, 34], which was delivered to a convention of the American Society of Mechanical Engineers in 1906 and published in the Society Transactions in 1907. The results were based on an estimated 30,000 to 50,000 cutting tests conducted over a period of 26 years, consuming 800,000 pounds of metal, and costing ten industrial sponsors an estimated $200,000. In conducting his tests, Taylor was fortunate to have at his disposal an essentially unlimited supply of consistent work material. Taylor's most important research contribution was his recognition of the importance of tool temperatures in tool life, which led to the development of his famous tool life equation (Section 9.7).

Taylor was careful to vary only one factor at a time in his experiments. This is sometimes cited as the reason for the utility and impact of his results [12, 35]. In fact, Taylor's work is a striking example of the inefficiency of this experimental approach; more accurate results could have been obtained with a fraction of the effort using factorial experimental designs.

Machine tools built after 1900 differed greatly from their predecessors. They were designed to run at much higher speeds to take advantage of high speed steel tools; this required the use hardened steel (rather than cast) gears and bearings and improved bearing lubrication systems. They were also fitted with more powerful motors and feed drives, and in

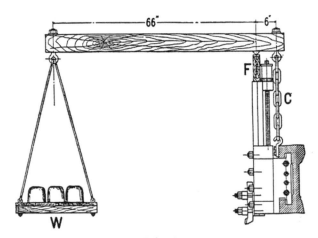

FIGURE 1.10 Balance beam dynamometer for measuring the feed force on a boring mill. (From F.W. Taylor [33].)

many cases, with recirculating coolant systems. Also, electric motors gradually began to replace steam engines as the common power source for machine tools after 1900.

The substantial capital invested in older tools prevented the immediate introduction of new tools in established industries. The new tools found a large market, however, in the emerging automotive industry (Figure 1.11). Automotive engine and transmission manufacture required reliably interchangeable parts in unprecedented volumes; because of this, the automotive industry had become the largest market for machine tools by World War I, a distinction which it still holds. It has consequently had a great influence on machine tool design. Due to accuracy requirements, grinding machines were particularly critical, and a number of specialized machines were developed for specific operations. Engine manufacture also required rapid production of flat surfaces, leading to the development of flat milling and broaching machines to replace shapers and planers. The development of the automobile also greatly improved gear design and manufacture, and machine tools were soon fitted with quick-change gearing systems similar in concept to automotive transmissions.

The automotive industry had its greatest impact on machine tool design in encouraging the development of dedicated or single purpose tools. Early examples included crankshaft grinding machines and large gear cutting machines, but the most striking examples were dedicated production systems. Linear systems evolved to produce engine blocks and heads prior to World War I (Figure 1.12). Part transfer in the early systems was manual. By the 1930s, however, linear transfer machines, first pioneered in the watchmaking industry in the 1880s [36], came into widespread use in the automotive industry. Transfer machines consist of multiple special purpose tools (or stations) connected by an automated materials handling system (Figure 1.13). A large system may be hundreds of meters long and employ dozens of stations. Transfer machines require very large capital investments but can produce parts at high rates, so that the cost per piece is usually lower than for general purpose

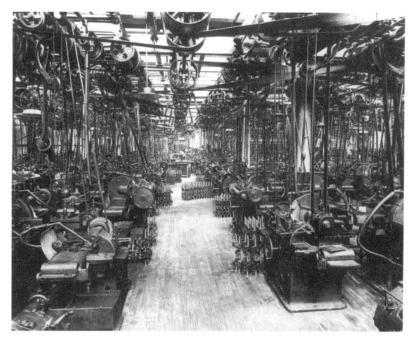

FIGURE 1.11 The crankshaft machine shop at the Ford Model T Plant in Highland Park, MI, photographed on January 2, 1915. Note the line shafting and manual part transfer. (Henry Ford Museum Collection.)

FIGURE 1.12 Multispindle drilling station on an engine block system at the Chalmers car plant, ca. 1920. Parts were transferred manually on rail cars in this case. (National Automotive History Collection, Detroit Public Library.)

FIGURE 1.13 A V8 engine block transfer machine at the Cadillac Livonia engine plant, photographed in 1981. In this and many much earlier systems, part transfer is completely automated. (National Automotive History Collection, Detroit Public Library.)

machines provided production volumes are in the hundreds of thousands for a period of years. The economics of transfer machines and other production systems are discussed in more detail in Chapter 3.

Another significant development in metal cutting in the twentieth century was the globalization of the machine and cutting tool industries. The Swiss and German machine tool industries had developed rapidly and become the design leaders in Europe by 1880, producing in particular, advanced types of gear cutting machines, automatic lathes, and jig boring and grinding machines. In the 1930s, the Krupps company introduced sintered tungsten carbide cutting tools, first in brazed form and later as a detachable insert. This material is superior to high speed steel for general purpose machining and has become the industry standard. The Soviet Union and Japan also became industrialized and developed machine tool industries early in the century; Japan, in particular, has developed into an important center for machine tool research, design, and production. Since World War II, there has been widespread development of machine and cutting tool industries in many countries. Top-quality machines and inserts may now be sourced from many parts of the world.

A great deal of research in metal cutting has been conducted since 1900. A bibliography of work published prior to 1943 was compiled by Boston [37]; much of this earlier work has been critically summarized by Shaw [38], King [39], and others. A number of significant research advances were made between 1940 and 1960. The shear plane theory of metal cutting was developed by Ernst [40] and Merchant [41] and provided a physical understanding of cutting processes which was at least qualitatively accurate for many conditions. Trigger and Chao [42] and Loewen and Shaw [43] developed accurate steady-state models for cutting temperatures. Somewhat later, a number of researchers studied the dynamic stability of machine tools, which had become an issue as cutting speeds had increased. This resulted in the development of a fairly complete linear theory of machine tool vibrations [44, 45]. Research in all of these areas continues to this day, particularly numerical analysis work made possible by advances in computing.

After World War II, the aircraft industry became an important market for machine tools. Since aircraft manufacture differs from automotive manufacture in that small batches of complex parts are required, it led to the development of more advanced general purpose machine tools. The most important innovation was the introduction of numerical control. Numerically controlled machines are the descendants of cam and template copying machines which move the cutting tool precisely along complicated paths in accordance with a stored program. There was considerable development of servomechanism technology for fire control and similar applications during World War II. Beginning shortly after the war, various attempts were made to apply this technology to automatic machine tool control [36]. Current numerical control (NC) technology developed largely from work initiated by John Parsons of the Parsons Corporation [46, 47]. Parsons developed early NC concepts for producing airfoil sections, and received an Air Force contract to develop a card-controlled machine. The Parsons Corporation subcontracted the servo system design to the Servomechanisms Laboratory at the Massachusetts Institute of Technology (MIT) [47, 48]. After considerable refinement of the initial concepts, the MIT group developed and demonstrated a tape-controlled milling machine in 1952 [48, 49] (Figure 1.14). (The complex funding and contractual arrangements generated by this project are described in melancholy detail in Reintjes' book [48].) A key contribution of the MIT group was Douglas Ross's development of the automatically programmed tools (APT) language for tool path code generation. This was one of the first higher level programming languages and served as the basis for many subsequent NC standards. Tape-controlled NC machining centers similar to the MIT machine became commercially available beginning about 1960. Tape control was replaced by direct computer numerical control (CNC) around 1980, and today all but the simplest machine tools are fitted with some kind of controller. There has been considerable interest in instrumenting machine tools so that the control computer can monitor the physical state of the process and take corrective action if required. Limited monitoring systems have been used in a number of applications and research into more general systems is being

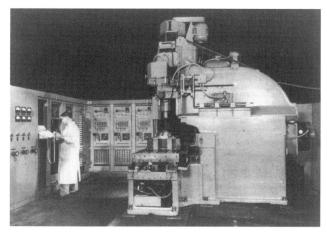

FIGURE 1.14 One of the first NC machine tools: the tape-controlled milling machine developed by MIT under an Air Force contract in 1952. The cabinets behind the machine house the controller, which used 292 vacuum tubes and had a clock speed of 512 Hz. (MIT Museum.)

conducted. Other current trends in machining development include movement toward more flexible, agile, or reconfigurable machining systems based on CNC machine tools, and the development of open-architecture, PC-based controls to replace proprietary controllers.

New tool materials have also greatly impacted recent metal cutting practice. A variety of ceramics are currently used for cutting tools, especially for hardened or difficult-to-machine work materials. Some, such as aluminum oxide, have long been used in grinding wheels, while others, such as silicon nitride, were initially developed for turbine blade applications. Ceramic and diamond tools have replaced carbides in a number of high-volume applications, especially in the automotive industry, but carbides (often coated with ceramic layers) have remained the tool of choice for general purpose machining. There has been a proliferation of grades and coatings available for all materials, with each grade containing additives to increase chemical stability in a relatively narrow range of operating conditions. For many work materials cutting speeds are currently limited by spindle and material handling limitations rather than tool material considerations. Dozens of insert shapes with hundreds of integral chip breaking patterns have been marketed; the further proliferation of insert shapes, however, is being opposed by standardization efforts in Europe. In the future, it is likely that tool materials and inserts will be specifically designed for critical operations.

1.3 TYPES OF PRODUCTION

Metal cutting practice has evolved to accommodate three basic types of production. The first is job shop production, or the production of small or medium lots of a variety of parts made of many materials. The second is the production of small volumes of a few complex parts, as in aircraft manufacture. The third is the mass production of large volumes of identical parts, as in automotive powertrain and component manufacture.

Job shop production is carried out using general purpose and CNC machine tools. Tool materials which cut a variety of work materials are employed. Tool geometries, cutting speeds, feed rates, and depths of cut are chosen based on experience, handbook recommendations, or tool manufacturer catalog recommendations. They are seldom economically optimum, but low production volumes do not justify tests to improve efficiency. CNC machine tools are used for the small volume production of complex parts. Tool materials can often be selected for optimum performance for a few work materials. Control and tool monitoring issues are important to ensure accuracy and to prevent damage to partially machined workpieces caused by tool wear or breakage. Mass production has typically carried out using dedicated, specially designed machining systems, although there is an increasing trend toward the use of CNC machine tools as discussed in Chapter 15. Tool materials effective for a narrow range of work materials are used. Tool geometries, cutting speeds, feed rates, and depths of cut must be selected to provide reliable, efficient operation. Machinability testing and process simulation can be used to choose optimum conditions; the cost of these activities can be offset by productivity improvements. Monitoring is less attractive in these applications; more emphasis is typically given to designing the process to ensure reliable operation, rather than to developing monitors which will periodically interrupt production.

REFERENCES

1. S. Malkin, *Grinding Technology*, Ellis Horwood, Chichester, UK, 1989
2. G. Benedict, *Nontraditional Manufacturing Processes*, Marcel Dekker, New York, 1987
3. J.E. Weller, *Nontraditional Machining Processes*, 2nd Edition, Society of Manufacturing Engineers, Dearborn, MI, 1984

4. G. Boothroyd and W.A. Knight, *Fundamentals of Machining and Machine Tools*, 2nd Edition, Marcel Dekker, New York, 1989, Chap. 14
5. J.F. Wilson, *Practice and Theory of Electrochemical Machining*, Wiley-Interscience, New York, 1971
6. G. Chryssolouris, *Laser Machining — Theory and Practice*, Springer Verlag, New York, 1991
7. A. Kobayashi, *Machining of Plastics*, McGraw-Hill, New York, 1967
8. S. Chandrasekar et al., Eds, *Machining of Advanced Ceramic Materials and Components*, ASME, New York, 1988
9. S. Smiles, *Industrial Biography*, John Murray, London, 1863, Chaps. 10–14
10. J.W. Roe, *English and American Tool Builders*, McGraw-Hill, New York, 1916; reissued by Lindsay Publications, Bradley, IL, 1987
11. R. Burlingame, *Machines that Built America*, Harcourt Brace, New York, 1953
12. L.T.C. Rolt, *Tools for the Job*, B.T. Batsford, London, 1965; reissued by The Science Museum, London, HMSO Press, 1986
13. R.S. Woodbury, *History of the Gear-Cutting Machine*, Technology Press, Cambridge, MA, 1958
14. R.S. Woodbury, *History of the Grinding Machine*, Technology Press, Cambridge, MA, 1959
15. R.S. Woodbury, *History of the Milling Machine*, Technology Press, Cambridge, MA, 1960
16. R.S. Woodbury, *History of the Lathe*, Society for the History of Technology, Cleveland, OH, 1961
17. R.S. Woodbury, *Studies in the History of Machine Tools*, MIT Press, Cambridge, MA, 1972
18. W. Steeds, *A History of Machine Tools 1700–1910*, Clarendon Press, Oxford, 1969
19. G. Lefebvre, *Le Tombeau de Petosiris*, Cairo, 1923/4, Plate 10
20. H. Hodges, *Technology in the Ancient World*, Barnes and Noble Books, New York, 1992
21. W.B. Parsons, *Engineers and Engineering in the Renaissance*, MIT Press, Cambridge, MA, 1932
22. B. Thomson (Count Rumford), An experimental inquiry concerning the source of the heat which is excited by friction, *Philos. Trans.* **88** (1798) 80–102; see also S.C. Brown, Ed, *Collected Works of Count Rumford*, Vol. 1, Harvard University Press, Cambridge, MA, 1968, 1–26
23. E.P. DeGarmo, *Materials and Processes in Manufacturing*, 5th Edition, Macmillan, New York, 1979, 18
24. D.A. Hounshell, *From the American System to Mass Production, 1800–1932*, Johns Hopkins University Press, Baltimore, MD, 1984
25. I. Finnie, Review of metal-cutting analyses of the past hundred years, *Mech. Eng.* **78** (1956) 715–721
26. N.N. Zorev, *Metal Cutting Mechanics*, Pergamon Press, Oxford, 1966, 1–2 (Translated by H.S.H. Massey)
27. E. Hartig, *Versuche Ueber Leistung und Arbeitsverbrauch der Werkzeugmaschinen*, B.G. Teubner, Leipzig, 1873
28. I.A. Time, *Resistance of Metals and Wood to Cutting*, St. Petersburg, 1870 (in Russian)
29. H. Tresca, Memoire sur le Rabotage des Metaux, *Bull. Soc. d'Encouragement l'Industrie Nationale* (1873) 585
30. A. Mallock, The action of cutting tools, *Proc. R. Soc. Lond.* **33** (1881) 127–139
31. F.W. Taylor, *Principles of Scientific Management*, Harper and Brothers, New York, 1913
32. J. Gies, Automating the worker, *American Heritage of Invention and Technology* **6**:3 (Winter 1991) 56–63
33. F.W. Taylor, On the art of cutting metals, *ASME Trans.* **28** (1906) 31–350
34. R. Kanigel, *The One Best Way: Frederick Winslow Taylor and the Enigma of Efficiency*, Penguin Books, New York, 1997, 387–389
35. E.G. Thomsen, F.W. Taylor A historical perspective, *On the Art of Cutting Metals — 75 Years Later*, ASME PED Vol. 7, ASME, New York, 1982, 1–12
36. D.F. Noble, *Forces of Production*, Oxford University Press, Oxford, 1984
37. O.W. Boston, *Bibliography on the Cutting of Metals 1864–1943*, ASME, New York, 1954
38. M.C. Shaw, *Metal Cutting Principles*, Oxford University Press, Oxford, 1984
39. R. I King, Historical background, *Handbook of High Speed Machining Technology*, Chapman and Hall, New York, 1985, 3–26
40. H. Ernst, Physics of metal cutting, *Machining of Metals*, American Society of Metals, Metals Park, OH, 1938, 24
41. M.E. Merchant, Mechanics of the metal cutting process. I. Orthogonal cutting of a type 2 chip, *J. Appl. Phys.* **16** (1945) 267–275

42. K. J. Trigger and B.T. Chao, An analytical evaluation of metal cutting temperatures, *ASME Trans.* **73** (1951) 57–68
43. E.G. Loewen and M.C. Shaw, On the analysis of cutting-tool temperatures, *ASME Trans.* **76** (1954) 217–231
44. S. Tobias, *Machine Tool Vibrations*, Wiley, New York, 1965
45. J. Tlusty, Machine dynamics, *Handbook of High Speed Machining Technology*, Chapman and Hall, New York, 1985, 48–153
46. John Parsons: The man behind numerical control, *Manuf. Eng.* **88**:1 (1982) 127
47. J. Pusztai and M. Sava, *Computer Numerical Control*, Reston Publishing Co., Reston, VA, 1983, 1–5
48. J.F. Reintjes, *Numerical Control: Making a New Technology*, Oxford University Press, Oxford, 1991
49. W. Pease, An automatic machine tool, *Sci. Am.* **187**:3 (1952) 101–115

2 Metal Cutting Operations

2.1 INTRODUCTION

This chapter describes the common machining operations used to produce specific shapes or surface characteristics. The primary focus is on process kinematics and equations for basic cutting parameters. Brief descriptions of the general purpose machine tools traditionally associated with each process are also included. More detailed information on machine tools is covered in Chapter 3. Additional descriptive information on basic cutting operations is available in the literature [1–10].

2.2 TURNING

In turning (Figure 2.1), a cutting tool is fed into a rotating workpiece to generate an external or internal surface concentric with the axis of rotation. Turning is carried out using a lathe, one of the oldest and most versatile conventional machine tools. The principal components and movements of a lathe are shown in Figure 2.2. The workpiece is mounted on a rotating spindle using a chuck, collet, face plate, or mandrel, or between pointed conical centers [2, 4]. Lathes may have a horizontal or a vertical spindle, with vertical spindle machines being used especially for large workpieces. The cutting tool is held on a translating carriage or turret or in the tailstock. The carriage travels along the bedways parallel to the part axis (Z-axis). Motion perpendicular to the part axis is provided by the X-axis or the cross slide, which is mounted atop the carriage. Contours, tapers, arcs, or other machine motions can be obtained by the X- and Z-axes.

In addition to the tool geometry, the major operating parameters to be specified in turning are the cutting speed, V, feed rate, f_r, and depth of cut, d. The cutting speed is determined by

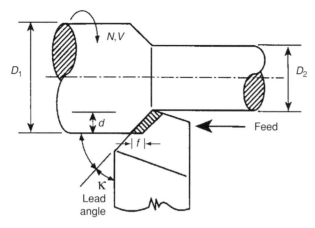

FIGURE 2.1 Turning.

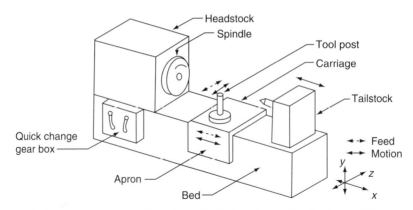

FIGURE 2.2 Principal components and motions of a lathe. (After E.P. DeGarmo [7].)

the rotational speed of the spindle, N, and the initial and final workpiece diameters, D_1 and D_2:

$$V = \pi N \frac{D_1 + D_2}{2} = \pi N D_{avg} \quad \text{for small } d: V \cong \pi N D_1 \tag{2.1}$$

The feed f is the tool advancement per revolution along its cutting path in mm/rev. The Feed rate (f_r) is the speed at which the tool advances into the part longitudinally in mm/min, and is related to f through the spindle rpm N:

$$f_r = fN \tag{2.2}$$

The feed influences chip thickness and how the chip breaks. The uncut chip thickness a is related to f through the lead angle of the tool, κ (Figure 2.1):

$$a = f \cos \kappa \tag{2.3}$$

The depth of cut, d, is the thickness of the material removed from the workpiece surface:

$$d = \frac{D_1 - D_2}{2} \tag{2.4}$$

The time t_m required to cut a length L in the feed direction is

$$t_m = \frac{L + L_e}{f_r}, \quad \text{where } L_e = \text{(approach allowance)} + \text{(overtravel allowance)} \tag{2.5}$$

The material removal rate per unit time, Q, is given by the product of the cutting speed, feed, and depth of cut:

$$Q = Vfd = \frac{\pi(D_1^2 - D_2^2)}{4} f_r \tag{2.6}$$

TABLE 2.1
Specific Cutting Power u_s for Various Materials (for Zero Effective Rake Angle Tools, 0.25 mm Undeformed Chip Thickness, and Continuous Chips Without Built-Up Edge (BUE))

Material	Unit Power (kW/cm³/min)	Unit Power (HP/in.³/min)
Cast irons	0.044–0.08	0.97–1.76
Steels:		
Soft	0.05–0.066	1.10–1.45
$0 < R_C < 45$	0.065–0.09	1.43–1.98
$50 < R_C < 60$	0.09–0.2	1.98–4.40
Stainless steels	0.055–0.09	1.21–1.98
Magnesium alloys	0.007–0.009	0.15–0.20
Titanium	0.053–0.066	1.16–1.45
Aluminum alloys	0.012–0.022	0.26–0.48
High temperature alloys (Ni and Co based)	0.09–0.15	1.98–3.30
Free machining brass	0.056–0.07	1.23–1.54
Copper alloys		
$R_B < 80$	0.027–0.04	0.59–0.88
$80 < R_B < 100$	0.04–0.057	0.88–1.25

Note: W = N m/sec.

The spindle power P_s is

$$P_s \cong Q u_s \qquad (2.7)$$

where u_s is the power required to cut a unit volume of work material. Typical values for u_s for common work materials are summarized in Table 2.1. These values may be multiplied by correction factors to account for changes in tool geometry, tool wear, and other factors [7].

If the efficiency of the drive system is η, the motor power P_m required is

$$P_m = \frac{P_s}{\eta} + P_t \qquad (2.8)$$

where P_t is the idling power. Finally, the efficiency of the drive system η is

$$\eta = \eta_m \eta_b \eta_{br1} \eta_{br2} \qquad (2.9)$$

where η_m, η_b, η_{br1}, and η_{br2} are the efficiencies of the motor, belt, front bearing, and rear bearing, respectively.

2.2.1 Hard Turning

A special case of turning is hard turning, in which hard metals are finish machined using ceramic or polycrystalline tools [11–18]. This process is sometimes used in place of rough turning, hardening, and finish grinding for parts made of tool steels, alloy steels, case hardened steels, and various hard irons. Very fine finishes and tolerances can be produced by this process, and in some cases part quality is better than can be obtained with grinding because intermediate chucking operations and associated setup errors are eliminated. In appropriate applications, hard turning also requires less capital investment, removes material more rapidly, and raises fewer environmental concerns than grinding. Hard turning became possible with the advent of hot-pressed ceramic and especially polycrystalline cubic boron nitride (PCBN) tools. Hard turning also requires high machine and toolholder rigidity, and strong insert shapes (negative rakes, large wedge angles, and special edge preparations such as chamfers discussed in Chapter 4). Modern CNC lathes usually have adequate rigidity for hard turning.

2.3 BORING

The boring operation (Figure 2.3) is equivalent to turning but is performed on internal surfaces. Boring is applied for roughing, semifinishing, or finishing cast or drilled holes. Finish boring in particular is a precision process characterized by small form, dimensional, and surface finish tolerances.

Boring can be performed on a number of machine tools, including lathes, drilling machines, horizontal or vertical milling machines, and machining centers. Most commonly, however, special boring machines of either the horizontal or vertical spindle type are used in high volume production. A horizontal boring machine, which can also be used for drilling and milling, is shown in Figure 2.4. Two- or three-axis machines are used depending on process requirements. The machine structure is similar to that of a milling machine with a precision spindle. High machine structural and spindle stiffness are required in order to generate quality bores. Very accurate work is often done using jig-boring machines, which are equipped with precision tables and spindles, stiff machine structures, and measuring devices built into the table.

Since boring is equivalent to turning, boring performance also depends on the cutting speed, depth of cut, and feed rate. The equations relating these parameters to part dimensions and machine variables reviewed above for turning are also applicable to boring. As will be discussed in Chapter 4, however, boring tools differ significantly from those used for turning due to their unique structural and dynamic requirements. Traditionally, moderate cutting speeds and small depths of cut and feed rates are used in boring to ensure accuracy, but in more recent practice higher cutting speeds have been used to reduce errors due to mechanical and thermal distortion. Heavier depths of cuts are used with multiflute boring tools. The hole

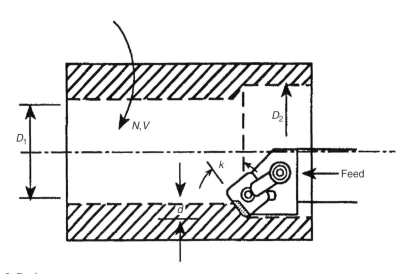

FIGURE 2.3 Boring.

Metal Cutting Operations

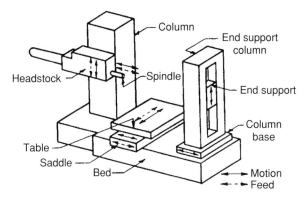

FIGURE 2.4 A horizontal boring machine, which can be used for boring, milling, and drilling. (After E.P. DeGarmo [7].)

depth (length) a boring bar can cut accurately is limited by the amount of bar deflection as will be discussed in Chapter 4.

2.4 DRILLING

Drilling (Figure 2.5), the standard process for producing holes, is the most common metal cutting process. Drilling is also often a bottleneck process in high volume manufacturing; in engine manufacture, for example, critical holes may be drilled in two or more passes, each penetrating to only a fraction of the final depth, to reduce cycle times.

Types of conventional drilling machines include upright machines, radial machines, and various specialized machines such as gang drill presses [2, 5]. The basic components of an upright drilling machine, shown in Figure 2.6, include the base, column, spindle, and work table. Radial drilling machines are designed for large workpieces and consist of a large

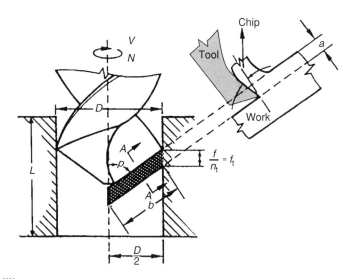

FIGURE 2.5 Drilling.

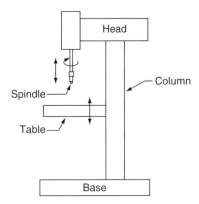

FIGURE 2.6 Principal components of an upright drill press. (After E.P. DeGarmo [7].)

horizontal arm extending from the column; both the height and angular orientation of the arm can be adjusted. A gang drill press is made up of two or more upright drilling machines placed next to each other on a common base. Often these machines operate sequentially with each spindle carrying a different tool, and workpieces moving from one spindle to the next. Drilling can also be performed on lathes, boring mills, and milling machines.

Geometrically, drilling is a complex process. The difficulty of producing drills with consistent geometries has traditionally limited accuracy, although drill consistency and repeatability has greatly increased recently with the advent of CNC drill grinders. Also, the complexity of the tool has inhibited the introduction of new tool materials, so that productivity gains in drilling have lagged those made in turning and milling over the past 30 years. A number of manufacturers have developed diamond or CBN tipped or coated drills to attempt to address this limitation.

Drilling performance depends on the materials involved, the drill geometry, the spindle speed, and the feed rate. The cutting speed V at the periphery of the drill is given by

$$V = \pi D N \quad (2.10)$$

where D is the drill diameter and N is the spindle rpm. As in turning, the feed rate f_r and feed per revolution f are related through Equation (2.2). The feed per tooth f_t depends on f and the number of flutes n_t

$$f_t = \frac{f}{n_t} \quad (2.11)$$

The effective depth of cut per flute d is

$$d = \frac{D}{2} \quad (2.12)$$

The width of cut b is given by

$$b = \frac{D}{2 \sin \rho} \quad (2.13)$$

where ρ is the half point angle of the drill. The uncut chip thickness a is

$$a = f_t \sin \rho \tag{2.14}$$

The metal removal rate Q is

$$Q = \left(\frac{\pi D^2}{4}\right) f_r \tag{2.15}$$

The time t_m required drilling a hole of depth L is

$$t_m = \frac{L + L_e}{f_r} \tag{2.16}$$

where the approach and overtravel allowance is

$$L_e = \frac{D}{2 \tan \rho} + \Delta L \tag{2.17}$$

and ΔL is the approach distance between the drill chisel edge and the entrance surface of the workpiece plus the overtravel of the drill in through holes.

In core drilling, in which a hole of initial diameter D_i is enlarged to diameter D, the width of cut is

$$b = \frac{D - D_i}{2 \sin \rho} \tag{2.18}$$

and the metal removal rate is

$$Q = \left(\frac{\pi (D^2 - D_i^2)}{4}\right) f_r \tag{2.19}$$

2.4.1 Deep Hole Drilling

A deep hole is a hole with a depth-to-diameter ratio of more than 5:1. Special machines are often required to drill deep holes with adequate straightness and to ensure efficient chip ejection and lubrication of the drill. Because the chips and heat generated by the process are confined, the deeper the hole, the more difficult it is to control heat buildup and remove the chips.

The traditional drilling operation, using standard or parabolic-flute twist drills, has been occasionally used for deep hole drilling but often requires "pecking" (drilling to intermediate depths and periodically withdrawing the tool to clear chips) unless a high-pressure coolant is used. Bushings with aspect ratios of 2:1 or 4:1 are also often used to support the drill at the entrance to improve location accuracy and to stabilize the drill.

Three distinct specialized deep hole drilling methods are in use: solid drilling, trepanning, and counterboring [2, 5, 19, 20] (Figure 2.7). Solid drilling is most common and can be further classified into four approaches: conventional twist drilling (discussed above), gundrilling, ejector drilling, and BTA (STS) (BTA, Boring and Trepanning Association; STS, single tube system) drilling (Figure 2.8). In gundrilling, the coolant is supplied through the center of the tool shank under high pressure, forcing the chips through the flutes. The BTA method uses

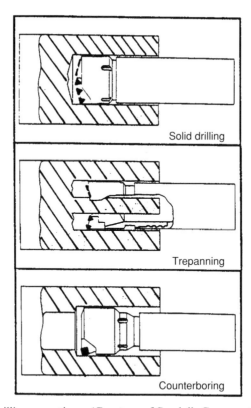

FIGURE 2.7 Deep hole drilling operations. (Courtesy of Sandvik Coromant Corporation.)

STS tools where the coolant is supplied under high pressure between the tool and hole surface. The coolant is then removed along with the chips through an opening above the cutting edge of the tool. The ejector or "two-tube" system feeds the coolant through a connector. Two-third of the coolant flows between the inner and outer tubes; the remaining coolant is drawn off through the nozzle, creating a vacuum which causes a backward ejection of chips through the inner tube. Tools for gundrilling, BTA drilling, and ejector drilling are discussed in Chapter 4.

Ejector drilling can be used for holes larger than 18 mm in diameter and is often the best choice for cost-efficient drilling of nonprecision holes. It does not require a pressure head for the coolant supply. The hole tolerance for ejector drilling is +0.075/−0.000 mm on the diameter; the hole straightness error is typically 0.05 mm/m. The BTA process can be used for holes over 10 mm in diameter. Holes with diameters between 3 and 20 mm can be drilled with a straightness error of 0.08 mm/m using the solid and gundrilling methods. The surface finish generated in deep hole drilling is typically between 1.0 and 1.5 μm R_a and is comparable to that obtained by reaming because the hole wall is burnished by the tool bearing surfaces. Deep hole drilling performance can sometimes be improved by supplying ultrasonic energy to the cutting zone [21].

The trepanning operation (Figure 2.7) removes material only at the periphery of the hole and leaves a solid core. Through holes with an aspect ratio of 2 to 4 can be trepanned with a hollow multitooth cutter and flood coolant. The depth of cut is equivalent to the width of the cutting edge and is smaller than for solid drilling. This operation, therefore, requires lower power and is well suited for through holes. Large blind holes can be trepanned using a tool

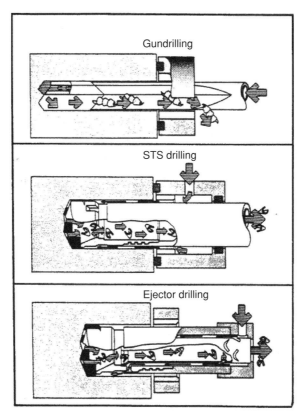

FIGURE 2.8 Solid deep hole drilling operations (STS, single tube system). (Courtesy of Sandvik Coromant Corporation.)

with a pivoting cutoff blade which separates the core from the base of the hole. Trepanning is less accurate than solid drilling due to misalignment caused by poor tool rigidity. It provides faster cycle times because there is no dead center area as with solid drilling.

Counterboring (Figure 2.7) enlarges an existing hole that is drilled or cast, normally for the purpose of improving its size, straightness, or surface finish. Often, a combination tool will perform a solid drilling operation while simultaneously counterboring the hole drilled ahead.

Deep hole drilling machines typically resemble horizontal turning or milling machines (Figure 2.9), although vertical spindle machines are also used. All three drill types (gundrill, ejector, and BTA) can be used, although special attachments for the bushing and the head are required for BTA setups. The tool, the workpiece, or both may rotate, depending on the size of the workpiece. The best hole quality is obtained when both the tool and workpiece rotate. Deep hole drilling machines are always equipped with high pressure, high volume coolant systems.

2.4.2 MICRODRILLING

Microdrilling is the drilling of small diameter (usually less than 0.5 mm) holes with a depth-to-diameter ratio greater than 10 [22]. Holes as small as 0.0025 mm have been successfully drilled. Microdrilling is similar to deep hole drilling but generally presents greater problems, since coolant fed drills cannot be used. High spindle speeds are required to generate sufficient

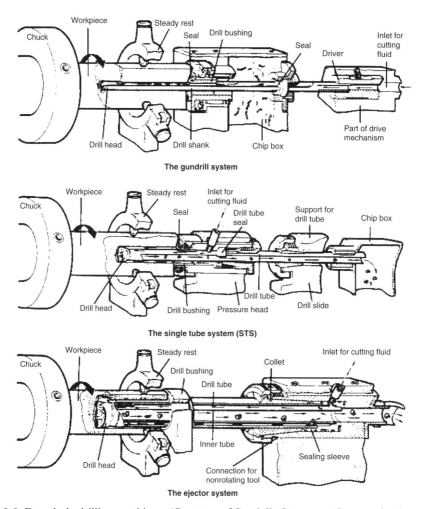

FIGURE 2.9 Deep hole drilling machines. (Courtesy of Sandvik Coromant Corporation.)

cutting speeds, especially when carbide drills are used. Low feed rates must also be used to avoid exceeding the buckling load of the drill. Feeds in the range of 0.00005 to 0.0005 mm/rev are common in microdrilling.

Microdrilling performance can sometimes be improved by supplying ultrasonic energy to the cutting zone [14]. High-frequency forced vibrations at frequencies between 15 and 30 kHz reduce friction and thus allow increased material removal rates. Ultrasonically assisted drilling is particularly effective for deep hole drilling and microdrilling in cases in which chip clogging limits the allowable penetration rate. The vibration tends to break chips into smaller sections while lowering forces. In these operations, vibrations can increase throughput by a factor of two while improving tool life and hole quality. The vibration frequency must be carefully controlled because vibration at frequencies outside the useful range reduces tool life.

When chips are not easily ejected, peck drilling (frequent withdrawals of the drill) is used to clear chips from the hole and to permit intermittent cooling of the drill. Peck drilling increases cycle time and drill wear (because the drill dwells at the bottom of the hole prior to retraction). Precise feed control is necessary to avoid excessive dwelling. Peck drilling may not be necessary if high spindle speeds are available, but may be preferred in drilling composites or layers of different materials.

Very small holes can also be produced by laser cutting and by electroplating a part in which a larger hole has been drilled. In some applications, small openings can also be produced through layering, that is, by drilling small holes in two thin sheets and then arranging the sheets so that the holes only partially intersect. These methods are usually slower and less accurate than drilling.

2.5 REAMING

Reaming (Figure 2.10) is used to enlarge a hole and improve its size control, roundness, and surface finish. Reaming is similar to boring, and in fact both processes can be used for some operations. The differences between reaming and boring result largely from tool design as discussed in Chapter 4. The choice of performing a boring or reaming operation depends on the hole diameter and length, interruptions within the hole due to internal cavities, and the required straightness, size tolerance, surface finish, and tool life. Reaming also resembles core drilling, and the equations and cutting parameters reviewed above for core drilling are also applicable to reaming. However, the stock removal (depth of cut) in reaming is generally between 0.25 and 0.7 mm, which is smaller than the stock removal in core drilling, counter-boring, and rough and semifinish boring.

Reaming machines are similar to drilling machines but have a less powerful motor and often a more precise spindle. Radial floating attachments (holders) are often used to ensure that the tool is aligned with the predrilled or cored hole. Special spindle heads with an attached, rotating bushing are used for squirt reaming. This improves machine capability and hole quality because the bushing travels with the spindle and supports the tool as it enters the hole.

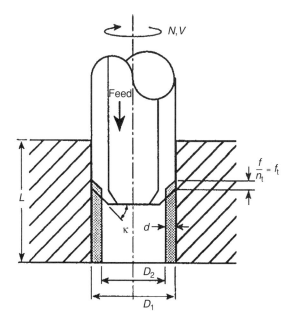

FIGURE 2.10 Reaming.

2.6 MILLING

In milling processes, material is removed from the workpiece by a rotating cutter. The two basic milling operations are peripheral (or plain) milling and face (or end) milling (Figure 2.11). Peripheral milling generates a surface parallel to the axis of rotation, while in face milling generates a surface normal to the axis of rotation. Face milling is used for relatively wide flat surfaces (usually wider than 75 mm). End milling, a type of peripheral milling operation, is used for profiling and slotting operations. Milling processes can be further divided into up (or conventional) milling and down (or climb) milling operations (Figure 2.12). If the axis of the cutter does not intersect the workpiece, the motion of cutter due to rotation opposes the feed motion in up (or conventional) milling but is in the same direction as the feed motion in down milling. When the axis of the cutter intersects the workpiece, both up and down milling occur at different stages of the rotation. Both up and

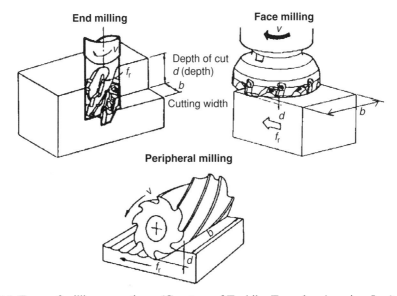

FIGURE 2.11 Types of milling operations. (Courtesy of Toshiba Tungaloy America, Inc.)

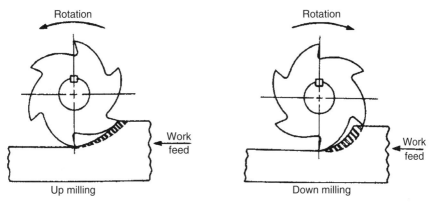

FIGURE 2.12 Schematic view of the up and down milling approaches. (Courtesy of Valenite Corporation.)

down milling have advantages in particular applications [2, 5]. Up milling is usually preferable to down milling when the spindle and feed drive exhibit backlash and when the part has large variations in height or a hardened outer layer due to sand casting or flame cutting. In down milling, there is a tendency for the chip to become wedged between the insert and cutter, causing tool breakage. However, if the spindle and drive are rigid, cutting forces in peripheral down milling tend to hold the part on the machine and reduce cutting vibrations.

The most common general-purpose milling machine is the knee and column milling machine or knee mill (Figure 2.13). The major components of the knee mill are the column, spindle, knee, saddle, and table. Both vertical and horizontal spindle milling machines are available. Universal mills are knee mills with a spindle head, which rotates at right angles to the table's longitudinal axis, so that the spindle can be either horizontal or vertical.

The uncut chip thickness varies continuously in milling. In up milling, the chip thickness is small at the beginning of cut and increases as cutting progresses, while in down milling the chip thickness is largest at the beginning of the cut. Cutting is also not continuous in milling, but rather is periodically interrupted as cutting edges enter and leave the part (Figure 2.14 and Figure 2.15). This leads to cyclic thermal and mechanical loads on the tool, which leads to a number of fatigue failure mechanisms not encountered in continuous cutting (Chapter 9).

The cutting action of each cutting edge on a milling cutter is similar to that of a single-point tool. The cutting speed is given by Equation (2.1), and the feed rate f_r, feed per revolution f, and feed per tooth f_t, are related by an equation similar to Equation (2.2):

$$f_r = Nf = n_t N f_t \tag{2.20}$$

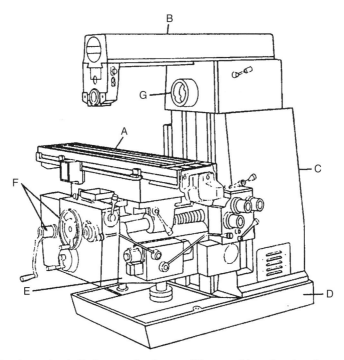

FIGURE 2.13 A horizontal spindle knee and column milling machine, showing the work table (A), ram (B), column (C), base (D), knee (E), and feed and height adjustments (F). (After M.C. Shaw [46].)

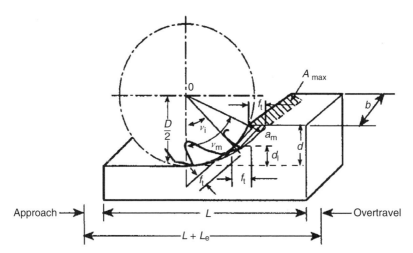

FIGURE 2.14 Characteristics of peripheral (or plain) milling.

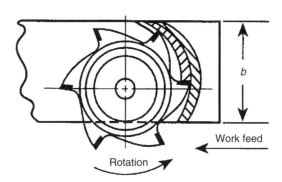

FIGURE 2.15 Face milling.

The variation of the uncut chip thickness in milling is complicated (Figure 2.14). Exact analyses [23–26] have shown that the uncut chip thickness varies trochoidally as the cutter rotates. For small feeds, however, a sinusoidal approximation is adequate. The uncut chip thickness a_i at an engagement angle of v_i, maximum uncut chip thickness, a_{max}, and average uncut chip thickness, a_{avg}, are given by

$$a_i = f_t \cos \kappa \sin v_i \qquad (2.21)$$

$$a_{max} = f_t \cos \kappa \sin v_m \qquad (2.22)$$

$$a_{avg} = f_t \cos \kappa \sin\left(\frac{v_m}{2}\right) \qquad (2.23)$$

where

$$\cos v_m = 1 - \frac{2d}{D} \qquad (2.24)$$

Metal Cutting Operations

and κ is the lead angle equivalent to the lead angle in turning (Figure 2.1). In peripheral milling, $\kappa = 0$; in face milling, ν_m is usually 90°.

If n_i teeth are engaged at a given instant, the maximum total undeformed chip area, A_{max}, is

$$A_{max} = f_t b \cos \kappa \sum_{i=1}^{n_i} \sin \nu_i \qquad (2.25)$$

The length of the arc of metal being cut, λ_c, which determines the chip length, in peripheral milling is

$$\lambda_c = (dD)^{1/2} \qquad (2.26)$$

In face milling, if the cutter is wider than the workpiece, λ_c is given by

$$\lambda_c = (bD)^{1/2} \qquad (2.27)$$

For end milling a slot (full engagement of the cutter),

$$\lambda_c = \frac{\pi D}{2} \qquad (2.28)$$

The metal removal rate Q is given by

$$Q = f_r b d \qquad (2.29)$$

where d and b are defined in Figure 2.11, Figure 2.14, and Figure 2.15. For example, d and b are the axial and radial depth of cuts for end milling, while radial and axial depth of cuts for peripheral milling. The time t_m required to mill a workpiece of length L is

$$t_m = \frac{L + L_e}{f_r} \qquad (2.30)$$

where the approach distance L_e in peripheral milling is given by

$$L_e = \sqrt{d(D-d)} + \text{(approach allowance)} + \text{(overtravel allowance)} \qquad (2.31)$$

while in face milling,

$$L_e = \frac{D}{2} \qquad (2.32)$$

The total travel length of the cutter is larger than the length of the workpiece due to the cutter approach and overtravel distances. The overtravel distance is normally very small, enough for the cutter axis to clear the end of the part.

When performing end milling or peripheral milling with a width of cut less than the cutter radius, chip thinning would occur and it is very important to increase the feed per tooth (using [Equations 2.21–2.24]) to keep the uncut chip thickness at the expected value.

When performing end milling, thread milling, and related operations on a machining center, internal or external circular interpolation is often required when generating a tool path

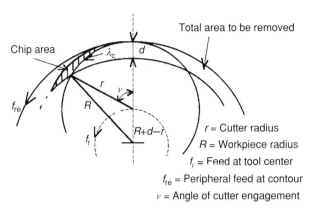
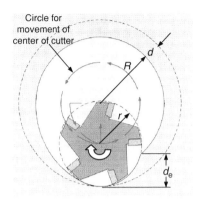

FIGURE 2.16 Tool path for circular interpolation in milling.

as shown in Figure 2.16. (When contour milling, the principles of straight end milling are applied, but additional checks on the feed rate, depth of cut, resultant chip thickness, and cutter edge density are required to avoid gouging when the cutter moves in a circular arc.) In these cases, the feed rate for the cutter center differs from the effective peripheral feed rate of the cutting edges, and somewhat different relations between cutting parameters are required.

The depth of cut engagement angle ν_i is given by

$$\nu_i = \cos^{-1}\left[\pm\frac{R^2 - r^2 - [R \pm (d-r)]^2}{2r[R \pm (d-r)]}\right] \quad (2.33)$$

where the plus and minus signs apply to internal and external milling, respectively. The plunge cut engagement is $2\nu_i$. The apparent length of contact of the cutting edges, λ_c, is given by

$$\lambda_c = (dD_e)^{1/2} \quad (2.34)$$

where D_e is the equivalent tool diameter given by

$$D_e = \frac{2r}{1 - [r/(d \pm R)]} \quad (2.35)$$

where the plus and minus signs apply to internal and external milling, respectively. λ_c can also be calculated from the relation

$$\lambda_c = r\nu_i \quad (2.36)$$

where ν_i is given in radians.

In a circular cut, the actual depth to which the edge penetrates is considerably greater than the radial depth of cut d. Therefore, it is important in circular cuts to determine the proper feed, depth of cut, and cutter diameter so that the cutter engagement is acceptable given the available horsepower, workpiece and fixture rigidity, and chatter limits. The feed per tooth, f_t, at the center of the tool is given by Equation (2.22) or (2.23) based on a known maximum or average chip thickness, respectively. The feed rate, f_r, at the center of the tool (used for programming the contour in most CNC machines) is given by Equation (2.20). The feed can be also obtained from Equation (2.24) using a modified depth of cut, d_e, that is not the same as the depth of cut value d. The real depth of cut is

$$d_e = \frac{d(R + d/2)}{[R + d \pm (-r)]} \quad (2.37)$$

The effective feed rate at the cutting edge is

$$f_{re} = f_r \frac{R \pm d}{R \pm (d - r)} \quad (2.38)$$

where the plus and minus signs correspond to internal and external circular interpolation, respectively.

The machining time can be calculated by dividing the total volume of material to be removed by the metal removal rate:

$$t_m = \pm \frac{\pi[(R \pm d)^2 - R^2]}{n_t N \lambda_c a_{avg}} + t_{feedin} \quad \text{or} \quad t_m = \frac{2\pi[R \pm (d - r)]}{f_r} + t_{feedin} \quad (2.39)$$

where t_{feedin} is the feed-in and feed-out times required during the ramping into and out of the cut. There are two approaches for the cutter entering and exiting the cut, either tangentially or radially. The tangential approach is preferable because the tool is ramping gradually into the cut and provides greater tool life, no marks are left on the workpiece and there is no tendency for tool vibration.

2.7 PLANING AND SHAPING

Planing and shaping are similar operations normally used to generate flat surfaces, although a variety of irregular or contoured surfaces can also be produced. In these processes the tool moves linearly and reciprocally with respect to the workpiece. In planing, the workpiece reciprocates while the tool is fed across the workpiece to provide the feed motion. In shaping, the tool reciprocates across the workpiece. Workpieces of any size can be machined with the planing operation while only small- or medium-sized parts can be machined by shaping. Neither process is used in mass production, since flat surfaces can be produced more rapidly by broaching or face milling. Planing is commonly used in machinery manufacture to produce flat surfaces on large castings or forgings.

Single-point tools are used in both operations and the basic elements involved, shown in Figure 2.17, are almost identical to those of turning. Therefore, the basic equations for planing and shaping are identical to those given for turning with the frequency of reciprocation or cutting stroke substituted for the rotational speed (rpm). Planing machines are generally simple in design; normally the tool is mounted on a gantry which spans the work table.

2.8 BROACHING

Broaching resembles shaping in that the tool translates linearly across the workpiece [2, 5, 27–29]. However, broaching employs a series of cutting edges ground or otherwise mounted on a tool body (Figure 2.18), rather than a single-point tool. By suitably varying the relative heights of successive cutting edges, a broach can be designed to perform roughing, semifinishing, and finishing cuts in a single stroke. Broaching is consequently a highly productive process. It is also precise and can produce very fine surface finishes. Accuracies of 0.025 mm with surface finishes of 0.8 to 2 μm are achievable.

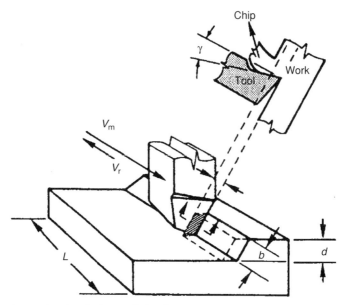

FIGURE 2.17 Cutting action in planing and shaping.

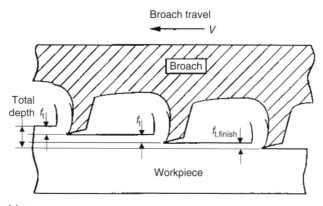

FIGURE 2.18 Broaching.

There are two basic types of broaching operations: external (surface) and internal broaching. External broaching is used to generate flat surfaces, contours, and various forms such as spur gears, splines, and slots. More than one surface on a workpiece can be produced in one pass. Internal broaching produces circular and noncircular holes, keyways, splines, serrations, and gear teeth. The internal broach requires a starting hole for insertion of the tool.

Broaching machines may be of the horizontal or vertical variety. Vertical machines are very popular because they require less space. Broaching machines may also be characterized by their driving mechanism as single or twin-ram, pull-down, pull-up, push-down, push-up, continuous, or rotary types [2, 5]. Broaching machines are designed with very high stiffness to support high cutting loads. The tooling is often large and heavy, so that special designs are required to ensure easy access for inspection and changing of broaches.

Metal Cutting Operations

An individual cutting tooth on a broach resembles a turning or a shaping tool, and the basic equations for turning reviewed above are applicable to broaching as well, except that the linear speed of the broach, V, is substituted for the rotational speed (m/min). The feed per tooth, f_t, is determined by the broach design (as illustrated in Figure 2.18) and cannot be altered by changing machine settings. The front teeth are used for roughing and the last ones for finishing as shown in Figure 2.18. The pitch of the broach is determined by the workpiece material and is coarser for roughing than for finishing. Each tooth removes a depth of cut that is equivalent to the feed per tooth and the total depth is removed by the summation of the individual depth of cuts (or feed per tooth). The machining time is estimated from the travel distance of the broach (which includes the roughing, semifinishing, and finishing teeth) over its cutting speed.

A process similar to broaching is turnbroaching or skiving, in which the broach is fed into a rotating workpiece. This process is used to machine crankshafts and camshafts in engine manufacture. Relations for the uncut chip thickness and other parameters are similar to those for milling.

2.9 TAPPING AND THREADING

Many workpieces require internal or external screw threads, which can be produced by a variety of cutting operations. The common cutting processes for producing internal threads are tapping, chasing, and thread milling. Cutting processes used to produce external threads include thread turning and die threading. Threads may also be produced by grinding and by various forming operations.

In *tapping*, a specially formed threading tool is fed into a hole drilled in a previous operation. The tap may either cut or plastically deform the hole wall material to form the thread. Roll form taps can be used in ductile materials such as free machining steels, soft carbon and alloy steels, austenitic stainless steels, and ductile aluminum, copper, zinc, and magnesium alloys. Roll and cut tapping produce different thread profiles as shown in Figure 2.19. In cut tapping, the minor diameter of the thread is determined by the diameter of the existing hole. The metal removal rate is governed by the tap's effective chamfer length, number of flutes, and rpm in addition to the minor diameter. The shear strength of the thread increases with the percent thread for rolled threads, but not significantly for cut threads. Roll taps produce no chips, but require consistent lubrication and control of the

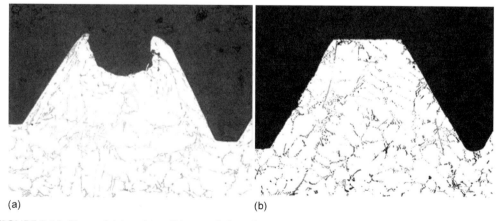

FIGURE 2.19 Formed (a) and cut (b) tapped threads.

pretapped hole diameter to prevent excessive tapping torque and tap breakage. The most common machines used for tapping are drilling machines, milling machines, and lathes. Rigid tapping, that is, CNC tapping without floating toolholders or self-reversing tapping units, requires proper synchronization of Z-axis feed to spindle revolution especially at higher speeds (>3000 rpm for a 6 mm tap).

Tapping is used for through holes and blind holes. Drilled through holes can be tapped along the entire length. Blind holes are tapped to a specified length and are sometimes called bottom holes. Blind-hole tapping requires accurate depth control to avoid ramming the tap into the bottom of the hole, causing tool failure or insufficient number of threads. Bottoming taps must pull chips up and out of the hole, as compared to through hole taps, which usually push chips out of the hole in front of the cutting edge. Roll form tapping can produce stronger threads with work-hardening materials (steels and stainless steels).

The thread height h_a or percentage thread η are important parameters in tapping. η is related to h_a, the theoretical thread height h_t, and the major and minor diameters of the thread form, D_m and D_μ (Figure 2.20) [30]:

$$\eta = \frac{\% \text{ thread}}{100} = \frac{h_a}{h_t} = \frac{D_m - D_\mu}{1.299p} \qquad (2.40)$$

where p is the thread pitch. h_t is the American National (AN) basic thread form height used in the standard method for describing the thread height. The torque required for tapping can increase by 200 to 300% when the percent thread increases from 50 to 75%. The manufacturing strategy should target a minimum percentage of thread if the thread strength is essentially insensitive to this parameter [30]. The unified (UN) form and ISO standard methods are equivalent to 83% of the AN thread depth. For cut tapping, η can also be calculated from the pitch diameter, D_p, and predrilled hole diameter, D_d:

$$\eta = \frac{0.7698(D_p + 0.6496p - D_d)}{p} \quad \text{(cut tapping)} \qquad (2.41)$$

The predrilled hole diameter for a target value of η is

$$D_d = D_m - 1.299\eta p \quad \text{(cut tapping)} \qquad (2.42)$$

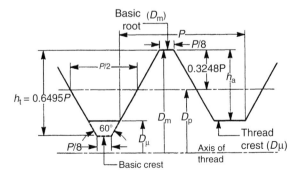

FIGURE 2.20 Thread profile parameters.

For roll form tapping, the equations corresponding to Equations (2.41) and (2.42) are

$$\eta \cong \frac{1.5396(D_d - D_\mu)}{p} \quad \text{(roll form tapping)} \quad (2.43)$$

$$D_d = D_m + 0.68 p \eta \quad \text{(roll form tapping)} \quad (2.44)$$

Generally, tap drill charts have been generated for $\eta = 75\%$, although the actual thread may deviate from the charted value [31, 32]. Form tapping requires a larger hole size than cut tapping because the minor diameter of the thread is produced by the extrusion inward from the drilled hole.

Figure 2.21 shows a plot of the ideal linear and rotational velocity versus time to generate a desired thread in a rigid tapping process. The cutting speed, feed rate, and depth of cut are given by

$$V = \pi D_m N \quad (2.45)$$

$$f_r = pN \quad (2.46)$$

$$d = \frac{D_m - D_d}{2} \quad (2.47)$$

The depth of cut per edge, d_i, is

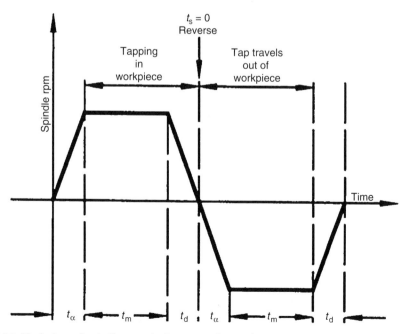

FIGURE 2.21 Variation of spindle rpm during a tapping cycle.

$$d_i = \frac{1.299 p \eta \cos \delta}{2 n_t n_f} \qquad (2.48)$$

where n_f is the number of threads along the chamfer, n_t is the number of flutes, and δ is the chamfer angle of the tap measured from the tap axis. The metal removal rate is

$$Q = \left(\frac{p}{4} + \frac{D_m - D_d}{\tan(\pi/3)} \right) \left(\frac{D_m - D_d}{4} \right) \frac{P}{\sin \lambda} N \qquad (2.49)$$

The time to cut a thread of length L, t_m, and thread helix angle, λ, are given by

$$t_m = \frac{L}{f_r} \qquad (2.50)$$

$$\tan \lambda = \frac{p}{\pi D_p} \qquad (2.51)$$

Thread turning, the oldest and most widely used threading operation, is a process for producing external or internal threads, usually using a single point tool [11, 33, 34]. This process has traditionally been used on soft materials, but with the availability of PCBN tools is now also used when turning hardened steels. The tool may be fed into the workpiece either radially or axially. Radial feed cutting generates higher cutting forces and leads to greater difficulty in chip disposal, and is used mainly on materials which produce short chips or with multitoothed inserts. In flank-infeed cutting (in which the tool is fed axially) the cutting action is more like conventional turning. There are many different flank-infeed sequences, which distribute the thread form between passes (Figure 2.22) [33]. The infeed sequence can be optimized to reduce the number of passes while keeping the chip load constant between passes as illustrated in Figure 2.23. The optimum number of passes depends on the tool geometry and edge strength. In some cases, the center portion of the thread is removed using radial infeed while the remaining stock is removed using flank infeed. In other cases, a significant amount of material is removed with a grooving tool, leaving only a small amount to be cut with a threading tool. The lead of the thread is determined by the longitudinal motion of the tool in relation to the rotation of the workpiece. The feed coincides with the pitch of the thread.

Multitoothed full-profile indexable inserts are also used to turn threads. Such inserts generate the full thread profile including the crest in a single pass, eliminating the multiple passes required to produce threads with a single point tool.

Thread milling is an operation used to generate internal or external threads using a milling cutter [11, 34, 35]. The cutter is fed along the axis of the workpiece as in thread turning to generate the threads in a single pass as shown in Figure 2.24. With a stationary workpiece, a rotating tool moves simultaneously along three axes to generate the helical thread (as compared to the two-axis motion used in circular interpolation). When cutting an external thread, the tool moves along the part's outside diameter; when cutting an internal thread, the tool moves inside a previously drilled hole. As in cut tapping, the feed rate is determined by the workpiece speed in a turning machine or by the helical path speed in NC machining centers or special machines. The accuracy of the thread is controlled by the accuracy of the axial and circular feed mechanisms of the machine, not by the cutting tool. It is preferable to start the thread milling operation at the bottom of blind hole so that the tool moves outward. In thread milling, the tool rotates at higher speeds and lower feeds than in tapping or thread

Metal Cutting Operations

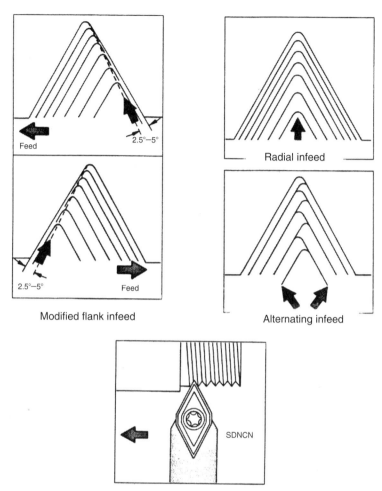

FIGURE 2.22 Schematic view of the radial, flank, and alternating infeed methods for threading. (Courtesy of Carboloy Corporation.)

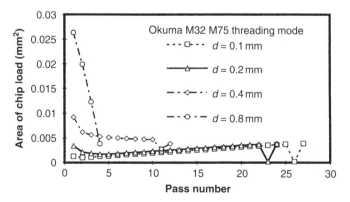

FIGURE 2.23 The effect of depth of cut on number of passes and on chip load area in thread turning.

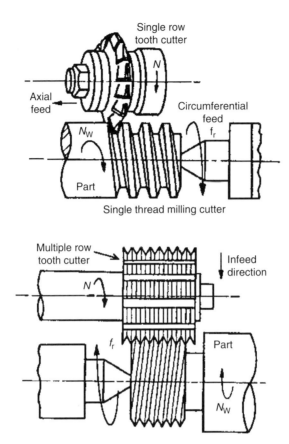

FIGURE 2.24 Thread milling operation performed in a turning center.

turning; the feed can be adjusted to generate the desired surface finish and is not constrained by the desired thread pitch as in other threading operations.

Compared to conventional tapping, thread milling allows more room for chip evacuation; chips are also much smaller. The power required for threading can be reduced considerably using thread milling. Percent threads approaching 100% can be generated, and tapered threads can be generated easily and accurately.

Thread milling is used primarily for large holes (diameter >30 mm) while tapping is used for smaller holes (diameter <40 mm). Recently, thread milling has been used for smaller holes with the introduction of a combined short-hole drilling and thread milling operation called thrilling or drill/thread milling (Figure 2.25) [34]. Thrilling uses a combined drill-threading tool rotating continuously at a high spindle speed to drill a blind or through hole and generates the thread through a helical retraction motion. Thread milling accuracy is dependent on the machine control system generating the helical interpolation including the machine motion accuracy. Thread milling tends to generate smoother and more accurate threads than tapping and is more efficient than thread turning. Thread milling also eliminates the spindle reversal at the bottom of the hole required in tapping. However, milled threads must typically be gauged much more carefully than tapped threads. Coolant requirements in thread milling are not as critical as those in cut tapping. Tooling costs are generally higher than for tapping.

Metal Cutting Operations 41

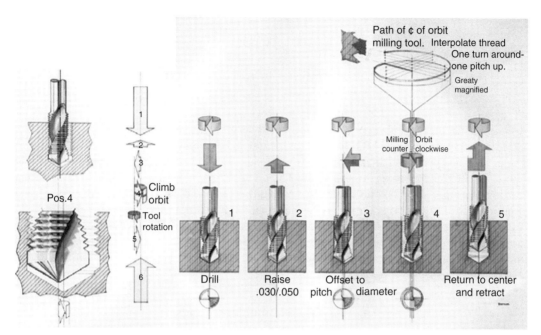

FIGURE 2.25 Drill/thread milling using a combination drill and thread milling tool.

Universal thread mills use a lead screw and can cut any thread. Production thread mills use a cam. NC machines capable of simultaneous three-axis motion are also used for thread milling. In some instances, special spindle units with an eccentric quill are used to generate helical motions of the tools. NC machines can be used for drilling and threading with a combination tool. A conventional lathe can also be used; in the case of a lathe, the tool is replaced by a milling cutter and milling head, or the milling cutter is set in a live spindle.

Thread milling machines usually employ a climb-milling motion, especially when producing internal right-handed threads, because the tool travels out of the hole during the cut, which reduces chip interference. However, a conventional up-milling mode is used when the part or fixture stiffness is low to reduce tool deflections and chatter. In addition, up-milling motion is also used with materials that are difficult to machine (e.g., stainless steels) to improve tool life.

Thread milling can produce high tool pressures when milling at full thread length and depth, which can result in excessive tool deflections and tool breakage. Machine requirements also limit the applicability of thread milling; the proper speeds and feed rates must be available, and the machine must be capable of producing an accurate circular motion at high speeds and feeds, especially with nonferrous parts. Thread milling can also only be applied when the ratio of the thread length to the major diameter of the tool falls within relatively narrow limits.

When thread milling a stationary workpiece, the tool feed rate at the cutting edge (true feed rate occurring at the perimeter) f_{rt}, and rotational speed N of the cutter are related through

$$f_{rt} = n_t f_t N \tag{2.52}$$

When thread milling a workpiece rotating at a spindle speed of N_w on a lathe, the feed rate is given by

$$f_{rt} = \pi D_m N_w \tag{2.53}$$

and the feed per tooth is

$$f_t = \frac{\pi D_m N_w}{n_t N} \tag{2.54}$$

In an NC machine, the feed rate is programmed so that the X-, Y-, and Z-axes are synchronized to generate the helical motion of the cutter. The chip load, which is proportional to cutter advance per tooth as in milling, can be changed by varying the tool speed, workpiece speed, or the number of teeth in the tool. The apparent contact length (chip length) for each cutting edge per tool revolution is given by either Equation (2.26) or (2.28). The cross-sectional area of the chip is trapezoidal. The uncut chip thickness does not correspond exactly to the feed per tooth because the programmed cutter feed is about its center and not at the tool periphery where the cut occurs (as explained in Equation 2.38 for circular interpolation milling applications). The centerline feed rate (on the orbiting diameter) for internal and external threads is, respectively,

$$f_r = f_{rt} \frac{D_m - D}{D_m} \quad \text{(internal threads)}$$

$$f_r = f_{rt} \frac{D_\mu + D}{D_\mu} \quad \text{(external threads)} \tag{2.55}$$

where D is the threading cutter major diameter. The metal removal rate for a tool with n_{tr} rows of teeth in contact with the workpiece is

$$Q = \left[\frac{p}{4} + \frac{D_m - D_d}{\tan 60}\right] \left[\frac{D_m - D_d}{4}\right] n_{tr} f_{rt} \tag{2.56}$$

The time required to cut a thread of axial length L in one pass is

$$t_m = 1.1 \frac{\pi D_p}{f_{rt} \cos \lambda} \tag{2.57}$$

When a single row tooth cutter is used, the machining time is

$$t_m = \frac{\pi D_p L}{p f_{rt} \cos \lambda} \tag{2.58}$$

Die threading is used to generate external threads using solid or self-opened dies (chasers). Materials with hardness lower than 36 HR_c can be threaded with a die. It is a slower operation than thread rolling but faster than thread turning. Die threading machines differ little from tapping machines.

Thread whirling [34] is a process used for internal and external threads. Whirling removes material in a manner similar to thread milling. This process uses a special head supporting the threading tool as illustrated in Figure 2.26 and Figure 2.27. The whirling head is mounted in a machine spindle that rotates eccentrically at high speed around the slowly rotating workpiece, or performs a planetary rotation along a stationary workpiece. The cutter uses inserts arranged along the inside or outside circumference of a ring for external and internal threads,

FIGURE 2.26 (a) External and (b) internal thread whirling operations. (Courtesy of Leistritz Corporation.)

FIGURE 2.27 An external whirling ring assembly using eight indexable inserts.

respectively. The cutting occurs when the whirling unit and whirling ring are off-center with respect to the workpiece (using a helically interpolated cutting path as in thread milling) so that the cutting edge is passed from the root diameter. This process is faster than thread milling and may generate better surface finish, lower lead error, and shorter chips. It has been successfully used on hardened steels in place of turning or grinding to reduce the cycle time. The most common applications are the machining of screws or worms. It is a preferred process when cutting very long threads or threads with high helix angles.

Thread rolling is an external cold forming process for producing threads on a cylindrical or conical blank. It is similar to internal roll form tapping. The functional principle of thread rolling is shown in Figure 2.28. The tool or die displaces or extrudes the metal from the part surface to form the threads. Generally, thread rolling is carried out using specially

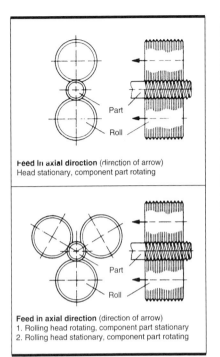
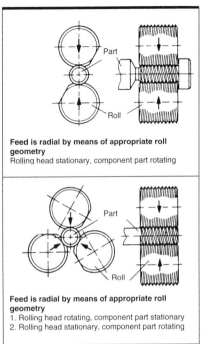
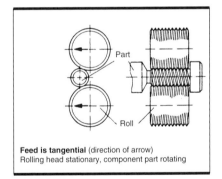

FIGURE 2.28 The functional principle of thread rolling operations. (Courtesy of Fette Tool Systems, Inc.)

designed vertical or horizontal machines. Thread rolling machines are designed with two flat die rollers in a side-by-side position. The threads on the blank, which is inserted between the rollers, are usually completed after eight revolutions. Other machines use two or three rollers, which form a radial-infeed cylindrical die. Thread rolling can be also carried out using lathes and automatic bar machines using special single- or double-roll attachments.

2.10 GRINDING AND RELATED ABRASIVE PROCESSES

Grinding is a general name for many similar processes in which a hard, abrasive surface is pressed against a work surface, resulting in the removal of material from both the workpiece and the abrasive. Grinding is a very common process traditionally used to produce parts with precise tolerances and fine surface finishes. It is also used in the general machining of hard or brittle materials and to produce complex contours. The major difference between traditional chip forming operations and grinding is the characteristics of the cutting tool. The individual grains on a grinding wheel are spaced randomly and have an irregular shape. The cutting speed of the wheel is very high while the feed is low compared to the traditional processes. Material removal in grinding takes place on a smaller scale than in conventional chip-forming processes; as a result, grinding requires different physical models for detailed analysis [2, 5, 8, 9, 36–48]. The chip thickness in grinding is very small (0.002 to 0.05 mm) compared to most other machining operations. Therefore, the power required for grinding is much higher than that of other machining operations. Another factor contributing to high energy is the random orientations of the individual grains in the wheel. Some grains are cutting while other grains are plowing or rubbing because they are either dull, generating a very high negative rake, or do not project into the workpiece and they rub the surface. The energy required to grind away a cubic centimeter of a given metal is typically 10 times that required to machine the same volume of material with a cutting tool given in Figure 2.1.

Grinding processes employ a variety of kinematic motions of the wheel and workpiece, some of which are shown in Figure 2.29. Processes can be broadly categorized as traverse or plunge grinding depending on the direction of feed of the wheel. In traverse grinding, the wheel moves parallel to the ground face, and only a portion of the workpiece surface is contacted at a time. The material is removed in several passes until the final depth of cut is achieved. Plunge grinding removes the entire depth of cut in a single pass. The width of the wheel is the limiting factor, and the wheel may have to traverse across or along the workpiece in order to grind the whole surface. Plunge grinding generates complex profiles very easily. Plunge grinding is faster than traverse grinding because the full wheel can be engaged; it is not usually possible to use traverse rates high enough to use the entire wheel. Grinding processes may be further classified as flat, cylindrical, or contour, depending on the shape of the ground surface, or as centerless, reciprocating surface, creep feed, microsizing, or honing depending on the kinematic motions of the workpiece and tool. The tool used may be a wheel, pad, disk, or belt made from a variety of bonded or loose abrasives.

Surface grinding is carried out either by reciprocating the workpiece or by rotating it about an axis perpendicular to the grinding surface while it is fed laterally in front of the rotating wheel. This operation produces flat, angular, and irregular shapes [43]. It uses either the periphery or face of the grinding wheel or a belt. This method competes with planing and milling operations, which generally require higher cutting forces.

Surface grinders may have either a horizontal or a vertical spindle with a table, which reciprocates and rotates [2, 4, 5].

Cylindrical grinding removes material from external cylindrical surfaces by rotating the workpiece and the wheel in opposite directions. The workpiece is supported between centers. The resulting surfaces can be straight, tapered, or contoured. It is used for hard or brittle workpieces or parts requiring fine surface finishes. The wheel or belt rotates at a much higher speed than the workpiece.

Cylindrical grinding machines resemble lathes since they are equipped with a headstock, tailstock, table, and wheel head. The workpiece is held either between centers or securely in a fixture mounted on the workhead spindle.

Centerless grinding is similar to cylindrical grinding, but the workpiece is supported by a blade between the grinding wheel and a small regulating (or feed) wheel [2, 4, 20, 49]. The regulating wheel holds the part against the grinding wheel and controls the cutting pressure and the rotation. The workpiece finds its own center as it rotates between the two wheels. Out-of-round material is pushed into the grinding wheel and ground away. The regulating wheel is usually made of a rubber-bonded abrasive and serves as both a frictional driving and

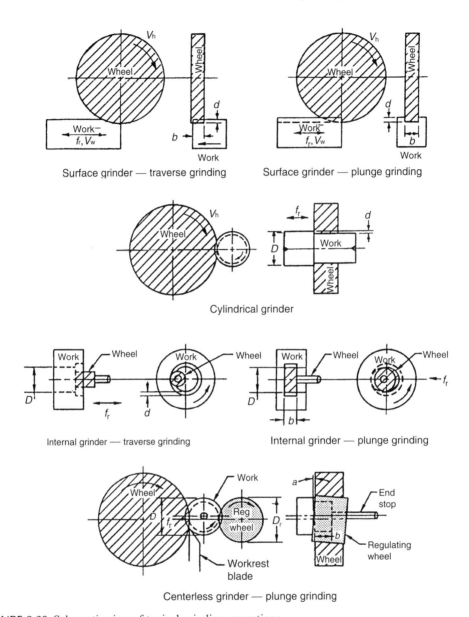

FIGURE 2.29 Schematic view of typical grinding operations.

braking element, rotating the workpiece at a constant and uniform surface speed. Centerless grinding is used primarily for external grinding.

Centerless grinding is used instead of cylindrical grinding when it is not possible to place centering holes in the part. Centerless grinding requires less grinding stock, can take heavier cuts, and is preferred for long, slender shafts. The centerless operation produces low cylindricity errors and consistent surface finish. It provides a better size control than cylindrical grinding. This process is necessary for parts which must rotate at high accuracy in a bearing or wear situation such as valves, camshafts, spindle shafts, etc.

Centerless grinders are similar to cylindrical grinders but without centers. The work is supported on shoes or a fixed blade under pressure applied by the regulating wheel.

Plunge grinding is a form of cylindrical grinding in which the wheel moves continuously into the workpiece rather than traversing.

Internal grinding generates internal cylindrical surfaces using very small wheels. It can be used to produce straight, tapered, blind, or through holes, holes with multiple diameters, contours, and flat sections. Stock removal rates are greater than for other abrasive hole-making processes such as honing or microsizing, which are used primarily to produce special surface finishes or close tolerances.

Creep feed grinding [50, 51] is a surface or external cylindrical grinding operation that removes a full depth of cut in a single pass at a very slow feed rate. Depths of cut are typically 2 to 6 mm, but may be as low as 0.5 mm for hard-to-machine materials. The speed of conventional wheels is generally 2000 m/min, while the feed rate is often only 25 to 400 mm/min. When properly applied, creep feed grinding can reduce the overall machining time by up to 50% with no loss of dimensional or geometric precision or surface quality. However, these results can be achieved only when the grinding machine is designed for creep feed applications to provide sufficient static and dynamic stability (two to three times that of conventional grinding machines), proper dressing capabilities, and adequate coolant control. It has been applied successfully to brittle materials such as ceramics and superalloys.

Input variables for grinding operations include the feed rate (down feed), wheel speed, workpiece speed, sparkout time and frequencies, and depth of wheel dressing. The wheel wear rate, the stock removal rate, and the wheel sharpness are also important parameters. The objective is to reach a steady state with a uniform wheel wear rate, steady power consumption, a constant pressure at the wheel–workpiece interface, and no apparent grinding burn and chatter. The workpiece surface is generated during the final contact between the wheel and workpiece; the grinding cycle starts with a heavy feed rate, which is reduced for the finish grind stage assuming that the same wheel is used. However, the finish wheel grade often differs from the rough wheel grade. The maximum usable stock removal rate usually remains approximately constant over a wide range of work speed and depth of cut combinations. It depends primarily on the grinding wheel characteristics, the work material, and the coolant type (oil or water). Oil coolants are often preferred because they reduce heat generation and residual stresses.

The equivalent diameter of the wheel-workpiece system is used to compare different grinding operations; this parameter provides the equivalent wheel diameter corresponding in a surface grinding operation and is defined as

$$D_e = \frac{D_h}{1 \pm (D_h/D_w)} \quad (2.59)$$

where D_h and D_w are the wheel and workpiece diameters and the + or − in the denominator is used for external (OD) or internal (ID) grinding, respectively. In the case of straight grinding, $D_e = D_h$ (or $D_w = \infty$). A large D_e usually results in higher contact area and higher threshold forces and power.

The apparent area of contact for the abrasives is calculated the same way as in milling processes. Therefore, the contact length for each abrasive grain per wheel revolution is given by Equations (2.26) and (2.27) with $D = D_e$. The chip thickness is dependent on the grinding method and can be calculated using relations similar to those for milling processes (Equations 2.21–2.25). However, the cross-sectional area of the chip is nonuniform and varies between abrasive grains. An extensive analysis of this subject is given in Ref. [44].

The volumetric rate of wheel wear Q_h is

$$Q_h = \pi D_h f_h b \quad (2.60)$$

where f_h is the rate at which the wheel diameter is decreasing. The stock removal rate from the workpiece Q_w for surface grinding is

$$Q_w = dV_w b \tag{2.61}$$

where V_w is the workpiece speed (equivalent to f_r in Figure 2.29). For cylindrical plunge grinding, Q_w is given by

$$Q_w = \pi D_w f_r b \tag{2.62}$$

where D_w is the diameter of the workpiece and f_r is the radial feed rate of the wheel into the workpiece. For cylindrical transverse grinding, Q_w is given by

$$Q_w = \pi D_w d f_r \tag{2.63}$$

where f_r is the feed rate of the wheel along the workpiece. The grinding power is proportional to the specific grinding energy (similar to the unit power discussed for turning):

$$P_m = u_s Q_w + P_t = F_t V_h \tag{2.64}$$

where F_t is the tangential force at the tool–workpiece interface and V_h is the wheel speed. Generally, the specific grinding energy, u_s, is higher than the unit energy for other processes. A performance index used to characterize wheel wear resistance in economic analyses is the *grinding ratio G* defined as

$$G = \frac{Q_w}{Q_h} \tag{2.65}$$

G varies from 0.018 to 60,000 [43, 52, 53]. Higher values of G indicate that the wheel is more productive. G generally increases with wheel speeds up to some level and then reduces again.

The processes described thus far use a hard abrasive wheel, and are intended to produce surfaces with fine finishes and close dimensional tolerances. Process conditions are selected to produce the required surface and dimensional properties as efficiently as possible without producing dynamic instability or surface damage or burn. Surface finish and integrity in grinding are discussed in Chapter 10. In troubleshooting grinding problems in practice, the grinding wheel is generally looked at first since it is relatively easy to change. Wheel characteristics of particular interest are grit size, hardness, and structure, which are described in Chapter 4. The manner and frequency of wheel dressing also has a significant impact on grinding performance and is also considered in troubleshooting. Once wheel issues have been resolved, the grinding fluid is examined for proper type, concentration, and delivery. Finally, the grinding infeed cycle is examined, especially when investigating dimensional or dynamic instability problems.

Microsizing (superabrasive bore finishing) is used to improve the accuracy of internal cylindrical surfaces. A preset, barrel-shaped tool fitted with bonded diamond or CBN abrasives, or an expandable tool with abrasive stones spaced around its periphery, is used. The tool, workpiece, or both may rotate (Figure 2.30) [54]. The tool is self-feeding and self-centering. The choice of coolant is very important, and honing type oils are generally used. Since the material is gradually removed by the thousands of superabrasive particles around the periphery of the tool, it generates very little heat or stress, enabling excellent control of bore geometry. Microsized holes have low roundness errors, limited taper, and no bell-mouth at the hole entrance.

Metal Cutting Operations

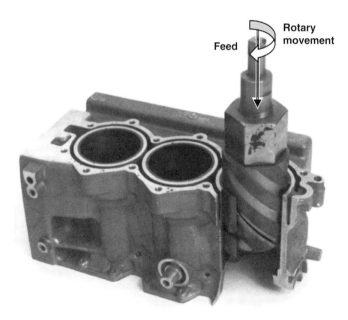

FIGURE 2.30 A microsizing tool used in engine bore manufacture. (Courtesy of Engis Co.)

Microsizing may be a single- or multipass operation. Multiple passes may be used when more than 0.05 mm of stock must be removed, when the part is weak structurally and would distort under heavy loading, or when a fine surface finish is required. The achievable surface finish and allowable depth of cut are functions of the abrasive grit size and coolant. To achieve 0.5 μm or better bore geometry in a single pass, three conditions must be met: (1) the tooling design should take into account variables such as the work material, hardness, bore size and length, bore shape (solid or interrupted), bore wall thickness, finish requirements, and incoming stock; (2) the bore must be aligned with respect to the tooling to minimize side forces; and (3) proper part fixturing must be used to prevent bore distortion. A floating holder can be used to improve tool alignment at conventional speeds but usually fails at higher speeds due to centrifugal forces. Therefore, the best results are usually obtained when a floating tool holder is used in conjunction with a radially and angularly floating type part fixture to allow for perfect alignment with low forces. Two- to four-pass microsizing processes are not uncommon in high volume production due to deviations from optimum setup conditions. Tool speeds range from 8 to 90 m/min and the feed rate can vary from 0.100 to 2.5 m/min, depending on the workpiece material. The maximum stock removal is a function of the workpiece material and the grit size of the abrasives and may be as high as 0.05 to 0.1 mm; in steel, it is typically 0.09 mm. Tolerances of better than 0.005 mm can be obtained with straightness errors less than 20 sec and roundness errors less than 0.001 mm. In many cases, microsizing has been found superior to honing, reaming, roller burnishing, and ID grinding.

Honing is a finishing operation similar to microsizing in which bonded abrasive sticks are pressed against the work surfaced using an expandable, reciprocating tool (Figure 2.31) [5, 55–61]. Honing corrects axial and radial distortions produced by previous operations (Figure 2.32) and imparts a controlled surface texture to the bore.

The expansion pressure is a critical parameter in honing since the wall temperature and bore thermal distortion increase linearly with pressure. The lowest expansion pressure and feed rate possible should be used to increase accuracy. Temperature is also proportional to

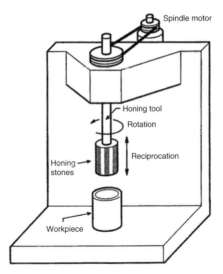

FIGURE 2.31 Schematic diagram of a honing machining system.

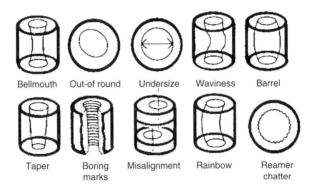

FIGURE 2.32 Bore form and size errors corrected by microsizing or honing.

honing speed, which must be as high as possible to maintain stone sharpness. The axial speed usually ranges from 12 to 25 m/min, while the circumferential speed varies between 15 and 100 m/min. Generally, a rough honing operation is required to create a geometrically correct bore, and a finish pass is used to generate the proper surface topography. Controlled temperature flood coolants are sometimes applied to the bore and honing tool to control thermal errors. The metal removal rate, Q_w, and depth of cut, d, in honing are related by

$$d \cong \frac{Q_w}{n_t \lambda_h V_h} \tag{2.66}$$

where n_t is the number of honing sticks, λ_h is the stick length, and V_h is the peripheral speed. Q_w depends on the expansion pressure and the abrasive grain size and sharpness. The combined rotary and reciprocating motion of the honing tool produces a cross-hatched surface pattern on the bore as shown in Figure 2.33; the cross-hatch angle κ is given by

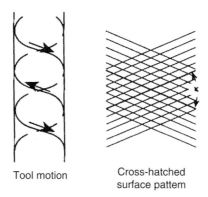

Tool motion   Cross-hatched surface pattern

FIGURE 2.33 The cross-hatched surface pattern generated by the tool motion in honing process.

$$\tan\frac{\kappa}{2} = 0.6366\frac{\lambda_s N_s}{DN} \qquad (2.67)$$

where λ_s is the stroke length and N_s is the stroke rate in strokes per minute.

Abrasive *brush honing* or flexible honing is a recently developed process using the same principle as the internal honing operation. Brush honing is carried out using flexible abrasive brush tools, which are equipped with compliant filaments which bend and press against the bore surface [62]. This process removes very little material from the surface using high production cycle times (less than 1 min); it removes the surface peaks and some of the smeared material left from previous honing processes. Unlike microsizing or honing operations, it cannot generally remove the tool feed marks left by prior turning or boring operations, and cannot produce a full plateau finish and cross-hatched pattern. It also differs from conventional honing in that the resilience of the abrasive brush results in a flexible cutting action, which does not alter the bore geometry with respect to concentricity, ovality, cylindricity, and squareness. Brush honing can be carried out with or without a coolant. Brush honing does not offer a significant economic advantage over microsizing and honing in high volume production.

External honing, which is also referred to as microfinishing, superfinishing, or superhoning, is a high precision, fine-grit abrasive process for removing imperfections on external surfaces which may result from conventional grinding. The tool is an abrasive stone that oscillates at a high rate (up to 2800 Hz) and is pressed against a rotating workpiece. A tool arm is used to position the stone and apply pressure. The stone oscillates in the direction parallel to the workpiece axis of rotation. There are four common microfinishing methods: centerless through-feed, plunge feed, plunge feed pivoting, and rotating tool/rotating workpiece. Parameters that affect part quality in microfinishing include the machine structure, work- and stone-holding methods, tool material, cutting fluid, and environmental control; requirements for these parameters are generally more stringent than for other abrasive processes [63]. External honing can be used in combination with turning on lathes by attaching a superfinishing arm behind the turning tool holder.

Stock removal is a function of stone cutting capability, pressure, oscillation amplitude and frequency, and rotational speed of workpiece. Use of a low peripheral work speed, moderate stone pressures, and wide workpiece contact lengths limit heat generation and produce fine surface finishes; the roughness may be as low as 0.025 μm R_a, compared to the range 0.1 to 0.15 μm for conventional grinding. Smaller tolerances and surfaces with a 100% bearing ratio can also be produced since this process removes the amorphous surface layer produced by turning or grinding operations. This process has achieved roundness errors of 0.12 μm. It has been applied to a variety of work materials, including high-strength steels, carbides, and ceramics.

In *belt grinding* (or *abrasive belt machining*), coated abrasive belts are used to remove material at a high rate. Due to its flexibility and the availability of stronger belts, abrasive belt machining has become increasingly popular, even replacing conventional turning, milling, grinding, and polishing operations [64, 65]. Belt grinding can be used in place of centerless grinding, surface grinding, and cylindrical grinding in roughing to fine finishing operations by replacing the wheel with a coated abrasive belt. Grinding can be performed using a slack belt, or with the belt riding over a contact wheel or platens which support the belt and apply a thrust force. The contact wheel can be hard or soft, and either smooth or serrated. Smooth hard contact wheels are used for precision abrasive machining and for maximum stock removal.

Belt grinding is more efficient than conventional grinding. The belt speed is usually between 800 and 2000 m/min, but speeds of up to 4000 m/min have been used to grind ceramics. The optimum speed is affected by the contact wheel or pressure, which supports the belt at the pressure point, regulates the cutting rate, and controls the grain breakdown as well as the grit size and work thickness. The feed is controlled either indirectly by adjusting the pressure or directly by actuating the contact wheel. Belt grinding is generally safer than standard grinding with brittle wheels. Adequate coolant flow and pressure must be used to reduce grinding temperatures and sparking.

Cycle times for belt grinding are comparatively low. It is particularly well suited for grinding complex profiles such as the oval-shaped lobes on cam shafts [64]. Belt grinders generally produce flatness values of 0.025 to 0.05 mm and surface finishes from 0.25 to 2.5 μm R_a with burr-free edges.

Disk grinding is a face grinding method in which all or most of the flat face of the wheel contacts the workpiece. Disk grinding offers advantages over other grinding methods in many applications because the wide face contact area generates surfaces with low flatness errors and fine finishes. Both single- and double-disk grinders (using single or double spindle types) are available. Face grinding is attractive on bimetallic surfaces where tool life in other operations may be limited.

Double-disk grinding is carried out using two opposing disks (wheelheads) between which a part is passed. Two opposite sides of the workpiece are ground simultaneously by opposing abrasive disks, producing flat, parallel surfaces. The process can be used for both thin and thick workpieces, and only lateral support of the work is required. The stock removal per side is usually between 0.25 and 0.70 mm for roughing, and between 0.1 and 0.25 mm for finishing. In single pass operations, a maximum of 0.5 mm stock per side can be removed. Double-disk grinding can produce surfaces with flatness errors of 0.0015 mm, parallelism errors of opposing surfaces of 0.0025 mm/25 mm, and squareness errors on the order of 0.0075 mm/25 mm.

2.11 ROLLER BURNISHING

Roller burnishing is a surface finishing operation in which hard, smooth rollers or balls are pressed against the work surface to generate the finished surface through plastic deformation [66–68]. Roller burnishing is used in combination with or as a replacement for lapping, honing, microsizing, grinding, or in some cases turning, milling, reaming, and boring. Burnishing is used to improve surface finish, control tolerance, increase surface hardness, and improve fatigue life.

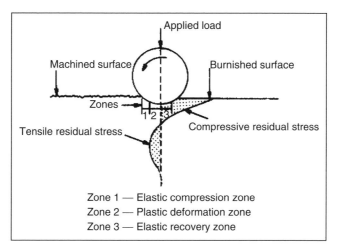

FIGURE 2.34 Schematic representation of the roller burnishing process. (Courtesy of Cogsdill Tool Products, Inc.)

As shown in Figure 2.34, the burnishing tool applies a pressure greater than the yield stress of the workpiece material. The rotating rollers or balls displace the surface asperities on the part and generate a plateau-like surface. Plastic deformation is induced in the surface layer of the workpiece, which work hardens from 5 to 20%. Burnishing produces a compressive residual surface stress, which often improves the wear resistance and surface fatigue life of the part. Surface finishes between 0.05 and 0.5 μm are generated in single or multipass operations; the surface bearing area can approach 100% when the preburnished surface has uniform asperities. Part dimensional accuracies can often be controlled within ±0.005 mm with proper part preparation. However, geometric errors produced in previous operations generally cannot be corrected by burnishing. Due to its dimensional consistency, burnishing is especially well suited for parts designed for press-fit assembly and face sealing.

Roller burnishing performance depends on the characteristics of the incoming surface. Turned, bored, and milled surfaces with roughnesses between 0.002 and 0.005 mm are suitable for burnishing because they have uniform asperities. Rough ground surfaces are also excellent for burnishing. Reamed surfaces are often not suitable due to surface and size inconsistencies. Special reaming tools with stationary carbide pads behind the cutting edge have been designed to ream and burnish a hole in a single operation. The surfaces generated by such reaming tools are not equivalent to those generated by roller burnishing process because the carbide pads apply a much lower surface pressure.

Roller or ball burnishing can be performed on flat surfaces, cylindrical ID and OD surfaces, tapered internal and external surfaces with taper angles up to 15°, spherical, radius, fillet surfaces, or a combination of these surfaces. The size reduction (of external dimensions) or increase (of internal dimensions) depends on the pre- and postburnished surface finish. The depth d to which the tool penetrates a surface is approximately

$$d = C_1(R_{a1} - R_{a2}) \qquad (2.68)$$

where R_{a1} and R_{a2} are the surface roughnesses of the preburnished and burnished surfaces, respectively. C_1 is a constant equal to 2 theoretically; in practice, however, C_1 is between 2 and 4. The burnished hole size tolerance is half of that for the preburnished hole. Many parameters, including the burnishing force, feed rate, roller or ball material and diameter, and lubrication, affect the final finish [69, 70]. The most important parameters are usually the burnishing force and the feed rate.

A knurling operation is occasionally required prior to roller burnishing to achieve a close tolerance on oversized bores or undersized shafts [71]. In these applications knurling raises the initial surface profile, providing enough material for the burnishing tool to roll down to the desired dimension.

2.12 DEBURRING

Burrs are undesired projections of material beyond the edge(s) of the workpiece due to plastic deformation during machining [72–78]. As discussed in Section 11.4 burr formation is a complex process that depends on the workpiece material, tool geometry, part design, degree of tool wear, and manufacturing process sequence. Burrs of excessive size often must be removed for the part to be handled safely, assembled, or function effectively. Deburring is generally a time-consuming and costly operation, and in many cases is difficult to automate.

Several deburring methods are widely used. These include barrel tumbling, centrifugal barrel tumbling, vibratory deburring, water-jet or abrasive-jet deburring, sanding, brushing, abrasive-flow deburring, ice blasting, liquid-honing, chemical deburring, ultrasonic deburring, electropolishing, electrochemical deburring, thermal-energy deburring, mechanical deburring, and manual deburring. Detailed descriptions of these operations are available in the literature [72–79]. The most widely used methods are manual deburring, brushing, and ultrasonic deburring.

Hole entrance and exit deburring is a very common process. Special deburring and chamfering tools have been developed to remove unwanted material created by drills or reamers. These tools generate a predetermined chamfer size at the entrance and exit of hole which remove burrs, so that a subsequent deburring operation (e.g., brushing) is not necessary. Robotic deburring strategies have also been successfully applied in holemaking [80, 81].

2.13 EXAMPLES

Example 2.1 A 50 mm diameter gray cast iron workpiece is rough turned with an indexable (throw-away) uncoated carbide insert tool. The feed for the tool is 0.4 mm per revolution (mm/rev), the depth of cut is 4 mm, and the recommended cutting speed is 70 m/min. Calculate: (a) the material removal rate, (b) power and torque required by the spindle, and (c) the main cutting force.

Solution:

(a) The material removal rate (MRR or volumetric rate of machining) is obtained from Equation (2.6)

$$\text{MRR} = Q = Vfd = (70 \text{ m/min})(0.4 \text{ mm/rev})(4 \text{ mm})(1000 \text{ mm/m})$$
$$= 112{,}000 \text{ mm}^3/\text{min} = 112 \text{ cm}^3/\text{min}$$

The cutting speed occurs at the average depth of cut diameter, which is $D_{\text{avg}} = 50 - 4 = 46$ mm.

(b) The specific cutting energy (unit power) is obtained from Table 2.1 as 0.065 kW/cm³/min (the average value). Hence, the machining power required in this turning operation can be estimated from Equation (2.7) to be

$$P = Qu_s = (112 \text{ cm}^3/\text{min})(0.065 \text{ kW/cm}^3/\text{min}) = 7.28 \text{ kW}$$

Metal Cutting Operations

The spindle torque can then be obtained from the knowledge of cutting power

$$P = T\omega = T(2\pi N) \Rightarrow T = P/(2\pi N) \text{ (W = J/sec = N m/sec)}$$

(c) The rpm (N) is calculated from Equation (2.1). This equation yields

$$N = \frac{2V}{\pi(D_1 + D_2)}$$

where $D_1 = 50$ mm, $d =$ depth of cut $= 4$ mm, and $D_2 = D_1 - 2d = 50 - 2 \times 4 = 42$ mm, Hence,

$$N = \frac{2(70 \text{ m/min})(1000 \text{ mm/m})}{\pi(50 + 42 \text{ mm})} = 485 \text{ rpm}$$

and

$$T = (7280 \text{ N m/sec})(60 \text{ sec/min})/(2\pi)(485 \text{ rev/min}) = 143 \text{ N m}$$

Example 2.2 Calculate the machining time in Example 2.1 if the length of the OD turning is 150 mm.

Solution: The distance that the tool travels along the workpiece is the length of cut plus the approach allowance to avoid the tool edge impact with the workpiece during the rapid travel. Let us assume the approach distance L_a is 3 mm. The machining time is equal to

$$t_m = (L + L_a)/f_r = (L + L_a)/(fN) = (150 + 3)/(0.4 \times 485) = 0.768 \text{ min} = 47.5 \text{ sec}$$

Example 2.3 For the turning operation in Example 2.1, evaluate the uncut chip thickness and the cutting edge engagement if the lead angle (SCEA) of the tool is 20°.

Solution: The uncut chip thickness is given by Equation (2.3)

$$a = f \cos \kappa = (0.4 \text{ mm})(\cos 20°) = 0.376 \text{ mm}$$

The cutting edge engagement is the product of the secant of the lead angle and the depth of cut (given by Equation 4.1) which is equal to

$$L_m = (\text{depth of cut}) \sec(\kappa) = d \sec(\kappa) = (4 \text{ mm}) \sec(20) = 4.257 \text{ mm}$$

Example 2.4 In a milling application, a four-fluted 75 mm diameter indexable end mill is used to clean a 200 mm wide sidewall into a 1035 steel workpiece as illustrated in Figure 2.35. Carbide coated inserts are used. The axial depth of cut is 80 mm and the radial depth of cut is 5 mm. The cutting velocity is 100 m/min with the suggested feed per tooth (or uncut chip thickness) is 0.2 mm/rev for the particular insert style and material. The cutting tool length extended out of the toolholder is 200 mm. Determine the taper on the sidewall of the workpiece generated due to tool deflection if the radial component of the force is estimated to be 30% of the cutting (tangential) force.

Solution: The wall of the part is cut by the periphery of the end mill that is similar to peripheral milling. The finish surface is perpendicular to the direction of feed. The end mill can be considered as an elastic cylindrical beam, cantilevered to the spindle by neglecting the

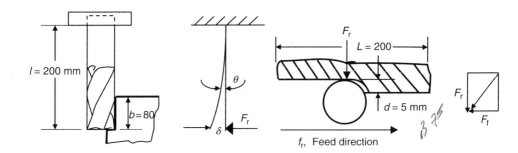

FIGURE 2.35 Schematic representation of Example 2.4.

compliance of the toolholder and spindle (see Chapter 5 for complete analysis of a toolholder–spindle interface). The static deflection due to the radial force (F_r) at the free end of the end mill is given by

$$\delta = \frac{F_r L^3}{3EI} \quad \text{where } I = \frac{\pi D_e^4}{64}$$

The effective diameter of the cutter is somewhat smaller than the outer diameter because the flute space is taken into consideration. The effective diameter has been found to be about 0.75 to 0.85 times the outer diameter. The radial cutting force is estimated to be 30% of the cutting (tangential) force. In addition, there is only one cutting edge in contact with the workpiece at any time. The radial force (F_r) is not constant because the chip thickness varies but let us assume that the maximum is $F_r = 0.3F$. The cutting force is estimated from the average cutting power.

From the information given, we can calculate the spindle rotational speed or the rpm of the milling cutter from Equation (2.10). Thus,

$$N = V/(\pi D) = (100 \text{ m/min})(1000 \text{ mm/m})/(3.14)(75 \text{ mm}) = 424 \text{ rpm}$$

The maximum uncut chip thickness (provided by the manufacturer of the cutter for roughing operations) is 0.2 mm. In peripheral and end milling application with the cutter diameter significantly larger than the radial depth of cut (d), the feed rate should be adjusted to avoid the chip thinning effect. Equation (2.22) gives the maximum uncut chip thickness with $\kappa = 0$ in this case (that is true generally in end milling) and ν_m is given by Equation (2.24)

$$\cos \nu_m = 1 - \frac{(2)(5)}{75} \quad \nu_m = 30°$$

$$a_{max} = f_t \cos \kappa \sin \nu_m$$

Hence,

$$f_t = a_{max}/\sin \nu_m = 0.2/\sin 30° = 0.4 \text{ mm}$$

The feed rate for the cutter is given by Equation (2.20)

$$f_r = n_t f_t N = (4)(0.4 \text{ mm/rev/tooth})(424 \text{ rev/min}) = 679 \text{ mm/min}$$

Metal Cutting Operations

The material removal rate, MRR, is obtained from Equation (2.29) to be

$$\text{MRR} = (\text{cross-sectional area of uncut chip})(\text{feed rate}) = dbf_r$$
$$= (5 \text{ mm})(80 \text{ mm})(679 \text{ mm/min}) = 271{,}600 \text{ mm}^3/\text{min} = 272 \text{ cm}^3/\text{min}$$

The unit specific energy for the steel material is estimated from Table 2.1 to be 0.08 kW/cm³/min. Hence, the power required for this operation is estimated from Equation (2.7) to be

$$P = \text{MRR} u_s = (271{,}600 \text{ mm}^3/\text{min})(0.08 \text{ W/mm}^3/\text{min}) = 21.7 \text{ kW}$$

The cutting force is estimated from the power

$$P = FV \Rightarrow F = P/V = (21{,}700 \text{ N m/sec})(60 \text{ sec/min})/(100 \text{ m/min}) \Rightarrow F = 13{,}020 \text{ N}$$

Hence, $F_r = 0.3F = 0.3 \cdot 13{,}020 = 3906$ N

The deflection is then

$$\delta = \frac{F_r L^3}{3EI} = \frac{(3{,}906 \text{ N})(200^3 \text{ mm}^3)}{3(400{,}000 \text{ N/mm}^2)(1{,}241{,}895 \text{ mm}^4)} = 0.021 \text{ mm}$$

$$I = \frac{\pi D_0^4}{64} = \frac{\pi (0.8 \times 75)^4}{64} = 1{,}241{,}894.5 \text{ mm}^4$$

The slope of deflection is 0.000105, which is the taper of the sidewall. The perpendicularity error of the wall due to cutter deflection only is 0.01 mm/100 mm.

Example 2.5 An end mill is used to clean up the surface of a 4140-forged steel part by performing face milling operation. An indexable 50 mm diameter cutter (shown in Figure 2.36) with three round inserts is selected. The IC size of the round inserts is 17 mm. The depth of cut is 3 mm. The suggested cutting conditions for this cutter are 150 m/min cutting speed and 0.2 mm/tooth feed. The specific energy for the workpiece material is 0.08 W min/mm³. Determine the metal removal rate.

Solution: The effective tool diameter (D_e) is not the actual cutter diameter because the axial depth of cut is smaller than the radius of the inserts

$$D_e = D - \text{IC} + \sqrt{\text{IC}^2 - (\text{IC} - 2d)^2} = 45.96 \text{ mm} = 46 \text{ mm}$$

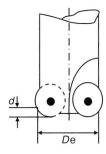

FIGURE 2.36 Schematic representation of Example 2.5.

Since the depth of cut is significantly smaller than the radius of the inserts, the chip thinning effect takes place.

Therefore, from Equations (2.22) and (2.24)

$$\cos \nu_m = 1 - \frac{(2)(3)}{17} \Rightarrow \nu_m = 50°$$

$$f_t = a_{max}/\sin \nu_m = 0.2/\sin 50° = 0.26 \text{ mm}$$

The feed rate for the cutter is obtained from Equation (2.20).

At smaller axial depth of cut, the spindle speed should be calculated based on the effective diameter in cut

$$f_r = n_t f_t N = (3)(0.26 \text{ mm/rev/tooth})(1038 \text{ rev/min}) = 747 \text{ mm/min}$$

The material removal rate, MRR, is obtained from Equation (2.29) to be MRR $= dbf_r =$ (3 mm) (46 mm) (747 mm/min) $= 103,136$ mm³/min.

Example 2.6 An indexable 25 mm two-flute ballnose end mill is used to clean up the surface of a 4140-forged steel part. The depth of cut is 6 mm. The suggested cutting conditions for this cutter are 150 m/min cutting speed and 0.2 mm/tooth feed. The specific energy for the workpiece material is 0.08 W min/mm³. Determine the metal removal rate.

Solution: The effective tool diameter (D_e) is not the actual cutter diameter because the axial depth of cut is smaller than the radius of the ballnose

$$D_e = 2\sqrt{d(D-d)} = 2\sqrt{6(25-6)} = 21.4 \text{ mm}$$

Since the depth of cut is significantly smaller than the radius of the nose, the chip thinning effect takes place. The thickness of the chip varies depending on the doc using Equations (2.21)–(2.24)

$$\cos \nu_m = 1 - \frac{(2)(6)}{25} \Rightarrow \nu_m = 58.7°$$

$$f_t = a_{max}/\sin \nu_m = 0.2/\sin 59° = 0.23 \text{ mm}$$

The feed rate for the cutter is given by Equation (2.20) and considering the effective diameter to calculate the spindle speed (rpm)

$$f_r = n_t f_t N = (2)(0.23 \text{ mm/rev/tooth})(2232 \text{ rev/min}) = 1027 \text{ mm/min}$$

The material removal rate, MRR, is obtained from Equation (2.29) with a radial depth of cut of 11 mm, which is about half the effective diameter of the cut

$$\text{MRR} = dbf_r = (6 \text{ mm})(11 \text{ mm})(2232 \text{ mm/min}) = 147,312 \text{ mm}^3/\text{min}$$

Example 2.7 A face milling operation is being carried out on a 150 mm wide by 700 mm long rectangular part of aluminum material as illustrated in Figure 2.37. A 200 mm diameter face milling cutter with 18 inserts is used to cut a 3 mm depth from the surface of the part. The lead

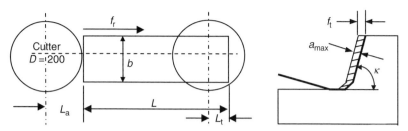

FIGURE 2.37 Schematic representation of Example 2.7.

angle of the cutter is 45°. The cutting speed is 300 m/min and the allowable uncut chip thickness is 0.3 mm. Calculate the machining time.

Solution: The cutter diameter is selected to be larger by 20 to 40% the width of the part and runs off center from the centerline of the part, as a rule discussed in Chapter 4. From the information given, we can calculate the spindle rotational speed or the rpm of the milling cutter from Equation (2.10). Thus,

$$N = V/(\pi D) = (300 \text{ m/min})(1000 \text{ mm/m})/(3.14)(200 \text{ mm}) = 478 \text{ rpm}$$

The feed per tooth is affected by the lead angle of the cutter as illustrated in the above figure and explained in Equation (2.21), where $\kappa = 45°$

$$f_t = a_i/\cos \kappa = 0.3/\cos 45° = 0.42 \text{ mm}$$

The feed rate for the cutter is given by Equation (2.20)

$$f_r = n_t f_t N = (18)(0.42 \text{ mm/rev/tooth})(478 \text{ rev/min}) = 3614 \text{ mm/min}$$

The cutting time is given by Equation (2.30). The approach and overtravel distances can be calculated from simple geometric relationships. The approach is a little larger than the radius of the cutter (see Equation 2.32) to avoid contact with the workpiece during the rapid travel motion. The overtravel distance varies depending on the surface requirements; there is a minimum negative distance (occurring when the cutter stops within the part) and a maximum distance if the surface texture is important. The selection depends on whether the trailing marks of the cutter (marks caused by a portion of the cutter's inserts that are not in the cut rubbing on the surface already milled) should show up across the full length of the surface or not. These marks are avoided in some specially designed machines for milling with their spindle tilted slightly so that the cutter is slightly off parallel from the part surface.

$$L_e|_{min} = L_a - L_t = \frac{D}{2} - \frac{1}{2}\left(\sqrt{D^2 - b^2}\right) = \frac{200}{2} - \frac{1}{2}\left(\sqrt{200^2 - 150^2}\right) = 34 \text{ mm}$$

$$L_e|_{max} = L_a + L_t = \frac{D}{2} + \frac{D}{2} = \frac{200}{2} + \frac{200}{2} = 200 \text{ mm}$$

The cutting time is calculated from Equation (2.16)

$$t_m = (L + L_e)/f_r = (700 + 34 \text{ mm})/(3614 \text{ mm/min}) = 0.20 \text{ min} = 12 \text{ sec}$$

Example 2.8 A 10 mm wide groove in a cast iron part is machined using a slotting milling operation as shown in Figure 2.38. The groove is made across the full length of the part that is 400 mm. The depth of the slot is 15 mm. A 200 mm diameter carbide brazed side milling cutter with 16 straight teeth is selected. Each insert cuts the full width of the slot. The cutting speed is selected to be 80 m/min and the maximum uncut chip thickness is 0.2 mm for carbide tool material. Calculate:

(a) The maximum feed rate (mm/min) possible if the machine spindle maximum available cutting power is 10 kW.
(b) The machining time to groove the part.
(c) What if an indexable staggertoothed slotting cutter is used with 16 inserts.

Solution:

(a) The maximum feed rate can be obtained from the maximum allowable material removal rate calculated based on the available power. Thus,

$$P = (MRR)u_s \Rightarrow MRR = P/u_s = (10 \text{ kW})/(0.07 \text{ W min/mm}^3) = 142.9 \text{ cm}^3/\text{min}$$

The specific energy for the workpiece material is estimated from Table 2.1 to be 0.07 W min/mm^3. The maximum MRR is given by Equation (2.29). In slotting, the radial depth of cut (d) is measured along the cutting diameter (the cutter radial portion engaged in the workpiece), whereas the axial width of cut (b) is the same as the width of the slot. Thus,

$$MRR = dbf_r \Rightarrow f_r = (MRR)/(db) = (142,900 \text{ mm}^3/\text{min})/(15 \text{ mm})(10 \text{ mm}) = 952 \text{ mm/min}$$

$$N = V/(\pi D) = (80 \text{ m/min})(1000 \text{ mm/m})/(3.14)(200 \text{ mm}) = 128 \text{ rpm}$$

The feed rate for the cutter is given by Equation (2.20)

$$f_t = f_r/(n_t N) = 952/(16)(128 \text{ rev/min}) = 0.465 \text{ mm/rev/tooth}$$

However, the maximum uncut chip thickness is 0.2 mm and based on Equations (2.22) and (2.24), the maximum feed per tooth allowed for this particular cutter is

$$\cos \nu_{max} = 1 - [(2d)/D] = (R-d)/R = (100-15)/100 \Rightarrow \nu_{max} = 32°$$

$$a_{max} = f_t \sin \nu_m, \quad \text{hence,} \quad f_t = a_{max}/\sin \nu_m = 0.2/\sin 32° = 0.377 \text{ mm/rev/tooth}$$

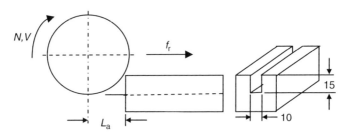

FIGURE 2.38 Schematic representation of Example 2.8.

Metal Cutting Operations

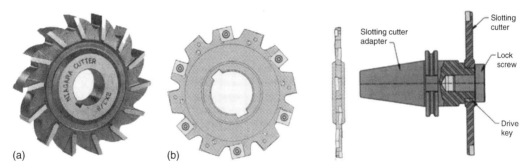

FIGURE 2.39 Comparison of slot milling cutters: (a) using an edge across the full width of the slot and (b) a staggertoothed cutter. (Courtesy of Kennametal and Carboloy.)

Therefore, the feed per tooth of 0.38 mm/rev/tooth is selected. This feed requires 8.2 kW or 82% of the available power.

(b) The cutting time is given by Equation (2.30). The approach distance L_a is given by Equation 2.31, while the overtravel distance is zero (or very small, 1 to 2 mm).

$$t_m = (L + L_e)/f_r = (400 + 52.7 + 10 \text{ mm})/[(16)(0.38)(128 \text{ mm/min})] = 0.595 \text{ min}$$
$$= 35.7 \text{ sec}$$

The total of 10 mm for the approach and overtravel distance is added to avoid the tool hitting the workpiece during the rapid travel.

(c) Figure 2.39 shows a comparison of the straight tooth carbide brazed cutter (a) with the staggertoothed cutter (b). If the cutter has 16 inserts in total, the effective number of teeth is eight because eight inserts on one side of the cutter and eight overlaping inserts on the other side are used to remove the full width of 10 mm. In this case, the feed rate is half of that for the carbide brazed cutter with 16 effective teeth. Hence, $t_m = (462.7 \text{ mm})/[(8)(0.38)(128 \text{ mm/min})] = 1.19 \text{ min} = 71.3 \text{ sec}$.

Example 2.9 The end face of a tube is face turned (a tube flange operation) on a lathe, using a depth of cut of 8 mm, and the inside and outside diameters of the flange are 100 and 200 mm, respectively. The tube is held with a hydraulic chuck at the other end as illustrated in Figure 2.40. The length of the tube is extended out 500 mm from the lathe chuck. The tube is made from aluminum material. The maximum rotational speed for this application is 400 rpm, and the maximum feed for the particular tool geometry is 0.25 mm/rev. The side

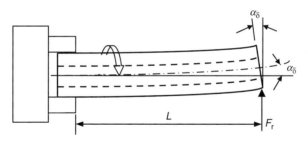

FIGURE 2.40 Schematic representation of Example 2.9.

cutting edge angle (lead angle) of the tool is 30°. Determine: (a) the maximum power, (b) the cutting time, and (c) the perpendicularity of the flange (face) to the part axis.

Solution:

(a) The power required to perform a face (turning) operation is given by Equation (2.7). The MRR is calculated from Equation (2.6) with $f_t = 0.25$ mm/rev and depth of cut = 8 mm.

The cutting speed varies as a function of diameter and the maximum occurs at the larger workpiece diameter or at the beginning of the cut which is

$$\text{MRR} = Vfd = \pi DNf_t d$$
$$= \pi(200 \text{ mm})(400 \text{ rpm})(0.25 \text{ mm/rev})(8 \text{ mm})$$
$$= 502,400 \text{ mm}^3/\text{min} = 502 \text{ cm}^3/\text{min}$$

The maximum power (P) required is

$$P = (\text{MRR})u_s = (502,400 \text{ mm}^3/\text{min})(0.025 \text{ W min/mm}^3) = 12.6 \text{ kW}$$

(b) The cutting time is given by Equation (2.5) with the length of cut equal to the thickness of the tube. Thus,

$$t_m = \frac{L}{f_r} = \frac{(D_o - D_i)}{2fN} = \frac{(200 - 100 \text{ mm})}{2(0.25 \text{ mm})(400 \text{ rpm})} = 0.5 \text{ min} = 30 \text{ sec}$$

(c) The perpendicularity error is equivalent to the deflection of the tube under the cutting load (or the radial force). Generally, the radial force is less than 50% than the cutting force in roughing turning operations, while they could be equal in finishing operations. Therefore, let us assume the worse case scenario where the radial force is equal to the cutting (tangential) force. The cutting force is estimated from the power

$$P = T\omega = T(2\pi)N \Rightarrow T = P/(2\pi N) \quad \text{and} \quad F_c = 2T/D \text{ or } F_c = P/(\pi ND)$$
$$F_c = (12,600 \text{ N m/sec})/[\pi(400/60 \text{ rev/sec})(200/1000 \text{ m})]$$
$$F_c = 3000 \text{ N}$$

Thus,

$$F_r \leq F_c \quad \text{or} \quad F_r \leq 3000 \text{ N}$$

The deflection at the tube at the end point (rigidly mounted to the machine spindle at the other end) is given by Equation (4.1)

$$\delta = \frac{F_r L^3}{3EI} = \frac{(3000 \text{ N})(500^3 \text{ mm}^3)}{3(50,000 \text{ N/mm}^2)(73.594 \times 10^6 \text{ mm}^4)} = 0.034 \text{ mm}$$

$$I = \frac{\pi(D_o^4 - D_i^4)}{64} = \frac{\pi(200^4 - 100^4)}{64} = 73.594 \times 10^6 \text{ mm}^4$$

The face of the tube has a taper (perpendicularity error in reference to the axis of the tube) if it deflects during cutting. The taper of the face angle is calculated from the deflection of the tube at the end of the face:

$$\tan a_\delta = \frac{x}{D_o} \Rightarrow x = D_o \tan a_\delta = D_o \left(\frac{\delta}{L}\right) = 200 \left(\frac{0.034}{1000}\right) = 0.007 \text{ mm}$$

Therefore, the taper or perpendicularity error is 0.007 mm.

Example 2.10 A slot 12 mm wide by 3 mm deep by 80 mm long in a cast iron workpiece material is rough broached. The broach has a step-per-tooth of 0.1 mm with 12.5 mm pitch. The cutting speed is 25 m/min. Determine: (a) the number of teeth on the broach, (b) the length of the broach, (c) the machining (cutting) time, and (d) the cutting force.

Solution:

(a) Figure 2.18 shows the characteristics of a broach tool. The feed per tooth is equivalent to the step-per-tooth in a broaching operation because it determines the chip thickness. If $f_t = 0.1$ mm and $d = 3$ mm, then the number of teeth are

$$n_t = d/f_t = 3/0.1 = 30 \text{ teeth}$$

(b) The length of the broach is $L_B = p n_t = (12.5)(30) = 375$ mm.
(c) The machining time is $t_m = (L_B + L_e)/V = (375 + 80)/(25 \times 1000) = 0.018$ min.
(d) The workpiece is 80 mm long and the cutter has a pitch of 12.5 mm. Hence, the maximum number of teeth engaging with the workpiece are

$$n_{te} = \text{int}[L/p] = \text{int}[80/12.5] = 7 \text{ teeth}$$

The maximum depth engage of the broaching tool is $d_e = n_{te} f_t = (7)(0.1) = 0.7$ mm. The MRR is given by Equation (2.29),

$$\text{MRR} = d_e w V = (0.7)(12)(25)(1000 \text{ mm/m}) = 210{,}000 \text{ mm}^3/\text{min}$$

The force can be obtained from the power, $P = F_c V$. Hence,

$$F_c = P/V = (\text{MRR} u_s)/V$$
$$= (210{,}000 \text{ W})(0.07 \text{ W min/mm}^3)(60 \text{ sec/min})/(25 \text{ m/min}) = 35{,}200 \text{ N}$$

Example 2.11 A 100 mm diameter hole is threaded for the first 30 mm depth using a 25 mm diameter solid carbide four-flute thread milling tool with three rows of teeth. The thread pitch is 3 mm. The allowable feed per tooth (chip load) in this case is 0.1 mm/rev., while the cutting speed is 100 m/min. Estimate the machining time.

Solution: Thread milling operation requires helical interpolation, which combines circular movement (interpolation shown in Figure 2.16) in one plane with a simultaneous linear motion perpendicular to it as illustrated in the Figure 2.41. Every tool orbit (on diameter D_x) represents a vertical movement of one pitch length.

There are two approaches for entering the cut in an internal hole along a tangential arc and radial. Radial approach is simple but the cutter should be entered to full depth of cut at 30 to 50% of the circular feed rate to avoid vibration. The tangential approach allows for the tool to ramp gradually into and out of the cut. It also eliminates the dwell vertical mark at the entry or exit point. The tangential approach requires more complex programming than the radial approach.

The feed rate at the cutting edge (true feed rate) is given by Equation (2.52)

$$f_{rt} = n_t f_t N = (4)(0.1 \text{ mm/rev/tooth})[(100)(1000)/(\pi 25)](\text{rev/min}) = 510 \text{ mm/min}$$

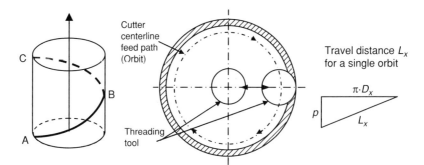

FIGURE 2.41 Schematic representation of Example 2.11.

The centerline feed rate is given by Equation (2.55). The major diameter of the thread is estimated from Equation (2.40)

$$\eta = (D_m - D_\mu)/(1.299p)$$
$$\Rightarrow D_m = \eta(1.299p) + D_\mu = 0.7(1.299 \times 3) + 100 = 102.7 \text{ mm}$$
$$f_r = f_{rt}(D_m - D)/D_m = 510(102.7 - 25)/102.7 = 386 \text{ mm/min}$$

The total travel length for the tool is estimated by calculating the time to complete a single orbit (as illustrated in Figure 2.41) during which three threads are cut and the number of orbits required to finish the full threaded depth in the hole. The length of a single orbit is

$$L_x = \sqrt{(p^2 + \pi^2 D_x^2)}, \quad D_x = D_m - D = 102.7 - 25 = 77.7 \text{ mm}, \quad L_x = 244 \text{ mm}$$

The number of orbits are

$n_{orbits} = L/(3p) = 30/[(3)(3)] = 3.33$, hence, the number of orbits will be either 3 or 4. The machining time is

$$t_m = L/f_r = n_{orbits}(L_x/f_r) = (3)(244/386) = 2.1 \text{ min}$$

for cutting which does not include the approach and exit distances for each orbit.

Example 2.12 A 30 mm deep hole is being drilled in a block of magnesium alloy with a 10 mm carbide drill at a feed of 0.3 mm/rev and at 200 m/min cutting speed. The drill point angle is 120°. Calculate: (a) material removal rate, (b) cutting time, and (c) torque on the drill (N m).

Solution:

(a) MRR = (cross-sectional area)(feed rate)

$$\text{MRR} = [\pi(10)^2/4](0.3 \text{ mm/rev})(200 \text{ m/min})(1000 \text{ mm/m})/(\pi 10)$$
$$= 150{,}000 \text{ mm}^3/\text{min}$$

(b) The cutting time is given by Equation (2.16), and let us assume that the approach and overtravel distance for the drill is $\Delta L = 3$ mm

$$t_m = (L + L_e)/f_r = (L + D/\tan \rho + \Delta L)/f_r$$

or

$$t_m = (30 + 10/\tan 60° + 3)/[(0.3 \text{ mm/rev})(6370 \text{ rpm})] = 0.02 \text{ min} = 1.2 \text{ sec}$$

(c) The torque can be estimated from the power required to drill the hole.

$$P = (\text{MRR})u_s = (150,000 \text{ mm}^3/\text{min})(0.008 \text{ W min/mm}^3) = 1.2 \text{ kW}$$
$$P = T\omega = T(2\pi N)$$
$$\Rightarrow T = P/(2\pi N) = (1200 \text{ N m/sec})(60 \text{ sec/min})/(2\pi \cdot 6370 \text{ rpm}) = 1.8 \text{ N m}$$

Example 2.13 A vertical surface grinding operation is being carried out with a 150 mm diameter wheel at 2000 m/min cuttung speed. The depth to be ground is 0.2 mm with a depth of grind of 0.01 mm per pass. The table traverse speed (feed rate of table or wheel) is 2200 mm/min. The length and width of the workpiece are 350 and 75 mm, respectively. Calculate the material removal rate and machining time.

Solution:

(a) The MRR for grinding is calculated using the same equations as in milling operations. In this surface grinding operation the width of the wheel covers the whole part width. The depth of cut per pass is generally very small in conventional grinding operations

$$\text{MRR} = bdf_r = bdV_{\text{work}} = (75 \text{ mm})(0.01 \text{ mm/pass})(2200 \text{ mm/min})$$
$$= 1650 \text{ mm}^3/\text{min}$$

where d is the depth of the layer of material removed during one cutting pass (stroke). The time to travel across the part length (or perform one pass) is

(b) $t_{mp} = (L_p + D + L_a)/f_r = (350 + 150 + 40)/1650 = 0.33 \text{ min/pass}$

where L_p is the part length, D is the wheel diameter, and L_a is the overtravel distance from both ends of the workpiece. The total time to grind the full depth is

$$t_m = \frac{(\text{depth of cut})_{\text{total}}}{d} t_{mp} + t_s = \frac{0.2 \text{ mm}}{0.01 \text{ mm}} 0.33 \text{ min/pass} + 0.33 = 6.93 \text{ min}$$

where t_s is the sparking-out time. In any grinding operation, the depth of cut during one cutting pass will initially be less than the nominal feed (d), in a direction normal to the work surface, setting on the machine. This infeed differential results from the deflection of the machine tool spindle, column, and workpiece under the forces generated during grinding.

Example 2.14 A 75 mm diameter cast iron workpiece is cylindrical ground using 150 mm diameter wheel at 2000 m/min cutting speed. The wheel thickness is 25 mm. The depth to be ground is 0.2 mm with a depth of grind of 0.015 mm per pass. The wheel (table) traverse speed (feed rate) is 1500 mm/min. The length of the bar to be ground is 200 mm. Calculate (a) the power required to grind and (b) the machining time.

Solution:

(a) $\text{MRR} = \pi D_w df_r = (75 \text{ mm})(0.015 \text{ mm/pass})(1500 \text{ mm/min})$
$$= 5299 \text{ mm}^3/\text{min}$$

The specific cutting energy for grinding should be much larger than that given in Table 2.1 based on turning tests. Therefore, the value of 0.08 W min/mm³ for cast iron from Table 2.1 should be at least three to five times higher for grinding operations. Hence, the power required is

$$P = (\text{MRR})u_s = (5299 \text{ mm}^3/\text{min})(5 \times 0.08 \text{ W min}/\text{mm}^3) = 2120 \text{ W}$$

(b) The machining time is

$$t_m = \frac{(\text{depth of cut})_{\text{total}}}{d} t_{mp} + t_s = \frac{0.2 \text{ mm}}{0.015 \text{ mm}} 0.155 \text{ min/pass} + 2(0.155) = 2.38 \text{ min}$$

The time to travel across the part length (or perform one pass) is

$$t_{mp} = (L_p + W_w + L_a)/f_r = (200 + 25 + 8)/1500 = 0.155 \text{ min/pass}$$

2.14 PROBLEMS

Problem 2.1 A 1020 steel bar 150 mm in diameter is turned on a 12 kW lathe at 100 rpm and a 0.5 mm/rev feed. The lathe has a 90% efficiency. What is the maximum depth of cut that can be used with this operation?

Problem 2.2 Estimate the machining time required in rough turning an 800 mm long, annealed 4240 steel round bar of 100 mm in diameter, using a carbide tool. Estimate the time for a ceramic tool.

Problem 2.3 Estimate the machining time required for rough turning the OD and facing the end of a 0.6 m long round bar 1040 steel, 100 mm in diameter using a carbide tool. The depth of cut for both operations is 1 mm. The maximum cutting speed allowed is 70 m/min, with a feed of 0.25 mm/rev.

Problem 2.4 The diameter of a 250 mm long by 95 mm diameter steel rod is being turned down to 90 mm on a lathe. The spindle rotates at 450 rpm, and the feed rate of the tool (traveling at an axial speed) is 250 mm/min. Calculate the cutting speed, material removal rate, time of cut, power required, and the cutting force.

Problem 2.5 A shaft of 1040 steel is grooved to a diameter of 80 mm from 100 mm. The width of the groove is 5 mm. The maximum cutting speed for the cutting tool material is 120 m/min with a feed of 0.25 mm/rev. Determine the chip area, material removal rate, power, and machining time.

Problem 2.6 A 200 mm long taper shaft is generated from 1040 steel 80 mm round bar stock in a lathe. The small diameter is 40 mm. The taper is 10° included angle. The maximum depth of cut for a roughing operation is 4 mm while for a finish cut it is 0.5 to 1 mm. The rapid travel feedrate is 12,000 mm/min, the rough feed is 0.35 mm/rev, and finish feed is 0.1 mm/rev. The suggested cutting speed is 100 m/min. Determine the following:

(a) number of passes from rough to finish part, and
(b) total machining time.

Problem 2.7 One thousand gray cast iron bars 100 mm diameter and 300 mm long must be turned down to 65 mm diameter for 200 mm of their length. The surface finish and accuracy requirements are such that a heavy roughing cut (removing most of the material) followed by a light finishing cut are needed. The available maximum power in the lathe spindle of 2.5 kW with an efficiency of 85% is used for the roughing cuts. The finish cut is selected at a feed of 0.13 mm, a cutting speed of 90 m/min, and at maximum power. Calculate the total production time in hours for the batch of work. Assume that the time taken to return the tool to the beginning of the cut is 3 sec, the tool index time is 1 sec, and the time taken to load and unload a workpiece is 2 min.

Problem 2.8 A 1 m diameter 6061 aluminum disk with a 300 mm diameter hole in the center is fixtured on the table in a vertical boring machine to perform a facing operation. The cutting tool starts to cut at the outside diameter and it is fed to the center (along its radius) for performing a facing operation. A constant table spindle rpm (rotational frequency) of 70 rpm is used, while the tool is fed at 0.25 mm/rev with a depth of cut of 6 mm. Calculate the following: (1) the machining time and (2) the power consumption at both the beginning and just before the end of the operation.

Problem 2.9 A hole is being drilled in a block of soft steel alloy with a 12 mm drill at 100 m/min cutting speed. The feed is 0.3 mm/rev and the hole depth (not including the drill point) is 25 mm. A standard solid carbide drill is used with a 120° point angle. Calculate the power and torque required drilling the hole and the machining time.

Problem 2.10 Calculate the time to drill a core hole through a cast iron workpiece material with a solid carbide three-flute drill that has a point angle of 120°. The thickness of the part at the drilling location is 50 mm. A drill diameter of 30 mm is used to enlarge a 15 mm hole. The recommended cutting speed for carbide drill is 80 m/min at a feed of 0.12 mm/rev/flute. What are the material removal rate (MRR) and the torque required by the spindle?

Problem 2.11 The hole generated in Problem 2.10 is reamed with a 30.5 mm eight-flute reamer. The feed for the reamer is 0.1 mm/tooth. The recommended cutting speed is 70 m/min. Calculate the machining time and power required to ream the hole.

Problem 2.12 A peripheral (slab) milling operation is being carried out on a 600 mm long, 50 mm wide steel (hardness of about 30 R_c) block at a feed of 0.15 mm/tooth and depth of cut of 10 mm. A 200 mm diameter staggertoothed side cutter is used with 20 inserts, and rotates at 250 rpm. The effective teeth are only one fourth the total number of teeth in the cutter (see Example 2.8). Calculate the material removal rate (MRR), the cutting time, and estimate the power and torque required by the machine tool spindle.

Problem 2.13 Face milling operation is being carried out on a 500 mm long by 50 mm wide rectangular stock of stainless steel material. A 200 mm diameter face milling cutter with 10 inserts is used to clean up the top 2 mm on the surface of the workpiece. The cutting parameters are 200 mm/min feed rate and 200 rpm on the spindle. Calculate the material removal rate (MRR), cutting time, and feed per tooth, and estimate the power required.

Problem 2.14 A face milling operation is being performed on an 80 mm wide aluminum part with a 300 mm diameter cutter. The feed per tooth is 0.2 mm and the depth of cut is 8 mm. The cutter center is offset 20 mm from the workpiece centerline. The cutter has 12 inserts.

(a) Determine the uncut chip thickness.
(b) Indicate in a graph the variation of uncut chip thickness on one tooth over two revolutions of the cutter.

Problem 2.15 Estimate the machining time required in tapping a hole at 20 mm depth with an M12 × 1 mm tap. A high speed steel cutting tap is used at 12 m/min cutting speed. What is the material removal rate?

Problem 2.16 A 100 mm diameter hole is threaded for the first 30 mm depth using a 25 mm diameter solid carbide four-flute thread milling tool with a single row of teeth. The thread pitch is 3 mm. The allowable feed per tooth (chip load) in this case is 0.1 mm/rev., while the cutting speed is 100 m/min. Estimate the machining time.

REFERENCES

1. M. Weck, *Handbook of Machine Tools*, 4 Volumes, Wiley, New York, 1984
2. T.J. Drozda and C. Wick, eds, *Tool and Manufacturing Engineers Handbook*, Vol. 1: Machining, SME, Dearborn, MI, 1983
3. A. Slocum, *Precision Machine Design*, Prentice Hall, Englewood Cliffs, NJ, 1992
4. S.S. Heineman and G.W. Genevro, *Machine Tools — Processes and Applications*, Canfield Press, San Francisco, CA, 1979
5. J.R. Davis Ed. *Metals Handbook*, Vol. 16: Machining, 9th Edition, ASM International, Metal Park, OH, 1989
6. O.W. Boston, *Metal Processing*, 2nd Edition, Wiley, New York, 1951
7. E.P. DeGarmo, *Materials and Processes in Manufacturing*, 5th Edition, Macmillan, New York, 1979, Chapters 16–28
8. E.J.A. Armarego and R.H. Brown, *The Machining of Metals*, Prentice Hall, New York, 1969
9. G. Boothroyd and W.A. Knight, *Fundamentals of Machining and Machine Tools*, 2nd Edition, Marcel Dekker, New York, 1989
10. Sandvik Coromant Corporation, *Modern Metal Cutting — A Practical Handbook*, Fair Lawn, NJ, 1996
11. G.T. Smith, *Advanced Machining — The Handbook of Cutting Technology*, Springer Verlag, New York, 1989
12. W. Konig et al., Machining of hard materials, *CIRP Ann.* **32** (1984) 417–427
13. D. Stovicek, Hard part turning, *Tooling Prod.* **57**:2 (1992) 25–26
14. C. Wick, Machining with PCBN tools, *Manuf. Eng.* **101**:1 (1988) 73–78
15. Y. Matsumoto, Review of current hard turning technology, *Abrasives* October/November (1996), 16–34
16. R. Sood, C. Guo, and S. Malkin, Turning of hardened steels, *SME J. Manuf. Process.* **2** (2000) 187–193
17. H.K. Tonshoff, H.G. Wobker, and D. Brandt, Tool wear and surface integrity in hard turning, *Prod. Eng.* **3**:1 (1996) 19–24
18. T.G. Dawson and T.R. Kurfess, Hard turning, tool life, and surface quality, *Manuf. Eng.* **126**:4 (2001) 88–98
19. H.J. Swinehart, *Gundrilling, Trepanning, and Deep Hole Machining*, ASTME, Dearborn, MI, 1967
20. *Machining Data Handbook*, 3rd Edition, Machinability Data Center, Cincinnati, OH, 1980
21. Sonobond Corporation, Ultrasonically assisted gun drilling, *Cutting Tool Eng.* September/October (1981), 26–27
22. A. Feifer, Drilling micro-size deep holes, *Tooling Prod.* October (1989) 58–60
23. A.J.P. Sabberwal, Chip section and cutting force during the milling operation, *CIRP Ann.* **10** (1961) 197
24. F. Koenigsberger and A.J.P. Sabberwal, An investigation into the cutting force pulsations during milling operations, *Int. J. MTDR* **1** (1961) 15
25. O.W. Boston and C.E. Kraus, Elements of milling, *ASME Trans.* **54** (1932) 71; see also *ASME Trans.* **56** (1934) 358
26. M. Martellotti, Analysis of the milling process, *ASME Trans.* **63** (1941) 677; see also *ASME Trans.* **67** (1945) 233
27. G. Augsten and K. Schmid, Turning/turn-broaching — a new process for crankshaft and camshaft production, *Ind. Prod. Eng.* **15**:2 (1991) 36–43

28. R.E. Roseliep, Advantages and applications in broaching, *Machining Source Book*, ASM International, Materials Park, OH, 1988, 117–120
29. L.J. Smith, Broaching: the fastest way to machine all kinds of external surfaces, *Machining Source Book*, ASM International, Materials Park, OH, 1988, 121–122
30. J.S. Agapiou, Evaluation of the effect of high speed machining on tapping, *ASME J. Eng. Ind.* **116** (1994) 457–462
31. R. Price, Don't gamble on thread depths, *Cutting Tool Eng.* April (1991) 40–44
32. B. Norton, Updating tap selection, *Cutting Tool Eng.* April (1992) 35–38
33. D.W. Yen, Turning of precision threads on heat treated alloy steel, *Trans. NAMRI/SME* **19** (1991) 90–95
34. F. Mason, New turns in thread cutting, *Am. Machinist*, November (1988) 137–144
35. G. English, Thread milling: a look at the basics, *Machining Source Book*, ASM International, Materials Park, OH, 1988, 123–127
36. K.B. Lewis and W.F. Schleicher, *The Grinding Wheel*, 3rd Edition, The Grinding Wheel Institute, Cleveland, OH, 1976
37. F.H. Colvin and F.A. Stanley, *Grinding Practice*, 3rd Edition, McGraw-Hill, New York, 1950
38. C. Andrew, T.D. Howes, and T.R.A. Pearce, *Creep Feed Grinding*, Holt Rinehart and Winston, London, 1985
39. C.P. Bhateja and R.P. Lindsay, *Grinding: Theory, Techniques, and Troubleshooting*, SME, Dearborn, MI, 1982
40. L.J. Coes, *Abrasives*, Springer-Verlag, New York, 1971
41. R.L. McKee, *Machining with Abrasives*, Van Nostrand Reinhold, New York, 1982
42. F.T. Farago, *Abrasive Methods Engineering*, Vols. 1 and 2, Industrial Press, New York, 1976 (Vol. 1) and 1980 (Vol. 2)
43. R.I. King and R.S. Hahn, *Handbook of Modern Grinding Technology*, Chapman and Hall, New York, 1986
44. S. Malkin, *Grinding Theory and Applications*, Ellis Horwood, Chichester, UK, 1989
45. G.C. Sen and A. Bhattacharyya, *Principles of Metal Cutting*, 2nd Edition, New Center Book Agency, Calcutta, 1969
46. M.C. Shaw, *Metal Cutting Principles*, Oxford University Press, Oxford, 1984
47. N.H. Cook, *Manufacturing Analysis*, Addison-Wesley, Reading, MA, 1966
48. J. Kaczmarek, *Principles of Machining by Cutting, Abrasion, and Erosion*, Peter Peregrinus, Stevenage, 1976
49. W.F. Jessup, Centerless grinding, *Handbook of Modern Grinding Technology*, Chapman and Hall, New York, 1986, Chap. 8
50. S.C. Salmon, Creep-feed grinding, *Handbook of Modern Grinding Technology*, Chapman and Hall, New York, 1986, Chap. 12
51. R.J. Fisher, Superabrasives and creep-feed grinding, *Machining Source Book*, ASM International, Materials Park, OH, 1988, 178
52. L.P. Tarasov, Grindability of tool steels, *ASME Trans.* **43** (1951) 1144
53. H.K. Tonshoff and T. Grabner, Cylindrical and profile grinding with boron nitride wheels, *Proc. 5th Int. Conf. Prod. Eng*, JSPE, Tokyo, 1984, 326
54. D.W. Bouchard, Single pass superabrasive bore finishing, *Superabrasives '85 Proceedings*, Chicago, IL, 1985, 5-37–5-41
55. H. Fischer, Honing, *Handbook of Modern Grinding Technology*, Chapman and Hall, New York, 1986, Chap. 13
56. T. Ueda and A. Yamamoto, An analytical investigation of honing mechanism, *ASME J. Eng. Ind.* **106** (1984) 237–241
57. A. Yamamoto and T. Ueda, Honing conditions for effective use of diamond and CBN sticks, *ASME J. Eng. Ind.* **109** (1987) 179–184
58. E. Salje, H. Mohlem, and M. von See, Comparison of grinding and honing process, *SME Manuf. Technol. Rev.* (1986) 649–653
59. E. Salje and M. von See, Process-optimization in honing, *CIRP Ann.* **36** (1987) 235–239
60. L. Jongchan, Fundamental Study of Honing, PhD Dissertation, Department of Mechanical Engineering, University of Massachusetts, Cambridge, MA, 1991

61. J. Gutowski, Honing of carbides and steel with diamond and CBN, *Superabrasives '85 Proceedings*, Chicago, IL, 1985, 5-26–5-36
62. F.J. Hettes, Which brush should you choose? *Manuf. Eng.* **107**:3 September (1991) 61–63
63. C.R. Nichols, External honing, microfinish, superfinishing, superhoning, *International Honning Clinic*, SME, Dearborn, MI, April 1992
64. J. Lee, E.E. Wasserbaech, and C.H. Shen, Camshaft grinding using coated abrasive belts, *Trans. NAMRI/SME* **21** (1993) 215-222
65. W. Konig, H. Tonshoff, J. Fromlowitz, and P. Dennis, Belt grinding, *CIRP Ann.* **35** (1986) 487–494
66. W.J. Westerman, An overview of roller burnishing as a surface conditioning, SME Technical Paper MR810401, 1981
67. J.A. Balcom, Bore to burnisher — robot to robot, SME Technical Paper MS830438, 1983
68. N.H. Loh, S.C. Tam, and S. Miyazawa, Ball burnishing of tool steel, *Precision Eng.* **15**:2 (1993) 100–105
69. N.H. Loh and S.C. Tam, Effect of ball burnishing parameters on surface finish — a literature survey and discussion, *Precision Eng.* **10** (1988) 215–220
70. N.H. Loh, S.C. Tam, and S. Miyazawa, A study of the effects of ball burnishing parameters on surface roughness using factorial design, *J. Mech. Working Technol.* **18** (1989) 53–61
71. W.J. Westerman, Salvage by knurling and burnishing, *Tooling Prod.* January (1987) 60–61
72. L.K. Gillespie, *Advances in Deburring*, SME, Dearborn, MI, 1978
73. L.K. Gillespie, *Deburring Technology for Improved Manufacturing*, SME, Dearborn, MI, 1981
74. A.F. Scheider and J.P. Gaser, Advanced flexible abrasive finishing tool technology, SME Technical Paper MR91-121, 1991
75. R. Frazier, Deburring update utilizing flexible radial wheel type processes, SME Technical Paper MR91-122, 1991
76. R.A. Tollerud, Deburring and surface conditions using nonwoven abrasives, SME Technical Paper MR91-124, 1991
77. C. Van Sickle and G. Flores, Mechanical deburring and honing in the automated environment, SME Technical Paper MR91-126, 1991
78. J.H. Indge, Edge round with flexible bonded abrasives, SME Technical Paper MR91-128, 1991
79. S. Kawamura and J. Yamakawa, Formation and growing up process of grinding burrs, *Bull. Japan Soc. Precision Eng.* **23**:3 (1989) 194–199
80. L.K. Gillespie, *Robotic Deburring Handbook*, SME, Dearborn, MI, 1987
81. E.G. Erickson, Automated robotic deburring produces quality components, *Automation* March (1991) 50–51

3 Machine Tools

3.1 INTRODUCTION

In addition to the workpiece and its fixture, there are three major components of a machining system: the machine tool, the cutting tool, and the toolholder. This chapter describes machine tools used to perform the various operations discussed in Chapter 2. Cutting tools are described in Chapter 4, and toolholders and fixtures are discussed in Chapter 5.

There are three principal types of machine tools. *Conventional machine tools* are designed to perform one or several operations on a variety of parts. These tools were developed early in the industrial revolution and are still found in every machine shop, where they are used for general purpose machining of small lots of parts, and for repair work. Their capabilities have been greatly enhanced by the advent of numerical control, which became available on most machine tools during the late 1970s. Conventional machine tools were described in connection with basic machining operations in Chapter 2, and will not be discussed further in this chapter.

Production machine tools are used in high-volume manufacturing systems to perform one or a sequence of operations repetitively. They can be adapted for more than one part of the same family but the changes required to switch from one part to another are usually time-consuming and uneconomical. They are composed of a series of simpler machines or mechanisms which resemble conventional machine tools, and which are connected by an automated materials handling system. Because of their lack of flexibility and large capital costs, they are only used when thousands of identical parts are required. Production machine tools are described further in Section 3.2.

CNC machine tools (machining and turning centers) are advanced types of numerically controlled machine tools used to produce a variety of complex parts. They are capable of moving cutting tools along complicated paths, often involving simultaneous motions of multiple axes, according to a stored program. Machining centers are flexible and can produce very complex parts in quantity with consistent quality and repeatability. They are generally economical for low or medium production volumes, but can be used in high-volume manufacturing if high spindle and feed speeds are available and the part and cutting tools are designed to minimize tool changes. Scalable manufacturing systems are easily designed using CNC machines. They are described further in Section 3.3.

Following descriptions of the principal types of machine tools, common designs of basic machine elements, including machine tool structures, slides and guideways, axis drives, spindles, coolant systems, and tool changers are described in Section 3.4 through Section 3.9. Only brief descriptions intended to acquaint end users of machine tools with the requirements of each element and the designs currently available are given. Machine tool builders will find more detailed descriptions of each component in books on machine tool design [1–6].

3.2 PRODUCTION MACHINE TOOLS

Most parts have a number of machined features and therefore must be produced on a group of machine tools arranged in a system. For parts required in high volumes, a number of specially designed machine tools may be grouped on a common base or connected by automated materials handling, coolant delivery, and swarf handling equipment to form a dedicated machining system. It is common to call these systems production machine tools, although in reality they are comprised of a grouping of simpler tools or stations.

The most common types of production machine tools feature automated part transfer between stations and are called transfer machines. There are two basic classes of such machines: *rotary transfer machines* and in-line or *conventional transfer machines* [7–9].

A common type of rotary transfer machine is the *rotary indexing system*, in which parts are mounted on a horizontal table or dial and transferred by rotation through various machining stations arranged around the table (Figure 3.1a). Stations are configured to perform specific operations repeatedly and are often numerically controlled. A trunnion or vertical drum can be used instead of a table if the parts require machining on opposite sides or at compound angles [1–4]. In a *dial system*, rotations are of a standard length. Dial systems provide two access planes and a limited number of stations, and cannot be expanded by adding additional stations. The "European" rotary type provides three access planes. A variation of the dial system is the *prism system* composed of fixtures that advance in a horizontal plane between workstations. Prism systems allow rotation of individual pallets so that parts can be machined on multiple surfaces.

Center column systems (Figure 3.1b) are similar to rotary indexing systems but have additional machining stations mounted on a central column. They can accommodate more operations than conventional rotary indexing systems. They provide three access planes. They are more difficult to maintain than rotary indexing systems since mechanisms in the center column are often difficult to access. Generally, rotary indexing systems are used for small or light parts, while center column machines are favored for heavy parts with greater machined content.

Conventional or in-line transfer machines (Figure 1.13 and Figure 3.2) are dedicated systems designed to produce a single part in large volumes (e.g., >25,000 per year). Since their output is high, the cost per part is relatively low. They consist of stations connected by an automated part transfer system. There are three conventional types of part transfer: sliding transfer, palletized transfer, and walking-beam or lift-and-carry transfer. In sliding transfer (Figure 3.3a), the part is moved on rails or rollers by an indexing bar, and is located and clamped at each station in turn. Sliding transfer is used especially for heavy iron or steel parts, such as engine blocks and motor housings. In palletized transfer (Figure 3.3b), the part is located and clamped on a traveling pallet, which is moved between stations by an indexing bar or moving chain. This method requires more investment and (frequently) floor space since some provision for pallet return must be made, but is preferred for relatively compliant parts such as aluminum transmission cases. In lift-and-carry transport the part is moved by a gantry or linkage; this method is particularly suited to small or irregularly shaped parts such as connecting rods and crankshafts. Other parts of a transfer machine include the center bed, which contains the transfer and swarf handling mechanisms, wing bases, tool driving heads, and load/unload equipment (Figure 3.4). Types of tool driving heads include milling heads, drilling and boring heads, and multispindle drilling heads as described in older books [8–11]. The dimensions and utility connections of these components have been standardized in both the United States and Europe, so that machines can be assembled modularly from basic components (Figure 3.5).

Conventional transfer machines are well suited for manufacturing products with long market cycle lives (greater than 10 years). They use well-established machine technologies but have little provision for adjustment. To operate economically, transfer machines require

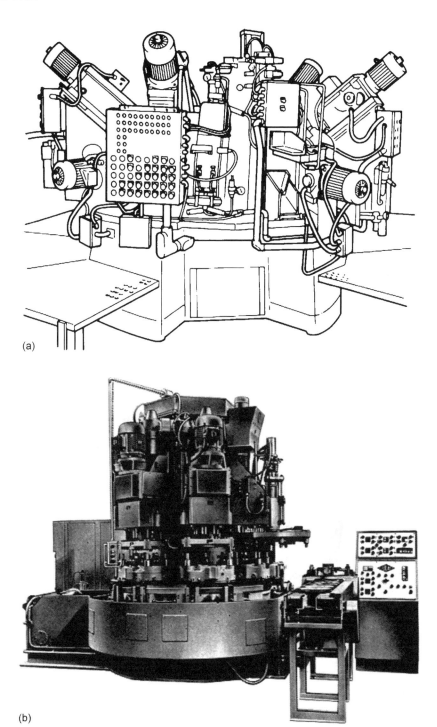

FIGURE 3.1 Rotary transfer machines. (a) Rotary indexing system. (After T.J. Drozda and C. Wick [2].) (b) Center column system. (After E.D. Lloyd [8].)

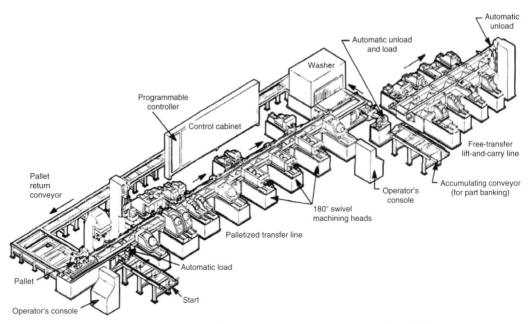

FIGURE 3.2 Conventional in-line transfer line. (After T.J. Drozda and C. Wick [2].)

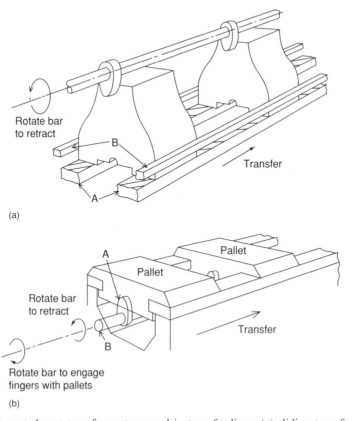

FIGURE 3.3 Automated part transfer system used in transfer lines: (a) sliding transfer; (b) palletized transfer. (After E.D. Lloyd [8].)

Machine Tools

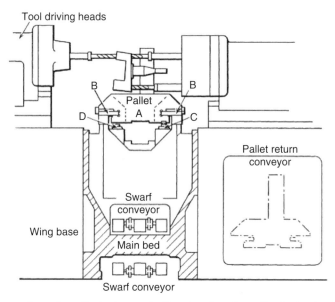

FIGURE 3.4 Cross section of a palletized transfer line showing the pallet, bed, wing bases, tool driving heads, swarf conveyer, and pallet return method. (After E.D. Lloyd [8].)

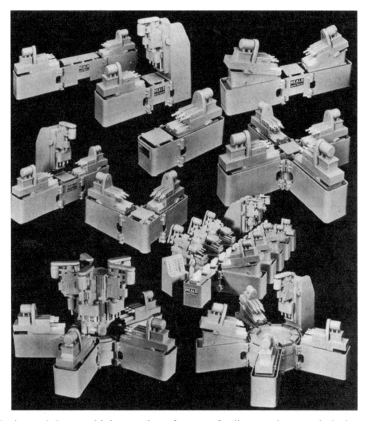

FIGURE 3.5 Basic modular machining stations for transfer lines and rotary indexing systems. (After E.P. DeGarmo [11].)

accurate preplanning of the product design and accurate forecasts of market demand. In addition to their lack of flexibility, which can lead to excessive new product lead times, a major disadvantage of conventional transfer machines is that they are serial systems, and thus do not run when any of the stations need repair or a tool change. Due to this limitation, large systems are often operating roughly 60% of the time. One strategy to improve machine utilization is to break large systems up into smaller subsystems with intermediate part buffers, but this increases in-process inventory and may reduce quality and increase repair costs if out-of-tolerance conditions are not noticed promptly in any subsection.

In recent years, a number of technologies have been applied to in-line automated production systems to improve flexibility and dimensional capability. A *convertible transfer line* operates like a dedicated line for one part, but is designed to accept a defined range or family of parts (e.g., iron and aluminum versions of an engine block, or six- and eight-cylinder version of an engine head). A convertible system allows flexibility in producing parts within a family based on changing demand with minimal loss of production time. Convertible systems can typically be switched over from one part to another in less than a month, compared to 6 to 12 for a conventional system (assuming the conversion is even feasible). Convertible transfer machines are designed for part volumes similar to those for conventional transfer machines, but have lower part life cycle times. A convertible system requires additional initial investment and accurate preplanning. It is especially important to use a common locating scheme for all parts to minimize fixture rework and to use common tooling when possible.

A *flexible transfer line* (FTL) is a production system designed for high-volume production that is capable of producing a family of similar parts with unplanned changes or additional machined content. The life cycle of individual parts can vary from a few years to several years as long as there is sufficient flexibility to fully utilize the system for 10 to 15 years. Such systems allow new products of the same family to be introduced quickly without major retooling. The changeover time between different products is usually a few hours and depends on the number of workstations involved and the available flexibility. Flexible transfer machines are well suited to applications in which a few similar parts are required in high volumes (e.g. >50,000 per year), as is often the case in automotive powertrain and component production. Flexible transfer lines require a significant initial investment premium as compared to conventional transfer machines, and still require accurate part planning and market forecasts to operate economically. Currently, flexibility is accomplished by using machining stations with indexable heads (turrets) or shuttle heads, each fitted with a number of different tools. CNC machines have also been used in FTLs, making them similar to the agile production systems described in Chapter 15. In addition to requiring a significant premium in initial investment, flexible systems increase the machine structure and fixture complexity and often required tooling and gaging inventories.

Product design is much more critical for FTLs than for conventional lines. Part features should be grouped and commonized so that a large number of features can be machined with single spindles. This means, for example, that the holes in the part should all be the same size so that they can be drilled using the same tool to reduce the number of tool changes. Other methods for increasing the production rate of an FTL include the use of multiple independent spindles for machining to minimize machining time or the use of multiple spindles with multiple part loading (e.g., twin spindles with dual part loading).

Automatic lathes, such as screw machines, bar chuckers, drum- and swiss-style automatics, and vertical turret lathes comprise another class of production machine tools still encountered in older operations. These machines, which are described in detail in the literature [2, 9, 10], have been superceded in most recent applications by the CNC turning centers described in the next section.

3.3 CNC MACHINE TOOLS AND CELLULAR MANUFACTURING SYSTEMS

Most machine tools built since 1985 are numerically controlled machines. The basic components of an NC control system are the program of instructions, machine control unit (MCU), servo drives for each axis of motion, and feedback devices for each axis of motion [12, 13]. The MCU moves the machine axes to drive the tool along the tool path specified in the stored programs. Both linear and rotary motions can be precisely controlled simultaneously.

NC systems can be classified by the method used to control machine slides, the number of axes, position information, the feedback mode, the interpolation method, or the data format. Machining centers with five, seven, or more axes, which can generate very complex surfaces that cannot be produced with conventional machines, are available. Naming conventions for axes and common structural configurations are described below. The methods available for controlling the relative motion of axes are: (a) point-to-point (usually for two-axis machines such as drilling machines with single or dual axis control); (b) straight cut (control along a path parallel to a linear or circular machine way); and (c) continuous path or contouring (continuous control along a path in two or more axes). These three types require an increasing level of control sophistication. The position information is either absolute (predetermined or fixed datum) or incremental (referenced from the current position).

Absolute control systems are closed-loop systems which rely on angular encoders or linear displacement encoders to determine absolute axis positions; feedback from the encoders is continuously compared to a reference value by a microprocessor, which adjusts the slide speed to eliminate deviation between the absolute position and the reference value. At lower speeds, the available motor torque is low, so that the adjustment may result in stalling and tool failure. Incremental control systems can be either closed- or open-loop systems. A closed-loop incremental control system uses an incremental displacement encoder which produces pulses corresponding to the smallest measurable unit of displacement, which are counted by the microprocessor and compared to the reference distance. The slide is stopped when the counter reaches the reference distance. An open-loop incremental control system operates without feedback; a stepping motor is used as an actuator that receives the number of pulses corresponding to a specified displacement directly from the controller. If power is lost, the operation may be resumed without rezeroing with an absolute control system; rezeroing is required with an incremental system. An open control system has better dynamic characteristics than a closed-loop system but does not provide position verification. Closed-loop systems with dynamic error compensation are required for high-speed contour milling. The communication rate for each individual axis processor for the servo interface can be the determining factor for the maximum cutting rate.

The MCU has historically been a proprietary or closed-architecture computer. Recently, however, control logic software has been developed which can be run on a number of platforms, including standard Windows-based PCs. Open controls are therefore becoming popular in machine tools as PC-based open-architecture alternatives are becoming attractive for major factory automation. Often, open control means using PC front ends and interfaces to proprietary machine tool controls that have full connectivity via standard networking and communications protocols. Open controls are enabling machine tools to take advantage of the latest software, networking, and operating system technologies, resulting in more flexibility. More detailed descriptions of these components and NC systems are available in the literature [1, 2, 5, 14–23].

Common types of machining centers are shown in Figure 3.6–Figure 3.15. *CNC lathes* or *turning centers* (Figure 3.6 and Figure 3.7) can perform turning, boring, facing, threading, profiling, and cut-off operations. If the turret is equipped with additional powered spindles (turn-mill NC turning machines), they can also be used for milling, drilling, and tapping.

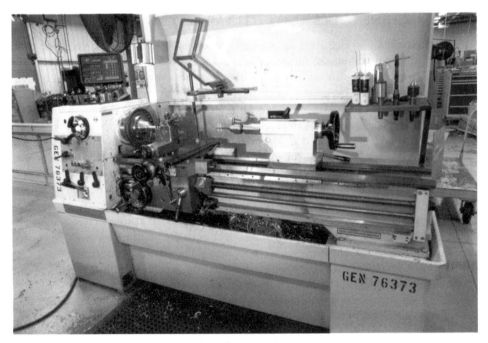

FIGURE 3.6 An NC turret lathe (a type of turning center).

FIGURE 3.7 A CNC turning center with a main spindle and a subspindle for second-side capability, a rotating tooling turret, and a contouring C-axis. (Courtesy of Cincinnati Milacron.)

Turning centers use quick-change modular tooling and may have two turrets (four-axis CNC lathe) that cut simultaneously. Lathes with multiple spindles (run simultaneously) can be used to reduce both the machining time and idle time for part setup and handling. *CNC automatics* are similar to turning centers but include more axes, rotating tooling (live tooling), and multiple slides and spindles. On CNC automatics, a job can be divided into segments so that many tools can work on different areas of a workpiece simultaneously. Cycle times can thus be shorter than on CNC lathes. However, they cannot be used to produce highly precise parts because they generally do not have fine controller resolution. As noted above, CNC lathes and automatics are modern equivalents of cam-driven screw and bar machines, which they have largely displaced.

Other types of *machining centers* are primarily milling and boring machines which can perform a variety of other operations such as drilling, reaming, and tapping without changing the part setup. They are sometimes referred to as *multitasking machines*. Common examples are shown in Figure 3.8–Figure 3.15. Machining centers are generally classified as horizontal or vertical, depending on which way the spindle is mounted. On a vertical machine, the workpiece is mounted on a horizontal bed; on a horizontal machine, the workpiece is usually mounted on a vertical fixture, which is less stable. Vertical or C-frame machines are preferred for large workpieces, flat parts, and especially for contoured surfaces in dies so that the thrust force is absorbed directly by the bed of the machine (see Figure 3.10). Gantry or bridge-type milling machines are used for very large workpieces because their two-column design gives greater stability to the cutting spindle(s). Horizontally configured machines are more versatile because four sides of the workpiece can be machined without re-fixturing if a rotary indexing worktable is available. Horizontals are finding increasing use in surface machining, since they provide access on larger complex parts and have less restriction on vertical height of the workpiece. Horizontal machines are preferred for untended use since they allow for easy chip and coolant evacuation. Pallet systems used to shuttle parts in and out of the workstation are better suited for horizontal machining centers as illustrated in Figure 3.20. Universal

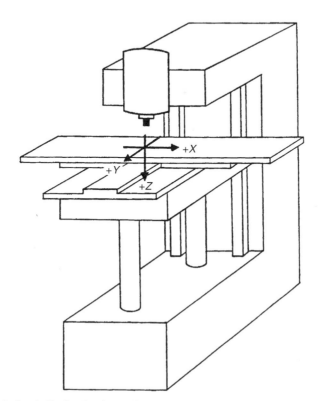

FIGURE 3.8 A vertical spindle fixed column (knee and column type) three-axis machining center.

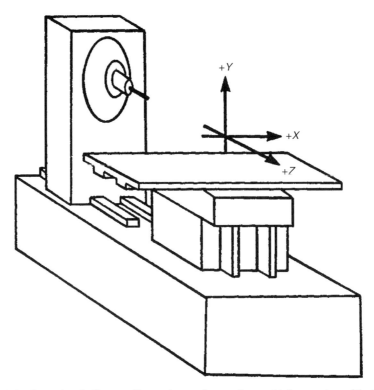

FIGURE 3.9 A horizontal spindle traveling column three-axis machining center with the lateral and vertical travels on the saddle.

FIGURE 3.10 Solid base, fixed column vertical spindle three-axis machining center. (Courtesy of Cincinnati Milacron.)

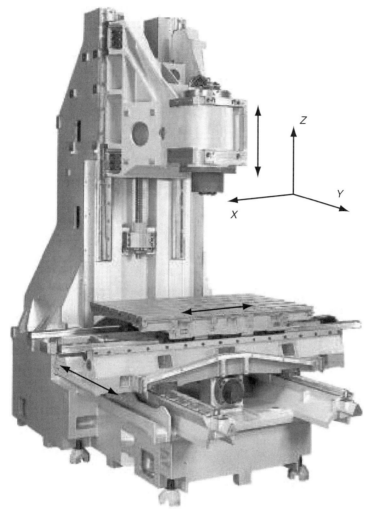

FIGURE 3.11 Solid base, fixed column vertical spindle three-axis machining center with Z-axis travel (vertical movement on spindle head), X-axis travel (longitudinal movement of table), and Y-axis travel (cross movement of saddle). (Courtesy of Mori Seiki Co., Ltd.)

machines have heads that rotate to act as a horizontal or a vertical machine. The combination of tilts and swivels available in the spindles and tables allows the workpieces to be addressed at various compound angles.

Conventional three-axis machines normally have three linear axes (X, Y, Z), but may have two linear and one rotational axis. Four-axis machines typically have three linear axes and a rotational axis on the work table. Five-axis machines have three linear and two rotational axes which may be provided by a rotating table, a spindle, or a combination of both (Figure 3.17 and Figure 3.18). The advantages of five-axis machines include: (1) their greater ability to maintain precision by minimizing the number of steps since most of the part is accessible to the cutting tool; (2) higher productivity due to fewer setups per part (using five-face machining); and (3) the ability to machine more complex shapes. There are several types of five-axis machine tools built on top of three-axis horizontal or vertical machines: those with a two-axis trunnion (as illustrated in Figure 3.17 and Figure 3.18), those with a swivel two-axis table, those

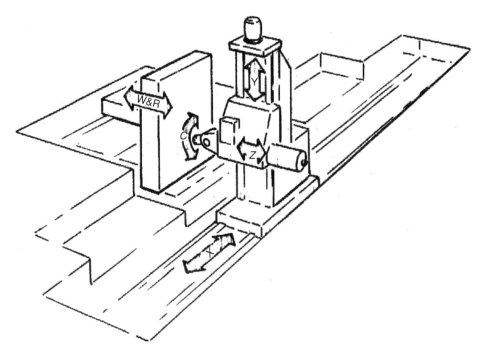

FIGURE 3.12 A horizontal spindle traveling column six-axis machining center with a column arranged as slide unit moving as the *X*-axis, a vertical slide moving as the *Y*-axis in the column, and quill for the *Z*-axis. The *C*-axis is on the spindle, while the *W*- and *R*-axes are provided on the table.

with a fork (two-axis swivel on the spindle), those with either an *A*- or a *B*-axis on the spindle and a rotary table (as illustrated in Figure 3.16 and Figure 3.19), and those with one rotary table on top of another (see *E*-axis table in Figure 3.18). Each configuration provides a range of spindle or workpiece orientations in the three orthogonal planes. For example, the fork or birotary head may have a *C*-axis travel of $+200°$ to $-200°$ and an *A*-axis travel of $+95°$ to $-110°$. Therefore, the selection or the particular fourth and fifth axis configuration depends on the geometry of the features to be machined and on the fixture design.

Hybrid machining centers combine the functions of both turning and machining centers and provide the ability to complete all machining operations in a single setup (Figure 3.16). A milling spindle is used that can tilt to make both horizontal and vertical machining operations possible as well as boring and milling on multiple faces. For example, such a machine can perform turning, milling, drilling, contouring with the *C*-axis, off-center machining with the *Y*-axis, milling of angled surfaces with the *B*-axis, grinding, etc. Such machines reduce cycle/lead times and work-in-process inventory, save setup and queue time, and possibly improve part quality by eliminating multiple fixturings.

The capabilities of machining centers are measured by maximum spindle rpm, power and torque versus speed curves, spindle taper size (which determines the size of the toolholder and tool), axis drive motor power, rapid feed rate, fastest cutting federate, machine construction (way style, stiffness, damping, etc.), workspace size, and support for networking.

High-speed machining centers (HSMCs) operate at spindle speeds of 20,000 to 40,000 rpm have high acceleration/deceleration (acc/dec) spindles (i.e., 1.5 sec from 0 to 20,000 rpm), high speed (200 m/min) and high acc/dec (2*g*) slides (with linear motors), and high-speed control systems. High-speed machines usually offer lower torque at high speeds than conventional machines at lower speeds (i.e., a 30,000 rpm/20 kW specification will yield 6.4 N m torque

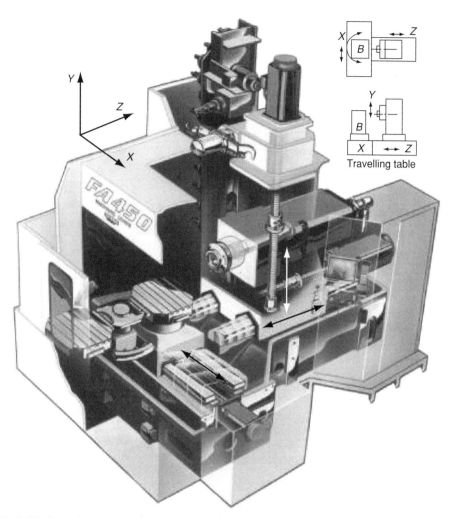

FIGURE 3.13 A horizontal spindle traveling column four-axis machining center with rotary table; Y- and Z-axis motions are performed by the column while the X-axis motion is provided by the table. (Courtesy of Toyoda Machinery USA, Inc.)

available at 30,000 rpm and 29 N m at 2500 rpm), and therefore the allowable depth of cut is often reduced at higher speeds. In HSMCs, the machining time for one part feature may be lower by as much as a factor of 20 when compared to a conventional machine. The machining time is typically one third of the total time, with the remainder being used for machine travel, tool changes, and pallet changes and rotations. As a result, high production rates are best achieved by reducing the noncutting time. Continuously raising the cutting speed proves cost-effective in just a few applications, such as aerospace applications in which parts are machined from billets.

As an example, a time study comparison of a standard CNC machine versus conventional (STD) HSMC versus an advanced HSMC in machining an aluminum automotive part is illustrated in Table 3.1. The cutting time for the advanced HSMC is improved significantly compared to STD HSMC because the process is changed somewhat so that different diameter holes are machined with single endmill and utilizing multifunctional tools. Actually the

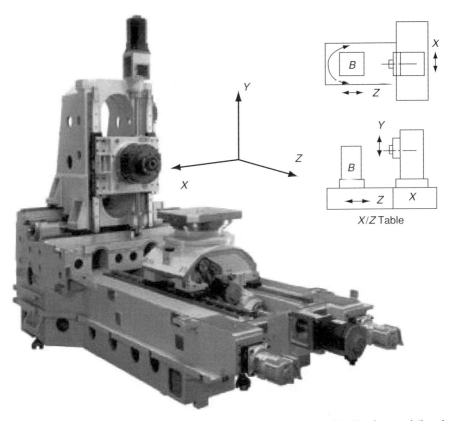

FIGURE 3.14 A horizontal spindle traveling column machining center with X-axis travel (longitudinal movement of the column), Y-axis travel (vertical movement of the spindle head), and Z-axis travel (cross movement of the table). (Courtesy of Mori Seiki Co., Ltd.)

number of tools was reduced from 17 to 12. The positioning time was found to be the largest contributor with the cutting time second to productivity improvement.

Drive dynamics and the dynamic characteristics of the machine structure are also important in HSMCs. High-speed machining or five-axis machine tool requires control with a look-ahead capability, and acceleration and deceleration control, and collision detection. Look-ahead capability allows for the CNC to read ahead a certain number of blocks in the program, in order to slow down the feed rate at anticipated sudden tool path direction changes. Nurbs interpolation has been used to curve interpolate a tool path so that the control system can change direction along the curve more gradually using a high average feed rate.

The selection of machining centers should be based on the total time required to finish a part, rather than on the machine spindle or slide speeds. The total time to finish a part depends on the machine motion time, the tool change time, the number of workpiece setups, the available number of axes, the available number of spindles, the pallet change time, and the workpiece fixturing approach. This time can be reduced by using machines with higher rigidity, accuracy, and flexibility to permit increased cutting speeds, automatic tool storage, transportation, and changing systems, modular workholding fixtures, head changers, universal spindle heads, and complex controllers. Head changers allow the use of horizontal spindles, vertical spindles, CNC universal angle heads, and multispindle heads. High-speed spindles of comparatively short design for small diameter tools can be automatically inserted

Machine Tools

FIGURE 3.15 A moving column high-speed machining center with the Z-axis on a horizontal ram. (Courtesy of Ingersoll Milling Machine Company.)

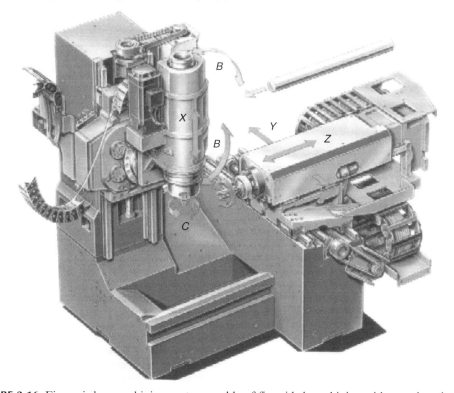

FIGURE 3.16 Five-axis bar machining center capable of five-sided machining with one clamping. The Y- and Z-axis motions are provided by the table while the X-axis motion is performed by the column. The workpiece carrier has a B-axis with a swiveling range of 270° and a C-axis with a swiveling range of 360°. (Courtesy of Hermle Machine Company LLC.)

TABLE 3.1
Comparison of Very High-Speed Machining for Manufacturing an Aluminum Part

Functions	STD CNC	STD HSM	Advanced HSM
Spindle (rpm)	12,000	15,000	20,000
Acceleration X-, Y-, Z-axes	0.7g	1.5g	2g
Rapid feed rate (m/min)	50	90	120
Reaches top slide speed in traveling (mm)	70	150	200
Table indexing time at 90° (sec)	3.0	1.9	0.8
Table indexing time at 180° (sec)	4.5	2.5	1.0
Table indexing time at 270° (sec)	5.0	3.1	1.3
Tool changing time (ATC), tool-to-tool (sec)	1.5	1.5	1.2
Tool changing time, chip-to-chip (sec)	4.5	2.8	2.4
Number of tool used to process all features	17	17	12
Results			
Cutting time (sec)	131	113	97
Positioning time (sec)	130	74	55
ATC time (sec)	25	25	17
Table indexing time (sec)	23	14	3.6
Total time (sec)	309	226	173
Productivity (%)	100	137	179

into the main machine spindle using the toolholder taper so that it operates from the same centerline. Either twin independent spindles or twin spindles with dual part loading may be used in certain applications to maximize machining time and increase the production rate. The attachment of a high-speed spindle with limited power on the side of a powerful low rpm main spindle balances power and rpm capabilities for a variety of applications. The number of machine axes can be increased by using a CNC rotary table in place of a standard table, an additional CNC rotary table on the pallet, or a two-axis swivel (fork) attachment (Figure 3.17–Figure 3.19). The selection and sequencing of machining processes is also important when all machining of a part has to be carried out using a single machine. It is often worthwhile to use machines with an increased number of axes of motion to reduce the machine setup time.

Ideally, machining centers should be designed to complete a workpiece in one chucking or setup. They should incorporate sensing and control strategies which compensate for tool wear, thermal expansion, workpiece material variation, vibration, and other variables in the machining process. They should offer a large range of spindle speeds (medium to high) with sufficient torque to handle a variety of workpiece materials (steel, cast iron, aluminum, etc.) and high traverse speeds, tool change speeds, and workpiece load/unload speeds. They should be capable of both wet and dry machining and include adequate ventilation and chip evacuation components. Finally, they should use sophisticated but easily programmed CNC controls. These considerations are discussed in more detail in Chapter 15.

NC machines allow for operations to be combined using combination tooling since variable speeds and feeds are available. Machining centers have traditionally been used for low to medium batch production. However, machining centers and turning centers are beginning to be used in place of transfer lines in the mass production of automotive engine and transmission components because they permit increased flexibility, allowing for a diverse product mix with frequent design changes, and because they can be operated profitably at reduced part volumes.

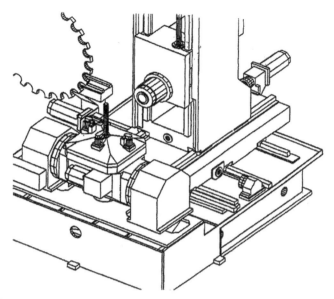

FIGURE 3.17 A five-axis machine using an integral tilt trunnion rotary table with 150° of motion on the A-axis (+30° to −120°) combined with a B-axis.

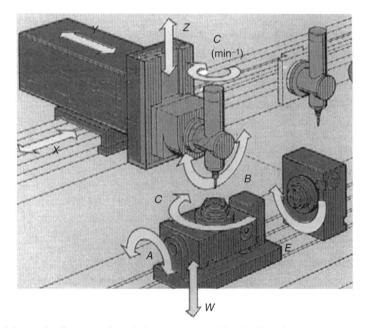

FIGURE 3.18 Schematic diagram of modular servo axes. The basic unit (X, Y, Z) carries the tool, the B- and C-axes are on the spindle, A/C axes are on an indexing head with a vertical rotational axis and a horizontal swivel head, and a horizontal E. (Courtesy of Klinx Hochgeschwindigkeitsbearbeitung GmbH, HSC.)

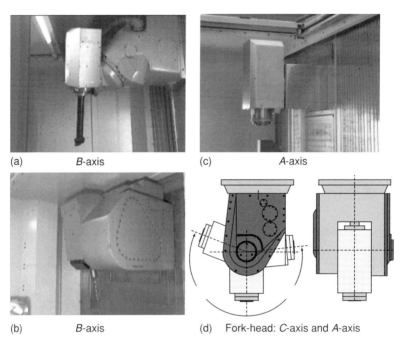

FIGURE 3.19 Spindle head styles for a five-axis system using a vertical spindle — fixed column machining center. (Courtesy of DMG Chicago, Inc. and Fidia Co.)

Machining centers can be used in sequence in cellular manufacturing systems or flexible manufacturing systems. A *cellular manufacturing system* (CMS) is designed to produce a family of parts of similar shapes. Machine tools are arranged in cells, which may consist of one or several stations linked by a common control system. A cell can continue to function regardless of the state of the other cells and systems, as long as it has the necessary parts and tools are available. A CMS is well suited to the multiproduct, small-lot-sized production requirements of a traditional machine shop. Various group technology concepts and schemes have been applied to manufacturing cells [24, 25]. Considerable research on optimizing CMS operations to minimize intercell part flow requirements has also been carried out [24–28]. In designing a CMS, it is often useful to start with a single machine (see Figure 3.20) and add capability as required. A vertical five-face machining center with a head changer or a turning center with either five- or six-sided machining capability can function as a basic cell. Additional machines can be added later as long as they have standard interfaces which allow additional software modules and material handling devices to be considered. Open architecture in cell controllers has been the major contributor to the implementation of cellular systems.

A *flexible manufacturing system* (FMS) [29–37] consists of a number of flexible machining cells or CNC machines. In principle, an FMS can handle a variety of similar or dissimilar part designs, and can enable new product designs to be introduced quickly without disturbing the production of other parts. An FMS should thus make possible improved machine utilization, part scheduling efficiency, and part quality, as well as reduced scrap, in-process inventory, and part setup and handling time.

FMSs are expensive and, to date, have not been widely applied. Most of the FMSs currently in operation have been specially designed for unique customer requirements, and are used to produce a family of similar parts. Several automotive companies have previously

Machine Tools

FIGURE 3.20 A single station palletized cellular manufacturing system. (Courtesy of Cincinnati Milacron.)

experimented with and been disappointed by FMSs for machining applications, because the promise of cost reductions for equipment reuse has not materialized as expected. An FMS requires significant additional investment and long lead times to convert or adapt to new unplanned products. In high-volume applications, they may also require dedicated fixturing and materials handling equipment at significant expense. There are many technical challenges to be considered in developing increasingly useful FMSs, such as tool condition monitoring, chip control, machine diagnostics, adaptive control, automated tool and pallet handling, flexible fixtures, and other untended machining concerns [38, 39]. A real FMS is today defined as a *reconfigurable manufacturing system* (RMS) by some researchers [40–43]. The argument is that FMS is configurable, but not reconfigurable after some years. An RMS utilizes modular system components that are reconfigurable machines and reconfigurable controllers, as well as methodologies for their systematic design and rapid ramp-up, which lead to reduced costs and times for retrofitting and conversion. Similarly, an *agile* manufacturing system conceptually can reallocate production line capacity to products that are in higher than expected demand, rapidly launch new products, and yet retain production ability for other products with lower expected demand [44–53]. The distinction between an agile and a flexible system is not rigorously defined in the literature, but it is generally understood that an agile system is a flexible system in which the machines can be rearranged and reused for a different product.

The design of a CMS or an FMS is very complex because several alternate configurations may be capable of producing a given part. For example, some configurations replacing long serial lines are multiple serial lines in parallel, parallel lines-with-crossover, hybrid or agile configurations (see Figure 3.21) [54, 55]. A cell controller can direct an automated work

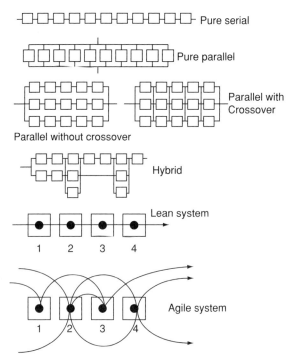

FIGURE 3.21 Differences between manufacturing system configurations and between lean and agile processes. The arrows illustrate the key distinction among the lean and agile systems (the part can take alternate routes in the agile process).

handling system to deliver the part to any one of a number of interchangeable machines, with each machine performing all of the necessary cutting for whatever workpiece it happens to see. Computer simulations are sometimes used to compare alternative designs. The input parameters for simulation include the part geometry, size, material, and tolerances, descriptions of each machining station, tool data, stand-alone availability, and mean times to failure and repair. In addition, constraints between operations such as precedence relations and contiguous allocations must be described. The process sequence can be specified or optimized. The available machine motions and tool data are used to determine the individual machine motions and cutting times. The simulation system should be capable of incorporating various machining centers, automatic pallet changers, material handling systems, buffers, and load/unload stations. A number of heuristic optimization algorithms have been proposed [56–67].

Several material handling methods for transferring parts between machines are available such as sliding or free transfer (roller conveyer systems), overhead crane systems, gantry transfer, robots, and walking-beam transfer. They provide three access planes and require relocation and reclamping of the part at each station. Palletized transfer (Figure 3.20) provides five access planes; parts are fixtured and clamped only one. In large systems, a number of pallets are used, and their dimensional variations must be tracked and compensated for to avoid impacting part quality. Pallet systems used to shuttle parts in and out of the machining stations are better suited for horizontal machines as illustrated in Figure 3.20.

CNC machines have also been integrated in automated flow line systems. An important technical problem in developing easily integrated manufacturing systems is the lack of standards which govern the type and form of information that manufacturing systems must exchange. In 1981 an Automated Manufacturing Research Facility (AMRF) was established to verify new concepts for automated manufacturing [68]. Later the Manufacturing Systems Integration (MSI) project at the National Institute for Standards and Technology (NIST) was created based on the AMRF technology for developing a system architecture that concentrated on the integration of manufacturing systems [69–71].

Lean machining systems represent a different philosophy in applying CNC and conventional machine tools to high-volume production [72, 73]. A lean system is similarly a dedicated transfer line in that they employ one-piece flow. Part transfer between machines is manual, in a sense, the operator provides the flexibility lacking in fixed automation. Lean systems work best when they operate to a fixed takt time and have carefully defined work standards. They also typically employ pull systems with minimal in-process buffers. Conscientious machine maintenance and operator training and motivation are required for optimum results.

3.4 MACHINE TOOL STRUCTURES

The machine tool structure supports the various parts of the machine tool, as well as the part and fixture, and provides rigidity to ensure accuracy. Some typical structures are shown in Figure 3.11–Figure 3.25; these figures illustrate traditional design variations in the base, column, slides, and spindle support. Newer hexapod and tripod structural configurations are shown in Figure 3.26–Figure 3.28, and Figure 3.30. Additional newer configurations not pictured include the open frame (G type), closed frame (portal type), tetrahedral (Lindsey's Tetraform), and spherical (NIST's M3) [6].

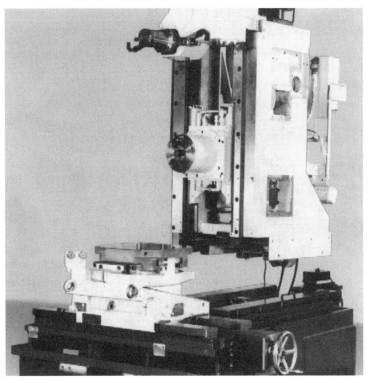

FIGURE 3.22 The structural components of the machining center in Figure 3.13. (Courtesy of Toyoda Machinery USA, Inc.)

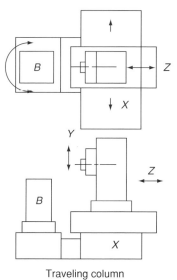

FIGURE 3.23 Structural components of a three- or four-axis horizontal spindle machining center with traveling column and either fixed or rotary table. All X-, Y-, and Z-axis motions are provided in the column. (Courtesy of Cellular Concepts.)

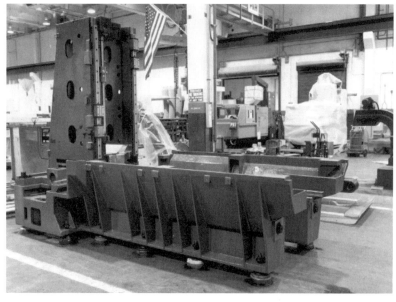

FIGURE 3.24 Cast bed and column components of a machining center with X- and Y-axis motions incorporated in the column and Z-axis motion provided by the table. (Courtesy of Cincinnati Milacron.)

Machine Tools

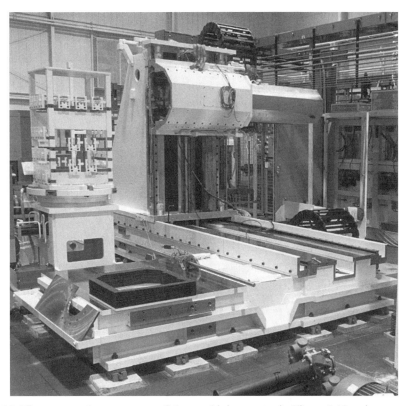

FIGURE 3.25 The structural components of the machining center in Figure 3.12, using magnetic linear motors for all axes. (Courtesy of Ingersoll Milling Machine Company.)

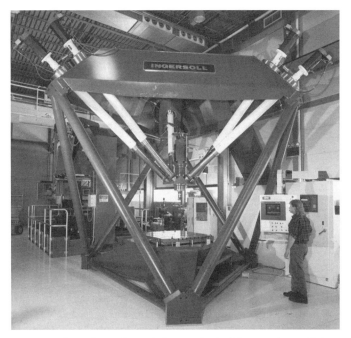

FIGURE 3.26 The structural configuration of the octahedral hexapod machining system (a 6-DOF PLM). (Courtesy of Ingersoll Milling Machine Company.)

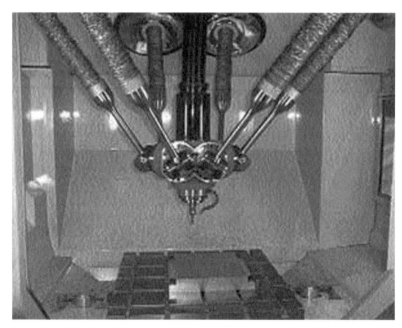

FIGURE 3.27 The structural configuration of the hexapod machining system (a six-axis control "Cosmo Center PM-600") (Courtesy of Okuma America Corp.)

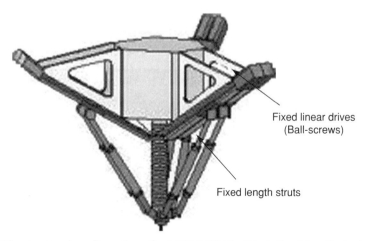

Fixed linear drives (Ball-screws)

Fixed length struts

FIGURE 3.28 The structural configuration of a 6-DOF PKM with six fixed length struts. This five-axis machine has noncoplanar linear joints fixed to the frame (www.toyoda-kouki.co.jp).

In designing the machine tool structure, system rigidity and inertia are the primary considerations. The machine structure must be rigid in order to resist deflections and vibrations due to cutting loads. The structural components should also be as light as possible to minimize the force required for acceleration or deceleration, to increase the maximum acceleration rate, to reduce jerk in machine motions, and to reduce stopping distances and increase machine accuracy. The damping characteristics of the machine structure are also very important because vibration energy is absorbed into the structure; the stability against chatter is determined by the product of modal stiffness and damping [74–76]. The static and dynamic loads including forces of acceleration, deceleration, and cutting must be analyzed. The hardness and elasticity of the material must be balanced in order to withstand impact and allow elastic deformations while preventing cracking or permanent deformations. Thermal expansions and distortions of the machine frame due to external or internal heat sources must be considered. A common current design approach is to start with an initial design based on experience, and refine this design through finite element analysis prior to machine building. Once a prototype is built, the design is further refined using experimental modal analysis and thermal mapping.

In general, a machine tool structure consists of a bed or base, column, ram, and saddle (carriage). Fixed components such as the base are most commonly made of cast iron, nodular iron, steel weldments, composites with polymer, metal, or ceramic matrices, or ceramics. Reinforced or polymer concrete and epoxy granite are sometimes used for machines subjected to high levels of vibration; their application has been limited mainly to grinding machines due to warping, thermal irregularities, and the absorption of coolant. Moving components are made of cast iron, steel, aluminum, and sheet metal. The design of moving columns is becoming increasingly critical as machine travel speeds increase, since the weight must be reduced to reduce inertia while maintaining high stiffness. Castings must be aged and stress-relieved. Steel weldments can reduce the structural weight significantly but require careful design to resist vibration and deformation. Welded bases can be designed with high stiffness and good control of damping because welds in the bases block vibration transmission. Machines using weldment structures are often classified as light-duty machine tools unsuitable for rough cutting or precision applications. Metal–matrix composites, ceramics, and reinforced concrete materials are most often applied in precision or high-speed applications [77–83]. Polymer or concrete filled bases are preferred for grinding machines because they increase system damping and stability. The advantages of ceramic materials include high stiffness and improved dimensional and chemical stability. The machine damping obtained from the structural material itself can be improved significantly by internal means, including: (1) filling structural cavities with oil, lead, sand, or concrete [1, 2, 6]; (2) circulating coolant through the machine structure; (3) allow for microslip at the joints; and (4) attaching a viscous material layer between joints. Damping can also be improved by using shear plates, tuned mass dampers, viscous shear dampers, and active dampers [6, 74, 84–86]. The ideal machine configuration is application dependent. The basics are explained in standards [87] and illustrated in Figure 3.11–Figure 3.18. Most of them have a serial kinematic architecture of the axes, with each axis of movement supporting the following axis and providing its motion. Traveling column designs (Figure 3.12, Figure 3.15, Figure 3.17, and Figure 3.23) have all three axes located on the column, with low rigidity on the cross slide, resulting in significant spindle droop during boring. X/Z table horizontal type or X/Y table vertical type machines (Figure 3.11) eliminate spindle droop; in this configuration the table is the weakest component. The traveling column design with a fixed table generally provides higher rigidity, while the traveling table design increases flexibility and reduces cost. Traveling column designs restrict access to the workzone when the column is moved forward toward the pallet. The traveling column design also simplifies chip control somewhat since a centrally mounted chip conveyor can be used. A horizontal spindle machine is generally more accurate than a vertical spindle machine because the spindle is not cantilevered off a large C-shaped support structure, which is subject to greater deformations. The loads on a spindle create a bending moment on the column of a vertical machine in contrast to the point force on the column of a horizontal machine; thus horizontal machines are more rigid than vertical machines. The spindle can be designed on a ram (quill), as shown in Figure 3.12, Figure 3.15, and Figure 3.25, to reduce the weight carried by the Z slide located on the ram; this improves the traveling speed and acceleration/deceleration of the slide. A fundamental drawback of serial architecture (one axis located on top of another) is that the structure of all axes must be heavy enough to provide the stiffness necessary to limit distortions that lower machine accuracy. This restricts dynamic performance and reduces operating flexibility.

The most common turret lathes are horizontal lathes using one or more turrets with several sides. The gravitational force on the workpiece is a concern in a horizontal lathe since it causes workpiece sag and uneven loading of the spindle bearings. Hence, vertical turret lathes are often used for heavy parts. Vertical lathes are becoming more common for smaller parts as well because they provide easier and more precise workpiece loading. Horizontals allow better chip removal for blind bores and more access for long shaft-type workpieces. The bed of a horizontal lathe can be horizontal, vertical, or slanted. The vertical and slant bed designs reduce thermal distortions and floor space requirements and simplify chip handling. Hybrid turret "turn-mill" NC lathes have milling or machining center capabilities. For milling operations, the workpiece is held fixed, or slowly rotated while rotating tools, or live tools, are brought to it. In these machines, rotary tools may be located on a separate slide or saddle that moves on the main ways or an auxiliary set, or rotary tools may be used in live spindles mounted in the turret itself. The powered turret design using both radially and axially mounted tools has been widely used. Multiple-spindle turning centers are also designed to perform operations on both ends of a part using either side-by-side or face-by-face machine configurations.

Grinding machines are usually similar to a lathe with a grinding wheel feed mechanism substituted for the turret. For heavy (i.e., creep feed) or ultra-precision applications, special designs are required to achieve adequate rigidity and dynamic stiffness.

On many systems, the workpiece is clamped to a pallet which is usually located on an indexing table. The clamping mechanism determines accuracy and repeatability; it should provide sufficient clamp pressure with minimum pallet deflection. A good clamping design is necessary for the effective use of the full vertical axis travel. Acceptable results can often be obtained by hydraulically clamping the table against a curvic coupling mechanism.

Some new machine structures use a parallel kinematic-link mechanism (PKM) in place of rectangular-coordinate serial link mechanisms, eliminating the machine slides [88, 89]. There are two classes of PKM structures: those having joints fixed on base and platform and extensible struts, and those having movable joints with fixed length struts. Fully parallel machines are called hexapod or Stewart platform (octahedron frame) machines [90]. The simple structure of the octahedral-hexapod machine (shown in Figure 3.26 and Figure 3.27) consists of a moveable spindle platform connected to a rigid base through six variable-length links (telescoping legs/struts providing 6 degrees-of-freedom [DOF] workspace) that control the position and orientation of the platform. The struts are connected to the platform and to the base using universal, spherical, or wobbling joints. Expanding and contracting the corresponding number of ball-screws control the attitude and position of the spindle. The interest of PKMs is largely due to their great flexibility in production using more than 3 DOF (resulting in more than three-axis machining and in some cases full five-side machining) and the force-to-weight ratio characteristics of these machines (potential for high dynamic capabilities). Some of the conceptual advantages for PKM relative to conventional machine tools are higher stiffness-to-mass ratio (due to closed kinematic loops and because struts are inherently stiff and light in weight), higher speeds, higher accuracy, modular design (reduced installation requirements), and mechanical simplicity (e.g., linear drives used for rotary movements) [91, 92]. These structures provide accelerations ranging from $1g$ to $3g$ and feed rates up to 100 m/min. The limiting factors for hexapods are [93]: (1) poor ratio of system size to workspace; and (2) requirement of 6-DOF passive joints having high stiffness and minimal backlash. Also, it is not clear that their enhanced performance will be fully utilized and provide a real benefit in many applications. The current systems seem best suited to mold and diemaking.

Structures with fixed length struts are less common for applications in machine tools due to their load characteristics [93–95]. Such nonhexapod fully parallel machines include the HexaM (Figure 3.28) and the Sprint-Z3 spindle (Figure 3.29) [91, 93, 96]. The HexaM has six fixed length struts and the Sprint-Z3 has three fixed length struts (three-axis spindle). The actuators (either ball-screws or linear motors) are fixed to the frame to reduce the maximum inertia effects. Single-DOF hinges have been used at the base end of fixed length struts to prevent any rotation of the platform from the outset.

Unlike a serial machine, the working envelope of a PKM is not cubic and depends on the geometry of the structural components (as illustrated in Figure 3.30a for the PKM in Figure 3.30b). However, five-sided machining (almost full five-axis) has been found possible with a Metrom Pentapod machine using five struts [97]. There are still many problems remaining concerning rigidity and accuracy. Rigidity requires highly rigid joints and accuracy requires proper calibration technology and a suitable control system. At present, these technologies are under development, and parallel–serial hybrid machines provide a better blend of the advantages of the PKMs and serial machines [92, 98–100]. Although several three- to six-axis commercial PKM machines have been available since 1994, especially for milling applications [91], the trend is toward hybrid solutions [89, 91, 93]. Two such systems are the Sprint-Z3 3-DOF PKM spindle system (that provides the Z-axis and two rotary A- and C-axes as illustrated in Figure 3.29) on top of a serial X- and Y-axis system to provide a 5-DOF machine [91], and a 5-DOF configuration using a tripod (3-DOF parallel) and two rotary axes (in serial) under the spindle head as shown in Figure 3.30b [100, 101]. The tripod mechanism consists of four kinematic chains, three variable length legs and one passive leg, connecting the moving platform to the fixed base that contains the bearings of the ball-screws. The platform is joined rigidly with the passive middle strut, which is connected with the base via a linear and universal joint. The struts are coupled via fixed Hooke joints to the base. At the lower end, the struts are joined via preloaded spherical or universal joints attached to the platform. The middle strut is used to constrain the rotary degrees of freedom of the platform so that the motion of the platform has only 3 DOF. A gimbal is used to connect the middle fixed length strut to the platform. A biaxial wrist (rotating head unit) is fixed to the platform to provide two additional wrist axes. The first rotary axis aligns to the middle strut; the second rotary axis is vertical to the first axis and carries the spindle. The two major design parameters affecting the performance characteristics are the selection of the appropriate geometric dimensions and kinematic topology (affecting the stiffness and accuracy). The

FIGURE 3.29 The structural configuration of the Sprint-Z3 spindle on parallel mechanism (3-DOF PKM with three fixed length struts). This 3-DOF head is used in a five-axis machine with X- and Y-axes with conventional drives on the head or workpiece carrier (www.DSTechnologie.com).

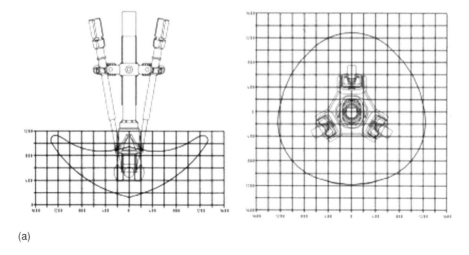

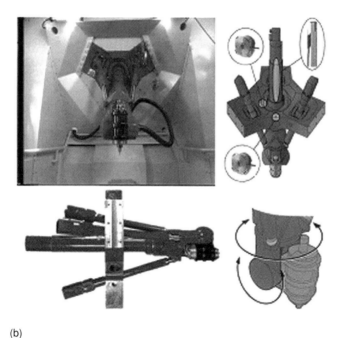

FIGURE 3.30 (a) Workspace sections of the Tricept PKM system. (b) The structural configuration of the Tricept machining system (5-DOF hybrid using a 3-DOF tripod and two rotary axes). (Courtesy of Tricept Inc.)

design of a PKM is very complex compared to serial kinematics and several approaches have been proposed [91]. The tool point stiffness and accuracy for hexapods are often found to be lower than conventional machines when using the same drive components because the strut stiffness is significantly reduced by the flexibilities of the joints at each end [76, 101–105]. Each extensible strut has a spherical joint at each end, which consists of a combination of pretensioned roller bearings ($k = 200$ N/μm), ball bearings, or gimbal ($k = 25$ N/μm). The stiffness

Machine Tools

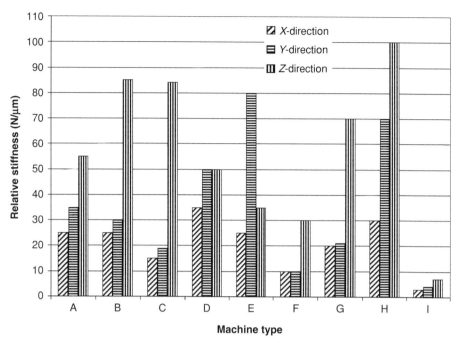

FIGURE 3.31 Relative static stiffness of different parallel and serial (PKM and SKM) machine tools. All machines are measured between table and a dummy tool. (A) Mid-size three-axis VMC. (B) Mid-size three-axis HMC 12,000 rpm spindle. (C) Mid-size three-axis HMC 24,000 rpm, 24 kW spindle. (D) Large-size four-axis HMC. (E) Mid-size five-axis VMC. (F) Large five-axis HMC. (G) Five-axis large hybrid PKM horizontal spindle on a PKM structure carried by serial X- and Y-axes. (H) Six-axis large PKM vertical spindle. (I) Five-axis hybrid PKM vertical spindle.

for a hexapod varies significantly (two to six times) across the workspace, while orthogonal three-axis machine tools have almost uniform stiffness over the workspace. Generally, the structural stiffness of hexapod machines (at least in one direction is less than the stiffness of one of the links) is significantly less than that of the three-axis classical machines but equal or better than the five-axis conventional machines as illustrated in Figure 3.31. In general, the stiffness in the Z-direction is greater than that of each strut itself as long as the tool is located within the space of the base joints. PKMs with fixed length struts could provide higher stiffness than machines with extensible struts.

Another very important parameter for machine tools is the chatter resistance criterion K (the product of the stiffness and damping, which should be greater than the spindle/tool stiffness [76]). The K term for a conventional machining center is generally greater than 6 N/μm, but for PKMs it varies between 0.5 and 4 N/μm. Generally, high-speed machine tools should have structural and drive stiffness $K > 3$ N/μm.

Building accuracy into a machining system is essential and requires building accuracy into both the machine and the process. The machine accuracy is usually expressed in terms of linear positioning accuracies of the individual axes. The linear positional accuracies of many CNC machines vary from 2 to 15 μm depending on the measurement standard. Even though the positioning accuracies of many CNC machines are similar, their volumetric positioning accuracies vary significantly (5 to 10 μm) due to compensation methods, geometric and kinematic behavior, static and dynamic behavior, thermal distortion, and workpiece fixture characteristics as illustrated in Figure 3.32. It is very difficult for a user to determine which

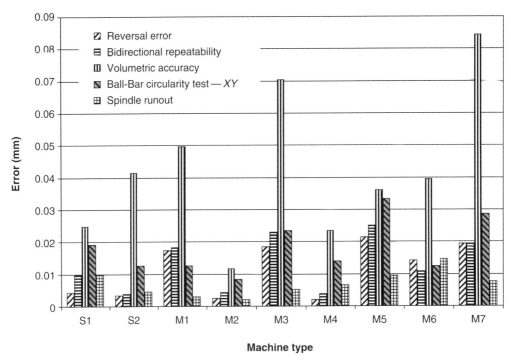

FIGURE 3.32 Comparison of accuracy for various small and medium horizontal CNC machines (the table sizes for the S and M machines are 500 and 630 mm diameter, respectively). All machines were measured with the same instrumentation as defined in ASME B5.54.

CNC machine has the volumetric positioning accuracy necessary for a particular precision part because the straightness and squareness or the volumetric accuracy are not specified in the standard specifications of machine tools. The three-dimensional (3-D) volumetric positioning accuracy is defined as the root-mean-square (RMS) of the sum of all errors in all axes of motion including the straightness errors, squareness errors, and angular errors. The machine's accuracy improves by calibrating and compensating volumetrically. The American Society of Mechanical Engineers (ASME) B5.54 [87] body diagonal displacement tests or the telescoping ball bar performance test can be used to evaluate the volumetric positional accuracy. If the machine is not accurate, however, it requires additional tests to define the straightness, squareness, and angular errors required for compensation.

The volumetric accuracy of parallel kinematic machines depends on how accurately the controller model describes the real kinematic behavior of the machine [106] because the position of the spindle nose is indirectly estimated from the rotational positions of the servo motors for the struts. In addition, the thermal effects are more difficult to model than for orthogonal structures. Due to the highly nonlinear kinematics and dynamics inherent to PKMs, the dynamic response errors are difficult to determine and cannot be assumed constant over the entire work volume. The interpretation of the data obtained when testing a PKM is often quite different than when the same equipment is applied to conventional machines [107]. The major concerns with PKM accuracy are the accurate identification of geometrical (kinematic) parameters, thermally insensitive strut length measurement, and deformation due to gravity [107]. A hexapod-type PKM machine may have hundreds of possible error sources. The volumetric accuracy could be as high as ±0.05 mm with reversal error of ±0.01 mm. Three-axis PKMs have achieved accuracy of 0.01 to 0.02 mm [91]. PKMs require further improvements to fulfill the high acceptance standards necessary for finishing operations in prismatic parts.

Passive means for temperature control should be considered in designing a machine structure. Minimizing and isolating heat sources, minimizing the coefficient of thermal expansion and maximizing the thermal diffusivity of the structural material, and isolating critical components can accomplish passive temperature control. Heat generated by the motors, by friction between moving components (spindle, ball-screws, etc.), and at the cutting zone must be contained and dissipated properly to minimize distortions in critical directions. Ceramics have good dimensional stability but low thermal conductivities. Polymer reinforced concrete conducts heat better than steel or cast iron. The thermal expansion of the machine structure components should be considered especially when dissimilar materials, such as composites and metals, are joined together. Active temperature control by circulating a temperature controlled fluid, air showers, or proportional control can be also used. Wet machining tends to minimize heat buildup in the tool and chips, especially when through-spindle coolant is used. Temperature sensors mounted on the saddle, base, column, and spindle provide feedback to assure machine accuracy and stability in spite of changes in machine temperature. The thermal analysis of the machine should be evaluated after the machine is anchored in the ground if anchors are used for the installation.

Machines should generally be fully enclosed by guards and safety devices. This ensures a clean machine environment and adequate protection of the operator from chip buildup or tool breakage. Optimum designs provide large gradient angles; the way telescoping covers and bed surfaces should be slanted as sharply as possible to allow free fall of chips and coolant to an oversized chip conveyer is shown in Figure 3.33. Single or dual swarf disposal conveyor systems should be designed to cover the tool's entire working range and should

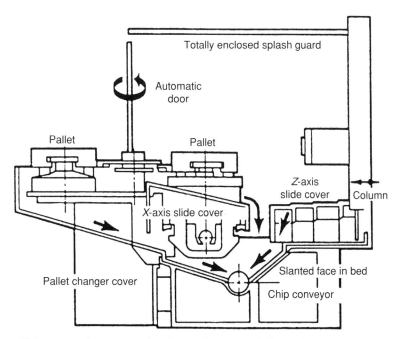

FIGURE 3.33 Chip evacuation approach using a single swarf disposal system. (Courtesy of Toyoda Machinery USA, Inc.)

ascend steeply to the ejection level to minimize coolant loss. In addition, noise from belts and motors can be reduced by isolating their guarding with foam.

Active or passive isolation of the machine from environmental vibrations through proper design of the machine foundation is also important, especially for high-precision machines [1]. Small machines with sufficient stiffness do not require special foundations. Larger machine tools usually require a separate concrete foundation to provide additional stiffness and to minimize the transmission of vibrations from neighboring equipment [108–110]. Separate foundations should be thick enough to provide adequate stiffness, support all machine elements, and be stable on the local soil. There are various types of mounting elements, such as machinery mounts and anchoring/alignment systems, designed for leveling, alignment, and damping of machines [1, 111]. Machines are commonly (a) anchored on a concrete or steel plate, (b) anchored on isolation pads which are located on concrete or steel plates, or (c) positioned on isolation pads without anchoring. Mounts which provide vibration isolation include pads of special materials, leveling mounts, press mounts, and inertia blocks [110]. Rubber mounts with enhanced electrorheological fluids [111] maybe used to reduce the settling time of high-precision, high-speed machines in which accelerations produce large inertia forces. Such mounts become stiffer upon the activation of an electric current during accelerations and decelerations.

3.5 SLIDES AND GUIDEWAYS

Slides are machine components that hold and move a workpiece or tool on guideways (or ways) to a specified position and at the same time absorb the forces generated during cutting. The dynamic and chatter behavior of the machine are affected significantly by the way system because they transfer the cutting forces into the mass of the machine structure. In addition to stiffness and damping, other important parameters considered in the selection and design of guides are machine speed and acceleration/deceleration requirements, load capacity, friction, thermal performance, material compatibility, environmental sensitivity, accuracy, repeatability, resolution, preload, and maintenance.

The directions of motion or the slides are identified by Cartesian coordinates designated as the X, Y, Z, or U and V directions (Figure 3.34) [87]. The Z-axis corresponds to the spindle

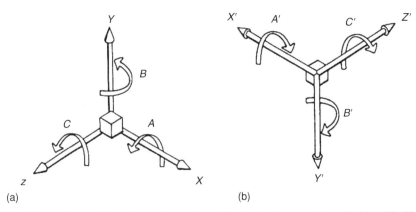

FIGURE 3.34 Nomenclature for axis motions of machine tool columns (a) and tables (b). The X'-, Y'-, and Z'-axes are sometimes alternatively referred to as the U-, V-, and W-axes. (After G. Boothroyd and W.A. Knight [7].)

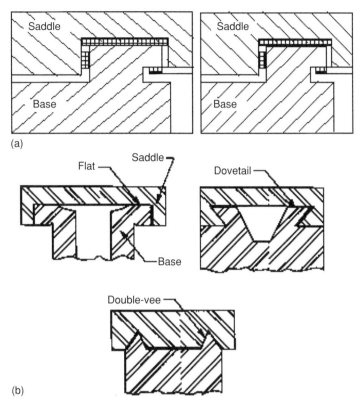

FIGURE 3.35 Common guideway designs.

axis, the X-axis is generally horizontal, and the Y-axis is perpendicular to the X- and Z-axes. Additionally, rotary axes designated as A, B, and C, are also used; their axes coincide with the X-, Y-, and Z-axes.

Common guideway (rail) designs include rectangular, cylindrical, vee, flat, vee and flat, dovetail, and circular-groove monorail (Figure 3.35). Guideways are commonly used in pairs or double tracks to ensure straightness. Dovetail and vee and flat guideways are accurate and have a high stiffness. Features which are easily machined in place are normally integrally cast in the structure, while more complex features, such as hardened steel cylindrical guideways, are bolted to the structure. Integrally cast flame-hardened ways are more difficult to manufacture and are usually hand scraped compared to the "bolt-on" type ways that are easily manufactured and reconditioned. Cast-in-place ways are less expensive but cannot be replaced and are generally used only in machines with short life expectancies. Traditionally, guideways have been made of steel or cast iron, but aluminum, ceramic, and composite guideways have been developed. Various low-friction coatings have been applied on ways to reduce friction and wear [112, 113].

The straightness and parallelism of the ways are very critical since they determine the pitch, roll, and yaw errors of each machine axis. The surface finish and flatness of the ways is controlled by the final manufacturing operation which may be scraping, super finish milling, or grinding plus etching or peening. Scraping a surface generates microgrooves for oil retention. Hardened and ground ways with epoxy-lined mating surfaces are often a good choice. The straightness error of the guideways is dependent on the final characteristics of the material used, the bearing type and its preload, and the roughness of contacting surfaces. For example, a light preload can lead to lower stiffness, error motion, and tool chatter, while high preload could increase bearing wear. In addition, the preload changes (decreases) with time due to the wear of the bearing elements. The ways should typically be parallel to each other and perpendicular to the bed within 0.01 mm, while the tops and bottoms of the ways should be flat and parallel within 0.005 mm [114]. Typical machine straightness accuracies are 0.05 to 0.1 mm/m for a single axis. The straightness accuracy of precise machines may be as low as 0.02 mm/m, and accuracies of 0.002 mm/m are becoming achievable.

The selection of the form for the guideways depends strongly on the type of bearing used to support and move the slide on the way [115]. Bearing systems can be classified as friction or antifriction types as illustrated in Figure 3.35a. Friction systems use a fluid film as a bearing media (e.g., hydrodynamic, hydrostatic, and aerostatic bearings), while antifriction systems use rolling element (linear) bearings.

Sliding contact (or plain sliding) bearings, sometimes called hydrodynamic or box-type bearings, are the oldest and most common type. They consist of two mating surfaces that have been machined, ground, or scraped to obtain the desired coefficient of friction; they provide the best combination of speed and load carrying capacity for general applications. They are used on flat (T way), vee–flat, double-vee, and dovetail guideways (Figure 3.35b). Flat guideways have the best load carrying capacity. They add to the rigidity of the base or column and provide very stable support and damping for large loads spread over a large area. They are insensitive to crashes. The vee shape is very accurate but the loads must be vertical. They use hydrodynamic lubrication supplied by oil under pressure. Sliding contact bearings usually exhibit significant static friction depending on the materials, surface finish, and lubricant; the static friction changes to the dynamic friction as motion begins. Typical static friction coefficients range from 0.03 to 0.3, while dynamic coefficients are usually between 0.01 and 0.1. Stick-slip or stiction action due to higher static than dynamic friction causes positioning errors. The friction is high as motion starts, then reduces and then increases with velocity.

Sliding contact and stick-slip friction can be reduced by coating the ways with a thin layer of low-friction material made of Teflon-impregnated sheets (polytetrafluorethylene — PTFE, Turcite, Rulon) or molybdenum disulfide and graphite poured into place (Diamond DWh310, Moglite, SKC) [6]. For example, Turcite has a friction coefficient of 0.04 compared to 0.2 for cast iron. Pulse-metered permanent lubrication is often used to reduce friction as well. The speed and acceleration of sliding contact bearings are limited to about 15 m/min and $0.1g$, respectively. Their accuracy and repeatability range from 6 to 10 µm and from 2 to 10 µm, respectively, in the axial direction depending on the drive system. Their straightness accuracy and repeatability fall in the range of 0.1 to 10 µm and are affected by the quality and alignment of the guideways. A preload of about 10% of the allowable load must be used to obtain good stiffness for bidirectional loading. Details of the design and analysis of sliding contact bearings and of material considerations for slide and guideway design are discussed in Refs. [1, 6, 112, 116 118].

Rolling element bearings (or linear guides shown in Figure 3.35a) have also been widely used in guideways (as shown on the column in Figure 3.24). The three major types of linear rolling element bearings are the nonrecirculating balls or rollers, recirculating balls, and recirculating rollers. Nonrecirculating systems are used where short travels and compact designs are needed. Cylindrical roller bearings provide higher stiffness and load carrying capacity than ball bearings, due to increased contact area, and are normally used for accurate machines. Recirculating roller bearings provide good load capacity and stiffness, but their alignment is critical.

Generally, rolling linear guides require lower power for motion and eliminate the stick-slip action characteristic of sliding contact bearings, but exhibit less stiffness and damping and lower load carrying capacity. The static and dynamic friction coefficients for rolling element bearings are roughly equal and are typically between 0.001 and 0.01, leading to improved resolution and controllability. Linear guides reduce axis reversal error. The upper limits for speed and acceleration are about 50 to 100 m/min and $1g$ to $2g$, respectively, for adequate bearing life. The upper speed limit for cylindrical roller bearings is generally 20 to 40% lower than for ball bearings. Their axial accuracy can be better than for sliding contact bearings and is in the range of 1 to 5 μm, while their straightness is similar and depends on the tolerance class of the rolling element and rails. Each machine axis typically has four bearing blocks, two on each rail, although many more may be used in high load applications. Rolling element bearings are sensitive to crashes. Generally, sliding contact bearings are used for heavy-duty applications with integrally cast ways (box-type ways), while linear bearing ways are used in lighter-duty and higher-speed applications. Correct linear guide selection allows a linear guide machine to approach slideway machine's rigidity. Details of the design, analysis, and applications of rolling elements for guides are given in Refs. [1, 6, 119–124].

Hydrostatic and aerostatic bearings are used for guideways in precision machines such as grinding and hard turning machines. These bearings have no mechanical contact between elements; the load is supported by a thin film of high-pressure oil or air that flows continuously out of the bearings. Static friction is eliminated and dynamic friction is insignificant at most speeds. Hydrostatic bearings are used at moderate speeds where high load capacity and stiffness are required, while aerostatic bearings provide a moderate load capacity and stiffness and are preferable at higher speeds. Hydrostatic bearings provide better vibration and shock resistance with superior damping characteristics. The dynamic stiffness is very high due to squeeze film damping. The damping is very good normal to the bearing surface but low along the direction of motion. A great deal of research and development work on analysis and development of hydrostatic and aerostatic bearings has been carried out [1, 6, 125–133]. Self-compensating or gap compensating bearings, which can be used with water as a bearing fluid, have recently been developed [6]. The use of water rather than oil as the sliding medium results in more stable temperatures, higher permissible speeds, and fewer fluid contamination problems. High performance hydrostatic slides have been made entirely from alumina ceramics. The Hydroguide and HydroRail are designs using a profile similar to linear rolling element bearings (Figure 3.36), and have been used in grinding and hard turning lathes [134].

Currently, slide speeds are generally restricted to below 20 m/min, although machines with slide speeds between 40 and 50 m/min are not uncommon and speeds between 70 and 100 m/min have been achieved on some high-speed machines. The maximum feed rate is typically between 40 and 50% of the maximum rapid travel rate.

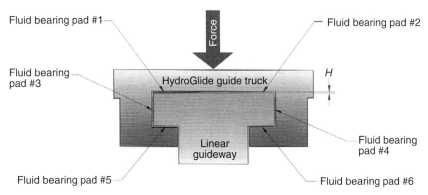

FIGURE 3.36 Example of the self-compensating HydroGlide^{TM} system with fluid gap H. (Courtesy of Aesop, Inc., and MIT.)

3.6 AXIS DRIVES

Relative motion between the tool and part in a machine tool is achieved by moving the machine table, spindle, or column separately or in combination. In any case, an axis drive (or servo) system is required. Desirable features in an axis drive system include accuracy, repeatability, and high dynamic stiffness. Other significant performance characteristics include the bandwidth or response time, peak drive current, servo gain, motion control, and torque-to-inertia ratio (acceleration/deceleration capability), as well as smoothness and accuracy.

An axis drive system (Figure 3.37 and Figure 3.38) consists of a drive motor or actuator, mechanical transmission elements, sensors, and a controller. The principal types of linear motion actuators are hydraulic or pneumatic pistons, conventional electric motors, and linear motors.

Hydraulic systems exhibit high dynamic stiffness and damping, can sustain high machine and cutting loads, and have good acceleration and power-to-weight ratios due to their low mass. They are most commonly used on single axis machines, especially on transfer lines. They are not widely used in machining centers because they are relatively inaccurate, generate excessive heat, and are often difficult to control.

Electric motors [135–141] have proven much more versatile and are the most common drive actuating systems. Both AC and DC drive motors are used. AC induction motors are more common because they require less maintenance and provide a greater bandwidth, higher gains, and better repeatability. AC motors, however, are less efficient and provide lower torque-to-inertia ratios. (Brushless, frequency-controlled AC drives, however, deliver considerably more power than comparable three-phase or DC motors.) The principal advantage of DC motors is that they generally produce higher torques at high speeds. A hybrid type of motor, the brushless DC motor, combines the advantages of both AC and DC brush motors. Brushless DC motors, commonly known as AC servos, are more durable than AC induction or synchronous motors, generate less heat and vibration, and are generally smaller in size and more efficient. Electric motors can be further divided into stepping and nonstepping motors. Stepping motors generate a precise motion increment in response to an electric pulse. They provide limited torque but are very accurate and more easily controlled than other types of

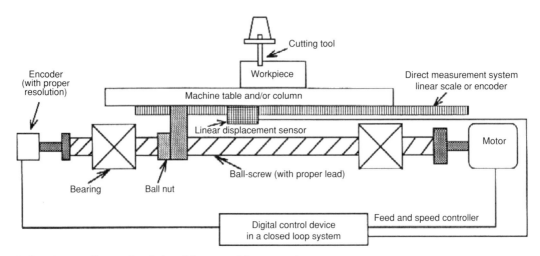

FIGURE 3.37 Conventional closed-loop machine control system.

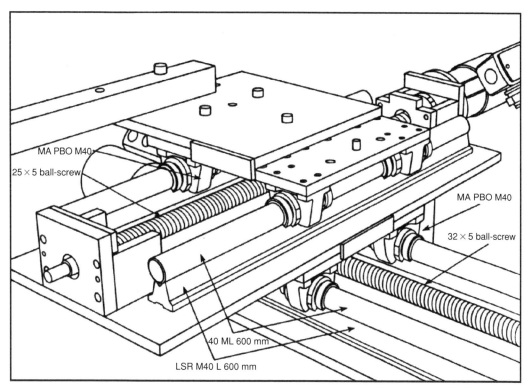

FIGURE 3.38 *X–Y* machine table system using linear motion systems for both rail bearings and the ball-screw.(Courtesyof Thomson Industries, Inc.)

motors. The best motor for a given application depends on the torques necessary for maximum acceleration.

Linear motors or *direct drives* (Figure 3.39) eliminate the mechanical transmission system required with rotary motors [6, 142–144] since they combine the functions of the ball-screw and motor. They are made with permanent or induction type magnets; induction magnets generally provide better performance and reliability. The force is transmitted through a magnetic field instead of a mechanical linkage. Because they are noncontact devices (except in the way system), they exhibit little wear and require less maintenance. Also, they can run at high speeds (up to 150 m/min) with high acceleration/deceleration characteristics ($>2g$) and eliminate deflections, backlash, and windup since bearings and ball nuts are not used. They can provide higher tool feed rates (almost twice that of ball-screws) with equivalent or increased accuracy when compared with traditional ball-screw systems since they can be controlled using many times the gain of their rotary counterparts [145]. The length of travel does not affect the performance of a linear motor as it does a ball-screw. However, they have a comparatively low dynamic stiffness and limited load capabilities. In addition, the utilization of linear drives is limited to applications with relatively small cutting forces (<5000 N) [145]. Higher forces can be achieved by placing separate drives in parallel. They require special machine designs and heat dissipation approaches due to their low mechanical efficiency. They are not yet a suitable replacement for ball-screws in all cases [146] but are used for high-speed machining applications not possible with other systems or for large machines made for very large dies, molds, and aerospace parts.

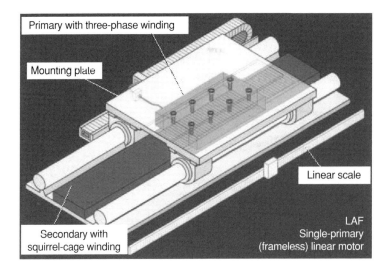

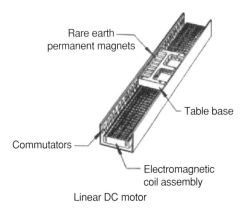

FIGURE 3.39 Linear axis drive motors. (Courtesy of The Rexroth Corporation, Indramat Division.)

When rotary electric motors are used to drive axes, a mechanical transmission is required to convert the motors' motion a linear slide motion. The most common type of transmission is the lead screw (Figure 3.37 and Figure 3.40), which is either directly coupled to the motor or connected to it by a toothed belt [1, 6, 115, 147–149]. Traditionally, the lead screw is driven and the lead screw nut is fixed rigidly to the table; the screw is supported with two bearings near the nose and two near the rear, with the rear bearings mounted on floating mounts to compensate for shaft expansion. There are various types of lead screw mechanisms including sliding contact thread screws (power screws), recirculating or nonrecirculating ball-screws, planetary roller screws, recirculating roller screws, traction drive screws, and hydrostatic nuts and screws. In each device, a nut is essentially sliding along a screw; the nut advances linearly with each rotation of the screw by an increment dependent on the screw pitch. It is essential that preload be established to eliminate backlas and increase stiffness. Preload can be provided by forcing the nut in one direction, using a doubled-nut design, or using oversized balls or shifting pitch. The preload is typically one third of the rated maximum load.

Special manufacturing processes are required for ball-screws used in high-speed machines to prevent errors and wear due to friction [150]. Ball-screws are generally better than lead

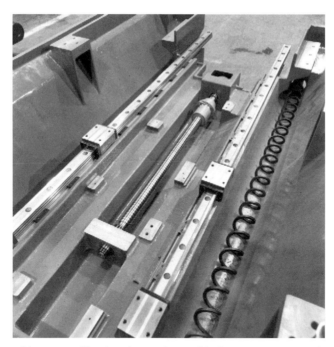

FIGURE 3.40 The guideways, rolling bearing packs, and ball-screw system used on the machine bed shown in Figure 3.24. (Courtesy of Cincinnati Milacron.)

screws because they are ground and can offer greater maximum speed, accuracy, and load capacity. The accuracy also depends on alignment, preload, and system stiffness. The location of the ball-screw is very critical to its performance because it affects both accuracy and durability. Generally, a ball-screw should be mounted as closely as possible to the guideway of the primary motion when only one of the ways is responsible for location and the other primarily supports the load. The ball-screw should be precisely centered when both guideways control the location and guide functions. In a horizontal machining center, two ball-screws may be used to drive the Z-axis so that the center of the base is left open for chip evacuation. Similarly, two ball-screws may be used for the Y-axis to eliminate the need to counterbalance the spindle carrier. In these cases, one of the two AC servomotors is used as the master and controls the axis movements and the other is the slave motor that follows. The ball-screw should also be closer to the table surface in this case. The ball-screw shaft can be hollow to reduce inertia. In addition, passing coolant through the core of ball-screw helps to maintain a constant temperature and prevents thermal expansion, especially during rapid feed, rapid acceleration and prolonged operation. For increased rigidity, the end brackets of balls-crews should be integrally cast with the base and column. With the present ball-screw materials, the upper limit of the rapid traverse speed is 50 to 60 m/min with an acceleration/deceleration capability of 8 to 10 m/sec^2. In recent development work [151], ball-screw systems with rigidly mounted screws and servo-driven nuts have reportedly achieved traverse speeds up to 120 m/min and accelerations up to 25 to 40 m/sec^2.

Less common types of transmissions include rack and pinion, ram and piston, and friction drive designs. *Rack and pinion* transmissions are used in large machines where long drives are needed (>4 m). *Ram and piston* drives are typically used for short displacements since they employ hydraulic cylinders. *Friction* or *traction* drives are used mainly in high-precision machines. They provide minimum backlash and low friction but have a low load capacity and moderate stiffness and damping.

The servo systems described above may exhibit backlash. Backlash can be eliminated when a precision gear-type speed reducer is used. Precision machines use harmonic drive gearing with elastic body mounts, rather than conventional gearing and rigid mounts. Lost motion due to backlash, shaft deflections, bearing looseness, and other sources is corrected for through software compensation. A feedback device at the opposite end of the ball-screw (see Figure 3.37) confirms that the CNC command of a certain number of revolutions (corresponding to a specified linear distance) has been executed.

The ratio of the load inertia to rotor inertia is an important factor in servo system performance. Inertial loads that are too large for the motor often result in sluggish response, causing overshoot, slow settling time, and stalling when the rotor position falls too far behind the drive signal. This problem is avoided by limiting the load-to-rotor inertia to a maximum of 10:1. For a large speed range, the motor and drive can be designed so that the torque is a function of the motor current at low speeds and a function of the motor voltage at high speeds; this approach optimizes motor performance. The damping characteristics of the servo are also important since they affect system stability. Ensuring proper damping, however, is difficult because it depends on several factors.

The acceleration/deceleration characteristics of the axis drives are especially important for high-speed machines used for rigid tapping, end-milling, and similar processes. There are two basic strategies in accelerating and decelerating an axis: the "acceleration rate" and "time to acceleration" methods. The "acceleration rate" method can improve precision at higher speeds while maintaining the desired path speed, but can result in machine shock at a corner in the workpiece. The "time to acceleration" method can reduce this shock but can cause path errors at corners. Therefore, the best approach is to combine both methods in high-speed, high-precision applications. The positioning distances and the value of the position controller's proportional velocity gain are the key parameters because they determine the following error and thus the time response of the system as it accelerates/decelerates [145]. High acceleration/deceleration coupled with high-resolution control systems that prevent overshoot during slide reversal should be used for threading at higher cutting speeds. The acceleration/deceleration over $1g$ and high proportional velocity gain are very important when machining parts with short positioning distances (≤ 50 μm) [145].

The natural frequencies of the system are affected by the gears and ball-screw as spring elements and the masses of the worktable and load. The natural frequencies of a linear motor drive system are significantly higher than a ball-screw system due to the lack of mechanical transmission elements.

Sensors and controls are very important components of the axis drive because they determine the positional accuracy of the drive system. The difference between a servo system and a simple motor is the fact that the position of the driven device for the servo is constantly being monitored. The positional accuracy depends on the feedback or closed-loop communication system, which compares the actual and programmed locations to determine the servo lag or following error. The controller is often the only intelligent segment of the servo system and can be a separate or integral part of the machine's controller.

Control of the table slide's position is accomplished by attaching a transducer to the motor or the table, slide, or other driven device. The most common approaches are to attach an encoder to the servo motor shaft or a resolver on the motor, and to attach linear sensors or linear encoders (scale feedback) to the moving machine elements. The latter method is more accurate because it measures the real motion produced by the motor and not just the number of shaft rotations. High-precision linear encoders use a graduated glass scale and an interferential measuring principle to measure distances (see Figure 3.39). For high-precision machines, both methods can be used to improve the axis feedback response.

3.7 SPINDLES

The spindle forms the interface between the machine tool structure and the cutting tool, and its properties determine how efficiently and accurately the motion capabilities of the machine tool are transferred to the cutting tool. The spindle is therefore one of the most important machine components, and must be designed to ensure accuracy and durability when subjected to the expected thrust and radial loads. Typical spindle requirements are running accuracy, speed range capability, high rigidity, minimal level of vibration, low and stable operating temperatures, long life, and minimal need for maintenance. Some of the critical factors to consider in designing a spindle are drive choices (belt driven, gear driven, motorized), bearing arrangement and mounting, bearing types (frictional characteristics and size), permissible operating conditions, external cooling conditions, weight, thermal growth, operating environment and bearing seals, method of lubrication and type of lubricant, resonant frequencies, allowable static overload, and tool retention requirements [152, 153]. Compromises must generally be made in order to provide the best combination of speed, power, stiffness, and load capacity. Spindles should also be designed for quick replacement to ensure rapid recovery from bearing failures or crashes.

Spindle performance is most often characterized by its torque and power versus speed characteristics. Spindle power–speed and torque–speed curves (Figure 3.41) are very important to machine performance and should be evaluated over the complete speed range for

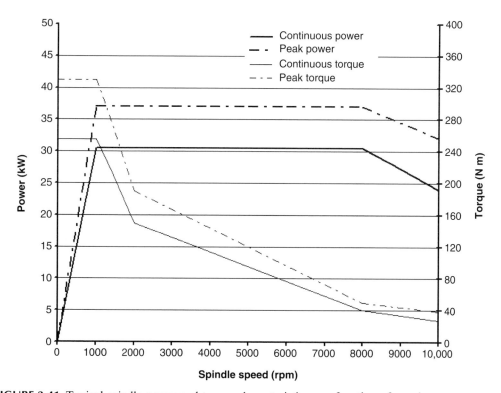

FIGURE 3.41 Typical spindle power and torque characteristics as a function of speed.

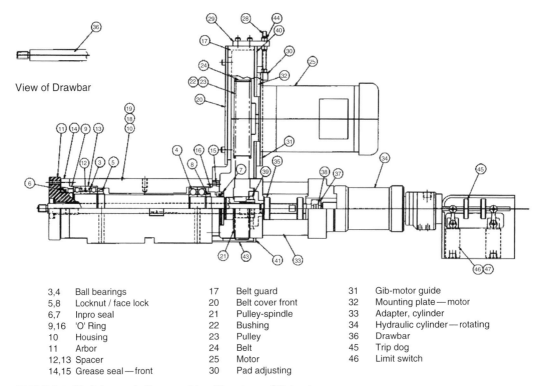

3,4	Ball bearings	17	Belt guard	31	Gib-motor guide
5,8	Locknut / face lock	20	Belt cover front	32	Mounting plate—motor
6,7	Inpro seal	21	Pulley-spindle	33	Adapter, cylinder
9,16	'O' Ring	22	Bushing	34	Hydraulic cylinder—rotating
10	Housing	23	Pulley	36	Drawbar
11	Arbor	24	Belt	45	Trip dog
12,13	Spacer	25	Motor	46	Limit switch
14,15	Grease seal—front	30	Pad adjusting		

FIGURE 3.42 A box spindle assembly. (Courtesy of Setco.)

continuous and intermittent duty. Proposed speeds and feeds should be checked against the spindle characteristics. Large diameter tools are typically run at low rpms to reduce the cutting speed, but on some machines this is not practical since the available torque drops sharply at speeds less than 20% of the maximum rating.

There are many types of spindles which may be broadly categorized as box spindles and motorized spindles. *Box spindles* (Figure 3.42) are driven externally by an electric motor; common drive mechanisms include belt drives, toothed belt drives, shaft couplings, and gear case speed reducers. Belt drives are most common on general purpose machines. Adjustable belts should be used on belt driven spindles to reduce bearing stresses and improve spindle accuracy. Gear drives are normally used for high power spindles and machining centers. Although gear trains may exhibit backlash and lost motion due to gear profile errors, poor gear tooth surface finish, and vibration, properly designed gear drives can increase drive rigidity and reduce susceptibility to malfunction when compared with toothed belts. Planetary gear systems are also used to provide a wide torque output. Variable-speed direct drives with shaft couplings are available for higher-speed applications. The spindle shaft incorporates the tooling system, including the tool taper, drawbar mechanism, and tool release system.

The motor is an important spindle component, which determines many performance characteristics, including the weight, inertia, size, acceleration/deceleration, torque, and power. The most common spindle drive motors are induction motors, variable-speed AC motors, and DC motors [154–157]. Generally, AC motors provide quiet, vibration-free rotation at controllable speeds, but have significant rotor and stator losses and comparatively low efficiencies and torque-to-inertia ratios. AC induction motors are commonly used in transfer machines and other constant speed applications because of their low cost, availability, and maintainability. Variable speed AC synchronous motors are common on NC machines. Their speed is controlled by input frequency changes; an input voltage change results in a torque change. Variable speed DC motors are used when high power and a very wide speed range are required. DC motors produce less rotor heat than AC motors. Permanent magnet (PM) brushless DC motors, switched reluctance (SR) brushless motors, and synchronous reluctance motors are the basic types of DC motors. PM brushless and SR motors are used in high end motor applications in which weight, size, and efficiency are critical. PM motors are generally smaller and more efficient than equivalent SR or AC induction motors. Brushless DC motors develop higher torques at lower speeds compared to AC motors (except vector-controlling AC induction motors), and are less sensitive to harmonics, which can create audible noise. Permanent magnet rotors result in substantially decreased rotor temperature, which improves bearing life.

A wide variety of spindle characteristics can be obtained since the spindle power, torque, and speed are largely dependent upon the driving motor, so that the final specifications can be modified for a particular application by using a different motor or belt ratio. High power and torque are possible because the spindle motor is mounted externally from the actual spindle shaft, and therefore it is often possible to use a very large motor compared to integral motor-spindles usually having limited space. Belt-driven spindles are limited in maximum rotational speeds due to the limitations of belts and gears at higher speeds. Typically, belt-driven spindles are used to a maximum rotational speed of 12,000 to 15,000 rpm and power ratings up to 30 HP.

In a *motorized spindle* (Figure 3.43), the electric drive motor is directly integrated into the spindle housing or body, eliminating the mechanical transmission. They are better suited to higher-speed applications because they eliminate vibrations associated with gear trains and drive belts. The rotor is usually placed directly between and supported by the spindle shaft

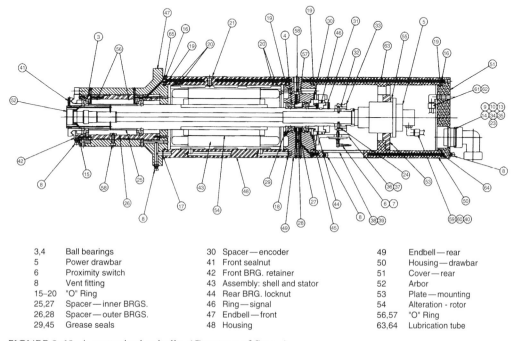

3,4	Ball bearings	30	Spacer — encoder	49	Endbell — rear
5	Power drawbar	41	Front sealnut	50	Housing — drawbar
6	Proximity switch	42	Front BRG. retainer	51	Cover — rear
8	Vent fitting	43	Assembly: shell and stator	52	Arbor
15–20	"O" Ring	44	Rear BRG. locknut	53	Plate — mounting
25,27	Spacer — inner BRGS.	46	Ring — signal	54	Alteration - rotor
26,28	Spacer — outer BRGS.	47	Endbell — front	56,57	"O" Ring
29,45	Grease seals	48	Housing	63,64	Lubrication tube

FIGURE 3.43 A motorized spindle. (Courtesy of Setco.)

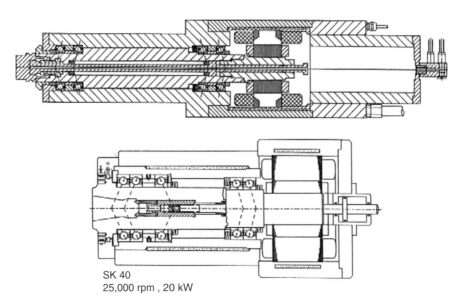

FIGURE 3.44 Motorized spindles with "flywheel" motor arrangement. (Courtesy of The Precise Corporation.)

bearings. This design provides comparatively short and compact spindles with adequate stiffness; the spindle natural frequency can be very high, allowing high subcritical (below the first natural frequency) operating speeds. Occasionally, the motor is placed behind the two bearing systems at the end of the shaft (Figure 3.44). This "flywheel motor" design can provide higher stiffness since the shaft diameter between the bearing systems is not limited by the rotor bore diameter; it also decreases the spindle clamping diameter and thermal expansion of the shaft. On the other hand, the rotor mass extending in the rear limits the natural frequency of the spindle. This frequency can be increased by adding a third bearing to support the very end of the shaft, but this requires extreme alignment accuracy and control of the angular thermal expansion between the three bearing positions. New motorized spindle designs nearly match the torque/speed characteristics of gear driven box spindles; consequently, they are finding increasing use in high-speed machining centers and machines designed for high precision and C-axis capability.

The spindle's torque and power characteristics are adjusted by a frequency converter. Typically, spindles are used in a constant torque range, where power and voltage increase linearly with frequency. However, greater power at lower speeds can be obtained by programming the constant torque range to reach maximum voltage before reaching maximum frequency; from this point up to the maximum frequency, the voltage remains constant and the torque decreases (see Figure 3.41). This design approach results in lower spindle power at top speed. Another important characteristic of the spindle is the control loop is open or closed (vector control). Closed-loop spindles incorporate an encoder for speed feedback. Closed-loop spindles are necessary for rigid tapping, supporting orientation, and high torque applications.

The spindle shaft is held in position by a set of high-precision bearings. In many cases and especially for high-speed spindles, *bearings* are the critical components of the spindle. As for slides and guideways, several types of spindle bearings are available including rotary bearings (angular contact ball and rolling element), hydrostatic bearings, hydrodynamic bearings, aerostatic bearings, aerodynamic bearings, and magnetic bearings. Table 3.2 summarizes

TABLE 3.2
Comparison of Various Types of Spindle Bearings

Characteristic	Rolling Element	Hydrodynamic	Hydrostatic	Aerostatic	Aerodynamic	Magnetic
Cost	Low	Medium	High	High/medium	Low/medium	Very high
Maintenance	Low	Medium	Medium/high	Medium/low	Low/medium	Low
Damping	Low/medium	High	Very high	Low	Medium	Medium/high
Stiffness	Medium/high	High	Very high	Medium/low	Low/medium	Medium/high
Load capacity	Medium/high	Medium/high	High	Low	Low	Medium/high
Running accuracy	Medium/high	High	High	High	High	High
Speed range	Medium	Low	Medium/high	High	High	Very high
Wear resistance	Medium	Medium	High	High	High/medium	Very high
Power loss	Low	High	High	Medium	Medium	Very low
Cooling capacity	Medium	Medium	High	Medium/low	Low/medium	Medium/low
Environmental factors	Low/high	Low/high	Low/high	Low	Low	Low
Reliability	Medium/high	High	High	Medium/high	Medium/high	Medium/high

the characteristics of the various bearing types. *Rotary bearings* for spindles are similar to linear bearings for slides (Section 3.4) and are subject to the same strengths and limitations. Similarly, *hydrostatic*, *hydrodynamic*, *aerostatic*, and *aerodynamic* spindle bearings are similar to corresponding slide bearing types. They can be installed parallel, perpendicular, or at an angle to the axis of rotation to resist radial, axial, or combined loads as shown in Figure 3.45. In *hydrodynamic* bearings, the shaft turns in a sleeve containing pressurized oil or water that generates a hydrodynamic film. True hydrodynamic bearings do not contain a prepressurized film; the film separating the sleeve and the shaft is established entirely by relative motion between the shaft and sleeve, which draws the fluid into the bearing gap. The motion of the fluid film creates pressure and lifts the shaft to create a gap. Most fluid film bearings, especially high-speed bearings, are initially hydrostatic at lower speeds so that frequent high-speed starts and stops do not damage the bearing surface, but develop a significant hydrodynamic component at higher speeds. These bearings are typically referred to as hybrid-hydrostatic or hydro-stato-dynamic bearings. They have very good damping properties and are well suited for high-performance and high-precision machining and may yield useful lives exceeding 20,000 h [152]. New hybrid-hydrostatic bearing designs have reduced power loss and can be used at speeds up to 100 m/sec, making them an attractive alternative for high-speed applications. Several hydrostatic bearing spindle systems have been used as illustrated in Figure 3.45 and Figure 3.46. The major concern with hydrostatic bearings is frictional loss, which is typically between 35 and 50%. The dampening characteristics of these spindles increase cutting tool life and overall part accuracy and improve part surface finish. *Magnetic* bearings have been used recently for rotary applications [153, 158–160] especially at very high speeds relative to their load capacity. Radial and axial magnetic bearings are used in the

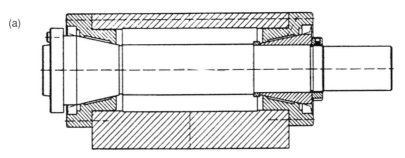
Spindle with hydrostatic sliding bearings in "O" arrangement

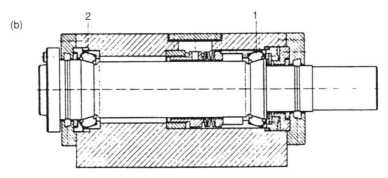
Spindle with hydrodynamic sliding pad bearings in "O" arrangement

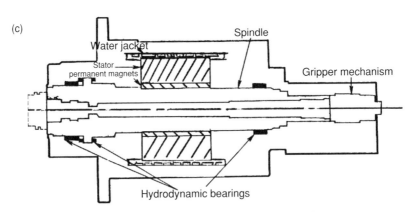
Spindle with hydrodynamic sliding axial and radial bearings

FIGURE 3.45 Spindles with hydrostatic/hydrodynamic bearings. ((a) and (b) courtesy of Pope Corporation/FAG Kugelfischer Georg Schafer KGaA; (c) courtesy of Ingersoll Milling Machine Company.)

spindle shown in Figure 3.47. There is no shaft wear because they are frictionless bearings. They can achieve the fastest shaft surface speeds (e.g., the highest DN; DN is the speed index and is defined later in the text) currently available. Catch bearings are used to hold the shaft during a power failure and prevent catastrophic contact. The electric fields of the bearings can be controlled to optimize stiffness and damping characteristics. With proper control, they can also be made self-balancing. Their benefits include: no mechanical contact (no wear or heat),

Machine Tools

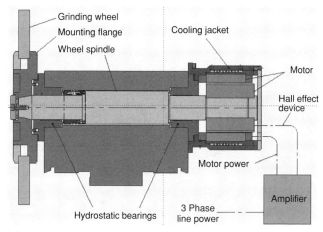

(a) Hydrostatic spindle for grinding machines

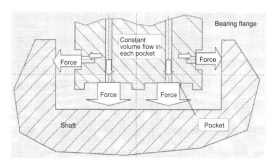

(b) Gyroscopic hydrostatic bearing construction with axial and radial capacity

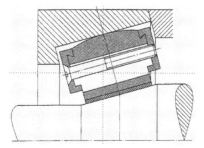

(c) Hydrostatic bearing construction with axial and radial capacity

FIGURE 3.46 Spindles and hydrostatic bearings. ((a) courtesy of Landis Gardner, a UNOVA Company; (b) courtesy of FAG Kugelfischer Georg Schafer KGaA; (c) courtesy of IBAG North America.)

very high accuracy, very high-speed capability (twice that of ball bearings), and provision for process monitoring. The limitations are: high cost compared to mechanical systems, limited load capability, moderate stiffness, and sensitivity to dynamic conditions. *Air bearings* are used at higher speeds, most commonly for grinding but also for milling, boring, and microdrilling. Their load capacity is low but they have excellent runout and long life and are suitable for ultra-precision machining. These bearings are extremely sensitive to contamination and moisture in the air. If the load capacity of the air film is exceeded, the film collapses and the bearing makes contact, causing catastrophic failure.

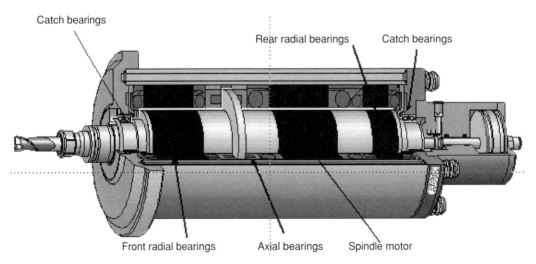

FIGURE 3.47 An active magnetic bearing spindle. (Courtesy of IBAG.)

There are two basic types of antifriction or rotary element bearings [161, 162]: ball or point contact bearings and roller or line contact bearings. *Ball bearings* typically provide lower stiffness compared with roller bearings but generate less heat; they are widely used in all types of spindles, but are especially common in high-speed, low-load applications. Angular contact ball bearings are most commonly used in machine tool spindles because they help to retain the preload at higher speeds. The contact angle determines the ratio of axial to radial loading possible, with radial load capability being the primary benefit. Typically, contact angles of 12°, 15°, and 25° are available. The higher the contact angle, the greater the axial load carrying capacity. Therefore, it may be desirable to use a bearing with a contact angle of 25° for a spindle that will be used primarily for drilling, and a contact angle of 15° for a spindle that will primarily be used for milling. The ball diameter becomes significant at high speeds due to centrifugal loads, which can damage the race and reduce bearing life. This problem is most acute at speed indices greater than 800,000 DN [163]. One response to this problem is to use smaller diameter balls to reduce ball mass and forces. Hybrid ceramic ball bearings are comprised of steel inner and outer races, ceramic (silicon nitride) balls, and a retainer cage. Ceramic ball bearings, being 60% lighter than steel with a higher modulus of elasticity and thus providing a smaller contact area and greater stiffness, are preferred at high speeds (ceramic balls allow up to 30% higher speeds for a given ball bearing size before the centrifugal force begins to deform the balls resulting in reduced bearing life). Ceramic ball bearings generate less heat and run about 10°C cooler than all-steel bearings with grease lubrication in comparable applications [164]. The thermal expansion of ceramic bearings is negligible, resulting in longer life and better accuracy due to lower friction and linear wear. Tests have shown that spindles using hybrid ceramic bearings exhibit higher rigidity and have higher natural frequencies, making them less sensitive to vibration. Standard dimensions of hybrid and steel bearings are the same, so a switch to hybrid bearings requires no design changes. *Roller bearings* may be of the cylindrical, spherical, tapered, or needle type [120]; they resist shock and impact loads better than ball bearings, but generally must be used at lower speeds and higher loads, and are more subject to misalignment. Straight and tapered roller bearings are commonly used in machine tool spindle applications. Line contact bearings are stiffer than point contact bearings by a factor of 6 to 8, and have higher load carrying capacities. Double-row cylindrical roller bearings are typically recommended to carry high

TABLE 3.3
Selection Criteria on Selecting Bearings for Spindles

Spindle Requirements	Bearing Selection	Spindle Design Impact
High torque	Larger bearings	Large shaft, lower speed
High speed	Low contact angle	Small shaft, low power
High stiffness	Roller bearings, larger size	Large shaft, lower speed
Axial loading	High contact angle	Lower speed
Radial loading	Low contact angle	Higher speed
High accuracy	ABEC 9, high preload	Lower speed

loads. Single-row bearings are used at higher speeds. Special bearing types such as hydra-rib [120] and hollow roller [165] bearings have also been investigated and appear to provide advantages over conventional bearings in some applications.

Generally, the selection of a bearing type depends on many factors such as diameter, thermal stability, stiffness requirements, load carrying capacity, speed, and lubrication as explained in Table 3.3. This table shows that many factors that determine the final spindle design and that compromises must generally be made to meet all requirements. Bearing sizes determine the spindle shaft diameter and therefore the stiffness of the shaft. Increasing the bearing diameter increases stiffness and load carrying capacity but reduces speed capability. The *speed index* DN, which is determined by multiplying the bearing bore diameter (mm) with spindle speed (rpm), is used as a measure of a bearing's relative reaction to speed. The bearing speed rating is also expressed in terms of D_mN (using the bearing pitch diameter [mm] multiplied by the rpm); the DN value is usually about 50 to 88% of D_mN. This factor is determined based on fatigue tests and is thus a function of the mechanical properties of the bearing materials. Actual limiting bearing speeds, however, are dependent on the bearing application (mounting arrangement and preload method) and lubrication as well as the speed factor. Typical speed ratings for oil jet, oil mist, or oil–air (ratio 1:2) bearings are 10^6 to 2.5×10^6 DN for angular contact bearings and 600,000 to 900,000 DN for roller element bearings; speeds below these levels can be achieved comfortably with grease-packed bearings. Current limits in standard products are about 1.5×10^6 DN for grease-packed ceramic ball bearings (average below 800,000) [152, 153]. The highest DN values are obtained with radial and angular contact single-row bearings with light preloads, oil lubrication and external cooling. When the DN value exceeds 800,000, the application is considered a high-speed application and cooling of the spindle becomes necessary. The accuracy of a bearing is defined by the ABEC (Annular Bearing Engineers' Committee) number; many machine tools are using ABEC-5 bearings, although high quality and precision spindles use ABEC-7 or -9 bearings. The bearings are typically combined and customized to meet a machine's specific operating requirements; as noted above, this requires a compromise between several performance characteristics such as load carrying capacity, rigidity, speed, preload, and tolerance.

The bearing service life at conventional speeds, at which the effects of load are dominant, is usually considered equivalent to the fatigue life, although the actual service life may be limited by wear due to environmental factors (contamination, corrosion, heat, or poor lubrication) and mounting problems. The major causes of bearing failures are contamination and lubrication breakdown due to inadequate sealing [152, 153]. High-speed spindles are typically designed to yield fatigue lives of 2000 to 3000 h under specified axial and radial

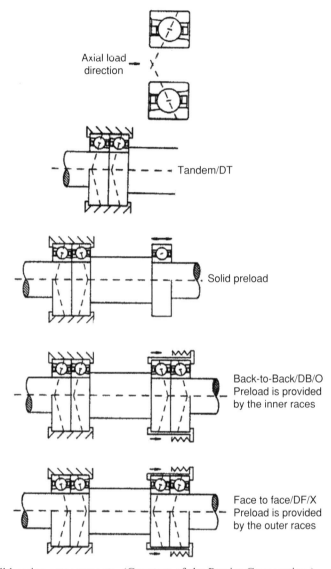

FIGURE 3.48 Ball bearing arrangements. (Courtesy of the Precise Corporation.)

loads. Under normal operating conditions, 3000 to 5000 h of service life can be expected, provided the bearings do not become contaminated and vibration remains within acceptable limits.

The bearing arrangement in a spindle has a strong influence on spindle performance. Typical angular contact bearing arrangements, shown in Figure 3.48, include back-to-back "DB" or tandem "DT" arrangements. Tandem mounting does not allow forces in both directions, unless another pair of bearings is used on the spindle shaft, facing in the opposite direction. Generally, two or three bearings are used at the front in a tandem setup and another tandem set is added at the rear of the spindle. Together, the bearing sets form an overall "DB" setup (as illustrated in Figure 3.44). Bearings should be positioned in an arrangement that does not allow loss of preload when the cutting forces are in the direction opposite the preload. The load capacity in radial and both axial directions is determined by the combination of bearings. The "O" arrangement (tandem "DB") is loaded against springs which limit its application to milling operations using either standard or high helix cutters. The "X" arrangement (tandem face-to-face "DF") is not subject to these limitations since the axial loads are absorbed directly in the bearings. Spindle growth with the "O" arrangement is directed out of the spindle, causing the shaft to grow longer and the tool to cut deeper. In contrast, bearings with an "X" arrangement yield inside, leaving additional stock on the part, which can be removed in a finishing pass. From one to three sets of bearings are used to support the spindle shaft at each end. In the case of a tandem set, most of the external radial load (70 to 80%) is carried by the first bearing. Since tandem sets are more difficult to lubricate, single bearing arrangements are preferred especially when the axial load is not of concern.

Most commonly, angular contact bearings are used at both ends of conventional or high-speed spindles. However, under high-speed, moderate load conditions such as those typical of drilling, a tandem "O" bearing arrangement with three bearings as locating bearings on the work end and a single row of cylindrical roller bearings as a floating bearing on the drive end may be used. Under high load, average speed conditions such as those typical of milling, a bearing arrangement with two double row cylindrical roller bearings as radial bearings and a double direction angular contact thrust ball bearing to support the axial force may also be used (Figure 3.49). Since tapered roller bearings can carry much higher loads than ball bearings, they save space because one tapered roller bearing is used at each end of the spindle (Figure 3.50), instead of two or three ball bearings. Standard tapered rolled bearings are normally used for high loads, lower speed applications, while moderate to high speeds (about 50 m/sec) are achieved either with modified cage designs to direct the oil to the roller-rib contact area or by providing a secondary lubricant source to the rib. The hydra-rib bearing is a self-contained tapered roller bearing specifically designed to provide the optimum bearing preload setting and thus the optimum dynamic stability for the spindle system. This bearing can maintain the optimum preload when the speed, load, or dimension changes because it allows for adjustment of the preload. Pressure changes controlled outside the machine by a single adjustment will change the preload in the spindle system. The compressibility of hollow roller bearings can be used to provide a better adjustability for preload. Hollow roller bearings prevent roller skidding, which is a persistent problem with cylindrical roller bearings. Finally, the physical operation of bearings is affected by the bearing and spindle runout (affecting accuracy), fitting practice, bearing adjustment and setting (which affect rigidity, accuracy, spindle runout, and optimum bearing life), and lubrication.

Bearing lubrication is a critical component of the spindle system. Depending on bearing size, type, and speed, bearing lubrication may be permanent grease pack or some type of oil system. The majority of conventional spindles have permanent grease pack bearings. These are preferred where practical because they are simple, relatively inexpensive, and are maintenance free for extended periods assuming the sealing is good. They result in very low temperatures if the correct type of grease is properly applied. The grease life usually determines the bearing life in high-speed applications. To preserve the grease at bearing speeds above 700,000 to 850,000 DN, temperatures of 35 to 40°C should not be exceeded. The run-in period for temperatures and torque has an important effect on the successful operation especially for bearings used at speeds exceeding 13 m/sec. Oil mist, pulsed oil–air (ratio 1:2), and occasionally oil-jet or oil circulation systems are also used to improve spindle stability, but are costly, pose environmental concerns, and require scheduled maintenance. Such systems require that clean, dry, and continuous air be supplied. Also, use of the correct type, quantity, and cleanliness of lubricating oil is critical [152]. Insufficient lubricant volume results in bearing burnout, but as the volume of lubricant increases, the temperature and

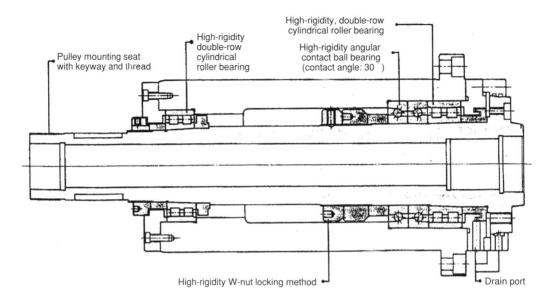

High-rigidity grease lubricated type for NC lathes
Spindle with rolling bearings at front and rear

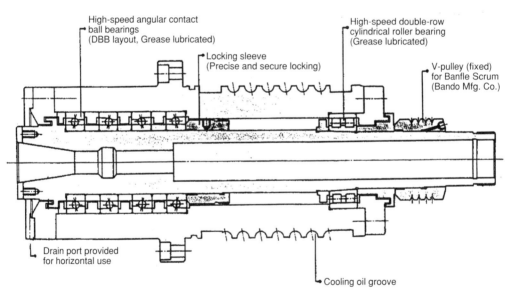

High-speed grease lubricated type for machining centers
Spindle with rolling bearings at rear

FIGURE 3.49 Machining center spindles with a combination of double-row roller bearings and angular contact ball bearings. (Courtesy of NSK Corporation.)

friction increase. Higher speeds up to about 2.5×10^6 DN can be accomplished with oil-jet lubrication even though bearing power consumption due to friction is very high. Oil circulation is regulated between 2 and 10 l/min.

The preload greatly affects bearing performance. Angular contact bearings are loaded in the axial direction. The preload maintains the contact points of balls at their original

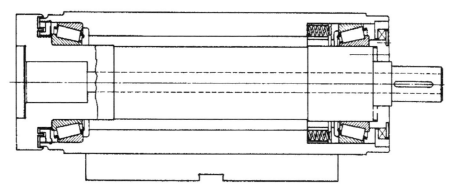

FIGURE 3.50 Spindle with tapered roller bearings (grease lubrication). (Courtesy of Timken Company.)

positions and raises the speed at which the effects of sliding become significant. Therefore, preload is increased as speed increases with external preloading arrangements. Static stiffness increases with preload, while the capacity to bear additional external loads and speed capabilities decrease. Ultimately, a load limit is reached which gives an unacceptable fatigue life even though the mechanism of failure in very high-speed applications is frequently wear rather than fatigue. Accuracy and repeatability increase with preload until the bearing gap is closed; after this point they decrease with increasing preload. The optimum preload setting gives the minimum compliance or the maximum dynamic stiffness. The static coefficient of friction increases with preload from 0.001 to about 0.01. Increasing the preload generally reduces the DN factor. The preload is determined by bearing specification, speed, and lubrication [74, 152, 153]. The preload is generally designed based on the highest speed and temperature condition, which determine the minimum preload and material DN. A solid or fixed preload (obtained with a "DB" arrangement) allows for loads in both axial directions. This design does not perform well if the shaft changes length due to thermal expansion. A variable preload bearing mechanism enhances the characteristics of high-performance bearing arrangements, enabling them to run under a range of conditions. Variable preload, provided by springs or hydraulic pressure, allows thermal growth. Heavier preload is used at lower speeds, which increases stiffness and allows higher cutting forces. Lower preload is used at higher speeds. However, it is often impossible to regulate a variable preload in an NC machine due to continuous changes in cutting conditions and loads. A controlled or variable preload is obtained with springs positioned at the rear of the bearing that allows the bearing set to move axially within the spindle housing. The preload must be taken into account when calculating load–life requirements because there is a trade-off between the preload and maximum load capacity.

Thermal stability is also an important concern in spindle selection, especially for finishing operations. ANSI standard B89.6.2 [166] addresses thermal stability and thermal errors due to axial drift, radial drift, and tilting. At present, spindles with a growth of 0.025 mm over a 3 h time span are very common. The use of warm-up cycles can reduce but not eliminate thermal errors. Further control of errors depends on lubrication and cooling of the spindle. The best approach for improving the spindle's thermal stability is to use more thermally stable materials for the spindle components and to cool the outer bearing races and motor with a water jacket or controlled temperature oil bath [167]. Steel has traditionally been used for most spindle components, but can be replaced in shafts and housings by fiber-reinforced metal or plastic composites [168]. Heavy metals and refractory materials with high modulus, such as molybdenum, tungsten, ceramics, and invar, can also be used but are expensive and difficult to fabricate. Sensors in the tailstock, saddle, and column have been used to measure the temperature of these units and automatically compensate for distortion. Thermal distortion is minimized by circulating temperature-controlled oil or coolant through the headstock, saddle, spindle, and cores of the ball. The coolant temperature is controlled to room temperature by a chiller unit to ensure machine temperature.

The spindle rigidity (the static stiffness or spindle deflection measured at the spindle nose) is determined by the number, arrangement, and stiffness (type) of the bearings, the stiffness of the shaft, the bearing spacing, the housing stiffness, the preload, and the overhang distance between the front bearing and the cutting tool. Thermal expansion also has an important influence on the static and dynamic stiffness of the spindle. Optimum bearing spacing is designed for individual maximum speeds. The bearings selected should have as large a bore as practical because a 19% increase in bore size results in 100% increase in the stiffness contributed by the bearings and shaft. Material selection for the shaft and housing can improve the spindle stiffness significantly since the bearings typically account for 30 to 50% of spindle deflection, while the shaft and spindle housing deflections are responsible for 50 to 70% . Elastic analyses of the spindle system have been effectively applied to determine the shaft stiffness as compared to the bearing stiffness and to optimize the housing design and heat transfer. Often a small stiffness improvement in a relatively stiff spindle is ineffective because the weakest, most elastic element in the system is the tool. Both the static and dynamic stiffness of the spindle are important; they differ significantly especially at higher speeds. The dynamic stiffness is influenced to a large degree by the damping characteristics and the static stiffness of the spindle. A load that could cause very little static deflection can result in very high dynamic deflections if the frequency of the dynamic load is the same as the natural frequency of the spindle system.

The rotary seals at the front and rear of the spindle prevent dirt, coolant, and chips from entering the bearings and spindle housing. Seals are even more critical when through-the-spindle pressurized coolant is used because the coolant splashes against the front of the spindle with significant energy. Inadequate sealing has been found to be one of the major failure modes of spindles in [153]. Spindles are usually easy to seal using lip seals, labyrinths, packing, face seals, and stuffing boxes. Some seals are designed dynamically so that they do not contact the shaft at running speeds, reducing heat generation. Other types of seals include noncontact types with air purging to prevent leakage. The mean time to failure for most seals is 500 to 1000 days [153]. The design of bearing seals is discussed in detail in Refs. [153, 169].

The tool retention system is also a critical spindle component for small or high-speed spindles. Figure 3.51 shows a power drawbar commonly used on NC machines. The drawbar passes through the center of the spindle shaft. The ID of the spindle nose is usually tapered to accept a tool holder; the drawbar pulls the holder securely into the spindle nose with an axial force of between 5 and 25 kN, depending on the application. (As discussed in chapter 5, HSK tooling couplings require higher retention forces than CAT-V tapers.) An improperly designed or adjusted drawbar can apply an excessive force to the front spindle bearings, leading to stresses and bulging which could reduce bearing life and performance. This is especially true for small spindles and spindles used in high power applications requiring high drawbar forces.

Spindle performance is also affected by balance characteristics, which determine limiting rotational speeds. Balance is affected by the characteristics of the rotor, the characteristics of the machine structure and foundation, and the proximity of resonant speeds to the service speed. Precision and high-speed spindles must be dynamically balanced. The balance grade G1, representing a maximum unbalance vibration velocity of 1 mm/sec, is normally sufficient

Machine Tools

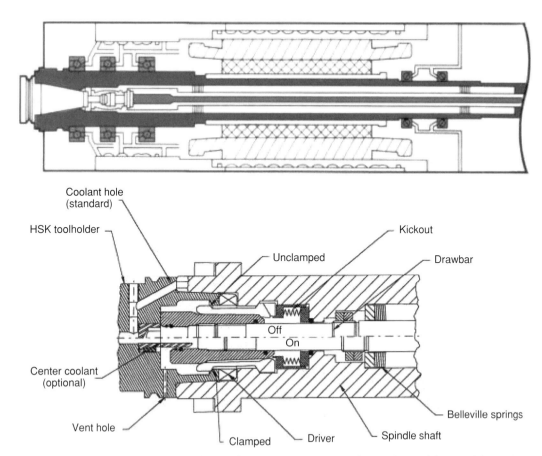

FIGURE 3.51 A power draw bar tool retention system. (Courtesy of Toyoda Machinery USA, Inc.)

for machine tool spindles, while a G0.4 may be needed for precision spindles. Details on the determination of the permissible residual unbalance over speed are given in ISO Standard 1940 [170].

Spindle vibration has a detrimental effect in spindle life, tool life, and part quality. The vibration can be detected in the displacement (amplitude), velocity, and acceleration in the axial and radial directions on the housing close to the bearings. Vibration of contactless bearings should be measured directly on the shaft or tool using noncontact methods. Vibration displacement and acceleration limits are spindle speed dependent; velocity limits are not and thus can be used to define one acceptance criterion for the entire range of spindle speeds. The velocity amplitude acceptance limit for gearless type spindles (box, cartridge, and multispindle/cluster type spindles) for up to 10,000 rpm is 0.01 mm/sec; it increases at frequencies greater than four times the spindle speed frequency, to 0.12 mm/sec for angular contact bearings and 0.18 mm/sec for roller bearings. The acceleration limits are $0.5g$ for angular contact bearings and $1g$ and $1.5g$ for roller bearings for speeds below 1000 rpm and above 1000 rpm, respectively. The displacement limit (peak-to-peak) for speeds below 700 rpm is 25 μm. The vibration velocity acceptance limit for gear-driven spindle assemblies is 0.2 mm/sec. The ASA S2/WG88 standard is being prepared to define acceptable vibration limits for high-speed spindles. Some general limits for high-speed spindles are 1 mm/sec RMS for the spindle, 2 mm/sec RMS for the spindle with tool, and 4 to 6 mm/sec RMS during cutting.

Most spindles exhibit axial and radial motion errors between 0.25 and 2 μm at the spindle nose, although precision spindles with total motion errors of less than 50 nm are available. Spindles with radial motions of up to 10 μm are common. The dynamic spindle runout is often larger than the static runout by a factor between 10 and 20 at high spindle speeds due to increased vibration. The radial runout measured at a specific gauging distance from the spindle nose using a precision arbor is a good indication of spindle runout quality because most tools extend 100 to 150 mm from the nose. Systems have been designed which incorporate counterweighed rotor assemblies fixed to the machine spindle nose for active real-time balancing of the entire spindle assembly after each tool change [171–174].

3.8 COOLANT SYSTEMS

The coolant is supplied to the spindle nose and to the tool either externally or internally through the spindle. The coolant is supplied externally using swiveling jets mounted on a plate at the top of the spindle, or more effectively around the spindle housing on the machine unit or on the machine top cover (Figure 3.52). These nozzles are freely adjustable to different distances between spindle and workpiece, so that they are effective over the entire workspace. External coolant can be also supplied through permanent nozzles located on the face of the spindle housing; this is especially effective for a ram/quill type spindle since the nozzles travel with the spindle in the Z-direction. In this case, the coolant is supplied through the spindle housing.

The coolant can be supplied either through the center or the perimeter of the spindle shaft (Figure 3.52). When the coolant is supplied through the shaft perimeter, so that it does not interfere with the drawbar system, it is routed through the flange of the tool holder. In the case of tools with internal coolant supply holes or when manufacturing precision bores, the coolant should be filtered to less than 40 μm, and down to 5 to 10 μm at high speeds to prevent wear and damage to the rotary union, tool holder/spindle interface, drawbar fingers, and tool, and to reduce stresses on seals. Filtering not only prevents blockage of the tool oil holes, but also ensures a constant surface finish in the bore and a longer tool and coolant life. Paper filters or a hydrostatic filter system with high filter capacity should be used as discussed in Chapter 14.

Remote control coolant nozzles may be used at the spindle head; these nozzles around the spindle are automatically controlled by the machine software according to the length and diameter of the tool to provide coolant at the cutting edge and workpiece.

Optimum cooling and rinsing is ensured by a combination of external cooling with annular spray nozzles and internal coolant through the tool. When high-pressure internal coolant is used, a mist collector may be necessary to meet mist exposure standards. Common coolant formulations, maintenance issues, and health and safety concerns are described in detail in Chapter 14.

3.9 TOOL CHANGING SYSTEMS

Machining centers are usually equipped with an automated tool change (ATC) system. The five basic designs used on various types of machining or turning centers are the swing arm, rack, multiple arm, wheel, and turret types shown in Figure 3.53. One or more *turrets* are always used in turning centers as shown in Figure 3.54. On machining centers, the *double arm*

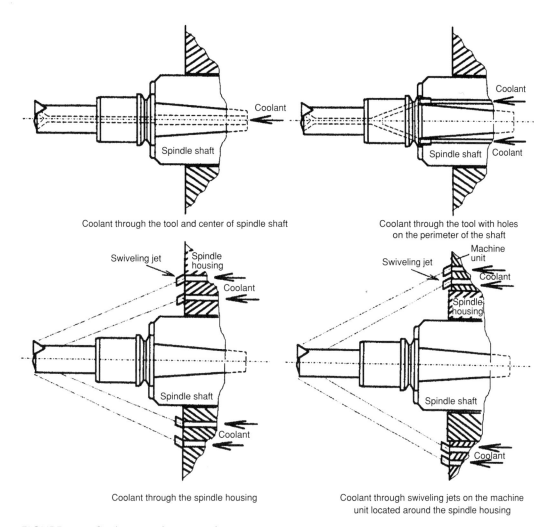

FIGURE 3.52 Coolant supply approaches.

swing system is the most common. *Single or multiple arm systems* use either a tool magazine or a tool runner for tool storage. A tool runner system which uses a monorail to convey tools from one or more magazines to a tool change station can be used to increase tool storage capacity. Some ATC systems can select tools at random (rather than in a predetermined sequence), and some can store information such as the accumulated tool life for a particular tool. The setup and organization of tools in the tool magazine is often important and is part of the design process especially when large batches of identical parts are to be machined [64].

The ATC cycle time varies depending on the class of machining center. The tool-to-tool change time is typically 1 sec for a multiple arm system, 2 to 5 sec for a swing arm system, and 5 to 10 sec for a rack system. The tool change time is typically 1.5 to 2.5 sec for a wheel and turret system, which carries a limited number of tools. Tool-to-tool change times are continuously decreasing, currently to below 0.6 sec in some cases, which is leading to reduced noncutting times. The ultimate goal is to change tools without stopping the spindle, especially when operating at higher spindle speeds.

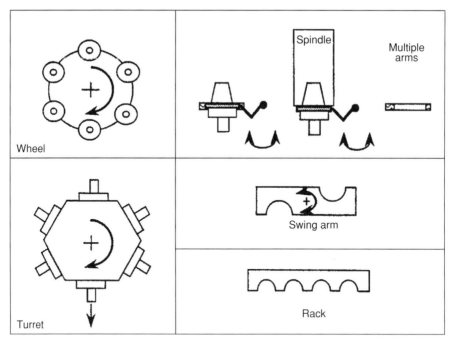

FIGURE 3.53 Types of automatic tool changers with corresponding tool change times.

FIGURE 3.54 The turret on a turret lathe.

3.10 EXAMPLES

Example 3.1 Estimate the stiffness of a strut used in a PKM machine tool if the stiffness of the ball-screw itself is 140 N/μm, the stiffness of a gimbal joint is 25 N/μm, and the stiffness of a preloaded roller ball/socket is 200 N/μm. Spherical joints at each end support the ball-screw and should be included in the estimation of the overall strut stiffness.

Solution: The strut and the joints at its ends are in series and their stiffness is

$$\frac{1}{k_s} + 2\frac{1}{k_g} = \frac{1}{k_T} \quad \text{hence} \quad \frac{1}{140} + 2\frac{1}{25} = \frac{1}{11.5} \quad \text{or} \quad k_T = 11.5 \text{ N/μm}$$

For gimbal joints $k_T = 11.5$ N/μm and for preloaded roller ball joints $k_T = 58$ N/μm. This is the concern with PKMs compared to serial machine tools assuming that the strut is constructed using the same ball-screw/nut assembly used for conventional machines.

Example 3.2 Consider a two-dimensional PKM system (2-D hexapod with variable strut length) with two struts (planar scissors-pair) 600 mm apart at the base as shown in Figure 3.55. Each strut is a ball-screw attached with a ball joint at the base and at the end where the spindle is mounted. Struts (A) and (B) swivel around their joints at the base and can stretch from a minimum length of 600 mm to a maximum of 1200 mm. The motions cover a large workspace. Estimate the stiffness of the PKM system at four different locations as a function of the ball-screw stiffness defined in Example 3.1.

Solution: The strut stiffness is estimated in Example 3.1 assuming the ball-screw is supported by spherical joints at each end. Let $p = \{px; py\}$ be the location of the tip. Let a and b be the location of the ground pivots A and B, respectively. These are all 2 × 1 vectors. For shorthand, let vectors from each point in the workspace to the ground pivots be represented by the vectors

$$pa = p - a \quad \text{and} \quad pb = p - b$$

Using k_s as the stiffness of the strut, the formula for the stiffness matrix is

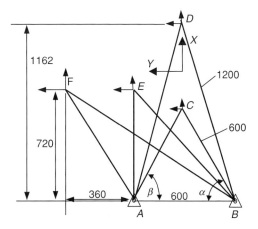

FIGURE 3.55 Illustration of several locations for a planar hexapod machine (PKM using two struts).

$$K \begin{bmatrix} k_{xx} & k_{xy} \\ k_{xy} & k_{yy} \end{bmatrix} = k_s \begin{bmatrix} \frac{pa \cdot pa^T}{pa^T \cdot pa} + \frac{pb \cdot pb^T}{pb^T \cdot pb} \end{bmatrix}, \quad pa = \begin{Bmatrix} pa_x \\ pa_y \end{Bmatrix} \quad \text{and} \quad pa^T = \{pa_x \ pa_y\}$$

Let us now calculate the parameters for the C point in the workspace

$$pa_x = l_s \cos(\alpha), \quad pa_y = l_s \sin(\alpha), \quad pb_x = l_s \cos(\beta), \quad pb_y = l_s \sin(\beta)$$

where α and β are the included angles of the struts at point C with the ground and l_s is the length of the strut. The stiffness matrix is equal to

$$K = k_s \begin{bmatrix} \cos^2 \alpha + \cos^2 \beta & \sin \alpha \cos \alpha + \sin \beta \cos \beta \\ \sin \alpha \cos \alpha + \sin \beta \cos \beta & \sin^2 \alpha + \sin^2 \beta \end{bmatrix}$$

The angles α and β for the location C are equal to 60° in which case the stiffness matrix is

$$K_C = k_s \begin{bmatrix} 0.5 & 0 \\ 0 & 1.5 \end{bmatrix}$$

Therefore, $k_{xx} = 0.5k_s$ and $k_{yy} = 1.5k_s$. Note that the off-diagonal entries are zero because point C has principal directions along X- and Y-directions (its stiffness matrix is already diagonal). This is also true for point D. The angles α and β for the location F are equal to 36.9° and 63.4°, respectively. The stiffness matrix for point F is

$$K_F = k_s \begin{bmatrix} 0.84 & 0.88 \\ 0.88 & 1.16 \end{bmatrix}$$

The principal stiffnesses for locations F and E are not in the X- and Y-directions and are computed using the eigenvalues which give the principal stiffnesses. They both have weak directions due to the small angle between the struts. The values of directional stiffness k_{xx}, k_{yy}, and those along the principal directions K_{p1}, K_{p2} are given in Table 3.4.

In summary, it is shown that the stiffness at the tip joint of the two struts of the hexapod machine varies throughout the workplace. In most places, it is much lower than the stiffness of a single strut. If it is assumed that the strut stiffness was calculated based on the stiffness of a classical machining center structure (where the ball-screw/nut/thrust bearing system effectively determines the stiffness in each direction), the stiffness of the 2-D hexapod is significantly lower in the weak directions. Generally, the closer the struts are to perpendicular, the better the stiffness.

TABLE 3.4
Stiffnesses in the Workspace of a Two-Dimensional Hexapod with Variable Strut Length

Location	k_{xx}/k_s	k_{yy}/k_s	K_{p1}/k_s	K_{p2}/k_s
C	0.5	1.5	0.5	1.5
D	0.125	1.875	0.125	1.875
E	0.41	1.59	0.232	1.768
F	0.84	1.16	0.106	1.894

Machine Tools

Example 3.3 A conventional machine tool with 10,000 rpm/10 kW rated belt-driven spindle is being replaced with a high-speed machining (HSM) center with a 40,000 rpm/20 kW rated spindle. The new machine will be used for machining aluminum workpieces, for which it seems suited based on its rated characteristics. The conventional CNC was used to rough and finish a part. The rough process included a face milling operation using a 100 mm diameter face milling cutter at 2000 rpm. However, the new machine stalls during the face milling operation. What is the reason for such a problem with the new machine?

Solution: The power and torque required by each of the operations in the new machine should be estimated and checked against the power and torque curves. The power and torque curves were obtained upon request from the manufacturer of the HSM center as illustrated in Figure 3.56. This graph will help us decide if the power required by the cutting tool is available in the machine. This information is estimated using the material in Chapter 2. The power and torque required for the cut (depth of cut = 3 mm, width of part 70 mm, feed per tooth = 0.13 mm, cutter with five inserts) are estimated to be 4.9 kW and 23 Nm, respectively. It can be seen that the belt-driven spindle can deliver the required power and torque at 2000 rpm while the motorized high-speed spindle does not have either the power or torque for this operation. The belt-driven spindle has three times the torque of the high-speed spindle because a larger motor drives it.

This illustrates:

1. The benefit of the high-speed spindle is obtained at higher speeds using lower depth of cut to avoid exceeding the torque and power available, in which case the machining time per pass is significantly lower for the high-speed spindle.

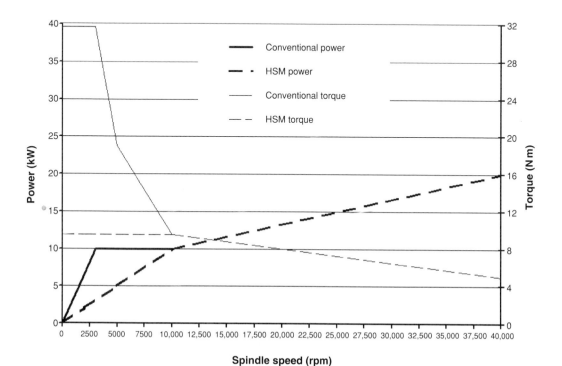

FIGURE 3.56 Machine tool spindle power and torque characteristics versus speed used in Example 3.3.

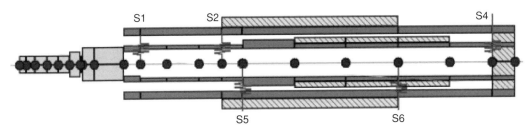

FIGURE 3.57 Two-dimensional axisymmetric finite element model of a spindle used to analyze static and dynamic characteristics in Example 3.4.

2. The specification of the spindle by a single number (i.e., 40,000 rpm/20 kW) is not sufficient to describe the spindle characteristics; graphs of power and torque versus speed are required to properly select the cutting conditions and optimize each operation.

Example 3.4 Evaluate and compare two integral-motorized spindles of the same design but different length. Spindle (A) has a 264 mm long stator with 26 N m output torque, while spindle (B) has a 100 mm shorter stator with 15 N m output at 20,000 rpm. The shorter spindle allows reducing the distance between the front and rear bearing by 100 mm, which affects the characteristics of the spindle. Analyze the characteristics for both spindles.

Solution: A finite element analysis (FEA) program was used to create a two-dimensional axisymmetric model that matched the geometry of the spindle shaft and nose and rotor. The FEA program is capable of simulating both the dynamic and static behavior of a spindle and tooling system. The spindle shaft and the housing were modeled as shown in Figure 3.57. Springs S1, S2, and S4 represent the spindle bearings and springs S5 and S6 represent the connections between the spindle housing and the machine tool structure. The parameters of springs S1, S2, and S4 were determined based on the stiffness provided by the bearing manufacturer and by matching the measured dynamic response and static deflections to those generated by the model.

The model provides estimates of the difference in natural frequencies. The first natural frequency for motor A is 672 Hz, while for motor B it is 836 Hz. The modal stiffnesses for motors A and B are 2.3×10^9 and 3.0×10^8 N/m, respectively. These frequencies correspond to 40,320 and 50,160 rpm, respectively, for motors A and B. Since the natural frequency for both spindles is more than twice higher than the maximum operating speed, both spindles are safe.

A 50 mm increase in the distance between bearings in a 40,000 rpm spindle resulted in reduction of the critical spindle speed from 51,198 to 44,445 rpm [154]. This brings the 40,000 rpm operating speed close to the first natural frequency, which is not generally acceptable due to higher vibration levels resulting in lower bearing and cutting tooling life.

REFERENCES

1. M. Weck, *Handbook of Machine Tools*, 4 Volumes, Wiley, New York, 1984
2. T.J. Drozda and C. Wick, Eds., *Tool and Manufacturing Engineers Handbook*, Vol. 1: Machining, SME, Dearborn, MI, 1983
3. M.P. Groover, *Automated Production Systems and Computer-Integrated Manufacturing*, Prentice-Hall, New York, 1987
4. D.J. Williams, *Manufacturing Systems*, Halsted Press, New York, 1988
5. J.R. Davis ed. *Metals Handbook*, Vol. 16: Machining, 9th Edition, ASM International, Materials Park, OH, 1989
6. A. Slocum, *Precision Machine Design*, Prentice Hall, Englewood Cliffs, NJ, 1992
7. G. Boothroyd and W.A. Knight, *Fundamentals of Machining and Machine Tools*, 2nd Edition, Marcel Dekker, New York, 1989
8. E.D. Lloyd, *Transfer and Unit Machines*, Industrial Press, New York, 1969
9. O.W. Boston, *Metal Processing*, 2nd Edition, Wiley, New York, 1951
10. C.R. Hine, *Machine Tools for Engineers*, 2nd Edition, McGraw-Hill, New York, 1959
11. E.P. DeGarmo, *Materials and Processes in Manufacturing*, 5th Edition, Macmillan, New York, 1979
12. C.T. Jones and L.A. Bryan, Programmable control, *Manufacturing High Technology Handbook*, Marcel Dekker, New York, 1987, 205–254
13. F.J. Hilbing, Data communications, *Manufacturing High Technology Handbook*, Marcel Dekker, New York, 1987, 255–298
14. G.E. Thyer, *Computer Numerical Control of Machine Tools,* Industrial Press, New York, 1988
15. J.G. Bollinger and N.A. Duffie, *Computer Control of Machines and Processes*, Addison-Wesley, Reading, MA, 1988
16. D. Gibbs and T.M. Crandell, *An Introduction to CNC Machining and Programming*, Industrial Press, New York, 1991
17. U. Rembold, C. Blume, and R. Dillmann, *Computer-Integrated Manufacturing Technology and Systems*, Marcel Dekker, New York, 1985
18. Y. Koren, *Computer Control of Manufacturing Systems*, McGraw-Hill, New York, 1983
19. P. Ranky, *Computer Integrated Manufacturing*, Prentice-Hall, New York, 1986
20. J. Pusztaai and M. Sava, *Computer Numerical Control*, Reston Publishing, Reston, VA, 1983
21. C. Tien-Chien, R.A. Wysk, and H.P. Wang, *Computer Aided Manufacturing*, 2nd Edition, Prentice-Hall, New York, 1998
22. T.O. Boucher, *Computer Automation in Manufacturing: An Introduction*, Chapman and Hall, London, 1996
23. W.W. Luggen and W.M. Luggen, *Fundamentals of Computer Numerical Control*, 3rd Edition, Delmar Publishers, New York 1994
24. I.Y. Alqattan, Systematic approach to cellular manufacturing systems design, *J. Mech. Working Tech.* **20** (1989) 415–424
25. R.L. Diesslin, Group technology, *Manufacturing High Technology Handbook*, Marcel Dekker, New York, 1987, 55–82
26. A. Mungwattana, Design of Cellular Manufacturing Systems for Dynamic and Uncertain Production Requirements with Presence of Routing Flexibility, PhD Thesis, Virginia Polytechnic Institute and State University, September 2000
27. I.A. Shahrukh, *Handbook of Cellular Manufacturing Systems*, Wiley, New York, 1999
28. P.A. Lewis and D.M. Love, The design of cellular manufacturing systems and whole business simulation, *Proceedings of the 30th International MATADOR Conference*, Manchester, April 1993, 435–442

29. G. Spur, Computer integrated manufacturing in Europe, *Proceedings of the European Conference on Flexible Manufacturing*, Dublin, 1988, 1–21
30. N. Alberti, U. La Commare, and S.N. la Diega, Cost efficiency: an index of operational performance of flexible automated production environments, *Proceedings of the 3rd ORSA/TIMS Conference on Flexible Manufacturing Systems: Operations Research Models and Applications*, Amsterdam, 1989, 67–73
31. W.N. Nordquist, Flexible manufacturing, *Manufacturing High Technology Handbook*, Marcel Dekker, New York, 1987, 105–144
32. Y.P. Gupta and S. Goyal, Flexibility of manufacturing systems: concepts and measurements, *Eur. J. Oper. Res.* **43** (1989) 119–135
33. H.T. Goranson, Agile manufacturing, in: A. Molina, J.M. Sanchez, and A. Kusiak, Eds., *Handbook of Life Cycle Engineering Concepts, Models and Technologies*, Kluwer Academic, Dordrecht, 1998, 31–58
34. R. Askin and C.R. Standridge, *Modeling and Analysis of Manufacturing Systems*, Wiley, New York, 1993
35. J.K. Shim and J.G. Siegel, *Operations Management*, Barron's Educational Books, New York, 1999
36. A.W. Hallmann, Flexible manufacturing for auto parts, *Tooling Prod.* **69**:1 (2003) 32–33
37. S.K. Das, The measurement of flexibility in manufacturing systems, *Int. J. Flexible Manuf. Systems* **8** (1996) 67–93
38. J.M. Fraser and F.F. Leimkuhler, Justification of flexible manufacturing systems, *Manufacturing High Technology Handbook*, Marcel Dekker, New York, 1987, 145–174
39. H.J. Warnecke and R. Steinhilper, *Flexible Manufacturing Systems*, Springer Verlag, New York, 1985
40. H.F. Lee, Optimal design for flexible manufacturing systems: generalized analytical methods, *IIE Trans.* **31** (1999) 965–976
41. U. Heisel and M. Mitzner, Progress in reconfigurable manufacturing systems, *CIRP 2nd International Conference on Reconfigurable Manufacturing*, Ann Arbor, MI, 2003
42. A. Urbani, S.P. Negri, and C.R. Boer, Example of measure of the degree of reconfigurability of a modular parallel kinematic machine, *CIRP 2nd International Conference on Reconfigurable Manufacturing*, Ann Arbor, MI, 2003
43. Y. Koren, U. Heisel, F. Jovanne, T. Moniwaki, G. Pritschou, G. Ulsoy and van Brussel et al, Reconfigurable manufacturing systems, *CIRP Ann.* **48**:2 (1999) 527–540
44. A. Gunasekaran, Agile manufacturing: enablers and an implementation framework, *Int. J. Prod. Res.* **36** (1998) 1223–1247
45. A. Gunasekaran, Design and implementation of agile manufacturing systems, *Int. J. Prod. Econ.* **62** (1999) 1–6 (editorial)
46. R.N. Nagel, R. Dove, S. Goldman, and K. Preiss, *21st Century Manufacturing Enterprise Strategy: An Industry-led View*, Iacocca Institute, Lehigh University, Bethlehem, PA, 1991
47. Y.Y. Yusuf, M. Sarhadi, and A. Gunasekaran, Agile manufacturing: the drivers, concepts, and attributes, *Int. J. Prod. Econ.* **62** (1999) 33–43
48. A. Gunasekaran, Agile manufacturing: a framework for research and development, *Int. J. Prod. Econ.* **62** (1999) 87–105
49. R. DeVor, R. Graves, and J.J. Mills, Agile manufacturing research: accomplishments and opportunities, *IIE Trans.* **29** (1997) 813–823
50. A. Gunasekaran, *Agile Manufacturing: The 21st Century Competitive Strategy*, Elsevier, Amsterdam, 2001
51. L.M. Sanchez and R. Nagi, A review of agile manufacturing systems, *Int. J. Prod. Res.* **39** (2001) 3561–3600
52. D.B. Sieger, A.B. Badiru, and M. Milatovic, A metric for agility measurement in product development, *IIE Trans.* **32** (2000) 637–645
53. R.E. Giachetti, L.D. Martinez, O.A. Saenz, and C.-S. Chen, Analysis of the structural measures of flexibility and agility using a measurement theoretical framework, *Int. J. Prod. Econ.* **86** (2003), 47–62

54. M. Shpitalni and V. Remennik, Practical number of paths in reconfigurable manufacturing systems with crossovers, *CIRP 2nd International Conference on Reconfigurable Manufacturing*, Ann Arbor, MI, 2003
55. P. Spicer, Y. Koren, M. Shpitalni, and D. Yip-Hoi, Design principles for machining system configurations, *CIRP Ann.* **51** (2002) 275–280
56. L.O. Morgan and R.L. Daniels, Integrating product mix and technology adoption decisions: a portfolio approach for evaluating advanced technologies in the automobile industry, *J. Oper. Manage.* **19** (2001) 219–238
57. Z. Xiaobo, Y. Jiancai, and L. Zhenbi, A stochastic model of a reconfigurable manufacturing system. Part 1. A framework, *Int. J. Prod. Res.* **38** (2000) 2273–2285
58. Z. Xiaobo, Y. Jiancai, and L. Zhenbi, A stochastic model of a reconfigurable manufacturing system. Part 2. Optimal configurations, *Int. J. Prod. Res.* **38** (2000) 2829–2842
59. H.F. Lee, Optimal design for flexible manufacturing systems: generalized analytical methods, *IIE Trans.* **31** (1999) 965–976
60. C.H. Fine and R.M. Freund, Optimal investment in product-flexible manufacturing capacity, *Manage. Sci.* **36** (1990) 449–466
61. S.M. Ordoobadi and N.J. Mulvaney, Development of a justification tool for advanced manufacturing technologies: system-wide benefits value analysis, *J. Eng. Technol. Manage.* **18** (2001) 157–184
62. D. Gupta and J.A. Buzacott, Models for first-pass FMS investment analysis, *Int. J. Flexible Manuf. Systems* **5** (1993) 263–286
63. B.P. Pang and K. Khodabandehloo, Modeling the reliability of flexible manufacturing cells, *Proceedings of the 29th International MATADOR Conference*, 1992, 183–190
64. J. Agapiou, Sequence of operations optimization in single-stage multifunctional systems, *J. Manuf. Systems* **10** (1991) 194–207
65. K. Hitomi, Analysis of optimal machining conditions for flow-type automated manufacturing systems: maximum efficiency for multi-product production, *Int. J. Prod. Res.* **28** (1990) 1153–1162
66. J. Agapiou, Optimization of multistage machining systems. Part I. Mathematical solution. Part II. The algorithm and applications, *ASME J. Eng. Ind.* **114** (1992) 524–538
67. R. Bakerjian, Ed., *Tool and Manufacturing Engineering Handbook, Design for Manufacturability*, Vol. 6, SME, Dearborn, MI, 1992, Chap. 11
68. J.A. Simpson, R.J. Hocken, and J.S. Albus, The automated manufacturing research facility, *J. Manuf. Systems* **1** (1982) 17–32
69. M.K. Senehi, S. Wallace, and M.E. Luce, An architecture for manufacturing systems integration, *Proceedings of the ASME Manufacturing International Conference*, Dallas, TX, 1992
70. M.K. Senehi, E. Barkmeyer, S. Ray, and E. Wallace, Manufacturing Systems Integration Initial Architecture Document, NISTIR 91-4682, September, 1991
71. M.K. Senehi, S. Wallace, E. Wallace, S. Ray and E. Barkmeyer. Manufacturing Systems Integration Control Entity Interface Document, NISTIR 91-4626, June, 1991
72. J.T. Black and S.L. Hunter, *Lean Manufacturing Systems and Cell Design*, SME, Dearborn, MI, 2003
73. C. Standard and D. Davis, *Running Today's Factory — A Proven Strategy for Lean Manufacturing*, SME and Hanser Gardner, Dearborn, MI, 1999
74. J. Lazan, *Damping of Materials and Members in Structural Mechanics*, Pergamon Press, London, 1968
75. E. Rivin, *Stiffness and Damping in Mechanical Design*, Marcel Dekker, New York, 1999
76. J. Trusly, J. Ziegert, and S. Ridgeway, Fundamental comparison of the use of serial and parallel kinematics for machines tools, *CIRP Ann.* **48** (1999) 351–356
77. Z. Li and R. Katz, Theoretical study on a reconfigurable parallel kinematic high speed drilling machine, *CIRP 2nd International Conference on Reconfigurable Manufacturing*, Ann Arbor, MI, 2003
78. Y. Furukawa, N. Moronuki, and H. Kubo, Application of a ceramic guideway in a ultra precision machine tool, *Bull. Jpn. Soc. Precision Eng.* **20** (1986) 197–198

79. Y. Furukawa, N. Morinuki and K. Kitagawa, Development of ultra precision machine tool made of ceramics, *CIRP Ann.* **35** (1986) 279–282
80. H. Sugishita, H. Nishiyama, O. Nagayasu, T. Shin-Nov, H. Sato and M.O-Hari, Development of concrete machining center and identification of dynamic and thermal structural behavior, *CIRP Ann.* **37** (1988) 377–380
81. D.G. Lee, H.C. Sin and N.P. Suh, Manufacturing of a graphite epoxy composite spindle for a machine tool, *CIRP Ann.* **34** (1985) 365–369
82. J.D. Suh and D.G. Lee, Composite machine tool structures for high speed milling machines, *CIRP Ann.* **51** (2002) 285–288
83. K. Takada and I. Tanabe, Basic study on thermal deformation of machine tool structure composed of epoxy resin concrete and cast iron, *Bull. Jpn. Soc. Precision Eng.* **21** (1987) 173–178
84. M. Tsutsumi and Y. Ito, Damping mechanism of a bolted joint in machine tools, *Proceedings of the 20th International MTDR Conference*, 1979, 443–448
85. S. Haranath, K. Ganesan, and B. Rao, Dynamic analysis of machine tool structures with applied damping treatment, *Int. J. Mach. Tools Manuf.* **27** (1987) 43–55
86. E. Rivin and H. Kang, Improvement of machining conditions for slender parts by tuned dynamic stiffness of tool, *Int. J. Mach. Tools Manuf.* **29** (1989) 361–376
87. ASME B5.54, Methods for Performance Evaluation of Computer Numerically Controlled Machining Centers, An American National Standard, 1992
88. V. Gopalakrishnan, D. Fedewa, M.G. Mehrabi, S. Kota and N. Orlandea. Parallel structures and their applications in reconfigurable machining systems, *Parallel Kinematic Machines International Conference*, Ann Arbor, MI, 2000, 87–97
89. M. Mandeli, T. Nagao, Y. Hatamura, M. Mitsuishi, and M. Nakao, New machine tools and systems, in: A. Dashchenko, Ed., *Manufacturing Technologies for Machines of the Future — 21st Century Technologies*, Springer Verlag, New York, 2003, 611–619
90. D. Stewart, A platform with six degrees of freedom, *Proc. Inst. Mech. Eng.* **180** (1965) 371–386
91. M. Weck and D. Staimer, Parallel kinematic machine tools — current state and future potentials, *CIRP Ann.* **50**:2 (2002) 671–684
92. B.S. El-Khasawneh and P.M. Ferreira, On using parallel link manipulators as machine tools, *Trans. NAMRI/SME* **25** (1997) 305–310
93. F. Pierrot and O. Company, Towards non-hexapod mechanisms for high performance parallel machines, *IECON 2000, IEEE Industrial Electronics Conference*, Nagoya, Japan, October 22–28, 2000
94. L. Molinari-Tosatti, G. Bianchi, I. Fassi, C.R. Boer and F. Jovane. An integrated methodology for the design of parallel kinematic machines (PKM), *CIRP Ann.* **47** (1998) 341–344
95. H.J. Warnacke, R. Neugebaver and F. Wieland, Development of hexapod based machine tool, *CIRP Ann.* **47** (1998) 337–340
96. F. Majou, P. Wenger, and D. Chablat, Design of a 3 axis parallel machine tool for high speed machining: the orthoglide, *IDMME 2002*, Clermont-Ferrand, France, May 14–16, 2002
97. M. Schwaar, T. Jaehnert, and S. Ihlenfeld, Mechatronic design — experimental property analysis and machining strategies for a 5-strut-PKM, *3rd Chemnitz Parallel Kinematics Seminar*, May 25, 2002, Verlag Wissenschaftliche Scriptien, Zwickau, 671–682
98. T.H. Chang, I. Inasaki, K. Morihara, and J.J. Hsu The development of a parallel mechanism of 5-DOF hybrid machine tool, *Parallel Kinematic Machines International Conference*, Ann Arbor, MI, 2000, 79–86
99. H.K. Toenshoff, G. Gunther, and H. Grendel, Influence of manufacturing and assembly errors on the pose accuracy of hybrid kinematics, *Parallel Kinematic Machines International Conference*, Ann Arbor, MI, 2000, 255–263
100. D. Zhang and C.M. Gosselin, Kinetostatic analysis and optimization of the tricept machine tool family, *Parallel Kinematic Machines International Conference*, Ann Arbor, MI, 2000, 174–187
101. S.A. Shamblin and G.J. Wiens, Characterization of dynamics in PKMs, *Parallel Kinematic Machines International Conference*, Ann Arbor, MI, 2000, 24–33
102. M. Weck and D. Stainer, Accuracy issues of parallel kinematic machine tools: compensation and calibration, *Parallel Kinematic Machines International Conference*, Ann Arbor, MI, 2000, 36–41

103. C.C. Vong, M.H. Perng, and C.T. Lee, On the inertial coupling effect and control of a parallel manipulator, *Parallel Kinematic Machines International Conference*, Ann Arbor, MI, 2000, 136–142
104. Y. Takeda, H. Funabashi, G. Sen, K. Ichikawa, and K. Hirosa, Stiffness analysis of a spatial six-degree-of-freedom in-parallel actuated mechanism with rolling spherical bearings, *Parallel Kinematic Machines International Conference*, Ann Arbor, MI, 2000, 264–273
105. C.M. Clinton, G. Zhang, and A.J. Wavering, Stiffness modeling of a stewart-platform-based milling machine, *Trans. NAMRI/SME* **25** (1997) 335–340
106. G. Wiens and D. Hardage, Dynamics and controls of hexapod machine tools, *Proceedings of the First European-American Forum on Parallel Kinematic Machines: Theoretical Aspects and Industrial Requirements*, Milan, Italy, 1998
107. A.J. Wavering, Parallel kinematic machine research at NIST: past, present, and future, *Proceedings of the First European-American Forum on Parallel Kinematic Machines: Theoretical Aspects and Industrial Requirements*, Milan, Italy, 1998
108. T.H.N. Brogden and F.M. Stansfield, The design of machine tool foundations, *Proceedings of the 11th International MTDR Conference*, 1970, 333
109. B.S. Baghshahi and P.F. McGoldrick, Machine tool foundations — a dynamic design method, *Proceedings of the 20th International MTDR Conference*, 1979
110. Unisorb, Machinery Installation Systems, Master Catalog and Engineering Guide, Michigan, 1989
111. T.C. Duclos, D.N. Acker, and J.D. Carlson, Fluids that thicken electrically, *Mach. Des.* **60**:2 (1988) 42–46
112. A. Devitt, Replication Materials in New Machine Architecture, SME Technical Paper MS90-408, 1990
113. M.W. Browne, New diamond coatings find broad application, *New York Times*, January, 1989
114. T. Subramanian and S.P. Rangosuami, Factors influencing the positioning accuracies of CNC machine tool slides, *Proceedings of the 12th AIMTOR Conference*, IIT, Delhi, India, 1986
115. M. Arnone, *High Performance Machining*, Hanser Garner Publications, Cincinnati, OH, 1998
116. K. Tanaka, Y. Uchiyam, and S. Toyooka, The mechanism of wear of polytetra-fluoroethylene, *Wear* **23** (1973) 153–172
117. B. Mortimer and J. Lancaster, Extending the life of aerospace dry bearings by the use of hard smooth surfaces, *Wear* **121** (1988) 289–305
118. K. Lindsey and S. Smith, Precision Motion Slideways, U.K. Patent 8,709,290, April 1988
119. B.J. Hamrock and D. Dowson, *Ball Bearing Lubrication*, Wiley, New York, 1981
120. *Hydra-Rib Bearing*, The Timken Company, Canton, OH, 1980
121. Z.M. Levina, Research on the static stiffness of joints in machine tools, *Proceedings of the 8th International MTDR Conference*, 1967, 737–758
122. M. Dolbey and R. Bell, The contact stiffness of joints at low apparent interface pressures, *CIRP Ann.* **19** (1979) 67–79
123. J.C. Bandrowski, Antifriction bearings for linear motion systems, *Mech. Eng.*, April 1987
124. A.G. Boim and Z.M. Levina, Effect of guideway type on machine tool positional accuracy, *Soviet. Eng. Research* 1(9), 1981, 71–75
125. W. Rowe, *Hydrostatic and Hybrid Bearing Design*, Butterworths, London, 1983
126. F. Stansfield, *Hydrostatic Bearings for Machine Tools*, Machinery Publishing Co., London, 1970
127. D. Fuller, *Theory and Practice of Lubrication for Engineers*, 2nd Edition, Wiley, New York, 1984
128. R.J. Welsh, *Plain Bearing Design Handbook*, Butterworths, London, 1983
129. J. Zeleny, Servostatic guideways — a new kind of hydraulically operating guideways for machine tools, *Proceedings of the 10th International MTDR Conference*, 1969
130. J. Powell, *Design of Aerostatic Bearings*, Machinery Publishing Co., London, 1970
131. K.J. Stout and E.G. Pink, Orifice compensated EP gas bearings: the significance of errors of manufacture, *Tribol. Int.* **13** (1980) 105
132. H. Yabe and N. Watanabe, A study on running accuracy of an externally pressurized gas thrust bearing (load capacity fluctuation gas to machining errors of the bearing), *JSME Int. J. Ser. III* **1** (1988) 114

133. N.K. Arakere, H.D. Nelson, and R.L. Rankin, Hydrodynamic lubrication of finite length rough gas journal bearings, *STLE Tribol. Trans.* **33** (1990) 201
134. A.H. Slocum Self Compensating Hydrostatic Linear Bearing, U.S. Patent No. 5,104,237, April 14, 1992
135. H.C. Town, Control of hydraulic transmission elements, *Power Int.*, March 1986, 65–67
136. R.L. Rickert, AC servos increase machine tool productivity, *Mach. Des.* **57**:4 (1985) 135–139
137. B.H. Carlisle, AC drives move into DC territory, *Mach. Des.* **57**:4 (1985) 61–64
138. W. Leonhard, Adjustable-speed AC drives, *Proc. IEEE* **76**:4 (1988) 455–471
139. S.J. Bailey, Step motion control 1985: direct digital incrementing with servo-like performance, *Control Eng.*, **32**:8 August 1985, 49–57
140. C. Mirra and R. Quickel, Drive for success, *Cutting Tool Eng.* **41**:4 (1989) 75–84
141. C. Mirra, What digital AC drives bring to machining, *Tooling Prod.* **57**:4 (1991) 45–46
142. D. Horn, Linear motors provide precise positioning, *Mech. Eng.* (November 1988) 70–74
143. B.L. Triplett, Linear motors combine muscle with a fine touch, *Mach. Des.* **59**:10 (1987) 94–97
144. W.E. Barkman, Linear motor slide drive for diamond turning machine, *Precision Eng.* **3**:1 (1981) 44–47
145. G. Pritschow, A comparison of linear and conventional electromechanical drives, *CIRP Ann.* **47**:2 (1998) 541–548
146. J.S. Agapiou and C.H. Shen, High speed tapping of 319 aluminum alloy, *Trans. NAMRI/SME* **20** (1992) 197–203
147. V.G. Belyaev, et al., How ballscrew diameter affects the speed of response of a machine feed drive, *Soviet Eng. Res.* **66**:5 (1986) 7–8
148. SKF, Transrol High Efficiency Ball and Roller Screws, Publication #3369/Z U.S.
149. Basics of design engineering: mechanical systems, *Mach. Des.* **64**:12 (1992) 293–300
150. Thyssen Corporation, Hueller Hille, Specht CNC High Speed Machining Center, 1993
151. H. Weule and T. Frank, Advantages and characteristics of a dynamic feeds axis with ball screw drive and driven nut, *CIRP Ann.* **48** (1999) 303–306
152. B. Popoli, Selecting the proper bearing system for your high speed spindle application, *Proceedings of the 1st Machine Tool Conference*, May 28–29, 2003
153. B. Hodge, Spindle designs for high-speed machining, *Proceedings of the 1st Machine Tool Conference*, May 28–29, 2003
154. P. Frederickson and D. Grimes, Optimizing high-speed spindle torque characteristics, *SME High Speed Machining Technical Conference*, Northbrook, IL, 2001
155. E.G. Korolev, Motor/spindles for NC machine tools, *Soviet Eng. Res.* **6**:12 (1986) 60–61
156. J.L. Reif, How to select cost-effective electric drive systems, *Proceedings of the 21st Annual Meeting Technical Conference*, Numerical Control Society, Chicago, IL, 380–395
157. J. Hendershot, Selection of motor types for high performance machine tool spindles, *Proceedings of the 1st Machine Tool Conference*, May 28–29, 2003
158. D. Weise, Present industrial applications of active magnetic bearings, *22nd Intersociety. Energy Conversion Engineering Conference,* Philadelphia, PA, August 1987
159. W.S. Chung, G.T. Gillies, C.H. Leyh, and R.C. Ritter, Ultra stable magnetic suspensions for rotors in gravity experiments, *Precision Eng.* **2**:4 (1980) 183–186
160. E.H.M. Weck and U. Wahner, Linear magnetic bearing and levitation system for machine tools, *CIRP Ann.* **47** (1998) 311–314
161. Y.C. Shin, K.W. Wang, and C.H. Chen, Dynamic analysis and modeling of a high speed spindle system, *Trans. NAMRI/SME* **18** (1990) 298–304
162. T.A. Harris, *Rolling Bearing Analysis*, Wiley, New York, 1966
163. C. Moratz, Contact angle and bearing selection for high speed spindle bearings, *SME High Speed Machining Conference*, Chicago, IL, 2003
164. C. Moratz, Barden/FAG *Ceramic Hybrid Spindle Bearings: Optimum Bearing for Machine Tools,* The Barden Corporation, CT, 2003
165. C.P. Bhateja and R.D. Pine, The rotational accuracy characteristics of the preloaded hollow roller bearing, *ASME J. Lubric. Technol.* **103** (1981) 6–12

166. ANSI/ASME B89.6.2, Temperature and Humidity Environment for Dimensional Measurement, ASME, 1973
167. W. Blewett, J.B. Bryan, R.R. Clouser, and R.R. Donaldson, Reduction of machine tool spindle growth, Report UCRL 74672, Lawrence Livermore Laboratory, 1973
168. D.G.S. Lee, H.C. Suh, and P. Nam, Manufacturing of a graphite epoxy composite spindle for a machine tool, *CIRP Ann.* **34** (1985) 365–369
169. Basics of design engineering: mechanical systems, *Mach. Des.* **64**:12 (1992) 208–237
170. ISO 1940/1, Mechanical Vibration — Balance Quality Requirements of Rigid Rotors, Part 1: Determination of Permissible Residual Unbalance, 1986
171. "BalaDyne" System by Balance Dynamics Corporation, Ann Arbor, MI.
172. P. Zelinski, Should you balance your tools? *Mod. Mach. Shop*, August 1997
173. S. Zhou and J. Shi, Optimal one-plane active balancing of a rigid rotor during acceleration, *J. Sound Vibr.* **249** (2002) 196–205
174. V. Wowk, *Machining Vibration: Balancing*, McGraw-Hill, New York, 1995

4 Cutting Tools

4.1 INTRODUCTION

Cutting tool design has a strong impact on machining performance. Properly designed tools produce parts of consistent quality and have long and predictable useful lives. An improperly designed tool may wear or chip rapidly or unpredictably, reducing productivity, increasing costs, and producing parts of deteriorating quality. Optimizing cutting tool technology has a major influence on the productivity and economics of a process. It is important to understand the cutting tool materials and geometries and their corresponding limitations with respect to the cutting conditions (speed and feed). On the average, the cutting tool cost represents about 3% the total component cost in high-volume production. Therefore, a 50% increase in tool life reduces the total cost per component by 1 to 2%. On the other hand, a 20% increase in the material removal rate could reduce the total cost per component by 15%.

Cutting tools may be broadly classified as single point tools, which have one active cutting edge, and multipoint, multifunctional, or multitasking tools, which have multiple active cutting edges. Single point tools are commonly used for turning and boring, while multipoint tools are used for drilling and milling. Multifunctional or multitasking tools are used to machine multistep holes or several features with one tool. Tools may be further classified based on the cutting edge material, geometry, and clamping method. The best choice of tool material and geometry in a given operation depends on the volume of parts to be machined, the workpiece material, the required accuracy, and the capabilities of available machine tools.

The objective of this chapter is to discuss conventional and advanced cutting tool technologies and to explain the properties and characteristics of tools which influence tool design or selection. The general properties of available tool materials and tool coatings are discussed in Section 4.2 and Section 4.3. Specific information on the design and selection of tools for turning, boring, milling, drilling, reaming, threading, grinding, honing, microsizing, and burnishing are discussed in Section 4.4–Section 4.13.

4.2 CUTTING TOOL MATERIALS

4.2.1 Tool Material

This section discusses the basic properties of tool steels and high-speed steels, carbides, cermets, ceramics, superabrasives, and the corresponding coatings for the above tool materials. Critical factors for performance and comparison are explained. The proper selection of a specific tool material and coating for a particular application is also discussed.

4.2.2 Material Properties

Cutting tools must be made of materials capable of withstanding the high stresses and temperatures generated during chip formation. Ideally, tool materials must have the following properties:

1. high penetration hardness at elevated temperatures to resist abrasive wear;
2. high deformation resistance to prevent the edge from deforming or collapsing under the stresses produced by chip formation;
3. high fracture toughness to resist edge chipping and breakage, especially in interrupted cutting;
4. chemical inertness (low chemical affinity) with respect to the work material to resist diffusion, chemical, and oxidation wear;
5. high thermal conductivity to reduce cutting temperatures near the tool edge;
6. high fatigue resistance, especially for tools used in interrupted cutting;
7. high thermal shock resistance to prevent tool breakage in interrupted cutting;
8. high stiffness to maintain accuracy; and
9. adequate lubricity (low friction) with respect to the work material to prevent built-up edge, especially when cutting soft, ductile materials.

The first three properties are required to prevent sudden, catastrophic failure of the tool. Properties 1, 4, and 5 are required for the tool to resist the high temperatures generated during chip formation; as discussed in Chapter 9, elevated tool temperatures can cause rapid tool wear do to thermal softening or diffusion and chemical wear [1–4]. Properties 3, 6, and 7 are required to prevent chipping of the tool, especially in interrupted cutting. Fracture toughness is usually characterized by the transverse rupture strength; fatigue resistance is characterized by the Weibull modulus. As shown in Figure 4.1, properties 1, 4, and 5 generally determine the maximum cutting speed at which a tool can be used, while properties 3 and 6 determine allowable feed rates and depths of cut. It should be noted that fracture and fatigue strength requirements vary with the cutting speed, since cutting forces usually decrease with increasing cutting speed.

The common tool materials currently in use include high-speed steels (HSS), cobalt-enriched high-speed steels (HSS–Co), sintered tungsten carbide (WC), cermets, ceramics, polycrystalline cubic boron nitride (PCBN), polycrystalline diamond (PCD), and single-crystal natural diamond. These materials provide a wide range of combinations of properties [5]. A relative comparison of their mechanical and physical properties, hardness, toughness, Young's modulus, thermal conductivity, specific heat, softening temperature, and Weibull modulus is given in Table 4.1 [1–9]; the magnitude of the property increases between the different materials moving from the bottom row to the top row in the table. Figure 4.2 shows a more specific comparison of hardness and toughness for various tool materials. Figure 4.3 shows a similar comparison of relative abrasion resistance, and Figure 4.4 shows typical

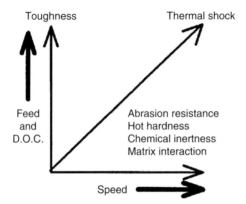

FIGURE 4.1 The influence of tool material properties on optimization of cutting conditions.

TABLE 4.1
Relative Comparison of Various Properties of Cutting Tool Materials

Microhardness	Fracture Toughness	Young's Modulus	Modulus of Rupture	Thermal Conductivity	Specific Heat	Softening Temperature	Chemical Inertness	Weibull Modulus
PCD	C2 carbide	PCD	C2 carbide	PCD	Al_2O_3	PCD	Al_2O_3	C2 carbide
PCBN	(SiCw)-Al_2O_3	PCBN	PCD	PCBN	Si_3N_4, HIP	PCBN	(SiCw)-Al_2O_3	(SiCw)-Al_2O_3
Al_2O_3 + TiC	ZrO_2	C2 carbide	Sialon	C2 carbide	Si_3N_4, RB	Al_2O_3	ZrO_2	Si_3N_4, HIP
C2 carbide	Sialon	Al_2O_3	Al_2O_3 + TiC	Si_3N_4, HIP	Al_2O_3 + TiC	C2 carbide	Sialon	Si_3N_4, RB
Si_3N_4, HIP	Si_3N_4, HIP	Al_2O_3 + TiC	Si_3N_4, HIP	Sialon	Sialon		Si_3N_4	Sialon
Sialon	Al_2O_3 + TiC	(SiCw)-Al_2O_3	ZrO_2	Al_2O_3 + TiC	ZrO_2		C2 carbide	Al_2O_3 + TiC
Al_2O_3	Al_2O_3	Si_3N_4, HIP	Al_2O_3	Si_3N_4, RB				Al_2O_3
ZrO_2	Si_3N_4, RB	Sialon	PCBN	Al_2O_3				
Si_3N_4, RB		Si_3N_4, RB	Si_3N_4, RB	ZrO_2				
		ZrO_2						

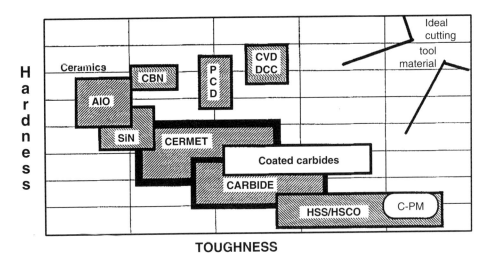

FIGURE 4.2 Comparison of hardness and toughness for various cutting tool materials.

allowable speed ranges for given combinations of tool and work materials, based largely on hot hardness and chemical inertness considerations. For specific grades, additional information on material properties, as well as application guidelines (e.g., allowable speed, feed, and depth of cut ranges) can be obtained from tool suppliers. The cutting speed selection is the most important parameter among the cutting conditions as explained in Chapter 9 and Chapter 13. The cutting speed is affected by the expected tool life curve as discussed in Chapter 9.

As is evident from Table 4.1, Figure 4.2, and Figure 4.3, no single material exhibits all of the desirable properties for a tool material. Some of the desired properties, in fact, are mutually exclusive. Very hard materials, for example, tend to be brittle and thus have poor fracture toughness. The best tool material for a given application depends on several factors; as shown in Figure 4.5, these include part requirements, constraints imposed by available machine tools and tool holders, and economic considerations. For new processes, the cutting speed regime is often selected first, as this has a strong influence on the metal removal rate and tool wear characteristics as discussed in Chapter 9. The tool material selection is then made from the restricted list of materials suitable for use within that range for the given work material (Figure 4.4). In troubleshooting existing operations to improve tool life or part quality, tool materials are often substituted based on the speed range and expected type of

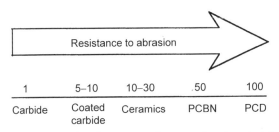

FIGURE 4.3 Relative comparison of abrasive wear resistance of tool materials.

Cutting Tools

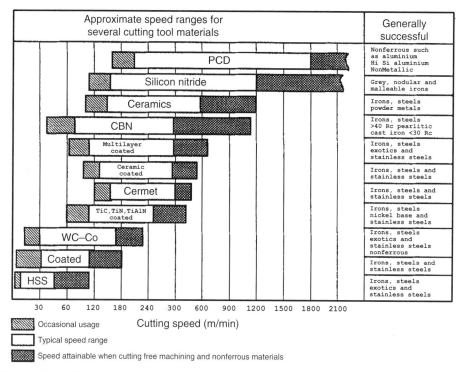

FIGURE 4.4 The effect of tool material on cutting speed.

tool failure as shown in Figure 4.6. In either case, an understanding of the properties and application ranges of specific tool materials is required to make sound choices. The properties of specific classes of tool materials are discussed in detail in the remainder of this section.

4.2.2.1 HSS and Related Materials

HSS are self-hardening steels alloyed with W, Mo, Co, V, and Cr. They exhibit "red hardness," which permits tools to cut at a dull red heat without loss of hardness or rapid blunting of the cutting edge. HSS is inexpensive compared to other tool materials, is easily shaped, and has excellent fracture toughness and fatigue and shock resistance. However, the hardness of HSS decreases rapidly at temperatures above 540 to 600°C, and HSSs have limited wear resistance, limited chemical stability, and a greater tendency to form a built-up edge than other tool materials. Their limited wear resistance and chemical stability makes HSS tools suitable for use only at limited cutting speeds. Plain HSSs are generally used at cutting speeds below 35 m/min, although special alloys such as the HSS–Co alloys discussed below, can be used at speeds up to 50 m/min. HSS is very commonly used for geometrically complex rotary tools such as drills, reamers, taps, and end mills, as well as for broaches. HSS tools are also widely used in multispindle machines (e.g., gang drill presses, screw machines, and older transfer machines), which have limited rigidity and speed capabilities.

Plain HSSs are broadly classified as T-type steels, which have tungsten as the major alloying element, and M-type steels, in which the major alloying element is molybdenum. The T-types are less tough than M-type, but are heat treated more easily. M-types are more widely used for rotary tooling, especially drills, end mills, and taps. Of the standard plain HSSs, M2 is most widely used for drills and taps, T42 has the best abrasion resistance, and M42 has the greatest hot strength.

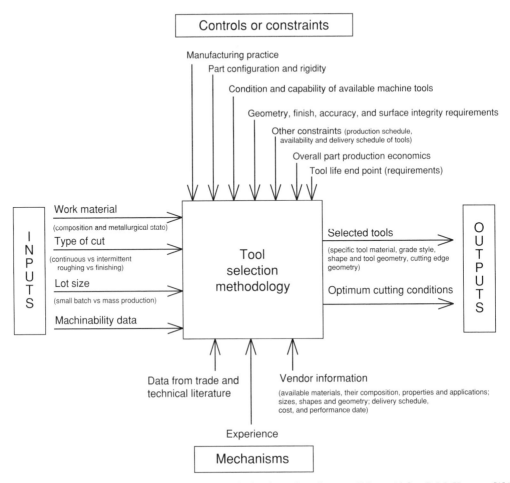

FIGURE 4.5 Procedure for tool selection and optimization of cutting conditions. (After B.M. Kramer [1].)

Sintered or powder metals (P/M) provides benefits in tool performance. P/M HSS material results in minimal distortion and improved wear resistance, hot hardness, and toughness compared with plain HSS. P/M HSS materials are especially beneficial in tapping and broaching but are also used for drills and end milling cutters because they have better toughness than carbide tools, which may chip and crack in interrupted cuts or when encountering hard spots. In proper applications, P/M tools can double tool life and remove material at twice the rate of HSS or even carbide [10].

In some applications, HSS is alloyed with cobalt (HSS–Co) or vanadium to produce grades with increased toughness, hot hardness, and wear resistance. HSS–Co grades are used especially for drills and taps. They are suitable for use at higher speeds, feeds, and depths of cut than plain HSSs, and have better fracture toughness than sintered carbides. Similarly, stellite, a cobalt-based alloy containing chromium, tungsten, and carbon, is sometimes used for tools. Most commonly, solid stellite tool bits are used to turn difficult to machine materials (e.g., weldments) which cause chipping of carbide tools.

4.2.2.2 Sintered Tungsten Carbide

Sintered tungsten carbide-based hardmetals are the most common tool materials for turning, milling, threading, and boring using indexable inserts and solid round tooling. Cemented WC

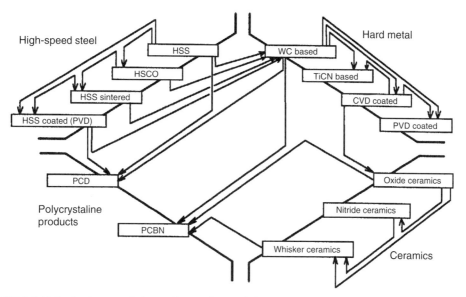

FIGURE 4.6 Substitution trends in cutting tool materials. (After I.E. Clark and J. Hoffmann [9].)

inserts and blanks are manufactured by mixing, compacting, and sintering WC and Co powders. The Co acts as a binder for the hard WC grains; as discussed below, the grain size and binder content largely determines the insert's physical properties. Characteristics of WCs include high transverse rupture strength, high fatigue and compressive strength, and good hot hardness. The modulus of elasticity and torsional strength are twice those of HSS. Carbides conduct heat away from the tool–chip interface well, and can be tailored to meet specific thermal shock requirements. By varying the cobalt content, the relative balance of hardness and toughness can be changed. Their main drawback is that they have only average chemical and thermal stability at high temperatures, which makes them unsuitable for machining steels at high cutting speeds. For nonferrous workpiece materials, WC tools will exhibit two to three times the productivity and 10 times the life of HSS tools; in steels two times the productivity and five times the life. The useful life of uncoated carbide at conventional speeds in hard materials is generally unacceptable. There is a trend away from HSS to carbides in most applications because new machining centers with higher speeds are stiffer and require more wear resistant tools.

In the United States, WC grades are usually classified into eight categories denoted C1 through C8. The grades are broadly divided into two classes (C1 through C4 and C5 through C8) according to the workpiece material (see Table 4.2). As the number increases within each class, shock resistance (toughness) decreases, hardness increases, high-temperature deformation resistance and wear-resistance increase, and carbide grain size decreases. Therefore, an increase in cutting speed or the feed load should be followed by an increase or decrease, respectively, of the classification number for the carbide grade. Hardness measurements primarily indicate the Co volume and WC grain size, and not the actual hardness of the carbide constituent. Similarly, a European classification has been adopted in ISO Standard 513 which consists of three categories designated as P- (heavily alloyed multicarbides), M- (low-alloyed multicarbides), and K-grades. A summary and comparison of the characteristics of C-, K-, P-, and M-grades is given in Table 4.2. This table can be used to determine appropriate application ranges for specific grades; for example, of K-grades suitable for use on cast iron, K01 is a wear-resistant, finishing grade suitable for finish boring

TABLE 4.2
Classification of Carbide Tool Materials According to Use [Per ISO 513-1975(e)]

C System	Materials to be Machined	Operation	Direction of Increase in Characteristic Of Cut	Direction of Increase in Characteristic Of Carbide	ISO System
C1 C2 C3 C4	Cast iron (all types) Hard steel Nonferrous Nonmetallics	Roughing Finishing	Increasing Speed ↑↓ Increasing Feed	Wear Resistance ↑↓ Toughness	K-40 K-35 K-30 K-10 K-01
C5 C6 C7 C8	Ferrous Carbon steel Alloy steel Stainless steel anneal Steel casting	Roughing Finishing	Increasing Speed ↑↓ Increasing Feed	Wear Resistance ↑↓ Toughness	P-50 P-40 P-30 P-20 P-10 P-01
	Low-strength steels Ductile cast iron High-temper. alloys Nonferrous	Roughing Finishing	Increasing Speed ↑↓ Increasing Feed	Wear Resistance ↑↓ Toughness	M-50 M-40 M-30 M-20 M-10 M-01

with no shock, while K40 is a tough grade suitable for rough milling. By convention, K25 represents general purpose milling.

Two basic classes of carbide materials are used for cutting tools: two-phase WC–Co (straight cemented tungsten carbide; grades K01–K40), and alloyed WC–Co grades, in which part of the WC is replaced by a solid solution of cubic carbides, such as titanium carbide (TiC), tantalum carbide (TaC), niobium carbide (NbC), or a combination of these materials (grades P01–P50 and M01–M50). Straight WC–Co grades are used for workpiece materials that cause primarily abrasive tool wear and generate a short, discontinuous chip; typical materials include cast iron, high-temperature alloys, high-silicon aluminum, other nonferrous alloys, and nonmetals. When used with steels, which generally yield longer continuous chips, straight WC–Co grades often fail due to crater wear on the rake face of the tool. As discussed in Chapter 9, crater wear is caused by diffusion of constituents of the tool into the chip at high temperatures. In straight grades, reducing the Co content and carbide grain size improves cratering resistance. Alloying also improves cratering resistance; alloyed grades containing TiC or TaC were developed for improved crater wear resistance when machining steels. Compared to straight grades, alloyed grades have increased heat resistance, compressive strength, and chemical stability. TaC alloyed grades have a higher hot hardness and better thermal shock resistance than TiC alloyed grades.

The hardness, fracture toughness, and heat resistance of carbide grades depend on the Co, TiC, and TaC contents and on the carbide grain size. Increasing the Co content decreases hot hardness and edge wear, crater wear, and thermal deformation resistance, but increases fracture toughness. The compressive strength is affected significantly by the Co content, and increases with Co percentage to a maximum at about 4%. The abrasive wear resistance of cemented carbides increases with increasing TiC content and decreases with increasing TaC content. Substrates with Co content less than 6% are attractive for ferrous machining at higher speeds with smaller depths of cut and limited interruptions.

The WC grain size affects the tool's hardness, toughness, and edge strength. Fine WC grains generate thin sections of Co binder which are constrained by the WC grains, resulting in harder and tougher materials with lower fracture strength. Typical WC grades have 2 to 5 μm grain size and a hardness between 89 and 93 Rockwell "A." Fine grain WC is used for inserts and for solid drills, reamers, end mills, and other rotary tools. Coarse grain WC (6 to 8 μm) is stronger but softer and less wear resistant than finer grades and is used primarily for roughing. Since hardness increases with decreasing grain size, micrograin carbides exhibit superior resistance to crater wear, notching, and chipping in semiroughing through finishing applications and perform well in applications where normal carbide grades tend to chip or break. Micrograin carbide grades are classified as fine grain (<1.5 μm), finest grain (<0.7 μm), and ultrafine grain (<0.5 μm). Ultrafine grain carbide is used in special cases such as rotary tooling for high-speed and high-throughput machining applications in ferrous materials and is becoming common for high performance rotary tooling. It improves cutting edge strength and stability, preventing premature chipping and material loading and allows sharper edges because ultrafine grain carbide is 15 to 20% tougher than common carbide grades (grain size 2 to 3 μm with 8% Co). Nanophase carbides are being developed for very small tools and for round high-performance tools for exotic materials. An improvement of carbide performance is obtained by providing a Co-enriched (binder phase) layer (about 13 to 25 μm thick) on the surface of the tool or the tool corner. This enriched layer contains two to three times the cobalt concentrations of the bulk material. This improves the toughness of the cutting edge and resistance to chipping.

Basic guidelines for selecting carbide grades are: (1) use the lowest Co content and finest grain size, provided edge chipping and tool breakage do not occur; (2) use straight WC grades when abrasive edge wear is of concern; (3) use TiC grades to prevent crater wear and/or both crater and abrasive wear; (4) use TaC grades for heavy cuts in steels. The greatest limitation of uncoated carbide, when used at higher cutting speeds in ferrous materials, is rapid wear due to abrasion.

4.2.2.3 Cermets

Cermets are TiC-, TiN- or TiCN-based hardmetals often described as ceramic or carbide composites. The physical properties and application range of cermets generally fall between those of WC and plain ceramics. Cermets are less susceptible to diffusion wear than WC, and have more favorable frictional characteristics. However, they have a lower resistance to fracture (lower strength and toughness) and a higher thermal expansion coefficient than WC, and are more feed sensitive. (A new generation of micrograin cermets provides bending strength closer to that of WC.) Cermets have a higher bending strength and fracture toughness than ceramics, higher thermal shock resistance than oxide-based ceramics, and lower hardness than all ceramics [11]. In general, cermets have excellent deformation resistance and high chemical stability, but relatively poor edge strength. They can be used with sharp cutting edges in many finishing applications, which enables them to produce exceptionally good surface finishes.

Cermets consist of TiC and TiN particles sintered with a refractory metallic binder, usually composed nickel (Ni), cobalt (Co), tungsten (W), tantalum (Ta), or molybdenum (Mo). Ni and Ni–Mo are the most commonly used binders, and the binder volume is typically between 5 and 15%. TiC provides hot hardness for wear resistance, oxidation resistance, chemical stability, and improved notch resistance, and reduces the tendency of the work material to adhere to the tool; TiN provides fracture toughness and thermal shock resistance. Cermets are generally available in three grades — hard, tough, and the relatively tough but hard. Hard cermets are used in applications requiring high resistance to wear and plastic deformation, such as semifinish and finish cutting of steels, stainless steels, free machining aluminum, and other nonferrous alloys (brass, zinc, and copper) and in some cast irons as shown in Table 4.3 Tough cermets are also used in semifinish and finish applications (especially milling) and in some rough continuous cuts in low alloyed steels, stainless steels, ductile irons, and relatively hard steels. The tough but hard grade is used for turning and boring and for finish milling operations. In appropriate applications, cermets provide 20 to 100% longer life than coated carbides. The principal advantage of cermets over carbide is their ability to operate at much higher surface cutting speeds, with longer cutting edge life. Cermets are generally used at semifinish to finish applications and especially high-speed finishing applications. The allowable cutting speed is usually lower than that attainable with ceramics. Suggested recommendations are given in Table 4.3. Cermets tend to be more shock resistant than ceramics and coolant is often recommended for finish turning, threading, and grooving with coated cermets. Micrograin cermets have much better thermal shock resistance and coolant can be used in all operations when necessary [12].

4.2.2.4 Ceramics

Ceramic tools are hard and chemically stable, and have replaced carbide tools in many high-speed machining (HSM) applications. Ceramics can withstand higher temperatures than carbides, allowing a three- to tenfold increase in cutting speed and a two- to fivefold increase in the metal removal rates. The mechanical properties of ceramics are superior to those of carbides only at higher temperatures (e.g., above 800°C). They retain excellent hardness and stiffness at temperatures from 1000 to 1500°C (carbides soften appreciably at temperatures above 850°C), and do not react chemically with most workpiece materials at these temperatures. They provide better size control due to lower tool wear rates, resulting in improved quality. They do, however, have several weaknesses: relatively low strength, poor resistance to thermal and mechanical shock, and a tendency to fail by chipping. Most ceramic monolith

TABLE 4.3
Cermet Machining Recommendations

Turning and Boring	Grade	Speed (m/min)	Feed (mm/rev)
Material			
Steels	Tough	60–300	0.15–0.35
Steels	Hard	60–340	0.10–0.35
Stainless steels	Tough and hard	45–270	0.10–0.30
Cast and nodular iron	Tough	60–250	0.15–0.45
Cast and nodular iron	Hard	60–360	0.15–0.45
Nickel	Hard	60–200	0.10–0.25
Milling			
Steels	Tough	60–250	0.05–0.15
Steels	Hard	60–340	0.05–0.15
Stainless steels	Tough and hard	60–230	0.05–0.15
Cast iron	Hard	100–360	0.05–0.15
Nodular iron	Hard	45–170	0.05–0.15

materials do not have predictable failure times and fail catastrophically in ways that may ruin workpieces. To address this limitation, ceramic composite materials are being developed. Ceramics are usually used without coolant to avoid thermal shock. Mechanical shock should be minimized by using stiffer machine tools and low frequency interrupted cuts. This makes selecting the proper speed and edge preparation essential when using ceramic tooling. In general, cutting speed should decrease as the workpiece material's hardness increases.

Ceramic cutting tools, mostly made of Al_2O_3- and Si_3N_4-based materials, can be divided into four categories:

1. Alumina, Al_2O_3, sometimes mixed with zirconium oxide, Al_2O_3–ZrO_2. These tools are yellow to gray/white in color. Tools made of Al_2O_3 and phase-transformation toughened Al_2O_3–ZrO_2 exhibit high chemical inertness and resistance to wear and thermal deformation. They are used for continuous shallow cuts (semifinishing and finishing operations) at relatively low feed rates. Typical applications include turning carbon steels, alloy steels, tool steels (<38R_c), and gray, nodular, or malleable cast irons (<300 BHN (Brinnel Hardness Number)) at speeds up 1000 m/min [7, 13].
2. Alumina–titanium carbide composites, Al_2O_3–TiC, containing 30 to 40% TiC. This material, which is black in color, has higher transverse rupture strength, thermal shock resistance, and hardness than conventional Al_2O_3, but still has a relatively low resistance to fracture. It is effective for uninterrupted cuts on alloy steels, chilled and malleable cast irons, hardened ferrous materials (35 to 65R_c), and exotic alloys [7, 13]. Coated Al_2O_3–TiC tools are used for finish turning of hardened steels and irons. All alumina-based (Al_2O_3, Al_2O_3–ZrO_2, and Al_2O_3–TiC) materials tend to crack and exhibit notch wear when machining steel. Also, chemically induced wear can occur, depending on the cutting temperature and the surrounding environment (air, humidity, coolant, etc.). Al_2O_3-based tools are unsuitable for machining aluminum alloys and titanium alloys because of their strong chemical affinity to these materials. They substitute for P01 to P05 or C8 carbide inserts.
3. Reaction-bonded silicon nitride (Si_3N_4, RB), hot pressed silicon nitride (Si_3N_4, HIP), and sialon (Si_3N_4–Al_2O_3), all of which are gray in color. Si_3N_4 RB and HIP are combinations of Si_3N_4 with yttrium, Al_2O_3, and TiC. Sialon includes silicon, aluminum, oxygen, and nitrogen. Compared to the materials in the first two categories, these tough ceramics exhibit superior wear and notch resistance, high red hardness, and resistance to thermal shock; tools made from them are consequently more reliable. Si_3N_4-based tools are extremely wear resistant when used to machine cast iron and superalloys, but are subject to excessive temperature-activated wear when machining steels at high speeds [14]. Sialon, conversely, has been applied successfully on both gray cast iron and steel at high speeds. Sialon is more chemically stable than Si_3N_4 but not quite as tough or resistant to thermal shock. Automobile engine blocks represent the largest single application for these materials, followed by nickel-based superalloys for aerospace and corrosion components, and hard steels for a variety of uses. They are not generally used for aluminum alloys due to the high solubility of silicon in aluminum. They substitute for K01 to K05 or C4 carbide inserts.

The wear rate of TiC and Ti(C,N)-coated Si_3N_4–(30%)TiC tools has been found to be significantly lower than Al_2O_3–TiC tools [15]. When a Si_3N_4–(15%)Al_2O_3–(30%)TiC tool is coated with TiC–Al_2O_3, or TiN–Al_2O_3, it outperforms Si_3N_4- and Al_2O_3-based ceramics for HSM of steels [15]. Si_3N_4 is the most appropriate ceramic tool material for machining cast iron at a speed up to 1200 m/min; for this work material, the speed limit for plain WC tooling is 100 m/min.

4. Silicon carbide whisker-reinforced alumina, (SiC_w)–Al_2O_3. This material was designed to combine reliability and superior resistance to fracture, thermal shock, and wear [13, 16, 17]. The tool life achievable with this material is not necessarily greater than that achievable with other ceramic tools; Sialon often outperforms (SiC_w)–Al_2O_3 when machining gray cast iron and 1045 and 4340 steels. It is most effective when used to machine high-temperature alloys at higher cutting speeds, especially those varieties that have a nickel base (e.g., inconel) [7]. Unlike other ceramic tools materials, it can be run with coolant. They substitute for K01 to K05 or C3 and C4 carbide inserts.

Figure 4.7 shows the hot hardness and Palmquist fracture toughness of these ceramic materials. Most ceramic cutting tools are presently used in the form of indexable inserts, which have been well proven over the past decade for higher speed machining. They have replaced carbides in many turning and milling operations. A limited number of ceramic inserts with grooved chip breakers are available. Operations such as drilling and end milling of ferrous materials and superalloys could be performed at higher speeds if the appropriate ceramic tools were available. Solid and indexable ceramic rotary tools are available in limited sizes and geometries, but because they are very sensitive to torsion and bending, they require cutting edge preparations which are difficult to produce by grinding. There is a need for ceramic tool inserts and heads brazed or bonded onto HSS or carbide drill and end mill bodies, but to date reliable tools of this type have not been developed.

When using ceramic tools, a chamfer at the entrance and exit of interrupted surfaces is recommended.

4.2.2.5 Polycrystalline Tools

Polycrystalline tools provide maximum tool life at high cutting speeds. Two materials have been developed: polycrystalline cubic boron nitride (PCBN) and polycrystalline diamond

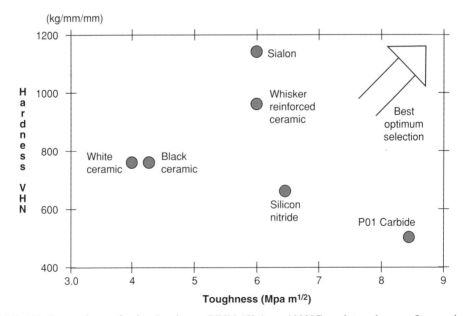

FIGURE 4.7 Comparison of microhardness (VHN-1Kg) at 1000°C and toughness of ceramic tool materials using P01 carbide as a baseline. (After P.K. Mehrotra [17].)

Cutting Tools 153

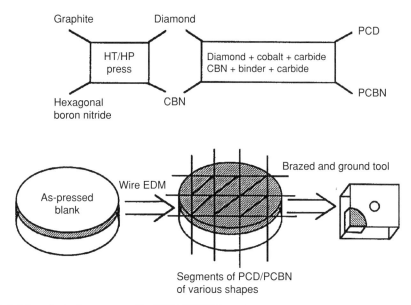

FIGURE 4.8 Manufacturing process for PCD/PCBN inserts.

(PCD). Both materials are manufactured using a high-temperature, high-pressure process in which individual diamond or CBN particles are consolidated in the presence of iron, nickel, or cobalt catalysts that promote grain consolidation into a solid mass (solid polycrystalline) or on a WC substrate (backed polycrystalline). The grain size is small. Once manufactured, the polycrystalline compact is cut by electrodischarge machining (EDM) into smaller pieces which are often brazed onto WC inserts as shown in Figure 4.8 [18]. A slightly different approach presses the PCD or CBN layer on the carbide insert or round tool and bonds the two materials together during sintering. This increases edge integrity and allows for more complex geometries [19].

4.2.2.5.1 Polycrystalline Cubic Boron Nitride
At elevated speeds, PCBN, the second hardest material known, remains inert and retains high hardness and fracture toughness. Although it has superior hardness, the fracture toughness of PCBN falls between that of WC and ceramics. PCBN has a high thermal conductivity and low thermal expansion coefficient, which makes it less sensitive to thermal shock than ceramics [20]. It is thermally stable at temperatures to 1400°C [20, 21]. PCBN can wear by diffusion when cutting ferrous alloys at high speeds, but outperforms WC in this respect. The use of PCBN tooling promotes self-induced hot-cutting because it can be used at cutting speeds sufficient to cause workpiece heating and softening.

There are several grades of PCBN insert materials and tools. The major types are listed in Table 4.4. The fracture toughness and thermal conductivity increase with increasing CBN content. However, inserts with a low PCBN content have greater compressive strength.

PCBN is well suited for HSM of ferrous materials with hardness between 45 and 65R_c; such materials include sintered (P/M) irons, hard and soft pearlitic cast irons, hardened steels, HSSs, and nickel-based alloys [22–24]. (Excessive wear is sometimes reported when cutting cast irons with high ferrite content, perhaps due to a chemical interaction with the binder.) PCBN can be used to machine hard steels and superalloys at speeds roughly equal to those attainable using conventional tools to machine soft materials. The high fracture toughness of PCBN tools makes them suitable for interrupted cutting operations such as milling. Ceramic

TABLE 4.4
Major Characteristics of PCBN Materials

CBN Content (%) (approximately)	Catalyst/Matrix/Binder	Format
90	Metal (Al)	Solid or carbide substrate
80	Metal/ceramic (Ti and Al)	Carbide substrate
<70	Ceramic (TiN or TiC)	Carbide substrate

tools fail in these applications at medium to high speeds. The straight PCBN grade can be used for most of the above applications for roughing and finishing operations. Composite PCBN grades can be used for high-speed finish machining of hardened steels, and for interrupted and continuous finishing operations on hard and soft cast irons. PCBN/ceramic grades are best for turning hard steels since they have less tendency to interact chemically with the chips. The operative speed range for PCBN when machining gray cast iron is 600 to 1400 m/min; speed ranges for other materials are as follows: hard cast iron (>400 BHN), 80 to 300 m/min; superalloys (>35R_c), 180 to 400 m/min; hardened steels (>45R_c), 70 to 300 m/min; sintered iron, 100 to 300 m/min. Case histories for cutting a variety of materials have been reported [25]. Materials which should not be cut with PCBN include soft steels, ferritic gray iron (as noted above), ductile and malleable iron <45R_c, stainless steels, inconels, and nickel-based high-temperature alloys.

In addition to cutting speed, the most important factor affecting PCBN insert performance is the cutting edge geometry or preparation. It is best to use PCBN tools with a honed or chamfered edge preparation, especially for interrupted cuts. As with ceramics, PCBN tools are available only in the form of indexable inserts. Development of brazed PCBN drill bits and end mills is proceeding, but extensive testing is required to identify the best geometries for high volume applications.

4.2.2.5.2 Polycrystalline Diamond
PCD, the hardest of all tool materials, exhibits excellent wear resistance, holds an extremely sharp edge, generates little friction in the cut, provides high fracture strength, and has good thermal conductivity. These properties contribute to PCD tooling's long life in conventional and HSM of soft, nonferrous materials (aluminum, magnesium, copper, and brass alloys), advanced composites and metal–matrix composites, superalloys, and nonmetallic materials (plastics, ceramics, fiberglass, wood, and graphite). PCD is particularly well suited for abrasive materials (i.e., drilling and reaming metal–matrix composites) where it provides 100 times the life of carbide. PCD is not usually recommended for ferrous materials or hardmetals because of the high solubility of diamond (carbon) in these materials. However, they can be used to machine some of these materials under special conditions; for example, light cuts are being successfully made in gray cast iron using chip loads between 0.025 and 0.075 mm.

PCD tooling requires a rigid machining system because PCD tools are very sensitive to vibration. In mass production operations, the attainable tool life may be over 1 million parts (e.g., for diamond-tipped drills or PCD milling cutters machining soft aluminum alloys). Tool lives of this length may never be achieved, however, because tooling breaks due to vibration or rough handling before wear becomes significant.

Various grades of PCD have been developed for specific applications [26–28]. The major difference between grades is the size of the individual diamond grains, which varies between 1 and 100 μm. Tool performance is affected by the microstructure of the PCD. Grades are grouped in several categories with average grain sizes of 1 to 4, 5 to 10, and 20 to 50 μm [25]. The abrasive wear resistance, thermal conductivity, and impact resistance increase with increasing grain size, but finer grained tools produce smoother machined surface finishes. For example, a coarse-grained PCD tool may provide 50% better abrasive wear resistance than a fine-grained tool, but produce a surface with 50% higher roughness. Because of their increased impact and abrasive wear resistance, coarse grades are preferred for milling and for machining high-silicon aluminum alloys and metal–matrix composites. Multimodal PCD grades (made with bimodal, trimodal, or quadrimodal distributions of PCD particles) provide the highly abrasiveness of unimodal grade of coarse particles with high toughness structure and superior edge sharpness of medium-size grain tools [28]. The PCD density increases with multiple particles sizes. Multimodal grade results in less chipping compare to a unimodal grade.

PCD-tipped HSS or carbide rotary tools (e.g., reamers, end mills, drills, etc.) are available in a limited range of geometries due to difficulties in grinding complex geometries, particularly on small diameter tools. More complex geometries can be used on carbide rotary tools by sintering the diamond into slots located at the point and along the flutes [19].

As with PCBN, the development of improved PCD drills remains an important goal for the tool industry especially for high-throughput applications. Issues to be resolved include identifying the optimal cutting edge geometry for the diamond tip and the best method of pocketing the polycrystalline blank for strength and manufacturability. The point geometry, flute geometry, and web thickness have not been refined sufficiently to allow use of polycrystalline brazed drills at penetration rates comparable to the feed rates attainable in turning and milling. Although methods of brazing the polycrystalline/carbide substrate tip to the main tool body have been improving steadily (with each manufacturer using its own proprietary procedures for surface cleaning and brazing), one of the major failure modes is still the detachment of the polycrystalline tip or the wear and erosion of the braze joints intersecting the cutting edge. Wear and erosion of brazed joints is avoided when the diamond is sintered into slots within the carbide tool.

4.3 TOOL COATINGS

HSS, HSS–Co, WC, and ceramic tools are often coated to increase tool life and allowable cutting speeds. The majority of inserts used in production are coated. Coatings act as a chemical and thermal barrier between the tool and workpiece; they increase the wear resistance of the tool, prevent chemical reactions between the tool and workpiece material, reduce built-up edge formation, decrease friction between the tool and chip or the tool and workpiece, and prevent deformation of the cutting edge due to excessive heating. Coated tools, therefore, can be used at higher cutting speeds, provide longer tool lives than uncoated tools, and broaden the application range of a given grade. Workpiece surface finish can be also improved with coated tools. A comparison of applicable cutting speed ranges for representative coated and uncoated tools is shown in Figure 4.9. The coatings are most effective in situations where crater or flank wear predominates, as is usually the case when machining steels and cast irons.

A number of factors affect coating performance, including the coating thickness, hardness, chemical compatibility and interfacial adhesion with substrate, crystal structure, chemical and thermal stability, elastic modulus, fracture toughness, wear resistance, thermal conductivity, diffusion stability, frictional properties, the tool geometry, and the intended application [29]. Postcoating treatment affects the interfacial adhesion with substrate and the smoothness of the coated surfaces.

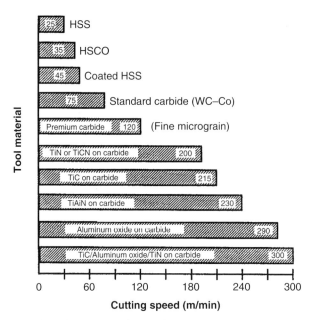

FIGURE 4.9 Comparison of cutting speeds of coated and uncoated tools.

4.3.1 Coating Methods

The two most common coating processes are chemical vapor deposition (CVD) and physical vapor deposition (PVD). Both are used for both single- and multilayer coatings. CVD coatings are generally applied at much higher temperatures (typically 1000°C) than those used for PVD coatings (typically 500°C). CVD coatings are typically between 5 and 15 μm thick; PVD coating thickness typically varies between 2 and 5 μm. There is a trend toward thicker coatings (>20 μm) on turning tools; milling tools have coating thicknesses up to 7 μm [30].

In general, the bond between a CVD coating and substrate (tool) is metallurgical and stronger than the mechanical bond produced by PVD. As a result, CVD coatings are harder than PVD coatings and provide longer tool lives when properly applied. However, the temperature requirements of traditional CVD techniques reduce the range of substrate materials to which these coatings can be applied. The CVD coatings on carbide substrates are in residual tension at room temperature because the coating materials have a higher thermal expansion coefficient than carbide; this may lead to transverse cracks resulting in tool failure in interrupted cutting. The high temperatures used in the CVD process can cause degradation of WC substrates due to the formation of an eta-phase (a thin, brittle glassy layer) at the interface between the coating and the substrate [31]. This process reduces the transverse rupture strength of the tool by as much as 30% and in particular embrittles sharp edges by breaking down the substrate's cobalt binder. Hence, CVD-coated tools require a honed edge. This has been overcome by reducing the process temperature to 700 to 900°C in the medium-temperature CVD (MT-CVD) process. This increases toughness and minimizes chipping and improves surface finishes of the coating. Most WC inserts are coated using a CVD or MT-CVD processes. A Co-enriched zone is sometimes used near the surface of WC insert to prevent the propagation of cracks from the coating into the core of the substrate to improve the toughness and fracture resistance. MT-CVD coatings are particularly well suited for interrupted and roughing cuts.

PVD coatings are essentially free of the thermal cracks that are common in CVD coating. Generally, PVD coatings are better suited for precision HSS, HSS–Co, brazed WC, or solid WC tools. In fact, PVD is the only viable method for coating brazed tools because CVD methods use temperatures that melt the brazed joint and soften steel shanks. PVD coatings are finer grained (effectively conform to the sharp edges of finishing tooling) and are generally smoother and more lubricious than CVD coatings, which build up on sharp corners. PVD coatings are preferred for positive rake and grooved inserts because they produce compressive stresses at the surface that reduce crack initiation and propagation. Several PVD coating methods are available such as ion bean evaporation, arc evaporation, and reactive sputtering.

Coating/substrate compatibility is improved by applying one or more intermediate layers between the surface coating and the substrate to balance chemical bonding and thermal expansion coefficients, resulting in a multilayer coating system which optimizes tool performance by providing resistance to several kinds of wear. Multilayer coatings may be produced by combined CVD, MT-CVD, and PVD methods; in such cases the CVD or MT-CVD process improves adhesion between the substrate and the first coating layer, while the subsequent PVD coating layers provide a fine-grained microstructure with better wear resistance and toughness or lower friction. Multilayer coatings are very common for turning and boring inserts because they provide the best combination of properties. Because machining processes result in many types of wear mechanisms, multiplayer coatings can add to a tool's multipurpose capability. CVD multiplayer coating has been also used in solid CBN inserts to improve the resistance to chemical and crater wear. The effect of the coating method and type in terms of withstanding different types of wear mechanisms is illustrated in Table 4.5.

There is considerable current development in nanolayer and nanocomposite or adaptive nanocrystalline coatings. Nanocomposite coatings may provide high hardness (4000 to 5000 HV (Vickers Hardness)) and high heat resistance (up to 1100°C), as well as toughness and hardness comparable to nanolayers [32].

4.3.2 Conventional Coating Materials

More than 50 combinations of coating and substrate materials are used for cutting tools; complete world-wide charts of coatings offered by most cutting tool manufacturers are available in the literature [33, 34]. Materials used for single layer coatings include titanium

TABLE 4.5
Indication of How Various Coating Methods Withstand Different Types of Wear (Courtesy of Sandvik Coromant)

Coating	Abrasion Wear	Adhesion Wear	Fatigue Wear	Plastic Deformation	Chipping/ Fracture	Try for a Sharp Edge
PVD	−	+	+	−	+	+
CVD	+	0	0	+	0	−
MT-CVD	+	0	0	+	0	−
TiAlN	+	+	0	+	−	0
TiCN	+	0	0	0	0	0
TiN	0	0	0	0	0	0
Al$_2$O$_3$	++	+	−	+	−	−

Note: "+" means positive impact; "−" means negative impact; "0" means neutral impact.

nitride (TiN), titanium carbide (TiC), titanium carbo-nitride (TiCN), titanium aluminum nitride (TiAlN), aluminum oxide (Al_2O_3), chromium nitride (CrN), hafnium nitride (HfN), titanium diboride (TiB_2), boron carbide (BC), and WC/C (amorphous diamond-like carbon) hard lubricant. Material combinations used in multilayer coatings include: Al_2O_3 on TiC or TiCN, TiN on TiC, TiN/TiC/TiN, TiN/TiCN/TiN, TiN/TiC/TiCN, Ti(C)N/Al_2O_3/TiN, Al_2O_3/TiC, TiN/TiC/Al_2O_3/TiN [29, 33, 35–46], and TiAlN+WC/C. Coatings containing Al_2O_3 are CVD grades.

A brief comparison of coating material properties is given in Table 4.6, and a comparison of coating hardness levels is given in Figure 4.10. TiC provides superior abrasive (flank or crater) wear resistance. TiN reduces friction and resists adhesive wear and built-up edge formation as well as increasing oxidation resistance (Figure 4.11). It also acts as a chemical barrier to diffusion wear when cutting ferrous materials with carbide tooling. TiCN provides low friction and good abrasive wear resistance. Al_2O_3 provides excellent thermal heating and oxidation resistance as well as abrasive wear and adhesion (built-up edge) resistance. TiAlN exhibits high hot hardness, ductility, and impact resistance.

PVD TiN, TiCN, and TiAlN, and AlTiN (high Al content TiAlN) coatings are very commonly used for rotary tooling. The performance of these coatings depends on the work material, the type of process, the cutter geometry, and the cutting conditions as shown in Figure 9.9 and Figure 9.10 in Chapter 9. All these coatings have performed very well in niche applications, but TiAlN is the most versatile of the three [37]. TiAlN has superior ductility and it is stable at higher temperatures than TiN and TiCN. TiAlN is not as hard as TiN and TiCN, but can be applied in thicker layers to compensate and provide equivalent tool life. TiAlN provides high resistance to oxidation, high thermal conductivity, greater hot hardness, and enhanced chemical resistance [31]. The oxygen with the aluminum in the coating forms an amorphous Al oxide layer at the interface with the chip (where temperature can reach more than 1000°C) that is thermodynamically stable and very protective and lubricious (as shown in Figure 4.12). TiAlN coatings are available in several forms such as low stress, hard, etc.

TABLE 4.6
Properties of Materials and Coating Materials [31, 35]

Material and PVD Coatings	Hardness HVN (kg/mm²)	Friction Coefficient versus Steel	Maximum Operating Temperature (°C)	Coating Thickness (μm)	Thermal Expansion (m/m K)
HSS	900	0.3–0.4	500		10–14
Oxides and nitrides	1,500–3,000	0.1–0.2	800		2.8–9.4
Carbides and borides	2,000–3,600	0.1–0.2	500		4.0–8.0
TiN	2,200	0.4–0.5	600	1–4	
TiCN	3,000	0.25–0.4	430	1–7	
TiAlN layered	3,300	0.3–0.65	800	2–5	
TiAlN monolayered	4,500	0.3–0.65	800	1–5	
AlTiN	3,800	0.5–0.65	900	1–5	
WC/C	1,000	0.1–0.2	300	1–4	
MoS_2-based	20–50	0.05–0.15	800	1	
TiAlN + WC/C	3,000	0.1–0.25	800	2–6	
CBN	4,700	NA	1400		4.7
Diamond	7,000–10,000	0.05–0.1			1.5–4.8

Note: Oxides and nitrides materials include Al, Si, Ti, Zr, Cr, Mo, Hf, and TiAl.

Cutting Tools

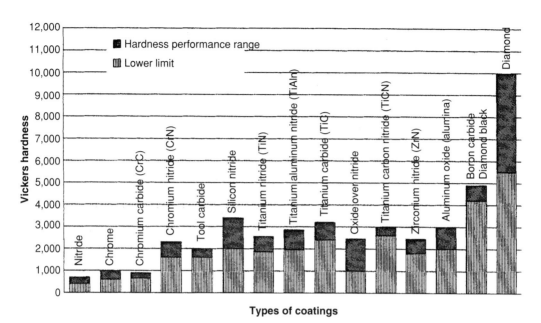

FIGURE 4.10 Comparison of microhardness for several types of coatings.

Their performance is altered by changing the aluminum content to suit specific applications. For example, the high content Al or AlTiN coating or hard TiAlN is ideal for HSM and dry HSM of hard and abrasive materials (i.e., ferrous, titanium, and exotic alloys). As with the carbon in TiCN, the aluminum in TiAlN can make a coating harder and more wear resistant.

TiN coatings perform best in cast, medium-alloy, and high-alloy steels. The TiN/TiAlN multilayer coating combines benefits of TiN, TiAlN, and TiCN such as high toughness, hardness, and heat resistance. Solid CBN inserts coated with a CVD Al_2O_3 coating provide good resistance to chemical and crater wear needed for roughing and finishing pearlitic grey cast irons, as well as chilled cast irons and hardened steels.

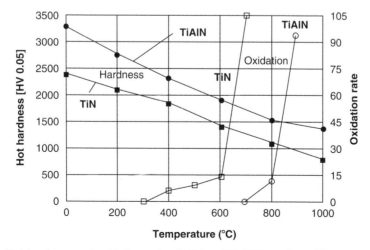

FIGURE 4.11 Hot hardness and oxidation rate of TiN and TiAlN coatings. (Courtesy of Kennametal, Inc.)

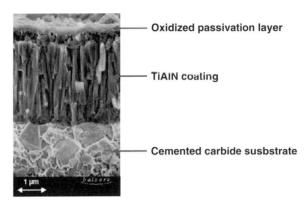

FIGURE 4.12 Photomicrograph (scanning electron microscopy) of a TiAlN coating on a carbide tool. (Courtesy of Balzers, Inc.)

Titanium diboride coatings are suitable for free machining aluminum and low silicon (hypoeutectic) aluminum alloys. They are hard, have low affinity to aluminum, and resist built-up edge formation. They are harder than TiN and TiAlN coatings and compete well with diamond coated carbide.

Lubricating coating materials, such as molybdenum disulfide (MoS_2) [47], tungsten disulfide (WS_2) [48], and WC/C (amorphous diamond-like carbon), are sometimes used in dry machining applications and also for drills and taps. Thin layers of these coatings are deposited on top of hard coatings, for example, MoS_2 over TiN or TiAlN and WC/C over TiAlN. These coatings improve chip flow and minimize built-up edge especially when drilling and tapping aluminum, cast irons, and steels. WC/C is also used as a single-layer coating to reduce built-up edge at lower speeds. The friction coefficient for these softer coatings is less than half that for TiN.

As noted above, multilayer coatings combine the best properties of single coatings, with each coating layer contributing a specific function and application [34, 38]. Al_2O_3 is usually deposited over TiC or TiCN since the TiC or TiCN layer provides a good base for the top layer to adhere. For example, the first layer of TiCN can be deposited at moderate temperature of 850°C to reduce the eta phase formation at the substrate/coating interface. A typical microstructure is shown in Figure 4.13.

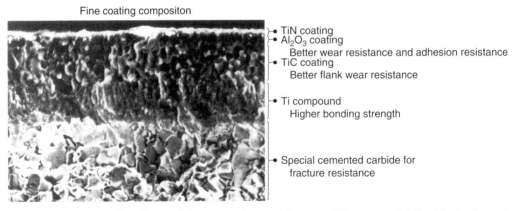

FIGURE 4.13 Schematic of a multilayer coating architecture. (Courtesy of Mitsubishi Materials Corporation.)

The primary concern in selecting a coating thickness is the tool edge strength, although tool wear is a significant secondary concern. Coatings reduce tool strength in proportion to their thickness since they act as stress risers. For example, a 2 μm coating could result in 15% reduction in tool strength, while a 10 μm coating could cause a 50% reduction. Therefore, the coating thickness is limited by the application, cutting conditions, and machining system.

4.3.3 Diamond and CBN Coatings

Pure crystalline diamond and CBN coatings [29, 49–53] provide low friction, good thermal conductivity, and excellent wear resistance. Since CBN coatings present similar technical challenges, the general concerns discussed for diamond coatings are applicable to both coating materials.

Diamond is deposited onto a substrate when carbon-based gases and hydrogen are disassociated at high temperature. Diamond coatings result in a fully dense layer of diamond at the tool surface, as compared to the porous layer containing cobalt phases at the surface of a PCD blank. Due to this higher density, coated diamond tools have a higher microhardness than PCD blanks, and can withstand higher cutting temperatures. A CVD thin-film diamond coating (a layer of diamond crystals less than 50 nm thick) can be deposited on either a ceramic or tungsten carbide substrate. Coating silicon nitride substrates has been successful because diamond adheres well to silicon nitride and because the thermal expansion coefficients of the coating and substrate are comparable [50, 51, 54]. These tools can also be reground and recoated after being worn out. However, the range of application of these tools is limited by the comparatively low toughness and impact strength of the substrate, the cost of the base material, and the difficulty in manufacturing ceramic tools with complex geometries [51, 52]. In addition, silicon nitride fails catastrophically after coating wear-through when machining aluminum workpieces, which is the most attractive application for such tools because CVD coatings cannot be used on sharp edges. CVD diamond thin films have not been yet fully developed for aluminum work materials, but have been successfully applied in the machining of phenolic resins, graphite, carbon graphite, fiberglass, and carbon graphite epoxy.

Thick CVD diamond films (0.1 to 1 mm thick) have also been explored to overcome the adhesion concerns with thin film coatings. Like PCD blanks, thick films are brazed onto a tool's cutting edge or corner. Therefore, brazed CVD diamond films cannot be applied easily to complex tool geometries. It has been reported [51] that thick film CVD diamond tools have wear characteristics surpassing PCD by 10 to 300% in selected applications. For example, CVD diamond coated inserts significantly outperform PCD inserts when turning abrasive materials because the cratering occurring on the PCD inserts induces softening of the cobalt phase, which is not present in CVD diamond coatings. In addition, CVD diamond films provide 30% higher hardness than PCD inserts. However, diamond-coated tools, especially those coated with thick diamond films, are avoided in interrupted cutting because they have lower fracture resistance than PCD tools.

Much development has also been directed toward developing coated rotary WC tools, since such tools cannot yet be made of solid ceramic due to their complex geometries. Rotary WC tools, usually containing between 5 and 6% cobalt binder, are difficult to coat using a CVD process because the cobalt reacts with the diamond. In addition, the thermal expansion coefficient of WC–Co is much higher than that of diamond. A number of methods have been explored to improve diamond/WC adhesion and to improve the reliability of diamond coatings; limited success has been obtained using WC inserts with low binder contents (<4%) [53]. Rotary tools (drills, end mills, taps) are available with fine diamond coatings (1 to 4 μm grain size) for use in nonferrous metals.

4.3.4 Cryogenic and Pulsed Magnetic Treatments

Processes using deep cryogenic or pulsed magnetic treatments are sometimes reported to improve the wear resistance of steel and WC tools.

In deep cryogenic tempering treatments, the tool is cooled to below 150 K for a number of hours using liquid nitrogen. Such treatments are used on steels to transform the retained austenite into martensite. This process reportedly improves the steel tool performance because it increases the tool's wear resistance by producing a tough surface; this helps to prevent particles from being removed from the surface by abrasion, and limits the penetration of the surface by hard particles in the environment.

In pulsed magnetic treatments, a high-strength, pulsed magnetic field with special characteristics is used to relieve residual stresses. Ferromagnetic materials expand and contract when placed in magnetic fields; this property is known as "magnetorestriction." Pulsed fields produced alternating expansions and contractions. Since dimensional changes do not occur uniformly throughout the thickness, internal vibrations that lead to relaxation of residual stresses are generated. This process reportedly can relax any grinding- and coating-induced stresses, increasing the toughness of the tool and its resistance to chipping or breakage and supposedly increasing tool life. Pulsed magnetic treatments can be used on tools made of HSS, HSS–Co, and powder metals, as well as solid WC and WC-tipped tools.

A cryogenic treatment combined with pulsed magnetic stress relief reportedly produces better results than a cryogenic treatment alone. The performance of both treatments depends on the manufactured quality of the steel tool, with the effects diminishing as the material quality and grinding procedures improve. These treatments are difficult to justify for good quality tools because the tool life improvement achieved is often not statistically significant.

4.4 BASIC TYPES OF CUTTING TOOLS

The six basic types of cutting tools, some of which are shown for drills in Figure 4.14, are solid tools, welded or brazed tip tools, brazed head tools, sintered tools, inserted blade tools, and indexable tools.

Solid HSS, HSS–Co, WC, and ceramic tools are made from a single piece of homogeneous material. P/M, carbide, and ceramic round tools are produced using a metal injection molding process. They can be resharpened after use, but this is a precision process requiring accurate grinding machinery.

WC, PCD, and PCBN-tipped tools consist of inserts or tips of hard tool materials brazed or welded to a steel or WC tool body. Tipped tools combine tool bodies that are tough, inexpensive, and easy to manufacture with tool points that are hard, wear resistant and typically brittle and difficult to manufacture. They can be reground, but as with solid tooling resharpening is a precision process. The methods of attaching the tip to the body affect the durability of the cutting edge(s). Welding produces the strongest bond; brazed bonds are weaker, but the brazing material (brass, bronze, or silver) acts as a cushion between the insert and tool body.

WC brazed head tools consist of a complete tool point made of WC brazed onto a steel tool body or shank. This method is used mainly for small tool sizes and in applications in which the joints of brazed inserts are not stable.

In *PCD and PCBN sintered tools*, PCD or PCBN powder is sintered into slots in WC tool bodies. This method is most commonly applied to drills, taps, and end mills.

Inserted blade tools consist of blades made of HSS, HSS–Co, WC, PCD, or PCBN, which are inserted into slots around the periphery of a tool body and locked in place mechanically. These cutters are normally used in high-production applications. They are ideal for

Cutting Tools

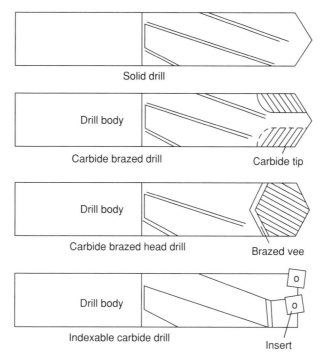

FIGURE 4.14 Types of rotary tooling.

close-tolerance finishing operations. Blades can sometimes be resharpened, but resharpening alters the tool size and (in some cases) geometry.

Indexable inserts are standard-sized wafers of hard tool material which are clamped into tool holders or cutter bodies. Inserts generally have more than one cutting edge; when a cutting edge is worn out, a new edge can be brought into use by rotating or "indexing" the insert. Inserts made of almost all tool materials are available. Many are made by direct pressing processes which allows production of complex shapes and eliminates the need for grinding. Indexable tooling offers advantages in both high- and low-production applications. Toolholders or cutter bodies may be made of steel, WC/heavy metal, or a combination of these materials to optimize performance. Inserts usually have uniform properties and are usually tougher and more wear resistant than other types of tools, so that they can be used at higher metal removal rates. They permit a reduction in tool inventory because a few standard inserts can replace a large number of complete tool bodies. They eliminate regrinding, are interchangeable, and are easily changed. They are less adaptable, however, than the brazed or inserted-blade tools for special cutters used to machine complex contours or holes with multiple diameters.

4.5 TURNING TOOLS

Turning is carried out primarily using single-point cutting tools, or tools with a single active cutting surface. Together with planing and shaping, turning is the most basic cutting process; an understanding of the cutting action and geometry of single-point turning tools therefore provides insight into the basic cutting action and geometry of more complex tools.

In the past, solid single-point tools ground from a HSS blank (Figure 4.15) [55] were most commonly used for turning. At present, however, turning is carried out using indexable

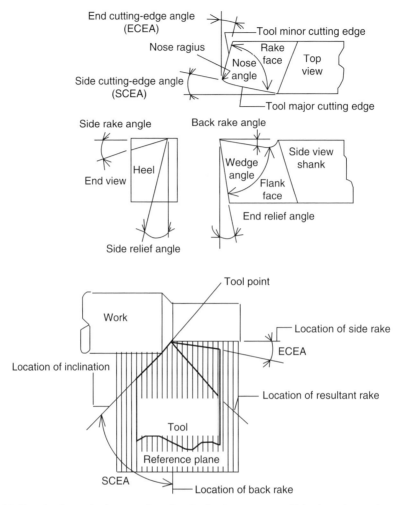

FIGURE 4.15 Standard terminology to describe the basic geometry of single-point tools.

inserts. This section discusses the design and selection of turning tools with emphasis on the proper selection of inserts and tool holders. The discussion of tool angles and insert selection (e.g., the strengths of various shapes of inserts, relative merits of different insert holding methods, etc.) is generally applicable to other cutting operations such as milling and boring.

4.5.1 Indexable Inserts

Figure 4.16 shows a representative sample of indexable inserts used for turning. The orientation of the insert to the workpiece is largely determined by the geometry of the toolholder. Selecting an insert and toolholder for a given operation, the following factors must be taken into account [4, 55–58]:

1. type of operation (roughing, finishing, etc.)
2. continuous versus interrupted cut
3. workpiece material and primary manufacturing process (casting, forging, etc.)

Cutting Tools

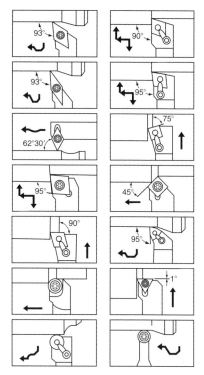

FIGURE 4.16 Representative inserts in several turning operations. (Courtesy of Mitsubishi Materials Corporation.)

4. condition of the machine tool
5. required tolerance
6. feeds and speeds (production rate).

Based on these factors, the following parameters must be selected:

7. insert material and grade
8. insert shape
9. insert size
10. insert thickness
11. nose radius
12. groove (chip breaker) geometry
13. edge preparation
14. edge clamping/holding method
15. lead, rake, relief, and inclination angles.

Most of these parameters are designated in the ANSI or ISO standard cutting insert nomenclature [56]. As discussed in Section 4.2 and Section 4.3, the selection of the insert material depends primarily on the work material, the required production rate, and the condition of the machine tool.

The shape of an insert is specified by the first letter of the insert designation; a TPGT-322 insert, for example, is triangular, while an SNGA-532 insert is square. Available insert shapes include diamond (C), triangle (T), square (S), octagon, round (R), and trigon (W). The shape

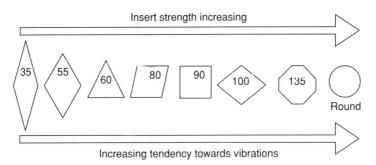

FIGURE 4.17 Relative insert strength.

of an insert largely determines its strength, its number of cutting edges, and its cost. As a general rule, an insert becomes stronger and dissipates heat more rapidly as its included angle is increased (Figure 4.17). The selection of the included angle is limited by the part configuration, the required tolerances, the workpiece material, and the amount of material to be removed. Therefore, the insert shape selection requires a trade-off between strength and versatility. For example, round inserts provide maximum edge strength and are therefore a good choice for roughing operations. They also provide a maximum number of effective cutting edges since they can be rotated (or indexed) through small angles when a given edge wears out. Round inserts thin the chip, however, and generate high radial forces; as a result, they should not be used when chatter or instability are expected, or when tight tolerances are required. Square inserts are common in general purpose applications because they provide good edge strength and a large number of cutting edges (8 for a negative rake tool, 4 for a positive rake). A 80° diamond insert is very versatile because it performs turning with 90° shoulder and facing operations. Generally, the largest included angle suitable for the workpiece geometry should be selected.

The insert size is determined by the largest circle that can be inscribed within the perimeter of the insert. The size of the inscribed circle (IC) determines the volume of tool material in the insert, and thus the insert cost. The IC size should be selected based on the depth of cut to be taken. Large IC inserts are used in heavy interrupted cuts or similar roughing operations. Smaller IC inserts perform more effectively in semifinish and finish operations.

As a rule, the cutting edge length should be two to four times the maximum cutting edge engagement (b or L_m), (Figure 4.18 and Figure 2.1). The allowable L_m is a function of the cutting insert shape and size.

The required insert size is determined based on: (a) selected depth of cut, and (b) effective cutting edge length in contact with part that determines the minimum edge length.

The thickness of an insert should also be selected based on edge strength requirements that relate to the selection of the IC. In general, larger IC inserts are thicker than a smaller IC inserts. The manufacturer determines the thickness based on the intended application (work material type and hardness) and its range with respect to depth of cut and feed.

A nose radius is usually provided on an insert to increase edge strength; a corner with a nose radius is much stronger than a sharp corner. The nose radius also serves to improve surface finish; at a constant feed, a larger nose radius generates a smoother finish (see Chapter 10). Hence, the largest nose radius allowed by the workpiece configuration and operating conditions (feed) should be used. Deflection and chatter are the primary concerns limiting the size of the nose radius, since an increase in the nose radius increases radial cutting forces. As a rule of thumb, the feed should not exceed 2/3 the nose radius. Some workpiece materials

Cutting Tools 167

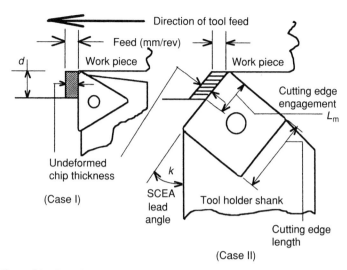

FIGURE 4.18 Effect of lead angle on undeformed chip thickness and cutting edge engagement.

(such as nickel-based alloys) require a nose radius larger than the depth of cut by as much as a factor of 2 to 4 for semifinish and finish operations, since the use of large radii allows the use of higher feed rates.

The insert IC, thickness, and nose radius are specified in the first three numbers in the insert designation. In the examples used above, a TPGT-322 insert has an IC size of 3/8 (0.375) in., a thickness of 2/16 (0.125) in., and a nose radius of 2/64 (0.03125) in.; and SNGA-532 insert has an IC of 5/8 (0.625) in., a thickness of 3/16 (0.187) in., and a nose radius of 2/64 (0.03125) in.

Standard inserts are available with varying tolerances (roughing inserts, semifinishing, and finishing inserts, etc.). Inserts can be used as sintered, ground on top and bottom faces only, or ground on all faces for precision. Ground inserts eliminate the need for manual adjustments of the tool position and hence maintain the required tolerance with greater ease on close tolerance work (less than 0.013 mm).

4.5.2 Groove Geometry (Chip Breaking)

The chip-breaking ability of an insert is important especially with ductile workpiece materials. The shape of the chip, cutting forces, and tool performance all are affected to a large extent by the groove geometry on the insert. There are several types of molded chip breakers (single-, two-, and three-stage chip breakers, vee-type chip grooves or land-angle designs, wavy and bumpy designs, and scalloped types as illustrated in Figure 11.6–Figure 11.9 in Chapter 11). The insert groove controls the chip flow direction, reduces cutting forces and edge wear, and controls vibration and temperature elevation. The groove geometry is dependent on the desired rake angle (positive, negative, or neutral) for a specific application. The geometry of a particular groove or chip breaker has a specific feed range that will produce acceptable chips; charts of effective feed and depth of cut ranges for a given insert design are available from the insert manufacturer (see Figure 11.10 in Chapter 11). Negative/positive inserts (with a negative side rake and a positive back rake) provide better chip control than either negative/negative or positive/positive inserts (see Figure 4.19). The mechanics of chip breaking is discussed in more detail in Chapter 11. Each insert geometry has a specific application area indicated by the feed versus depth of cut diagram (Figure 11.7).

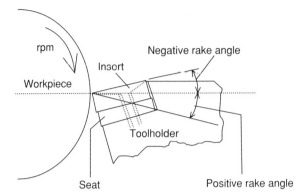

FIGURE 4.19 Grooved tool inserts which could provide a positive rake even when used in a negative rake toolholder.

4.5.3 Edge Preparations

Proper edge preparation is very important to the performance of a tool because it strengthens and protects the cutting edge and delays or eliminates chipping or breakage of sharp edges [57, 59, 60]. It has a significant effect on cutting force, surface finish, residual stresses, burr formation, and wear rate [61]. It also removes edge imperfections (cracked crystal layers) and prepares the edge for coating. Edge preparations are especially important for tools made of brittle materials. There are three basic preparations (Figure 4.20): hones (honed radius), chamfers, and negative lands, as well as combinations of these three. The hone is either a radius hone ($T = F$) or oval/elliptical hone ($T > F$). Most standard inserts have a light hone as part of their manufacturing process. An up-sharp or light hone is sufficient for machining soft, nonferrous materials such as aluminum. Nickel alloys and exotic aerospace alloys also usually require a light hone because they are cut at low feeds and cutting speeds. Steels and nodular or gray irons often require tools with medium-honed edges. Lightly honed edges are not recommended for cermet, ceramic, or CBN inserts, which are brittle and are normally provided with chamfered edges. A hone or small chamfer is generally required for the high throughput drills. The size of the hone varies with insert size may range from 12 μm (light

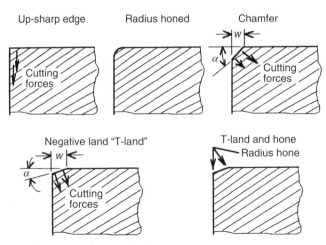

FIGURE 4.20 Cutting edge preparation methods.

hone) to 130 μm (heavy hone). Variation of the hone geometry is expected because it is made by manual stoning, brushing, tumbling, slurry honing, or abrasive blasting [62]. The uncut chip thickness and feed should be greater than the edge hone radius. Effective hones are approximately 1/3 to 1/2 the feed.

The bevel angle of chamfered edges ranges from 20° to 45°. The negative land (T-land) is similar to the chamfer but its bevel angle is smaller (usually between 5° and 20°). Chamfered and negative land edge preparations provide an effective negative rake angle to strengthen the edge and reduce chipping with brittle tool materials. However, chamfers or T-lands increase cutting forces and could have a potentially detrimental effect on both tool life and the machined surface finish. The T-land width can be smaller, equal, or larger than the feed per tooth. Generally, the chamfer or T-land width should be at maximum 70% of the feed; it should usually be kept below 30% to prevent chip jamming, high tool pressures, and heat build-up, all of which promote tool failure. T-lands are normally used on negative rather than positive rake inserts. However, a short negative land 25° × 0.05 mm can be used with positive rake tools to redirect the forces through the tool and provide good support for the cutting edge; in this case the T-land should be about 60% of the feed.

Edge preparations affect the edge strength, tool life, and chip control. Even though tool life reliability improves by avoiding microchipping or breakage, the actual life could be reduced because it acts as a pre-wear feature at the cutting edge. Edge preparation is a trade-off between edge strength versus edge wear and feed versus size of edge preparation. For example, the recommended edge preparation for turning gray cast iron with PCBN is a 20° × 0.2 mm chamfer for roughing and 20° × 0.1 mm chamfer for finishing. In more severe applications, an additional 0.025 mm radius hone is used to achieve additional strength. In the lightest cuts, a 0.025 to 0.05 mm radius hone is often acceptable.

4.5.4 Wiper Geometry

Wiper tools have additional radii behind the tool nose at the leading edge that are kept in contact with the surface to remove the peaks of the grooves made by the leading edge as shown in Figure 10.10 in Chapter 10. Even though wiper inserts are designed for better finishes, they are not used much for light finishing operations because they require semifinishing depth of cut. They improve surface finishes by 100% (as explained in Chapter 10 and Figure 10.11) but their main benefit is to reduce the cycle time by increasing the feed (50 to 100% higher feeds of conventional inserts) with maintaining surface finish.

4.5.5 Insert Clamping Methods

The insert is usually set on a square or rectangular shank holder with adjustable screws or buttons to control the insert position. Qualified tool holders provide precise control using indexable inserts since constant adjustment of the cutting tool or machine tool compensating devices are not necessary. The insert may be set in a pocket which is either machined directly into a holder or into a cartridge and adjustable head, which are mounted on the holder. In the former case, the rake and lead angles on the insert are designed into the pocket of the holder as shown in Figure 4.21. The cartridge style is more versatile, since various insert styles can be accommodated using different heads. It is used mainly for internal turning (or boring) operations, since it permits small tool size adjustments. The manufacturing tolerances of the insert pocket are very critical in providing proper support and interchangeability between inserts.

There are a few basic insert clamping methods such as top-clamp for plain inserts, cam pin or screw clamp for inserts with center hole, and a combination of cam pin with top-clamp

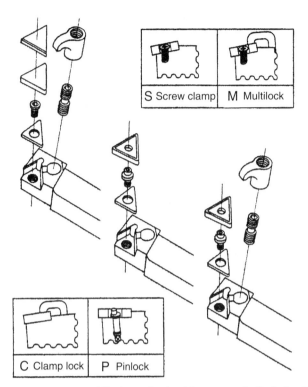

FIGURE 4.21 Basic insert locking (holding) methods. (Courtesy of Carboloy/Seco and Sumitomo Electric Corporation.)

(Figure 4.21). Top-clamp methods vary from manufacturer to manufacturer, but as the name implies generally hold the insert with a top clamp. One limitation of top-clamping methods is that they often require molded features on the insert surface, rendering inserts one-sided and reducing the number of available edges. The cam pin method is a very common insert locking approach in which the insert is clamped by a pin through the center hole; when the pin is rotated into the locked position, it pulls the insert down and back into the toolholder pocket.

In general, a combined top-clamp and cam pin locking method provides the most rigid clamping condition, while a cam pin clamping method alone provides the least rigid condition. An even better system is the wedge top-clamp and pin because the rear edge of the clamp locates against the rear angled surface of the insert pocket which ensures alignment. This clamp ensures accurate positioning both vertically and horizontally. Conventionally, indexable inserts are fixed in pockets in the holder, although freely rotating round inserts are used to improve tool life [63–65] and reduce cutting forces in some applications.

4.5.6 TOOL ANGLES

The lead angle, κ, is equivalent to the side cutting edge angle (SCEA) for turning and to the end cutting edge angle (ECEA) for boring and facing in the English system. The lead angle is sometimes called the bevel reciprocal angle and the cut-entering angle. The bevel angle, $k_r = 90° - \kappa$, is used as the lead angle in the metric system. An increase of SCEA reduces the chip thickness and increases the chip width (see Figure 2.3 in Chapter 2). The largest lead angle allowed by the workpiece geometry should be selected because: (a) it thins the chip and reduces the unit pressure on the insert, allowing the feed per revolution to be increased; (b) it protects the nose radius as the insert enters a workpiece, reducing the nose wear and chipping; and (c) it reduces depth of cut notching because the scale or work hardened surface of a workpiece is spread over a larger area of the cutting edge. The chip thickness decreases with increasing lead angle. A zero lead angle results in sudden loading and unloading of the cutting edges it enters or exits the workpiece. Geometric considerations aside, the maximum lead angle which can be used in a given operation depends on the rigidity of the machining system. An increase in the lead angle results in higher radial forces and lower axial (feed) forces, and can lead to increased tool or workpiece deflections and the occurrence of chatter. Lead angles of 15° to 30° are normally used in general purpose machining, while 0° lead angles may be used for long, slender workpieces. The lead angle often determines the size of the insert for a given depth of cut. Typical insert setups with the corresponding lead angle are shown in Figure 4.18. The ECEA must be positive to reduce the tool contact area with the metal being cut and to prevent vibration and chatter, which result in poor surface finish and tool life. The effect of SCEA and ECEA on surface finish is discussed in Chapter 10.

Other major features of the tool geometry are the relief angles and rake angles shown in Figure 4.15. The end and side relief angles protect the tool from rubbing on the cut surface. Both these angles are a function of the workpiece material hardness and the tensile strength of the tool material; and they are usually kept to a minimum to provide sufficient support on the cutting edge especially for positive rake tools. Under certain conditions with weak setup, relief angles can cause the onset of chatter. The relief angles for tools used in hard materials are low and approach 5° when small feeds are also used. However, higher angles are required with soft materials and higher feeds. Angles of 12° to 15° are not uncommon with soft materials such as aluminum and occasionally, cast iron.

Both the back and side rake angles, which provide inclination between the face of the tool and the workpiece, have a strong influence on tool performance. The rake angles can be positive, neutral, or negative as illustrated in Figure 4.19. The rake angles affect the magnitude and direction of cutting forces, edge strength, chip formation, chip flow direction, chip-breaking characteristics, and surface finish. A positive angle moves the chip away from the machined surface of the workpiece while a negative angle directs the chip toward the workpiece. The combination of the rake angle and relief angle determines the wedge angle. A positive rake angle reduces the wedge angle which results in smaller shear cross-sectional area of the cutting edge, reducing edge strength. However, positive rakes increase the shear angle of the chip, reducing the chip thickness and resulting in lower cutting forces. Positive rake inserts therefore require less cutting power and exert less stress on the workpiece, tool, and the machine spindle and ways. They also generate less vibration and help control chatter. Negative rake inserts alter the direction of forces in a manner that places the cutting edge under compression, which is desirable since the compressive strength of most cutting tool materials is more than twice as large as the transverse rupture strength. Traditionally, negative rake tooling is used whenever it is allowed by the workpiece and the machine tool. In general, a 1° increase in the rake angle results in 2% decrease in power consumption. Therefore, a tool with +5° rake requires about 20% less power than a tool with −5° rake angle. A positive geometry should be considered when it can reduce production cost while improving the quality of the work. Negative side rake angles are preferred for interrupted cuts, heavy feeds, hard metals, and some nonferrous metals. Positive side rake angles are often necessary for work-hardening alloys, when using slow surface speeds, soft metal and nonmetals and when the rigidity or strength of workpiece and fixture is limited.

The rake angle of a flat-top surface insert is determined solely by the rake angle of the toolholder. The rake angle can also be pressed directly into the insert by using chip groove geometries which provides a positive rake cutting action when used in a negative rake toolholder as illustrated in Figure 4.19. Even though the range of rake angles is generally

−5° to +15°, higher positive angles are slowly being used to improve productivity. Positive-geometry inserts are most commonly made of WC; ceramic materials are very brittle and commonly available only in negative rake varieties. The rake angle of an insert is specified in the second letter of an insert designation. An SPE-422 insert can be used in a neutral or positive (P) rake holder, while a TNG-333 insert must be used in a negative (N) rake holder.

4.5.7 Thread Turning Tools

Thread turning can be performed in several ways. The three popular ways to machine thread vees (using several passes) were discussed in Chapter 2 and were illustrated in Figure 2.22. The cutting speeds for the tool material recommended in ordinary turning are usually reduced by 25% to keep the cutting temperature below critical values.

Typical threading type inserts are shown in Figure 4.22. Multitoothed inserts cut the vee in a single pass, using the radial infeed method, because the succeeding teeth cut deeper than the teeth preceding them. Full-profile inserts cut a complete thread including the crest. V-profile inserts are used for various pitches with same thread angle but the crest is generated by a preturning operation [66]. The insert is inclined an amount equal to the thread's helix angle (λ) to best match the lead angle of the thread, which varies along the thread height; this ensures uniform edge wear of both flanks and therefore longer tool life as well as better surface finish. The inclination angle of an insert is between 0° and 4°; a value of 1° generally produces a satisfactory thread. The relief angles on the flank of the insert depend on the inclination angle; the side relief angle in the direction of the feed must be greater than the thread helix angle to prevent the insert from rubbing. The effective rake angle varies along the cutting edge; it is positive at the leading edge and negative at the trailing edge. Standard triangular turning inserts (60° included angle) are often used for roughing; they generate a thread form of slightly more than 60° when the tool is set with a negative rake angle. Cermet cutting tools produces better surface finish and size control than WC tools. Ceramic inserts are generally not as effective in thread turning as in hard thread turning.

4.5.8 Grooving and Cutoff Tools

Grooving and cutoff tools are similar to threading tools in that the tool is surrounded on three sides by the workpiece, so that the cutting forces and heat are concentrated on the weakest part of the insert. Grooving inserts and tool holders are narrow and must be relieved on all the three sides (as shown in Figure 4.23), and therefore do not provide as much support as

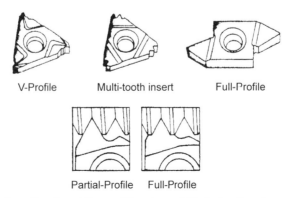

V-Profile   Multi-tooth insert   Full-Profile

Partial-Profile   Full-Profile

FIGURE 4.22 Typical threading insert forms. (Courtesy of Carboloy/Seco.)

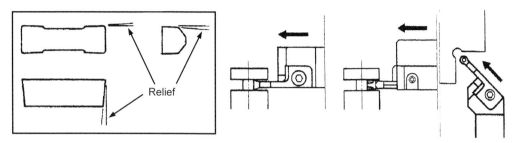

FIGURE 4.23 Multiple relief weakens grooving or cutoff inserts. (Courtesy of Valenite Corporation.)

other types of turning inserts. The lead angle can be neutral, right, or left (of a few degrees). The ratio of radial depth to insert width should be less than 8 to 10 to maintain stability. The insert width is generally affected by the workpiece material. Thin inserts are less stable than the wider ones. The lead angle provides more stable cutting. Inserts with lead are used to reduce the small diameter protrusion left at the center in cutoff operations. Special inserts with and without chipbreaking grooves are available for these operations.

4.5.9 FORM TOOLS

Form tools are used to cut a specific profile on a rotating workpiece in a single plunge cut (Figure 4.24). Grooving and cutoff tools are types of form tools. A form tool is fed perpendicular to the surface of the workpiece and contacts the workpiece at multiple points. Form tools produce shapes with less machining time than single-point turning tools can because multiple passes must be used with single-point tools to maintain accuracy. The surface finish obtained with form tools is usually about four times rougher than the finish on the form tool itself.

Form tools are designed in various sizes and can range in width from 4 to 200 mm. The maximum effective width of a form tool depends on its geometry, the workpiece material, infeed rate, and the available machine power. Combined grooving and form tools are used for wide grooves or for generating wide shapes. One form tool can replace several single-point turning tools and can provide increased dimensional accuracy and repeatability for complex profiles. Form tools are produced by electodischarge machining (EDM), or by grinding in the case of thin inserts. Tool materials used for form tools include brazed or clamped WC, HSS, and P/M tool steels. Form tools can be manufactured with 0.005 mm dimensional repeatability using EDM.

4.6 BORING TOOLS

Boring tools have the same insert or point geometry as turning tools, and most of the information reviewed in the previous section is applicable to boring as well as turning. One exception concerns the nomenclature of cutting angles; the end (rather than the side) cutting edge is the major cutting edge in boring, so that the ECEA in boring corresponds to the SCEA in turning and is generally negative.

The boring bar, however, is long and compliant compared to a turning toolholder. In fact, due to their length to diameter ratios, boring bars and end mills are the least rigid of all cutting tools. The rigidity or compliance of the boring bar is often the primary factor determining boring process performance; deflections of the bar constrain its reach, affect the bored hole diameter and accuracy, and may lead to vibration and chatter. Displacement

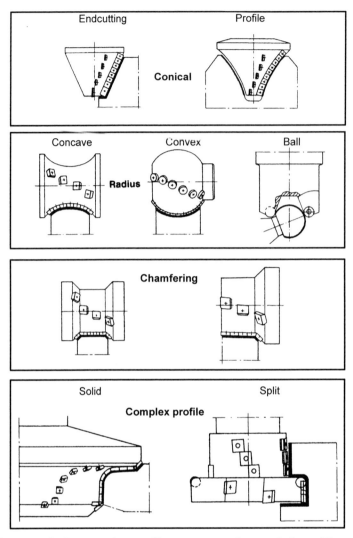

FIGURE 4.24 Form tools for complex profiles on a rotating workpiece. (Courtesy of Sandvik Coromant.)

of the boring bar due to cutting forces, especially in interrupted cuts, may result from: (a) bending deflection of the bar structure; (b) deformation in any joint(s) between the parts of a composite bar; (c) deformation in the joint between the spindle and the bar flange; and (d) deformation of the spindle and its bearings. Factors affecting boring system rigidity therefore include the basic machine design; the drive mechanism design; the spindle stiffness; the boring bar diameter, length, overhang, and material; and the workpiece material and fixturing.

In many general purpose applications, the deflection of the bar itself is larger than the other components. The deflection of a simple bar, rigidly mounted (cantilevered) to the machine structure, with an overhang L and radial force F_r, can be calculated using the equation

$$\delta = \frac{F_r \cdot L^3}{3 \cdot E \cdot I} \tag{4.1}$$

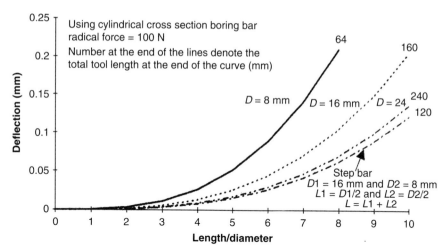

FIGURE 4.25 Deflection of a tool/toolholder system versus length/diameter ratio.

where the E is the modulus of elasticity of the material and I is the cross-sectional moment of inertia. For a solid round bar,

$$I = \frac{\pi \cdot D^4}{64} \qquad (4.2)$$

The deflection under 100 N load for a cylindrical cross section single or step diameter bar is illustrated in Figure 4.25; this figure underlines the importance of minimizing tool overhang and L/D ratio. As a rule of thumb, deflections should be less than 0.025 mm to ensure adequate system performance.

Two major problems often encountered with boring bars which are related to bar rigidity are springback and chatter. Bar springback is caused by insufficient stiffness and the resulting deflection or deformation of the bar due to cutting forces. Excessive springback (or elastic recovery) of the boring bar will result in a smaller bore or scratch marks during tool retraction. Spingback and chatter during tool retraction can be eliminated by designing the tool to permit a minor (0.02 to 0.04 mm) shift in the cutting edge position during the tool retraction; the shift should not change the rotational axis of the tool. Due to the small size of the tool, the shifting mechanism is difficult to package and may cause balancing problems at high rpms.

Chatter can occur either during the feeding or retracting motions. The resonant frequencies and chatter resistance of a bar depend strongly on the bar length to diameter (or overhang) ratio; overhang ratios greater than 4 or 5 (4 for steel bodies) are especially susceptible to chatter. On a more basic level, chatter resistance is strongly affected by the value of the quantity $k \cdot \zeta$, where k is the effective dynamic stiffness and ζ is the damping ratio of the system. The stiffness of the boring bar can be increased by using the largest diameter and shortest shank length bar permitted by the bore geometry, and by using bar materials with a high elastic modulus. Deflections can also be minimized by reducing cutting forces through the use of tip geometries designed to ensure free cutting. The damping characteristics of a boring bar system depend on the structural damping of the bar (a material property) and the interfacial frictional damping between the bar, the tool holder, and their connection or mounting technique. In a conventional setup the boring bar is mounted in a holder (chuck) which is subsequently mounted to the spindle. An improved but less flexible approach is to bolt the tool directly to the spindle via a flange. In designing a boring bar system, a trade-off between stiffness and damping characteristics may be required. Solid bars made of rigid materials enhance stiffness, but often have relatively poor damping properties which reduce chatter resistance. Composite extended bars provide increased damping but often reduced stiffness. In summary, the stiffness and damping (chatter-resistance) characteristics of a boring tool are enhanced primarily by: (1) using anisotropic mandrels with directionally enhanced stiffness axes [67]; (2) using bar materials with high Young's modulus and/or high damping materials; and (3) using passive (dynamic vibration absorbers, DVA, for L/D ratio <10) [68, 69] and active [70] vibration control methods.

The use of optimal materials and geometries is often the preferred approach to enhancing the stability of boring bars, and cantilever tools is general [71–73]. Variations of composite boring bar structures using tool steels, heavy metals, hard carbides, and light alloy materials can be used effectively as shown in Figure 4.26. Composite bars generally consist of a heavy metal root segment (which is machinable) with a relatively high Young's modulus. The overhang segment at the end of the tool, which holds the tool bit or insert, is designed to be light, and may be made of aluminum, titanium, or similar materials to increase the natural frequency of the cutting point without significantly affecting bar stiffness. The middle section is often made of carbide. These designs are compromises between economical solid steel and costly solid carbide bars (see Example 4.4 to Example 4.6). Generally, solid steel boring bars are used for $L/D<4$, composite bars are used with $L/D<6$ or 7, and composite bars with DVA are used for $L/D>6$ [68, 69].

Boring tools may be fixed or adjustable as shown in Figure 4.27. Fixed boring tools generate a particular diameter predetermined by the radial distance of the cutting tip from the tool axis. Such tools may be made of solid HSS, WC, PCD, or PCBN tips brazed to a solid bar, solid WC, or indexable inserts of various materials set in pocketed bars. They may have one or more cutting edges depending on the bore size and bore quality and surface finish requirements. As discussed below, adjustable boring tools are equipped with heads containing precision mechanisms to offset the cutting edge to compensate for tool wear or deflections.

Edge clamping methods are similar to those discussed in the previous section for the turning tools. Boring bar with rotary inserts can be used to improve tool life [74, 75]. Insert

FIGURE 4.26 Combination structure boring bar.

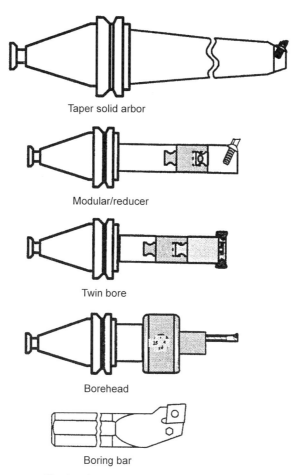

FIGURE 4.27 Various types of boring bars with and without diameter adjustments.

size tolerances and boring bar manufacturing tolerances often make it difficult to control the size of resulting bored holes. Therefore, a size adjustment capability is often necessary. The conventional approach for adjusting the size of boring bars is to use manually or automatically adjustable boring heads. These heads are mounted at the end of the boring bar. The manual boring head uses a precision micrometer adjustment or a cartridge. The former uses a dial (micrometer spindle) that offsets the insert or the boring bar in 0.0025 to 0.01 mm increments per division on the diameter. The latter design does not provide any indication for size adjustment and uses a cam underneath the cartridge which permits a small range of adjustments (Figure 4.28). Manual adjustable tooling is set up by trial and error and is unsuitable for mass production. The dial system has a graduated head (Figure 4.29) attached on a set screw or on a ring around the head. To set the boring diameter, the dial is turned to the desired setting. In cartridge-type systems, adjustments are made using a preloaded screw without any indication of tool offset; a gage is therefore required for any subsequent adjustments.

The automated approach requires a very complex tooling system to actuate the drawbar in the spindle to automatically perform small adjustments in the boring bar size. The boring bar requires a built-in mechanical or hydraulic tool compensating system for offsetting the tool to adjust size as shown in Figure 4.30. The tool positioner is very complex and can be fit

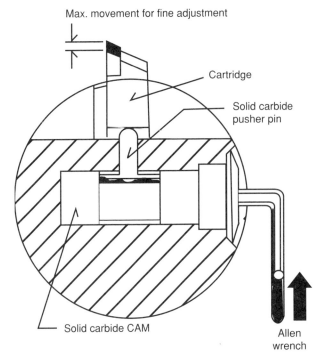

FIGURE 4.28 Manual adjustable cartridge. (Courtesy of Master Tool Corporation.)

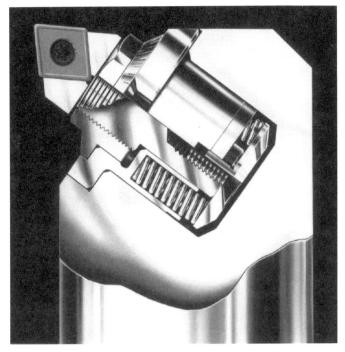

FIGURE 4.29 Single-point boring bar with a diameter adjustment dial cartridge. (Courtesy of Valenite Corporation.)

Cutting Tools

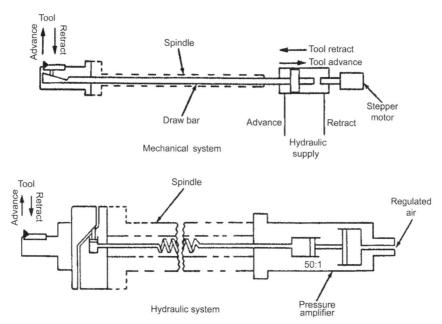

FIGURE 4.30 Schematic diagram for two automatic adjustable boring systems.

only in boring bars with diameters larger than 50 mm. Most of the available designs have a weak link between the boring bar and the tool adjustment mechanism and therefore have significantly lower static and dynamic stiffness (20 to 50%) compared to non-adjustable solid boring bars. As an alternative, an adjustable boring bar assembly in which the cutting radius can be altered by predetermined increments simply by releasing and turning the bar within a sleeve can be used. The holder sleeve is located slightly eccentric relative to the spindle axis, so that turning of the bar effectively changes the working radius of the cutting point as shown in Figure 4.31.

Multiple-step tool bars using several cutting edges, as shown in Figure 4.32, are designed to generate bores with varying diameters and to bore and counterbore in one pass. Similarly, a boring bar with a roughing or semifinishing tool ahead of a finishing tool (axially staggered inserts and radial spacing) can be used to rough and finish a bore in one pass.

Draglines (withdrawal marks) generated during the reversal of the bar from the bottom of the bore, due to springback of the bar, can be eliminated using either an automatic compensation system or through proper programming on CNC machines. A compensation system pivots the insert or the head of the boring bar at the bottom of the bore so that the effective tool diameter becomes smaller during retraction. In the CNC approach, the bar is offset radially for clearance during retraction.

4.7 MILLING TOOLS

Milling is carried out using a rotary tool with multiple cutting edges. Both solid cutters with ground cutting edges and pocketed cutter bodies fitted with inserts are used as shown in Figure 4.33. Design criteria for milling tools are similar to those for turning tools in many ways, but milling involves additional concerns because it is an interrupted cutting process. The cutting edges on a milling cutter enter and leave the cut each rotation, and are actually cutting for less than half of the total machining time.

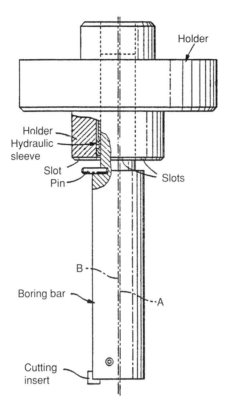

FIGURE 4.31 Manual adjustable boring bar with improved accuracy.

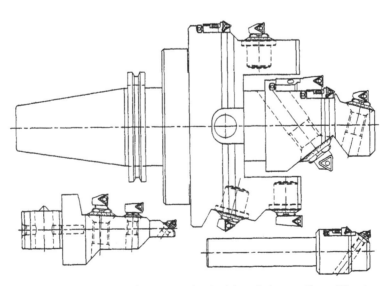

FIGURE 4.32 Multiple operation boring/counterboring/chamfering tooling. (Courtesy of Rigibore Tooling Systems.)

Cutting Tools

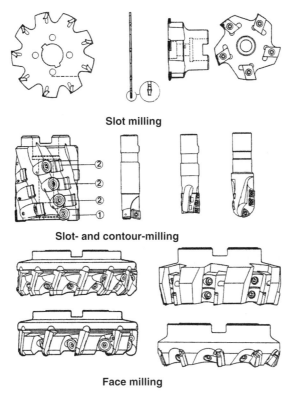

FIGURE 4.33 Indexable milling cutters. (Courtesy of Carboloy/Seco.)

This section discusses common cutter designs and geometric factors which affect cutter and insert selection [76–81]. The nomenclature used follows the American National Standard for face and end mills [82].

4.7.1 Types of Milling Cutters

Several types of milling cutters are used for different operations. *Face milling cutters* are used to produce flat surfaces. They are usually fitted with indexable inserts or inserted blades. Indexable face mills are the most common and versatile type because they can be used with a variety of insert materials. Conventionally, indexable inserts are fixed in pockets in the cutter body, although freely rotating round inserts are used in rotary cutters. The geometry and nomenclature of a face milling cutter is shown in Figure 4.34.

Slot milling cutters are used for grooving, slitting, and side and face milling. Slot cutter designs and manufacturing methods vary depending on their size and application. Solid HSS, solid WC, and WC-tipped designs are often used for small diameter cutters (<100 mm), while indexable cutters are preferred for larger diameters. Several insert geometries are available with or without chip breakers for indexable cutter bodies. The cutting edges of such cutters are similar to those used on circular saws or hole saws. Both full and single side slot mills are available. Full side cutters may have continuous edges across the width of the cutter, or located alternatively around the cutter periphery. The cutting edges of full side cutters can be square, trapezoidal (chamfered on both sides), pointed, or radiused. The triple-chip ground-tooth form (in which the first tooth cuts a narrow chip from the center of the cut while the second tooth, which follows the first, cuts two equal-size chips from the sides of the cut) has

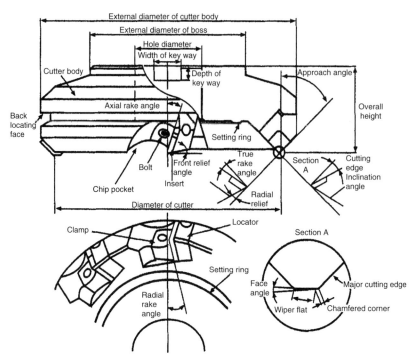

FIGURE 4.34 Milling cutter nomenclature. (Courtesy of Sumitomo Electric Corporation.)

become popular because it removes the material much more easily. The edges are generally designed with positive or neutral rake angles.

End milling cutters generate two workpiece surfaces at the same time; cutting edges are located on both the end face and the periphery of the cutter body. They are usually used in facing, profiling, slotting, shoulder, slabbing, and plunging operations and are the most versatile milling tools. When they are used for plunging, the teeth must cut all the way to the center of the tool. They are produced in solid HSS, HSS–Co, WC, ceramic, PCD/PCBN brazed or vein construction, inserted blade, and indexable insert designs. Inserted blade end mills produce better finishes than indexable WC end mills. Much of the previous discussion about stiffness and deflection of boring bars is also applicable to end mills. Shank-type end mills are available in a wide variety of sizes, flute configurations, and lengths. The geometry of the end mill is shown in Figure 4.35 [81]; its diameter, flute length, and corner radius must match the dimensions of the pocket to be machined. Generally, the two- and three-flute end mills are center cutting, while end mills with larger number of flutes are noncenter cutting. The end mill's core diameter, number of flutes, and flute space determine its rigidity in the manner discussed below for drills. The tooth edge geometry, radial and axial rakes, relief angles, and flute forms are important, and for solid end mills should be designed according to the same guidelines used for single-point turning tools, indexable milling cutters, and drills.

Rotary milling cutters can be used in certain cases to improve productivity. Rotary milling cutters are equipped with round inserts which are secured in interchangeable cartridges and which are free to rotate as they cut [63]. The insert rotates on a bearing attached on the cartridge. The rotation of the insert alters chip formation on the rake face by increasing the effective inclination angle. The insert rotation is caused by the action of the cutting forces and chip flow over the insert face; a separate drive mechanism is not required. The contact edge length between the chip and the tool changes during cutting, leading to reduced wear

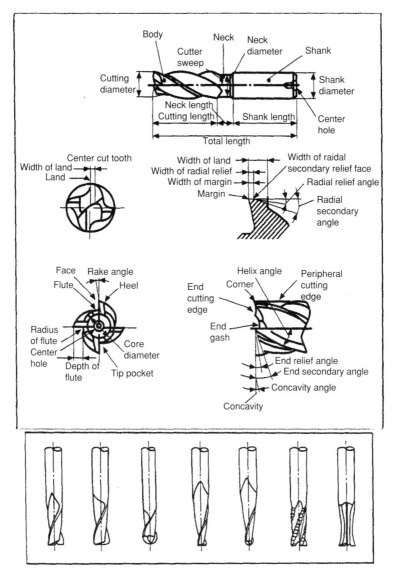

FIGURE 4.35 End mill geometry. (Courtesy of Sumitomo Electric Corporation.)

distributed evenly around the periphery of the insert, and ultimately to increased tool life. They can also produce smoother finishes due to the large radius on the inserts.

4.7.2 Cutter Design

Factors to be selected in designing a milling cutter for a specific operation include the cutting edge geometry (rake and lead angles), cutter density, cutter diameter, entry and exit angles, and cutter construction (indexable, inserted blade, etc.) [58, 83–85]. For slot and end mills, the helix angle also strongly affects cutter performance.

The standard angles used to describe rotary tools such as milling cutter geometry are the radial (side) and axial (back) rake angles (Figure 4.34). The axial rake angle affects the chip flow direction and the thrust force, while the radial rake angle has a strong effect on the cutting power and tool life. The radial (α_f) and axial (α_p) rakes determine two additional angles, the

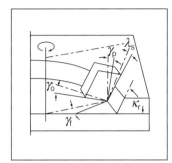 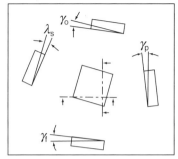

FIGURE 4.36 Radial and axial rake angle in a rotary tool. (Courtesy of Sandvik Coromant.)

inclination (λ_s) and orthogonal (top) rake (α_0) angles (Figure 4.36). The orthogonal rake is the true top rake which influences cutting forces and power requirements. The true rake angle is measured from, and perpendicular to, the lead angle. The inclination angle is similar to the helix angle and is measured from the cutting edge face to the reference plane on a line parallel to the lead angle cutting edge. This angle determines the chip flow direction and significantly affects cutting forces and tool life. These angles are related through equations:

$$\tan \alpha_0 = \tan \alpha_f \cdot \cos \kappa_r + \tan \lambda_p \cdot \sin \kappa_r \tag{4.3}$$

$$\tan \lambda_s = \tan \alpha_p \cdot \cos \kappa_r - \tan \lambda_f \cdot \sin \kappa_r \tag{4.4}$$

where κ_r is the bevel angle, which is related to the lead angle as discussed below. For example, a combination of $\alpha_p = 7°$, $\alpha_f = 0°$, and $\kappa_r = 15°$ results in $\alpha_0 = 2°$ and $\lambda_s = 7°$. Three common combinations of these two angles are used on standard milling cutters as shown in Figure 4.37. Chip formation for each of the configurations is shown in Figure 4.38.

1. Cutters with positive axial and radial rake angles are called *double-positive* cutters. The positive axial rake lifts the chip and curls it away from the finished workpiece surface and toward the inside of the cutter body. The positive radial rake provides a sharper cutting edge that tends to pull the tool into the work (free cutting). Double positive cutters reduce the cutting pressure, consume less power, create less heat, reducing deflection and result in less strain on the machine bearings, ways, and spindle. This geometry creates a true shearing action which is suitable for finish milling of free-cutting steel, aluminum, and brass; for materials which form a continuous chip; and for work-hardening and nonferrous materials and many soft stainless steels.
2. Cutters with negative axial and radial rake angles are called *double-negative* cutters and are the most common type of cutters. The double-negative geometry offers several advantages: first, negative rake inserts with twice as many cutting corners as positive inserts are used; second, the strongest part of the insert, away from the cutting edge, enters the work first, whereas with positive rake cutters the edge enters first. Also, the high edge strength permits use of harder, more wear resistant insert materials. Double-negative cutters tend to push the work away from the tool, and therefore require greater power and higher system rigidity than positive cutters. Moreover, with a double-negative cutter, the chip is bent forward and downward under pressure, which can cause chip evacuation problems for soft steels. Double-negative cutters are effective in rough and finish milling of steel and cast iron, including hard and high-strength grades.
3. Cutters with positive axial and negative radial rake angles are called *shear-angle cutters*. These cutters combine some of the advantages of both negative and positive rake

Cutting Tools

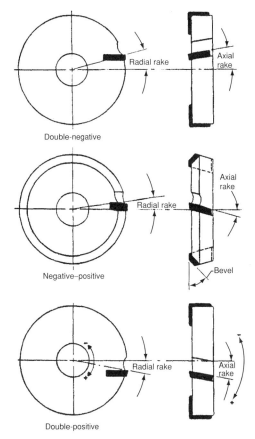

FIGURE 4.37 Three milling cutter geometries with differing radial and axial rake angles. (Courtesy of Valenite Corporation.)

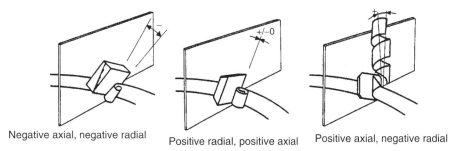

FIGURE 4.38 Chip formation for the milling cutter geometries shown in Figure 4.39. (Courtesy of Carboloy/Seco.)

cutters. The negative radial rake provides a strong cutting edge, while the positive axial rake angle, combined with the bevel on the cutter body, lifts chips up and directs them away from the surfaces being machined. They are most often used for rough and finish milling of the tougher grades of aluminum and other nonferrous materials, and for free machining of cast iron, steels, stainless steels, and most high-temperature alloys which are milled with difficulty using double-negative cutters. They can provide a finished surface in a single pass.

The lead angle, shown in Figure 4.34, is equivalent to the SCEA or approach angle in turning, and is equal to 90° minus bevel angle (κ_r) shown in Figure 4.36. The best lead angle for a specific operation depends on several factors including the part configuration, the distribution of cutting forces, and the chip thickness. An increase in the lead angle results in an increase in the axial force and a reduction of the radial force. A proper lead angle also allows a cutter to enter and exit the cut more smoothly, reducing the shock load on the cutting edge. Proper selection of the lead angle is especially important for shear-angle cutters. A 15° lead is often used for face milling. Larger lead angles are used in shear-angle cutters because they provide a large inclination angle. A 45° lead provides the strongest cutting edge for heavy duty milling; in this case the radial force will be approximately equal to the axial force. Large lead angles should also be used when the spindle is weak radially (e.g., due to a long spindle overhang), and when the radial stiffness of the part or fixture is limited.

The cutter density or tooth pitch is determined by the number of teeth in the cutter body. Cutters are classified as having fine (3 to 5 inserts per 25 mm of diameter), medium, or coarse (1 to 1.5 inserts per 25 mm of diameter) pitch. A fine-pitch cutter has about half the chip clearance of a coarse pitch cutter. The milling cutter with the highest number of teeth allowed by the workpiece and machine tool is most efficient. However, coarse pitch cutters allow higher chip thickness or feed per tooth, resulting in fewer teeth in the cut to prevent machine overloading. Therefore, a coarse pitch is usually used for large diameter cutters and for heavy-duty applications, while fine pitch cutters are often used in high production rate applications, when machining thin wall sections, and when the depth of cut does not exceed 7 mm. The amount of chip space required for a specific operation is determined by the workpiece material and the width of cut. Two essential requirements must be considered in selecting the cutter density: (1) at least one (and preferably two) cutting edge (insert) must be in the cut at all times to prevent instability of the tool as it enters and exits the workpiece; and (2) adequate chip space and cutting edge support must be provided. Cast iron and titanium require little chip space, but steel and aluminum require relatively large chip slots. A larger number of flutes on an end mill reduces chip loads and improves the workpiece finish.

Staggered or white-noise cutters, with unequal spacing of cutting edges around the cutter, can reduce or eliminate chatter by varying the chip load on each insert and reducing the effective tooth pass frequency. They result in changes in the cutting forces which interrupt harmonic vibrations. Differential helix angles on successive flutes can be used for long edge end mills to reduce vibration during cutting. Roughing shell end mills may have a spirally staggered arrangement of indexable cutting edges to provide smooth cutting action and quiet operation.

The diameter of a face, end, side or slot mill depends on the radial depth or width of cut (or part), axial depth of cut, and feed, which also determine the required number of cutting edges per millimeter of diameter (e.g., deep slots require low-density cutters). The cutter diameter should allow for optimum feed per tooth and depth of cut based on the available power. The cutter diameter for a face mill should generally be at least 1.3 to 1.6 times the width of the part. This allows the cutter to be offset from the centerline of the workpiece and to overhang it on the tooth entry side. Symmetrical positioning of the cutter should be avoided especially when the engagement width is considerably smaller than the cutter diameter in order to prevent vibration due to pulse loading of the machine tool especially for light-duty machines and positive tools. Pulse loading could lead to excitation of various structural vibration modes that may cause transient vibrations during the entry and exit of the cutter. Coarse pitch cutters are normally the first choice for diameters larger than 120 mm. The bending moment on the cutter is affected by its diameter and overhand and it should be below the maximum allowed by the spindle interface to avoid cutter vibration and deflection. When a long tool overhang must be used, the smallest diameter cutter possible should be selected. Therefore, it may be preferable to use a smaller diameter cutter and to cover the width of the workpiece in two or more passes, unless this arrangement produces an uneven surface finish.

The diameter of the cutter also affects the chip thickness. As discussed in Chapter 2, chip thinning effect is very important in milling. The maximum radial chip thickness is affected by the cutter geometry, the lead angle, and (most significantly) by the position of the cutter on the workpiece (see Figure 2.11 and Equation 2.12). The average chip thickness should be greater than 0.07 mm, and if it becomes less than 0.025 mm, a mixture of rubbing and cutting will occur resulting in greater heat generation and excessive flank wear. More details on chip thinning effect are discussed in Example 2.5 to Example 2.8.

The selection of the cutter density and diameter in contour or pocket milling is more complex than in straight line milling because the cutter engagement during convex and concave arcs must be considered. This is true, for example, when traveling around a 90° corner in pocket milling.

Cutter designs and feed rates that produce narrow chips with a heavy cross section can be used to improve metal removal rates and tool life when machining relatively deep cuts using machines with limited power and rigidity. Either step cutters or chip-breaker cutters can be used for this purpose. Inserts are stepped radially and axially in step cutters as shown in Figure 4.39. This arrangement increases the chip thickness at reduced feedrates by breaking up each chip among several inserts in a sequence. The use of a step cutter reduces the effective number of cutting edges [86]. Chip-breaker cutters are designed with grooves around the body that are either semicircular or sinusoidal in shape. They are also called hogging cutters. Such designs are used in solid and ball nose end mills (Figure 4.40) and in indexable face, slot, and helical end mills equipped with scalloped inserts (Figure 4.41). The design of the chip-breaker grooves, specifically coarse or fine pitches, controls the chip size. The grooves are arranged so that every other edge removes the ridges left on the part by the preceding insert as shown in Figure 4.40 and Figure 4.41. Use of chip-breaker cutters lowers cutting forces and reduces tool deflection caused by long cutter overhangs, and therefore permits use of high feed rates in roughing operations.

The insert contact angles during entrance (especially in down milling) and exit have a strong influence on tool life and burr formation. These angles are a function of the relative

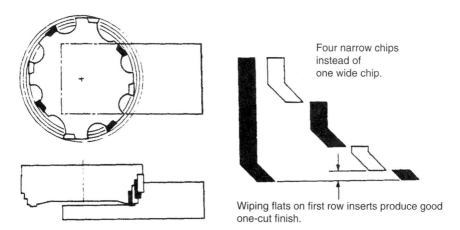

FIGURE 4.39 Milling cutter with four radial and axial steps and the corresponding chip cross sections. (Courtesy of Ingersoll Cutting Tool Division.)

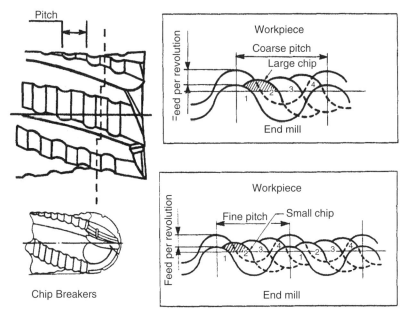

FIGURE 4.40 End mills with chip breakers. (Courtesy of OSG Tap & Die, Inc.)

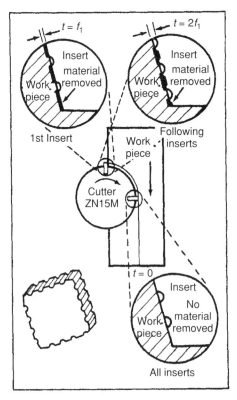

FIGURE 4.41 Sequential cutting action of a milling cutter with scalloped inserts. (Courtesy of Carboloy/Seco.)

dimensions of the cutter and workpiece, the cutter overhang on the entry/exit side, and the radial rake angle of the cutting edge. The insert contact angle is

$$\varepsilon_c = 90° - \alpha - \varepsilon \tag{4.5}$$

where the insert entry angle, ε, is defined by a line through the cutter axis and perpendicular to the entrance surface and a line through the cutter axis passing through the cutting point:

$$\varepsilon = \tan^{-1}\left[\frac{\sqrt{(2 \cdot R - w_s) \cdot w_s}}{R - w_s}\right] \tag{4.6}$$

where R is the cutter radius and w_s its overhang on the entry side (Figure 4.42). The sign of the insert contact angle determines the location of the initial impact between the tool and the workpiece in down milling. The initial contact occurs on the cutting edge (the weakest section of tool) when ε_c is positive and on the rake face when ε_c is negative as shown in Figure 4.43. A positive ε_c (defined in the direction of the + sign in Figure 4.3) should be avoided unless the work material is soft and ductile. The cutter position relative to the workpiece should be controlled so that insert contact angle is negative (defined in the direction of the − sign in Figure 4.43). However, ε_c is often positive at the beginning of the workpiece cut while negative at exit and re-entry (see Figure 4.42). Tool entry angle problems are often encountered when the cutter re-enters the workpiece after crossing cavities or holes in the workpiece, as is often the case in face milling automotive powertrain components. Generally, having the centerline of the cutter well inside the workpiece width is considered a good practice for face milling. Re-entry in peripheral shoulder and through-slot end milling operations often results in severe entry angle conditions which can lead to tool failure. Various tool routing approaches can be used to reduce tool entry or re-entry problems. Improper exit angles can cause burr formation as discussed in Chapter 11. Burr formation is a major concern in milling and has been the limiting factor preventing milling operations from providing part quality equivalent to grinding in many applications.

Indexable and inserted blade milling cutters can be designed for roughing and finishing in one cut. This can be done either by using a combined rough and finish cutter, or by using

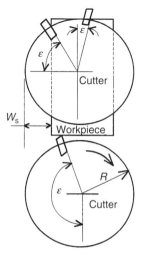

FIGURE 4.42 Schematic view of initial cutting edge contact at the entrance of a workpiece. (Courtesy of Carboloy/Seco.)

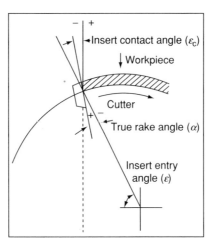

FIGURE 4.43 Schematic view of cutting edge contact during milling at the entrance and exit of a workpiece.

roughing and finishing inserts on the periphery of the tool only as shown in Figure 4.44. Combination cutters use roughing inserts around the tool periphery and finishing inserts mounted on the tool face. In some cases, finishing inserts with wiper flats are substituted for some of the roughing inserts. Finishes equivalent to grinding can be produced when milling with wipers, especially when wipers are located precisely with respect to the cutter's axial plane. These cutters perform best at light depths of cut and medium-to-high feed rates. The feed per revolution of these tools must be about 1 to 1.5 mm less than the length of the flat or wiper flat. The flat wiping surface of the insert is often crowned with a large radius to prevent the formation of a sawtooth workpiece surface profile, which may occur when the flat land is not exactly parallel to the direction of feed. Surface roughnesses of 1.6 μm or less can be produced with these cutters.

In end and slot milling, the helix angle of the cutting edges or flutes determines how rapidly cutting forces increase and decrease as the cutting edges enter and exit the cut. Higher helix angles increase the length of the cutting edge engaged in the cut and allow for coarser tooth spacings because they insure continuous cutter–workpiece contact; they often improve the surface finish and permit use of increased speeds and feed rates (see Examples in Chapter 8). Helix angles up to 60° may be used in order to reduce cutting forces for materials such as stainless steel, titanium, and inconel. However, lower helix angles provide higher edge strength in corner areas and reduce edge chipping or flaking.

Both square and ballnose end mills are used. Ballnose end mills require the consideration of the effective diameters affected by the depth of cut as explained in Problem 2.6 in Chapter 2. The radial depth of cut affects the surface finish because the tool leaves behind scallops on the surface that vary in height according to the stepover distance. The stepover is given by

$$\text{Stepover} = 2[(D/2)^2 - (D/2 - h)^2]^{0.5} \qquad (4.7)$$

where h is the height of peaks.

4.7.3 MILLING INSERTS AND EDGE CLAMPING METHODS

The criteria for selecting inserts in milling are very similar to those used in turning, although additional considerations also apply. There are several insert shapes available for milling that

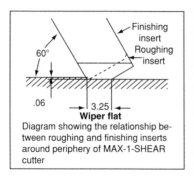
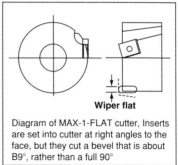
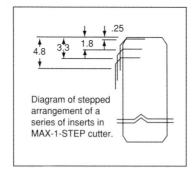
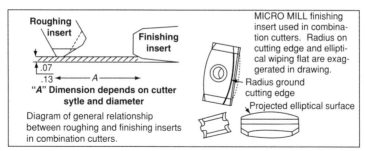

FIGURE 4.44 Diagram showing the relationship between roughing and finishing inserts in a combination milling cutter. (Courtesy of Ingersoll Cutting Tool Division.)

are not often used for turning due to the nature of the operation. The cutting insert nomenclature is described in Figure 4.45. The insert corner configuration can be a radius, a chamfer, a double chamfer, or a flat. The proper selection of the edge preparation (up-sharp, hone, chamfer, or combined hone plus chamfer) has a strong influence on tool life as discussed in Section 4.5.

The methods of holding the insert in the cutter body are very critical to cutter performance especially at higher speeds due to centrifugal force of the clamping components. The most common methods are: (1) nonadjustable insert pockets, (2) axially or radially adjustable designs using a pocket, a precision rail and wedge lock, or a cartridge, and (3) brazed tips. Some of these systems are illustrated in Figure 4.33. An on-edge insert held in fixed pockets by a single screw provides secure locking against movement, permits simple, accurate indexing, and allows for the strongest section of the carbide to support the cutting forces. The adjustable designs generally lack insert strength and do not provide security against insert movement during cutting. Brazed teeth are usually used on smaller cutters and provide the lowest edge height and roundness error.

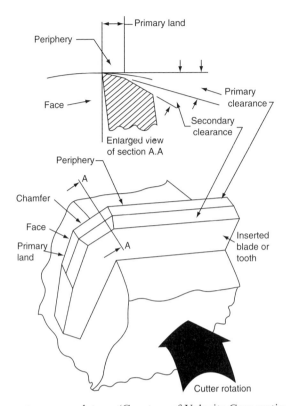

FIGURE 4.45 Insert geometry nomenclature. (Courtesy of Valenite Corporation.)

4.8 DRILLING TOOLS

A drill is an end-cutting tool which has one or more straight or helical flutes, and which may have a hollow body for the passage of cutting fluid and chips during the generation of a hole in a solid or cored material. Drills vary widely in form, dimension, and tolerance. Drills are classified according to the materials from which they are made, their lengths, shapes, number of flutes, point characteristics, shank style, and size series [87]. The best type of drill for a given application depends on the material to be drilled, its structural characteristics, the hole dimensions, whether the material to be drilled is cored or solid, whether a through or blind hole is required, the entrance and exit characteristics of the workpiece (Figure 4.46), the expected hole quality, the characteristics of the machine tool and fixture, and the cutting conditions. Selecting the proper style of drill for a given application requires consideration of all these factors. Drill manufacturers offer the same style drill with slight variations in both configuration and metallurgy. These slight variations strongly affect drill life and hole quality, especially for small diameter drills.

Three types of conventional drills are widely used: regrindable drills, spade drills, and indexable insert and tipped drills. As shown in Figure 4.47, there are several types of regrindable drills including twist (or regular) drills, gundrills, counter drills, and pilot drills. Twist drills (Figure 4.48) differ widely in the number of flutes they contain, and in geometric characteristics such as the helix angle, relief (clearance) angle, point style, flute shape, web taper, web thickness, and margin width. A standardized system of identifying or classifying twist drills made by different manufacturers has not yet been developed.

Cutting Tools

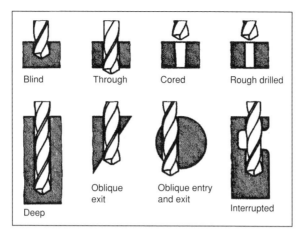

FIGURE 4.46 Influence of workpiece and hole form characteristics. (Courtesy of Guhring Inc.)

This section describes the basic types of drills and factors affecting drill design and selection. Much of the material pertains specifically to twist drills, which are the most widely used drills in practice, but other types of drills are discussed as well.

4.8.1 TWIST DRILL STRUCTURAL PROPERTIES

The structural properties of the drill have a direct bearing on drilling performance. From a structural viewpoint, several features of a twist drill are significant as illustrated in Figure 4.48:

(a) *Shank or drive type.* The butt end of the drill is generally held in a holder or the spindle and driven. Driver design is important because it determines the roundness accuracy and stiffness of the drill-holder system, as well as the limiting speeds and coolant

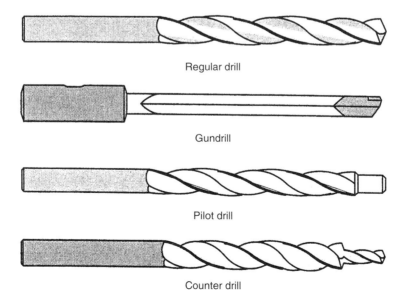

FIGURE 4.47 Regrindable drill types.

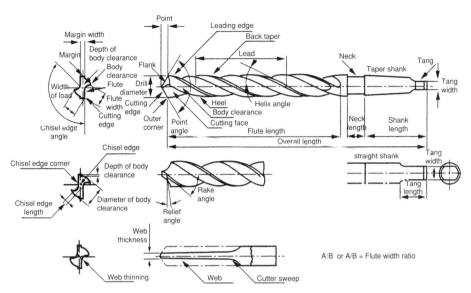

FIGURE 4.48 Drill nomenclature. (Courtesy of Sumitomo Electric Corporation.)

pressures which can be used in some applications. Maximizing the driver length improves rigidity and concentricity.

Straight shank drills with or without tapered flats are mounted in end mill holders, collets, chucks, or special hydraulically or mechanically clamped holders described in Chapter 5. *Tapered shank* drills are mounted directly into the spindle with or without intermediate sleeves or adapters.

(b) *Helix angle.* The helix angle is the angle between the leading edge of the land and the drill axis.

Standard helix drills have a helix angle of approximately 30° and are used for drilling malleable and cast irons, carbon steels, stainless steels, hard aluminum alloys, brass, bronze, Plexiglass, and hard rubber. *Low (slow) helix* drills have helix angles of approximately 12°. They have increased cutting edge strength and are used for drilling high-temperature alloys and other hard-to-machine materials. They are also used for brass, magnesium, aluminum alloys, and similar materials, since they provide for quick ejection of chips at high penetration rates, especially for shallow holes. *High (quick) helix* drills have helix angles of approximately 40°, as well as wide, polished flutes and narrow lands. They are used for drilling low strength nonferrous materials such as aluminum, magnesium, copper, zinc, plastics, and for low-carbon steels. *Zero helix (straight flute)* drills have a 0° helix angle. They are used for materials which produce short chips such as brass, other nonferrous materials, and cast iron. They are especially common in horizontal drilling applications. Low or zero helix drills can be used for holes with length to diameter ratios exceeding four provided pressure-fed through the tool coolants are used to evacuate chips.

The helix angle affects not only the chip ejection capability of a drill, but also its cross-sectional strength, area moments of inertia, rigidity, and rake angle. Some evidence indicates that a spiral flute counters the tendency of a straight flute drill to whip. A left-hand spiral on a right-hand cutting drill resists the winding up of the shank at higher feed rates, especially for long drills. Left-hand helix is also used for taper drills to prevent chatter occurring with right-hand drills under right-hand rotation. The torsional stiffness of a drill varies parabolically with the helix angle and reaches a maximum at a helix angle of approximately 28° [88]. The radial stiffness of a drill decreases with increasing helix angle, and reaches a minimum at a helix angle of approximately 35° [89]. The axial stiffness also varies parabolically with the helix angle, with a minimum occurring at approximately 20° [89]. The allowable thrust and critical cutting speed are also affected by the helix angle, especially for small diameter drills. An increase in helix angle results in increased rake angles and lower torque and thrust. The helix angle influences drill life in a complex fashion depending on other parameters and the workpiece material [90, 91].

(c) *Hand or direction of rotation.* The vast majority of drills are right-handed and rotate in a clockwise direction when viewed from the shank end.

(d) *Number of flutes.* The number of flutes may vary from one to four, with two being the most common choice. The optimum number of flutes on a drill depends on the drill diameter, the work material, required hole quality, and hole exit conditions. Generally, one-flute drills are used for deep hole drilling, two-flute drills are used for most general purpose applications, and (as discussed below) three- and four-flute drills are used for close tolerance work and for drilling interrupted holes or through holes in workpieces with inclined exit surfaces.

(e) *Coolant hole(s).* Solid drills without coolant holes are used for shallow holes (up to two diameters deep) and for conventional tool penetration rates (using feeds in the range of 0.008 to 0.011 mm per cutting edge per millimeter of drill diameter). Solid drills with coolant feeding hole(s), called coolant-fed or oil-hole drills, have passages that run through the drill body. Cutting fluid is fed through these passages to improve chip ejection and cool the cutting edges, which permits use of higher cutting speeds and penetration rates.

(f) *Web and flute geometry.* The strength of a drill is largely determined by its web and flute sizes. The two main conflicting parameters in drill body design are adequate flute area for efficient chip disposal and high drill rigidity to reduce deflection and increase dynamic stability. The ratio of the web thickness to the drill diameter directly affects the drill's torsional and bending strength. For conventional two-flute drills, this ratio is usually about 0.21:1. The flute-to-land ratio also significantly affects the drill's strength. Conventional two-flute drills have a flute-to-land ratio of about 1.1:1, which provides a flute space area between 45 and 55% of the total cross-sectional area. This amount of flute space is sufficient for general purpose applications with most work materials. The diameter of the inscribed circle (on each side of the thin web section) tangent to the drill radius ending at the two intersections the flute forms with the drill periphery and the web of the drill has been found to be related to the measured torsional stiffness of the drill section [92]. These ratios can be optimized for specific work materials and hole depths as shown in Figure 4.49. On the other hand, the inscribed circle diameter of the flute cross section is also a critical design parameter since the drill chip has a conical shape [91]. The area moments of inertia of the drill cross section are very important because they affect drill deflection; the moments of inertia for two-flute drills are 50 to 65% of the value for a solid shaft in one principal direction, and 10 to 15% of the solid shaft value in the other principal direction. Two-flute drills therefore tend to deflect significantly in the weaker principal direction. A comparison of the two-, three-, and four-flute drills with respect to the area moments of inertia in both principal directions is given in Table 4.7 [93, 94]; in this table, the ratio A_f represents the ratio of the flute area over the drill diameter area. The stiffness of the three- and four-flute drills in both principal directions is the same because the principal area moments of inertia are equal.

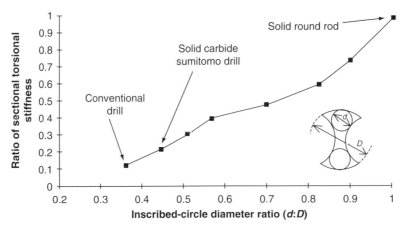

FIGURE 4.49 Effect of web thickness on drill's torsional stiffness. (After Ref [92] and J. Skoglund [97].)

Parabolic (rolled-heel) flute forms increase the chip space and enhance chip ejection and are therefore widely used for dry and deep hole applications. The use of parabolic flutes with high helix angles (greater than 30°) further improves chip ejection. Parabolic drills have a heavier core, approximately 40% the diameter compared to 20% of standard twist drills; the heavier core adds rigidity and increases stability when drilling deep holes and harder materials [95]. Drills of this type are used not only for soft materials, such as plastics, aluminum, copper, and low-carbon steel, but also for stainless steel, cast iron, and nodular iron.

(g) *Material.* Twist drills are most commonly made of HSS, HSS–Co, solid WC, or with WC tips or heads brazed on a steel body. Twist drills made of solid ceramics, PCD edges or tips brazed on a steel body, PCD heads brazed on a WC body, PCBN and ceramic tips, and PCD veined on a WC body are also used in specialized applications.

The greatest improvements in productivity have resulted from the acceptance of solid carbide drills [96, 97]. Compared to HSS drills, carbide drills permit an increase in productivity by a factor of 2 to 10, and increase the hole quality. Solid carbide drills are especially well suited for high-throughput precision hole manufacturing.

Solid ceramic and cermet and CBN- and PCD-tipped drills are used at higher speeds than carbide drills. Ceramic drills can be used for fiber-reinforced composites, but their application

TABLE 4.7
Comparison of Area Moments of Inertia between Two-, Three-, and Four-Flute 21 mm Diameter Drills

Number of Flutes	Web Diameter (mm)	Land Width (mm)	Inscribed Circle Diameter (mm)	Ratio A_f	Moments of Inertia Principal Direction I	Principal Direction II
2	7.00	10.4	6.96	0.45	5180	1224
3(I)	8.73	11.2	6.10	0.40	4100	4100
3(II)	12.73	5.0	4.10	0.36	4005	4005
4	11.5	5.2	5.00	0.40	4329	4239

in ferrous materials (e.g., cast iron) has been limited by a lack of machine tools with sufficient speed capability and acceleration/deceleration rates for the spindle slide reversal in blind holes. PCD-veined drills eliminate concerns about the integrity of the braze interface with the carbide blank [19]. PCD drills have been used extensively to drill aluminum alloys, other nonferrous alloys, and fiber-reinforced composites at conventional and high speeds.

Generally, improved drill rigidity has a positive effect on most aspects of the drilling operation. A stiffer drill exhibits less of vibration and deflection, which allows the use of higher speeds and feed rates, and produces better hole quality and longer tool life. This property of solid carbide drills usually eliminates the need for a guide bushing, which is often required with HSS drills to maintain hole location accuracy. Short carbide stub drills are invariably used without bushings. The use of bushings with long solid carbide drills is usually avoided because any misalignment between the drill and the bushing can result in drill breakage. Carbide head drills or brazed carbide drills should be used for deep hole applications which require a bushing.

Three-flute drills produce holes with precision hole size, finish, and roundness tolerances. The three-flute design is a compromise between the conventional two- and four-flute drills [94]. A larger number of flutes provides additional guidance for the drill. The flute cross section is the critical design parameter because there are three flutes compared to two flutes for standard drills. Only conical and planar (three-facet chisel) points can be ground on three-flute drills because of the odd number of cutting edges. A planar point (Figure 4.50) has three V-shaped chisel edges. Notch web thinning is used because the web (core diameter) at the point is a large percentage (usually 40%) of the drill's diameter. The residual length of the three chisel edges after web thinning is usually 5% of the diameter, which nonetheless results in much higher thrust than that obtained with a conventional two-flute drill. The residual chisel can be thinned further when drilling soft, nonferrous materials, assuming that chips do not pack in the notched areas. In some cases a thick chisel at the center is avoided by continuing only one or two of the three edges all the way in to center [98]. However, a three-flute drill can be ground with inverted center point to which eliminate the chisel edges as discussed above and shown in Figure 4.50 [94]. The inverted point reduces the drill thrust force by up to a factor of two compared to the planer notch web-thinned point. In an alternative design, all three cutting edges are brought together into a pyramid-shaped "spur" at the center [99]. This design reduces thrust but has two drawbacks: the center spur is too weak for many hard materials, and the axial rake angle of the cutting edge shifts from positive to negative at the chisel edge.

Four-flute drills generate even better hole quality than the three-flute drill because they have four margins guiding the tool. The additional margins act as a bushing and increase drill stability during interrupted cuts. The best four-flute design has two flutes that cut to center and two flutes separated from the chisel edge by an undercut as shown in Figure 4.50 [93, 100]. The chisel edge is generated by the intersection of the major flanks from only two cutting edges 180° apart. Also, a wider range of point geometries can be ground on a four-flute drill than on a three-flute drill; this range includes two-facet, four-facet, and helical points. The flute space is more restricted than on a three-flute drill, however, making four-flute drills most attractive for drilling materials such as cast iron which form short chips. Either two or four coolant holes can be used in a four-flute drill to improve chip evacuation.

4.8.2 Twist Drill Point Geometries

One very important feature of a twist drill is its point geometry. The geometry of the point determines the characteristics of the drill's, three cutting edges: the main cutting edges or cutting lips, the chisel edge, and the marginal cutting edges. The geometry of the main cutting

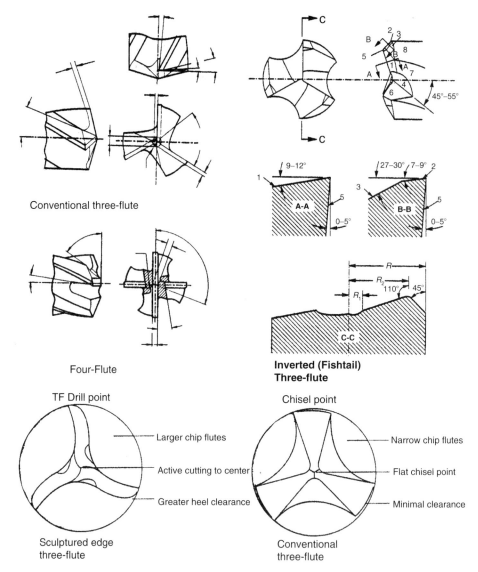

FIGURE 4.50 Configuration of three- and four-flute drills. (Courtesy of Guhring Inc.)

edges affects the drilling torque, thrust force, radial forces, power consumption, drilling temperature, and entry and exit burr formation [101, 102]. The chisel edge positions the drill before the main cutting edges begin to cut, and stabilizes the drill throughout the cutting process; it also affects the drill's centering characteristics (skidding and wandering at entry) and the thrust force. The margins guide and locate the drill and affect the hole straightness and roundness errors and the drilled surface finish.

The best point geometry for a given application depends most strongly on the drill and workpiece materials, hole depth and size, required hole quality, and expected chip form. Other factors to be considered include the entrance and exit surface orientation with respect to hole axis, hole interruption, burr formation and tool life concerns, and the presence or absence of a bushing (Figure 4.46). The principal geometric features of the point are the point angle, the chisel edge angle, and the relief angle.

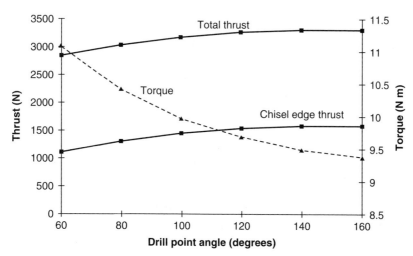

FIGURE 4.51 Effect of point angle on thrust force and torque (considering the main cutting lips only).

The cutting edge length is inversely related to the point angle. An optimum point angle which yields maximum drill life and hole quality exists for every work material. A standard 118° point is used for general purpose drilling of readily machined materials. Point angles smaller than 118° are preferred for many cast irons, copper, fiber aluminum alloys, die castings, and abrasive materials. Point angles greater than 118° are used for hard steels and other difficult materials. Generally, a lower point angle reduces the thrust force while increases the torque (Figure 4.51); the thrust force varies parabolically with the point angle and reaches a minimum value at roughly 118°, the point angle used on standard drills. The cutting edge is formed by the intersection of the flute face and the flank face; the shape of lip is determined by the point angle, helix angle, and flute contour. The flute contour and helix angle are usually designed to provide a straight lip with a 118° point angle. A straight lip is desirable because it generally provides maximum tool life. Specialized drill designs with concave lips (e.g., racon point drills) are used for some steels, since a concave lip induces more strain in the chip and improves chip breaking.

Photographs of the point geometries of several solid carbide drills used for machining aluminum are shown in Figure 4.52. The corresponding relief and rake angles for the drills in Figure 4.52 are shown in Figure 4.53. Similarly, photographs of the point geometries of several solid carbide drills used for machining cast iron are shown in Figure 4.54. The corresponding relief and rake angles for the drills in Figure 4.54 are shown in Figure 4.55. The rake angle distribution across the main cutting edges depends on the flute helix angle. The flute helix reaches its maximum at the margin and decreases to zero at the center; it can be calculated at any radius using Equation (2.51). Similarly, the rake angle decreases near the web; it is typically negative at the center of a drill and roughly equal to the helix angle at the outer corner as shown in Figure 4.53 and Figure 4.55. Lip correction can be used to reduce the rake angle and increase edge strength along the main cutting edge; it generates a constant rake (helix) angle along the entire length of the cutting edge. Lip correction is used especially for inhomogeneous materials such as cast iron, and when small, discontinuous chips are desired. A 0° to 5° positive rake angle produced by lip correction or the use of straight flute drills provides a strong edge for general purpose drilling of hard and brittle materials such as cast iron, metal–matrix aluminum composites, stainless steel, steel alloys, nickel-chrome steel, titanium alloys, and high-temperature alloys. A small or neutral rake angle will not help chip evacuation and may cause build-up at the point in softer materials. The strength of a positive

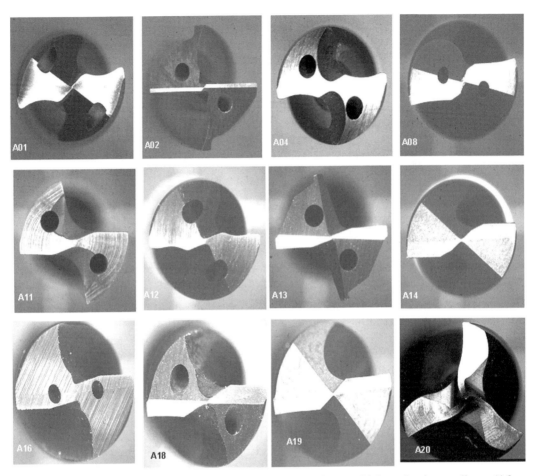

FIGURE 4.52 Photographs of 8.5 mm solid carbide drills for machining aluminum alloys. (After J. Agapiou [101].)

rake lip can be increased by grinding a 0.025 to 0.1 mm hone on the edge (needed for high-throughput applications) [101, 102].

The corresponding cross-section profiles and the characteristics of the chisel edge for the solid carbide drills in Figure 4.52 are shown in Figure 4.56. The chisel edge is the blunt cutting edge at the center of a center-cutting drill. It is formed by the flank surface ground on the drill web. The ratio of the web or core diameter to the drill diameter is usually large. The optimal web thickness depends primarily on the work material. The web thickness is usually about 15 to 20% of the drill diameter for large drills, but may reach 50% of the diameter for small drills, which require a proportionately heavier web to maintain stiffness.

Because it cuts slowly and has a large negative rake angle, the chisel edge produces a chip by an extrusion or smearing action, rather than by cutting (Figure 4.57). Because chisel edge chips have a less direct path to the flutes, they are more likely to pack and build up in the hole. The chisel edge contributes substantially to the thrust force; the size of the contribution depends on the relative lengths of the chisel and main cutting edges. The chisel edge contributes roughly 50% of the thrust for a drill with a typical drill with a web thickness equal to 20% of the diameter (Figure 4.56). If the web thickness to diameter ratio is increased to 30%, the chisel edge thrust doubles; if this ratio is further increased to a 40%, it will increase by an additional factor of 2 (or to a factor of 4 total as compared to a 20% web drill). The three

Cutting Tools

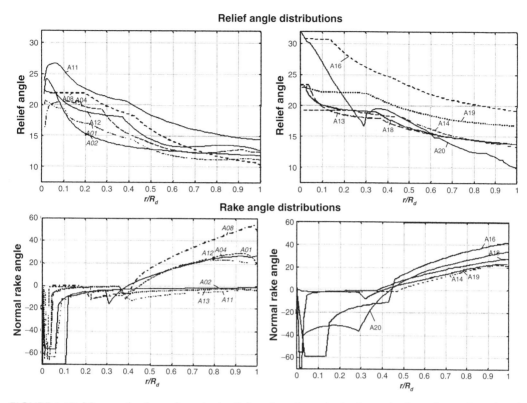

FIGURE 4.53 Measured values of nominal relief angle and nominal rake angle along the cutting edge of the aluminum drills in Figure 4.52. (After J. Agapiou [101].)

common approaches to reducing problems associated with the chisel edge are: (1) reducing the chisel edge length by thinning or splitting the drill point, (2) changing the shape of the chisel edge to improve its cutting action, and (3) eliminating the chisel edge altogether.

Edge preparation follows guidelines similar to those used for turning and milling as shown in Figure 4.58 for the drills in Figure 4.54. The edge treatment is very critical to drill performance. The cutting forces will be excessive if the edge treatment is excessive (even if the rake angle is large). For example, drill C01 (in Figure 4.54) with a 29° helix angle and chamfered and honed edge (shown in Figure 4.58) requires 40% larger thrust than drill C18 with a 14° helix angle and honed edge (also shown in Figure 4.58). In general, cutting forces depend mainly on the three parameters: the relief angle distribution (Figure 4.53 and Figure 4.55), edge treatment, and chisel edge length.

A wide variety of drill point and body configurations have been developed (as shown in Figure 4.52, Figure 4.54, and Figure 4.56) to improve aspects of drill performance such as the drill's centering ability, thrust force, and rigidity. The effect of drill point geometry on drill performance and cutting forces is discussed in detail in Refs. [82, 101, 102]. Some common drill point geometries and their range of application are shown in Figure 4.59. Briefly, the common types of points include the following.

Conventional point: The conventional or conical point is the most common type of drill point ground on standard, 118° point drills. The chisel edge is usually either conventional (with conical relief) or two faceted (which results in a flat or blunt chisel) [91], and has a high negative rake angle (−50° to −60°). Conventional drills tend to "walk" or drift during entry and thus often require a centering hole. Conventional point drills are most often used in operations which

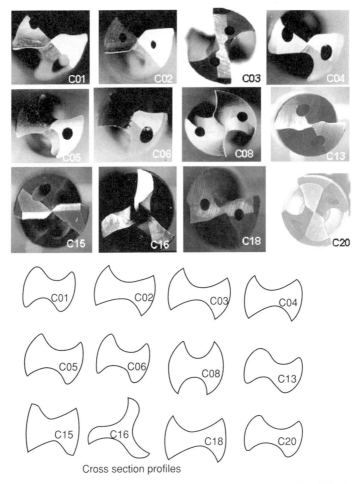

FIGURE 4.54 Photographs and cross-sectional profiles of 8 mm solid carbide drills for machining cast irons. (After J. Agapiou [102].)

do not require high precision or production rates. The conical point can be ground to provide a small crown (0.07 to 0.2 mm depending on the drill diameter) along the chisel edge, which results in significant improvement in the chisel edge cutting action and centering characteristics.

Radial/racon point: This type of drill has an arch-shaped radiused point, resulting in a more positive rake angle at the center. The radial lip provides a self-centering effect; it can therefore drill more accurate holes than a conventional drill. The cutting edge is longer than on a conventional point, resulting in slightly higher torque and thrust, but also lower edge temperatures and stresses since the heat and forces generated during drilling are spread out over a larger area. Radial points thin the chip at the outer corner, protecting the corner and margin from wear, reducing burr formation, and improving drilled surface finish and tool life.

Web thinned point: In this point, the chisel edge is thinned by grinding a notch at the chisel edge corner with a radiused wheel. There are several variations of web thinning such as that described in DIN-Standard 1412 Form A. Web thinning reduces the chisel edge length, reducing thrust and improving chip evacuation from the center of the drill. Reducing the chisel edge length also improves drill centering properties. The web is typically thinned to a diameter between 8 and 12% of the drill diameter. Lip correction can be used to thin the chisel edge as described in DIN-standard 1412 Form B.

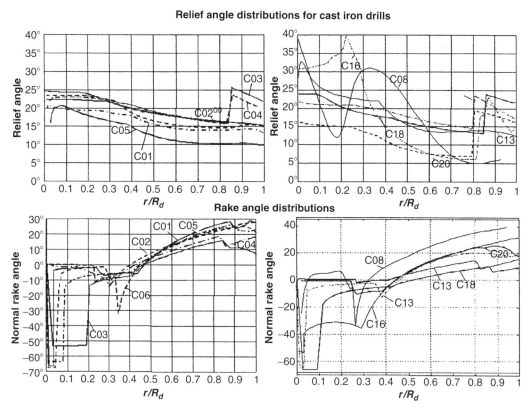

FIGURE 4.55 Measured values of nominal relief angle and nominal rake angle along the cutting edge for the cast iron drills in Figure 4.54. (After J. Agapiou [102].)

Split point: The split point is often referred to as a crankshaft drill. It is produced by notch type web thinning as described in DIN-standard 1412 Form C. There are two or three similar variations of this point style. The most common split point type is a special case of the web-thinned point with a much smaller residual chisel edge length (typically 2 to 3% of the diameter). There are however a number of drill point designs, produced by the Hosoi grinding technique, which use a modified split with an S-shaped secondary cutting edges; these include the Dijet-Hosoi, Sandvik Delta, OSG Ex-Gold, Mitsubishi New Point, Sumitomo Multi Point, Tungaloy Spiral Jet DSC, and Guhring RT 80, RT100, and Precision Twist Drill points [103]. The S-form split reduces secondary edge wear and drill failures when drilling hard materials. The notches in split point drills are prone to build-up when drilling soft materials; the split point is also prone to edge chipping when drilling tough alloys.

The notches on the crankshaft or true split point often do not reduce the chisel edge length, but generate two small cutting edges, one on each side of the chisel edge passing ahead of center. This point is self-centering and also reduces the thrust force, especially when drilling work-hardening materials. The split point is especially common on long drills, such as crankshaft oil hole drills, and on small diameter (<12 mm) drills. It is also a preferred point configuration for drilling titanium, stainless steel, and high-temperature alloys.

Helical/spiral point: This point has an "S" contour with a radiused crown chisel that reduces the thrust force and makes the drills self-centering [104]. It eliminates the need for web thinning. In general, a spiral point drill has a thicker web than a conventional drill because a thin S-shaped chisel limits its effectiveness when drilling soft materials. The main advantage of

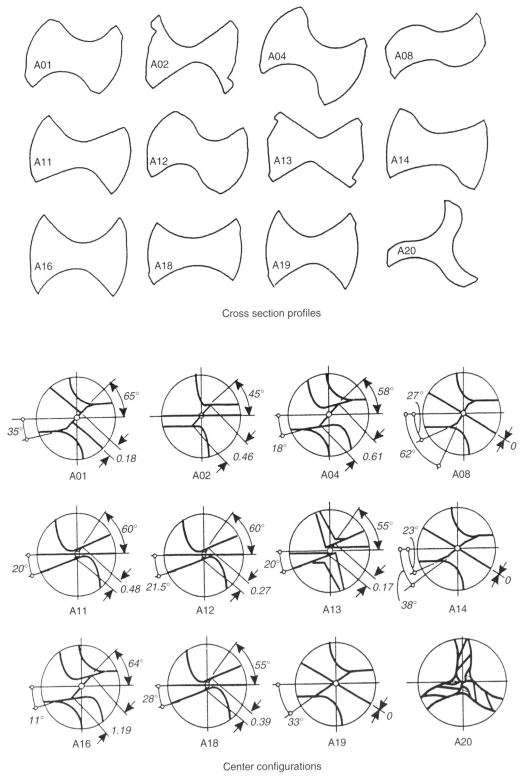

FIGURE 4.56 Cross-sectional profiles and center configurations for the aluminum drills in Figure 4.52. (After J. Agapiou [101].)

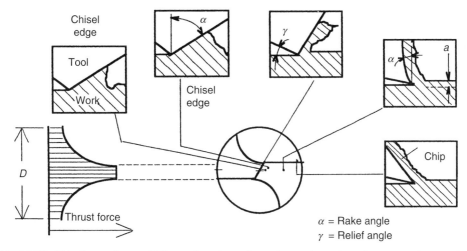

FIGURE 4.57 Chip formation at different locations along the cutting edge of a conventional drill.

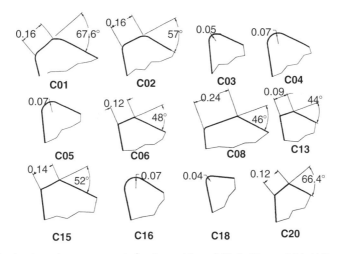

FIGURE 4.58 Edge treatment measurements for the cast iron drills in Figure 4.54. (After J. Agapiou [102].)

this point is that it reduces burr formation at drill exit. Helical points are weaker than split points and require a special drill grinder. The Hertel SE point is a special type of the helical point with concave lips.

Bickford point: The Bickford point geometry is a combination of the helical and racon point geometries. The helical point is ground on the center of the drill, while the racon point is used for the outer portion of the drill. This point combines the benefits of both the helical and racon points.

Four-facet chisel point: Also called the bevel ground point, this point has a chisel edge formed by the intersection of primary and secondary relief planes ground on the flank, producing a less negative rake angle as compared to the conventional point. The more favorable chisel edge geometry reduces the thrust force, improves centering accuracy, and increases drill life. This is the most common point on microdrills and can be used successfully with most workpiece materials.

Double angle points: These points are ground with a corner break (chamfer) to reduce the included point angle at the drill periphery, resulting in four-facet lips. This is the

DIN-standard 1412 Form D point. The reduced peripheral point angle reduces corner wear and burr formation at breakthrough and improves size accuracy; the abrupt change in the point angle also serves to split or break the chip. This point is particularly effective when drilling brittle ferrous or hard materials with severe break-through conditions.

Multifacet point: Several types of multifacet points are available [105]; they are most easily classified by the number of facets, or the number of primary and secondary relief surfaces ground on the flank. The six-facet Avyac point resembles a four-facet chisel point with an additional web thinning notch at the chisel edge [106]. A number of eight- and ten-facet drills

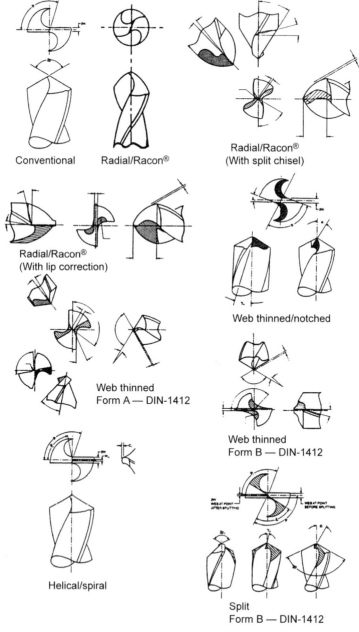

FIGURE 4.59 Various drill point geometries. (Courtesy of Guhring Inc.)

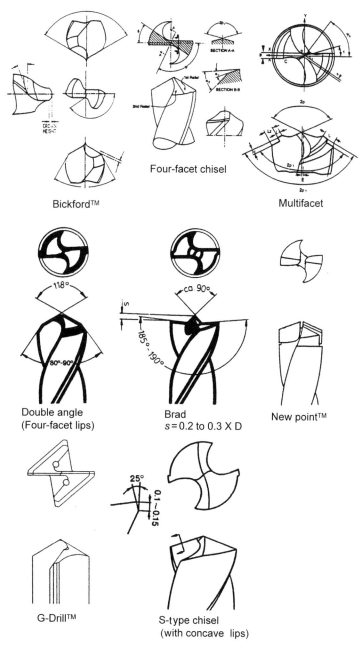

FIGURE 4.59 *Continued*

were developed by Wu and his colleagues for different workpiece materials [107]. As the number of facets is increased, the point becomes increasingly difficult to grind consistently.

Brad point: The brad point (DIN-standard 1412 Form E) has a web-thinned center point ground at an acute point angle (less than 120° and usually 90°) and slightly concave main lips. The outer corner functions as a trepanning tool. The center point length is usually equal to 20 to 30% of the drill diameter. The brad point is designed for drilling accurate, round holes in sheet metal with minimal burr formation. A disk or slug of material is produced at exit.

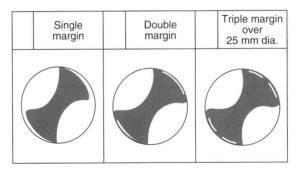

FIGURE 4.60 Cross sections of single and multiple margin drills. (Courtesy of Metcut Tools.)

Nonchisel edge drills: Drills can be ground without a chisel edge to reduce the thrust force. This type of point is sometimes called an inverted point because the point angle near the drill's center is inverted (greater than 180°). Grinding the inverted center section, splits the cutting lips into two segments and serves to split the chip. Similar points include: (a) the *new point* (Figure 4.59), which produces a small diameter core at the drill center that breaks off as drilling proceeds; (b) *three-flute inverted* or *fishtail point* (Figure 4.50, inverted three-flute) which produces less than half the thrust force of a standard three-flute drill.

Multimargin drills: These drills have two or more margins per flute, depending on their diameter (Figure 4.60), in contrast to the conventional single margin drills. The most common double margin drill, at least for diameters less than 25 mm, resembles a multiple flute version of the pressure-fed gundrill used for deep hole drilling (described below under multitip drills). They provide twice the number of contact points on the hole surface, which serves to guide the drill and results in higher drill stability and stiffness in the hole; the drill deflection therefore is significantly reduced, especially for interrupted holes. The secondary margins also burnish the hole and produce a smoother finish. Very accurate holes can be produced with diameter close to the drill size. The drill point geometry is generally independent of the number of margins. The *G-drill* (Figure 4.59) is a double margin drill with straight flutes. This design has been successfully applied to several materials, especially aluminum and cast iron. The back taper of the margins is required to be two to four times that of single margin drill to maximize drill life and to prevent early margin build-up and cracking. Flood coolant is usually adequate for shallow holes.

4.8.3 Spade and Indexable Drills

Spade and indexable drills [108] have fixed lengths, which can reduce setup time. They are similar to brazed drills in that they employ a steel body. They are available in diameters from 12 to 75 mm for L/D ratios from 2 to 10. Many point geometries which can be ground on a twist drill are not available on spade and indexable drills.

Spade drills consist of a body and a removable (throw-away) cutting blade or bit, which is precisely located and clamped in a special slot at the end of a steel drill body (Figure 4.61). The blade may be clamped with either one or two screws, with the two screw system usually being more stable. Spade drills are available in diameters greater than 9 mm. Spade drills can be used at high penetration rates and are comparatively rigid. In general, spade drills are not used for finishing operations requiring tolerances better than 0.08 mm on the diameter unless special care is used to set the blade in the drill body. Spade drills increase the range of possible cutting edge materials as compared to conventional twist drills; the blade may be made of solid HSS, HSS–Co, WC, cermets, or ceramic, or may be PCD or PCBN

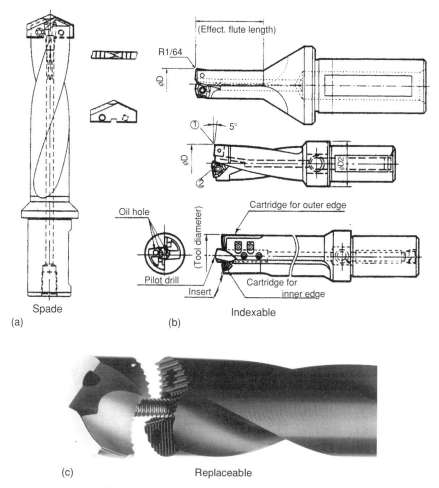

FIGURE 4.61 Indexable drills — spade, indexable insert, and replaceable head drill designs. ((a) Courtesy of Allied Machine & Engineering Corporation; (b) Courtesy of Carboloy/Seco.; (c) Courtesy of Toshiba Tungaloy America, Inc.)

tipped. This drill type offers the economic benefit of throwaway inserts, and often eliminates geometric variation due to regrinding.

Indexable insert drills consist of a steel body and shank with indexable inserts replacing the cutting edges at the end of the body (Figure 4.61). They also eliminate the need for a pilot or center hole, which may be required with conventional twist drills. These drills have a point angle between 170° and 200° and generate a round slug during breakthrough for through holes. Indexable drills can be used for solid drilling, core drilling, counterboring, and back boring (i.e., drilling a pilot hole and boring either a straight or taper hole on the backstroke).

Indexable drills can be used with most workpiece materials because several insert geometries can be applied. As in turning and milling, inserts can be coated and provided with formed-in chip breakers. Indexable drills are used mainly for holes less than two or three diameters deep. The feed per revolution used is typically half that for multiflute twist drills because the inserts on the two opposite edges do not overlap across the radius. However, the inserts can withstand higher speeds than solid twist drills because a larger selection of tool materials and coatings is available. One concern when using indexable drills is the reliability of the center insert; the cutting speed drops to zero at the center, which may result in insert

TABLE 4.8
Comparison of Hole Tolerance from Nominal Diameter for Different Drill Types Based on Drill Size Tolerances

Drill Diameter (mm)	Solid Carbide X3	Brazed Carbide Tipped $L/D < 3$	Brazed Carbide Tipped $L/D < 5$	Indexable Insert[a]	Replaceable Head X1	Replaceable Head X2	Replaceable Head
<10		+0.035	+0.055				
<17	+0.030	+0.042	+0.070	+0.10	+0.042	+0.070	0.0
<25	+0.040	+0.050	+0.080	+0.15	+0.050	+0.080	+0.40

Note: X1 replaceable heads are of a higher quality than the X2 heads.
[a] These have the largest tolerance depending on L/D ratio.

breakage due to increased cutting forces and poor chip removal. To prevent such failures, the center insert should have a stronger edge geometry and be made of a tougher material than the outer corner insert.

Indexable head (*replaceable tip or crown*) *drills* consist of a steel body and shank with a replaceable carbide head (Figure 4.61). They are very similar to the carbide brazed head drills (see Figure 4.14). They provide a head with a complex point geometry in an indexable design. The head is replaced by rotating the holding clamp; the alignment repeatability is typically within 0.05 mm.

Replaceable head drills offer greater accuracy (as shown in Table 4.8) than indexable drills because the head is ground to a precise diameter; they are also more stable in large L/D applications and run at higher feeds than the indexable insert drills. They provide faster tool changes and good coolant flow compared to carbide head brazed drills.

4.8.4 Subland and Step Drills

Subland drills are special tools for drilling multidiameter holes [109]. Each diameter has its own flute and land as shown in Figure 4.62; this results in a complex flute geometry, which is necessary for what is effectively two or more tools sharing a common axis and core. In a

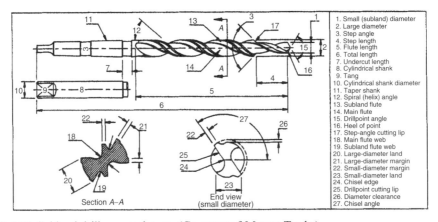

FIGURE 4.62 Subland drill nomenclature. (Courtesy of Metcut Tools.)

Cutting Tools

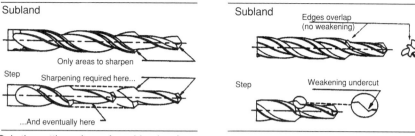

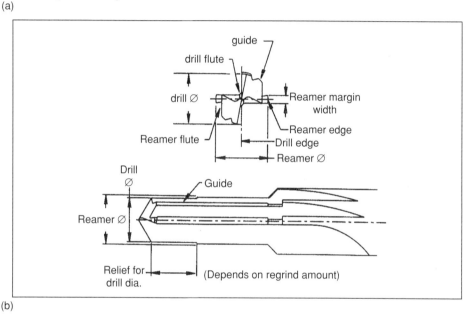

FIGURE 4.63 Comparison between subland and step drills. ((a) Courtesy of Metcut Tools; (b) Courtesy of Kennametal Inc.)

conventional *multistep drill* (Figure 4.63), the smaller diameter ends at the larger diameter's cutting lips, and both share common flutes, lands, and margins; it is thus a modified standard drill. The advantage of the subland drill over the step drill is the preservation of the geometry for all diameters after regrinding and therefore the larger number of regrinds possible.

Subland and step drills are usually used for manufacturing blind multistep holes in one pass as shown in Figure 4.64, but they can also be used as a combined drill and reamer (shown in Figure 4.65) to generate close tolerance through holes in a single pass. A combined drilling and reaming tool ("drillmer") can be used for through holes. In this case, a straight or helical two-flute drill is followed by a reamer section with several straight or helical flutes; when the length of the drill section is a little longer than the part thickness to be drilled, the reamer can generate its own hole location (i.e., does not follow the drilled hole). Drill/reamers generally are used to produce relatively shallow through holes (with depths less than 2 to 2.5 times the reamer diameter). Though the tool coolant is usually necessary for drill/reamers to improve chip evacuation and prevent the damage of the hole surface by chips trapped between the lands and the wall.

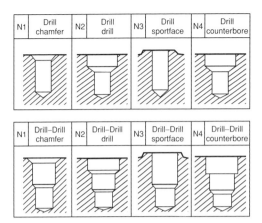

FIGURE 4.64 Hole styles generated with subland or step drills. (Courtesy of Metcut Tools.)

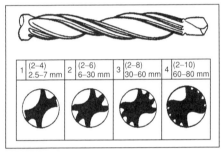 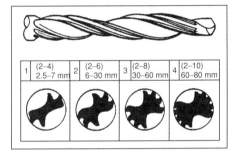

Step drill/reamer design. Cross section shows drill and reamer lands sharing the same diameter.

Subland drill/reamer design. Cross section shows independent diameters for drill and reamer lands.

FIGURE 4.65 Comparison between step and subland drill/reamer designs. (Courtesy of Metcut Tools.)

4.8.5 MULTITIP (DEEP HOLE) DRILLS

Multitip drills have segmented cutting edges as illustrated in Figure 4.66. Types of multitip drills include indexable insert drills (described above), gundrills, and BTA (Boring and Trepanning Association) and ejector drills. Most are essentially two-fluted drills with one or more tips arranged on each flute; the number of tips is a function of the drill diameter. For sizes less than 35 mm, two tips, one cutting at the center and one at the periphery, are usually used. For larger diameters, a third tip cutting in the gap between the first two may be added. Multitip drills generally use either brazed or indexable WC tips. Multitip designs are classified as either balanced or unbalanced. The tip of a balanced drill is symmetric, resulting in minimal radial loads and unbalanced torque. Unbalanced drills generate larger radial forces and require guide pads to support the resultant force. Some indexable drills are nominally balanced like spade or twist drills, but even in this case radial forces may result due to manufacturing tolerances in the inserts and insert pockets. Most of the gundrills, BTA and ejector drills are unbalanced.

BTA and ejector drill heads [110, 111] are used in single- and double-tube deep hole drilling systems (respectively) with high pressure coolant. The difference between these two drilling heads is illustrated in Figure 4.66. Multiple-lip heads provide better chip control than single-lip heads and reduce power requirements due to a reduction in friction on the bearing pads. The unbalanced forces generated by twist drills, gundrills, and BTA drills are shown in Figure 4.67. A large hole is present above the cutting edge(s) to accommodate chip disposal. The pads

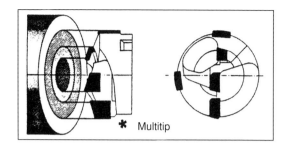

(a)

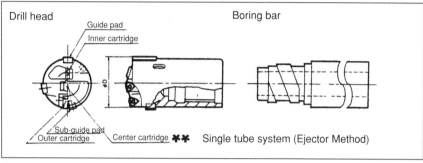

(b)

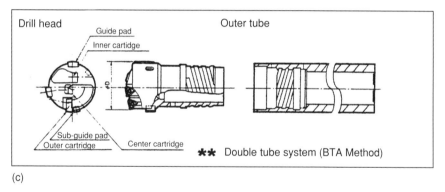

(c)

FIGURE 4.66 Diagram of deep hole drill heads. ((a) Courtesy of Sandvik Coromant; (b) and (c) Courtesy of Toshiba Tungaloy America, Inc.)

burnish and work-harden the hole wall surface, which acts as a pilot bushing. These tools are ideal when hole straightness is critical or when fine surface finishes are required. BTA heads are preferable for materials with poor chip formation characteristics such as low-carbon steels and stainless steels. They can be used at much higher penetration rates than twist drills for holes more than five diameters deep. Ejector drills are only available in sizes larger than 18 mm.

Gundrills [112] traditionally have a single lip with carbide wear (bearing) pads to support and guide the tool as shown in Figure 4.68. Two-fluted double margin or multiple-lip high-pressure coolant tools, which are becoming increasingly common, are also often called gundrills. The two-fluted drill is guided by margins approximately 90° from each cutting edge as shown in Figure 4.67 and Figure 4.68. The maximum practical hole depth depends on the torsional stiffness of the drill body, which made either of single- or double-crimped 4130 aircraft quality steel tubing bodies (heat treated to 35–40 R_c) or solid single- or double-milled style bodies (Figure 4.69). The crimped tubing design provides for larger flute space and coolant holes than the milled flute type. Single- or double-crimped tube designs are used in

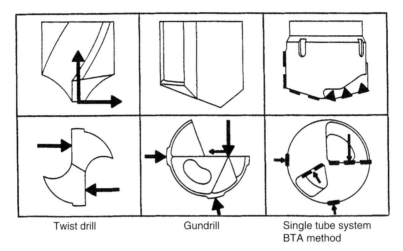

FIGURE 4.67 Comparison of the radial cutting force for balanced and unbalanced drills. (Courtesy of Sandvik Coromant.)

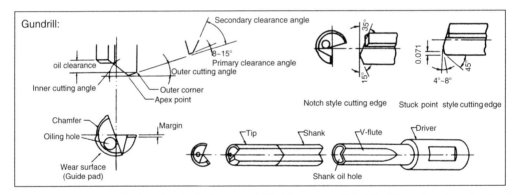

FIGURE 4.68 Characteristics of a single-flute gundrill. (Courtesy of Toshiba Tungaloy America, Inc.)

severe applications where high coolant volume and large flute cross section are required to improve chip-removal efficiency. However, the milled style body provides higher torsional stiffness. The drill tip is supported by the drilled hole wall since it cannot be fully supported by its body. Tool life is affected by the body design because it controls the volume of the coolant, the flute space, and the drill stiffness. Although gundrills usually have straight flutes, a spiral flute may be used to reduce the drill's tendency to whip.

Two types of gundrill construction are common. In the first, a molded solid WC head is brazed onto a tubular steel body; in the other, WC tips are brazed onto a steel head attached to a tubular body. Solid carbide heads are normally used for holes less than 37 mm in diameter; for larger holes the insert-carbide tips are often used for cost reasons.

Due to the nature of the deep-hole drilling process, gundrills often have distinctive point geometries. A multifacet point is often ground on single-lip gundrills in order to break the chips in two or more segments. While notch style (double ground) cutting edges are very common, a triple ground edge (shown in Figure 4.69) is often used to distribute cutting forces more evenly at the cutting edge and to permit use of higher penetration rates. The triple-ground design is used especially for tough work materials such as stainless steel and high-temperature alloys. Single-flute drills are designed for materials with long and stringy chips

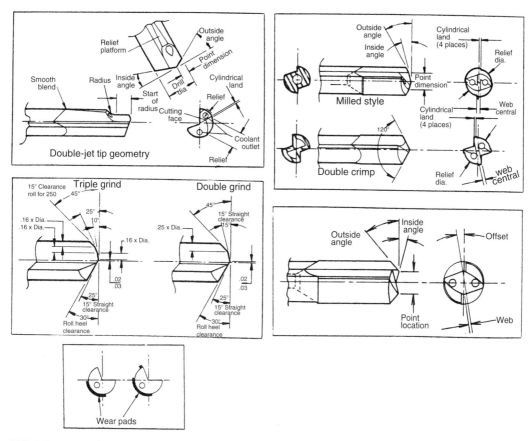

FIGURE 4.69 Point geometry of typical gundrills. (Courtesy of Star Cutter Co.)

such as tool steels. Double-flute drills are preferred for materials with small or soft chips such as cast iron and aluminum.

Double-jet gundrills (U.S. Patent No. 4,092,083, shown in Figure 4.69) have two holes for coolant through the carbide head instead of the single hole in conventional gundrills. This design improves chip ejection by increasing and properly directing the coolant to the cutting edges.

The practical cutting speeds of gundrills are similar to that of conventional drills with comparable tool materials. However, gundrills are used at lower penetration rates than the conventional two-flute drills since they are often a single-flute drills and their tubular body may twist at higher feeds. Common feeds for gundrills are 0.05 to 0.125 mm/rev for single-flute designs and 0.125 to 0.25 mm/rev for two-flute designs.

4.8.6 OTHER TYPES OF DRILLS

Half-round drills are constructed from a round rod with roughly half the diameter ground away to provide the flute area; they are normally tipped with a conical point. The ground flat is of approximately the same length as the flutes of a conventional drill, and results in a zero rake angle at the cutting edge. The conical point is ground on the centerline, resulting in no chisel edge. This drill works on the same principle as the gundrill and requires a bushing for guidance at entry unless the drill is very short or has a very small diameter (i.e., for micro-drilling). This drill produces reamed hole surface quality due to its gundrilling-like support.

However, its range of applicability is limited due to the difficulty of ejecting chips through the straight flat flute with only flood coolant. The practical depth-to-diameter ratio is usually below 10:1 in horizontal drilling and 3:1 in vertical drilling.

Trepanning drills cut only an annular groove at the hole periphery and leave a solid core or slug at the hole center. Trepanning drills cut more efficiently than conventional and gundrills because they cut less material overall and no material at low velocities near the center of the hole. Trepanning drills are most often used for holes larger than 40 mm even though they are available in drills down to 12 mm. Trepanning drills produce less chips and therefore require less power and thrust capability. Types of trepanning drills include single-edge with wear pads (with a solid WC head, brazed carbide tips, or indexable tip and wear pads; Figure 4.70) and multiple-edge. A *Rotabroach* drill is a trepanning drill (Figure 4.70) with several teeth depending on the tool diameter; the cut is distributed over a greater number of cutting edges resulting in longer tool life.

Microdrills are small-diameter drills (<4 mm) [113]. The design of microdrills is especially critical in applications such as printed circuit board manufacturing in which accurate, high-quality holes must be drilled in composite materials. Hole damage such as delamination, fiber pull out, strain, and phase change must be avoided in these applications.

Both the drill point and body geometry are critical to microdrill performance. Complex point designs such as the spiral point, split point, notched point, etc. cannot be ground on

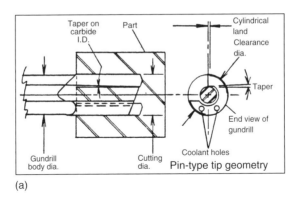

(a)

(b)

FIGURE 4.70 Point and body geometry of trepanning type drills. ((a) Courtesy of Star Cutter Co.; (b) Courtesy of Hougen Manufacturing, Inc.)

small drills. The most common microdrill designs include conventional twist drill designs with a four-facet point [114], half-round designs, and pivot (spade) drill designs. Proper flute space is required for adequate chip ejection so that such drills can be used at comparatively large penetration rates. The microdrill should have a point angle smaller or equal than that of the pilot drill.

4.8.7 Chip Removal

Chip control in drilling is critical because often determines hole quality and drill wear and reliability. The importance of chip control increases with hole depth. Chips should be rapidly ejected out of the hole to prevent heat build-up and flute packing, which can cause accelerated tool wear, plastic deformation of the cutting edge, or drill breakage.

Several techniques are used to expedite chip ejection. Proper design of the drill point geometry and flute shape can facilitate chip breakage and efficient chip flow up the flutes. For example, a parabolic flute drill, made by modifying the heel shape on the land of a high-helix twist drill, has larger flute space than standard high-helix drills, providing significantly improved chip flow and coolant access. Chip breakers or chip splitters (Section 6.5), or convex cutting edges can be used to break chips in smaller segments (if this cannot be accomplished by adjusting the feed and speed) to improve chip ejection, especially for mild and alloy steels which often generate long chips.

Chips can also be flushed from holes by properly applied coolant. Either flood coolant or internal (through the tool) coolant can be used. Internal coolant is more effective; this method cools and lubricates the cutting edge and reduces thermal shock in addition to removing chips, and therefore often results in improved tool life and productivity. Flood coolant is usually sufficient for drilling holes less than two diameter deep at moderate feeds and speeds. However, three- and four-flute drills may require through the tool coolant for holes deeper than 1 to 1.5 diameters depending on spindle orientation (vertical or horizontal) and penetration rate. A safe drilling length for the indexable drills with flood coolant is less than 0.75 diameters. Spade drills usually require internal coolant. High-pressure coolant (300 to 1500 psi) is often required with soft and alloy steels or for high-speed drilling of aluminum in order to clear chips. Through the tool coolant is required for high penetration rate and high-speed drilling.

Internal coolant can be introduced either through the spindle or through a rotary coolant inducer attached between the tool holder and the spindle. The latter approach reduces the coolant pressure, the system stiffness, and (generally) the allowable rotational speed. The coolant holes on the drill may extend to the flank faces on the drill point, or may exit on the flutes. Some drills have a central coolant hole through the web and which branches near the flank, while others have one spiral coolant hole per flute. Coolant holes on the flank should be located as close to the outer diameter as possible so that the outer portion of the drill is cooled during exit for through holes, and to minimize coolant spray into the air during exit. Drill margin damage during breakthrough is also a concern when using coolant-fed drills. Additional flood coolant or coolant through holes exiting in the flutes during breakthrough could help reduce or prevent outer corner wear and welding of chips on the margins.

Flood coolant is applied from nozzles located above the spindle, around the spindle housing, on the tool holder, or around the tool shank through grooves in the holder collet. The nozzles should be directed so that some apply a coolant stream to the drill point while others direct a stream behind the point; this results in more effective cooling as the drill travels into the hole.

The pressure and especially the volume of the coolant have a strong impact on drill performance. Increasing volume is required as the drill diameter increases. Larger diameter drills require increased coolant volumes or flow rates because the size of the coolant holes increases with diameter; for deep holes or vertical spindle setups, coolant pressure should also be increased. The required pressure and volume depends on the drill geometry, the size and shape of the chips produced for the particular workpiece material, and the number and size of the coolant holes through the drill [115]. The amount of coolant is important to the performance of the cut. As a general rule a 0.5 gpm coolant per horsepower (0.746 W) is good [116]. The force of the liquid exerted on the chips (through the impulse–momentum equation) is proportional to the coolant mass flow rate and velocity in the cutting area. The velocity is proportional to the square root of pressure from Bernoulli equation. This means that a 100% increase in pressure will result in 40% increase in the force pushing the chips. On the other hand, a 100% increase in the flow rate increases the force by 100%. Therefore, it is preferable to increase the flow rate rather than the pressure. Figure 4.71 can be used to estimate the required coolant pressure and volume as a function of diameter for coolant fed two-flute solid carbide, brazed tipped, and replaceable head drills. The low limit is used for $L/D < 2$ while the high limit for $L/D > 3$. Indexable insert and spade drills require flow rates near the high limit and pressures closer to the lower limit. The required coolant flow rate and pressure for BTA and ejector type drills can be estimated from Figure 4.72. The maximum conveyable flow rate as a function of pressure for different diameters and type drills can be estimated from Figure 4.73. Figure 4.73a shows that the two-flute solid carbide helical-flute drills with two large coolant holes can deliver higher flow rates than the three smaller coolant holes in the three-flute drills. However, straight-flute carbide drills have larger coolant holes than the corresponding helical-flute drills; this means that the straight-flute drills can deliver higher coolant flow rate than helical-flute drills as shown in Figure 4.73b. Indexable drills can deliver higher flow rates at lower pressures than solid carbide drills as shown with

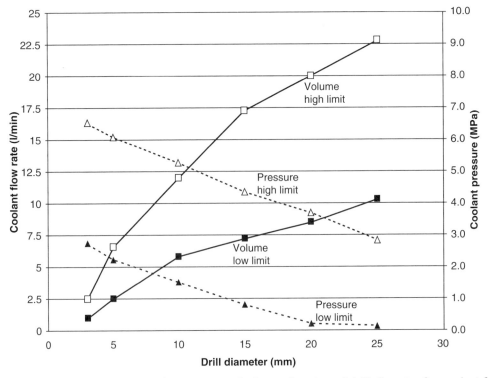

FIGURE 4.71 Coolant volume and pressure required as a function of drill diameter for coolant-fed drills.

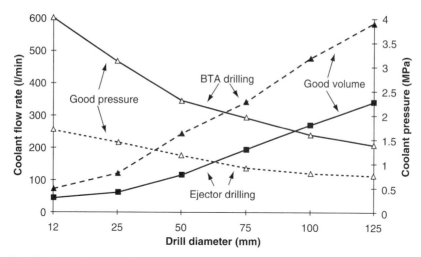

FIGURE 4.72 Coolant volume and pressure required as a function of drill diameter for gundrills.

the replaceable head drills in Figure 4.73b. Indexable insert drills having similar coolant supply characteristics as replaceable head drills. However, the spade drill could provide higher flow rate at lower pressures than the replaceable head or indexable insert drills especially at the larger sizes (>20 mm). The required flow rate as a function of the drill diameter for different hole depths (applicable especially to gun drilling) can be estimated from Figure 4.74.

Spray mist systems that provide small amount of lubricant at low pressure can be used successfully on machines lacking a coolant recovery system. Compressed air can also be used to clear chips on such machines even though the noise level increases due to air flow.

An interrupted feed cycle can be used if the above approaches do not evacuate chips effectively. The most common method is peck drilling, or periodically retracting the drill to break and clear chips from the hole. Peck drilling is often necessary when drilling deep holes with flood coolant. Clearing all chips before the drill re-enters is critical in peck drilling to ensure that chips do not become wedged beneath the cutting edges. An alternative method is to allow the drill to dwell for an extremely short period every two or three revolutions to break chips into short segments; this is not an optimum approach, however, because it increases drill wear.

4.8.8 Drill Life and Accuracy

From a drill life viewpoint, through holes are easier to drill than blind holes because they eliminate drill dwell at the bottom of hole when the drill reverses direction. However, drill chatter or breakage can occur during breakthrough when drilling through holes, due to feed surging (i.e., the occurrence of higher effective feed rates due to the tendency of the drilling machine to spring back because of a sudden reduction in the thrust force). In such cases, slowing the feed during exit, coolant fed drills, or a drill point geometry change may be required. A conventional practice with certain work materials which generate long, continuous chips (e.g., high strength and stainless steels) is to retract the drill before breakthrough to clear chips and relax the strain on the system; this prevents feed surging at breakthrough. Whirling vibrations may also occur during initial penetration, especially for high spindle speeds or long, slender drills [117, 118].

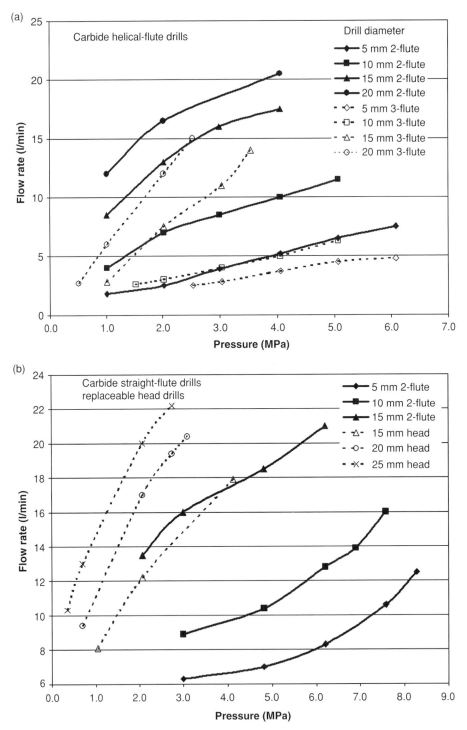

FIGURE 4.73 Relationship between coolant volume and pressure required as a function of depth of hole.

Cutting Tools

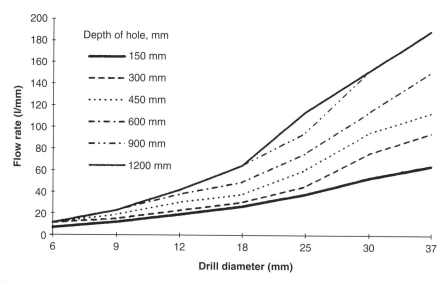

FIGURE 4.74 Relationship between flow rate required and drill diameter required as a function of depth of hole.

As with boring bars, a drill's body geometry and material determine its rigidity. The drill geometry and its length should be designed to provide maximum stiffness and minimum deflection under the expected thrust loads and interruptions. A solid WC drill is superior to a brazed or tipped drill with a steel body in this respect. The toolholder and machine stiffness are also critical to tooling performance.

Precision holes can be produced in a single pass (i.e., without subsequent boring and reaming) when a proper drill design is used. Most drilling operations are carried out with HSS or HSS–Co drills. Although conventional drilling has become more accurate over the last decade, only a small percentage of manufacturers use the proper drills to produce tight tolerance holes. The concentricity/runout tolerance of HSS and HSS–Co drills is a function of the drill diameter, size, and length (e.g., a 6 mm drill with 20 mm length has 0.06 mm runout error while a 200 mm long drill of the same diameter has 0.3 mm runout error specification). Although the typical diametrical tolerance for drilled holes is between 0.05 and 0.13 mm, tolerance of 0.015 mm can be achieved if the drill geometry, tool holder, and machine are properly designed. Some of the important parameters controlling the accuracy of the drill are the roundness accuracy or TIR, which should be within 0.005 mm, the lip height error, which should be less than 20% of the feed per flute, and the symmetry of the cutting edges (flute spacing, web centrality, and chisel centrality). Hole location accuracy is a function of drill point geometry (especially the chisel edge), the drill stiffness, and the rigidity of the toolholder, fixture, spindle, and machine structure. Of all the drill types, solid carbide drills have the tightest tolerances and produce the best holes as summarized in Table 4.8. This makes them the best choice for extremely close tolerance holes. Brazed tipped drills and replaceable head drills produce hole tolerances about twice those of solid carbide drills. Indexable drills have the largest tolerances and hole variations, especially at high L/D ratios.

4.8.9 Hole Deburring Tools

There are several types of general deburring tools as discussed briefly in Chapter 2. For deburring holes, special chamfering tools are used either independently or in combination on drill or reamer bodies as illustrated in Figure 4.75 and Figure 4.76. These tools chamfer the entrance and exit of the hole. There are several variations of such tools.

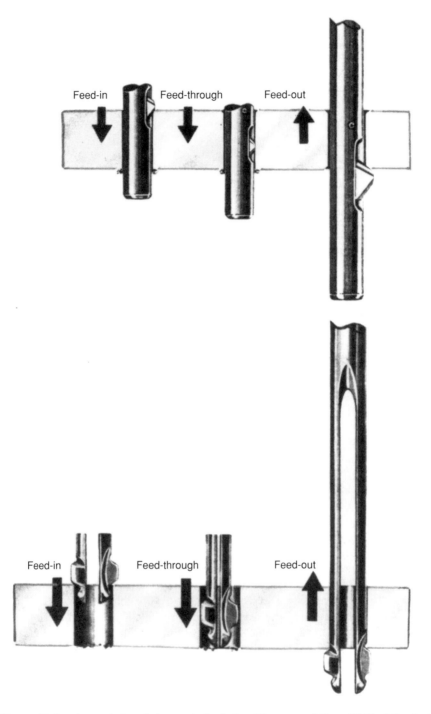

FIGURE 4.75 Deburring and chamfering tools for holes. (Courtesy of Cogsdill Tool Products, Inc.)

Cutting Tools

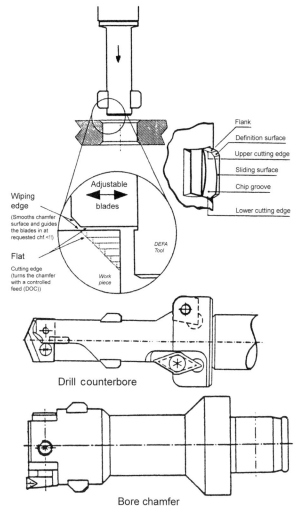

FIGURE 4.76 Deburring and chamfering tool used in combination with other hole making tooling. (Courtesy of Heley.)

4.9 REAMERS

Reamers are rotary single- or multiedged cutting tools used to enlarge holes and improve hole quality. The geometry of a typical reamer is shown in Figure 4.77 [119]. Reamers are similar to drills without a chisel edge, and specifically to core drills; they are also quite similar to boring tools. Reaming tools generally do not affect the location and straightness of the starting hole since they usually have a chamfer angle and follow the centerline of the existing hole. In addition, the reamer is supported primarily by the workpiece during cutting through circular margins which are not present on boring tools. The cutting edge geometry of reamers is specified in the same way as for boring, milling, and drilling tools. The major grinding parameters are the edge preparation at the outer corner (i.e., the dimensions of the chamfer), the axial and radial rake angles, the primary and secondary relief angles, the helix angle and direction, and the margin design.

The major difference between a reamer and a boring tool has traditionally been the way in which they are applied. The reamer traditionally is held in a radial floating holder that allows

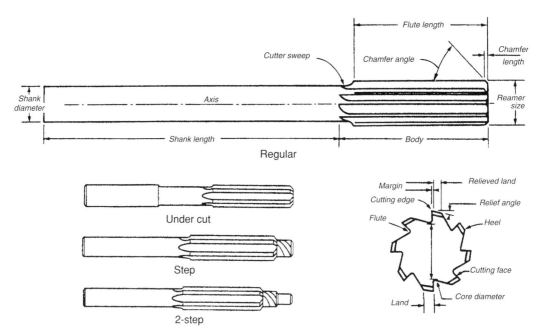

FIGURE 4.77 Reamer nomenclature. (Courtesy of Precision Twist Drill Co.)

it to locate in the existing bore; it follows the centerline of the hole being reamed and does not correct the centerline of the hole in relation to the centerline of the machine spindle. However, rigid reamers are also common. Boring tools are held rigidly in order to correct the centerline of the hole being enlarged. A reamer usually operates at lower speeds and higher feed rates than a boring tool because the cutting speed must be kept low for a reamer used in a floating holder to improve hole roundness, and because reamers are usually designed with more cutting edges than boring tools.

The quality of reamed holes depends primarily on the condition of the prereamed hole, the geometry of the reamer, the stiffness of the reamer, machine and fixture, and on the cutting conditions, coolant application, and compatibility of the coolant with the workpiece. Reamers can produce holes with a size and roundness variation of less than 0.025 mm; a gun reamer can generate holes with size tolerances less than 0.012 mm.

Most reaming tools are of solid or brazed construction. Indexable inserts and heads have also been used in reaming tools to reduce tooling cost and eliminate regrinding. Inserted-blade reamers are also used when size control is critical, since they allow for size adjustments. As with indexable boring tools, the indexable blade is mounted directly into the tool body for small diameter tools (<15 mm) and in a cartridge for larger tools.

4.9.1 Types of Reamers

Different types of reamers include single- and multiflute reamers, straight and tapered reamers, single- and multistep (multidiameter) reamers, and expansion or adjustable reamers (Figure 4.78) [120]. This section describes mainly single-flute, multiflute, and multistep straight reamers; design concerns for tapered reamers are similar.

A single-flute reamer is self-guiding and follows its own centerline. It therefore corrects straightness, angularity, and location errors in the initial hole within narrow limits. The WC wear pad, guide pad, and peripheral tip determine the diameter of the reamed hole and also

Cutting Tools 225

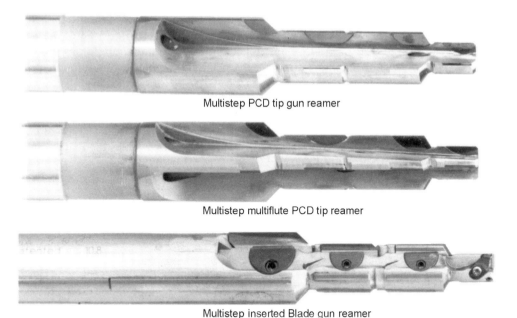

FIGURE 4.78 Configurations of reamers for short and deep holes.

could burnish the hole surface to improve roundness. A two-flute reamer with secondary support margins (Figure 4.79) provides similar cutting performance.

Single-flute reamers, often called gun reamers or gun bores, generally produce bores with lower roundness errors than multiflute reamers. Gun reamers can often be used to produce tight tolerance holes without microsizing. A gun reamer acts as a reamer when mounted in a floating holder, as a fine boring tool when it is held in a rigid holder, and as a gun reamer when a bushing is used to control motion at hole entry.

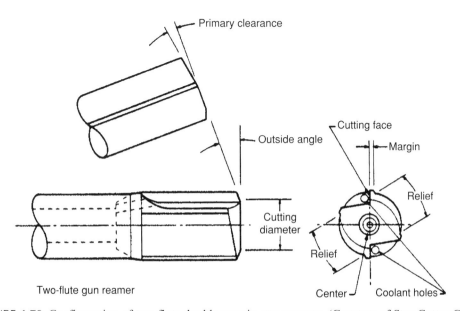

FIGURE 4.79 Configuration of two-flute double margin gun reamers. (Courtesy of Star Cutter Co.)

Multiflute reamers remove metal faster and have a longer effective tool life than single-flute reamers, but usually produce bores with higher roundness errors. A multiflute reamer generally produces a multicornered "lobed" hole profile due to radial vibrations of the tool axis which result in deviations in the initial hole diameter, misalignment of the cutting portion of the reamer with respect to the hole axis, and differences in the widths of tool margins [121, 122]. A reamer is often guided through a bushing or pilot surfaces to follow the desired path; when this approach is not feasible, a radially and axially floating holder is used to align the reamer in the hole and allow the reamer to follow the prereamed hole. The use of a short entering taper on the front of the reamer, or a short pilot section, guides the reamer into the hole and reduces chatter and vibration. Increasing the number of flutes on a reamer produces a hole profile with a larger number of corners or lobes; this in turn reduces the out-of-roundness errors of the reamed hole.

Multiflute reamers usually run at higher feed per revolution than a drill due to larger number of flutes. Occasionally, reamers with staggered (irregularly space) teeth are used to either prevent or reduce chatter.

Reamers are also used to produce a hole with stepped diameters concentric to each other. Step or subland designs can be used for multidiameter reamers; the design selection depends on the required number of flutes for each step. Subland reamers provide fewer flutes per diameter than step reamers because step reamers use a single set of flutes for all diameters (Figure 4.59 and Figure 4.60). The design of multidiameter reamers is also influenced by the length of each step, depending on the configuration of the prereamed hole [123].

Multistep reamers can be also used to bore and ream a hole or to double ream a through hole in order to improve hole quality, especially when the stock remaining for reaming is excessive due to misalignment concerns between machining stations (see Figure 4.80). A drill-reamer (dreamer) is used mainly for through holes to drill and ream holes in a single pass as explained in Section 4.8.

4.9.2 Reamer Geometry

Generally, a reamer has a chamfer at the outer corner of the cutting edges to guide it into the hole. The chamfer angle, shown in Figure 4.81, is complimentary to the lead angle used in turning and milling tools. A standard chamfer of 45° is used for most applications. However, chamfers between 30° and 45° can be used for steel, while chamfers up to 20° are used for cast iron and aluminum. When the reamer is being used as a boring tool (with a fixed holder),

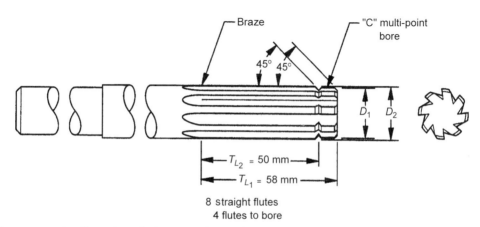

FIGURE 4.80 Configuration of a bore-reaming tool.

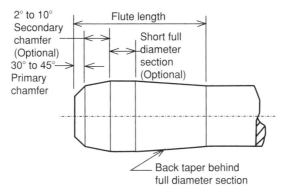

FIGURE 4.81 Several features of the point of a reamer to improve hole quality and surface finish.

an initial chamfer angle greater than 15° followed by a secondary angle smaller than 20° is used so that the tool acts as an end cutting tool (as shown in Figure 4.79) which further improves hole size and surface finish by a scraping action. A reamer with a 45° or smaller primary chamfer will not completely correct hole location and straightness errors. Higher chamfer angles result in reduced tool life when cutting ferrous materials, especially for interrupted cuts, due to rapid tool corner wear and a higher susceptibility to corner chipping. The chamfer angle also affects the rake angle. Left-hand reamers are usually designed with a 30° chamfer, which results in a positive rake angle. The length of the secondary chamfer is very critical to the tool performance; it can generate chatter when it is too long depending on its relief angle, the workpiece material, and part–fixture stiffness. The chamfer width along the cutting edge should be larger than the depth of cut. The smaller secondary and primary chamfer angles and the longer secondary length result in better surface finish.

The chamfer relief angle has a strong impact on tool life and can vary from 6° to 12° and from 10° to 20°, respectively, for the primary and secondary relief angles. Reamers usually have a narrow circular land with a width which varies with the reamer diameter and the workpiece material; the width increases with diameter (i.e., 0.17 to 0.6 mm, respectively, for a tool diameter ranging from 6 to 50 mm).

Reamers have axial and radial rake angles defined in the same way as for end mills (Figure 4.34, Figure 4.35, and Figure 4.37). The axial rake is determined by the flute helix. A left-hand helix on a right-hand cut results in a negative axial rake; conversely, a right-hand helix on a right-hand cut provides a positive axial rake. A straight flute reamer has a zero axial rake angle. Reamers with a negative axial rake angle require higher feed forces and produce rougher hole surfaces. Reamers with a positive axial rake angle are more susceptible to chatter, especially when the machine tool is in poor condition or a floating holder is used. However, a freer cut is generated with a positive axial rake reamer. In a right-hand cut, a right-hand helix pushes the chips to the top of the hole through the flutes, while a left-hand helix pushes the chips ahead of the tool. Left-hand helixes are therefore used for through holes. Tapered reamers should be designed with a helix direction opposite the rotation to prevent the tool from pushing material ahead of the edge.

A back taper is a small taper (or longitudinal relief) on the diameter along the flutes. Generally, a back taper of 0.005 mm/mm of flute length is sufficient, but larger values can be used when binding is a problem.

Reamers can produce precision holes only if they run true on the spindle and are squared to the workpiece, which is most easily accomplished through the use of axially and radially adjustable holders. Radially floating holders are often used (especially on transfer lines) to ensure alignment with the prereamed hole. The holder for a fixed reamer may have radial and/or axial adjustments to zero any misalignment or roundness errors between the spindle axis and the reamer.

4.10 THREADING TOOLS

Types of threading tools include cut taps, roll form taps, thread mills, thread turning inserts, thread chasers, and dies. This section describes taps, chasers, and thread mills; thread turning inserts were discussed in the Section 4.5.

4.10.1 TAPS

A tap is a rotary tool with a geometry similar to that of a screw. The size and geometry style are the two major considerations in selecting the correct tap for a particular material and machine/part setup. The tap style is defined by the number of flutes, rake or hook angle, chamfer length, land, and helix angle [124, 125] as shown in Figure 4.82. *Cut and roll forming taps* (Figure 4.83) are the major tap styles used for internal threading. Cut taps cut and remove metal to produce threads; roll form taps displace or deform metal to form the thread profile.

A cut tap has a series of single point cutting edges arranged linearly and radially on the tool periphery as shown in Figure 4.84. Only the chamfered teeth on the front end portion the tap contribute to the cutting action as shown in Figure 4.84; the uniform threads, behind the tapered portion, guide the tap by bearing on the threads already generated. The equivalent of one revolution of full thread form is produced with each revolution of the tap after all of the chamfered teeth are engaged in the workpiece. However, the actual work is spread out along the full length of the chamfer. The chamfered teeth and the tap's first full thread cut or deform the material. Each succeeding chamfered tooth makes a deeper cut until the full thread is generated. Commonly used chamfers include taper, plug, and bottoming, which are illustrated in Figure 4.85. The chamfer has a direct relationship to the chip load as shown in Figure 4.86. Therefore, it is important to use the proper chamfer depending on the thread pitch, workpiece material, percentage of thread, and type of hole. The taper chamfer is designed for difficult-to-machine materials because it provides a better distribution of chip load per tooth, resulting in better thread quality. The chamfer should not exceed three to five

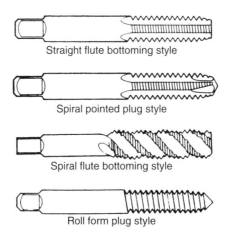

FIGURE 4.82 Configuration of cut and roll form taps.

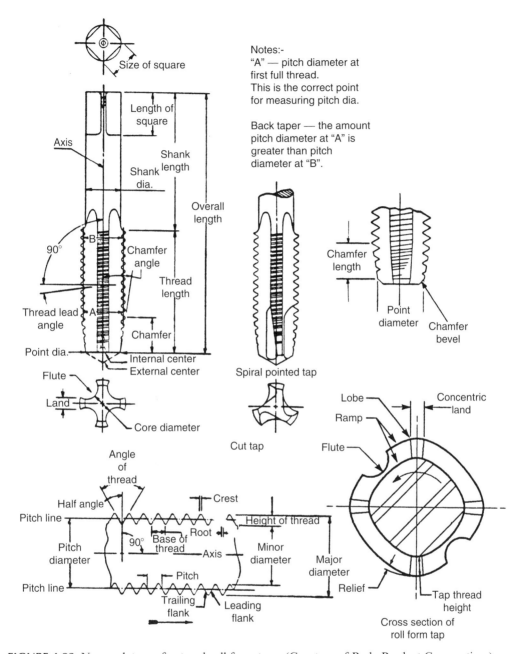

FIGURE 4.83 Nomenclature of cut and roll form taps. (Courtesy of Besly Product Corporation.)

threads for taps used in work-hardening materials such as superalloys and stainless steels, so as to produce chips that are thick enough to allow the cutting edges to undercut any previously work hardened surface. The plug chamfer, with an average chip load per tooth, is used for through holes in most materials at conventional or higher speeds. The bottoming chamfer results in a large chip load, which requires less torque, and is used mainly for blind holes. The longest chamfer possible should be used to improve the tap's efficiency, dimensional accuracy, and life, even though a long chamfer increases the cycle time and requires a

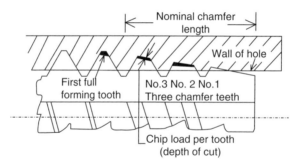

FIGURE 4.84 Chip load per individual tooth for one flute of a tap.

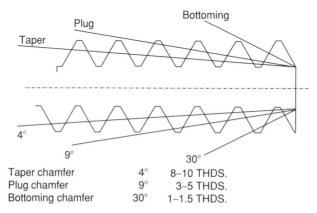

FIGURE 4.85 Influence of chamfer on tap's end geometry.

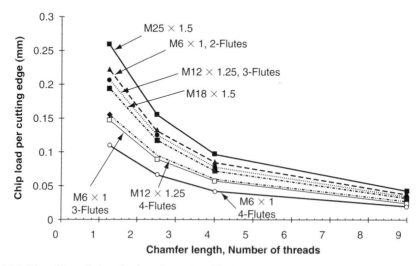

FIGURE 4.86 The effect of chamfer length on the chip load per cutting edge for several sizes of taps.

deeper drilled hole. Titanium alloys often require longer chamfers to prevent galling on the chamfer relief surfaces. The geometry of the transition from the chamfer to the tap's full diameter also affects tap performance.

The number of flutes generally varies from two to four and affects the chip load and the available chip and coolant space. As with drills and reamers, several factors affect the optimum number of flutes. The chip load per tooth is reduced either by increasing the flute number or increasing the chamfer length as shown in Figure 4.86; a four-flute tap removes about 20% less chips per flute than a three-flute tap. However, larger number of flutes produces a larger core diameter, reduces the land width, and reduces the chip space, which increases the likelihood of tap breakage due to chip clogging. Nevertheless, when chips are manageable (easily broken or powdery), an increase in the number of flutes should be considered; too wide a land will produce excessive friction, so that it is better to use a larger number of narrow lands on larger taps. Deeper holes require more chip space on the tool, especially when tapping blind holes or when a vertical setup is used; in these cases a three-flute tap is often the best choice. As in drilling, the nature of the chips produced often dictates the number of flutes. High-speed tapping requires wider flutes, narrower lands, and a smaller core diameter than conventional tapping.

The radial rake or hook angle, defined on the face of the teeth in the flutes of the tap (Figure 4.87), is equivalent to the side rake angle on a single-point tool; the rake angle may be positive at the outer diameter, but decreases to negative values near the center. It should be selected based on the characteristics of the workpiece material. A positive rake angle provides the best cutting action in soft and ductile materials, but results in a weak cutting edge. A positive rake reduces torque and forms chips which are more easily disposed of. Zero or negative rake taps should be used for hard and brittle materials to prevent chipping of the thread crests, which usually occurs during tap reversal. The flute shape and rake at thread crest of conventional taps are often modified to provide increased support to the thread crest edges to reduce chipping. A spiral point at the end of the tap along each flute face as shown in Figure 4.82 and Figure 4.83 is used to provide a positive rake along the secondary face which covers the whole taper length. Spiral-pointed taps form the chip into a tight curl and eject chip ahead of the tap; this keeps the flutes as clean as possible, allows adequate coolant access to the cutting edge, and eliminates chip interference during tap retraction or backing out. Special shear grinds have been developed to provide a progressively changing rake angle (variable rake) over the entire thread and to improve tap performance as compared to conventional spiral points. Spiral point taps are designed primarily for tapping through holes. Fluteless taps with a spiral-point result in maximum body strength; they are designed to tap holes less than one diameter deep in sheet metal and in soft materials.

The flute helix is important especially when tapping blind holes. Spiral fluted taps are designed to draw chips smoothly out of blind holes, resulting in faster chip removal and less clogging. Spiral flutes reduce torque requirements and breakage. Spiral flute taps with multiple chamfered teeth may result in poor chip control because they produce multiple

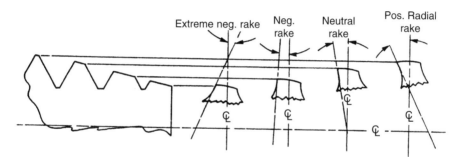

FIGURE 4.87 The effect of chamfer on the rake angle at the cutting teeth of a tap. (Courtesy of T.M. Smith Tool Int. Corporation.)

chips with different radii of curl, which can become entangled in the flutes or (eventually) the tool, resulting in damage to the threads during tap reversal.

The flute helix angle varies with the workpiece material. Slow helix angles of about 15° with low rake angles between 3° and 5° are used for titanium alloys and monel; the same helix with higher rake angles between 14° and 18° are used for tapping short holes (<1.5 × D) in structural, carbon and alloyed steels, nodular and malleable cast irons, brass, bronze, and copper. Faster helix angles between 35° and 45° with 7° to 13° rake angles are used for deeper holes (<2.5 × D) in structural, carbon and alloyed steels and for all stainless steels. A medium helix angle of 25° with a small rake angle between 3° and 5° is used in hardened and tempered alloy steels, nickel-based alloys, and inconel. Finally, fast spirals of about 40° with high rake angles between 15° and 25° are used for low-silicon aluminum, long chip brass, and thermoplastics. Spiral flute taps with as few flutes as possible and a short chamfer length to increase the cutting volume per tooth may alleviate some tapping problems, especially when tapping brittle materials such as cast iron, work-hardening materials such as stainless steel, or soft materials such as aluminum.

The thread relief corresponds to side clearance angle on single point cutting tools. Several types of thread relief have been developed, as shown in Figure 4.88, to provide radial clearance in the thread form at the pitch diameter of the flanks in order to reduce the contact

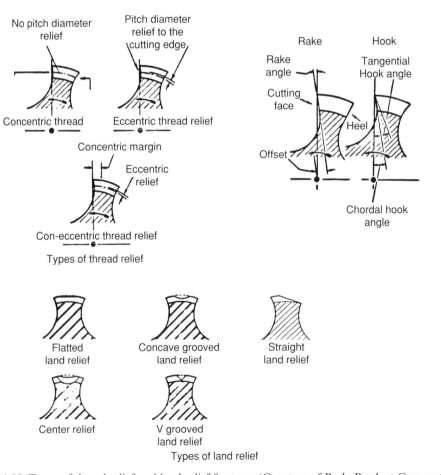

FIGURE 4.88 Types of thread relief and land relief for taps. (Courtesy of Besly Product Corporation.)

between the tap's land and workpiece, thereby lessening the forces and heat generated. The actual thread size and quality are strongly affected by the amount of thread relief. Concentric relief provides zero relief; it supports the tap well during cutting and produces threads which are not oversized. It is used for soft materials. Eccentric relief provides radial clearance in the threads from the cutting edge to the heel of the land; this type tends to cut oversized threads and is used mainly on tougher materials. Con-eccentric type thread relief is a combination of the above two types; the first third of the land has concentric relief, while the remaining two thirds of the land has the eccentric relief. This design provides support to prevent oversize and results in a freer cutting tap. Double eccentric thread relief is a combination of a slight radial relief starting at the cutting edge and continuing for a portion of the land width, and a greater radial relief for the balance of the land.

There are also several types of land relief (Figure 4.88), which provide clearance on the thread crest. Flatted land relief truncates the crest between the cutting edge and heel so that there is contact of the crest only at the cutting edge and the heel. Concave grooved land relief is similar to flat relief but provides a groove in the center of the land which does not quite reach to the pitch diameter. The center relief design has a wider and deeper groove (going down to the root of the thread form) than grooved land relief; it supplies more coolant to the cutting area. V-grooved land relief is similar to the other grooved types but uses a longitudinal V-notch at the center of the land. Chamfer or straight land relief provides a gradual decrease in land height from some distance of the cutting edge to heel.

The combination of the rake angle, thread relief, and land relief largely determine the cutting action and life of the tap. As with other rotary tools, back taper is also used on tap threads to minimize drag between the material being threaded and the tap. Increased back taper is required for materials with large elastic moduli.

In some cases, negative relief is substituted for the more common radial (positive) relief to effectively form the thread's final dimension. This design has a lobe or concentric land at the center of the conventional land, which results in a combined cutting and roll forming action in a single operation that is completed during the first full thread. The roll form component tends to eliminate any tap unbalance, which is often present due to radial forces imposed by grinding errors between the teeth of different flutes. In addition, it reduces the tendency of chips to re-enter the hole and damage the cut threads or the tap. A cut tap with roll forming action requires higher torque than a plain cut tap.

An interrupted thread design, produced by removing every other tooth from a standard tap, can be used to break chips in smaller sections and assist in chip ejection. In addition, it improves coolant supply to the cutting edges as compared to standard designs.

Coolant-fed cut taps have coolant exit holes in each of the tap's flutes or a center hole through the body that exits at the front end face and are used for through and deep blind holes to improve chip ejection. Coolant-fed taps are also used to improve lubrication in soft materials such as low-carbon steels.

External or internal threads can also be cut using collapsible or solid adjustable taps with blades or chasers as shown in Figure 4.89. Collapsible taps with blade or circular chasers are made in several designs and are used for cutting threads in stationary or rotating workpieces. The chasers are mounted on a die head. A single collapsible tap die head can be designed to accommodate both multiple blades or circular chasers. The circular shape of the chaser permits only enough rubbing action immediately behind the cutting edge to ensure proper lead control. Blades and chasers are designed using principles similar to those used for cut taps. They are used for both conventional threading applications and for deep hole tapping. A single chaser threading tool cuts threading time in half compared to single point threading, while its multiple cutting edges increase the tool life proportionately. More than five chasers are generally used when the workpiece has either an unusually wide flat or more than one

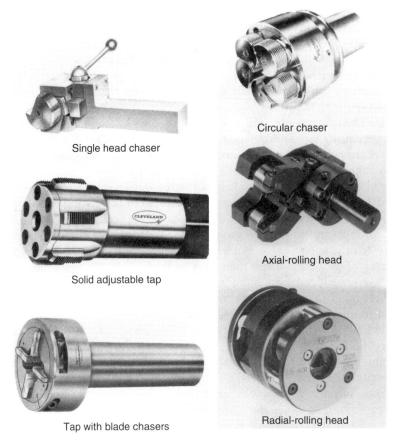

FIGURE 4.89 Configuration of collapsible and solid adjustable taps with blades or chasers. (Courtesy of Cleaveland Twist Drill Co.)

keyway or groove on the diameter to be threaded, to provide a steady threading action and produce round, accurate threads. Figure 2.28 shows two die heads with two and three circular chasers. A range of thread diameters and pitches may be cut with each die head simply by changing chasers.

Roll form taps have neither flutes nor cutting edges. Some of the important geometric features are the number and geometry of lobes located around the periphery of the threads and vent groove (Figure 4.82 and Figure 4.83). The torque required for roll taps is generally up to a factor of 5 higher than that required for cut taps. The design of roll taps significantly affects their torque requirements [126–128]. The frictional heat generated with roll taps is higher than that for cut taps, so that the lubricity of the cutting fluid becomes a critical concern, particularly at higher tapping speeds. Some medium- to high-carbon steels including alloy steels lead to severe tool galling and high torques when threaded using roll taps. Roll form taps are best suited for use on ductile materials such as aluminum, brass, tellurium, armco iron, and die-cast zinc.

A roll tap requires a tighter size control of the pretapped hole than cut tap because an oversize roll tap will create an error in both the minor and major diameters in the threaded hole. The chamfer geometry for form taps is the same as for cut taps. The tap diameters are radially relieved to form a lobed structure as shown in Figure 4.83. The tap diameter determines the number of lobes; there are always at least two. Taps can have either a cylindrical or polygonal geometry. The lobe shape can be cylindrical or spiral. The concentric land width is critical because it contacts the formed threads and may lead to galling. One or more vent grooves are generally used to provide a path for lubrication and to act as a vent for blind hole tapping. Extra vents may be required in deep hole tapping to increase lubricant flow to the chamfer portion of the tap. There is approximately a 2 to 5% variation on the generated thread by any roll tap as compared to only 2% variation by most of cut taps. Generally, the taps with either higher thread relief or lower thread length require lower torque.

Excessive torque is the most common cause of tap failure. The torque depends on the tool geometry and the tapping speed. In general, the torque is influenced more significantly by the tap design in roll form tapping than in cut tapping.

HSS is the most common material for taps, followed by HSS–Co, brazed WC, and solid WC. Brazed WC taps are used for holes larger than 10 mm in diameter. Solid micrograin WC taps were initially developed for smaller holes, but are now used in a large range of sizes. Solid micrograin WC taps perform well in abrasive workpiece materials such as sintered metals, laminated materials, cast iron, wrought and cast aluminum, brass, and titanium, but should not be used in cast materials with hard spots. Solid WC taps also perform best in high-speed applications. Finally, Solid WC taps cut cleaner, last longer, and achieve higher metal removal rates than taps made of other materials. Diamond taps have been manufactured either by brazing the diamond along the flute for larger size taps or sintering the diamond into a slot cut in a WC body; in the latter method, the thread is formed on the diamond using EDM.

Tap class specification is based on the thread fit and the thread tolerance. A tap's thread tolerance depends on its basic pitch diameter. The letters H and L or D and DU are used, respectively, for unified inch or metric screw threads for high and low limits. The number next to letters H or D represents an even multiple of 0.0005 in. or 0.0127 mm, respectively, above basic diameter and likewise for L and DU below basic pitch diameter. The class fit determines the total tolerance zone of the thread for the pitch diameter and the minor diameter (drilled hole size for cut taps) in the case of internal thread [129], which control the percent thread. For each class of thread, there are a maximum of 12 different "D" tolerance limits (D0 to D12). If no tolerances are specified, the D5 or D6 are used to meet the common 6H class. Improved control of the tap tolerances can significantly reduce tapping difficulty. The use of full thread tolerance allows: (1) an increase in tap life by specifying a higher H or D number, and (2) the selection of the proper drill size for generating the correct percent thread (thread height). For example, use of a lower "D" number in soft materials without chip ejection limitations can relax the class specification for the minor diameter. A wide tolerance range within specifications [129] can reduce the thread and therefore the tapping torque; the percent change in torque is normally approximately equal to the percent change in thread. This, in turn, can increase tool life and eliminate tap breakage and galling problems, which are generally associated with chip ejection limitations and high thread percentages.

4.10.2 Thread Mills

Thread mills for internal or external threading may be solid or indexable. Indexable thread mills (Figure 4.90) have one or more inserts peripherally depending on the tool diameter; each insert may have one or multiple teeth. More than one insert can be used, one above the other and offset symmetrically, to mill longer threads. Solid thread mills are similar to end mills with either a single row or multiple rows of teeth [130]. The geometry of solid multithread milling tools is similar to that of cut taps from the point of view of the helix angle, thread relief, land relief, and rake or hook angle. However, there is not backtaper and chamfer at the

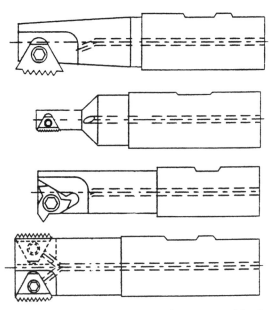

FIGURE 4.90 Indexable thread milling tools. (Courtesy of Kennametal Inc.)

end point as in cut taps. Indexable internal thread mills are used for holes larger than 12 mm, and provide the same flexibility as indexable boring or drilling tools. However, solid thread mills reduce machining cycle times because they have more flutes and because their teeth generally cover the full threaded length, so that the threads are completed in one revolution or orbiting cycle. The number of orbiting cycles for an indexable tool is equal to the ratio of the threaded workpiece length to the length of the teeth on the insert. The smallest possible cutter diameter should normally be selected to minimize the cutting time and the chip length; shorter chips reduce vibration and thus may improve tool life. On the other hand, a larger tool diameter is sometimes used to minimize tool deflection and produce an equal thread height from all teeth. The optimum tool diameter depends on a trade-off between these factors and can be estimated the same way as an end mill. The tool deflection can be also reduced by using interrupted (staggered) thread design on a two-fluted geometry as in cut taps.

Combined drilling and thread milling tools (Figure 4.91) can be used to drill and thread a hole in a single stroke. In this case, the drill point is designed based on the principles discussed in the drilling section, while the thread mill is added behind the drill point. The diameter of the thread mill is slightly smaller than the drill diameter, and an undercut is generated at the bottom of the hole during thread milling. A drill/threadmill is weaker than a plain thread mill because it has deeper flutes for proper chip ejection during the drilling phase; the tool diameter in relation to the hole size is more critical especially for small holes. The tool body stiffness can be optimized by varying the diameter, web back taper along the flute length, number of flutes, and chip load. When used at high speeds in nonferrous materials, either coolant through the tool or an effective flood coolant is required.

4.11 GRINDING WHEELS

Grinding wheels are composed of abrasive grains, which provide cutting edges, and a bond material, which holds the abrasive grains and thus acts as a toolholder. Factors to consider in selecting grinding wheels include the abrasive (type, properties, particle size, distribution, and

Cutting Tools 237

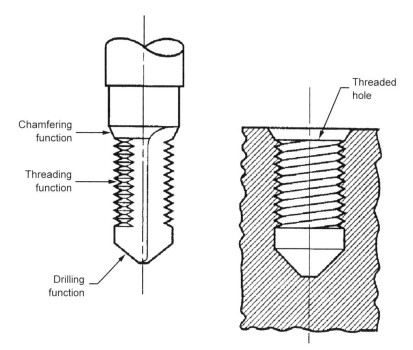

FIGURE 4.91 Configuration of a drill/threadmill tool.

content/concentration), bond (hardness/grade, stiffness, porosity, and thermal conductivity) and wheel design (shape/size and core material). The wheel matrix exhibits porosity, which is important for the effectiveness of the coolant and chip disposal. Production speed and cutting efficiency are influenced by the abrasive type, grain size, hardness and brittleness, and dressing or sharpening properties, and by the bonding material type and its grain retaining and renewal properties. Grinding wheels must be hard and tough to withstand the cutting forces while the abrasive grains must be capable of resharpening themselves by gradual break down to expose sharp new cutting edges. Grinding wheels range in size upto to 1 m in diameter and from 1 to 525 mm in thickness. Several wheel shapes, some of which are shown in Figure 4.92, are available.

4.11.1 ABRASIVES

Practically important properties of the abrasive grains include their hardness, abrasion resistance, crystal structure, and size/shape, which affect the relative friability and durability of the abrasive. Friability defines the ability of an abrasive grain to fracture under certain grinding conditions. The friability index increases with increasing hardness. A grain with good toughness can sustain wear and high cutting pressures but may result in excessive heat generation and (eventually) surface, or wheel damage. On the other hand, easily friable grain tends to wear away very rapidly. Therefore, the objective should be to use grains with sufficient toughness that are not too friable.

In order of increasing hardness, the available abrasive grain materials include aluminum oxide (Al_2O_3), silicon carbide (SiC), zirconia alumina, cubic boron nitride (CBN), and diamond; the last two are often referred to as superabrasives, while the first three are called conventional abrasives [131, 132].

Aluminum oxide is the most widely used abrasive and is used to grind carbon steel, alloy steel, malleable iron, and superalloys. There are various types of aluminum oxide grain types

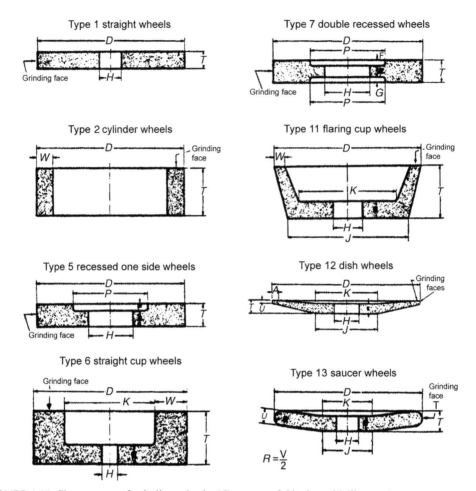

FIGURE 4.92 Shape types of grinding wheels. (Courtesy of Cincinnati Milacron.)

using several impurities (such as titanium, sodium, and chromium) which affect the friability of the grains. The most recently developed types are "sol–gel" and "seeded-gel" abrasives (ceramic aluminum oxide — a high purity grain manufactured in a gel sintering process), which consist of aluminum oxide grains with a randomly oriented microcrystalline structure [133]. They are used for precision grinding for steels particularly difficult alloys. Zirconia alumina (mixture of aluminum oxide and zirconium oxide) grains are used for rough grinding, particularly of ferrous metals. Silicon carbide is used to grind low tensile strength materials such as aluminum, copper, bronze, gray and chilled iron, cemented tungsten carbide, and nonmetallic materials such as ceramics. Silicon carbide grains are generally black, but green in purer forms; green SiC is used for heat-sensitive nonferrous materials.

The superior hardness, abrasion resistance, thermal conductivity, and compressive strength of the superabrasives make them the best candidates for grinding hard metals, carbides, ceramics, and many other hard, tough materials, especially at higher cutting speeds.

Aluminum oxide is often replaced by the CBN for hardened steels (>45 HR_c), superalloys (nickel, cobalt, or iron base with hardness greater than 35 HR_c), HSSs, and cast iron. CBN has four times the abrasion resistance of aluminum oxide. The high thermal conductivity of CBN prevents heat build-up and associated problems such as wheel glazing and workpiece metallurgical damage.

CBN abrasives are available in several types to match the bond system and grinding application. Medium and high toughness uncoated CBN abrasives are used in vitrified, metal, and electroplated bond grinding wheels. Nickel-coated CBN abrasives are used for resin bonds, and are available in two types: medium and high toughness single-crystal (monocrystalline) and tough microcrystalline. Monocrystals (consisting of single CBN crystals) tend to fracture macroscopically after dulling under high grinding forces, which provide a continuous supply of sharp cutting edges. Microcrystalline CBN particles fracture on a microscopic scale after dulling, and thus produce lower wheel wear rates.

Diamond has three times the abrasion resistance of silicon carbide and is used effectively in grinding cemented carbide tools, ceramics, and glass. Superabrasive wheels provide repeatability, surface speeds, and metal removal rates not attainable with other wheels. Diamond and silicon carbide are not effective for grinding steel due to the high chemical solubility of carbon in iron, which leads to rapid wear. Superabrasive wheels can be used at peripheral speeds up to 15,000 m/min. They can be used at material removal rates one to two orders of magnitude greater than those achievable with conventional abrasives without overheating the workpiece. Diamond and CBN superabrasives are much harder than the other abrasives and can be used to cut a variety of materials faster than aluminum oxide, silicon carbide, and zirconia alumina with slower wheel wear rates [134–137].

Three types of diamond crystals are available: RVG(Resenoid vitrified grade), MBG,(metal bonded grade) and MBS(metal bonded saw grade). RVG, medium friability diamond, is used in resin and vitreous wheels primarily for grinding tungsten carbide. MBG is used in metal bonded wheels, which are tougher and less friable than RVG wheels; MBG wheels are used for glass and ceramic materials. MBS is used in metal bonded saws for cutting stone and concrete. RVG diamond abrasives may be used in an uncoated state, with a nickel coating to improve grain retention in resin bonded wheels, or with a copper coating to reduce resin temperatures in dry grinding. Copper-coated wheels generally contain 50% copper by weight; nickel-coated grains contain between 30 and 56% nickel by weight; the lower concentration is used for steel–carbide composites, while the 56% concentration is used for wet grinding tungsten carbide.

4.11.2 Bonds

Bonds hold the abrasive grains together in the wheel. Bonds should have sufficient rigidity and the ability to retain sharp abrasive grains during cutting, yet release dulled grains. The bond must withstand grinding forces and temperatures, and resist chemical attack by the cutting fluid. The bond type also determines the wheel's maximum safe speed.

There are three major types of bonds: vitrified, organic, and metal [4, 132, 138]. Vitrified bonds are made of inorganic materials, generally glass or silicates. Organic bond materials include resin, shellac, rubber, and oxychloride. Organic bonds are flexible on a microscopic scale, which allows grains to translate and rotate during grinding. Metal bonds are more heat resistant than organic bonds and more impact resistant than vitrified bonds. Bronze, nickel, and iron are used as metal bonds; sintered bronze is the most common material. Vitrified and organic bonds are used with the conventional abrasive grades wheels. Vitrified wheels are used mainly for precision grinding, while organic (resinoid) wheels are used in high stock removal operations and all dry applications because they have a much higher resistance to thermal shock and can sustain high pressures. High porosity bond structure can be used for aggressive removal of material without burning because high porosity absorbs copious quantities of coolant and diffuses it readily in the cutting zone, facilitating heat removal resisting clogging of the cutting edges with swarf [139]. Resinoid, vitrified, and metal bonds are used with superabrasive wheels. Electroplated metal bonds, usually composed of a matrix of nickel or nickel alloy, produce single-layer superabrasive wheels by attaching individual grains of diamond or CBN to a steel preform; the layer of abrasive particles is only partly submerged in the metal matrix, rather than being fully encapsulated, so that they cannot be dressed or trued on a grinding machine. This eliminates friction between the bond material and workpiece and therefore reduces grinding temperatures. Electroplating technology produces complex-shaped wheels more easily and less expensively. These wheels perform well in high-speed grinding operations (with peripheral wheel speeds between 5000 and 10,000 m/min) since grinding speeds can be increased by 20 to 50% compared to other bonded wheels. Detailed discussions of the selection of abrasives and bond types are available in the literature [4, 136, 140]. Vitrified wheels traditionally can be used safely at wheel speeds up to 2000 m/min. The maximum safe speed for resinoid wheels is typically approximately 3000 m/min. Speeds between 4000 and 5000 m/min have become possible with recent advances in vitrified bonds and reinforced bonds that combine glass and ceramic materials.

4.11.3 Wheel Grades

The strength of the bonding holding the abrasive grains in the wheel is defined as the wheel grade. The grade depends on the percent of grain and bond in the wheel. The wheel is characterized as hard or soft depending on the strength of the bond and its ability to withstand cutting forces. Hard wheels retain grains more strongly, while soft-grade wheels lose their grains easily. Several factors, such as wheel speed, work speed, workpiece hardness and ductility, grinder structural condition, and the contact area between the wheel and the workpiece, determine whether the wheel acts as a hard or soft wheel. A decrease or an increase of the infeed rate, traverse speed, work speed, and through-feed rate can make a wheel act harder or softer, respectively. An increase in the workpiece or wheel diameter causes the wheel to act harder because the larger contact area distributes the stock removal over a larger number of grains, which reduces the specific force on individual grains. Soft grades are preferred for rough grinding, vertical-spindle surface grinding, and for machines that are relatively free for vibration. Hard grades are used in internal grinding, peripheral surface grinding, and peripheral cylindrical grinding.

The strength of a bond is controlled by the number and size of the microscopic flaws that are inherent in the bond material. The strength of vitrified or organic bond wheels sometimes decreases over time as the wheel is placed under stress due to stress-corrosion mechanisms that act on the bond when water in the coolant meets the bond material at the exposed edges of the flaws.

Wheels with a finer grit and a harder bond are preferred for profile grinding of intricate shapes and finely detailed workpieces. Harder bond wheels are required to obtain tight workpiece size tolerances. Resinoid or rubber bonded wheels, rather than more commonly used vitrified bond wheels, may be needed when excessive variations in the size and condition of the workpiece are present in centerless grinding.

Conventional grinding wheels are characterized by their grit size; for example, 80-grit size means that the average size of the abrasive grains is approximately 80 particles per inch. Grit sizes below 50 are considered coarse, those between 50 and 90 are considered medium, and grit sizes above 90 are classified as fine. Generally, the percentage of grains coarser than the specified average grain size is smaller than the percentage of smaller grains.

Coarse grains generally remove more stock per unit time than fine grains. However, this is not always true when grinding very hard work materials; in this case a medium or fine grain may be preferred because they provide more cutting edges on the wheel. In addition, coarse grains produce scratches on the surface in which it may be impossible to remove with finer grade wheels in subsequent processing.

The grain compaction, or grain spacing, density, or concentration, is a measure of the number of grains per unit volume of the grinding wheel. The compaction is not as critical for conventional wheels as for diamond and CBN wheels. Single-layer wheels offer a higher density of abrasive grains and more grain exposure than other bonded superabrasives.

The factors to be considered in selecting the proper wheel grade are the workpiece material, diameter, and size; the size of grinding wheel; the grinding contact area; the stock removal; the required surface finish; the coolant; the wheel speed; power requirements; and wheel safety. Detailed discussions of the impact of these factors on wheel selection are available in the literature [4, 138, 140, 141]. A chart of the standard marking system for conventional wheels is shown in Figure 4.93. The maximum stock removal rate depends primarily on the wheel characteristics, work material, and the coolant type (oil or water); variables such as the wheel speed, truing method, and coolant application method, also affect wheel capability to a lesser degree. Wheel selection becomes less critical when grinding, using a high G-ratio. Some general rules are: wet grinding is preferable to dry grinding, especially in heavy stock applications; a coarser grained wheel with a more open structure and a less friable abrasive can be used to increase the material removal rate; a harder grade with a coarser grain wheel is preferred for soft metals, while a softer grade and a finer grained wheel should be used for hard metals; a finer grain wheel with a denser structure and a less friable abrasive generates smoother finish; and finer wheel dressing results in better surface finish on the workpiece.

Small workpiece diameters require finer and harder wheels than larger diameters. In internal grinding, wheel diameters close to that of the hole which allow the proper coolant application should be selected for increased wheel life, particularly for small holes (<60 mm diameter); a larger wheel diameter produces better part size and form tolerance.

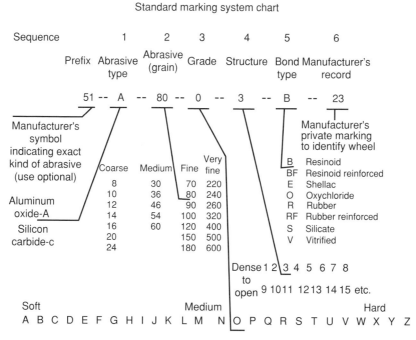

FIGURE 4.93 Standard marking system for aluminum oxide and silicon carbide grinding wheels. (After S. Malkin [143].)

Organic bonded (Al_2O_3 and SiC) wheels are very effective in internal grinding because they effectively withstand the high temperatures often generated in such operations; organic bonded wheels are also used when fine finishes are required. Vitrified bonded wheels are used primarily in peripheral surface grinding. Resinoid bonds are most frequently used in disc grinding because they can withstand higher speeds and greater shocks than the most rigid vitrified bond. Vitrified bond Al_2O_3 and SiC wheels are used in conventional creep-feed grinding.

4.11.4 Operational Factors

Operational factors such as wheel balancing, truing/dressing, grinding cycle design, and coolant application also affect and are affected by grinding wheel selection. Wheel truing and dressing in particular have a strong impact on wheel performance and life. The truing operation removes material from the cutting face of the wheel to generate a particular profile (or to maintain a flat profile), to minimize wheel runout, and remove lobes or irregularities. Truing is necessary if the wheel runout exceeds 0.013 mm after mounting. The dressing operation conditions the wheel surface to maintain particular cutting characteristics; it removes loaded material and glaze from a dull wheel face [4, 142] to control the surface finish of workpiece. A smooth wheel face containing dulled grains is called glazed. In some cases, the same wheel is used for rough and finish grinding a part; in these cases the coarse wheel is dressed to generate dull grains so that it can produce a finish normally produced by a fine-grained wheel.

Truing can be carried out either when the wheel is mounted on the machine spindle or on a fixed wheel-holding device. The wheel should be trued every time it is removed from its holder or, when used without a holder, removed from the spindle. For conventional wheels, truing and dressing are commonly carried out in a single process using the same tool; in this case the combined process is simply referred to as dressing. Superabrasive wheels usually require separate truing and dressing operations.

Conventional wheels may be trued using several methods, including those employing an abrasive stick, brake controlled truing device, soft steel block, or crush rolls manufactured with conventional abrasive grains. Diamond truing tools such as single- or multipoint diamond tools, diamond blocks, and diamond rolls are also common. In general, dressing tools are similar to truing tools. Dressing tools should be softer and finer than the wheel being dressed. Crush dressing can increase the free-cutting properties of the grinding wheel and therefore allows the use of a harder bond, finer grain, or denser structure than recommended for similar applications when dressing by conventional diamond methods.

Truing and dressing of diamond and CBN wheels are critical operations which strongly impact grinding effectiveness and wheel cost. The most common devices for truing and dressing superabrasive wheels are brake-controlled silicon carbide, aluminum oxide, or diamond-impregnated nibs. The truing frequency is controlled by the grinding G-ratio. Continuous-dress is used in some aggressive applications, such as creep-feed grinding using a diamond-roll dressing wheel, so that the wheel is dressed continuously as it grinds to maintain consistent edge sharpness. The traverse rate for a diamond truing tool is 250 to 500 mm/min for rough grinding and 100 to 180 mm/min for finish grinding; the rate of travel for metallic truing tools is 1000 to 1500 mm/min for rough grinding and 250 to 500 mm/min for finish grinding. Wet truing is often necessary to remove heat and abrasive grits. The depth per pass should be approximately 0.012 mm on average and less than 0.025 mm across the face of the wheel. In side truing, the traverse travel distance per wheel revolution should be approximately 0.005 mm. Further information on truing and dressing is available in the literature [136–138, 143].

4.12 MICROSIZING AND HONING TOOLS

Microsizing tools for bore finishing (Figure 4.94) hold a fixed diameter during a single stroke operation with manual or automatic size compensation. The tool is usually allowed to float on a floating holder, although two universal joints attached to the spindle can also be used. Types of microsizing tools include one-piece arbor, adjustable arbor, and auto-expansion arbor tools [144]. They generally consist of two major components: a cast iron or steel sleeve, and a corresponding tapered arbor. The sleeve, which is the cutting portion, has a barrel shape with electrochemically plated abrasives. The sleeve also has a helical slot for expansion of its diameter. Once the sleeve is mounted on the corresponding tapered arbor, the tool can be brought to size and adjusted for wear by moving the sleeve up the arbor. An auto-expansion arbor with a mechanism similar to those found in automatic adjustable boring bars (Figure 4.30) can also be used. One-piece arbor designs are recommended when other types are not available, such as for very small diameter bores. Various styles of coolant flutes, tool lengths and size adjusting features can be incorporated to improve tool performance in a particular application as illustrated in Figure 4.94. The sleeve is replaced by a multiple-stone tool (carrying a number of abrasive stones around the periphery of the arbor) for larger bores between 25 and 300 mm in diameter.

Unlike microsizing tools, honing tools are generally multistroke tools which are passed through the bore repeatedly. Common honing tool designs [145] include single-stone, multistone, and Krossgrinding tool designs. Single-stone tools have a single stone, as illustrated in Figure 4.95, and are commonly used for small bores (diameter <25 mm). Multistone tools are

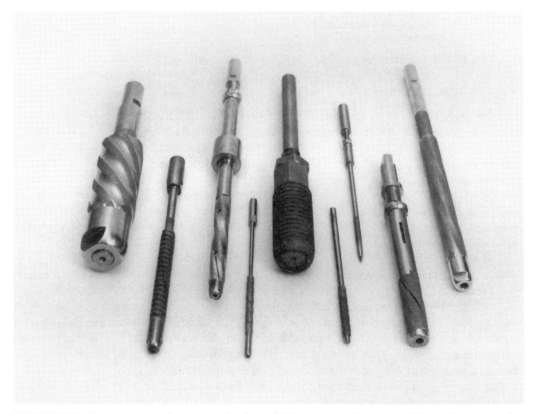

FIGURE 4.94 Various styles of superabrasive bore finishing tools. (Courtesy of Engis Corporation.)

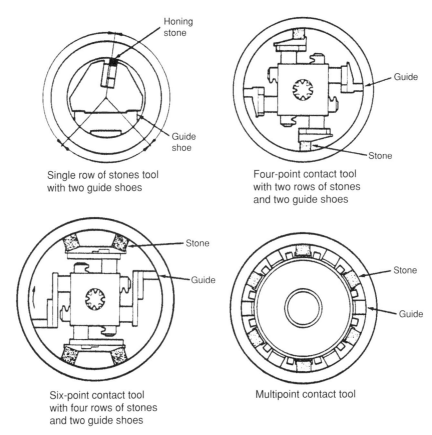

FIGURE 4.95 End view of honing tool designs. (After H. Fischer [145].)

used for larger bores (15 to 150 mm diameter) and have multiple contact points in order to provide better force distribution, better geometry, longer tool life and (possibly) faster stock removal; they are similar to large microsizing tools. Multistone tools are comprised of a number of guide shoes and stones. The arrangement of the shoes and stones on the tool periphery is very important to tool performance (Figure 4.96); they are used for large bores from 50 to 300 mm in diameter. Cone-fed expansion mechanisms (Figure 4.97) are commonly used. The angular spacing of the contact lines for the guide shoes and stones is very important. A three-point tool (two fixed guide shoes and one abrasive stone) is usually better than the four-point contact design (two fixed guide shoes and two sticks) [145, 146].

The Krossgrinding tool [145] is a multistroke superabrasive design (for small bores 10 to 32 mm diameter), which combines the best features of multistroke honing tools and single-stroke microsizing tools; it is equipped with a plated superabrasive sleeve.

Successful honing depends not only on the tool design but also on the use of the proper honing stone. The selection of the stone is based on the same principles as the selection of a grinding wheel: grit size, hardness, bond type, and abrasive material [4, 138, 146]. In honing, unlike grinding, the abrasive stones cannot be dressed, and so must be self-dressing; this requires a close control of both the reciprocation and rotational speed.

Coolant-fed tools have been used to improve the lubrication, flushing, and temperature control characteristics of the coolant. Vitrified and diamond honing abrasives in particular benefit from coolant-fed tooling.

Cutting Tools 245

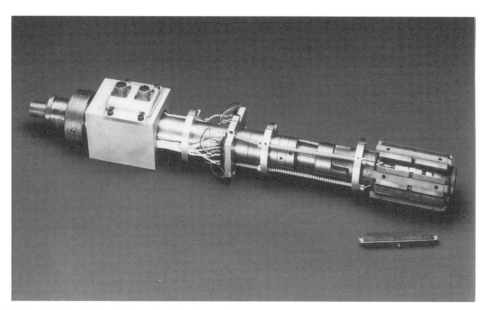

FIGURE 4.96 Typical multiple-point honing tool.

Finally, two types of abrasive brushes are used for brush honing [147]. The first is the ball type designed with clusters of abrasives attached to the ends of flexible wire filaments as shown in Figure 4.98. This tool is readily adapted to any machine tool with a rotating spindle. The second is the bristle type, made of nylon bristles impregnated with abrasives (Figure 4.98). The bristles are mounted on the same kind of honing shoes and tools as abrasive sticks; they are designed to be adjustable either manually or automatically for interference setting and tool compensation.

4.13 BURNISHING TOOLS

Roller burnishing tools consist of a series of hardened, highly polished rollers positioned in slots in a hardened mandrel or cage; they resemble roller bearings in construction. They are made either with fixed diameter, which remains constant during the operation ("interference tools"), or with a diameter that can be changed by exerting a predetermined burnishing pressure ("expander tooling") [148]. Fixed diameter tools provide tolerance control, but produce surface finishes which depend on the preburnished surface dimension and quality. Typical designs for various applications are shown in Figure 4.99.

4.14 EXAMPLES

Example 4.1 A 20 mm deep hole is drilled and tapped with a M10 × 1 mm cutting tap in a cast iron workpiece material. A solid carbide drill and a HSS tap are used. Estimate the machining time required in drilling and tapping the hole. What is the material removal rate for both the drill and tap?

Solution: The cutting speed for both tools can be selected from Figure 4.4. The cutting speed of 100 m/min for an uncoated carbide drill is typical for cast iron. The cutting speed for the HSS tap is assumed to be 30 m/min. The depth of the drilled hole must be deeper than the tapping length to provide clearance for the chips and the front taper of the tap (the first few

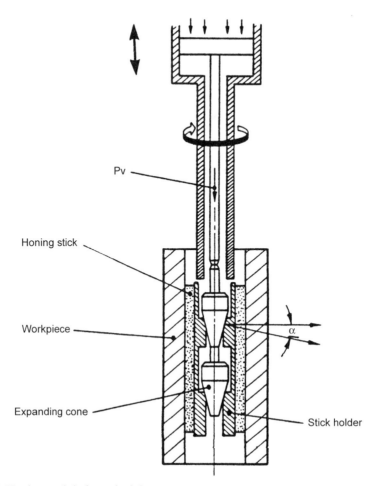

FIGURE 4.97 Honing tool design principle.

taper threads) at the bottom of the blind hole, and is assumed to be 27 mm. The spindle rpm for drilling and tapping is

$$N_d = V_d/(\pi \cdot D) = (100 \text{ m/min})(1000 \text{ mm/m})/(3.14)(9 \text{ mm}) = 3839 \text{ rpm}$$

(the drill diameter for a 10 mm tap is about 9 mm), and

$$N_t = V_t/(\pi \cdot D) = (30 \text{ m/min})(1000 \text{ mm/m})/(3.14)(10 \text{ mm}) = 955 \text{ rpm}$$

The feed for drilling cast iron is assumed to be 0.15 mm/tooth. The feed rate is then

$$f_r = n_t \cdot f \cdot N = (2)(0.15 \text{ mm/rev/tooth})(3839 \text{ rev/min}) = 1152 \text{ mm/min}$$

and for tapping, the feed rate is determined by the pitch of the tap (given from Equation 2.46). The pitch for the M10 × 1 mm tap is 1 mm

$$f_r = p \cdot N = (1 \text{ mm})(955 \text{ rev/min}) = 955 \text{ mm/min}$$

FIGURE 4.98 Brush honing tools: (a) ball, (b) bristle. (After Y.T. Lin [147].)

FIGURE 4.99 Internal and external roller burnishing tools. (Courtesy of Cogsdill Tool Products, Inc.)

The cutting time for drilling with 120° point drill is

$$t_{md} = (L_d + L_e)/f_r = (L + D/\tan \rho + \Delta L)/f_r$$
$$= (27 + 9/\tan 60° + 2)/(1152 \text{ mm/min}) = 0.030 \text{ min} = 1.78 \text{ sec}$$

and for tapping is

$$t_{mt} = (L_t + L_e)/f_r = (L + D/\tan \rho + \Delta L)/f_r$$
$$= (23 + 2)/(955 \text{ mm/min}) = 0.026 \text{ min} = 1.57 \text{ sec}$$

L_e is assumed to be 2 mm since a hole is made. The total machining time to drill and tap the hole is the sum of the drilling and tapping times plus several other tool travel times (i.e., retract the tool from the bottom of hole, approach time, etc.) discussed in Chapter 13. Let us assume that the total time is $t_T = 1.78 + 1.57 + \Delta t = (3.35 + \Delta t)$ sec.

The material removal rate for the drill, MRR, is obtained from Equation (2.15) to be

$$\text{MRR} = (\text{cross} - \text{sectional area})(\text{feed rate}) = (\pi \cdot D^2/4)f_r$$
$$= [\pi(9)^2/4](1152 \text{ mm/min}) = 73{,}287 \text{ mm}^3/\text{min}$$

The MRR for tapping is given by Equation (2.57)

$$\text{MRR} = \left(\frac{p}{4} + \frac{D_m - D_d}{\tan 60°}\right)\left(\frac{D_m - D_d}{4}\right)\left(\frac{p \cdot N}{\sin \lambda}\right) \quad \text{where } \tan \lambda = \frac{p}{\pi \cdot D} = \frac{1}{\pi \cdot 10}$$

The helix of the thread $\lambda = 1.82°$

$$\text{MRR} = \left(\frac{1}{4} + \frac{10 - 9}{\tan 60°}\right)\left(\frac{10 - 9}{4}\right)\left(\frac{1 \cdot 955}{\sin 1.82°}\right) = 3732 \text{ mm}^3/\text{min}$$

Example 4.2 Estimate the change in machining time for drilling and tapping the hole in Example 4.1 if both the drill and tap are made from HSS.

Solution: The cutting speed for the tap is the same. The cutting time for the drill will be different because the maximum cutting speed allowed for HSS drills is much lower than carbide drills as shown in Figure 4.4. A cutting speed of 30 m/min for an uncoated HSS drill is common for cast iron.

The spindle rpm for drilling is

$$N_d = V_d/(\pi \cdot D) = (30 \text{ m/min})(1000 \text{ mm/m})/(3.14)(9 \text{ mm}) = 1152 \text{ rpm}$$

The feed for drilling cast iron with HSS drill can be the same as that with carbide drill, i.e., 0.15 mm/tooth. The feed rate is then

$$f_r = n_t \cdot f \cdot N = (2)(0.15 \text{ mm/rev/tooth})(1152 \text{ rev/min}) = 346 \text{ mm/min}$$

The cutting time for drilling with 120° point drill is

$$t_{md} = (L + L_e)/f_r = (L + D/\tan \rho + \Delta L)/f_r$$
$$= (27 + 9/\tan 60° + 2)/(346 \text{ mm/min}) = 0.10 \text{ min} = 6 \text{ sec}$$

Cutting Tools 249

The total machining time to drill and tap the hole is then

$$t_{T2} = 6 + 1.57 + \Delta t = (7.57 + \Delta t) \text{ sec}$$

The change in machining time between HSS and carbide drill is

$$t_{TX} = t_{T2} - t_{T1} = (7.57 + \Delta t) - (3.35 + \Delta t) = 4.22 \text{ sec}$$

Therefore, the total machining time is reduced by 53% when a carbide drill is used instead of HSS.

Example 4.3 Evaluate the effect of using CBN inserts instead of aluminum oxide CVD-coated carbide inserts on a boring bar to finish bore a 25 mm diameter hole 50 mm deep in cast iron.

Solution: A good cutting speed for Al_2O_3 multicoating carbide inserts is 220 m/min, while for CBN inserts it is 1000 m/min for cast iron. Therefore, the carbide inserts will be run at 2800 rpm and the CBN inserts at 12,750 rpm. The *cutting time will be 5.46 times lower* for CBN inserts since the feed rate ($f_r = fN$) for the CBN bar ($f_r = [f(12,750 \text{ rpm})]$ mm/min) is 5.46 times higher than that of the carbide bar ($f_r = [f(2,800 \text{ rpm})]$ mm/min).

However, the material removal rate (MRR) with the CBN boring bar will be 5.46 times higher than that for the boring bar using carbide coated inserts. This means that the power required for boring with CBN inserts is about 5.46 times higher than the boring bar with carbide coated inserts even though the torque and force for both boring bars will be about the same. Generally, the torque required to remove the same material at higher speeds tends to be smaller than at lower speeds because the chip removal is more efficient at higher speeds. This illustrates the importance of checking the spindle power and torque availability as a function of speed as explained in Chapter 3.

Example 4.4 A 55 mm diameter hole is being bored to 59 mm diameter bore with a single point boring bar. The workpiece material is gray cast iron. A carbide insert is used in the indexable boring bar made of steel. The cutting speed for the bar is 100 m/min and the feed is 0.2 mm/rev. Two different boring bar designs (A) and (B) are used as shown in Figure 4.100.

(a) Estimate the deflection at the tool point for bar (A) due to radial force in order to determine the effective bore diameter.
(b) Check the effect of cutting conditions on improving hole size control.
(c) Calculate the hole size improvement by changing the boring bar (A) material from steel to heavy metal.
(d) Calculate the improvement on bore size by optimizing the boring bar design to (B).

The material properties are: $E_{steel} = 206,700$ MPa and $E_{heavy\ metal} = 330,000$ MPa.

Solution: In order to understand the effect of the bar design and cutting conditions on deflection, a rigidly clamped beam with a force acting at the free end is considered.

(a) In order to find the tool point deflection to determine the hole size quality, the radial cutting force acting on the boring bar should be estimated. It is known from previous work that in roughing and semifinishing operations the feed and radial forces are smaller than half of the cutting force along the speed vector ($F_r \leq 0.5 F_c$). The radial force in this problem can be equated to half the cutting force since the depth of cut is 2 mm (the worse case scenario is $F_r = F_c$). The cutting force is estimated from the torque and power required for the cut as explained in Chapter 2. Thus,

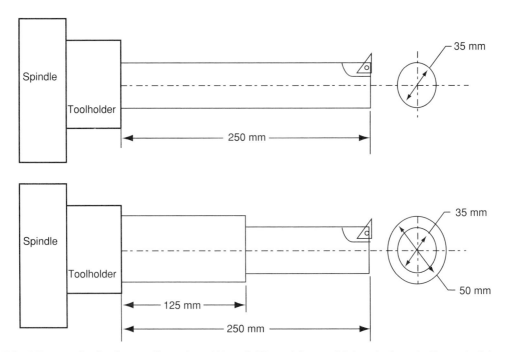

FIGURE 4.100 Boring bar configurations (A) and (B) used for machining the bore in Example 4.4.

$$T = F_c R = 2 F_r R$$

and

$$P = T\omega = 2F_r R(\pi N) = (\text{MRR})u_s \Rightarrow F_r = (\text{MRR})u_s/(4\pi RN)$$

$$N = V/(\pi D_{\text{avg}}) = (100 \text{ m/min})(1000 \text{ mm/m})/(3.14)[(59+54)/2] = 559 \text{ rpm}$$

$$\text{MRR} = \frac{\pi(D_o^2 - D_i^2)}{4} f_r = \left[\frac{\pi(59^2 - 55^2)}{4}\right](559 \text{ rpm})(0.2 \text{ mm/rev})$$
$$= 40{,}040 \text{ mm}^3/\text{min} = 40 \text{ cm}^3/\text{min}$$

$$F_r = \frac{(\text{MRR})u_s}{4\pi RN} = \frac{(40{,}040 \text{ mm}^3/\text{min})(0.06 \text{ W min/mm}^3)(\text{N m/W sec})}{4\pi\left(\frac{59+55}{4}\text{ mm}\right)(559 \text{ rpm})\left(\frac{\text{min}}{60 \text{ sec}}\right)\left(\frac{\text{m}}{1000 \text{ mm}}\right)} = 720 \text{ N}$$

The deflection at the cutting tool point for bar A is calculated using Equation (4.2)

$$\delta_A = \frac{F_r L^3}{3EI} = \frac{(720 \text{ N})(250^3 \text{ mm}^3)}{3(206{,}700 \text{ N/mm}^2)(73{,}624 \text{ mm}^4)} = 0.246 \text{ mm}$$

$$I = \frac{\pi(D_o^4 - D_i^4)}{64} = \frac{\pi(35^4)}{64} = 73{,}624 \text{ mm}^4$$

The calculated deflection is an approximation because the clamping is never absolutely rigid and it is impossible to predict the cutting and radial forces exactly. The contribution of the flexibility from the toolholder–spindle interface connection(s) is discussed in Chapter 5. The radial deflection can be compensated by offsetting the turning machine at a cutting depth equal to $D_O+2\delta$. However, the retraction marks could be a concern especially if the spring back of the bar is significant as explained in Section 4.6.

(b) The deflection of 0.246 mm is very large and the hole diameter will be smaller by 0.492 mm and will generate the scratch marks during tool retraction. Therefore, the deflection should be reduced by reducing the radial force F_r, which is proportional to the power or MRR. Hence, the cutting conditions, depth of cut, or feed, must be changed in order to reduce the force by reducing MRR. Since a large reduction on the deflection is required, both depth of cut and feed will be reduced by 50%. However, the reduction of the depth of cut by 50% will require two passes to remove the 2 mm full depth from the bore. The reduction of the feed from 0.2 to 0.1 mm/rev is acceptable but the machining time will be doubled. The reduction of the cutting conditions results in lower productivity

$$d_2 = \frac{1}{2}d_1 \quad \text{and} \quad f_2 = \frac{1}{2}f_1, \quad \text{hence } \text{MRR}_2 = \frac{1}{4}d_1 \cdot f_1 \cdot V = \frac{1}{4}\text{MRR}_1$$

Hence, the power required at the above lighter cut is 25% of the power at the original conditions. Likewise, the radial force is reduced by 75%. Thus,

$$F_{r2} = 0.5 F_{c2} = F_{r1}/4 = (720/4) = 180 \text{ N} \quad \text{and} \quad \delta_2 = \delta_1/4 = (0.246/4) = 0.062 \text{ mm}$$

(c) The effect of heavy metal boring bar on deflection is

$$\delta_{hm1} = \delta_1(E_{steel}/E_{heavy\,metal}) = 0.246(206{,}700/330{,}000) = 0.155 \text{ mm}$$

or using the 50% reduced feed and depth of cut, the deflection is reduced to

$$\delta_{hm2} = \delta_{hm1}/4 = 0.155/4 = 0.039 \text{ mm}$$

Therefore, the substitution results in a significant reduction on deflection and in better hole size control and quality.

(d) The boring bar geometry can be changed from (A) to (B) to reduce its deflection by increasing its stiffness. Bar (B) has two different diameter cross sections along its length. The front diameter is the same as bar (A) to have an effective chip clearance between the bar and the hole since it is a horizontal boring operation. The deflection of bar (B) is

$$\delta_B = F_r\left[\frac{L_1^3}{3 \cdot E_1 \cdot I_1} + \frac{L_2^3}{3 \cdot E_2 \cdot I_2} + \frac{L_1 L_2}{E_2 \cdot I_2}(L_1 + L_2)\right] \quad \text{where } L_1 = L_2 = 125 \text{ mm}$$

$E_1 = E_2 = 206{,}700$ MPa, $F_r = 720$ N, and the moments of inertia for the two cross sections are

$$I_1 = \frac{\pi D_1^4}{64} = \frac{\pi(35^4)}{64} = 73{,}624 \text{ mm}^4 \quad \text{and} \quad I_2 = \frac{\pi D_2^4}{64} = \frac{\pi(50^4)}{64} = 306{,}796 \text{ mm}^4$$

hence, the deflection for boring bar B is

$$\delta_B = 720[4.28 + 1.03 + 6.16]10^{-5} = 0.083 \text{ mm}$$

which is much lower than the deflection for bar (A). If the bar is made from heavy metal the deflection is calculated to be 0.052 mm.

Example 4.5 Consider the boring bar (A) in Example 4.4 and discuss all the factors influencing the deflection due to forces acting on the free end.

Discussion: Example 4.4 was a simple case of a real problem because the parameters considered were the bar material, diameter and overhang, and the radial force. However, the radial force (and therefore the tool/bar deflection) is also dependent on the geometry of the insert (lead angle, rake angle, corner radius, and edge preparation).

The cutting (tangential) force, F_c, will push the tool downwards and away from the center line by δ_T (vertical deflection) as shown in Figure 4.101. The radial force, F_r, will push the tool away from the workpiece δ_R in a radial direction. The tool point deflection in the direction of the cutting force will reduce the clearance angle γ (side relief angle in Figure 4.16) on the flank face of the insert as shown in Figure 4.102. Therefore, the clearance angle of the tool should be large enough to avoid contact between the flank face of the tool and the wall of the hole; this becomes very important with small diameter holes due to the small curvature of the internal diameter. As explained in Section 4.5, the selection of the rake angle α, wedge angle w, and the relief angle γ have a decisive influence on cutting efficiency and tool strength. The deflection δ_T can be compensated by positioning the cutting edge above the centerline of the workpiece in a turning machine. However, the change of the bore radius due to the δ_T is small and it is estimated for boring bar (A) in Example 4.4(a) to be

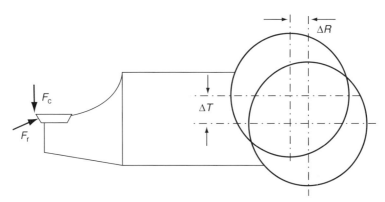

FIGURE 4.101 Free body diagram of the forces and corresponding deflections at the cutting edge of a boring bar.

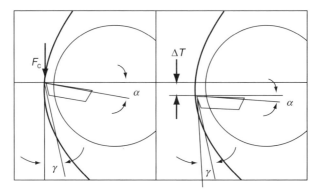

FIGURE 4.102 Free body diagram of the cutting (tangential) force and the corresponding deflection at the cutting edge of a boring bar. (After Ref [83].)

$$\delta_{Tr} = R - (R^2 - \delta_T^2)^{0.5} = \frac{59}{2} - \left[\left(\frac{59}{2}\right)^2 - (0.493)^2\right]^{0.5} = 0.004 \text{ mm}$$

The δ_{Tr} of 0.004 mm is very small compared to the radial deflection of 0.246 mm due to the radial force. Therefore, the effect of the δ_T in the bore diameter is generally neglected. In addition, this error cannot be compensated in a milling machine because the tool rotates. However, the clearance angle is reduced by an amount equal to $\gamma = \sin^{-1}(\delta_T/R)$ that is equivalent to 1° for Example 4.4(a). This means that the 7° clearance angle of the tool is reduced to 6° during cutting.

The radial force effects the radial (horizontal) deflection, which reduces the depth of cut (since the tool moves and affects the diametrical accuracy of the hole as shown in Figure 4.103 and analyzed in Example 4.4). In addition, the chip thickness will change with the varying size of the cutting forces which causes vibration and may lead to chatter. Adjusting the depth of cut to be δ_r greater than the designed depth of cut, compensates for the deflection. Likewise, the diameter of the boring bar should be adjusted to be $2\delta_r$ larger than the desired diameter. The diameter of the boring bar in Example 4.4(a) should be adjusted to 59.493 mm in order to generate the desired 59 mm bore.

The lead angle affects the feed and radial components of the cutting forces as explained in Section 4.5. If the lead angle is zero, the radial force is a function of the corner radius of the insert and the depth of cut when it is smaller than the corner radius. As the lead angle increases, the radial force component increases while the feed force decreases. Therefore, a small lead angle is better in boring even though it affects the chip thickness and the direction of the chip flow, a compromise often has to be made. It is suggested that a lead angle of 15° or less is used [83]; a 15° lead angle will generate twice the radial force that a 0° lead angle produces.

The nose radius of the insert also affects the radial force; the greater the nose radius, the greater the radial and cutting forces. The deflection of the tool in the radial direction is affected by the relationship between the nose radius and the depth of cut. The radial force increases with increased depth of cut as long as the nose radius is smaller than the depth of cut. The effect of the lead angle is present on the radial deflection only when the nose radius is smaller than the depth of cut. Therefore, it is better to select a somewhat smaller nose radius than the depth of cut.

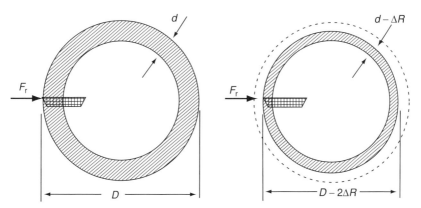

FIGURE 4.103 Free body diagram of the radial force and the corresponding radial deflection at the cutting edge of a boring bar.

Edge rounding on the primary cutting edge of the insert has a significant influence on the size of the all forces (cutting, radial, and feed forces).

Example 4.6 Optimize the static stiffness and the dynamic characteristics of the boring bar in Example 4.4(a).

Solution: The static stiffness of a solid boring bar can be increased by maximizing its diameter or by reducing its length. The static stiffness is also dependent upon the modulus of elasticity and moment of inertia as shown in Equation (4.1). The dynamic characteristics of the bar relate to its first natural frequency, the dynamic stiffness, and damping. Considering forced vibration, the amplitude of force vibration depends upon the static stiffness and the natural frequency of the bar. If the tooth passing frequency is near the natural frequency of the bar, resonance will occur and the amplitude of vibration tends to be very high. The natural frequency of vibration of a cylindrical cantilevered bar with a center hole is approximated by the equation [61]:

$$\omega = \left(\frac{3EI}{0.23mL^3}\right)^{0.5} = C_1 \left(\frac{D_o^2 + D_i^2}{L^3}\right)^{0.5}$$

The natural frequency of vibration for the boring bar is affected by the same parameters as the static stiffness and the mass of the bar. Therefore, a hollow bar could improve the dynamic characteristics since it reduces the mass of the bar. The stiffness at the end of the bar is defined by Equation (4.1) as

$$K = \frac{F}{\delta x} = \frac{3EI}{L^3} = C_2 \frac{D_o^4 - D_i^4}{L^3}$$

The rate of change of static stiffness between the solid and hollow bars is estimated as

$$\Delta K = \frac{K_s - K_h}{K_s} = \left(\frac{D_i}{D_o}\right)^4$$

The rate of change of static stiffness between the natural frequency of the solid bar and that of hollow bar is

$$\Delta \omega = \frac{(D_o^2 + D_i^2)^{0.5} - D_o}{D_o}$$

There is a trade-off between the static stiffness and the natural frequency when selecting the diameter of the center hole in the boring bar. Therefore, the solution is obtained by equating the above two equations ($\Delta K = \Delta \omega$). If the equations are plotted, the solution indicates that $\Delta K = \Delta \omega$ when the hole diameter (inner diameter) is approximately 67% of the bar diameter. Therefore, the 35 mm diameter bar in Example 4.4(a) will require a 23.5 mm hole in which case the static stiffness will be reduced by 20%:

$$\Delta K = \left(\frac{D_i}{D_o}\right)^4 = \left(\frac{23.5}{35}\right)^4 = 0.20$$

and the corresponding natural frequency of the bar will be increased by 20%. However, it is important to mention that the dynamic stiffness is also imortant in designing a boring bar.

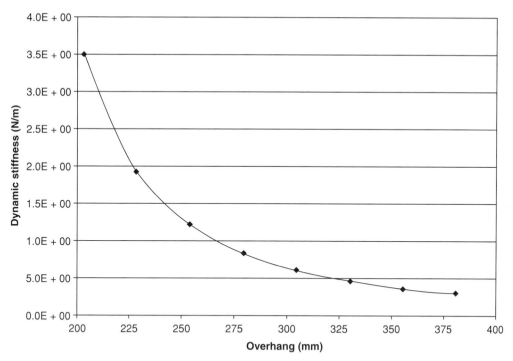

FIGURE 4.104 Dynamic stiffness of a 38.1 mm diameter boring bar. (After A. Alev and W. Eversole [149].)

For example, the dynamic stiffness of a boring bar with an internal chatter suppression system [149] is shown in Figure 4.104 for different overhangs of a 38.1 mm diameter boring bar. This boring bar has a center hole with an internal device that consists of several spring-loaded high-inertia disks as illustrated in Figure 4.105. The internal device suppresses chatter by increasing the damping of the bar. The dynamic stiffness is a function of static stiffness, natural frequency, and damping characteristics. In addition, the moment of inertia of the boring bar at the end can be increased by placing the cutting edge of the insert above the neutral axis of the bored hole without affecting the cutting tool geometry [150].

Example 4.7 A 25 mm hole is made in solid aluminum with a required size tolerance of ± 0.025 mm, location tolerance of ± 0.025 mm, and a surface finish of $R_a = 0.002$ mm. The hole depth is 50 mm. Define the process and the corresponding cutting tool(s) required to manufacture this hole.

Solution: Generally, the hole is roughed with a drill and if the required tolerances are tight, a finish operation is used such as boring, reaming, or circular interpolating (with an end mill). The L/D is 2:1 and the tolerance requirement allows any one of the above operations to be

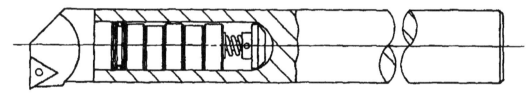

FIGURE 4.105 Illustration of a boring bar with an active vibration control. (Courtesy of Kennametal Inc.)

used. Note that, in general, reaming is applied to any L/D hole, boring for $L/D < 10$, and circular-interpolation for $L/D < 4$. In addition, boring and reaming can hold tighter tolerances than circular interpolation.

Generally, using the largest drill with the proper stock allowance for the finish pass is the fastest approach (compared to end milling) as a first operation. The second operation could be either a semifinish or a finish operation depended upon the hole quality generated by the drill. The semifinish operation is eliminated if a solid carbide drill is used with the proper chisel edge (web thinned) for aluminum. However, the semifinish operation could be either end milling (with 0.75 to 1 mm depth of cut) or reaming (with 0.3 to 0.5 mm depth of cut). The finish pass could be any one of the above three suggested operations depending on the finish hole requirements.

End milling cannot produce surface finishes equivalent to boring and reaming operations. The surface finish marks for milling are around the perimeter of the bore, while for boring/reaming they are along the bore length as explained in Chapter 10. A 15 to 20 mm three-flute end mill will generate an ideal surface finish of about 0.0012 mm. However, the contribution of the milling tool runout in the spindle is very significant because it directly affects the surface finish. If the tool runout is 0.003 to 0.010 mm, the surface finish will be at best equivalent to the runout value.

The surface finish requirement of 2 μm rms can be obtained by boring but the reaming and end milling operations are questionable. Only multiflute reamers with double margins or gun reamers can obtain consistently 2 μm finish (see Equation 10.8 in Chapter 10).

Finally, the location tolerance of ±0.025 mm can be obtained by all three operations, although meeting tolerance with a reamer will require the proper toolholder and cutting edge geometry unless the hole is semifinished with an end mill because conventional reamers tend to follow the predrilled hole location.

Example 4.8 A thin-walled aluminum part is being machined in a 50 mm thick plate using a 10 mm solid carbide end mill. The wall height and thickness are 30 and 0.4 mm, respectively. Select the process and the corresponding cutting conditions.

Solution: The wall of the part is machined from both sides using an alternating approach as shown in Figure 4.106 to reduce the deflection of the wall. The axial depth of cut of the first cut on one side is 1 mm. The second is done on the other side with an axial depth of cut of 2 mm. Cutting proceeds with 2 mm axial depth of cut until the whole 30 mm is machined. Climb (down) milling is preferable for this application with a square-shoulder end mill. Higher speeds are also preferable to reduce the contact time of the cutting edge with the wall. Therefore, the maximum spindle speed (rpm) is selected that is stable for the particular mill geometry, overhang, and cutting conditions as is explained in Chapter 12. Let assume that a 20,000 rpm spindle is used and 18,000 rpm is acceptable for the rough end mill which corresponds to a 565 m/min cutting speed, which is acceptable for aluminum with carbide tooling. A feed of 0.12 mm/tooth is selected for the two-flute mill in the roughing passes and 0.1 mm/tooth for the four-flute mill in the finishing cuts.

The axial depth of cut of 5 mm with a radial depth of cut of 6 to 10 mm is selected for the roughing cut based on the spindle capability and tool stability. The analysis of Example 2.4 is used here to evaluate the spindle power and tool deflection. The 10 mm radial depth of cut is used to initiate a groove through the solid material during the initial contact with the part. During the full cut the tool deflection is about 0.12 mm while for the remanding cuts with a 6 mm radial depth of cut the deflection is 0.07 mm.

A 2 mm axial depth of cut and 1 mm radial depth of cut are selected for the finish cut. The tool deflection in this case using a four-flute end mill is about 0.008 mm. The axial depth of cut can be increased as long as the deflection of the end mill and the wall is acceptable and the

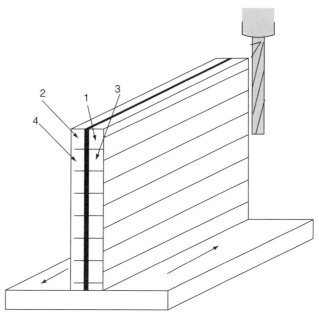

FIGURE 4.106 Illustration of the end milling tool path (process) for machining a thin wall in a part. (After Ref [84].)

tool does not chatter. The cutter diameter for the finish cut should be also reduced to 8 mm to reduce the chip thinning effect (estimated by Equation 2.21 in Chapter 2).

Example 4.9 Machine a groove that is closed at both ends. The dimensions of the finished groove are: 25 mm wide by 200 mm long by 6 mm deep. The workpiece material is low-alloy steel with hardness 180 HB. Select the process and the corresponding cutting conditions.

Solution: This operation requires an end mill rather than a side and face mill because the groove is closed at least at one end. An end mill with center cutting (having overlapping cutting edges on the axial end face) allows the end mill to be fed axially (as a drill) into the workpiece for a small depth of 6 mm in this case. A solid carbide or indexable end mill can be selected for this operation. There are three alternatives with respect to the cutter pitch or number of cutting edges: two-flute (coarse pitch), three-flute, and four-flute (dense pitch). The two-flute cutter has only one tooth in cut, while the three- and four-flute cutters have two cutting edges in cut. The general rule is to use less flutes for deeper cuts, with four or higher flutes for light cuts (depth $<0.2D$). The estimation of the cutting forces, power, and stability will help in the decision of the cutter selection (more detailed discussion on this is given in Example 8.1 to Example 8.3 in Chapter 8). In this case, a three-flute end mill should be used in order to increase the cutting edge engagement to avoid chip thinning that may cause vibration especially when extended tool overhang is used to reach in the slot. In addition, the three-flute cutter ensures good chip evacuation out of the groove.

The cutting edge geometry is either a zero or 30° helix indexable end mill with a corner radius of 0.75 mm in the inserts, or an equivalent solid carbide end mill. The down milling process is used to achieve the most favorable cutting action. A feed of 0.12 mm/tooth and the cutting speed of 200 m/min are selected based on tooling manufacturer recommendations for coated inserts or a coated solid carbide end mill. The cutter diameter is 25 mm for a conventional slot milling operation. If the slot is generated through circular interpolation, a 13 mm end mill with three flutes should be used instead.

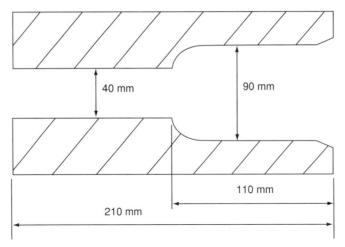

FIGURE 4.107 Illustration of the cross section of a step hole for boring the 40 mm hole in Problem 4.2.

4.15 PROBLEMS

Problem 4.1 A 40 mm diameter by 100 mm long hole (shown in Figure 4.107) is being bored to 42 mm diameter with a single-point boring bar. The bar should enter from the 90 mm cavity due to the complexity of the part. The workpiece material is medium-carbon steel. A carbide insert is used in the indexable boring bar made of steel or heavy metal. The boring bar is integral to a CAT-50 toolholder and extends 220 mm. The cross section of the solid boring bar is cylindrical with 30 mm diameter. The cutting speed for the bar is 80 m/min while the feed is 0.2 mm/rev. Design a boring bar and minimize the deflection as much as possible. What should be the size of the boring bar in order to obtain a bore with tolerance of ± 0.05 mm?

Problem 4.2 A 100 mm × 100 mm × 35 mm deep cavity is machined in steel with a hardness of 46 HRC. It is suggested to use rough, semifinish, and finish passes to finish the cavity. A 25 mm diameter two-flute ballnose end mill is used for the rough cut with an axial depth of cut of 3 mm and a radial depth of cut of 10 mm. A 20 mm diameter two-flute ballnose end mill is used for the semifinish cut with an axial depth of cut of 0.8 mm and a radial depth of cut of 0.05 mm. A 12 mm diameter four-flute ballnose end mill is used for the finish cut with an axial depth of cut of 0.3 mm and a radial depth of cut of 0.002 mm. Select the cutting conditions and estimate the machining time for each tool.

REFERENCES

1. B.M. Kramer, On tool material for high speed machining, *ASME J. Eng. Ind.* **109** (1987) 87–91
2. S.J. Burden, J. Hong, J.W. Rue, and C.L. Stromsborg, Comparison of hot-isostatically-pressed and uniaxially hot-pressed alumina–titanium–carbide cutting tools, *Ceram. Bull.* **67** (1988) 1003–1005
3. R. Komanduri and J.D. Desai, Tool Materials for Machining, Technical Information Series, Report No. 82CRD220, General Electric, August 1982
4. T.J. Drozda and C. Wick, *Tool and Manufacturing Engineers Handbook*, Vol. I — Machining, SME, Dearborn, MI, 1988, Chapter 1
5. R. Edwards, *Cutting Tools*, Institute of Materials, London, Ashgate Publishing Co., Brookfield, VT 1993

6. K.H. Smith, Whisker-reinforced ceramic composite cutting tools, *Carbide Tool J.* **8**:5 (1986) 8–11
7. W.W. Gruss, Ceramic tools improve cutting performance, *Ceram. Bull.* **67** (1988) 993–996
8. P.N. Tomlinson and R.J. Wedlake, The current status of diamond and cubic boron nitride composites, *Proceedings of the Efficient Metal Forming and Machining Symposium*, Pretoria, South Africa, November 1982
9. I.E. Clark and J. Hoffmann, PCD tooling in the automotive industry, *Diamond & CBN Ultrahard Materials Symposium*, IDA, September 29–30, 1993, 115–130
10. M. Deren, P/M edge — powder-metal end mills: the tougher roughers, *Cutting Tool Eng.* **55**:1 (2003) 52–55
11. W.W. Gruss, Turning and milling of steel with cermet cutting tools, *Proceedings of the ASM Tool Materials for High-Speed Machining Symposium*, Scottsdale, AZ, 1987, 49–62
12. A. Richards, Cermets: a machinist's tooling solution, *MoldMaking Technol.* **May** (2004) 19–22
13. D. Bordui, Hard-part machining with ceramic inserts, *Ceram. Bull.* **67** (1988) 998–1001
14. J.G. Baldodi and S.T. Buljan, Silicon Nitride Based Ceramic Cutting Tools, SME Technical Paper MR86-912, 1986
15. V.K. Sarin and S.T. Buljan, Coated ceramic cutting tools, *Proceedings of the ASM High Productivity Machining Materials Symposium*, New Orleans, 1985, 105–111
16. E.D. Whitney and P.N. Vaidyanathan, Engineered ceramics for high speed machining, *Proceedings of the ASM Tool Materials for High-Speed Machining Symposium*, Scottsdale, AZ, 1987, 77–82
17. P.K. Mehrotra, Productivity improvement in machining by applying ceramic cutting tool materials, *A Systems Approach to Machining*, ASM, Materials Park, OH, 1993, 15–20
18. D. Novak, A better braze, a comparison of two processes for brazing PCBN onto carbide, *Cutting Tool Eng.* **56**:6 (2004) 34–37
19. K.R. Pontius, New veins of application, *Cutting Tool Eng.* **54**:8 (2002)
20. P.K. Bossom and J. Hoffmann, Turning and milling with PCBN in the automotive industry, *Diamond & CBN Ultrahard Materials Symposium*, IDA, September 29–30, 1993, 99–114
21. P.T. Heath, Structure, properties and applications of polycrystalline cubic boron nitride, *Proceedings of the Superabrasives '85 Symposium*, Chicago, IL, 1985, 10–47
22. T.A. Notter, P.J. Heath, and K. Steinmetz, Amborite for machining hard ferrous materials, *Proceedings of the Advances in Ultrahard Materials Application Technology*, Volume Two, 1983, 51–70
23. Y. Kohno, T. Uchida, and A. Hara, New Applications of Polycrystalline CBN, SME Technical Paper MR85-283, 1985
24. T.J. Broskea, High speed machining of cast iron with polycrystalline cubic boron nitride, *Proceedings of the Tool Materials for High-Speed Machining Symposium*, Scottsdale, AZ, 1987, 39–47
25. M.S. Deming, B.A. Young, and D.A. Ratliff, "Turning Gray Cast Iron with PCBN Cutting Tools," *Diamond & CBN Ultrahard Materials Symposium*, IDA, September 29–30, 1993, 131–142
26. T.J. Broskea, New Applications of Polycrystalline Diamond in the Automotive Industry, SME Technical Paper MR91-175, 1991
27. T.J. Broskea, Superabrasives use in engine block production, *Diamond & CBN Ultrahard Materials Symposium*, IDA, September 29–30, 1993, 143–152
28. A. Richter, A mixed grade — the latest grades of PCD mix coarse particles to increase packing density and yield a more continuous cutting edge, *Cutting Tool Eng.* **56**:6 (2004) 28–33
29. C. Subramanian and K.N. Strafford, Review of multicomponent and multilayer coatings for tribological applications, *Wear* **165** (1993) 85–95
30. Anon Tooling and workholding, *Manuf. Eng.* **133**:2 (2004) 95–110
31. M.J. McCabe, How PVD coatings can improve high speed machining, High Speed Machining Conference, SME, Northbrook, IL, September 20–21, 2001
32. M. Olbrantz, Composition matters — new equipment allows deposition of hard and tough nanocomposite tool coatings, *Cutting Tool Eng.* **55**:12 (2003) 39–41
33. K.J.A. Brooks, *Word Directory and Handbook of Hard Metals and Hard Materials*, 5th Edition, International Carbide Data, United Kingdom, 1992/1993
34. J. Destefani, Cutting tools coatings, *Manuf. Eng.* **129**:4 (2002) 47–56
35. D.G. Bhat and P.F. Woerner, Coatings for cutting tools, *J. Met.* **February** (1986) 68–69

36. B.M. Kramer, Requirements for wear-resistant coatings, *Proceedings of the International Conference on Metallurgical Coatings Symposium*, San Diego, CA, April 18–22, 1983.
37. R. Horsfall and R. Fontana, TiAlN coatings beat the heat, *Cutting Tool Eng.* **45**:1 (1993) 37–42
38. J.R. Coleman, Make money with multicoats, *Manuf. Eng.* **104** (1990) 38–42
39. W. Koenig, R. Fritsch, and D. Kammermeier, New approaches to characterizing the performance of coated cutting tools, *CIRP Ann.* **41** (1992) 49–54
40. T. Leyendecker, O. Lemmer, S. Esser, and J. Ebberink, The development of the CVD coating TiAlN as a commercial coating for cutting tools, *Surf. Coat. Technol.* **48** (1991) 175–178
41. W. Koenig and D. Kammermeier, New ways towards better exploitation of physical vapour deposition coatings, *Surf. Coat. Technol.* **54/55** (1992) 470–475
42. W. Koenig and R. Fritsch, Performance and wear of carbide coated by physical vapour deposition in interrupted cutting, *Surf. Coat. Technol.* **54/55** (1992) 453–458
43. S.E. Franklin and J. Beuger, A comparison of the tribological behavior of several wear-resistant coatings, *Surf. Coat. Technol.* **54/55** (1992) 459–465
44. O. Knotek, F. Loffler, and G. Kramer, Multicomponent and multilayer physically vapor deposited coatings for cutting tools, *Surf. Coat. Technol.* **54/55** (1992) 241–248.
45. O. Knotek, F. Juhgblut, and R. Breidenbach, Magnetron-sputtered superhard coatings within the system Ti–B–C–N, *Vacuum* **41** (1990) 2184–2186
46. R.R. Irving, How ion implantation is extending tool life, *Iron Age*, March 5 (1984) 28–33
47. A. Richter, Top coat — molybdenum disulfide tool coatings improve the machining of aluminum, titanium and nickel-base alloys — even when dry, *Cutting Tool Eng.* **55**:12 (2003) 36–38
48. R. Dzierwa, Slippery when blue — a coating system that combines a patented surface impingement process and tungsten disulfide prolongs cutting tool life, but does it promote faster speeds? *Cutting Tool Eng.* **56**:1 (2003) 36–41
49. K.E. Spear, Diamond — ceramic coating of the future, *J. Am. Ceram. Soc.* **72** (1989) 171–191
50. K. Ito, Diamond coated cutting insert by using carbon monoxide as source gas, *Gorham Advanced Materials Institute International Conference on Diamond and DLC Coatings*, Marco Island, FL, October 15–17, 1989
51. Norton DiamaTorr Diamond Coated Cutting Tool Catalog, 1992
52. P.M. Noaker, Diamond thin film, *Manuf. Eng.* **July** (1993) 63–65
53. C.H. Shen, Machining performance of thin diamond coated inserts on 390 Al, *Trans. NAMRI/SME*, 1994
54. C.H. Shen, Performance and applications of CVD thin film diamond coated tools, in: I.M. Low, Ed., *Advanced Ceramic Tools for Machining Applications — III*, Trans Tech Publications, Switzerland, 1998, Chapter 2
55. *Basic Nomenclature and Definitions for Single-Point Cutting Tools*, American National Standard ANSI B94.50, 1975
56. *Indexable Inserts — Identification System*, American National Standard ANSI B212.4, 1986
57. D.O. Wood and R.E. King, *Modern Metal Cutting: Turning Theory*, Chapter 1: Turning and Boring Angles and Applications, SME, Dearborn, MI, 1985, 3–26
58. J. Destefani, Cutting tools geometries, *Manuf. Eng.* **133**:5 (2002) 41–49
59. G.T. Smith, *Advanced Machining: The Handbook of Cutting Technology*, IFS Publications Ltd., Kempston, UK and Springer Verlag, New York, 1989
60. J.D. Agnew, Edge preparation: its importance and how, *Tooling Prod.* **38**:11 (1973) 42–44
61. B. Kennedy, A better edge — proper edge preparation improves tool performance, *Cutting Tool Eng.* **56**:2 (2004) 54–59
62. R.J. Schimmel, J. Manjunathaiah, and W.J. Endres, Edge radius variability and force measurement considerations, *ASME J. Manuf. Sci. Eng.* **122** (2000) 590–593
63. L. Briese, Rotary Tool Cutting Cartridge, European Patent Number 0038923, Publ. 6-26-85
64. Komanduri R.H. Ettinger, M.P. Casey, and W.R. Reed., Dynamically Stiffened Rotary Tool System, U.S. Patent Number 4,515,047, May 7, 1985
65. Takama and T. Hayase., Rotary Cutting Tool, U.S. Patent Number 5,505,568, April 9, 1996
66. J. Rowe, The lowdown on laydown inserts — understanding the benefits of a laydown threading system, *Cutting Tool Eng.* **54**:10 (2002) 45–48

67. M.D. Thomas, W.A. Knight, and M.M. Sadek, Comparative dynamic performance of boring bars, *Proceedings of the 11th International MTDR Conference*, Pergamon Press, Oxford, 1970
68. A. Ruud, R. Karlsen, K. Sorby, and C. Richt, Minimizing vibration tendencies in machining, *MMS Online*, March 2004
69. Tuned Boring Bars, Kennametal Inc., Catalog 1010, 745
70. M. Kemmerling, Active damping of a boring bar, *Industrie-Anzeiger* **107**:63 (1985) 42–43 (in German)
71. E.I. Rivin, Chatter-resistant cantilever boring bar," *Proc. NAMRC* **11** (1983) 403–407
72. E.I. Rivin and H. Kang, Improving cutting performance by using boring bar with torsionally compliant head, *Proc. NAMRC* **18** (1990) 230–236
73. E.I. Rivin, Structural optimization of cantilever mechanical elements, *ASME J. Vibr. Acoust. Stress Reliab. Des.* **108** (1986) 427–433
74. H.M. Weiss, et al., Metal Boring With Self-Propelled Rotary Cutters, U.S. Patent Number 6,073,524, June 13, 2000
75. P. Szuba, et al., Boring Tool With Staggered Rotary Cutting Inserts, U.S. Patent Number 6,135,680, October 24, 2000
76. E.L. Sorice, Selecting milling cutters, in: B.K. Lambert, Ed., *Milling Methods and Machines*, SME Dearborn, MI, 1982, 62–63
77. V.J. Tipnis, The working face mill, in: B.K. Lambert, Ed., *Milling Methods and Machines*, SME, Dearborn, MI, 1982, 64–67
78. V.J. Tipnis, The versatile end mill, in: B.K. Lambert, Ed., *Milling Methods and Machines*, SME, Dearborn, MI, 1982, 68–73
79. E.M. Sautel, How to select and use tools for milling slots. Parts I and II, in: B.K. Lambert, Ed., *Milling Methods and Machines*, SME, Dearborn, MI, 1982, 90–95
80. D. Smith, Tips on how to do a better job of finish milling. Parts I and II, in: B.K. Lambert, Ed., *Milling Methods and Machines*, SME, Dearborn, MI, 1982, 96–103
81. P.C. Miller, Insert update: the right tool boosts profits, *Tooling Prod.* **February** (1989) 50–53
82. Milling Cutters and End Mills, American National Standard, ANSI B94.19, 1985
83. *Modern Metal Cutting — A Practical Handbook*, Sandvik Coromant, Fair Lawn, NJ, 1996
84. *Productive Metal Cutting*, Sandvik Coromant, Fair Lawn, NJ, 1998
85. Tooling University Geometries at work — optimize edge and insert placement in facemill and endmill cutters, *Cutting Tool Eng.* **55**:5 (2003) 56–63
86. More chips, less chatter, in: B.K. Lambert, Ed., *Milling Methods and Machines*, SME, Dearborn, MI, 1982, 104–105
87. *Twist Drills*, American National Standard, ANSI B94.11, 1979
88. K. Narasimha, M.O.M. Osman, S. Chandrashekhar, and J. Frazao, An investigation into the influence of helix angle on the torque–thrust coupling effect in twist drills, *Int. J. Adv. Manuf. Technol.* **2** (1987) 91–105
89. T.R. Chandrupatla and W. Webster, Effect of drill geometry on the deformation of a twist drill, *Proceedings of the International MTDR Conference*, 1985, 231–235
90. M.C. Shaw and C.J. Oxford Jr., On the drilling of metals 2 — the torque and thrust in drilling, *ASME Trans.* **79** (1957) 139–148
91. D.F. Galloway, Some experiments on the influence of various factors on drill performance, *ASME Trans.* **79** (1957) 191–231
92. National Twist Drill and Tool Company, Rigidity of twist drills, *Met. Cuttings* **10**:3 (1962)
93. J.S. Agapiou, Design characteristics of new types of drill and evaluation of their performance drilling cast iron — I. Drills with four major cutting edges, *Int. J. Mach. Tools Manuf.* **33** (1993) 321–341
94. J.S. Agapiou, Design characteristics of new types of drill and evaluation of their performance drilling cast iron — II. Drills with three major cutting edges, *Int. J. Mach. Tools Manuf.* **33** (1993) 343–365
95. M. Plankey, When the chips are down, *Cutting Tool Eng.* **55**:6 (2003)
96. J.S. Agapiou, An evaluation of advanced drill body and point geometries in drilling cast iron, *Trans. NAMRI/SME* **19** (1991) 79–89

97. J. Skoglund, Carbide drills: the answer to high-productivity steel drilling, *Cutting Tool Eng.* **February** (1990) 35–37
98. A. Maier, Multigroove Drill Bit with Angled Frontal Ridges, U.S. Patent Number 4,594,034, June 10, 1986
99. A. Maier, Multiple-Tooth Drill Bit, U.S. Patent Number 4,645,389, February 24, 1987
100. J.S. Agapiou, Four Flute Center Cutting Drill, U.S. Patent Number 5,173,014, December 22, 1992
101. J. Agapiou, High speed drilling of aluminum workpiece material, *High Speed Machining 2003 Technical Conference*, Chicago, IL, April 7–9, 2003
102. J. Agapiou, High speed drilling of gray cast iron workpiece material, *High Speed Machining 2003 Technical Conference*, Chicago, IL, April 7–9, 2003
103. F. Mason, Whatever happened to the chisel edge, *Am. Machinist Automat. Manuf.* **February** (1988) 49–52
104. H. Ernst and W.A. Haggerty, The spiral point drill — a new concept in drill point geometry, *ASME Trans.* **80** (1958) 1059–1072
105. H.T. Huang, C.I. Weng, and C.K. Chen, Analysis of clearance and rake angles along cutting edge for multifacet drills (MFD), *ASME J. Eng. Ind.* **116** (1994) 8–16
106. F. Fiesselmann, High-performance drilling, *Cutting Tool Eng.* **February** (1988) 61–64
107. S.M. Wu, Multifacet drills, in: R.I. King, Ed., *Handbook of High-Speed Machining Technology*, Chapman and Hall, New York, 1985, Chapter 13
108. M. Deren, Check the index — indexable drills allow higher speed and feed rates, *Cutting Tool Eng.* **54**:9 (2002)
109. M. O'Donoghue, The simple subland solution, *Cutting Tool Eng.* **February** (1994) 35–39
110. A. Alongi, BTA systems — finish what they start, *Cutting Tool Eng.* **October** (1993) 32–36
111. T. Yakamavich, Deep hole drilling — now it is faster, easier, *Automation* **January** (1991) 30–31
112. H.J. Swinehart, *Gundrilling, Trepanning, and Deep Hole Machining*, ASTME, Dearborn, MI, 1967
113. B. Kennedy, The skinny on microdrills — the correct application of microdrills requires attention to numerous operational parameters, *Cutting Tool Eng.* **55**:1 (2003)
114. C. Lin, S.K. Kang, and K.F. Ehmann, Planar micro-drill point design and grinding methods, *Trans. NAMRI/SME* **20** (1992) 173
115. L. Jung and J. Ni, Prediction of coolant pressure and volume flow rate in the gundrilling process, *ASME J. Manuf. Sci. Eng.* **125** (2004) 696–702
116. G.S. Antoun, The science of high-pressure coolant, *Prod. Machining* **July/August** (2004) 36–39
117. S.J. Lee, K.F. Ehmann, and S.M. Wu, An analysis of the drill wandering motion, *ASME J. Eng. Ind.* **109** (1987) 297
118. H. Fuji, E. Marui, and S. Ema, Whirling vibration in drilling. Parts 1, 2, and 3, *ASME J. Eng. Ind.* **108** (1986) 157
119. *Reamers*, American National Standard ANSI B94.2, 1993
120. M. Rubemeyer, Reaming right — a guide to reaming holes cost-effectively, *Cutting Tool Eng.* **56**:1 (2004)
121. K. Sakuma and H. Kiyota, Hole accuracy with carbide-tipped reamers (1st report) — behavior of tool and its effect on multicornered profile of holes, *Bull. Jpn. Soc. Precision Eng.* **19**:2 (1985) 89–95
122. K. Sakuma and H. Kiyota, Hole accuracy with carbide-tipped reamers (2st report) — effect of alignment error of pre-bored hole on reaming action, *Bull. Jpn. Soc. Precision Eng.* **20**:2 (1986) 103–108
123. J.S. Agapiou, Cutting tool strategies for multi-functional part configurations. Part II. Discussion of experimental and analytical results, *Int. J. Adv. Manuf. Technol.* **7** (1992) 70–78
124. W. Verfurth, Taps and tapping, *Modern Machine Shop*, October 1980
125. *Taps — Cut and Ground Threads*, ASME/ANSI Standard B94.9, 1987
126. J.S. Agapiou and C.H. Shen, High speed tapping of 319 aluminum alloy, *Trans. NAMRI/SME*, **20** (1992)
127. G. Lorenz, On tapping torque and tap geometry, *CIRP Ann.* **29** (1980) 1–14
128. G. Lorenz, A study on the effect of tap geometry, *Mech. Eng. Trans. Inst. Eng. Aust.* **29** (1978) 1–4
129. ISO General Purpose Metric Screw Threads — Tolerances — Deviations for Constructional Threads, International Standard ISO 965/III, 1973 (E)

130. A. Richter, Down to size — there is an alternative to threading small-diameter holes with taps using a thread mill, *Cutting Tool Eng.* **55**:2 (2003)
131. J. Sullivan, Choosing the right grinding wheel, *Modern Machine Shop*, December 2000
132. E. Galen, Superabrasive grinding: why bond selection matters, *Manuf. Eng.* **126**:2 (2001) 80–88.
133. M.A. Leitheiser and H.G. Sowman, Non-fused Aluminum Oxide-based Abrasive Mineral, U.S. Patent Number 4,314,827, February 9, 1982
134. J. Jablonowski, Enter the vitrified CBNs, *Machining Source Book*, ASM International, Materials Park, OH, 1988, 171–173
135. J. Cleason, CNC + CBN = formula for high grinding productivity, *Machining Source Book*, ASM International, Materials Park, OH, 1988, 174–177
136. G.G. Rooney, Choosing wheels for today's jig grinding, *Machining Source Book*, ASM International, Materials Park, OH, 1988, 184–188
137. S.F. Krar and E. Ratterman, *Superabrasives: Grinding and Machining With CBN and Diamond*, Glencoe/McGraw-Hill, Westerville, OH, 1990
138. *Machining Data Handbook*, 3rd Edition, Vol. 2, Machinability Data Center, Cincinnati, OH, 1980
139. High Porosity Grinding up Output, Avoids Burning, Saint-Gobain Abrasives, Manufacturingtalk, December 8, 2004
140. R.L. Mckee, *Machining With Abrasives*, Van Nostrand Reinhold Company, New York, 1982
141. J. Schwarz, Precision grinding: tips on selecting the right wheel, *Machining Source Book*, ASM International, Materials Park, OH, 1988, 168–170
142. Making for Identifying Grinding Wheels and Other Bonded Abrasives, American National Standard ANSI B74.13, 1982
143. S. Malkin, *Grinding Technology — Theory and Applications of Machining With Abrasives*, Ellis Horwood Limited, Chichester, U.K., 1989
144. R. Marvin, Achieving maximum effectiveness with single pass superabrasive bore finishing, *International Honing Clinic*, SME, Dearborn, MI, April 1992
145. H. Fischer, Tutorial, *International Honing Clinic*, SME, Dearborn, MI, April 1994
146. P.P. Bose, Honing with CBN, *Am. Machinist Automat. Manuf.* **January** (1988) 66–69
147. Y.T. Lin, Honing with abrasive brushes, *International Honing Clinic*, SME, Dearborn, MI, April 1992
148. C. Wick and R.F. Veilleux, Roller and ball finishing/burnishing, *Tool and Manufacturing Engineers Handbook*, Vol. III: Materials, Finishing and Coating, SME, Dearborn, MI, 1980, Chapter 16
149. A. Alev and W. Eversole, Design and Devices for Chatter Free Boring Bars, ASTME Technical Paper MR69-266, 1969
150. B. Berdichevsky, Method of Increasing Rigidity of Boring Bars for Boring Small Holes, SME Technical Paper TE81-187, 1981

5 Toolholders and Workholders

5.1 INTRODUCTION

The design and structural properties of toolholders and workholders have a strong influence on machining cost, accuracy, and stability.

In considering accuracy, the entire tooling structure must be taken into account. The tooling structure is composed of the tooling itself and the tool–machine interface. The tooling may be a solid structure or a composite structure composed of jointed elements; cutting tools may either be mounted directly to the spindle/turret, or may be connected through an adapter (arbor or toolholder). The spindle connection for a toolholder (or integral cutting tool) is the foundation that supports the cutting edge in any rotating or stationary tool machining system, and may take a variety of forms. The toolholder is often the weakest link in the machining system which has limited, in some cases, full utilization of the potential of advanced cutting tool materials [1, 2].

There are several factors influencing the design of a connection as indicated in Figure 5.1. No single style of toolholder or cutting tool–toolholder interface is superior for all applications. Each one of the interfaces can perform well in particular applications depending on the performance requirements. Great care must be taken to ensure that the right toolholder is chosen for a particular job. All three toolholder–device components (the tool, toolholder, and tool–machine connection) must be given equal consideration.

Similarly, the fixture supports, locates, and constrains the part during machining and has a strong influence on the rigidity and dynamic characteristics of the part–fixture–machine tool structure.

This chapter discusses toolholding and fixturing methods. Section 5.2 describes toolholding systems. Section 5.3 and Section 5.4 describe toolholder–spindle connections and tool clamping systems. Section 5.5 discusses balancing toolholders. Fixtures are described briefly in Section 5.6.

5.2 TOOLHOLDING SYSTEMS

5.2.1 General

Tool–toolholder and toolholder–spindle interfaces are the keys to achieve high-performance and high-throughput machining operations in the advanced machines and cutting tools. Toolholder quality, dimensional tolerances, and axial alignment vary broadly from manufacturer to manufacturer. This is unacceptable in many cases; if these interfaces are manufactured improperly or worn (e.g., out of tolerance or improperly fitting, resulting in toolholder tilting and out-of-roundness), the performance of the cutting operation degrades, resulting in poor accuracy, repeatability, rigidity, and tool life.

The important structural and dynamic characteristics of a tooling structure interface are the manufacturing tolerances, static and dynamic runout, radial and axial positioning

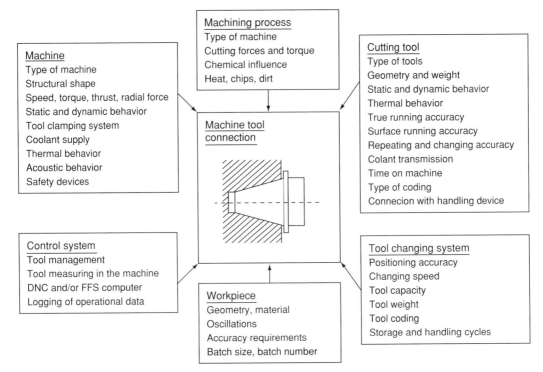

FIGURE 5.1 Factors influencing the design of the toolholder and spindle connection.

accuracy and repeatability, connection rigidity (static and dynamic stiffness), force transmission capability, momentum and torque characteristics, clamping forces, balance requirements, fatigue life and durability, retention force requirements, safety, locking/unlocking forces, coolant capability, ease of connection and disconnection, chemical and thermal stability, maintenance requirements, sensitivity to contamination, and cost. Other aspects to be addressed include the efficiency of the connection over a long period, tool presetting requirements, and location of a data-carrying module.

The cutting tool body and toolholder are made either in one solid piece (a monolithic tool) or as a mechanically connected modular system (Figure 5.2). The tooling structure is composed of attachment devices for the cutting inserts or the tooling itself (Figure 5.3a and b). Integral toolholders are used: (a) in dedicated machines and transfer lines which produce components that will not change; (b) for tools that recur in several tooling setups, such as face mill arbors and end mill holders of fixed gage length; (c) when runout is very critical; and (d) for tooling packages for which cost is the primary consideration. Integral tooling is not versatile because it can be used only for specific applications; it is preferred when the part features in a family of components are commonized so that such tools can be applied effectively. In addition, in the event of a crash, the whole tool and holder may need to be replaced. Composite or modular systems are preferred in small lot production and other applications requiring flexibility without excessive inventory. In the event of a crash, composite/modular systems require replacement of only the damaged component or adapter. However, modular tooling provides a weaker mounting system than an integral tool, since each additional joint or interface reduces stiffness. It is also extremely difficult to achieve good alignment with multiple interfaces, especially when the tolerance requirement is high. The more joints or interfaces in a system, the weaker and less accurate it becomes.

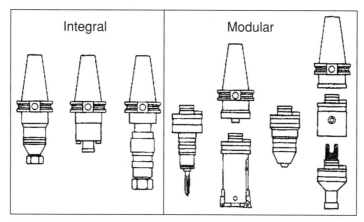

FIGURE 5.2 Comparison between monolithic (solid) and modular toolholding systems. (Courtesy of Sandvik Coromant.)

Toolholders for turning and machining centers have undergone great changes in the past few decades. The range available includes conventional square-shanked toolholders (Figure 5.4), round-shanked boring bars, and Morse taper round tools loaded manually into a turning center, as well as quick-change toolholders that are loaded either manually or automatically. Changes in toolholders for machining centers have been even more pronounced because additional design requirements have to be met for rotary toolholders. Today, the end user is confronted with an enormous challenge in choosing an optimum tooling system, since there are currently more than 30 different modular and quick disconnect systems on the market.

5.2.2 Modular and Quick-Change Toolholding Systems

Modular holding systems, shown in Figure 5.2–Figure 5.6, consist of stationary or rotating adapters in a variety of configurations to fit various machines with a common coupling. There are several major types of connections with respect to centering and locating characteristics as shown in Figure 5.7. These include different types of cylindrical shafts including single and multiple cylinders, face and nonface contacts, as well as different types of tapers and taper/face contact systems. There are also several designs of connections with respect to torque transmission, such as polygon, straight and spiral gear designs, the more conventional key and pin drive methods (Figure 5.8), and mounting bolt patterns or draw bars. Cylindrical shaft forms and both straight and spiral gear designs or pin configurations for torque transmission are primarily used as interfaces in manual or semiautomatic holding systems because the male and female portions of these connections are difficult to assemble automatically. Some toolholders are designed with a flange ahead of the toolholder shank to provide high positioning accuracy.

Modular components can be assembled in different configurations to make a variety of tooling systems that can be shared by multiple machines. Several assembled modular toolholder systems for boring, drilling, reaming, tapping, etc., tooling are shown in Figure 5.9. Rotating toolholders, including collet chucks, tap chucks, various milling adapters, and rotating boring tools with a corresponding coupling, can be used on any machining center. Extension and reduction adapters facilitate assembling tools to the required gage length. Modularity reduces lead-time due to the speed with which new tools can be built from standard components, and improves tool utilization, management, and standardization. The interfaces should be designed/selected so that the decrement in stiffness due to an additional connection is not so significant as to counterbalance the benefits of such designs.

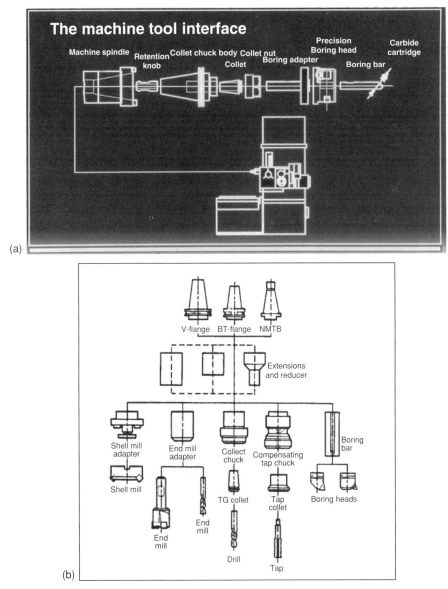

FIGURE 5.3 (a) Modular/flexible tooling system. (Courtesy of TSD.) (b) Modular machining center tooling system. (Courtesy of Kennametal, Inc.)

Generally, modular toolholding connections can provide precision equivalent to the H8/H9 ISO-tolerance class. The standard industry tolerances for male and female adapters are +0.0 to −0.013 and +0.013 to −0.0 mm, respectively. However, some manufacturers guarantee +0.0 to −0.004 and +0.004 to −0.0 mm tolerances, respectively for the male and female adapters, which results in a worst case eccentricity of 0.008 mm at the front of the adapter with total indicated runout (TIR) repeatability of ±0.008 mm considering a single connection. The runout is a function of the tool axis misalignment with reference to

FIGURE 5.4 Manual tool system used in lathes. (Courtesy of Valenite, Inc.)

the spindle axis of rotation and is dependent on the spindle accuracy, the squareness of the connecting faces (for face contact systems) with respect to their centerline of rotation and mating surfaces for the male pilot. The runout is also a function of the total tool length, which may be comprised of one or more spacer adapters. Tilting errors are much more critical than eccentricity errors for extended tools (made of several spacer adapters). Modular tooling systems tend to lose accuracy with extended use due to dirt and chip contamination and wear, scratching, or other damage to their interfaces produced during tool change, cleaning, and storage.

Radial and angular errors, caused by the machine spindle itself, the toolholder–spindle nose interface, or toolholder–tool interface, can be corrected by an adapter located between the cutting tool and the toolholder as shown in Figure 5.10. Four bolts, 90° apart, are used for each of the radial and angular adjustments on individual adapters or a single, integrated adapter. The importance of accurate angular adjustment increases with increasing L/D ratio (and especially when $L/D > 4$ as explained in Example 5.1).

An important outgrowth of modular tooling has been *quick-change tooling*, which has led to significant improvements in productivity. A quick-change tooling system (Figure 5.4–Figure 5.6) consists of a tool or holder that can be changed for another as quickly as possible. It allows the correct length tool to be built offline to maintain maximum performance. The presetting capability of quick-change tooling, in conjunction with the repeatability of the coupling between the cutting unit and the toolholder, ensures that the cutting edge is properly positioned in relation to the workpiece.

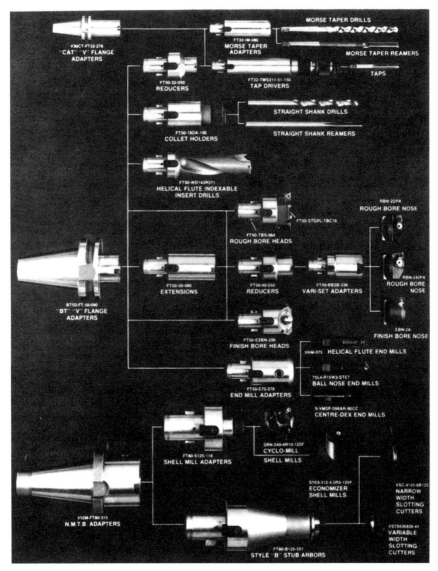

FIGURE 5.5 Automatic modular quick-change tooling system using the VTS (European FTS) adapter. (Courtesy of Valenite, Inc.)

5.3 TOOLHOLDER–SPINDLE CONNECTION

This section reviews the most common designs for the connection between the toolholder and the spindle [3].

5.3.1 General

A cutting tool is most commonly mounted in the spindle using a toolholder. In some cases, a cutting tool is mounted directly into the spindle nose to provide accurate location and high stiffness; in such cases, the spindle nose designs may include straight or tapered pilot holes with drive keys, pins (buttons), or gear teeth, and mounting bolt patterns or drawbars

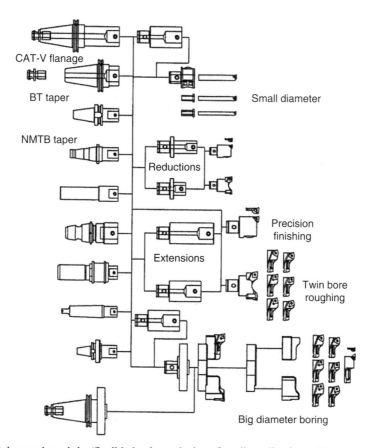

FIGURE 5.6 Advanced modular/flexible boring solutions for all applications. (Courtesy of Parlec, Inc.)

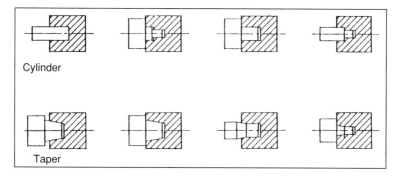

FIGURE 5.7 Configuration of centering and axial locating shaft forms available in the market under various trade names. (Courtesy of Valenite, Inc.)

(Figure 5.7 and Figure 5.8). Manually changed tooling systems are generally bolted onto the spindle face with one or more screws, supported with side screws, held hydraulically, or shrink-fit as illustrated in Figure 5.11. Toolholders or cutting tools with *cylindrical shanks* (end mills, drills, reamers, etc.) are mounted directly into the spindle shaft using a Weldon connection (Type A). In other cases, the toolholder shank is flattened or has a whistle notch and one or two mounting setscrews used to lock the cutter (Type B). Toolholders with a centering (taper) shank style adaption (Types F, G, and H) are very common; they are usually driven

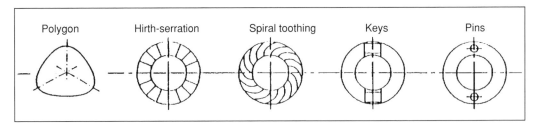

FIGURE 5.8 Configuration of tooling interface forms for torque transmission available in the market under various trade names. (Courtesy of Valenite, Inc.)

by a radial key on the spindle nose or the drawbar and are secured to the spindle with or without face contact by a drawbar in the spindle head. They can also be bolted to the spindle through their flange, in which case they have a short tapered shank; the bolts may transmit the torque in this case. The Type H interface provides much higher bending stiffness than Type G due to face contact. The flat back Type E or F interfaces have higher bending stiffness than the taper shank Type G.

The *American National Acme threaded (automotive) shank* with a whistle notch and a key drive is used for drilling, reaming, and tapping holders in transfer line applications. The whistle notch on automotive shanks or cutting tools does not provide preloaded face contact. However, a preloaded face contact can be obtained using adaption Type C, in which the screw in the spindle shaft is slightly behind the centerline of the seat on the shank of the tool or holder; the radial force generated by the fine pitch screw results in axial force and pushes the faces of the tool or holder and the spindle together under high pressure. Variations of this type of adapter have often been used in modular quick-change tooling systems.

The standard *Morse taper* shank with self-locking characteristics (Type D) is one of the oldest connections without face contact, a drawbar, or mounting screws. It has been replaced in many applications by the *CAT-V taper* discussed below. Simultaneous fit of the shank and face can be obtained by using a tapered or cylindrical pilot for centering the connection and a flat flange to provide the desired stiffness. This is illustrated in designs E, F, and H.

Standard face milling and boring cutters use the flat back or *flange mount* design with a straight or tapered pilot center (Types E and F). The centering plug is often an integral part of the spindle shaft in which case the milling cutter or adapter is secured to the spindle with lockscews or mounting bolts. The stiffness of the toolholder interface with flange connection is a function of the contact pressure applied by tightening of the flange bolts; the axial force (F) and contact pressure (P) acting on the contact surfaces can be calculated approximately using the formulas

$$F = \frac{5 \cdot n \cdot T}{D} \quad (5.1)$$

$$P = \frac{F}{A} \quad (5.2)$$

where T is the tightening torque, D is the bolt diameter, A is the contact area (flange surface area minus the bolt holes), and n is the number of bolts. The flange thickness and the waviness of the joint surfaces affect the form of the interface pressure distribution [4]. A thick flange requires a larger tightening force to provide the same interface pressure as the thin flange.

Figure 5.12 shows a milling cutter with a *curvic coupling* interface (QCS) (consisting of a set of mirror-image flat precision helical gears permanently attached to the spindle and to the

Toolholders and Workholders

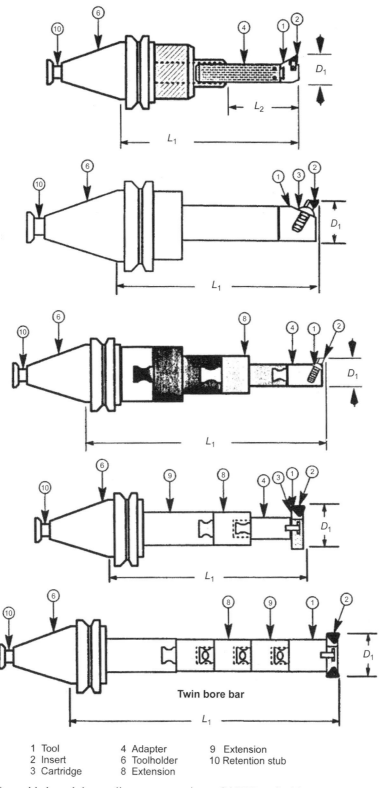

1 Tool
2 Insert
3 Cartridge
4 Adapter
6 Toolholder
8 Extension
9 Extension
10 Retention stub

FIGURE 5.9 Assembled modular tooling systems using a CAT-V toolholder.

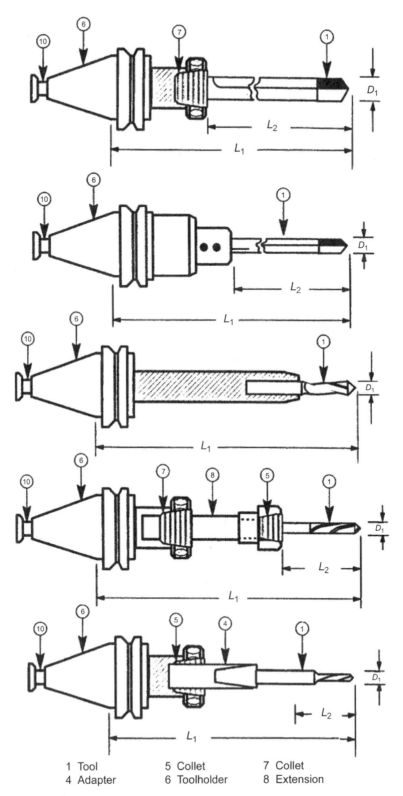

FIGURE 5.9 *Continued*

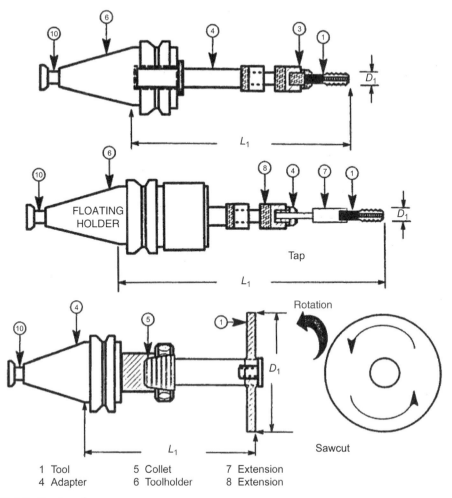

FIGURE 5.9 *Continued*

toolholder), mounted directly to the spindle shaft through a bolt-on tool connection; their engagement assures a unique position of the toolholder relative to the spindle [5]. The curved gear design makes tool mounting self-centering, highly rigid, and capable of transmitting as much torque as a steel tube of equivalent diameter. This toolholder interface can be mounted in a semiautomatic mode using a knob at the back of the holder and a retention system in the spindle that clamps the two halves together hydraulically (Type I). A *curvic flange* mount provides good radial and axial accuracy (better than 2 µm) with repeatability of 1 µm. A comparison of the curvic coupling with the CAT-V and flat joint is shown in Figure 5.13 (in which the curvic couplings A and B have, respectively, 20 and 24 teeth [5]). A flat back joint provides higher bending stiffness than an equivalent curvic coupling connection (Figure 5.11). A static axial preload markedly increases the joint stiffness of the curvic coupling, which in turn increases the natural frequency. A curvic coupling has a natural frequency lower than that of a flat joint but higher than that of a tapered connection. On the other hand, a curvic coupling has higher damping capacity than either a flat joint or a tapered connection.

A straight shank holder can be mounted in a *hydraulic expansion sleeve* in the spindle nose that is activated manually or semiautomatically (Type J). This provides 360° uniform pressure

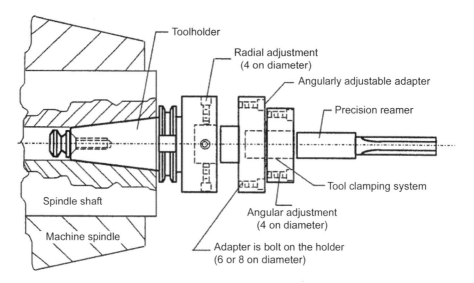

FIGURE 5.10 Adjustable adapter for radial and angular adjustments of a cutting tool in the spindle.

which clamps the toolholder concentrically, and uniform contact is achieved over the full length of engagement; the clearance between the toolholder shank and the hydraulic sleeve should be smaller than 0.015 mm for proper gripping. The normal amount of contraction for a 25 mm diameter arbor is 0.01 to 0.03 mm. Contraction is proportionately higher or lower for larger or smaller diameters. It is characterized by very low runout (TIR of 3 to 5 μm, 100 mm from the front face of the spindle) and high repeatability (0.0013 mm or better); it is balanced by design, and is used for high-precision operations or higher rotational speeds. It also provides better damping than most of the other interfaces. The clamping force for a 70 to 85 MPa expansion pressure can transmit a torque of 60 and 500 N m, respectively, for 10 and 32 mm tool shanks. The torque capability can be calculated from the available pressure using Equation (5.3). The design of the hydraulic compression sleeve could significantly affect the tool stiffness based on its geometry.

Another system comparable in performance to monolithic toolholders uses a straight shank that is secured in the spindle nose using a *shrink-fit* (Type K). The advantage of this method is that the shrinking process is reversible. Standard straight shank carbide tools are shrunk into the spindle nose shaft with a high interference fit. In this case, the spindle nose is induction heated; with the proper match of materials, the shank expands less than the spindle shaft, eliminating the interference and allowing for removal of the toolholder. A shrink-fit provides the best possible TIR, inherent balance, and gripping force. However, shrink-fit requires the spindle shaft to be extended out of the spindle housing by an amount equal to the tool shank length shrunk into the spindle shaft pilot.

The amount of interference can be optimized based on application requirements, but generally should be less than 0.01 mm to avoid excessive compressive stress on the tool shank and to increase the holding ability and safe operating stress in the spindle nose section. The magnitude of the shrink-fit or the unit pressure and tangential stress generated by the fit can be calculated from Lame's equations [6],

$$\delta = \frac{b \cdot P}{E_h}\left[\frac{b^2 + c^2}{c^2 - b^2} + \nu_h\right] - \frac{b \cdot P \cdot \nu_t}{E_t} \qquad (5.3)$$

Toolholders and Workholders

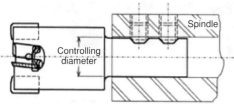

A: (Poor) Straight shank without face contact "Weldon type"

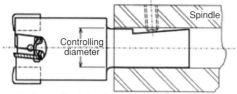

B: (Poor) Straight shank without face contact "whistle notch type"

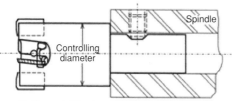

C: (Better) Simultaneous fit straight shank

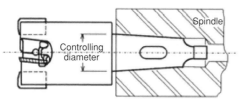

D: (Poor) Std. Morse taper shank (ANSI B5.10)

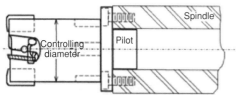

E: (Better) Flat back (flange mounting) with a straight pilot

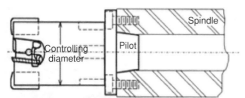

F: (Very good) Flat back (simultaneous fit) with paper pilot

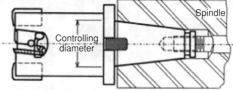

G: (Better) Taper shank without face contact

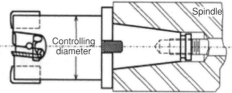

H: (Best) Simultaneous fit with taper shank

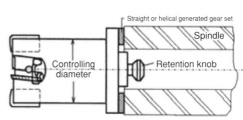

I: (Very good) Curver flange mount with a retention system

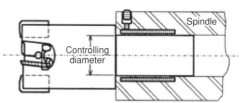

J: (Very good) Hydraulic toolholder with a straight shank

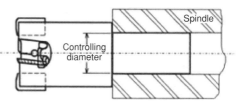

K: (Very good) Shrink-fit toolholder with a straight shank

FIGURE 5.11 Spindle–toolholder connections.

FIGURE 5.12 Spiral toothed curvic coupling interface. (Courtesy of Valenite, Inc.)

The tangential stress at the inner surface of the ID of the holder due to the shrink is

$$\sigma_t = \frac{P[b^2 + c^2]}{c^2 - b^2} \tag{5.4}$$

In these equations, δ is the diametrical interference, P is the compressive stress (pressure) between the toolholder and the tool shank, E_h and E_t are, respectively, the moduli of elasticity for the holder and tool shank materials, ν_h and ν_t are, respectively, Poisson's ratios for the toolholder and tool shank materials, and b and c are, respectively, the inside and outside diameters of the toolholder nose where the tool shank is inserted.

The torque T which can be transmitted without relative displacement between the tool and toolholder is

$$T = \frac{\pi \cdot P \cdot L \cdot b^2 \cdot \mu}{2} \tag{5.5}$$

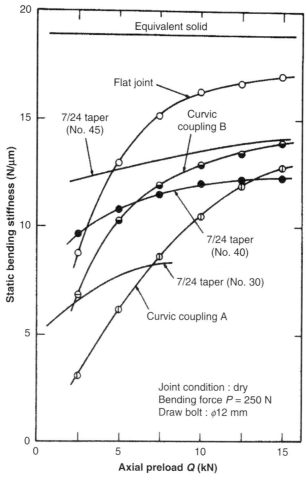

FIGURE 5.13 Effect of axial preload on static stiffness of various interface systems. (After S. Hazem et al., [5].)

where L is the contact length of the shrink-fit joint and μ is the coefficient of friction. The static coefficient of friction between steel and carbide is 0.25 to 0.30 with a surface finish of between 0.0003 to 0.0004 and 0.0002 mm, respectively, for the toolholder and tool shank materials. The use of kinetic coefficient of friction, which varies between 0.12 and 0.20, in this equation results in a high safety factor. The torsional holding power is proportional to the applied pressure at the joint, which is proportional to the diametrical interference. Torque capability between 50 and 400 N m, respectively, for 8 and 25 mm tool shank diameters can be achieved.

The amount of interference has a significant influence on the joint tangential displacement [4]. The displacement decreases with increasing interference. The relationship between the tangential displacement and the applied load is linear for a large interference, but deviates from linearity for small interference values [7]. The joint tangential displacement and microslip increases with increasing surface roughness at the same interference value. The tangential stiffness of shrink-fit joints increases with increasing interference and decreasing surface roughness. The stiffness is approximately 75 to 93% of the stiffness of and equivalent solid system for large interference values [7]. Under dynamic loading the observed tangential displacement is considerably larger than that observed under static loading.

Toolholder dimensions have a strong influence on toolholder performance. In cases without good face contact, the diameter of the toolholder at the front of the flange or the tool itself should not be any larger than the gage diameter of the toolholder shank from a bending stiffness viewpoint. The diameter of the tool or toolholder body is the controlling dimension when proper face contact at the interface is used (so that the radial and tangential loads do not separate the unit), and it can be as large as the flange diameter of the toolholder.

Toolholders with a centering (tapered) shank style adaption, as shown by the tapered shank (Types F, G, and H in Figure 5.11), are very common; they are usually driven by a radial key on the spindle nose or the drawbar and are secured to the spindle with or without face contact by a drawbar in the spindle head. They can also be bolted to the spindle through their flange, in which case they have a short tapered shank; the bolts may transmit the torque in this case. The Type H interface provides much higher bending stiffness than Type G due to face contact. The flat back Type E or F interfaces have higher bending stiffness than the tapered shank Type G as shown in Figure 5.11.

A flowchart of the requirements for selecting a machining toolholder interfaces is given in Figure 5.14. The procedure provides only guidelines of the different steps that should be used to review system components critical to the successful application of a particular toolholder interface with the machine tool. If the machine tool is new, the toolholder size and machine tool characteristics should be selected based on the application. For an existing machine tool, the cutting conditions should be selected based on the machine tool and spindle characteristics including the size of the toolholder–spindle interface.

5.3.2 Conventional Tapered "CAT-V" Connection

The Caterpillar V-flange "CAT-V" steep (7/24) taper design (Figure 5.11, Type G) is the most common toolholder interface system for CNC machines. CAT-V tapers are designated by size as 30, 40, 45, 50, and 60 tapers; the shank dimensions for each size are given in Figure 5.15. The ISO 1947 standard for conical surface accuracy provides for classification of the taper with respect to rated accuracy (based on the "AT" number), surface finish, roughness, and roundness of any section on a conical surface. The CAT-V flange design is included in ANSI Standard B5.50, ISO Standard 7388/1, "SK" DIN 69871-A and -B, DIN 2080, JIS B6339 BT Japanese standard (MAS BT 403), and the NMTB standard (Figure 5.15). Some of these standards are very similar and interchangeable, while others appear similar but are not interchangeable.

Conventional 7/24 taper interfaces have two major shortcomings: (1) radial clearance in the back part of the tapered connection, due to taper tolerancing, reduces stiffness and increases runout, and potentially generates micromotions which cause fretting corrosion; and (2) the mandated axial clearance between the flange of the toolholder and the face of the spindle creates 25 to 50 μm uncertainty in the axial positioning of the toolholder. However, there has been a product consistency and worldwide acceptance with the steep taper. It is ideal for most applications including high-speed operations up to 15,000 rpm.

The main advantages are: (1) it is not self-locking and is secured by tightening the holder taper in the tapered hole of the spindle; and (2) the design of the 7/24 steep tapers is relatively simple and requires precision machining of only one dimension — the taper angle. The taper plays two important roles simultaneously: precision location of the toolholder relative to the spindle, and clamping to provide adequate rigidity.

The allowable tolerance for the interface taper is split between the shank (toolholder taper) and the socket (machine spindle taper). The standard tolerances specify a minus deviation of the hole angle and a plus deviation of the toolholder angle, resulting in clearance between the back of the taper and the spindle hole. The ANSI/ASME tolerance has been

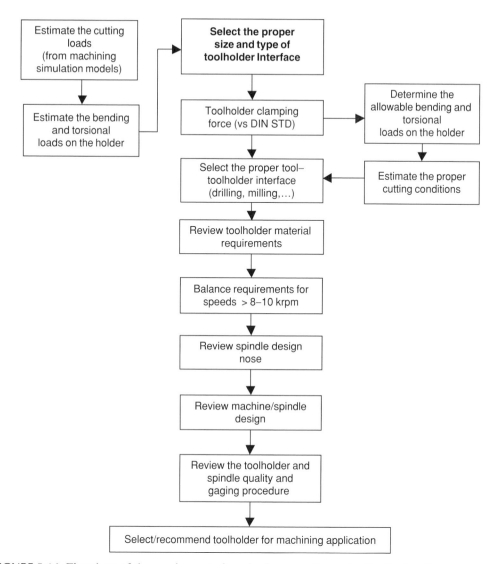

FIGURE 5.14 Flowchart of the requirements for selecting a machining toolholder interface.

revised to AT4 from the middle of AT4 and AT5 (ISO 1947 STD, see Figure 5.16), which is equivalent to BT holders. However, a large number of standard commercial toolholders were measured with an air gage and several of them found between AT5 and AT6. This leads to mobility of the taper under heavy cutting forces and to runout if the draw bar force is not perfectly symmetric. The radial repetitive accuracy of #40 and #50 tapers is within 0.02 mm for new tools, but may be two or three times higher with worn tools. The repeatability of the #30 taper is generally worse than that of #40 and #50 tapers. The maximum possible runout effect from the "AT3" accuracy as compared to the ANSI specification is shown in Figure 5.17. However, if the toolholder and spindle taper tolerance are AT2 to AT3, the radial repetitive accuracy will be better than 0.006 mm.

A steep taper connection is often not suitable for high rotational speed applications (30 to 40 m/sec peripheral speed) because the taper may expand (at speeds >10,000 rpm) due to centrifugal forces, substantially reducing the contact area between the toolholder and the

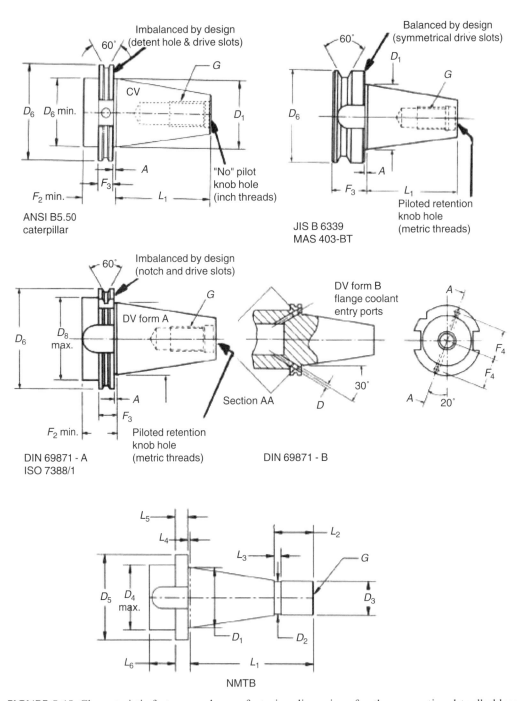

FIGURE 5.15 Characteristic features and manufacturing dimensions for the conventional toolholders with a 7/24 taper. (Courtesy of Kennametal, Inc.)

spindle [8]. The front end of high-speed spindles expands as much as 4–5 μm at 30,000 rpm for taper #30; German researchers have demonstrated that the front end of a high-speed spindle expanded by as much as 0.2 μm at 10,000 rpm and 2.8 μm at 40,000 rpm [1]. As a consequence, the dynamic runout and dynamic stiffness of a 7/24 steep taper system progres-

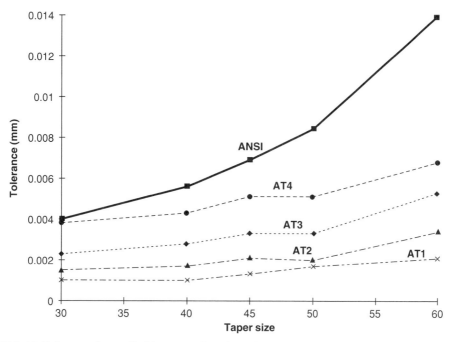

FIGURE 5.16 Tolerance for toolholder tapers based on the ISO 1947 standard.

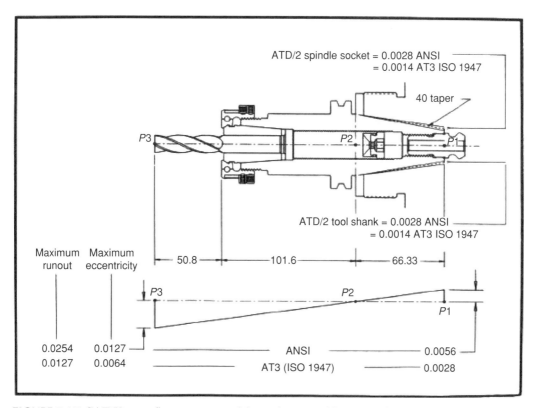

FIGURE 5.17 CAT-V taper fit versus eccentricity and runout. (Courtesy of Kennametal, Inc.)

sively worsens as tool speed increases. As the female taper expands, the male taper is pulled further into the shaft (by as much as 0.05 mm depending on the drawbar force), changing the axial position of the tool. The border between high- and low-speed applications, by consensus of several spindle manufacturers, is 8000 to 15,000 rpm for a #50 taper, 12,000 to 20,000 rpm for a #40 taper, and 16,000 to 25,000 rpm for a #30 taper.

Many tool and machine manufacturers realize the importance of improving the accuracy of the toolholder–spindle combination from AT4/AT5 to AT3/AT3 for low- to medium-speed, high-precision spindle systems (Figure 5.16) and AT2/AT2 for high-speed, high-performance spindle systems.

Another concern in heavy machining (as in high-speed machining) is fretting corrosion (tarnishing) in the taper portion of both the toolholder and the spindle. Fretting corrosion is a frictional oxidation phenomenon due to the relative movement of the tapers under vibration or heavy cutting. For this reason, black oxide or other treatments are often not suitable for high speed and heavy load applications [9].

The axial, bending, and torsional stiffness of the toolholder–spindle interface are important for both static and dynamic performance characteristics. The axial movement of the holder taper in the spindle is about 0.02 mm [8, 11] due to elastic deformation at the taper interface under clamping forces. The saturation of the axial displacement and a better repeatability is achieved at the suggested clamping forces (drawbar preload) given in Table 5.1. For example, the CAT-40 results in 6 μm movement in the spindle nose while the drawbar force increased from 5 to 15 kN. The maximum torque that the CAT-V interface will carry is provided in Table 5.2.

The static bending and torsional stiffness (based on bench tests) for the conventional interface at three drawbar forces is given in Figure 5.18 and Figure 5.19, respectively [10]. It was observed that there was not a clear trend between the clamping force or the class of fit on the test tool taper and the static bending stiffness or torsional stiffness in the range of the applied cutting loads tested. This indicates that the static stiffness saturates at the drawbar preloads (7 and 12 kN, respectively, for the CAT-40 and CAT-50 interfaces) [11]. Other researchers found that stiffness is very sensitive to minute differences in the taper angle and to axial preload [12]. In another research, the drawbar's pull force had a strong impact on stiffness as illustrated in Figure 5.13. It shows that a 400 to 800% increase in clamping force results in 20 to 50% higher bending stiffness [5, 12, 13].

The normalized bending stiffness as a function of bending moment can be obtained from the static bending stiffness measured at very high radial loads. The rotation (tipping) angle of the toolholder flange is measured using a special setup [10]. The measurement of the rotation

TABLE 5.1
Drawbar Forces for CAT-V and HSK Toolholders

Size	CAT-V Flange Force (kN)			Size	HSK-A Force (kN)	HSK-B Force (kN)
	Conventional	High Speed	Max. Allowable			
30-T	5–6	3–6		40	6.8	
40-T	8–15	5–10	35–42	50	11	6.8
45-T	14–18	—	56–62	63	18	11
50-T	18–40	12–20	78–88	80	28	18
60-T	60–80	—		100	45	28
				120		45

TABLE 5.2
Maximum Torque (N m) Carried by the Various Toolholder Couplings

Size	CAT-V (N m)	Size	HSK-A (N m)	HSK-E (N m)	KM (N m)	Size	HSK-B (N m)	HSK-F (N m)	Size	Capto (N m)
		32	45	10–19	155	40			C3	
30		40	81	15–35	325	50		20–40	C4	
		50	185	30–75	780	63	800	45–85	C5	1,600
40	800	63	360	70–150	1,530	80	1,200	100–175	C6	3,600
45	1,800	80	710		2,800	100	1,800		C8	7,000
50	7,000	100	1,420		4,200	125	7,000		C10	13,000
55		125	2,850			160				
60		160	5,700							

at the interface under an applied bending moment (as shown in Figure 5.20) provides a normalized measurement of static bending stiffness of the interfaces [13, 14] (see Figure 5.21):

$$k = \frac{M_b}{\theta} = \frac{FL}{\theta} \quad (5.6)$$

The normalized stiffness is given in Figure 5.22 and Figure 5.23 [11, 13]. This normalized stiffness can be used in the selection of the interface based on the expected requirements of the cutting operations (i.e., light or heavy operations, short or long tooling, etc.). CAT-50 provides about four to five times greater bending stiffness that CAT-40 interface. The values for the CAT-V interfaces are small but still sufficient to support the bending moments for many conventional cuts since the bending moment is generally less than 200 N m. The normalized stiffness was found to be the same for all the clamping forces (7 to 17 kN for

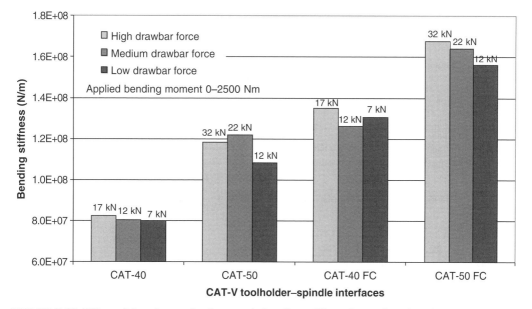

FIGURE 5.18 Effect of drawbar preload on static bending stiffness for various interfaces.

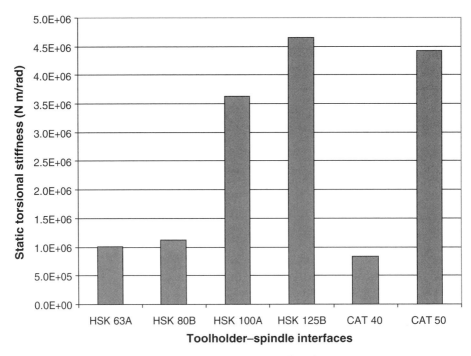

FIGURE 5.19 Static torsional stiffness for various toolholder interfaces.

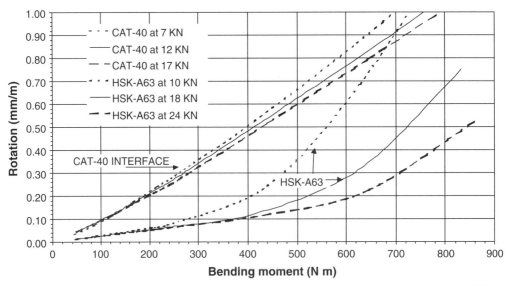

FIGURE 5.20 Comparison of toolholder tilt for CAT-40 versus HSK-A63 interfaces at three different clamping forces (bench tests).

CAT-40 and 12 to 32 kN for CAT-50) for bending moments less than 200 N m. On the other hand, the normalized stiffness increased by 10 to 15% between the high and low clamping forces when higher than 200 N m bending moment was applied at the interface.

The results from the dynamic bending stiffness and torsional stiffness indicate that the toolholder–spindle interfaces are dynamically more flexible as the drawbar preload increases

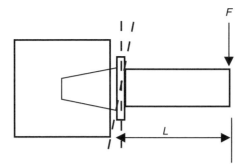

FIGURE 5.21 Normilized bending stiffness of an interface.

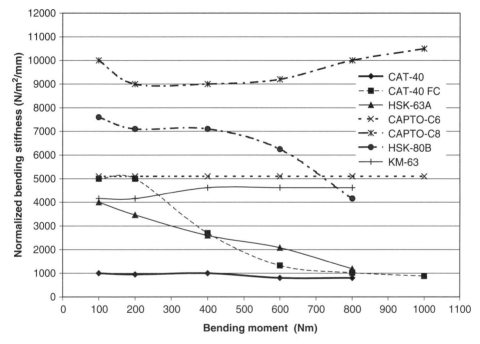

FIGURE 5.22 Normalized static bending stiffness for various small interfaces.

[10]. In addition, the modal stiffness increases and the damping ratio decreases with the application of higher clamping forces (drawbar loads). With increasing clamping forces, the damping ratio decreases at a rate that is greater than the rate at which the modal stiffness increases thus resulting in a dynamically more flexible system. The average static, modal, and the dynamic bending and torsional stiffness are given in Figure 5.24 and Figure 5.25, respectively. CAT-50 provides significantly better dynamic characteristics than CAT-40 interface. The torsional stiffness for the CAT-50 is superior to CAT-40.

Power drawbars for CAT-V style holders consist of several components, which may vary depending on the retention knob used. A clamping device extends through the spindle to grip the holder at the knob as shown in Figure 3.51 in Chapter 3. For each of these different standards it is necessary to have a gripper, which is designed to conform to the specific retention knob [10].

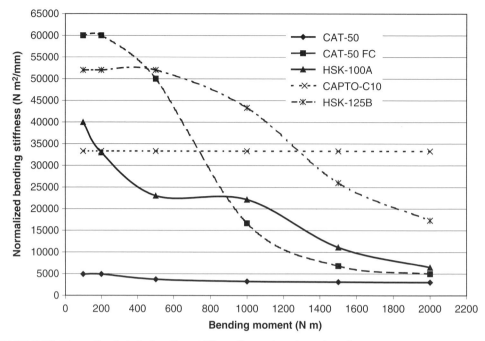

FIGURE 5.23 Normalized static bending stiffness for various large interfaces.

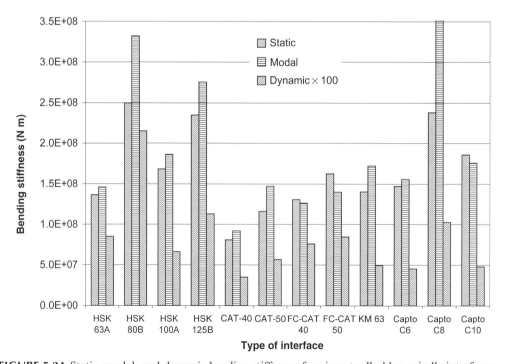

FIGURE 5.24 Static, modal, and dynamic bending stiffness of various toolholder–spindle interfaces.

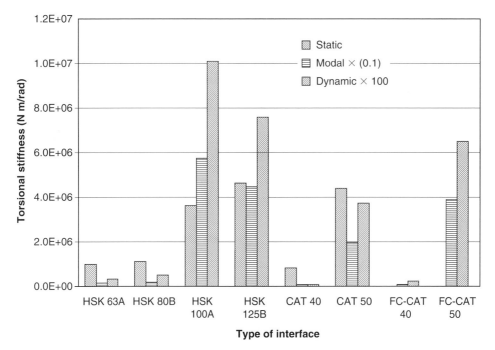

FIGURE 5.25 Static, modal, and dynamic torsional stiffness of various toolholder–spindle interfaces.

5.3.3 FACE-CONTACT CAT-V INTERFACES

We have seen the undesired effects of using a conventional 7/24 taper toolholder in high-speed or large bending moments applications. The advantages of the face contact "CAT-V FC" interface versus the CAT-V are: (1) interchangeability with existing toolholders, (2) applicability to higher speed machining, (3) reduction or prevention of fretting corrosion (in heavy cuts), (4) elimination of Z-axis movement at higher speeds, (5) Z-axis repeatability within one or a few microns, (6) good repeatability among a family of spindles, and (7) minimal deflection for maximum machining accuracy and better finish. There are several approaches aiming to provide both taper and face contact in the CAT-V interface to improve the conventional 7/24 taper design without major changes in spindles and with full compatibility with conventional toolholders. Two main approaches are: (a) metal-to-metal contact (rigid systems) by holding very close tolerances on both halves of the coupling, and (b) using flexible elements either on the taper or flange face of the toolholder or spindle.

The *Big-Plus* interface system offers simultaneous contact on the taper and flange face using the rigid system approach [15]. This system is based on the standards for CAT-V, JIS-BT, and DIN69871. Dual contact is achieved with the Big-Plus system by eliminating the gap between the machine spindle face and the toolholder flange face (as illustrated in Figure 5.26). By having tight tolerances on the gage diameter (for both toolholder and spindle) and flange face, about 20 μm axial clearance exists when the toolholder is inserted in the spindle nose, which is closed by the clamping forces given in Table 5.1. One of the important features of the Big-Plus system is interchangeability with existing toolholders as shown in Figure 5.26. A spindle with Big-Plus interface (combination A in Figure 5.26) is required to benefit from the technical advantages that the Big-Plus interface offers.

The static bending stiffness of the Big-Plus CAT-40 and CAT-50 interfaces at three clamping forces is shown in Figure 5.17 (defined as face contact "FC") [11]; the stiffness

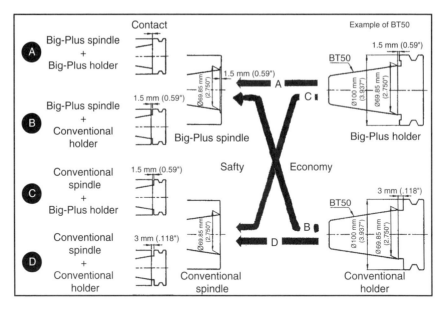

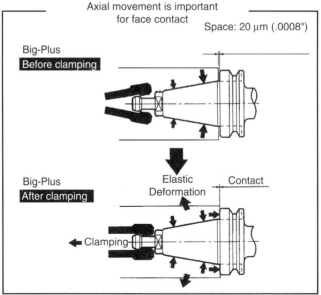

FIGURE 5.26 Big-Plus interface illustrating the clamping–unclamping position of the toolholder and interchangeability with existing CAT-V interface. (Courtesy of Daishowa Seiki Co. and Stanley Sheppard Co.)

for the face contact (Big-Plus) interface is 40 to 60% greater than that of the conventional CAT-V interface. The stiffness does not change significantly with increasing clamping force. The normalized static bending stiffness is superior for the face contact interface for bending moments less than 200 to 300 and 500 to 600 N m, respectively, for the CAT-40 FC and CAT-50 FC as shown in Figure 5.22 and Figure 5.23. The static and dynamic bending stiffness for the CAT-V FC interface is as good or better than that of the CAT-V without FC interface (Figure 5.24). The dynamic torsional stiffness for the face contact interface is better than that without face contact as shown in Figure 5.25.

Similar characteristics of simultaneous contact have been obtained using spacers located between the toolholder flange and the spindle face. The spacers have to be precisely ground to the needed thickness. This approach will provide the advantages of a rigid face contact (discussed above) but will result in somewhat lower static stiffness than the Big-Plus due to the additional contact with the spacer. However, the dynamic stiffness could increase. The toolholder must be fit individually to a particular spindle and is not interchangeable with other spindles unless the gage diameter of the spindle is controlled to a few micrometers.

The rigid approach for simultaneous contact allows for spindle nose expansion of about 6 μm per diameter. This will affect the front spindle bearings unless the spindle nose female taper is extended ahead of the front bearings; this will result in lower stiffness and natural frequency.

There are several interface systems which use flexible elements to obtain simultaneous fit. The flexible elements create a dampening effect reducing cutting vibration, thus extending the cutting tool life and applicable cutting regimes. The *3-Lock* taper interface provides simultaneous fit (see Figure 5.27) [16] using a spring loaded taper cone. The term 3-Lock refers to the 3 locking points, which consist of the two internal "wedging or locking points" and the face contact between the machine tool spindle face and the toolholder flange. When the toolholder is inserted into the spindle, taper contact is first achieved. As the machine tool continues to pull the tool holder up into the spindle the taper is pushed down on top of the Bellville springs located below the base of the taper cone to preload the taper as illustrated in Figure 5.27. It counteracts the centrifugal forces generated at higher speeds. The toolholder taper is preloaded in the spindle nose [16]. The 3-Lock contact ratio at taper and flange is

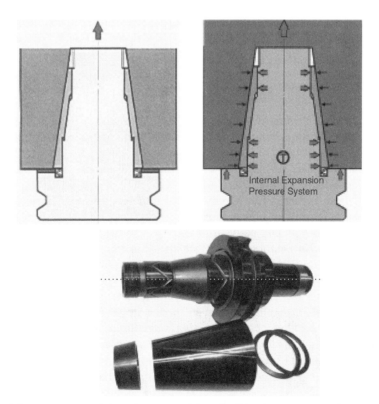

FIGURE 5.27 CAT-V FC 3-Lock interfaces containing flexible elements on their taper. (Courtesy of Nikken Kosakusho Works Co.)

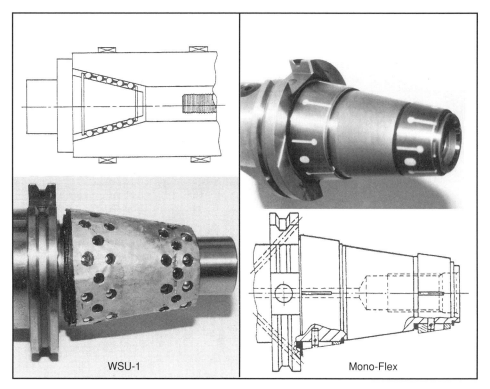

FIGURE 5.28 CAT-V FC WSU-1 and Mono-Flex interfaces containing flexible elements on its taper (After J.S. Agapiou, et al., [17]) (Mono-Flex: Courtesy of Rigibore Tooling Systems.)

claimed to be 90%:10% compared to a solid double contact with a taper contact of 10 to 50% and flange contact of 90 to 50%.

Some other flexible element interfaces are the *WSU-1* and *Mono-Flex* shown in Figure 5.28 [17]. The results of an evaluation of these toolholders are given in Figure 5.29 for two clamping forces. Both systems require somewhat higher clamping force to outperform the conventional interface. The stiffness values in Figure 5.29 cannot be compared to those in Figure 5.17 because the test parameters are different. The clamping force was saturated at about 44 kN for both systems. The Mono-Flex toolholder provides improvements over the conventional toolholder only at higher clamping forces. The WSU-1 system provided higher stiffness than the Mono-Flex and conventional CAT-V system.

Another interface with flexible elements on the flange face is illustrated in Figure 5.30. The flange face of the toolholder is made in two pieces with bevel washers in between; this provides the preload on the flange face during clamping. The dynamic bending stiffness of this face contact interface is much better than the nonface contact CAT-V interface as shown in Figure 5.31. However, the static bending stiffness is not as high as that of a rigid simultaneous interface but somewhat better than the conventional CAT-V interface at bending moments less than 50 to 100 N m. In general, the simultaneous interfaces using flexible elements will provide better dynamic bending stiffness than the conventional CAT-V interface; however, their static stiffness is improved compared to that of conventional CAT-V only at higher clamping forces.

The *CAT-V DFC* shank is another double face contact 7/24 taper interface using flexible elements to provide the simultaneous fit [18]. The toolholder consists of two pieces, the main

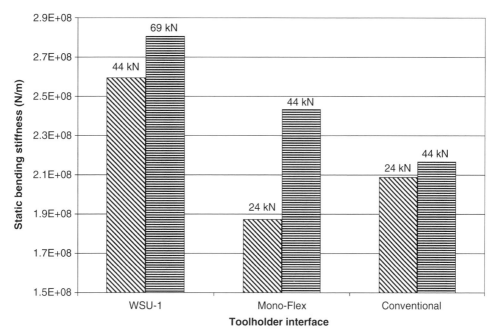

FIGURE 5.29 Static bending stiffness comparison of two CAT-50 interfaces with flexible elements.

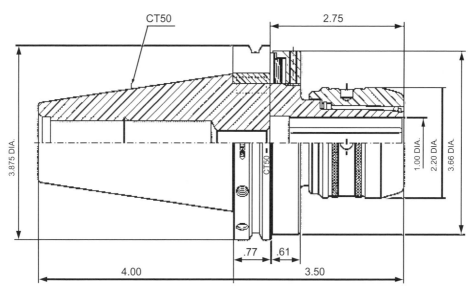

FIGURE 5.30 CAT-V FC toolholder–spindle interface with flexible elements on the flange face. (Courtesy of Richmill Manufacture Co.)

toolholder body and a cylindrical sleeve with the 7/24 taper on the outside. Drawing the pull stud of DFC tooling results in face contact (see Figure 5.32). The clamping force requirement is similar to conventional 7/24 taper (given in Table 5.1) at the higher range. The static and dynamic bending stiffness and concentricity of DFC interface is claimed to be better than the conventional 7/24 interface.

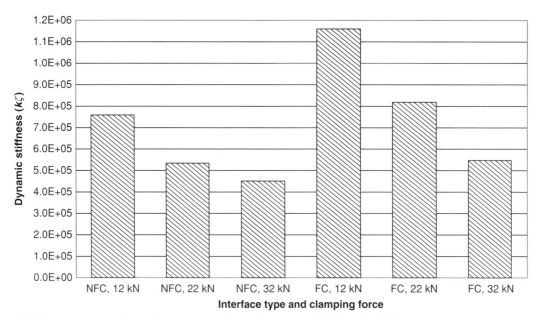

FIGURE 5.31 Comparison of dynamic bending stiffness for CAT-V with (FC) and without (NFC) face contact interface shown in Figure 5.30.

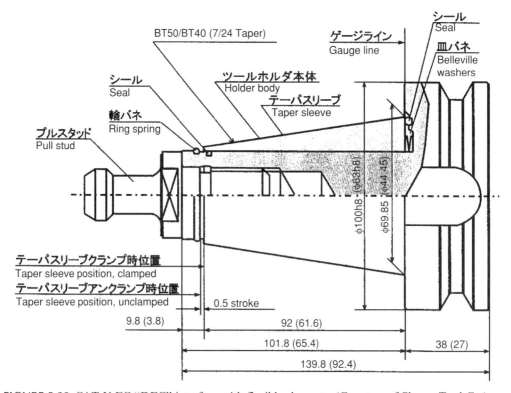

FIGURE 5.32 CAT-V FC "DFC" interface with flexible elements. (Courtesy of Showa Tool Co.)

Toolholders and Workholders

5.3.4 HSK Interface

The HSK (Hohl Schaft Kegel, or hollow taper shank) system has become a common toolholder–spindle connection in industry in recent years; it was developed to overcome the limitations of the steep 7/24 tapers. The shallow 1/10 taper angle and resilient thin-walled design (Figure 5.33) allow for expansion of the taper shank during clamping and results in simultaneous taper and face contact [13, 14, 19–21]. The HSK interface is intended to outperform CAT-V and other toolholder connections at high rotational speeds (>10,000 rpm) with respect to stiffness, radial and axial positioning accuracy and repeatability, and short tool exchange time [14, 19–21].

The hollow taper HSK shank must be manufactured at very tight tolerances, and particular care must be exercised in the selection of materials, heat treatment, and gauging [22–26]. The tooling shanks and spindle receivers are described by DIN69893 (ISO/DIS 12164–1) and DIN69063 (ISO/DIS 12164–1), respectively. Other very critical parameters for the HSK system are the cleanliness of the interface, the drawbar clamping device and its draw force magnitude, the design and manufacture of the spindle nose taper, dynamic balance at higher speeds, and allowable bending and torsional loads [25, 27].

The HSK design uses a 1/10 taper and is short compared to the standard 7/24 taper, resulting in faster tool exchange time. Tolerances on critical dimensions of both the

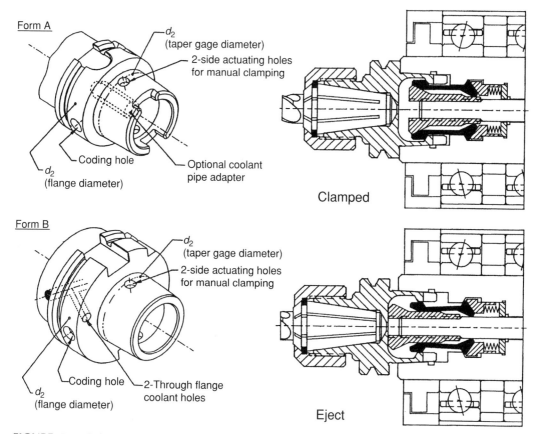

FIGURE 5.33 HSK toolholder DIN 69893 taper and face fitting and toolholder–spindle interface illustrating the clamping–unclamping position of the drawbar fingers and the knock out of the toolholder during ejection. (Courtesy of Precise Corporation.)

toolholder and spindle are between 2 and 6 μm. The slow taper reduces the force required to produce a given amount of elastic deformation in the hollow shank to bring it into contact with the spindle hole. However, a slow taper requires a higher ejection force to separate the male and female tapers. Due to such stringent tolerances, the shallow taper angle, and resilient thin-walled design, application of an axial force results in simultaneous contact between the taper and face surfaces (simultaneous fit with face contact as illustrated in Figure 5.33). The design includes a mechanical preload on the flange face to ensure full diametrical contact. The friction resulting from the preload also helps to transmit torque. Tests have shown that positioning accuracies of 1 μm can be repeatably attained when the same toolholder is clamped and unclamped in a specific clamping unit. In applications in which more than one toolholder is used, the accuracy of each toolholder must be considered. The TIR at 100 mm from the spindle face is typically less than 5 μm.

HSK holders come in six versions (Forms A, B, C, D, E, and F) [22–26]. Forms A and B are designed for automatic tool changing; Forms C and D are for manual tool changing, and Forms E and F are for very high-speed applications. The ratio of the V-flange diameter to the gage diameter for Forms A, C, and E is approximately 1.35:1 as compared to 1.7:1 for Forms B, D, and F. Form B provides about 25% larger load capacity than Form A with the same gage line diameter; therefore, if the flange diameter of the interface represents the limiting factor for application, Form A holders will provide higher bending load capacity than Form B holders. If the shank diameter is limited, a Form B with the next highest nominal size offers a higher bending capacity than a Form A. A comparison of the corresponding sizes of CAT-V and HSK interfaces is given in Table 5.2 together with their torque capability.

The drawbar in the HSK system is more complicated and expensive than that in a CAT-V system (see Figure 3.51 in Chapter 3). The toolholder clamping force suggested by DIN [23, 27] is given in Table 5.1. During rotation the centrifugal force acting on the fingers adds to the clamping force. The positive locking action and the centrifugal clamping enhancement greatly increase the inherent safety of the HSK interface for high-speed applications. The HSK interface provides full-face contact and partial contact along the taper as explained using finite element analysis (FEA) in Ref. [25].

The static bending at the tip of a tool and torsional stiffness (based on bench tests) for the HSK interface at three drawbar forces are given in Figure 5.34 and Figure 5.19, respectively. As with the CAT-V interface, the HSK interface shows no clear trend between the clamping forces and the static bending stiffness or torsional stiffness in the range of the applied cutting loads tested. This indicates that the static stiffness saturates at the lower clamping forces. The HSK-B80 is about 80% stiffer (in bending) than the HSK-A63 automatic change interface that is 50% stiffer than the HSK-C63 manual change interface. The HSK-B125 is about 40% stiffer (in bending) than the HSK-A100 interface. The large and small sizes interfaces cannot be compared in Figure 5.34 because of the difference in the tooling sizes and the distance of the applied load from the spindle face. The static bending stiffness for the HSK-A63 and HSK-A100 interfaces is about 60 to 70% and 40 to 45%, respectively, greater than that of the CAT-40 and CAT-50 interfaces (see Figure 5.24). The static torsional stiffness for the HSK-B80 interface is not significantly different (about 10 to 15% greater) than that of the HSK-A63 interface (see Figure 5.19). However, the static torsional stiffness for the HSK-B125 interface is about 25 to 30% greater than that of the HSK-A100 interface. On the other hand, the HSK-A63 provides equal or about 20% higher static torsional stiffness than the CAT-40 interface, while the CAT-50 shows better torsional stiffness than the HSK-A100 interface.

The normalized bending stiffness as a function of bending moment at the interface for the HSK interfaces is given in Figure 5.22 and Figure 5.23 [21]. The HSK-B80 provides about 100% higher stiffness than the HSK-A63 that is 200 to 300% better than the CAT-40 interface at bending moments lower than 200 N m. CAT-50 provides about four to five times greater

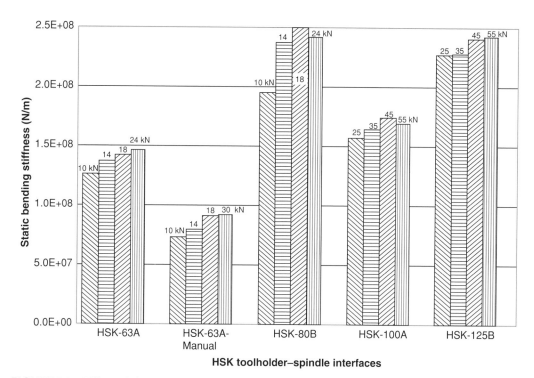

FIGURE 5.34 Effect of clamping force on static bending stiffness for HSK interfaces.

bending stiffness than the CAT-40 interface. The HSK-B125 provides about 30 to 50% higher stiffness than the HSK-A100, that is, 400 to 700% better than the CAT-50 interface at bending moments lower than 500 N m. The normalized bending stiffness decreases with increasing bending moment for the HSK interfaces. The normalized stiffness was found to be the same for all the clamping forces for bending moments less than 200 N m for the smaller sizes and 500 N m for the larger-sized interfaces.

The maximum bending moment that a toolholder can support is a function of the clamping force [10, 21, 25, 27] and it should always exceed the bending moment generated during cutting as explained in Example 5.9. The maximum allowable bending moment before the face joint is separated at the toolholder flange is given in Figure 5.35 [10, 27] for a range of clamping forces for different sizes of HSK toolholders. For example, 320 N m is the maximum bending moment that the HSK-63-A interface can support before complete separation of the face contact occurs when the clamping force is 18 kN (see Figure 5.35). The boundary bending moment for an HSK-A63 is about 30 to 35% greater than that of HSK-B63. However, the style B can be loaded at 30% higher bending moment than the same gage diameter style-A interface.

The maximum allowable torque that the HSK coupling can carry is relatively low, as shown in Table 5.2, since it is carried by a single key (located at the foot of the thin cross section shank) and the friction generated between the taper and face contact areas. The HSK standard specifies two drive keys, but with the tolerances involved, it is difficult to get both drive keys in contact at the same time unless severe deformation is present [21, 25, 27].

The dynamic bending stiffness and torsional stiffness were better for the style-B HSK interfaces than for the style-A interface as illustrated in Figure 5.24. The dynamic bending stiffness for HSK-A63 interface is also much better than the CAT-40, but the HSK-A100 was

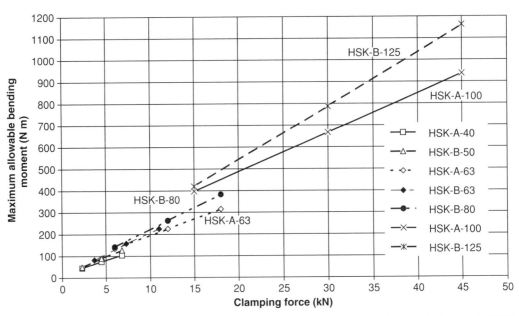

FIGURE 5.35 Maximum allowable bending moment for HSK toolholders. (After I. Schubert et al., [21].)

marginally better than the CAT-50 interface (see Figure 5.24). The static and dynamic torsional stiffness of the larger sizes of HSK interfaces is superior to the smaller HSK sizes. In addition, the dynamic torsional stiffness for the HSK interface is better than the CAT-V interface. In addition, generally, the dynamic stiffness increases with a decrease in the clamping force.

The dynamic runout of HSK systems (due to the expansion of the taper spindle hole caused by the increased centrifugal forces) does not significantly increase with spindle speed as it does with 7/24 steep taper toolholders (Figure 5.36); the expansion ratio of the female to

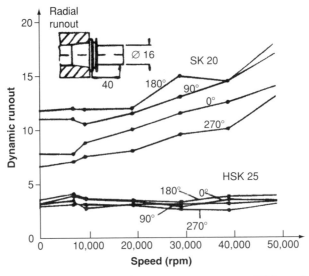

FIGURE 5.36 Dynamic runout of HSK interface versus steep taper (7/24) toolholders. (Courtesy of Precise Corporation.)

hollow tool shank for HSK is much lower than that of the 7/24 taper due to the high compliance of the hollow taper shank. The centrifugal force, being exerted on the hollow shaft of the HSK toolholder and on the fingers of the clamping system, prevents total separation between the hollow shank and the spindle hole, and therefore the holder maintains its radial (centerline) accuracy. The stress on the toolholder increases by 30 to 40% while the speed increases from 1,000 to 16,000 rpm [25]. Finally, several features on the toolholder are nonsymmetrical, complicating balancing in high-speed applications. Balancing the HSK holder can be challenging because its shank is deformed to the geometry of the spindle cone. In addition, balancing the drawbar is difficult because it contains moving parts.

Through numerous tests, collected plant data, and theoretical modeling and analysis [25, 27], the following technical and operational issues on the HSK toolholder system were identified:

1. The material of the toolholder is critical when used on machines with frequent automatic tool changes or used under heavy loads. A minimum requirement of 1380 MPa yield strength, 1500 MPa tensile strength, 50 HR_C shank hardness should be strictly followed.
2. The clamping force for the toolholder should at least be equal to the DIN standard recommendation, but should not be larger than 130% of this value. In aggressive operations, maintaining a high clamping force in the drawbar mechanism is critical in maintaining face contact between the toolholder and the spindle. When face contact is lost there is a dramatic decrease in interface stiffness, and a corresponding rise in toolholder stress, possibly leading to shear failure in the hollow toolholder shank.
3. The toolholder tapered shank and spindle tapered bore should be gauged to conform to the DIN standard using the proper gauging equipment. Typically there is one ring of contact near the spindle nose, and another ring of contact towards the back of the taper separated by a middle region with taper-to-taper clearance. However, slightly out of tolerance taper combinations could lead to a relatively unstable situation where there is only one ring of taper-to-taper contact near the spindle nose. This will have significant effects on the interface stiffness and stress.
4. The bending and torsional cutting loads must be simulated for a specific tool geometry to avoid overloading the HSK interface. Depending on the particular loading condition, the region of maximum stress in the toolholder is typically either in the root of the long keyway or in the thin walled segment of the toolholder shank near the edge of the 30° land gripped by the drawbar fingers. Toolholder failures in bending or in torsion will generally originate in one of these two areas.
5. Toolholder imbalance should be determined at higher speeds (>10,000 rpm) before any machining is performed. For a very well balanced system, there is minimal stress rise in the toolholder as rpm alone is increased, even up to 15,000 rpm for the HSK-A63. Excessive imbalance, however, could cause much higher stresses.
6. The toolholder shank and the spindle nose should be maintained chip-free and clean.

Other proprietary advanced toolholder systems include the Kennametal KM, NC5, and Sandvik Capto systems.

The *Kennametal KM (Krupp Widia — Widiaflex/UTS)* system [28] uses a 1/10 taper connection (Figure 5.37) very similar to HSK system. It is a common system for transfer equipment and stationary tooling (lathes). The gage diameter of KM toolholders is almost equal to HSK holders, but the shaft length is 20% longer. The KM interface is designed to allow for larger taper interference (about two to four times) than that of the HSK interface. The KM system generally uses large flange contact, equivalent to HSK Form B, and a

FIGURE 5.37 Hollow shank taper and face fitting KM toolholder system. (Courtesy of Kennametal Corporation.)

ball-track clamping mechanism with locking balls that is significantly different from that used in the HSK system (Figure 5.33). The clamping force for the KM interface is higher than that of the HSK (about 35 to 45 kN for the KM-63). Kennametal specifies the toolholder material in order to sustain the high stresses and deformation under high clamping forces.

A comparison of KM sizes with sizes of other style interfaces and their corresponding allowable torque is given in Table 5.2. The radial and axial accuracy and the repeatability of the KM system are similar to those obtained with the HSK system. The static and dynamic bending stiffness of KM interface falls between that of styles A and B of the corresponding HSK size as illustrated in Figure 5.24. The limit of bending moment is considered equivalent to the HSK interface. However, KM is designed to produce a higher clamping force than HSK resulting in higher bending load resistance because of its higher clamping force. For similar clamping forces, the properties are similar. A higher clamping force for KM can delay face separation for large bending moments. The dynamic torsional stiffness for the KM-63 is also equivalent to HSK interface as shown in Figure 5.25. The torque transmission stiffness of the KM system is larger than that of the HSK system since KM holders carry torque both with the locking balls and a key. The KM connection requires cleanliness practices equivalent to those of the HSK connection.

The *NC5 connection* is a 1/10 short taper interface with double contact obtained by flexible elements [29]. The NC5 taper toolholder is designed similar to the 3-Lock 7/24 system discussed above in Figure 5.27 using a drawbar similar to 7/24 taper, but its taper angle and length are the same as the HSK. The big advantage NC5 has over HSK is that the Bellville springs act as a shock absorber. Any vibrations created due to hard milling, unbalance, or extended length are absorbed by the Bellville springs. The clamping force requirement is similar to conventional 7/24 taper (given in Table 5.1) at the higher range. The static bending stiffness of NC5 is somewhat lower (about 10 to 15%) than that of the corresponding sizes of KM and HSK interfaces. The normalized bending stiffness for the NC5 is about 10 to 20% lower than that of HSK interface (which is given in Figure 5.22 and Figure 5.23) and the limit of bending moment for NC5 is 30% lower than the HSK system.

TABLE 5.3
Evaluation of NC5 Interface Against the Conventional 7/24 Taper and HSK (Courtesy of Nikken Works, Ltd.)

Tool Shank	BT-40	HSK-A63	NC5-63
Static stiffness, K	1	1.04	1.03
Damping ratio, ζ	1	0.77	1.13
Chatter criterion, $K\zeta$	1	0.80	1.16

However, the NC5 has better dynamic stiffness than the HSK and conventional 7/24 taper as shown in Table 5.3.

The *Sandvik Coromant Capto* system (Figure 5.38) uses a 1/20 tapered tri-polygon connection; it provides simultaneous taper and face spindle contact by allowing a relatively small elastic deformation of the shallow hollow taper shank in a manner similar to the HSK and KM systems. It provides self-centering and self-aligning properties and high torque stiffness due to the harmonic tri-polygon shape. The gage diameter of the Capto system is 8% smaller while its shaft length is 20% longer than the HSK system. The polygon transmits torque through the taper shank and eliminates the need for drive keys or balls and wedges (as required with all other taper systems), whose presence may create balancing problems.

The clamping force for Capto is about 75 to 100 kN for the three sizes C6, C8, and C10, which is higher than that of all other interfaces (see Table 5.1). The triangular form ensures that the forces applied to the coupling, particularly when rotating tools are used, will be evenly distributed in a symmetrical manner. The static and modal bending stiffness of Capto

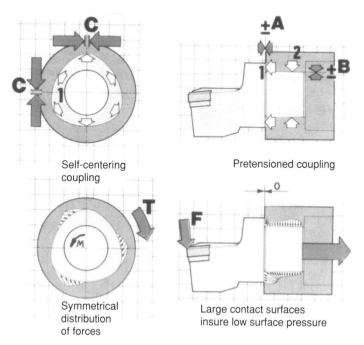

FIGURE 5.38 Hollow shank tapered polygon and face fitting Capto toolholder system. (Courtesy of Sandvik Coromant.)

interface is equivalent (or 10 to 20% higher) to that of HSK interface for the corresponding sizes as shown in Figure 5.24. The dynamic bending stiffness for the Capto system is significantly lower (30 to 50%) than that of HSK interface also shown in Figure 5.24. The Capto interface also transmits higher torque than any other automatic change coupling as long as a female adapter is not used (Table 5.2). When an adapter is used, as in manual systems, the bolts holding the adapter on the spindle or turret transmit the torque. The Capto system can potentially carry a larger torques than KM, HSK, and CAT systems, but we have not observed torque-carrying capability to be a problem with any of the connections. The dynamic torsional stiffness of the Capto system is somewhat less than that of HSK. The radial and axial accuracy and repeatability of the Capto system are equivalent or better than those obtainable with the HSK and KM systems.

The Capto system and, to a lesser extent, the KM system are most commonly used for stationary and rotary manual or semiautomatic applications (e.g., turning centers, transfer lines, etc.) because they provide higher bending stiffness than the HSK and CAT systems.

All the above advanced designs employed a "virtual taper" surface on the toolholder and provide face contact between the toolholder and the spindle and thus assure enhanced bending stiffness and high axial accuracy. A comparison between several toolholder coupling interface systems is given in Figure 5.39.

5.3.5 QUICK-CHANGE INTERFACES (TOOLHOLDERS/ADAPTERS)

A variety of quick-change manual toolholding systems are also available. The most significant differences between systems are generally in the design of the connection between the adapter head (male) and the base receiver (female) unit, which is either an integral part of the spindle shaft or an adapter that is usually removed only in case of catastrophic tool accidents.

Toolholder style	Bending stiffness	Torque stiffness	Accuracy	Tool change properties
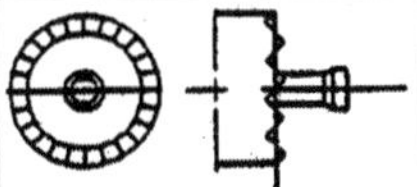	Poor	Very good	Good	Good
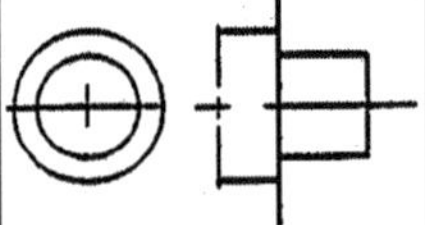	Very good	Poor	Good	Poor
	Good	Poor	Poor	Good
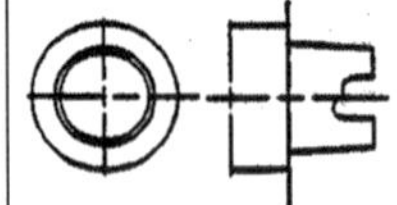	Very good	Poor	Good	Good
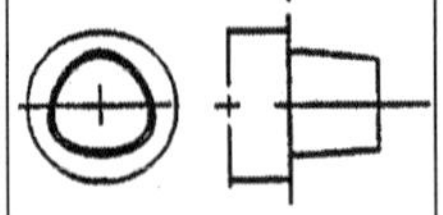	Very good	Very good	Very good	Good
	Very good	Poor	Good	Very good

FIGURE 5.39 Comparison between toolholder coupling principles.

Toolholders and Workholders

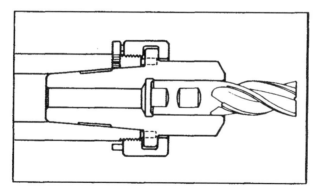

FIGURE 5.40 Quick-change toolholder taper connection. (Courtesy of Scully Jones Corporation.)

The configuration of male/female connections varies between manufacturers. In the Pawl-Lock design, shallow angle self-locking tapers are used in some connections for manual tool changing machines which can hold an average positional accuracy within 0.015 mm with maximum runout of 0.03 mm (Figure 5.40, e.g., Kwick-Switch, X-Press, Tru-Taper-Smith Lock, Smith Super Taper); to change tools, the operator merely gives a quarter turn on the locknut which secures the secondary holder in the spindle during cutting. The TIR for two such systems A and B using a precision single angle collet is listed in Table 5.4.

In other systems, a cylindrical male pilot in the back of the toolholder is inserted into a female bore and one, two or three set screws, placed symmetrically around the female bore wall, hold the mating cylindrical parts in place (Figure 5.41); the set screws often have a conical end or ballnose that seats in the rear of sockets or a notch (keyway) in the tool/holder pilot shank. When the socket is machined slightly below the setscrew centerline, the end of the setscrew pushes against the socket. This provides face contact and applies pressure to the flange face and on the periphery of the adapter and receiver interface; this creates a rigid connection. The preload of the connection at the face contact is a function of the geometry and tolerances of the setscrew end and the socket in the male pilot, as well as the tightening torque of the setscrew. Multiple setscrews are used in order to increase the preload at the face contact of the connection and possibly the torque capability.

A similar design uses a floating pin (ABS) that is axially offset from the thrust and receiving screws as shown in Figure 5.42; when the floating pin is compressed by tightening the thrust screw, forces F_1 and F_2 are compounded to $2F_A$ in the axial direction acting at the face of the connection. The axial force at the face of the interface and allowable transmission torque for the ABS system are given in Table 5.5 for a specified torque on the thrust screw.

TABLE 5.4
Total Indicator Runout (TIR) for Quick-Change Toolholders

Holder Type	Test Arbor Diameter (mm)	Test Arbor Projection in Diameters	Average TIR (mm)	Total Range TIR (mm)
A	9.5	5.3	0.014	0.030
A	9.5	9.3	0.021	0.046
B	7.9	5.3	0.015	0.025
B	7.9	8.6	0.021	0.041

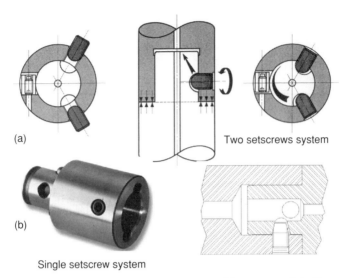

FIGURE 5.41 Quick-change interface with one and two side screws. ((a) Courtesy of EPB, Inc.; (b) Courtesy of Parlec, Inc.)

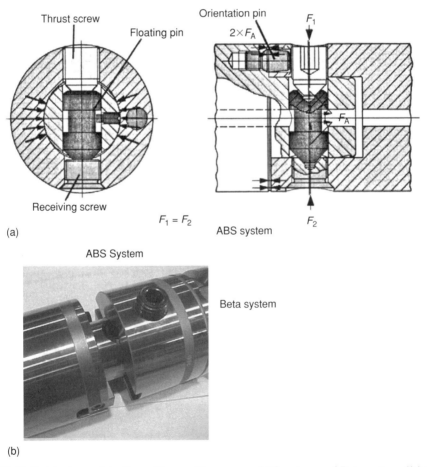

FIGURE 5.42 Quick-change interface with one side screw — ABS system and Beta system. ((a) Courtesy of Komet, Inc.; (b) Courtesy of Command Tooling Systems.)

TABLE 5.5
Specifications for the ABS Modular System Connection

ABS Size	Torque on Thrust Screw (N m)	Axial Force (kN)	Transmission Torque (N m)
20	2.8	9.8	51
25	4.5	19.6	73
32	18	29.4	192
40	28	36.0	294
50	36	42.7	700
63	56	48.9	1,491
80	68	55.6	2,994
100	90	75.6	5,999
125	102	91.1	10,959
160	113	104.5	14,687
200	124	113.4	17,512

A drive pin or key in such connections is used for locating and torque transmission. Other designs similar to this are the Beta interface (see Figure 5.41); their differences are in tolerancing and actuation pressures.

There are several other similar systems with side lock screws on the market. The important design and manufacturing characteristics of systems with side lock screws are the clearance tolerance between the straight pilot (adapter) and bore (receiver) and the perpendicularity of the flange face and pilot and mounting face and bore. The female bore is slightly larger than the male cylinder and should be eccentric to the spindle rotational axis, so that the male cylinder is concentric to the spindle axis after the set screws push the tool shank against the bore; however, this is not true with most systems. The clearance should be as small as possible and usually varies between 0.004 and 0.025 mm; increasing this clearance increases the TIR of the cutting edge. Generally, modular toolholding connections can provide precision equivalent to the H8/H9 ISO-tolerance.

The performance of all front side-locking systems suffers due to the extra length needed to incorporate the locking mechanism, which brings the cutting edge further from the gage line. Typically, a penalty of 30 mm can be expected. However, indexable tip tools with integral adapters reduce the total tooling length.

Some of these interfaces were compared and the results are given in Figure 5.43–Figure 5.45. Some of these interfaces can provide equal or better static bending stiffness than the HSK interface of similar size as shown in Figure 5.43. The static torsional stiffness for some of these modular connections is as good as for the HSK interface as illustrated in Figure 5.44. The dynamic stiffness for some of the modular connections was better than HSK interface as shown in Figure 5.45. Some of the simultaneous fit straight shank (side lock or side screw) interfaces perform as well or better than the simultaneous fit expandable taper shank interfaces.

5.3.6 TOOLHOLDERS FOR TURNING MACHINES

Tool blocks are generally used in turning centers, although conventional square-shanked toolholders, round-shanked boring tools (Figure 5.4), and quick-change tooling systems are also common.

Tool block systems have taken several forms which are combined in different machine-adaptable clamping units such as the VDI system, the standard MTP adapter plate, standard

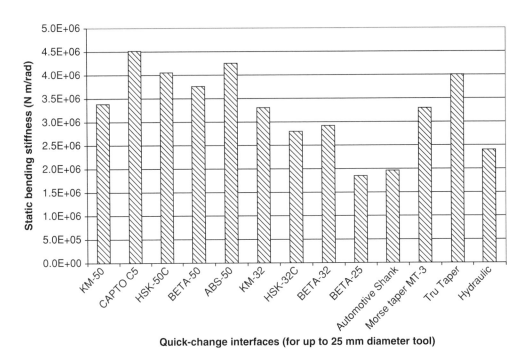

FIGURE 5.43 Comparison of static bending stiffness for quick-change interfaces using 25 mm tool.

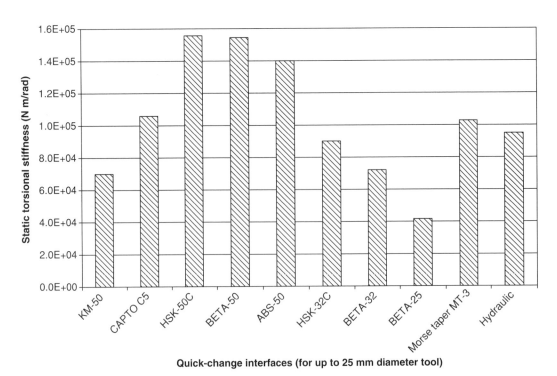

FIGURE 5.44 Comparison of static torsional stiffness for quick-change interfaces using 25 mm tool.

Toolholders and Workholders

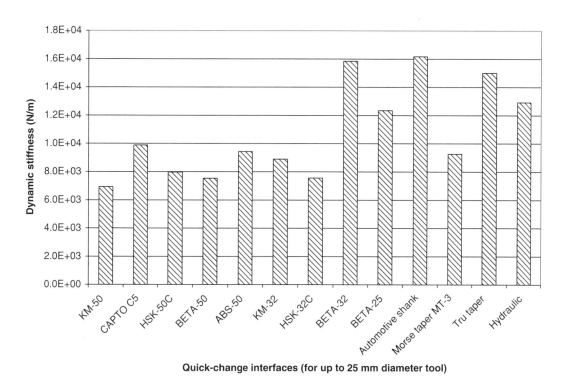

FIGURE 5.45 Comparison of dynamic bending stiffness for quick-change interfaces using 25 mm tool.

or custom bolt-on blocks, custom one piece turret disks, as shown in Figure 5.46. CAT-V style toolholders have been used widely as a tool adapter system (MATS) in the United States but have limitations at heavy bending loads, since they do not provide face contact. The VDI 3425/2 (DIN 69 880) toolholder system with a round shank and rack machined on the shank (see Figure 5.47) is widely used in Europe. A mating rack in the turret clamps the adapter in place. The KM and Capto modular toolholding systems were initially designed for lathes and provide good performance (Figure 5.22–Figure 5.24). HSK toolholders are also used for turning but have less bending load capacity than the Capto and KM systems. A modular lathe tooling system enables users to perform a wide variety of external, internal, and rotating tool operations with a minimum number of components.

The design features of quick-change tooling systems vary, but the systems available can be divided into two categories: cutting unit systems, and tool adapter systems (tool blocks) as shown in Figure 5.47. The cutting unit system uses a replaceable clubhead designed for square shank holders for turning operations, and less effectively for boring and especially rotating applications. In this system, turning, boring, and rotating applications are not interchangeable. The precision coupler delivers excellent repeatability when clamping the same unit with good rigidity.

5.3.7 EVALUATION AND COMPARISON OF TOOLHOLDER–SPINDLE INTERFACE

A comparison between several toolholder–spindle interfaces is given in Table 5.6. The static and dynamic bending stiffness and the torsional stiffness were measured from a bench test [10]. The reliability is based on evaluation matrix using factors such as fracture fatigue, dirt/nick sensitivity, wear/corrosion sensitivity, sticking extra force, torque slippage, temperature sensitivity, maintenance lubrication, coolant feed, and gauging. The ease of tool change is

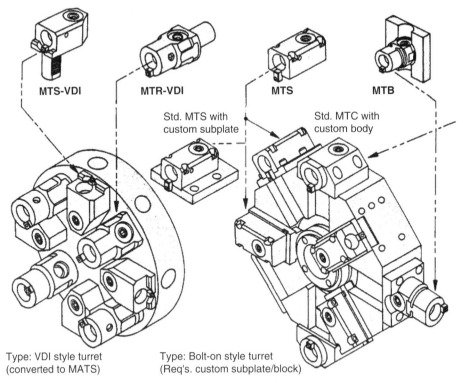

FIGURE 5.46 Tool blocks and turret designs (MATS) for quick-change tooling in turning machines. (Courtesy of Carboloy Corporation.)

based on an evaluation matrix with the factors of holder weight, length engagement, surface cleanliness, drawbar force, critical location, and coolant through ease.

Toolholder–spindle interfaces can be evaluated and compared based on experimental and analytical/finite element results. The bench test results of the static and dynamic stiffness, shown above among the different interfaces (Figure 5.18–Figure 5.45), indicate that some toolholders are superior to others (e.g., HSK-A63 interface is superior to CAT-40 interface). However, the bench test results (an approach always used in the past for comparing different sets of interfaces) to characterize toolholder–spindle interfaces do not represent properly the interfaces in a machine tool spindle. Bench tests provide a relative comparison between toolholders, but the results will be different when the same toolholders are evaluated on a spindle because the static and dynamic stiffness seen at the tip of the cutting tool also depend on the stiffness of the tool, the spindle geometry and bearings, the housing, and the overall machine structure. For example, a comparison of the static bending stiffness between the HSK-A63 and CAT-40 interfaces in identical machine tool spindles is shown in Figure 5.48. The tool stiffness is about equal for both interfaces especially below 600 N radial force at the tool end. In contrast, the results from the bench tests are also shown in Figure 5.48, which indicate that the HSK interface is much stiffer than the CAT system (as explained in Figure 5.22 and Figure 5.24). The natural frequencies for the two interfaces from the machine tool spindle tests were in good agreement, with a difference of less than 5%. The modal stiffness for the CAT-40 was somewhat larger than that of the HSK-A63.

A group of researchers [30–33] proposed the characterization of the toolholder–spindle interface using the joint stiffness parameters. A methodology for estimating the joint stiffness

Toolholders and Workholders

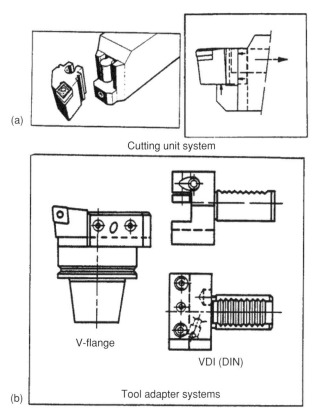

FIGURE 5.47 Quick-change tooling systems for turning machines. ((a) Courtesy of Kennametal, Inc.; (b) Courtesy of Valenite, Inc.)

parameters of a toolholder–spindle interface using one linear spring and one rotational spring (see Figure 5.49) has been found very effective [34]. A flow chart of the methodology is shown in Figure 5.50. It is based on the FRF (Frequency response function) of the interface system. It involves the FEA and experimental measurements of a bench fixture. The most important assumption is that the behavior of the test system is linear. After the joint stiffness parameters are estimated, they can be used in any machine tool spindle FEM Finite element model (as illustrated in Figure 5.51) to estimate the static and dynamic characteristic at the tool tip.

The Spindle Analysis program (SPA) from Manufacturing Laborites, Inc. was used to create a two-dimensional (2D) axisymmetric model of a spindle for evaluating the CAT-40 and HSK-A63 interfaces. The spindle shaft, the housing, and the toolholder structure were modeled as shown in Figure 5.51. The SPA program can model a two degrees of freedom (2 DOF) spring to include a linear direction and rotational direction to represent the joint between the spindle and the toolholder as illustrated by spring S3 in Figure 5.49. The SPA FEA program is traditionally capable of simulating both the dynamic and static behaviors of a spindle and tooling system. The taper joint stiffness parameters for the 2 DOF spring S3 are given in Table 5.7 and obtained using the methodology in Figure 5.50 [33]. Springs S1, S2, and S4 represent the spindle bearings and springs S5 and S6 represent the connection of the spindle housing to the machine tool structure. The static and modal analysis results are given in Table 5.8 and Table 5.9, respectively. The toolholder deflection and natural frequencies or mode shapes were similar for both interface styles. The first two natural frequencies from the SPA FEA (in Table 5.9) were within 5% those from the machine tool spindles. In addition, the

TABLE 5.6
Comparison of Several Interfaces With Respect to Reliability, Ease of Change, Relative Cost, Balance, and Stiffness

Toolholder Type	Reliability[a]	Ease of Change[b]	Relative Cost	Balance[c]	Static Bending Stiffness[d] ($\times 10^8$ N/m)	Dynamic Bending Stiffness[e] ($\times 10^5$ N/m)	Static Torsional Stiffness[f] ($\times 10^6$ N m/rad)
HSK 63A	1	2	5	8	1.44	5.12	1.01
HSK 80B	1	2	2	8	1.94	2.51	1.14
HSK 100A	1	1	3	5	1.65	9.92	3.25
HSK 125B	1	1	1	5	1.99	8.05	3.62
CAT-40	10	10	10	10	0.73	5.35	0.67
CAT-50	10	10	9	7	1.15	4.26	2.35
FC-CAT-40	9	6	5	10	1.26	6.36	NA
FC-CAT-50	8	4	6	7	1.64	8.52	NA
KM 63	6	5	6	8	1.42	4.56	NA
Capto C6	6	6	6	NA	1.47	4.51	NA
Capto C8	5	5	5	NA	2.47	10.20	NA

Note: 10 = best; 1 = worst.
[a] Reliability based on evaluation matrix.
[b] Ease of tool change based on evaluation matrix.
[c] Balance of holder as made.
[d] Static bending stiffness from measurements (ratio of the applied force and the resulting deflection).
[e] Dynamic bending stiffness from measurements (product of modal stiffness and the damping ratio).
[f] Static torsion stiffness from measurements (ratio of the applied torque and the resulting rotation).

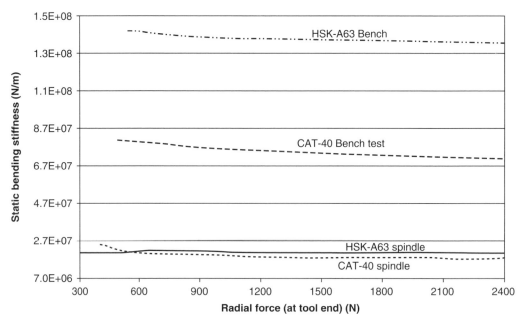

FIGURE 5.48 Comparison of static bending stiffness between CAT-40 and HSK-A63 toolholders on bench fixture and in machine tool spindle.

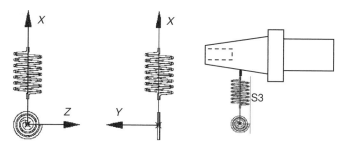

FIGURE 5.49 2 - DOF (linear and rotational) spring orientations used in SPA FEA model.

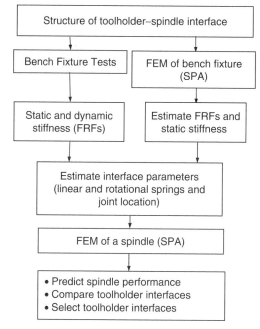

FIGURE 5.50 Flow chart of methodology for evaluating toolholder–spindle interfaces.

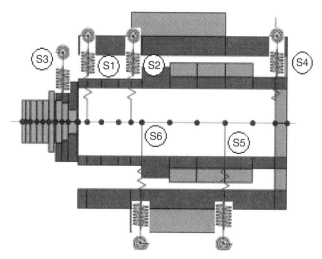

FIGURE 5.51 SPA model for CAT-40 spindle nose and test tool.

TABLE 5.7
Average Joint Characteristics for the CAT-40 and HSK-A63 Interfaces

		Spring Constants	
Interface	Joint Location (m)	Linear (N/m)	Rotational (N m/rad)
CAT-40	0.02	1.375×10^9	8.108×10^6
HSK-A63	0.02	2.335×10^9	1.587×10^7

TABLE 5.8
Comparison of Estimated Static Stiffness (SPA) in the Machine Spindle

Applied Load (N)	CAT-40 (N/m)	HSK-A63 (N/m)
1000	2.622×10^7	2.706×10^7

TABLE 5.9
Comparison of Estimated Modal (SPA) Parameters in the Machine Spindle

	CAT-40		HSK-A63	
Mode	ω_n (Hz)	K (N/m)	ω_n (Hz)	K (N/m)
1st	517	1.39×10^8	517	1.45×10^8
2nd	671	4.10×10^7	675	4.22×10^7
3rd	1521	3.12×10^8	1532	3.33×10^8

natural frequencies are much lower when the toolholder is in the spindle than in the bench fixture [33]. These results emphasize the importance of evaluating the toolholder system on the actual production spindle instead of a bench test.

The static bending deflection of the tool can also be defined analytically using the joint stiffness parameters; The bending deflection of the tool includes the elastic deflection of the bar itself, the deflection of the toolholder–spindle joint, and the spindle deflection at the front bearing(s). The stiffness of the spindle–toolholder interface is decomposed into a rotational spring and a translation (linear) spring:

$$\delta_{total} = \delta_b + \delta_h + \delta_r + \delta_s \tag{5.7}$$

where δ_b is the deflection for a cantilever (rigid body) beam, δ_h is the translation of the taper interface, δ_r is the deflection due to rotational spring of the taper interface, and δ_s is the translation of the spindle shaft due to bearing stiffness. If the tool is not monolithic and there is an additional connection between the tool and the toolholder, the toolholder interface should be also considered as a summation of the deflection due to linear and rotational springs. This is further discussed in Example 5.8.

The receptance coupling substructure analysis (RCSA) method [35, 36] can be used to predict the tool point dynamic response for machine tool applications. This method considers the toolholder and tool as an assembly of two substructures. The tool interface receptance is determined analytically, while the toolholder–spindle interface receptances are determined experimentally using a standard test holder and finite difference calculations. The tool point dynamics is predicted from the RCSA by coupling rigidly the two substructures. This method includes the following features: (a) experimental identification of the toolholder–spindle interface translational and rotational receptances using a finite difference approach; (b) analytical determination of the toolholder–tool interface receptances; and (c) rigid coupling of the toolholder–spindle and toolholder–tool interfaces to determine the tool point response. This method eliminates the need to measure the tool–toolholder–spindle dynamics for each tool and holder combination in a particular machine and, therefore, significantly reduces the number of impact tests required for stability lobe diagram generation.

The material removal rate that a machining center can achieve is strongly dependent upon the static and dynamic characteristics of the machine–tool–workpiece system as seen at the tip of the tool. The selection of the best spindle–toolholder interface is not simple because it depends on the intended use of the machine. The stiffness is one of the important parameters to consider in the selection. The static stiffness measures the deflection at the end of the tool in response to a static force. It provides some indication of the ability to create a surface or hole with the tool in the intended location. In milling, drilling, reaming, and boring with a relatively low spindle speed, the error in location of the machined surface or hole is related to the static stiffness. The higher the static stiffness, the more accurately the surface or hole will be located with smaller form error. If static stiffness is the performance criterion, then the data show that the face contact connections perform a little better than the nonface contact connections in a machine tool spindle, as opposed to the results from a bench fixture showing that the HSK interface is superior to the CAT interface. Therefore, the static stiffness of the tool should be estimated by considering the spindle stiffness in order to properly select an interface style. The static stiffness can be estimated using the 2D FEA or analytically assuming the joint stiffness parameters are available.

In other cases, the aims are high metal removal rate (in milling, boring, etc.), or accurate surface location at higher spindle speeds. In both of these cases, the better selection criterion is dynamic stiffness (or chatter criterion as discussed in Chapter 12) as seen at the tip of the tool. The dynamic stiffness is a combination of stiffness and damping in a particular mode of vibration (at a particular frequency). These are the "modal" stiffness and modal damping. The damping is estimated from the FRFs while the natural frequencies and modal stiffness can be estimated using several programs. For dynamic cutting performance at the tip of the tool, the 2D FEA can also be used assuming that the joint parameters are available.

There are cases where the bore quality or surface flatness are very critical, in which case the quality of the interface taper and face including the quality of the tool point (runout) with respect to the taper interface are very important. In addition, the static bending stiffness becomes important assuming it is not a high-speed application. It was observed that the rigid face contact interfaces provide higher static stiffness or normalized bending stiffness at higher moments assuming: (1) the taper quality for the nonface contact interfaces is worse than AT3 tolerance, and (2) the clamping force is high enough to sustain larger bending moments without losing contact between the toolholder and the spindle face.

5.4 CUTTING TOOL CLAMPING SYSTEMS

Cutting tools are generally held in the machine spindle by toolholders. Depending on the tool or cutter type, toolholders such as those shown in Figure 5.11, modular toolholders, standard collet chucks, or end milling holders, may be used. There are also several other approaches to hold round shank tools that deviate from the traditional collet chuck and eliminate the threaded locknut. These include hydraulic chucks, milling chucks, and shrink fit mountings. Toolholder dimensions outside the flange area and toolholder concentricity and tolerancing are not covered by ANSI standards and are left entirely to the manufacturer. Therefore, all toolholders are not alike.

Coolant-fed rotary toolholders are usually supplied through a center hole in the toolholder and less frequently from the flange of the toolholder (Figure 5.52). However, external coolant can be supplied through the tool using a rotary gland mounted either on the toolholder or the tool itself as shown in Figure 5.52.

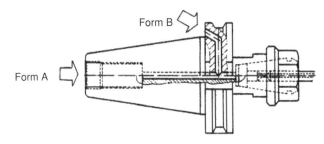

V-flange straight shank holder
Coolant: end entry through the spindle

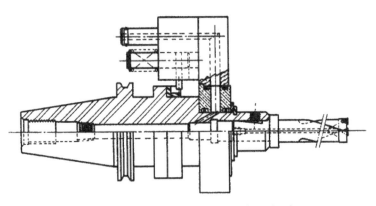

V-flange automatic tool change coolant gland
Coolant entry options:
1. End entry through the spindle
2. Coolant gland mounting for automatic tool change applications

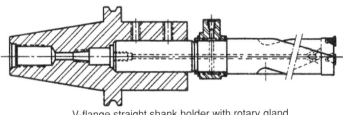

V-flange straight shank holder with rotary gland
Coolant: side entry

FIGURE 5.52 Typical coolant toolholder adapters for coolant-fed rotary cutting tools. (Courtesy of The George Whalley Co.)

Toolholders and Workholders

5.4.1 MILLING CUTTER DRIVES

Milling cutters can be connected directly into the spindle if quick-change adapters (e.g., CAT-V, HSK, KM, Capto, or curvic coupling shanks) are attached as an integral part of the cutter body (Figure 5.53). Milling cutters with diameters up to 160 mm are generally mounted on a "stub arbor" toolholder and held in place by a single arbor screw (Figure 5.53). Arbors are available to fit all different spindle sockets. This method is very popular, especially for aluminum cutter bodies which must be mounted on steel toolholders. Face milling cutters with diameter over 160 mm generally mount directly to the face of the machine spindle. The two most popular methods of mounting face mills are the centering plug (flat back drive) method and the National Standard (mounting ring) method. In the centering plug method, a centering plug is used to locate the cutter on the spindle face. In the National Standard method, there is either a recessed locating diameter on the back of the cutter or a mounting ring fitted to a groove on the rear face of the cutter which ensures proper location on the spindle face (Figure 5.53). The cutter is bolted directly to the spindle in both methods.

5.4.2 SIDE-LOCK-TYPE CHUCKS

Weldon chucks (Figure 5.11A) are used with round shank cutters with one or two Weldon flats and beveled corners. Weldon chucks are simpler and stronger but less versatile than collet chucks (described below). They were specifically designed to drive end-milling cutters,

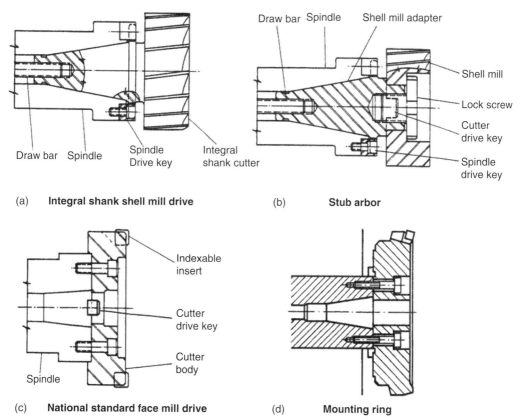

FIGURE 5.53 Types of milling cutter holding devices. ((a–c) Courtesy of Valenite, Inc.; (d) Courtesy of Sandvik Coromant.)

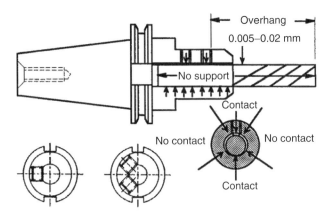

FIGURE 5.54 Clamping contact area for Weldon-type holders.

but have also been used with indexable drills and boring tools. The standard Weldon chuck has one side-mounted clamping screw; the modified Weldon chuck has two clamping screws. The resulting close fit between the screw and the flat prevents end mill slip or pullout while providing maximum driving power. The cutter is held only by a portion of its periphery against a thin holding surface (the cylinder inside a cylinder design means that there is only one line of contact) as shown in Figure 5.54. Runout is often unavoidable and is equivalent to the sum of the manufacturing tolerances of the tool shank and the pilot bore; it may exceed 0.01 mm. Runout can be minimized by machining the pilot bore about 0.008 mm eccentrically with respect to the axis of rotation of the toolholder in the direction of the side-screw. Using two clamping screws instead of one increases the static and dynamic stiffness of the system by as much as 50%.

Another type of side lock toolholding arrangement is the standard whistle-notch (Figure 5.11A), a 5° inclined flat ground on the tool shank; the side set screws on the holder are not perpendicular to the tool axis, but inclined by 5°. This allows the tool to be pushed against a back stop. Back stop contact is not always achieved, which can result in loosening of the tool at high force levels. The whistle-notch method is sometimes used with drills.

Side-lock-type holders are unsuitable for high speed operations because the pilot expands under centrifugal forces, resulting in a loosening of the tool–toolholder joint. The dynamic stiffness is low, particularly in the direction 90° offset from the set screw. Therefore, Weldon and whistle-notch system toolholders may cause high vibration, resulting in poor surface finish and reduced bearing life. The modified Weldon system using two set screws on the same diametrical plane provides increased dynamic rigidity and it is generally suitable for higher speed milling, but is not suitable for maximum speed and power applications.

5.4.3 Collet Chucks

Collet chuck toolholders are commonly used for drills and reamers and may also be used with boring bars and end mills (Figure 5.55). These toolholders consist of a front socket, a collect, and a locknut as shown in Figures 5.56. Collets are cylindrical fixtures with one or more slots to generate flexible fingers designed to grip smooth cylindrical elements (Figure 5.57). The collet is pushed into a tapered socket, which causes the fingers to deflect and grip. "Nonpull" style end mill collets have a plug that contacts a flat on the end mill shank (i.e., a Weldon connection) to provide both positive drive and positive axial tool retention (Figure 5.57). Extended length positive pull-collets have been used in extended nose toolholders; in this arrangement, a draw nut or draw bolt is used to open and close the collet as shown in Figure 5.58.

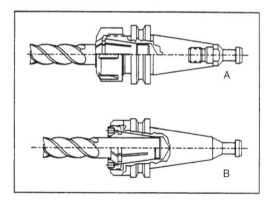

FIGURE 5.55 Single-angle collet chuck toolholders. (Courtesy of The Precise Corporation.)

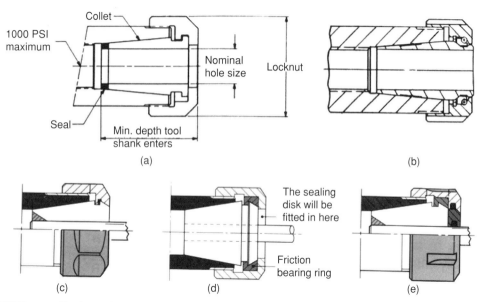

FIGURE 5.56 Single-angle collet systems: (a) STD nut, (b) ball bearing, (c) DIN 6499-D and -E, and (d, e) friction bearing system. ((a, b) Courtesy of TSD DeVied-Bullard, Inc.; (c–e) Courtesy of Rego-Fix Tool Corporation.)

Collets come is a variety of IDs to accept the different tool shank sizes. The cutting tool shank is inserted into the collet fingers which are compressed by tightening a threaded cap over the collet to secure the tool. Holding power around the shank is provided by compressive stress in concentric wedge-shaped collet sections. Holding power is increased by imposing axial motion on the tool shank as the collet is tightened. Gripping power is maximized when there is full length engagement between the tool shank and the collet bore. The grip power drops significantly when engagement becomes less than two thirds of the full length of the collet bore.

The collet envelope size is a function of the maximum bore size and the outside shape designation. The envelopes for Erickson collets of increasing gage diameter are 50TG, 75TG, 100TG, and 150TG, which have maximum bore sizes of 12.5, 19.05, 25.4, and 38.1 mm, respectively. The minimum collet envelope relative to bore size should be used for high-speed

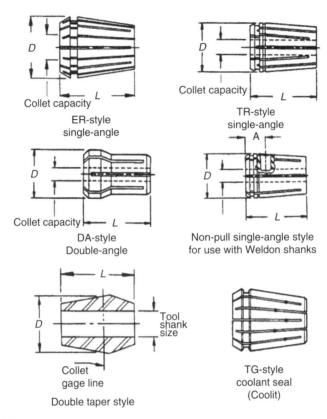

FIGURE 5.57 Collet types.

applications because it balances the collet grip force and mass of the chuck. However, the maximum bore size in the collet envelope should be selected in general-purpose applications to obtain maximum grip torque (e.g., a 12.5 mm collet provides only 60% the grip torque efficiency than that of the 19 mm shank in an Erickson 75TG envelope [37]). Using the largest envelope size collet allowed by the part and fixturing maximizes gripping power and stiffness (see Figure 5.61).

Types of collets include single and double angle collets (Figure 5.57). Single angle collets are generally superior. There are several single angle collets with different cone angles; the most popular are the Rego-Fix ER style 16° (DIN 6499) included angle and the Erickson TG style 8° included angle. A lower included angle results in higher grip power (rigid tool grip) for the collet (e.g., a TG 8° collet provides twice the holding power of an ER 16° collet). The number and width of slots around the periphery determines the collapsibility or size range for the collet. A larger number of slots around the collet results in a larger collapsible range, better accuracy, and higher gripping power. When possible, a shank size equal to the nominal size of the collet should be used [37]. Collets with fewer slots designed to clamp only nominal diameter tools provide the highest concentricity. A light film of oil on the collet chuck socket cavity can increase grip power by as much as 40 and 50% using a steel and carbide tool shank, respectively. Carbide shanks generally grip slightly tighter than steel shanks. A double taper collet with one slot which does not extend through the wall thickness has low collapsibility (roughly 0.07 mm as compared to 0.3 to 1.0 mm of standard collets) but very high gripping power.

Toolholders and Workholders

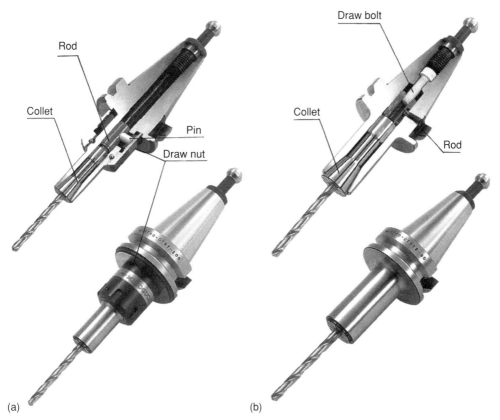

FIGURE 5.58 Extended positive pull-collet designs for extended toolholders with a 2 mm chucking range. (Courtesy of Tecnara Tooling Systems, Inc.)

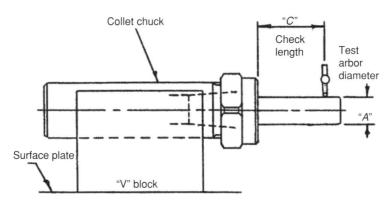

FIGURE 5.59 Collet chuck assembly total indicator runout measurement as properly noted in the DIN 6499 document. (Courtesy of TSD DeVlieg-Bullard, Inc.)

Collet accuracy is dependent on the taper fit and the roundness and surface finish of the collet and toolholder cone. The collet accuracy, concentricity tolerance, or TIR allowance is measured as an assembly as shown in Figure 5.59. The concentricity tolerance requirements for DIN 6499 are given in Table 5.10 measured in a chuck collet extension located on a "V" block or a precision spindle as shown in Figure 5.59. Class I and Class II generally represent

TABLE 5.10
Single Angle "ER" Assembled Collet Chuck Accuracy (Using the Procedure from DIN 6499)

"A" Rated Diameter (mm)		"C" Tool Overhang (mm)	Concentricity Tolerance (mm)			
Over	To		STD ER[a]	ER[a] Ultra Precision	Class I DIN 6499	Class II DIN 6499
1	1.6	6				
1.6	3	10	0.010	0.005	0.008	0.015
3	6	16				
6	10	25				
10	18	40	0.010	0.005	0.010	0.020
18	26	50				
26	34	60	0.015	0.010	0.015	0.025

[a] Specified by Rego-Fix.

high precision and standard collets, respectively. The runout of standard single angle collets is normally between 0.01 and 0.02 mm, but can be easily doubled or tripled when the holder is taken into account. The accuracy of precision single angle collets is about 0.008 to 0.010 mm. Gripping accuracy decreases as the shank size deviates further from the nominal collet bore; the accuracy of the extended collet in Figure 5.58 at an L/D ratio of 4 is between 0.01 and 0.02 mm, increasing as the tool shank deviates further from the nominal collet shank size. Growth during rotation affects slotted collets. For this reason, precision collets (accurate in terms of diameter and concentricity) are required for high-speed toolholders. The appropriate recommended nut tightening torque range should be used (100 to 135 N m for 75TG, 150 to 200 N m for 100TG, and 200 to 270 N m for 150TG). As the collet is tightened on the tool by the nut, the collet must be forced into the tapered socket, which can cause the tool to move from 0.2 to as much as 0.7 mm. The grip force for the 100TG collet chuck is between 300 and 600 N m. A 40% loss of grip can occur with a metal backup screw; much lower losses occur with plastic backup screws.

Double angle (DA) collet systems are used when maximum rigidity and close tolerance are not required (Figure 5.57). Their runout ranges from 0.025 mm for larger collets (>3 mm) to 0.1 to 0.2 mm or higher for smaller collet sizes. A DA collet produces roughly half the grip force of a TG collet.

Several uniquely designed coolant-fed collet chucks which seal the adapter and cutting tool are available. These include coolant-control collets, such as (1) collets with slots filled with silicon rubber, (2) Coolit collets using rubber composite plugs in the slots (Figure 5.57), and (3) collets with slots which are open at the small end and do not intercept the groove at the front of collet. Special washers can also be used to seal the collet; typical arrangements include (4) a nylon back-up screw at the cutter end, (5) a nylon seal at the back of the collet (Figure 5.56A), and (6) a metal–nylon washer seal at the front of collet in the nut (Figure 5.56E). Types 1, 2, and 3 work well, with small leakage if any, when the collet size matches the tool diameter and minimum collet collapse is required. Type 1 is often avoided because the tool concentricity can be substandard if the silicon in the slots is not perfectly deposited. Type 4 seals the back end of cutting tool. The cutting tool should be designed to match the seal; a common center oil-coolant hole or multiple oil-coolant holes without grooves should be used. Similarly, an O-ring assembled at the tip of the backup screw can

Toolholders and Workholders 321

be used to seal the end of tool shank. Type 5, which compresses the nylon against the collet (Figure 5.56A), may reduce the grip power of the tool by 20 to 30%.

Coolant can be supplied through the collet to the cutting area in four different ways (Figure 5.60): (1) through an oil hole in the cutting edge, (2) through adjustable nozzles mounted on the coolant cap, (3) through the space between tool shank and spacer, and (4) through the slits of collet. Methods 2, 3, and 4 are used for tools without oil holes.

The clamping locknut design also has an important impact on collet chuck performance. The clamping locknut provides the preload for the collet, and thus determines the chucking pressure. A standard nut is attached at the front of the collet by engaging the extractor spiral in the collet groove (Figure 5.56A). Another standard design described in DIN 6499-D and -E is shown in Figure 5.56C. A ball bearing clamping locknut (Figure 5.56B) is often used to provide increased chucking pressure [38, 39], which can result in a 40% increase in gripping capacity and better concentricity compared to standard locknuts with the same tightened torque. Roller bearing clamping nuts are also used to increase gripping power. A friction bearing system (Figure 5.56D and E) has also been used to improve clamping power. A comparison of the transfer torque for these nuts is shown in Figure 5.61. In high-speed applications, high-performance locknuts without an extractor spiral are used to eliminate unbalance. Although collet chucks generally provide good accuracy and dynamic stiffness, their utility at very high speeds is limited. The external collet nut expands under centrifugal force, so that the collet loses some gripping torque and may vibrate and loosen when the spindle is decelerated quickly. Better designs employ an internal nut (Figure 5.55B) [9] or no

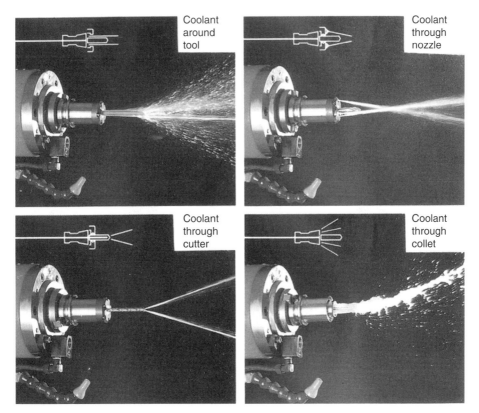

FIGURE 5.60 Coolant through systems for collet type toolholders. (Courtesy of Tecnara Tooling Systems, Inc.)

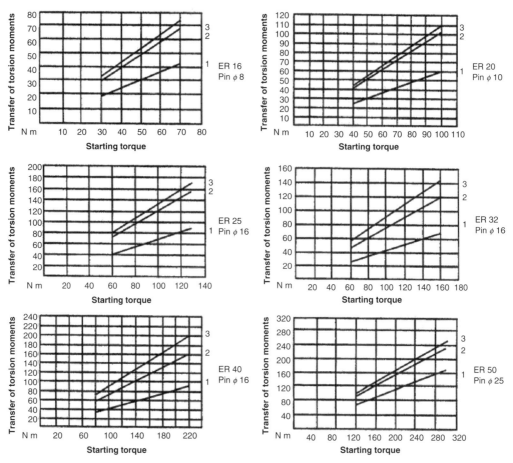

(1) Standard nut. (2) Ball bearing nut. (3) Friction bearing nut.

FIGURE 5.61 Torque transmitted by three different collet locknut designs: standard nut, ball bearing nut, and friction bearing nut. (Courtesy of Rego-Fix Tool Corporation.)

nut (Figure 5.58); at high speeds, the internal nut expands into the thread, tightening rather than loosening the connection. These designs are suitable for comparatively high power and speed. The clamping torque for a 13 mm tool shank in the extended positive pull-collet without nut, shown in Figure 5.58, is 95 N m.

The chucking pressure of the collet at the nose depends on the locknut type. A standard nut generally provides low pressure near to the nose, as compared to zero chucking pressure at the nose for the ball bearing locknut. The maximum chucking pressure for the ball bearing locknut occurs at about one fourth of the collet length from the nose [38]. The collet shown in Figure 5.58 should provide significant pressure at the nose.

5.4.4 Hydraulic Chucks

In hydraulic chuck toolholders, the wall of the inside diameter of the toolholder (an internal membrane) expands when a hydraulic pressure is applied using grease or oil, then returns to its original size when the pressure is released as explained in the toolholder–spindle connection section (Figure 5.11, Type J and Figure 5.62) [40]. An Allen screw is turned a few

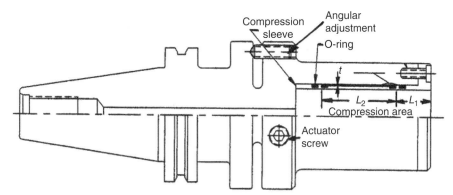

FIGURE 5.62 Hydraulic expansion toolholder with angular adjustment. (Courtesy of Hydra-Lock Corporation.)

revolutions to actuate a piston and force hydraulic fluid from the piston chamber into the expansion sleeve at 70 to 140 MPa, which clamps the tool shank with uniform pressure on its geometric centerline over the full length of engagement.

Hydraulic chucks provide very good concentricity (0.005 mm or better TIR at $2 \times D$ to $3 \times D$ from the nose of the chuck) and high repeatability (0.0013 mm or better). The runout can increase by 50 to 100% when precision sleeve collets are used to step down to smaller diameters with the same chuck. The normal amount of contraction for a 25 mm diameter toolholder is 0.01 to 0.03 mm. Contraction is proportionately higher or lower for larger or smaller diameters. The hydraulic bore is made 0.005 mm larger than the nominal diameter, while the toolholder shank has a recommended tolerance of H6 and at maximum $D/1000$. A closer fit between the cutter shank and the sleeve increases the gripping power. This arrangement provides a balanced system for medium to higher speeds.

Hydraulic toolholders provide sufficient pressure to drive the tool shank without a flat or a tang; the clamping force for a 70 to 85 MPa expansion pressure can transmit a torque of 60 and 500 N m, respectively, for 10 and 32 mm toolholder shanks. The torque capability is calculated from the available pressure using Equation (5.5). In several designs, the tool shank is clamped at the top and bottom of the sleeve while the middle part provides stiffness. The hydraulic fluid in the sleeve acts as a damping agent and impact cushion and, therefore, hydraulic holding systems are free of vibration and chatter. However, hydraulic chucks provide relatively low radial stiffness in the hydraulic membrane since the tool practically "floats" in the compression sleeve; they are thus not good for heavy side-milling applications. The compression sleeve (length L_2 in Figure 5.62) is some distance L_1 behind the nose of the chuck, which also reduces the stiffness of the cutter. Tools with Weldon shanks or whistle-notch flats are not acceptable because they would destroy the bore sleeve, unless they are used in a sleeved collet toolholder. There is not axial drawback of the tool shank as the chuck is tightened.

5.4.5 Milling Chucks

Milling chucks use a broken or unbroken clamping surface principle which permits uniform expansion and contraction of the clamping surface. Typical designs feature a Roll-Lock mechanism using a single or double set of needle rollers and an outer guide ring (Figure 5.63). The needle rollers are located between the bore, which acts like a master collet, and the movable guide ring. When the chuck is tightened, an inward uniform force is created, squeezing the chuck body inward and wrapping it around the tool shank.

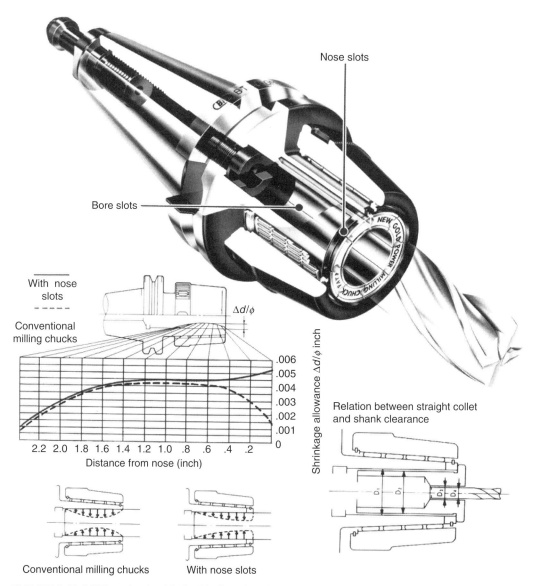

FIGURE 5.63 Milling chuck with double fine slots for better accuracy and high shrinkage allowance at nose. (Courtesy of Stanley Sheppard Company, Inc.)

The gripping power is dependent on the design characteristics and the ID size of master sleeve as shown in Figure 5.64. The chucking force is also affected by the design of the master sleeve, which is often slotted or grooved annularly to provide increased shrinkage of the bore and to increase the frictional resistance in the chuck by eliminating the effect of any oil film on the tool shank. The gripping force increases linearly with the guide ring travel distance. The addition of an extra row of roller bearings, five instead of four, increases the gripping power by as much as 30%. The gripping power for a double roller bearing chuck is in principle 1.5 to 2 times larger than that of a single roller bearing chuck, according to manufacturer, although some evidence indicates the actual gripping force may be lower [39]. It is preferable to use the smallest clearance between the tool shank and the bore of the master sleeve (<0.030 mm) to

Toolholders and Workholders

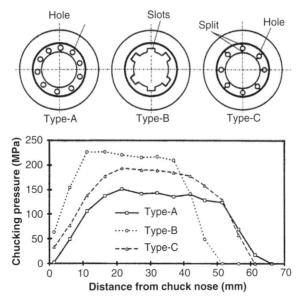

FIGURE 5.64 Various nose slot styles for milling chucks [39].

obtain sufficient clamping force at the nose of chuck. The clamping range is 0.2 to 0.3% of the relevant clamping diameter. The maximum shrinkage can be extended closer to the nose by using special nose slots as shown in Figure 5.63, which is especially important when using small diameter end mills. Various kinds of milling chuck toolholders are available which differ in the collet chuck design as shown in Figure 5.64 [39]; Type B provides the highest pressure while Type A provides the widest distribution as shown in Figure 5.64.

The allowable tool runout is given in Table 5.11. Milling chucks have less than 2 μm tool length movement in the chuck during tightening.

5.4.6 Shrink-Fit Chucks

A highly reliable type of toolholder, comparable in performance to a monolithic toolholder (Figure 5.65A), is the shrink-fit chuck (Figure 5.65B). In shrink-fit chucks, standard straight shank tools are shrunk into the toolholder body with a high interference fit as explained in the toolholder–spindle connection section (Figure 5.11, Type K). The advantage of this method is that the shrinking process is reversible. To insert the tool into the holder, the holder is rapidly heated (using induction or electromagnetic field) until the tool shank will slide into the holder. As the toolholder cools, the resulting thermal contraction exerts uniform pressure around the

TABLE 5.11
Milling Chuck Runout Allowance and Gripping Torque Values

	Maximum Allowable Runout (μm)		Gripping Torque (N m)	
Chuck Size	At Nose End	At 100–200 mm Out	Maximum	Hand Tightened
19			1000–3000	350–400
32	5	15	2500–4500	500–800
50			3000–5000	600–1100

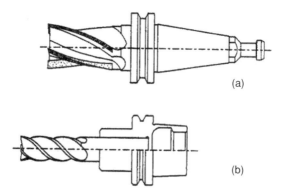

FIGURE 5.65 Advanced toolholder designs: (a) monolithic toolholder, and (b) shrink-fit toolholder. (Courtesy of Precise Corporation.)

entire surface of the tool shank. With a proper match of materials, the tool expands more slowly than the toolholder, eliminating the interference, and allowing the tool shank to be removed. A difference in the coefficients of thermal expansion in excess of a factor of 2 between steel and carbide is often an advantage in allowing for convenient removal of the tool shank. These toolholders are typically designed and tested to last 20,000 to 50,000 tool-change (shrink–unshrink) cycles. The amount of interference can be optimized based on application requirements, and the magnitude of the shrink-fit, the unit pressure, tangential stress generated by the fit, and the torque transmitted can be calculated from Equations (5.3)–(5.5). Generally the interference should be less than 0.01 mm to avoid excessive compressive stress on the tool shank and to increase the holding ability and safe operating stress in the toolholder nose section. In continuous dry cutting (end-milling long travels or drilling several holes consecutively) of ferrous metals, which produces a build-up of temperatures in the tool and holder, the difference in the coefficient of thermal expansion for the tool and holder materials will cause a decrease in the effective amount of interference; this must be considered in the design of the joint. A change in shrink pressure due to high centrifugal forces should also be considered for high-speed toolholders because the transmitted torque could drop by 70 to 80% as the speed increases from zero to 40,000 rpm with an H5/H6 tool shank tolerance. Therefore, tolerance H5 instead of H6 is recommended to provide a safer minimum clamping torque.

The shrink-fit interface effectively provides an integral shank tool, virtually eliminating tool deflection, vibration, and slippage at the interface [6, 7]. Because the thermal expansion and contraction in the toolholding system is symmetric, it provides the best possible runout, balance, and gripping torque combination characteristics. It provides inherent balance due to its symmetric design without nuts, collets, etc. Grooves around the toolholder ID can be provided for coolant access around the tool. Shrink-fit toolholders seem to offer their greatest advantage in extended length or reach applications.

5.4.7 Proprietary Chucks

There are also proprietary solutions to precision chucking, such as the Tribos, powRgrip, and CoroGrip, as well as various other interfaces using ultra precision collets with special nuts designs (e.g., HP collet chuck shown in Figure 5.66, XT precision collet system, etc.). Several of these chucks were developed to overcome the concerns with the shrink-fit systems because they do not require dangerously high temperatures to work correctly and no cool-down period is necessary as with shrink-fit chucks. They usually employ unique mechanical

Toolholders and Workholders

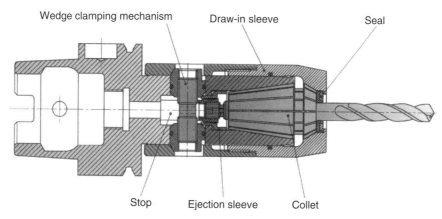

FIGURE 5.66 High-performance "HP Series" collet chuck. (Courtesy of Parlec, Inc.)

activating designs that develop significantly more clamping pressure than conventional collet chucks while providing the high accuracy of shrink-fit chucks. The *Tribos* system [41] is a tri-polygon clamping system (shown in Figure 5.67). Its ground polygon (three-lobed) shaped bore rigidly clamps the tool shank in three places. A clamping device mechanically deforms this bore (generating precise force at three points on the toolholder) to make it round, allowing for the tool shank to be inserted; unclamping the device will return the clamping diameter to its initial three-lobed shape and clamp the tool shank at three locations under very high forces. The Tribos system provides about the same accuracy as a shrink-fit chuck (TIR of 0.003 mm at $2 \times D$ to $3 \times D$ from the nose of the chuck) because the tri-polygon bore is ground as a round bore under proper loads with the same accuracy as the shrink-fit bore.

The *powRgip* system [42] consists of a toolholder, precision collet (with 1:100 taper angle), and a hydraulic setup device to insert the collet and tool into the holder. A press is used to

FIGURE 5.67 Precision polygonal mechanical actuated "Tribos" toolholder chuck. (Courtesy of Schunk, Inc.)

expand the holder nose slightly and push the collet with the tool into the toolholder nose. The maximum TIR is expected to be 0.003 mm at $4 \times D$ out from the nose [42]. The collet has a built-in set screw for presetting the tool length. The tool shank should have an H6 tolerance. This system is available for clamping diameters between 3 and 20 mm.

The *CoroGrip* system [43] consists of a thin-wall nose with a movable sleeve on a 2° taper angle. A high hydraulic pressure (10 kpsi) is used to push the outer sleeve up on the taper when the tool is clamped, and down on the taper when it is unclamped as shown in Figure 5.68. This system is similar to milling chuck or shrink-fit chuck but uses sliding surfaces to collapse the inner body of the nose onto the tool shank. The TIR is expected to be 0.002 to 0.005 mm at $4 \times D$ from the nose [43].

The *SINO-T* system [41] is a universal toolholder design more slender than the standard Weldon and collet chuck holders. Its concept is similar to the hydraulic toolholder because it consists of an expansion sleeve and an expansion chamber with elastic medium (made from a solid body for stiffness). It clamps mechanically by tightening a clamping sleeve axially, which squeezes the elastic medium that transfers a clamping force uniformly over the clamped tool shank. Torques up to 95 and 290 N m can be transferred to tool shanks with quality H6, respectively, for 12 and 20 mm chucks. Its runout accuracy at the clamping bore is 0.005 mm. Its advantage over standard Weldon and collet chucks is a result from the vibration damping characteristics of the elastic medium; it provides good stiffness for end milling. In addition, the front cup is tightened to a dead stop and cannot be overtightened, as the nuts in collet chucks can be. The axial drawback of the tool shank is a few micrometers as the chuck is tightened.

5.4.8 TAPPING ATTACHMENTS

Solid toolholders with no axial or radial movement are used in synchronous tapping in NC machining centers; the tap is clamped using collet chucks, Morse taper split sleeve drives, Whistle-Notch solid-drives, or quick-change adapters. Quick tap replacement heads reduce

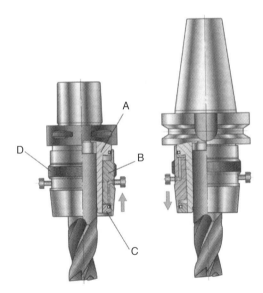

FIGURE 5.68 Precision mechanical (hydraulic actuated) "CoroGrip" chuck. (Courtesy of Sandvik Coromant.)

tool change times but are best suited for slow to moderate rpm due to lack of synchronization capability in the machine feed control.

Tapping attachments are used to assist synchronization of the machine feed (lead screw) with the spindle rpm to match precisely the pitch of the thread. The tapping device has the greatest influence on tapping performance. Tapping attachments can be held directly in the spindle shaft or toolholder or can be integrated into the toolholder body. Tapping heads are designed with features such as axial compression, axial tension, radial float, torque control, and provision for quick replacement of dull taps (Figure 5.69) [44]. Axial compression and tension cushions the tap at the entrance and bottom of a hole, while preventing double threading, and compensates for excessive feed rates when entering the hole.

Radial float can be designed into a chuck to account for location misalignment between the tap and the hole. Torque control assists in the bottom tapping of holes and prevents breakage of dull taps. The threaded depth is controlled to within one tenth of a revolution.

Self-reversing toolholders eliminate spindle reversals; they reverse the tap without reversing the machine spindle. The advantages of this chuck are pronounced with smaller tap sizes when the spindle rpm is high, and, therefore, the acceleration/deceleration time for the spindle is large. Self-reversing systems are currently operating at under 5000 rpm.

5.4.9 REAMING ATTACHMENTS

Reaming attachments are used to assist the alignment of a reamer in a hole. Floating reamer holders are used to compensate for radial, parallel, and angular movements (centerline

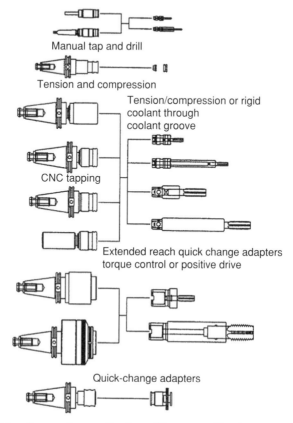

FIGURE 5.69 Toolholders for tapping applications. (Courtesy of Parlec, Inc.)

deviations between the machine spindle and the workpiece hole) necessary for the reamer to be positioned parallel and perpendicular to the existing hole axis; otherwise, the reamed hole squareness and angularity can be changed by the reamer.

5.4.10 Comparison of Cutting Tool Clamping Systems

One of the main criteria when selecting the toolholder is the runout error at the cutting edges. The radial runout affects the equality of the radial cutting forces among the cutting edges especially in a peripheral milling cutter. It also affects tool life directly as illustrated in Figure 5.70 and Figure 5.71. The graph in Figure 5.70 shows that precision chucks with runout at the tool point below 0.01 mm are desirable for high-accuracy and high-speed applications. Figure 5.71 shows tool life reduction as a function of runout given as a percentage of the uncut chip thickness.

It is important to note that the runout values provided in catalogs for different toolholders represent the runout of the chuck with a perfect tool and toolholder shank or spindle. It is usually measured in a precision setup (e.g., air bearing spindle, etc.). For example, a measurement of 11 identical HSK-A63 precision toolholders (using a straight collet on a 10 mm carbide end mill with $3 \times D$ overhang) with a runout specification of 0.003 to 0.005 mm at $4 \times D$ was performed by the manufacturer using the proper setup and the tool runout range was found to be 0.002 to 0.006 mm. The same toolholders were measured in a presetting machine (using a low clamping force for the HSK shank) and the runout range was 0.002 to 0.024 mm. The nominal runout of a complete tool–toolholder–spindle system is estimated using the best/worse case scenario approach as explained in Example 5.1 and illustrated in Figure 5.17. There are several different types of toolholder–tool interfaces within each style of interface. For example, every manufacturer has toolholder with standard collet quality and several of them also have precision collet types (as shown in Table 5.10). The accuracy of

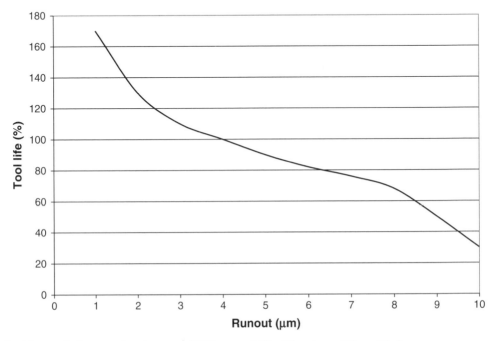

FIGURE 5.70 Influence of tool runout (TIR) on tool life. (Courtesy of Rego-Fix.)

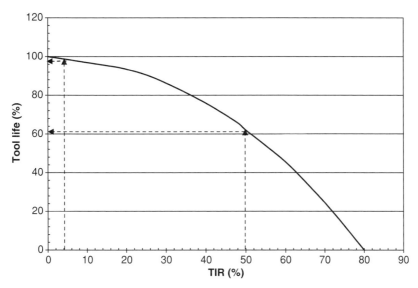

FIGURE 5.71 Influence of tool runout (TIR) as a percentage of uncut chip thickness on tool life. (Courtesy of Rego-Fix.)

precision single angle collets is generally specified as 0.005 mm at $4 \times D$ from nose but on the spindle CAT-V toolholders (with AT3 taper) the error can be as high as 0.013 to 0.020 mm.

The roundness of a 15.5 mm diameter carbide rod mounted in four different toolholding chuck systems, listed in Table 5.12, was compared against a gage mandrel using the HSK-80 spindle interface with all toolholders. Two toolholders were evaluated for 100 repetitions with each chuck style; the spindle was started and stopped between each trial. The minimum, maximum, range, and standard deviation are given in Table 5.12 for each toolholder. The best roundness was obtained with the hydraulic chuck.

The runout for several toolholders was evaluated with a solid tool at $3 \times D$ from the nose and the results are shown in Table 5.13. These toolholders were measured by four different manufacturers, three times each, and the minimum, maximum, and average values for each toolholder–tool interface are given in Table 5.13. The runout of some of these toolholders is higher than expected based on the manufacturer's specifications.

The toolholder quality was evaluated based on the concentricity and roundness of the bore ID to toolholder taper measurements for several end mill and collet toolholders. The results are shown in Table 5.14. The quality of the CAT-40 toolholders was 50% better than that of CAT-50 toolholder. However, the quality for the HSK-A63 was about the same as that of an HSK-A100 toolholder. In addition, the quality of the CAT-40 toolholders was better than the HSK toolholders.

The static and dynamic bending and torsional stiffness measured from the bench fixture tests with a CAT-50 as the basic toolholder and using 6.35 mm HSS and carbide tooling with an overhang length of 38.9 mm ($L/D = 6$) are shown in Figure 5.72 and Figure 5.73, respectively. The static and torsion bending for the collet type holders (using either 6.35 or 7 mm tool size collets) was better than the other type holders (i.e., Weldon, hydraulic, milling chuck, etc.). In addition, the TG collet type had better stiffness than the ER type collet. The dynamic (bending and torsional stiffness defined as the product of modal stiffness with damping ratio) performance of the hydraulic type holder is not good as expected compared to the other type holders, but its static bending stiffness is better than expected, even though the tool practically

TABLE 5.12
Roundness of Several Tool Chucks Using an HSK-80 Toolholder and a 15.5 mm Carbide Diameter Rod (Courtesy of Alfing Corporation)

Toolholder Style	Roundness (μm)			
	Min.	Max.	Range	Std Dev.
Gage mandrel (1)	1.1	2.8	1.7	0.25
Gage mandrel (2)	0.9	2.6	1.7	0.25
Gage mandrel (3)	0.5	2.2	1.7	0.29
Standard collet, ER 40/16	8.8	31.2	22.4	3.47
Standard collet, ER 40/16	22.2	82.2	60.0	8.85
Precision collet, ER 40/15.5	1.5	10.3	8.8	1.62
Precision collet, ER 40/15.5	4.1	28.1	24.0	4.36
Hydraulic chuck	1.5	3.3	1.8	0.35
Hydraulic chuck	5.2	8.7	3.5	0.55
Whislow-Notch in	6.7	9.4	2.7	0.54
ABS-FWD Adapter	0.9	6.2	5.3	0.93

"floats" in the expansion sleeve. The milling chuck type holder performed unexpectedly poorly compared to the collet and Weldon type holders due to collet size reduction characteristics. The stiffness of the carbide tooling is at least twice that of HSS as expected due to carbide's higher modulus of elasticity.

Performance for 12.7 mm toolholders with 108 mm overhang ($L/D = 8$) is shown in Figure 5.74 and Figure 5.75, respectively, for HSS and carbide tooling. Toolholders for a 12.7 mm tool performed somewhat similar to the 6.35 mm toolholders. The torsional characteristics of the shrink-fit and hydraulic holders are better than the other holder types. The wedge type end-milling holder provides better dynamic characteristics than the Weldon type holder. The Morse taper holder performed better than expected. The larger size milling chuck tends to perform better than the smaller size as also observed with the 6.35 mm tooling. The performance of the tooling in hydraulic holders with collets is worse than without collets. In addition, the modal stiffness decreases while the dynamic stiffness increases with a decrease in the clamping force for both styles of interfaces.

TABLE 5.13
Evaluation of Runout at $3 \times D$ from the Nose for Several Toolholder–Tool Interfaces

Toolholder–Tool Interface	Runout (mm)			Toolholder Accuracy (mm)	
	Minimum	Maximum	Average	Concentricity	Roundness
Milling chuck	0.0025	0.023	0.0010		
Hydraulic	0.0005	0.0076	0.0029		
Shrink-fit	0.0013	0.0076	0.0040		
Collet DA-180	0.0023	0.1524	0.0386		
Collet ER-25	0.0005	0.0330	0.0087	0.00384	0.00106
Collet TG-100	0.0005	0.0127	0.0054	0.00480	0.0016
Weldon	0.0005	0.0064	0.0046	0.00450	0.00115
Collet precision	0.0005	0.0152	0.0052		

TABLE 5.14
Evaluation of the Concentricity and Roundness of the Bore ID to Toolholder Taper from Several Toolholders Measured in a Roundness Instrument

	CAT-40		CAT-50		HSK-A100		HSK-A63	
	Concentricity (mm)	Roundness (mm)	Concentricity (mm)	Roundness (mm)	Concentricity (mm)	Roundness (mm)	Concentricity (mm)	Roundness (mm)
1 EM	0.00119888	0.0016256	0.00856361	0.0012014	0.00191897	0.0006896	0.00306324	0.000428
1 EM	0.00324993	0.0010541	0.0024892	0.0006922	0.00691769	0.0013576	0.00828548	0.0024346
1 EM	0.00058928	0.0004699	0.00531114	0.0008344	0.00218567	0.0019164	0.00313563	0.000635
1 EM	0.00174879	0.0006528	0.0094996	0.0012484	0.00300609	0.0012408	0.00522986	0.0008141
1 EM	0.0033274	0.0006096					0.00180213	0.0011697
0.5 EM	0.00154051	0.0012027	0.00784225	0.0007976	0.00747903	0.0005258	0.00209677	0.0016688
0.5 EM	0.0022098	0.0018783	0.00217805	0.000616	0.00630555	0.006731	0.01399921	0.0013729
0.5 EM	0.00221996	0.0004458	0.00077851	0.0005525	0.00246507	0.0014999	0.0053086	0.0008103
0.5 EM	0.00524929	0.000997	0.01011682	0.0005893	0.00378206	0.0014719	0.00276225	0.0009258
0.5 EM	0.0042926	0.0004724					0.002159	0.0008865
TG 100	0.00395605	0.0008992	0.00083185	0.000823	0.00254	0.0011925	0.00159385	0.0012916
TG 100	0.0015494	0.0012281	0.00155321	0.001632	0.00427863	0.0016764	0.0065405	0.0011341
TG 100	0.00568833	0.0036386	0.00704342	0.0006058			0.00294894	0.0007836
TG 100	0.00279908	0.0013411	0.004953	0.0006033				
TG 100	0.0013716	0.0006858						
ER 16	0.00536829	0.0029477	0.00477279	0.0012014	0.00296672	0.0011811	0.00401447	0.0004458
ER 16	0.00309753	0.0009068	0.0021971	0.0009182	0.00247523	0.0011887	0.00797941	0.0006693
ER 16	0.00056515	0.0003988	0.00202565	0.0015685	0.00507873	0.0008446	0.00299339	0.0010325
ER 16	0.00463169	0.0009982	0.00523875	0.0010579	0.00238633	0.0012573	0.00379857	0.0007404
ER 16	0.004064	0.0009525	0.00672592	0.0010998			0.00257302	0.0007722
1 SF					0.00838581	0.0013475		
0.5 SF					0.00824865	0.0008331		
Average	0.00011559	4.607E-05	0.00031725	8.526E-05	0.00020966	6.944E-05	0.00027505	6.73E-05

Note: 1 EM = 25.4 mm end mill; 0.5 EM = 12.7 mm end mill; TG 100 = TG collet; ER 16 = ER collet; 1 SF = 25.4 mm shrink-fit; 0.5 SF = 12.7 mm shrink-fit.

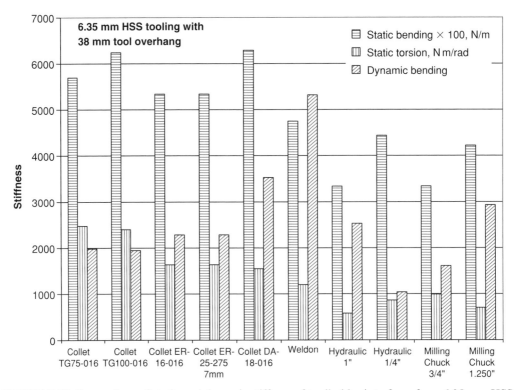

FIGURE 5.72 Comparison of static and dynamic stiffness of toolholder interfaces for a 6.35 mm HSS tool.

The results for 25.4 mm tooling interfaces are given in Figure 5.76 and Figure 5.77, respectively, for the HSS and carbide tooling with $L/D = 5.6$. Again the TG collet type holders performed as well as most of the other type holders. The modular interfaces with both cylindrical and taper face contacts (i.e., HSK, CAPTO, KM, ABS, BETA, Tru-Taper) of size 50 perform somewhat better than the other interfaces but in some cases not significantly so. In addition, the expected superior performance for the milling chuck was not observed and the hydraulic holder exhibited the highest dynamic and lowest static stiffness as shown in Figure 5.77.

The transmitted torque for four different chucks as a function of tool shank diameter is shown in Figure 5.78. The clamping transmitted torque increases with shank diameter, but there is a large scatter among different manufacturers due to differing design characteristics among similar interfaces and the tolerances of the bore and the tool shank. Generally, all high-performance interfaces (shrink-fit, hydraulic, etc.) require an H6 shank tolerance to achieve optimum results. The standard collet chucks and some of the high-performance collet chucks accept H6 to H11 tool shank tolerances. The collet chucks provide higher clamping torque than some of the high-performance interfaces for tool shank diameters smaller than 12 to 14 mm. In addition, the clamping transmitted torque for many high-performance chucks drops by about 5% for every 0.002 mm drop in tool shank diameter.

A comparison between toolholder–tool interfaces is given in Table 5.15 using several criteria. The accuracy results are based on a CAT-40 (AT3 taper) while the transmitted torque capability is based on 20 mm tool shank. The runout accuracy of shrink-fit, hydraulic, milling chucks, and high-performance collet systems (using super precision collets) is

Toolholders and Workholders

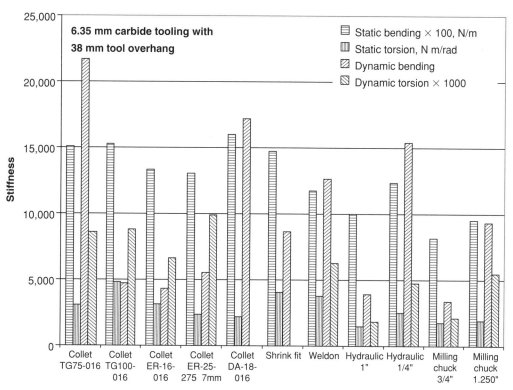

FIGURE 5.73 Comparison of static and dynamic stiffness of toolholder interfaces for a 6.35 mm carbide tool.

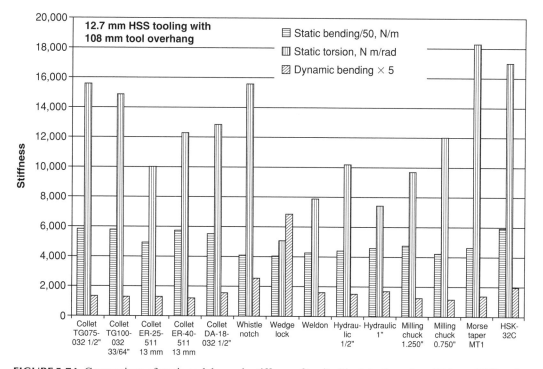

FIGURE 5.74 Comparison of static and dynamic stiffness of toolholder interfaces for a 12.7 mm HSS tool.

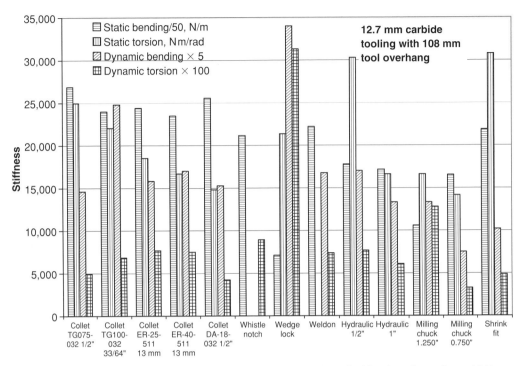

FIGURE 5.75 Comparison of static and dynamic stiffness of toolholder interfaces for a 12.7 mm carbide tool.

comparable and is 2 to 10 times better than that of standard collet and side-lock chucks. The clamping pressure and torque for shrink-fit and milling chucks are comparable, up to twice that of hydraulic expansion chucks, and up to 2.5 to 3 times that of collet chucks for large tool shank diameters.

A comparison of four tool–toolholder clamping systems tested on a CAT-V 50 spindle [38] showed that the parameters investigated do differ considerably between different kinds of locking systems as shown in Table 5.16. Collet chuck locking system performance is directly proportional to the tightening torque of the locknut. A collet chuck with ball bearing locknut system provides higher gripping capacity compared to standard locknut collet chucks. The static stiffness response for different systems for a 2000 N radial force is given in Table 5.16; damping characteristics, which affect chatter and forced vibrations, are also given in Table 5.16.

A comparison of a hydraulic with a precision collet chuck tool–toolholder interface is given in Table 5.17, based on dynamic tests using an HSK-63 spindle system and an L/D ratio of 5.9 [45]. The response of these two tool–toolholder connections is similar. The results of a second comparison of five tool–toolholder connections with an L/D ratio of 7.2 are given in Table 5.18. The static stiffness of the hydraulic toolholder was similar to that of the Weldon one-screw toolholder and lower than the other toolholders. The deflection response for the shrink-fit, Weldon two-screw, and collet chuck toolholders was linear with respect to overhang and had the same slope; the other two toolholders had a larger slope. The particular tested collet chuck performed exceptionally well statically and dynamically. The addition of a second screw on the Weldon type toolholder significantly improved its static and dynamic response.

Toolholders and Workholders

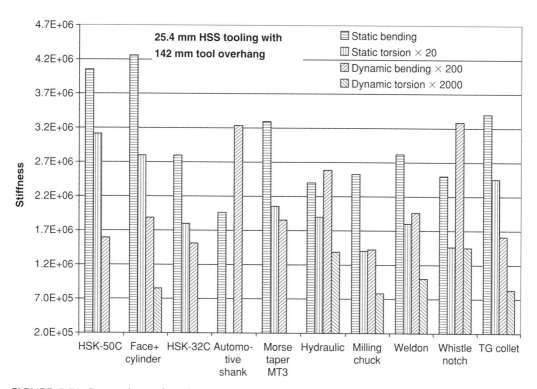

FIGURE 5.76 Comparison of static and dynamic stiffness of toolholder interfaces for a 24.5 mm HSS tool.

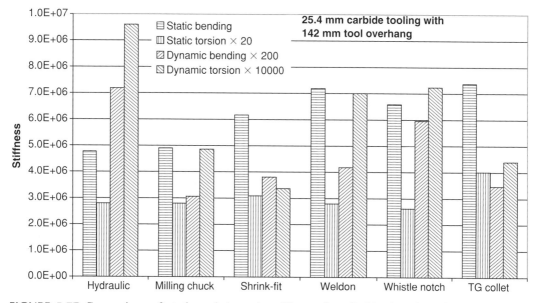

FIGURE 5.77 Comparison of static and dynamic stiffness of toolholder interfaces for a 25.4 mm carbide tool.

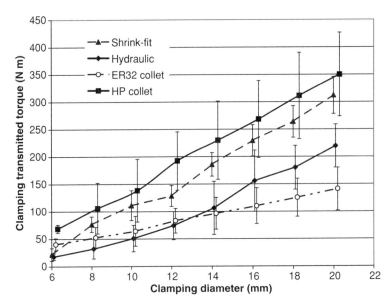

FIGURE 5.78 Transmitted torque by hydraulic (h6 tolerance), shrink-fit (h6 tolerance), standard collet (h6–h11 tolerance), and precision collet (h6–h11 tolerance) chucks. (This graph incorporates data from several manufacturers such as Parlec, Rego-Fix, Sandvik, and Schunk.)

Another comparison between standard collet chucks and milling chucks [39] found that the chucking pressure (and therefore the torque capability) of milling chucks is 3 to 20 times that of collet chucks. The maximum chucking pressure is generally only effective over the middle 40 to 60% of the chuck's clamping length. Still another comparison of a 31 mm milling chuck (using an unbroken clamping surface) with a hydraulic chuck of the same size and geometry indicated that the hydraulic chuck is stiffer by 20 to 30%.

The transfer functions for a 19 mm diameter tool with 50 mm overhang clamped in four different CAT-V 40 toolholders (a collet, Weldon with one screw, hydraulic chuck, and shrink-fit chuck) are summarized in Table 5.19 [46]. This comparison clearly indicates the importance of using a short holder with stiff clamping. A shrink-fit chuck provided the stiffest connection between the tool and spindle, while the hydraulic chuck provided the weakest connection. The Weldon type was weak due to its long overhang from the spindle face and its single clamping screw. The hydraulic toolholder had a high modal frequency. The static stiffness of the CAT-V 40 taper is much lower than that of the CAT-V 50 taper for all chucks, as can be seen by comparing Table 5.16 and Table 5.19.

Two sets of cutting tool clamping systems were evaluated with a CAT-V 50 toolholder using 19 and 25 mm diameter steel bars as cutting tools. The first 19 mm diameter set includes a hydraulic toolholder and three different collet chucks. Two 100TG collet chucks were used with 10 slits, one with a ball bearing nut (CNBB) and the other with a standard spring nut (CNSS); the third 150TG collet had three slits with a standard spring nut (CNSSL). The collet size was equal to the tool shank diameter. The second 25 mm diameter set included one end mill Weldon style holder with one set screw and a hydraulic chuck. The overhang of the tool point from the spindle face, the tool L/D from the toolholder face, and the overhang of the toolholder were not exactly the same for all the tooling systems as noted in Table 5.20. The end mill toolholder was evaluated for two orientations of the set screw: (a) the set screw was displaced 90° from the applied load, and (b) the set screw was in line and on the same side as

TABLE 5.15
Comparison between Tool–Toolholder Interfaces in a CAT-40 (AT3 Taper) for a 20 mm Bore

	Side Lock	Collet Type DA	Collet Type ER TG	"HP" Collet Type	Hydraulic Chuck	Milling Power Chuck	Shrink-Fit	Pow-Rgip Chuck	Tribos Chuck	CoroGrip Chuck
Application/type of operation	Heavy rough semifinish	Rough semifinish	Rough semifinish finish	Rough finish	Semifinish and finish	Heavy rough finish	Heavy rough finish	Heavy rough finish	Heavy rough finish	Heavy rough finish
Accuracy TIR $4 \times D$ (mm) Chuck Only	0.005–0.018				0.005	0.005–0.012	0.005		0.005	0.005
Chuck and Collet		0.025–0.050	0.010–0.040	0.005–0.020	0.010	0.010–0.025		0.005		0.006
Rigidity	+	--	--	+	+	+++	+++			
Transmitted torque[1] (N m)	Lock screw	65–100[N2]	ER32–170[N3]	295	150–280	1000	200–500	360	150–240	400–580
Max krpm	12–15		30–40	30–40	30–40	20	40	40	40	40
Range of collapse (mm)	None	0.3	0.2	0.05–0.2[N4]	0.008	0.012	None	None	None	0.012
Clamping range (mm)	5–50	0.5–50	1–50	3–25	5–31	6–50	6–38	3–20	6–32	6–32
Length adjustment accuracy (mm)	None or some	0.2–0.7	0.2–0.7	0.02–0.2	±0.005	>0.005	±0.005	>0.005	±0.005	±0.005
Relative cost[5]	1	1	2	4–5	10	6–10	5–7			
Balance quality as standard	--	--	--	--	+	--	+++			
Tool clamping	Allen key	C-spanner[8]	C-spanner[8]	C-spanner	Allen key	C-spanner	Device[9]	Device[10]	Device[10]	Device[10]
Maintenance	None[6]	Yes[7]	Yes[7]	Yes[7]	None[6]	Yes[7]	None[6]	Yes[7]	None[6]	Yes[7]

Notes:

[1] For tolerance h6 (values indicate minimum values measured at minimum size h6 tolerance).

[N2] 1:1 ratio of tightening torque versus grip.

[N3] 1:2 ratio of tightening torque versus grip.

[N4] High accuracy is obtained with at shank size collet.

[5] 1 = lowest price; 10 = highest price.

[6] None required (wipe bore for residual dirty coolant).

[7] Requires cleaning and changing collets, nuts, etc.

[8] C-spanner may need device to control nut tightening torque.

[9] Heat shrinking device

[10] Special clamping device.

-- = poor; - = fair; + = good; +++ = very good.

TABLE 5.16
Comparison of Tool–Toolholder Connections Using a 20 mm Diameter Round Shaft Extended 80 mm from the Toolholder Nose [38]

Toolholder System (CAT-V 50 Spindle Nose)	Static Stiffness (N/mm)	Damping Ratio
Spindle nose alone	200,000	
Direct mounting of the tool with conical shank	9,090	
Weldon-type chuck	6,800	0.045
Collet chuck with ball bearing locknut	6,060	0.123
Clarkson-type chuck	5,850	0.091
Standard collet chuck	5,320	0.081

TABLE 5.17
Comparison of Tool–Toolholder Connections Using a 10 mm Diameter Carbide Rod Extended 59 mm from the HSK-A 63 Toolholder Nose Based on an Impulse Test [45]

Toolholder System	Hydraulic Chuck	Precision Collet Chuck
Static stiffness (N/mm)	945	867
Min. dynamic stiffness (N/mm)	41.2	38
Resonance frequency (Hz)	1530	1490

TABLE 5.18
Comparison of Tool–Toolholder Connections Using a 12.7 mm Diameter Carbide Rod with 90 mm Overhang Based on Static Loading and Impulse Test [45]

Toolholder System (Clamped on Rigid Block Using a Flange Mount)	Static Stiffness (N/mm)	1st Natural Frequency (Hz)	Dynamic Stiffness (N/mm)
Shrink-fit chuck (0.1 mm interference fit)	2500	1184	40
Hydraulic chuck	1430	810	140
Weldon chuck — one screw	1540	845	94
Weldon chuck — two screws	2170	1029	160
Standard collet chuck	2440	1161	151

the applied load. A comparison of the static stiffnesses of all the above tooling systems is given in Table 5.20. The stiffness for all collet chucks was similar and about 50 to 60% higher than for the 19 mm hydraulic chuck. The stiffnesses of the 25 mm hydraulic chuck and the end mill holder were similar. The natural frequencies and damping ratios were also measured and are given in Table 5.20. The damping ratio of the hydraulic chuck was larger than that of the collet chucks. The ball bearing nut collet had a higher damping ratio than the standard spring nut collets. The hydraulic chuck had higher damping than the end mill chuck with the set screw along the applied load direction. The damping increased for the end mill chuck when the set screw was oriented 90° from the applied load. The static stiffness results correlate well with those summarized in Table 5.16. In contrast, the damping ratios in Table 5.16 are much higher than those in Table 5.20; the difference was attributed to higher drawbar forces since

TABLE 5.19
Comparison of Tool–Toolholder Connections Using a 19 mm Diameter Steel Rod With 50 mm Overhang on Standard 40 Taper Toolholder Based on Dynamic Tests [46]

Toolholder System	Modes	Frequency (Hz)	Stiffness (N/mm)
Standard collet chuck (TNL = 66 mm)	1	799	1000
	2	5328	1111
Weldon chuck — one screw (TNL = 92 mm)	1	772	690
Hydraulic chuck (TNL = 18.5 mm)	1	3700	625
Shrink-fit chuck (TNL = 33 mm)	Rather stiff modes	—	3333

Note: TNL = 18.5 mm toolholder nose from spindle face.

the 36 kN used on the toolholders evaluated in Table 5.20 is much higher than that used in most current machine tools.

The measured deflection at the end of the tool bar and the toolholder, and the predicted deflection at the end of the tool bar based on the measured toolholder deflection and based on a monolithic tool–toolholder using 500 N load at the end of the cutting tool, are shown in Figure 5.79. The spindle deflection was 0.0025 mm. The deflection at the end of the toolholder for the above seven CAT-V 50 toolholders was between 0.005 and 0.013 mm. There was generally some permanent deflection of the CAT-V toolholder due to the toolholder–spindle taper fit (a maximum of 0.007 mm) that was not recovered after the load was released. The difference between the actual deflection at the tool end and the predicted deflection based on the measured toolholder deflection reflects the contribution of the tool–toolholder connection to the tool deflection response. The ratio of the actual to predicted deflections is 2.5 to 2.8 for the hydraulic chuck, 1.2 to 1.7 for the collet chucks, and 1.1 for the end mill chuck. The difference between the hydraulic tooling system and a monolithic tooling system is very significant. In contrast, the difference between a collet or end mill chuck and a monolithic tooling system is not as pronounced.

Two face-mount toolholders, one a hydraulic chuck (HYD13/F) and the other a radially adjustable end mill toolholder (SL14/F), were also evaluated; the end-mill style had a toothed clamping slit sleeve that is compressed onto the cutting tool with a set screw. The contribution of the tool–toolholder connection of the face-mount toolholders to the total deflection is similar to that of the CAT-V toolholders as shown in Figure 5.79. The toolholder chuck style and the cutting tool diameter and L/D ratio are all important parameters when considering the tool stiffness. The decrease of the tooling system stiffness as a function of the cutting tool L/D is shown in Figure 5.80 for both face-mount toolholders. The selection of the proper toolholder chuck style is much more important for short cutting tools.

The static bending deflection of the tool can be also defined analytically using the joint stiffness parameters. The bending deflection of the tool includes the elastic deflection of the tool itself, the deflection of the toolholder–tool joint, the deflection of the toolholder, deflection of the toolholder–spindle joint, and the spindle deflection at the front bearing(s) as explained in Example 5.8.

5.5 BALANCING REQUIREMENTS FOR TOOLHOLDERS

High-speed machining requires improvements in the quality and design of toolholders, specifically improved precision and rigidity and better balance. Balance is particularly important in high-speed applications since oscillating radial cutting forces and centrifugal forces

TABLE 5.20
Comparison of Tool–Toolholder Connections Using a CAT-V 50 Toolholder Based on Static (at 500 N Load at End of Tool Bar) and Dynamic Tests

Toolholder System	Symbol	Diameter (mm)	L/D	TOL (mm)	TNL (mm)	Stiffness (N/mm)	Frequency (Hz)	Damping Ratio
Hydraulic chuck	HYD19	19	4.2	171	91	1980	800	0.024
100TG Collet with ball bearing nut	CNBB19	19	4.6	170	79	3400	955	0.021
100TG collet with standard spring nut	CNSS19	19	4.6	170	87	3500	908	0.014
150TG collet with standard spring nut	CNSSL19	19	4.8	170	87	3000	875	0.015
End mill Weldon style (set screw 90° to the load direction)	EMW25/HS	25	3.9	175	102	7800	858	0.035
End mill Weldon style (set screw along the load direction in the same side)	EMW25/VS	25	3.9	175	102	8130	830	0.0285
Hydraulic chuck	HYD25	25	3.1	178	100	8140	818	0.032

Static stiffness of spindle nose along = 350 kN/mm. TOL = total overhang (from spindle face to the end of the tool where the load is applied). TNL = toolholder nose overhang from spindle face.

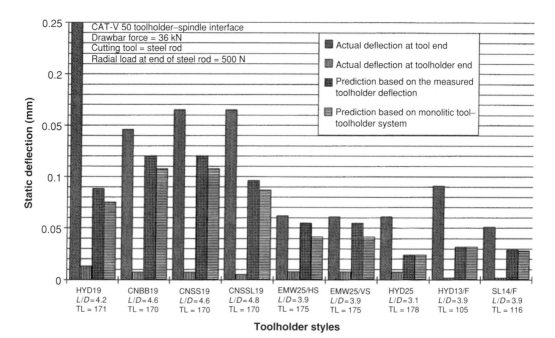

FIGURE 5.79 Comparison of several tool–toolholder connections.

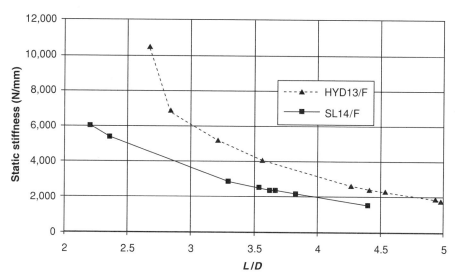

FIGURE 5.80 The effect of tool L/D ratio on the static stiffness of a tooling system.

(which depend on the magnitude of the unbalance) can result in a safety hazard, as well as premature spindle bearing wear or failure, holder–spindle interface fretting, poor tool life, and poor part quality or surface finish due to chatter.

Unbalance is caused by several factors [47]:

1. Expansion of the spindle nose at high rpm, resulting in diminished contact between the spindle and toolholder taper. This reduces rigidity and may lead to tilting or radial movement of the toolholder within the spindle nose (due to the absence of face contact) which generates unbalance.
2. Uneven mass distribution in the toolholder.
3. Poor tolerance in the toolholder which results in eccentric features with respect to the rotational axis.
4. The use of asymmetrical components, for example, spacers, drivers, set screws, cutting tools, or even retention knobs.

Toolholders must be balanced when used at spindle speeds greater than 8,000 to 10,000 rpm. Unbalance is defined as the condition which exists when the principal mass axis (axis of inertia) of a tool–toolholder assembly does not coincide with the rotational axis. An unbalanced tool–toolholder assembly results in oscillating forces and movement of the cutting tool. Balancing the rotating assembly, including the toolholder and the tool, distributes the mass correctly so that the rotary vibration frequencies stay within acceptable limits at operational speeds. There are standards (ISO 1940/1 and ANSI for Balance Quality of Rotating Rigid Bodies S2.19-1975 or -1989) defining the permissible residual unbalance relative to its maximum service speed.

There are three principal types of unbalance: static, coupled, and dynamic [29]. *Static "single plane" unbalance* occurs when the mass axis does not coincide with the rotational axis but is parallel to the rotational axis. *Coupled unbalance* occurs when the mass axis does not coincide with the rotational axis, but does intersect the rotational axis at the center of gravity of the toolholder; the force vectors created are equal in magnitude but 180° apart. *Dynamic "two plane" unbalance* occurs when the mass axis is not coincident or parallel to the axis of rotation and does not intersect this axis; it is a combination of static and coupled unbalances.

Generally, single plane balancing is applied to toolholders that operate at speeds below 15,000 rpm, which are prebalanced by the manufacturer, and which have a length less than two times the gage line diameter (e.g., CAT V-flange and HSK side-lock holders). Two plane balancing is considered for toolholders operating at speeds greater than 20,000 rpm with lengths more than two times the gage line diameter (e.g., boring bars) [48, 49].

The general equation for the allowable unbalance for a specific toolholder operating at a known rpm is

$$U = \frac{9549 \cdot G \cdot W}{\text{rpm}} \tag{5.8}$$

where U is the unbalance in g mm, W is the total tool weight in kg, rpm is the rotational speed of the tool, and the "G" number represents the quality-grade as defined in ANSI/S2.19-1975 or -1989. G can be calculated from

$$G = \frac{U \cdot \omega}{W} \tag{5.9}$$

where ω is the rotational speed in rad/sec. The units for "G" are mm/sec. Toolholders with unbalance of 250 g mm or more are not uncommon, although many manufacturers are reducing the unbalance to below 50 g mm. Balance requirements are determined based on the maximum radial force F_r tolerated for a given spindle or operation, given by

$$F_r = U \left[\frac{2\pi \cdot \text{rpm}}{60} \right]^2 \tag{5.10}$$

For example, 50 g mm of unbalance produces a continuous radial force of 123 N at 15000 rpm

The balance quality-grade "G" is very important. Once it is known and its physical significance is understood, the allowable unbalance can be calculated. The "G" number varies from case to case depending on the mass of the tool–toolholder assembly, the spindle speed to be used, and the quantity of unbalance present. A tool can meet any balance specification as long as it is heavy enough or turned slowly enough. $G6.3$ is appropriate for machine tools and general machinery parts, $G2.5$ is used for machine tool drives, and $G1.0$ is specified for precision spindles, especially for grinding machines (Figure 5.81). $G0.4$ is also used for spindles. In general, toolholder balance requirements vary between $G1.0$ and $G6.3$.

Fine balance to a specific "G" tolerance is performed on balancing machines by rotating the toolholder at lower speed (about 1000 rpm). Unbalance is detected and material is drilled or milled opposite to the unbalance to reduce it. This could mean redrilling the toolholder every time a tool is changed, which could result in the total destruction of the toolholder after several tool changes. To avoid this, a variety of toolholder balancing systems are available including prebalanced systems, axial, radial, or angled screws equally spaced, and weighted balls and balancing rings integrated into the toolholder body [50]. For example, a set screw can be inserted into the threaded holes which can be adjusted radially to correct for unbalance. The set screws can be held in place by a noncementing type of Loctite or by using a Nylok type of screw. All these innovative methods permit adjusting to minimum unbalance using either a balancing machine or balancing charts. Active balancing and vibration control of the spindle can be done in real time by mounting a balancer on the spindle [51]. In one case, two rings held in place by a permanent magnetic force are mounted on the spindle and rotate with the spindle and can also be controlled to rotate with respect to the spindle. This device is used to balance the spindle and toolholder as a system [52, 53].

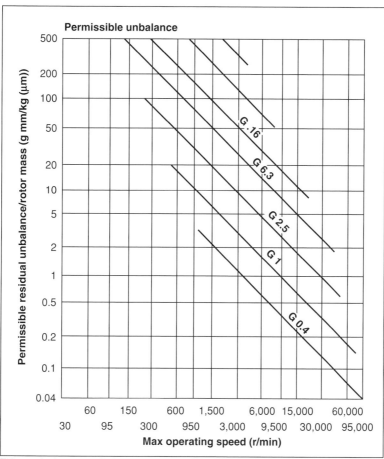

This graph shows permissible shaft unbalances as a function of operating speed and balancing grade.

FIGURE 5.81 Maximum permissible residual specific unbalance value corresponding to various quality grades G. The grade number (G) represents the unbalance vibration velocity in mm/sec. $G1$, a grade often required for precision spindles, allows a maximum rest vibration of 1 mm/sec. Typically, using components of grades $G1$ to $G6.3$ leads to acceptable system balance. (Courtesy of Precise Corporation.)

The tool–toolholder assembly must be balanced to a degree of accuracy that matches the machine spindle requirements. The requirements are determined based on general machinery vibration data, and are often defined in terms of the velocity of the out-of-balance condition or the vibration displacement. The permissible amount of unbalance decreases with increasing speed in rpm (Figure 5.81). Generally, a balancing operation (or at least checking the balance condition) is required each time a new tool setup occurs as spindle speeds increase [37, 48, 54]. The cutting tool itself can cause unbalance even if the toolholder is balanced. For example, a Roller Lock milling chuck requires rebalancing after each tool change due to the repositioning of the small needle bearings in the collet actuator [54]; while these parts are small in mass, they are distant from the rotational axis and thus have a strong effect on balance. The balance consistency for a side-lock end mill holder was found to be excellent while several end mills of same size were reassembled and remounted [54]. The balance consistency of a collet chuck was not as good as that of an end mill holder but the reassembly and remounting did not cause large balance variations. A $G2.5$ tolerance at 20,000 rpm is possible with precision collet chucks. However, as much as 30 g mm of unbalance can be introduced by simply loosening and retightening the nut of standard collets. The balance variation of the Roller Lock milling chuck was much higher than the side-lock end milling and collet chuck holder. If higher speeds are required (above 25,000 rpm) the chuck should be precision balanced with the cutting tool.

Mass symmetry by design is an effective approach to toolholder balance. There are controllable (fixed) and uncontrollable (variable) sources of unbalance [48, 54]. Controllable sources, such as different depths of drive slots, set screws on end mill holders, retention knobs, unground bases of V-flanges, and other geometric characteristics, can be eliminated either through proper design or through balancing of the toolholder by the manufacturer. From a balance viewpoint, drive slots should be positioned symmetrically, although slots with different depths are sometimes used for errorproofing (e.g., to prevent insertion of a tool into the spindle backwards). Compensation must be made by adding heavy metal screws to balance geometrical unevenness. The balance corrections are made by matching the displaced and unequal masses on opposite sides of the toolholder's main body. Balancing holders for high speeds require grinding all possible exposed surfaces after heat treatment using the ground shank taper as the base reference. Repeatable balance with collet holders is attainable using a clamping nut with a fine thread with or without a coaxial extractor ring. Uncontrollable sources of unbalance generally occur after the completion of the initial balance, every time the cutting tool is changed or the collet is loosened and reclamped. Figure 5.82 shows toolholder design principles used for a high-speed collet chucks. The possible source of unbalance is the eccentricity between the toolholder mass axis and the rotational axis, which can be minimized by using tight taper angle tolerances (AT3 per ISO 1947) for the spindle cone and for the toolholders. The toolholder should have a symmetrical design, and its ID features should be concentric to the taper within 0.005 mm TIR. Set screws in side-lock end mill holders require

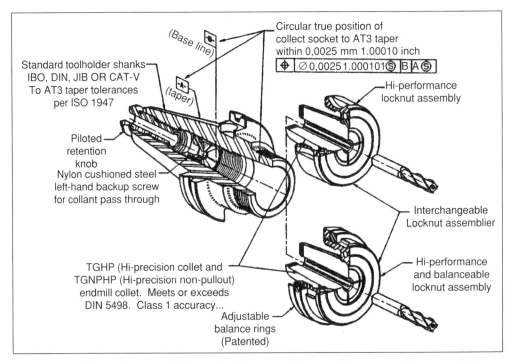

FIGURE 5.82 High-performance/high-speed collet system features. (Courtesy of Kennametal, Inc.)

built-in balance compensation. Selecting the correct toolholder having excellent taper contact, accurate (very low runout) and symmetry by design is the main goal. Toolholder bodies (usually without tool, retention knob, collet, or nut) balanced by design in the neighborhood of $G6.3$ at 15,000 rpm are available and should be selected because they often represent the heaviest part of the toolholder assembly. Prebalanced toolholders should be selected for speeds above 10,000 rpm.

5.6 FIXTURES

5.6.1 GENERAL

Fixtures are workholding devices used to locate, clamp, and support parts accurately and securely. Fixtures which include a build-in tool guidance device (such as bushing for drilling or reaming) are sometimes called *jigs*. A fixture or jig must accurately locate and position (clamp) the workpiece with respect to the tool in order to maintain the specified tolerances under the prevailing cutting forces. The part is supported to prevent its deflection due to cutting and clamping forces; it must also not interfere with machine motions during cutting or loading/unloading. Therefore, fixtures should be given the same attention as the rest of the machining and tooling system in designing the process.

The various elements and parameters involving in the fixture design process are shown in Figure 5.83. It is important for fixture planning to be integrated with process planning, providing the link between design and manufacturing [55, 56]. Generally, manufacturing practice requires an iterative process between fixture planning and process planning.

The major elements of a fixture are the *supporting structure*, *locating points*, and *clamps*, which provide positional accuracy. It is also important to consider a fixture's *operation efficiency* (see Figure 15.10 in Chapter 15).

Generally, as a basic *supporting structure*, subplates with a standard grind pattern are mounted on machine tools. Other structural elements include angle-plates, sine plates, vise-jaws, chucks, universal indexing heads, and standard clamps. The details of the structure of a

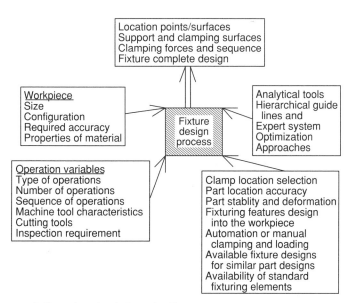

FIGURE 5.83 Factors influencing the design of a fixture.

fixture are typically part dependent; moreover, as discussed below under types of fixtures, different approaches are used to construct the structure for fixtures used in mass and batch production.

Clamps are used to hold parts against locating fixtures and to withstand secondary forces. The clamp type and clamping method have a strong influence on part quality [57–61]. Efficient clamping requires an understanding of basic actions in the workpiece. A variety of manual and power clamps are available. Power clamping provides better performance than manual clamping because the clamping force is more controlled and consistent and the clamping sequence is automated and thus also consistent. Hydraulic workholders (clamps and vises) are efficient and provide consistent pressures/forces; positive-locking cylinders (hydraulically activated and mechanically locked) can be used in palletized fixtures so that the hydraulic connections can be removed while the clamps are positively locked. Another way to provide clamping action is to make clamps from shape-memory alloys (SMAs) which respond to temperature changes and mechanical deformation. An SMA clamp can be deformed under low forces and returned to its original shape by heating; the heat for actuation can be supplied by several means (e.g., electrical-resistance heating). The clamping force should not exceed the level above which the part deformation is significant compared to the part tolerance. Therefore, clamping forces for finishing operations should be much lower than the forces used for rough machining operations.

Locators must be strong and rigid enough to resist the cutting forces exerted on the workpiece. For best locating repeatability and stability, locators should contact the workpiece on a machined surface; such surfaces can be a plane (internal machined surface), a concentric internal diameter, a precision hole (using a pin), or a combination of a surface and a pin. The most repeatable and reliable approach is to locate from an internal bore or a hole. For example, a plane surface supports/restrains the part in one direction while a surface with a round pin supports and locates the part because it restrains a motion in three directions.

The locating points are determined by the degrees of freedom of a part. Each part has 12 possible motions, six linear motions ($+X, +Y, +Z, -X, -Y, -Z$) and six rotational motions (clockwise and counterclockwise around each of the three axes). The fixture must position the workpiece in each of the three axial planes in order to fulfill a positive location criterion, the 3–2–1 (three–two–one or six-point locating principle) being the most common method of location in practice. In the 3–2–1 fixture principle (shown in Figure 5.84), the primary locating plane is provided three supporting points restricting five possible movements, a

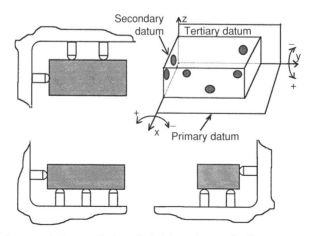

FIGURE 5.84 Six points restricting workpiece (3–2–1 locating method).

second plane is provided two supporting points restricting three possible movements, and the third plane is assigned one point to restrict one final possible movement [57–66]. The three planes must be mutually perpendicular. The 3–2–1 method, using six locators, constraints 9 out of 12 motions; the remaining three possible movements should be restricted by clamps. The locators are usually selected as far apart as possible in order to ensure maximum stability. In practice, additional redundant supports may be necessary to avoid vibration and deflections due to cutting forces. The function of a clamping device is to apply and maintain sufficient counteracting holding forces to a workpiece so that it is rigidly fixed against all external force fields during cutting. The part setup is generally determined based on relations between features and critical tolerances.

Operation efficiency is a basic fixture requirement because it involves the convenience and cleanness of the fixturing operation. The fixture should be designed for easy loading and unloading of workpieces, as well as easy chip disposal since the accumulation of chips around the fixture and its elements can affect part quality.

5.6.2 Types of Fixtures

Fixtures can be classified as dedicated (permanent), general-purpose, and reconfigurable workholders. Dedicated fixtures (made from tooling plates) are developed for a specific workpiece geometry and manufacturing operation, especially in mass production [57–59, 64]. Dedicated workholders are expensive because they are complex, one-of-a-kind devices (e.g., power operated locating and clamping fixtures are used in automated manufacturing systems as shown in Figure 15.8 and Figure 15.9). General-purpose fixtures, which have been described extensively in the literature [65–67], include plates, angle-plates, sine plates, vise-jaws, power chucks with quick jaw-change systems, universal indexing heads, and standard clamps. They are inexpensive and reusable and thus well suited to small lot production. Reconfigurable or flexible fixtures can be adjusted to accommodate a variety of parts.

Workholding devices for rotational parts involve various chucks (which grip rotational parts on the outside diameter) and expanding arbors (which grip rotational parts on the inside diameter). The performance of the chuck is affected by the eccentricity between spindle and chuck and random variation caused by conditions built into the chuck (internal friction, sliding fits, etc.).

Modular fixturing fills a significant gap between general-purpose workholders and special purpose workholders. They can be used for one-time jobs, infrequent production runs, and prototype and development work. They require more space than the permanent fixtures and are not well suited to mass production. Dedicated fixtures with modular versatility can be constructed by mounting the clamps and locators directly to a subplate or tombstone using standard bolts and T-nuts (see Figure 15.9 and Figure 15.13); manufacturer's specifications indicate they can hold workpieces securely with a positional accuracy between 0.01 and 0.035 mm.

Most flexible fixtures are modular fixtures which use standard components and assemblies arranged differently for different parts [68, 69]. A variety of standard components is shown in Figure 5.85. Modular fixtures feature either a gridwork of accurate locating holes or precision T-slots. Grid hole designs use some combination of precisely positioned dowel holes along with tapped holes to accurately align, locate, and secure fixturing elements; some systems alternate bushing and threaded holes, while other systems have an alignment bushing plus a threaded insert in every hole. Grid hole designs provide better spatial resolution and higher clamping stiffness than T-slots. Tooling cubes or tombstones provide a reasonably rigid framework that can be used in many variations. Fixtures with their own base plates can be mounted to them, or a fixture can be built directly on them. These

FIGURE 5.85 Modular fixturing systems. (Courtesy of Carr Lane Manufacturing Co.)

systems, however, lack sensor feedback capability and programmability. A fourth-axis fixture quick change [65] can be used which accepts faceplates or tombstones in the horizontal or vertical direction. Research on phase change flexible fixtures has also been reported [70–72]. In these fixtures, the part is partly immersed in a special fluid which can be made to change rapidly to a solid. Before removal of the part from the fixture, it is changed to liquid again. The phase change can be triggered by cooling/heating (using low melting point alloys as an encapsulating medium), an electric current (using electrically active polymeric materials), or removal of air pressure (in a particulate fluid bed). Conformable fixture surfaces, consisting of an array of discrete elements brought in contact with the workpiece surface, are used to machine turbine blades. Programmable fixtures are used to increase flexibility and are supported by a computer-controlled device or a robot to perform various tasks such as assembly from a CAD description, setup of the workpiece in the fixture, application of the clamps in the proper sequence, and control of the clamping pressure (see Chapter 15, Figure 15.14–Figure 15.16).

Several magnetic types of fixtures are used when grinding flat parts that would be distorted if held in other types of fixtures [73]. These fixtures are not usually very accurate and cannot be used in high-precision grinding to tolerances less than 0.003 mm.

5.6.3 Fixture Analysis

A *kinematic analysis* of the fixture–workpiece configuration should be performed to develop a good fixture design and limit trial-and-error in fixture implementation. Such an analysis ensures that the workpiece is easy accessible and detachable but does not account for its deformation or friction with the fixture elements. The cutting forces are not considered in a kinematic analysis. The linear motion of the workpiece is restricted by the selection of the reference surfaces, while its rotational movement due to the cutting forces is restricted by the positions of the locators. Kinematic and force analyses have been used to determine optimal locating and clamping points/surfaces and the positions of support points to minimize workpiece deflection [74–78].

Finite element analysis has been also used for fixture design; these analyses can take into account the deformation of the workpiece. The latter method requires a finite element model of the workpiece and the proper boundary constraints to represent support, locating, and clamping elements [79]. It can evaluate the clamping and cutting distortion including the contact deformation, fixture element stiffness/compliance, and dynamic response of the fixture–workpiece system. Specific examples are given in Chapter 8.

Expert systems have also been developed for assisting the fixture design and selection. Expert systems are built on rule based systems which consist of strategic rules, synthesizing rules, constraining rules and heuristics (rules of thumb) for fixtures for a given family of parts [80–83].

There are several more detailed approaches to fixture analysis. *Screw theory* has been used to model the fixture–workpiece system, with linear programming used to determine clamping forces [84]. *Nonlinear programming* has been used to derive a quadratic model for verification of the fixture configuration [85]. A *dynamic model* for the fixture–workpiece system has also been reported; this model was used to determine the minimum clamping forces [86]. Proper locating and clamping points can also be determined using *equilibrium equations* [87]. Slippage conditions for a clamped part can be determined by constructing limit surfaces in the force/moment space based on frictional assumptions [88].

The external forces and torques exerted on the workpiece by the cutting tool should be mathematically modeled together with the clamping and friction forces [37, 40, 60]. The objective of such analysis is to ensure that the clamping pressure is sufficient to keep the workpiece fixed during machining while not producing stresses that exceed limiting values, which lead to unacceptable deformations. Large clamping forces are often applied during roughing operations, but the clamping forces should be significantly reduced during finishing operations to meet tight part tolerances. The static force and moment equilibrium equations involving worst-case forces can be used to determine the clamping force requirements. The correlation between cutting force and clamping moment is very important to keep the workpiece stable during the metal cutting process [89]. The matrix form of the equations for the fixtured workpiece shown in Figure 5.86 is

$$\begin{Bmatrix} F_x \\ F_y \\ F_z \\ M_x \\ M_y \\ M_z \end{Bmatrix} = \begin{bmatrix} 0 & 0 & 0 & 1 & 1 & 0 \\ 0 & 0 & 0 & 0 & 0 & 1 \\ 1 & 1 & 1 & 0 & 0 & 0 \\ a_{41} & a_{42} & a_{43} & a_{44} & a_{45} & a_{46} \\ a_{51} & a_{52} & a_{53} & a_{54} & a_{55} & a_{56} \\ a_{61} & a_{62} & a_{63} & a_{64} & a_{65} & a_{66} \end{bmatrix} \cdot \begin{Bmatrix} R_{11} \\ R_{12} \\ R_{13} \\ R_{14} \\ R_{15} \\ R_{16} \end{Bmatrix} \qquad (5.11)$$

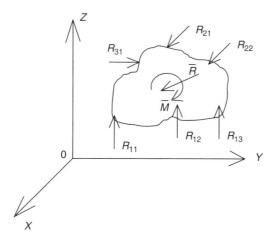

FIGURE 5.86 Free-body diagram for a fixtured workpiece.

where F_i and M_i are the Cartesian components of the resultant force and moment vectors and a_{ij} are the moment arms of R_{ij} contributing to moment M_i. The unknown parameters in the matrix system are the non-negative reactions R_{ij}, which are determined by solving the matrix system. These reactions are required for the design of the fixture because they form the constraints for the determination of the support points based on the 3–2–1 rule. The reaction forces also determine the deformation and resistance to machining forces as functions of applied clamping forces. The results of such an analysis may dictate the redesign of the fixture to include redundant supports and reduced clamping forces so that the proper location restraints are used and clamping loads do not deform the part significantly.

The number of clamps required is determined based on the applied resultant forces by the clamp type's rated holding capacity, which is the maximum force the clamp will sustain in the locked position. The force resisting motion in a direction perpendicular to the applied clamping force is determined using the Coulomb friction model,

$$F_i \leq \mu_s \cdot P_i \quad \forall i \tag{5.12}$$

where μ_s is the static coefficient of friction between the workpiece and the surface of the fixture in contact and P_i and F_i are, respectively, the exerted normal clamping force and the resultant motion resisting force. P_i and F_i are related by

$$F_i = n_c \cdot P_i \tag{5.13}$$

where n_c is the number of clamps for a given part.

The part deformation due to the clamping force can be calculated using the slab method. The average normal stress p_{avg} is given approximately by

$$p_{avg} \cong \frac{2}{\sqrt{3}} Y \left[1 + \frac{\mu_s \cdot \lambda_h}{h} \right] \tag{5.14}$$

where Y is the yield stress of a perfectly plastic material or the flow stress of a strain-hardening material, λ_h is the clamp half-width, and h is the part thickness. The load to cause part deformation is

$$P_i = 2 \cdot \lambda_h \cdot p_{avg} \cdot w \tag{5.15}$$

where w is the width of the clamp.

For a solid cylindrical clamp of radius r, the average normal stress is

$$p_{avg} \cong Y \left[1 + \frac{\mu_s \cdot r}{3 \cdot h} \right] \tag{5.16}$$

and the maximum force before deformation is

$$P_i = \pi \cdot r^2 \cdot p_{avg} \tag{5.17}$$

Part quality is impacted by the flow of variation through a manufacturing system governed by fixture locaters, sequencing, orientation to the machine, and other geometric relationships. Variation simulation techniques provide a means of predicting process capability by emulating the possible errors. Variation simulation analysis assesses the impact of locator variation, workpiece datum feature variation, orientation errors, and locator and clamping sequencing [90].

Toolholders and Workholders

5.7 EXAMPLES

Example 5.1 Estimate the nominal runout of a complete tool–toolholder–spindle system for various cases. A 10 mm diameter end mill is used with an overhang of 40 mm. Consider a precision chuck in an HSK-A63 toolholder with and without a collet, with and without an extension as shown in Figure 5.87, and a precision collet chuck in a CAT-40 toolholder.

Solution: The nominal runout for the cutting point of a tooling system can be estimated using the best/worst case scenario approach illustrated in Table 5.21. The runout of a precision collet itself is about 0.002 to 0.003 mm compared to the 0.005 to 0.010 mm range when the collet and the female taper in the toolholder are considered with a 30 mm tool overhang as shown in Table 5.21. The female taper is considered in the basic holder in Table 5.21. The contribution of the CAT-40 toolholder–spindle interface is evaluated using the approach illustrated in Figure 5.17 while for the HSK interface the spindle and toolholder face runout errors are used. The spindle contribution with the extension (in the fourth column) is twice that without extension because the runout at the end of the tool is a function of the total length of the toolholder as illustrated in Figure 5.17. In addition, the upper value of the total runout in Table 5.21 is higher than expected because it represents the worst case scenario in which the maximum tolerances of each component are summed.

However, in practice, some of the tolerance of the different components will cancel each other so that the system runout is expected to be lower than the worst case.

Note that the worst case stack-up tolerance analysis (linear stack-up of errors) is popular because of its simplicity and ease of computation. It also assumes that practically under no situation will there be a defect condition, as long as the errors of each component in the system have been considered within its specified error tolerance band. However, worst case tolerancing makes no assumptions on how different errors are distributed within their respective tolerance bands

Example 5.2 Evaluate the runout of two modular tooling assemblies with long overhangs as illustrated in Figure 5.88. The difference between the two modular systems is the clamping method. One has a center threaded joint (bolt clamping) and the second system has a front clamping (rack and pinion joint). The extension spacer is 120.7 mm long by 80 mm diameter and the basic toolholder is 79.4 mm long. Note: front clamping uses a differential screw and opposite sets of serrated clamping jaws to grasp and pull the adapter/tool back into the coupling. Front clamping provides for a fast tool change. Center bolt clamping provides a stiffer joint than front clamping and is considered the optimum solution for heavy machining with long overhangs.

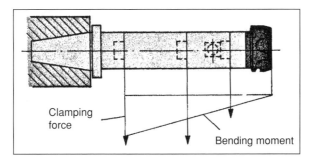

FIGURE 5.87 Standard and extended toolholder runout evaluation using the HSK-A63 and CAT-40 interfaces.

TABLE 5.21
Nominal Runout Values in Micrometers (μm) for Different Assemblies

	HSK-A63 Toolholder		CAT-40	
System Components	Solid Holder Without Collet	Solid Holder With Collet	Holder With Extension	Solid Holder With Collet
Spindle	3–5	3–5	6–10	6–25
Basic holder	2–6	2–6	2–6	2–6
Extension			2–4	
Clamping adapter			2–6	
Collet		2–3	2–3	2–3
Tool (end mill)	3–10	3–10	3–10	3–10
Total TIR (μm)	8–21	10–24	17–37	12–40

Solution: The runout as a function of the number of extension spacers connected together (as illustrated in Figure 5.9 and Figure 5.88) is shown in Figure 5.89 for two modular tooling systems using a CAT V-flange #50 toolholder system. The runout of the center threaded joint was higher than that of the rack and pinion joint. The difference between the two joints increases with an increasing number of spacers (or length). Large runouts may result with asymmetric key slots (slots located on one side of the contact surface only). Due to interrupted grinding conditions caused by the presence of the key slot, segments of the contact

FIGURE 5.88 Bending moment and clamping force for a combination of classic and front clamping systems, which performs better for extension tooling than either system by itself. (Courtesy of Sandvik Coromant.)

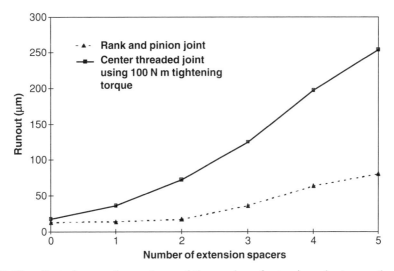

FIGURE 5.89 The effect of connection system and the number of extension adapters on the cutting tool runout.

ring-shaped surface near the key slot protrude roughly 2.5 to 5 μm above the surface. Key slot asymmetry also results in asymmetry in contact deformation between the mating surfaces. The runout of the center threaded joint increased linearly with increasing tightening torque. The minimum runouts occur at tightening torques between 90 and 110 N m. For heavy milling applications, however, higher torques should be used if runout is not a critical concern.

The static and dynamic stiffness response of the tool was measured in both the longitudinal (x) and transverse (y) directions while the tool was mounted in the spindle of a machining center. The static stiffness as a function of the number of extensions is shown in Figure 5.90. An exponential drop in stiffness occurred as the number of joint extensions

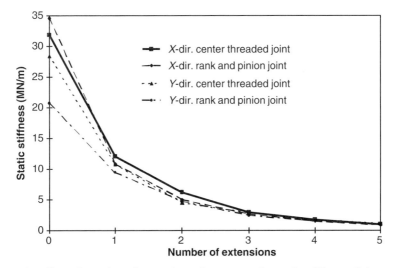

FIGURE 5.90 The effect of number of extension adapters on the static stiffness of the cutting tool at X- and Y-directions for both the classic and front clamping systems.

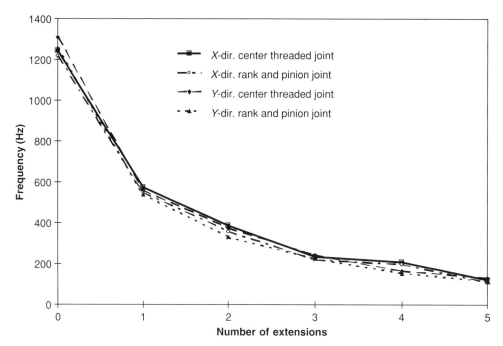

FIGURE 5.91 The effect of number of extension adapters on the frequency of the cutting tool at X- and Y-directions for both the classic and front clamping systems.

increased. There was no significant difference between the two tooling systems in either direction. The resonant frequency response (which determines the cutting speed range which should be avoided) was the same for both tooling systems and similar to that corresponding to the static stiffness as shown in Figure 5.91. In addition, it was found that the stiffness (and natural frequency) was stable at tightening torques over 120 N m.

Example 5.3 Calculate the total balancing grade of an assembled spindle–toolholder–tool system. The specified grade for each component is: spindle $G1$, toolholder $G2.5$, and tool $G6.3$. The mass of the spindle is 20 kg, of the toolholder is 1.27 kg, and of the tool is 0.35 kg and the tool rpm is 25000.

Solution: The total allowable unbalance for the system is equal to the sum of the allowable unbalances for each subsystem. Hence,

$$U_{tot} = U_{spindle} + U_{holder} + U_{tool}$$

Equation (5.8) is used to calculate unbalance for each component

$$U_{tot} = 7.639 + 1.213 + 0.842 = 9.694 \text{ g mm}$$

The G number is calculated from Equation (5.9) based on the total mass of the system $W = 21.62$ kg and the spindle rpm. Hence, G_t is calculated to be 1.17.

Example 5.4 Estimate the unbalance force reduction by selecting a CAT-40 prebalanced toolholder for a 15,000 rpm application.

Toolholders and Workholders

Solution: A standard CAT-40 toolholder without prebalance has about 30 to 75 g mm unbalance. The unbalance produces a centrifugal force given by Equation (5.10):

$$F_r = U \left[\frac{2 \cdot \pi \cdot \text{rpm}}{60} \right]^2 = 30 \left[\frac{2 \cdot \pi \cdot 15,000}{60} \right]^2 10^{-6} = 74 \text{ N}$$

The radial force for 30 g mm unbalance at 15,000 rpm is 74 N and for 75 g mm unbalance the force becomes 185 N. Hence, the force reduction is significant if a prebalanced toolholder is used.

Example 5.5 A 267 mm long, 31 mm diameter solid extended integral boring bar will be operated between 10,000 and 14,000 rpm. Estimate the allowable unbalance for a balance quality grade of G2.5. In addition, determine the radial force and boring bar deflection as a function of toolholder unbalance from 0 to 55 g mm.

Solution: The allowable unbalance of the boring bar is estimated from Equation (5.8), with the weight of the bar estimated as 3.63 kg

$$U = \frac{9549 \cdot G \cdot W}{\text{rpm}} = \frac{9549 \cdot 2.5 \cdot (3.63 \text{ kg})}{10,000 \text{ rpm}} = 8.66 \text{ g mm}$$

So, the allowable unbalance is calculated to be 8.7 and 6.2 g mm, respectively, for 10,000 and 14,000 rpm for balance quality grade of G2.5. The radial force at the point of the boring bar is calculated using several unbalance values in the range of 0 to 55 g mm in Equation (5.10) as in Example 5.4. The force as a function of unbalance for three different speeds (5,000, 10,000, and 14,000 rpm) is shown in Figure 5.92. The force increases significantly with unbalance and cutting speed. The tool deflection is calculated based on the deflection of the boring bar due to its weight plus the unbalance force:

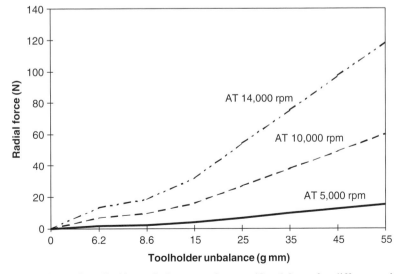

FIGURE 5.92 The effect of toolholder unbalance on the centrifugal force for different spindle speeds.

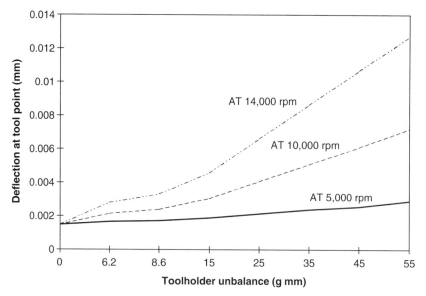

FIGURE 5.93 The effect of toolholder unbalance on the tool point deflection for different spindle speeds.

$$\delta = \frac{qL^4}{8EI} + \frac{FL^3}{3EI} = \frac{(7800 \text{ kg/m}^3)(267 \text{ mm})^4}{8EI} + \frac{F(267 \text{ mm})^3}{3EI}$$

The deflection is calculated for several values of radial forces estimated above for the corresponding unbalances for three different speeds; the results are shown in Figure 5.93. The deflection increases significantly at higher speeds.

Example 5.6 Determine the effectiveness of the toolholder clamping length on the deflection of a boring bar. Two different boring bar designs (A) and (B) can be used to bore the same hole as shown in Figure 5.94. The cutting force on the boring bar is 80 N. The clamping length of the tool shank in the holder is $2 \times D$ for bar (A) and $4 \times D$ for bar (B). The extended length of the boring bar out of the toolholder nose L_1 for both bars is equal to $7 \times D$.

Solution: The cutting force F_1 will create a reaction force at both ends of the holder in contact with the bar as shown above. The forces are calculated by the lever rule.

The lever rule for bar (A) is

$$F_1 \times L_1 = F_3 \times L_3$$

Hence, the force acting on the back end of the boring bar is

$$F_3 = (F_1 \times L_1)/L_3 = (80 \times 7)/2 = 280 \text{ N}$$

For equilibrium: $F_1 + F_3 = F_2$. Therefore, the force acting at the front of the chuck is $F_2 = 80 + 280 = 360$ N. This force tends to deform the nose of the chuck at the free end. The overhang of the boring bar increases if a deformation is present.

Toolholders and Workholders

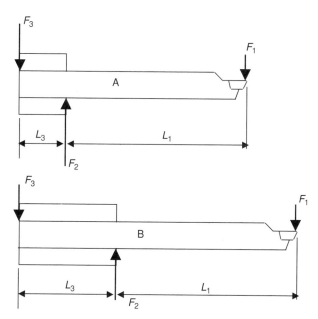

FIGURE 5.94 The effect of tool shank engagement with toolholder bore (tool clamped length) on the tool performance. (After J. Tlusty [46].)

The lever rule for bar (B) is

$$F_3 = (F_1 \times L_1)/L_3 = (80 \times 7)/4 = 140 \text{ N}, \quad F_1 + F_3 = F_2, \quad F_3 = 80 + 140 = 220 \text{ N}$$

Hence, the reaction force acting at the free end of the chuck is smaller than that in bar (A). This will reduce the deformation of the chuck at the free end.

This example illustrates how the clamping length of the boring bar shank in the holder, L_3, affects the force, F_2, which acts on the outer end of the clamping unit as a result of the cutting force, F_1, which bends the bar. Therefore, increasing the clamping length in the shank reduces the force which acts on the rear end of the boring bar. This, in turn, means that the stress at the point of clamping is reduced, which results in higher stability. The holder–tool clamping interface is exposed to a load which can deform the surface asperities or the bore in the holder under heavy loads; therefore, the internal surfaces of the tool clamping unit should have a high level of surface finish and hardness. It is possible that the overhang of the bar (L_1) may increase due to the deformation at the outer end of the holder.

Example 5.7 An OD turning operation is performed at 3000 rpm. The diameter of the cast iron part diameter (the chucking diameter) is 100 mm as shown in Figure 5.95. The front section of the part is turned down to 95 mm using 0.3 mm/rev feed. A 20 mm diameter hole is also drilled at the center of the part using 0.3 mm/rev feed at 2000 rpm.

Calculate the required gripping force of a power chuck.

Solution: The gripping force required depends on the type of cutting is performed. Assuming the two operations are performed separately, each operation is examined independently.

1. The cutting forces in turning are the tangential (cutting), radial, and feed forces. The cutting forces are absorbed by the jaw faces in contact with the OD of the workpiece. The main cutting force F_c produces a moment ($F_c d_z / 2$) which must be absorbed by the chuck and transmitted by friction on the jaws through the clamping contact surfaces. The clamp force or total jaw force is the algebraic sum of the individual radial forces applied by the top jaws on

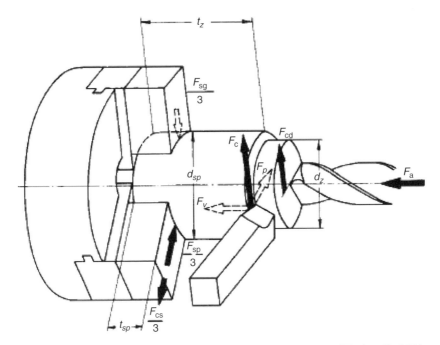

FIGURE 5.95 Clamping force consideration on a lathe chuck. (Courtesy of Rohm GmbH.)

the workpiece. In this case, there are three jaws and the force at each jaw is $F_{sg}/3$. The gripping force required by the jaw should be larger than the moment produced by the main cutting force during OD turning:

$$\left.\begin{array}{l} T = F_f \dfrac{d_{sp}}{2} \\ \mu = \dfrac{F_f}{F_{sg}} \\ T = F_c \dfrac{d_z}{2} \end{array}\right\} \Rightarrow F_{sg} = \dfrac{F_c \cdot S_z}{\mu} \dfrac{d_z}{d_{sp}}$$

where F_{sg} is the required total gripping force by the jaws without considering the effects of angular speed; d_{sp} is the chucking diameter; d_z is the machining diameter; S_z is the safety factor; l_z is the distance between machining and clamping points; and l_{sp} is the chucking length.

The calculation of the torque or cutting force during a turning operation was discussed in Example 2.1 (Chapter 2)

$$P = T \cdot \omega = T(2\pi N) = (MRR)u_s \Rightarrow F_c = P/(2\pi \cdot R \cdot N)$$

$$T = (MRR)u_s/2\pi N = [\pi \cdot (100^2 - 95^2)/4](0.3 \text{ mm/rev})N(0.065 \text{ W/mm}^3/\text{min})/2\pi N$$
$$= T + 142.6 \text{ N m}$$

Hence, $F_c = 2T/d_z = 2 \times 142.6 \times 1000/100 = 2852$ N

$$F_{sg} = \dfrac{F_c \cdot S_z}{\mu} \dfrac{d_z}{d_{sp}} = \dfrac{2852 \times 1.5}{0.20} \dfrac{95}{100} = 20{,}320 \text{ N}$$

2. Drilling the 20 mm diameter hole ($d_z = 20$ mm) at the center of the part requires MRR = 141,000 mm³/min and $T = 58.5$ N m

$$F_{cd} = 2T/d_z = 258.5 \times 1000/100 = 1170 \text{ N}$$

$$F_{sg} = \frac{F_{cd} \cdot S_z}{\mu} \frac{d_z}{d_{sp}} = \frac{1170 \times 1.5}{0.20} \frac{20}{100} = 1755 \text{ N}$$

This gripping force will determine the operating power for the chuck.

The gripping force of the chuck is largely influenced by the centrifugal forces (F_{cs}) of the jaws at higher speeds because all rotating parts tend to move away from the axis of rotation. The centrifugal forces are given by Equation (5.10) and the unbalance in this case is determined by how much the jaws distributed masses deviate radially from the rotating axis. This requires the knowledge of the mass (m) and the distance of the center of gravity for the jaws to the rotating axis (R). In this case, let us assume that $m = 4$ kg and $R = 0.075$ m. The centrifugal force by each jaw is

$$F_{cs} = U \left[\frac{2 \cdot \pi \cdot \text{rpm}}{60} \right]^2 = 4 \times 0.075 \left[\frac{2 \cdot \pi \cdot 2000}{60} \right]^2 = 13{,}146 \text{ N}$$

This force must be taken into account when determining the initial gripping force F_{sg}. The initial gripping force with the chuck stationary is the sum of the gripping force without considering the effects of speeds and the centrifugal force of the jaws (loss of gripping force) for external gripping:

$$F_{sp} = F_{sg} + F_{cs} = 20{,}320 + 13{,}146 = 33{,}466 \text{ N}$$

The centrifugal force constitutes a large percentage (65% of the gripping force required without considering the effects of speeds) of the total force required on the jaws.

Example 5.8 Consider the boring bar in Figure 5.96 for finishing a 30 mm diameter hole to 31 mm diameter with a single point. The workpiece material is gray cast iron. A carbide insert is used in the indexable steel boring bar. The cutting speed for the bar is 100 m/min and the feed is 0.15 mm/rev. The boring bar is integral to a CAT-40 toolholder. The spindle stiffness at the front (by the toolholder interface) is 2×10^8 N/m. Estimate the deflection at the tool point.

Solution: The two tapers at the toolholder–spindle interface (female taper in the spindle nose and the male taper on the holder) do not have a perfect fit because are made with certain tolerances. Therefore, a relative motion occurs between the two tapers under cutting loads at the tool point.

The static bending deflection is the first check someone can perform to avoid large deflection at the tool point. The bending deflection of the tool includes the elastic deflection of the bar itself, the deflection of the toolholder–spindle joint, and the spindle deflection at the front bearing(s). The stiffness of the spindle–toolholder interface is decomposed into a rotational spring and a translation (linear) spring

$$\delta_{\text{total}} = \delta_b + \delta_h + \delta_r + \delta_s$$

where δ_b is the deflection for a cantilever beam as discussed in Chapter 4, δ_h is the translation of the taper interface, δ_r is the deflection due to rotational spring of the taper interface, and δ_s is the translation of the spindle shaft due to bearing stiffness. If the tool is not monolithic and there is an additional connection between the tool and the toolholder, the toolholder interface should be also considered as a summation of the deflection due to linear and rotational springs. The joint location between the tapers is 20 mm from the spindle face inside. The

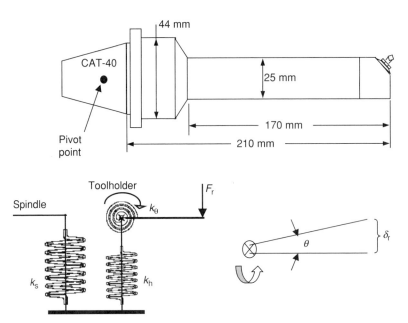

FIGURE 5.96 Analytical model of the toolholder–spindle interface with 2 DOF (linear and rotational) spring.

spindle stiffness is $k_s = 3 \times 10^8$ N/m. The toolholder–spindle interface has a linear spring $k_h = 1.375 \times 10^9$ N/m and a rotational spring $k_\theta = 8.108 \times 10^6$ N m/rad (from Table 5.7). The radial cutting force at the tool point was calculated based on the cutting conditions as explained in Example 4.4 (Chapter 4) to be $F_r = 185$ N.

The total linear stiffness in the direction of the radial force is equivalent to

$$\frac{1}{k} = \frac{1}{k_s} + \frac{1}{k_h} = \frac{1}{1.375 \times 10^9} + \frac{1}{3 \times 10^8} \Rightarrow k = 2.463 \times 10^8 \text{ N/m}$$

The deflection of the bar structure with two cross sections is

$$\delta_b = F_r \left[\frac{L_1^3}{3 \cdot E_1 \cdot I_1} + \frac{L_2^3}{3 \cdot E_2 \cdot I_2} + \frac{L_1 L_2}{E_2 \cdot I_2}(L_1 + L_2) \right]$$

where $L_1 = 170$ mm and $L_2 = 40$ mm.

$E_1 = E_2 = 206{,}700$ MPa, and the moments of inertia for the two cross sections are:

$$I_1 = \frac{\pi D_1^4}{64} = \frac{\pi (25^4)}{64} = 19{,}165 \text{ mm}^4 \quad \text{and} \quad I_2 = \frac{\pi D_2^4}{64} = \frac{\pi (44^4)}{64} = 183{,}890 \text{ mm}^4$$

Hence, $\delta_b = 0.078$ mm.

The spindle and the holder linear deflection is

$$\delta_1 = \frac{F_r}{k} = \frac{185 \text{ N}}{2.463 \times 10^8 \text{ N/m}} = 0.00075 \text{ mm}$$

The toolholder deflection at the tool point is affected by its shank rotation in the spindle nose taper. The toolholder rotation at the spindle nose is determined by the rotational spring at the interface located 20 mm behind the spindle face:

$$\theta = \frac{M_\mathrm{h}}{k_\theta} = \frac{F_\mathrm{r} \cdot L}{k_\theta} = \frac{185 \cdot (210 + 20)}{8.108 \times 10^6 \times 1000} = 5.3 \times 10^{-6} \text{ rad}$$

$$\delta_\mathrm{r} = L \tan \theta = (210 + 20) \cdot \tan(5.3 \times 10^{-6}) = 0.0012 \text{ mm}$$

The total deflection is the sum of the above three deflections:

$$\delta_\mathrm{total} = \delta_\mathrm{b} + \delta_\mathrm{l} + \delta_\mathrm{r} = 0.078 + 0.00075 + 0.0012 = 0.080 \text{ mm}$$

The contribution of the toolholder–spindle interface compliance in the boring bar deflection is only 2.5% in this case. However, for a shorter boring bar (e.g., $L = 75$ mm), the contribution of the toolholder–spindle interface compliance is increased to 29%.

Example 5.9 Evaluate the effect of the tool geometry on the performance of the toolholder–spindle interface considered in Figure 5.21.

Solution: The maximum bending moment that a toolholder can support is a function of the clamping force. The level of the clamping force is very important, especially in milling and boring operations, because it determines the limit of the torsional/bending forces at the cutting edge(s) and defines the stability and rigidity of the interface. The maximum allowable bending moment before the face joint is separated at the toolholder flange is given in Figure 5.35 for a range of clamping forces for different sizes of HSK toolholders. For example, 300 N m is the maximum bending moment that the HSK-63-A interface can support before complete separation of the face contact occurs when the clamping force is 18 kN. The torsion/bending forces at the flange must be estimated as explained in Chapter 2 and Chapter 4 (or simulated with software) based on the cutting conditions for semi and rough milling and boring applications to prevent face contact separation. The maximum value of tangential forces is a function of the two critical dimensions "R" and "L" that must be considered when designing turning tools and especially boring bars and milling cutters (Figure 5.97). The bending moment generated during cutting can be calculated from the equation:

$$M = \frac{9.55P}{RN}(L^2 + c_1^2 L^2 + c_2^2 R^2 \pm 2c_1 c_2 LR \cos\theta \pm 2c_2 LR \sin\theta)^{1/2} \qquad (5.18)$$

where M is the bending moment (N m), P is the cutting power (W), R is the radius of the milling cutter or the distance of the cutting edge to the center axis of the bar for boring tools (mm), L is the length of the cutting edge from the spindle nose (mm), and N is the spindle rpm. The + or − in the parenthesis is used for the axial force being away and into the spindle direction, respectively. In addition, c_1 and c_2 are the ratios of the tangential force to the radial force and axial force, respectively:

$$c_1 = \frac{F_X}{F_Y} \quad \text{and} \quad c_2 = \frac{F_Z}{F_Y} \qquad (5.19)$$

The ratio of the axial and radial force is primarily dependent on the lead angle with smaller influence by the rake of the cutting edge. In general, the axial force is the smallest of the three forces in milling, while in boring the axial (feed) force could be either higher or lower than the radial force, depending on the feed, depth of cut, lead angle, and corner radius of the insert. Equation (5.18) can be reduced to

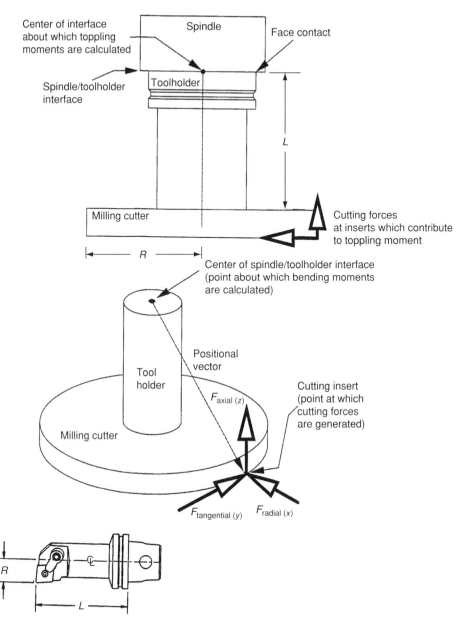

FIGURE 5.97 The maximum allowable tangential force on a milling cutter or boring bar or turning tool is a function of dimensions L and R, and the type and size of toolholder interface.

$$M = \frac{9.55\,P}{RN}L \qquad (5.20)$$

for small diameter (<100 mm) and small lead angle (<15°) milling cutters. Rake angles greater than 10° will contribute significantly on the cutting force ratios. All three forces and especially the tangential and radial forces (estimated through models) should be considered for all other cases in Equation (5.18) (Figure 5.97).

Toolholders and Workholders

Example 5.10 Consider three different shrink-fit holders for 6, 12, and 20 mm diameter tool shanks. Their nose characteristics are given in Table 5.22. The clamping length of the tool shank is considered the maximum in the toolholder. Estimate the clamping pressure and torque, the chucking axial force, and the temperature required to expand the toolholder bore to insert the tool shank for an interference fit of 0.005, 0.010, and 0.020 mm.

Solution: The analysis for the shrink-fit interface to evaluate the clamping pressure, torque and the interfacial stresses is discussed by Equations (5.3)–(5.5). The interference fit is the initial differences in diameters between the inner radius of the toolholder nose and the tool shank diameter. The clamping pressure is calculated from Equation (5.3) using different interferences and the toolholder characteristics given in Table 5.22. The pressure and hence

TABLE 5.22
Nominal Runout Values in Micrometers (μm) for Different Assemblies

	Interference (mm)	Tool Size			
		6	12	20	20
ID (mm)		6	12	20	20
OD (mm)		23	27	38	51
L (mm)		26	37	42	42
P-HSS (MPa)	0.005	150	57.7	29.3	38
T-HSS (N m)		33	72	116	150
F-HSS (N)		11	12	11.6	15
P-Carb (MPa)		134	52.8	27	34.4
T-Carb (N m)		30	66	107	136
F-Carb (N)		9.9	11	10.7	13.6
$\Delta\theta$ (°C)		107	44	19	19
P-HSS (MPa)	0.010	301	116	59	76
T-HSS (N m)		66	145	232	300
F-HSS (N)		22	10624	23	30
P-Carb (MPa)		268	106	54	69
T-Carb (N m)		59	132	215	272
F-Carb (N)		20	22	22	27
$\Delta\theta$ (°C)		168	76	38	38
P-HSS (MPa)	0.020	602	231	117	152
T-HSS (N m)		133	290	463	600
F-HSS (N)		44	48	46	60
P-Carb (MPa)		536	211	109	138
T-Carb (N m)		118	265	429	544
F-Carb (N)		39	44	43	55
$\Delta\theta$ (°C)		293	137	75	75

Notes:
ID and OD: are used at the toolholder nose.
P-HSS: clamping pressure with HSS tool shank.
ID and OD are used at the toolholder nose.
P-Carb: clamping pressure with carbide tool shank.
T-HSS: transmitted torque with HSS tool shank.
F-HSS: transmitted axial force with HSS tool shank.
L: length of contact in the holder bore.
α: thermal expansion coefficient for steel 0.0000075 mm/mm/°F.
E_{steel} = 207,000 MPa and $E_{carbide}$ = 400,000 MPa.
Poisson's ratios for steel and carbide material are considered as 0.29.
The coefficient of friction is considered as 0.15.

the stresses increase as the size of the tool shank decreases and as the wall thickness of the toolholder nose increases. The chucking torque is lower for carbide tools than for HSS tools. The chucking axial force is high for all sizes and interferences. Comparing the results in Table 5.22 with those in Figure 5.78 for shrink-fit indicates that the manufacturers are expecting lower interference fit for the smaller sizes of tools. This is the main reason for controlling the tool shank diameter with h6 tolerance or better (h5 for small size tooling). For example, the h6 tolerance specifies a variation of 0 to -0.008 mm for a 6 mm tool shank diameter; this means that the chucking pressure and torque would vary significantly by comparing the results between the 0.005 and 0.010 mm interference-fits since the toolholder nose is designed to accept tool shanks in the range of 5.992 to 6 mm. Therefore, there is a concern with small size tooling using the shrink-fit chuck. The variation of the toolholder bore size among different manufacturers results in different chucking torques. Note that the coefficient of friction of 0.15 used in Table 5.22 has a good safety factor as discussed in Equation (5.5). The toolholder nose is heated by a temperature ΔT. Assuming a uniform temperature field, the initial shape of the nose remains the same after heating. The result of the heating is an increase of the inner bore by an amount $\delta r = \alpha \Delta T r_i$ and the outer diameter of the nose by an amount $\delta r = \alpha \Delta T r_o$. The temperature difference ΔT required to expand the bore by 0.005 mm larger than the interference (between the tool shank and the toolholder nose) specified in Table 5.22 is calculated and is also given in Table 5.22. The temperature is higher for the smaller diameter tool shank interfaces as expected.

5.8 PROBLEMS

Problem 5.1 Consider the boring bar (B) in Example 4.4. The boring bar is integral to a CAT-50 holder. The toolholder–spindle CAT-50 interface stiffness is modeled by a linear spring and a torsional spring. The linear spring constant is 3×10^9 N/m, while the torsional spring constant is 4×10^7 N m/rad with the joint location (pivot point) 40 mm in the spindle nose. The length of the toolholder taper in the spindle is 100 mm. The spindle stiffness at the front (by the toolholder interface) is 2×10^8 N/m. Estimate the deflection at the tool point.

Problem 5.2 A 40 mm diameter hole is being bored to 44 mm diameter with a single point boring bar. The workpiece material is medium-carbon steel. A carbide insert is used in the indexable steel boring bar. The boring bar is integral to a CAT-50 toolholder and extends 200 mm. The cross section of the solid boring bar is cylindrical with 30 mm diameter. The cutting speed for the bar is 80 m/min while the feed is 0.2 mm/rev. The toolholder–spindle interface stiffness is characterized by a linear spring and a torsional spring. The linear spring constant is 3×10^9 N/m, while the torsional spring constant is 4×10^7 N m/rad with the joint location 40 mm in the spindle nose. The length of the toolholder taper in the spindle is 100 mm. The spindle stiffness at the front (by the toolholder interface) is 2×10^8 N/m. Estimate the deflection at the tool point in order to determine the bore size quality. The material properties are: $E_{steel} = 206{,}700$ Mpa

REFERENCES

1. E.I. Rivin, Tooling structure: interface between cutting edge and machine tool, *CIRP Ann.* **49**:2 (2000) 591–634
2. E.I. Rivin, Trends in tooling for CNC machine tools: tool–spindle interfaces, *Manuf. Rev.* **4** (1991) 264–274
3. S. Baier, Part I: Spindles and their relationship to high-speed toolholders, *Moldmaking Technol.* August 2003

4. M. Tsutsumi, A. Miyakawa, and Y. Ito, Topographical Representation of Interface Pressure Distribution in a Multiple Bolt–Flange Assembly — Measurement by Means of Ultrasonic Waves, ASME Paper 81-DE-7, 1981
5. S. Hazem, J. Mori, M. Tsutsumi, and Y. Ito, A new modular tooling system of curvic coupling type, *Proceedings of the 26th International Machine Tool Design and Research Conference*, MacMillan, New York, 1987, 261–267
6. *Designing With Kennametal*, Kennametal, Inc., Latrobe, PA, 1978
7. R.H. Thornley and I. Elewa, The static and dynamic stiffness of interference shrink-fitted joints, *Int. J. Mach. Tools Manuf.* **28** (1988) 141–155
8. E.I. Rivin, Tooling structures — a weak link in machining centers and FMS: some ways for improvement, *Proceedings of the International Conference on Manufacturing Systems and Environment*, Vol. 21, JSME, Tokyo, Japan, May 28, 1990
9. J.S. Zenker, Design and application of high speed spindles for high speed machining, *SME Hi-Speed Machining Seminar*, Minneapolis, MN, February 15, 1994
10. J.P. Thomas, Characterization of the Machine Tool Spindle to Toolholder Connection, MS Thesis, University of North Carolina at Charlotte, 1999
11. J.S. Agapiou, Selection of tool holding system for a machine tool spindle, *Proceedings of the First International Machine Tool Conference: The Dominance of Spindle Performance*, SME, Dearbon, MI, May 2003
12. M. Tsutsumi, Y. Anno, and N. Ebata, Static characteristics of 7/24 tapered joint for machining center, *Bull. JSME* **26** (1983) 461–467
13. M. Weck and I. Schubert, New interface machine/tool: hollow shank, *CIRP Ann.* **43** (1994) 345–348
14. M. Tsutsumi, M. Ohya, T. Aoyama, S. Shimizu, and S. Hachiga, Deformation and interface pressure distribution of 1/10 tapered joints at high rotational speed, *Int. J. Jpn. Soc. Precision Eng.* **30** (1996) 23–28
15. Stanley Sheppard Co., Inc., Big-Plus Spindle System, Big Daishowa, Catalog EX-48, 1998
16. Nikken Kosakusho Works, Ltd., 3-Lock System, Catalog 3341
17. J.S. Agapiou, E. Rivin, and C. Xie, Toolholder/spindle interfaces for CNC machine tools, *CIRP Ann.* **44** (1995) 383–388
18. Showa 2, Showa D-F-C Shank, Showa Tool Co., Ltd., Catalog DF-9808
19. W. Kelch, HSK tooling system for high speed machining, *Proceedings of the Sixth International Machine Tool Conference (IMTC)*, Osaka, 1994, 126–148
20. M. Tsutsumi, T. Kuwada, T. Aoyama, S. Shimizu, and S. Hachiga, Static and dynamic stiffness of 1/10 tapered joints for automatic changing, *Int. J. Jpn. Soc. Precision Eng.* **29** (1995) 301–306
21. I. Schubert, M. Weck, et al., Interface Machine/Tool: Testing and Optimization, Final Report on the Research Project, WZL Laboratory for Machine Tools and Applied Economics, Aachen University, March 31, 1994
22. Kegel-Hohlschafte Form A and Form C AnschluBmaBe, DIN 69893-1, 1996
23. Kegel-Hohlschafte nach DIN 69893 Form A AnschluBmaBe, DIN 69063-1, 1995
24. Kegel-Hohlschafte Form A and Form C nach DIN 69893-1 AnschluBmaBe (Spindle), Entwurf DIN 69063-1, 1996
25. H. Hanna, J.S. Agapiou, and D.A. Stephenson, Modeling the HSK toolholder–spindle interface, *ASME J. Manuf. Sci. Eng.* **124** (2002) 734–744
26. E. Kocherovsky, *HSK Handbook*, Intelligent Concept, West Bloomfield, MI, 1999
27. J.S. Agapiou, P. Bandyopadhyay, C.H. Shen, and D.A. Stephenson, Operational practices for using HSK tool-holder–spindle systems in machining applications, *Trans. NAMRI/SME* **30** (2001) 231–236
28. KM Modular Quick-Change Tooling Systems for Lathes and Machining Centers, Kennametal Catalog A93-49 (105) E3
29. Nikken Kosakusho Works, Ltd., NC5 Tooling System, Catalog 8205
30. T.R. Kim, S.M. Wu, and K.F. Eman, Identification of joint parameters for a taper joint, *ASME J. Eng. Ind.* **111** (1989) 282–287
31. D.M. Shamime and Y.C. Shin, Analysis of no. 50 taper joint stiffness under axial and radial loading, *Trans. NAMRI/SME* **28** (1999) 111–116

32. D.M. Shamime, S.W. Hong, and Y.C. Shin, An *in situ* modal-based method for structural dynamic joint parameter identification, *J. Mech. Eng.* **214**:C5 (2000) 641–653
33. J.H. Wang and S.B. Horng, Investigation of the tool holder system with a taper angle 7:24, *Int. J. Mach. Tools Manuf.* **34** (1994) 1163–1176
34. J. Agapiou, A methodology to measure joint stiffness parameters for toolholder/spindle interfaces, *Trans NAMRI/SME* **33** (2004) 503–510
35. T.L. Schmitz, G.S. Duncan, et al., Improved milling capabilities through dynamics prediction: three component spindle-holder-tool model, *Proceedings of the 2005 NSF DMII Grantees Conference*, Scottsdale, AZ, 2005
36. T.L. Schmitz, M. Davies, and M. Kennedy, Tool point frequency response prediction for high-speed machining by RCSA, *ASME J. Manuf. Sci. Eng.* **123** (2001) 700–707
37. D.L. Lewis, Factors for successful rotating tool operation at high speeds, Presented at the SME Hi-Speed Machining Clinic and Tabletop Exhibit, Schaumburg, IL, May 10–11, 1994
38. A. Mannan and B. Lindstrom, Investigations in the static and dynamic performance of different end mill locking systems, *CIRP Ann.* **30** (1981) 265–268
39. M. Tsutsumi, Chucking force distribution of collet chuck holders for machining centers, *J. Mech. Working Technol.* **20** (1989) 491–501
40. F. Mason, Tools run true in hydraulic toolholders, *Am. Machinist* **July** (1994) 59–61
41. M. Koch, Toolholders: an important connection between spindle and cutting tool, *Moldmaking Technol.* September 2003
42. J. Lorincz, Mechanical tool holder provides a gripping solution, *Tooling Prod.* November 2003
43. *Die & Mold Making — Application Guide*, Sandvik Coromant, Fair Lawn, NJ, 2000
44. D. Moore, Driving the tap, *Cutting Tool Eng.* **August** (1995) 52–60
45. E. Lenz, J. Rotberg, R.C. Petrof, D.J. Stauffer, and K.D. Metzen, Hole location accuracy in high speed drilling influence of chucks and collets, *CIRP Manufacturing Systems Symposium*, Ann Arbor, MI, May 1995
46. J. Tlusty, High speed milling, *Proceedings of the Sixth International Machine Tool Conference (IMTC)*, Osaka, 1994, 35–60
47. S. Baier, Part II: Spindles and their relationship to high-speed toolholders, *Moldmaking Technol.* September 2003
48. M.H. Layne, Detecting and correcting the unbalance SME High Speed Machining Clinic and Tabletop Exhibit, Chicago, IL, April 25–26, 1995
49. M.H. Layne, On balance, *Cutting Tool Eng.* **August** (1991) 36–41
50. D.W. McHenry, Keeping your balance, *Moldmaking Technol.* August 2004
51. S. Zhou and J. Shi, Active balancing and vibration control of rotating machinery: a survey, *Shock Vibr. Digest* **33**:4 (2001) 361–371
52. S.W. Dyer, B.K. Hackett, and J. Kerlin, Electromagnetically Actuated Rotating Unbalance Compensator, U.S. Patent Number 5,757,662, Baladyne Corporation, 1998
53. S. Zhou and J. Shi, Optimal one-plane active balancing of a rigid rotor during acceleration, *J. Sound Vibr.* **249**:1 (2002) 196–205
54. D. Chartier and G. VanWaes, Toolholders the critical interface, Presented at the SME High Speed Machining Clinic and Tabletop Exhibit, Chicago, IL, April 25–26, 1995
55. J.Y.H. Fuh, C.H. Chang, and M.A. Melkanoff, An integrated fixture planning and analysis system for machining processes, *Robot. Computer-Integr. Manuf.* **10**:5 (1993) 339–353
56. Y. Rong and Y. Zhu, *Computer-Aided Fixture Design*, Marcel Dekker, New York, 1999
57. E.G. Hoffman, *Jig and Fixture Design*, 2nd Edition, Delmar Publishers, Inc., 1985
58. W.E. Boyes and R. Subrin, *Low Cost Jigs and Fixtures & Gages for Limited Production*, SME, Dearborn, MI, 1986
59. W.E. Boyes and R. Bakerjian, *Handbook of Jig and Fixture Design*, SME, Dearborn, MI, 1989
60. B. Shirinzadeh, Issues in the design of the reconfigurable fixture modules for robotic assembly, *J. Manuf. Systems* **12**:1 (1993) 1–13
61. K. Nyamekye and S.S. Mudiam, A model for predicting the initial static gripping force in lathe chucks, *Int. J. Adv. Manuf. Technol.* **7** (1992) 285–291

62. M.V. Gandhi and B.S. Thompson, Automated design of modular fixtures for flexible manufacturing systems, *J. Manuf. Systems* **5**:4 (1986) 1–13
63. R.J. Menassa and W.R. DeVries, Locating point synthesis in fixture design, *CIRP Ann.* **38** (1989) 165–169
64. T.J. Drozda, R.E. King, and J.B. Creutz, *Jigs and Fixtures*, 3rd Edition, SME, Dearborn, MI, 1989
65. P.C. Miller, Workholding for CNC efficiency — Parts I and II, *Tooling Prod.* **December** (1994) 29–32; **January** (1995) 23–28
66. R. Okolischan and J. Camp, Modular fixturing slashes leadtime, *Cutting Tool Eng.* **June** (1992) 99–102
67. J.L. Colbert, R. Menassa, and W.R. DeVries, A modular fixture for prismatic parts in an FMS, *Proc. NAMRC* **14** (1986) 597–602
68. B. Benhabib, K.C. Chan, and M.Q. Dai, A modular programmable fixturing system, *ASME J. Eng. Ind.* **113** (1991) 93–100
69. J.H. Buitrago and K. Youcef-Toumi, Design of active modular and adaptable fixtures operated by robot manipulators, *Proceedings of the USA–Japan Symposium on Flexible Automation*, Minneapolis, MN, July 1988, 467–474
70. M.V. Gandhi and B.S. Thompson, Phase change fixturing for flexible manufacturing systems, *J. Manuf. Systems* **4**:1 (1985) 29–38
71. M.V. Gandhi, B.S. Thompson, and D.J. Mass, Adaptable fixture design: an analytical and experimental study of fluidized-bed fixturing, *J. Meachanisms* **108**:15 (1986)
72. J. Abou-Hanna and K. Okamura, Mechanical properties of steel pellets in particulate fluidized bed fixtures, *J. Manuf. Systems* **10**:4 (1991) 307–313
73. G.A. Phillipson, Electropermanent magnetic fixtures: an attractive alternative, *Cutting Tool Eng.* **June** (1993) 85–89
74. H. Asada and A. By, Kinematic analysis of workpart fixturing for flexible assembly with automatically reconfigurable fixtures, *IEEE J. Robot. Automat.* **RA-1** (1985)
75. Y.C. Chou, V. Chandu, and M.M. Barash, A mathematical approach to automatic configuration of machining fixtures, analysis and synthesis, *ASME J. Eng. Ind.* **111** (1989) 299–306
76. M. Mani and W.R.D. Wilson, Automated design of workholding fixtures using kinematic constraint synthesis, *Proc. NAMRC* **15** (1988) 427–432
77. R.J. Menassa, Synthesis, Analysis and Optimization of Fixtures for Prismatic Parts, PhD Thesis, Rensselaer Polytechnic Institute, Troy, NY, 1989
78. R.J. Menassa and W.R. DeVries, A design synthesis and optimization method for fixtures with compliant elements, in: P.H. Cohen and S.B. Joshi, Eds., *Advances in Integrated Product Design and Manufacturing*, ASME PED Vol. 47, ASME, New York, 1990, 203–218
79. J.Q. Xie, J.S. Agapiou, D.A. Stephenson, and P. Hilber, Machining quality analysis of an engine cylinder head using finite element methods, *J. Manuf. Processes* **5** (2003) 170–184
80. A. Markus, Z. Markusz, J. Farkas, and J. Filemon, Fixture design using Prolog: an expert system, *Robot. Computer-Integr. Manuf.* **1**:2 (1984) 167–172
81. P.M. Ferreira and C.R. Liu, Generation of workpiece orientations for machining using a rule based system, *Robot. Computer-Integr. Manuf.* **4**:3/4 (1988) 543–555
82. W. Jiang, Z. Wang, and Y. Cai, Computer-aided group fixture design, *CIRP Ann.* **37** (1988) 145–148
83. B. Bidanda and P.H. Cohen, Development of a computer aided fixture selection system for concentric, rotational parts, in: P.H. Cohen and S.B. Joshi, Eds., *Advances in Integrated Product Design and Manufacturing*, ASME PED Vol. 47, ASME, New York, 1990, 151–162
84. Y.C. Chou and M.M. Barash, A mathematical approach to automatic design of fixtures: analysis and synthesis, *Proceedings of the ASME Winter Annual Meeting*, Boston, MA, 1986, 11–27
85. J.C. Trappey and C.R. Liu, An automatic workholding verification system, *Proceedings of the International Conference on Manufacturing Technology of the Future (MSTF'89)*, Stockholm, 1989, 23–24
86. R.O. Mittal, et al., Dynamic modeling of the fixture–workpiece system, *Robot. Computer-Integr. Manuf.* **8** (1991) 201–217
87. S. Nnaji, et al., A framework for a rule-based expert fixturing system for face planer surfaces on a CAD system using flexible fixtures, *J. Manuf. Systems* **7**:3 (1988) 193–207

88. M.R. Cutkosky and S.H. Lee, Fixture planning with friction for concurrent product/process design, *Proceedings of the NSF Engineering Design Research Conference*, University of Massachusetts, Amherst, MA, 1989, 613–628
89. S.L. Jeng, L.G. Chen, and W.H. Chieng, Analysis of minimum clamping force, *Int. J. Mach. Tools Manuf.* **35** (1995) 1213–1224
90. J.S. Agapiou, E. Steinhilper, F. Gu, and P. Bandyopadhyay, Modeling machining errors on a transfer line to predict quality, *J. Manuf. Processes* **5** (2003) 1–12

6 Mechanics of Cutting

6.1 INTRODUCTION

The forces generated in metal cutting operations have long interested engineers. These forces determine machine power requirements and bearing loads, cause deflections of the part, tool, or machine structure, and supply energy to the machining system which may result in excessive cutting temperatures or unstable vibrations. Measured cutting forces are also sometimes used to compare the machinability of materials, especially in cases in which tool life tests cannot be performed due to time constraints or limited material supplies. It is also used for real-time sensor-based control in monitoring a cutting process and tool wear and failure.

This chapter gives a broad overview of cutting mechanics. Cutting force and chip thickness measurement methods are discussed in Section 6.2. Force components, empirical force models, and specific cutting power are discussed in Section 6.3–Section 6.5. A theoretical investigation of cutting mechanics requires consideration of deformation and frictional conditions, which are reviewed in Section 6.6 and Section 6.7. Shear plane, shear zone, and finite element models for cutting forces are reviewed in Section 6.8–Section 6.12. Most results are applicable only to continuous chip formation without a built-up edge, although the mechanics of discontinuous chip formation and built-up edge formation are discussed briefly in Section 6.13 and Section 6.14.

6.2 MEASUREMENT OF CUTTING FORCES AND CHIP THICKNESS

Cutting forces are often measured in machinability testing and research. Measurements are usually made using specially designed dynamometers. Early researchers used a variety of hydraulic, pneumatic, and strain gage instruments [1, 2]. More recently, however, piezoelectric dynamometers employing quartz load measuring elements have become standard for routine measurements [3–6]. Compared to other designs, quartz dynamometers have a high stiffness and broad frequency response; depending on mounting arrangements, bandwidths from 0 to over 1 kHz are common. They are also very stable thermally and exhibit little static crosstalk between measurements in different directions.

Most commonly dynamometers are mounted between the tool or workpiece and a nonrotating part of the machine tool structure (Figure 6.1). This arrangement simplifies the wiring required to route signals to amplifiers, as well as the interpretation of measured data. In some applications, however, it is desirable to mount the dynamometer in the spindle so that it measures forces in a rotating coordinate system [7, 8] (Figure 6.2). This is most often true in boring and milling tests, since forces on individual inserts can be resolved without geometric transformations of the signals, and in tests on irregular parts which present fixturing difficulties. Spindle-mounted dynamometers are also used when milling large workpieces; in this case, the weight of the workpiece on a table-mounted dynamometer can reduce the natural frequency of the measuring system to a value comparable to the tooth pass frequency [9]. The chief difficulty in using spindle-mounted dynamometers is that

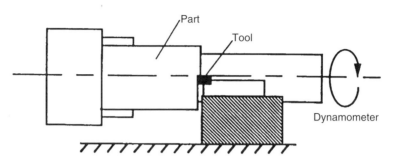

FIGURE 6.1 Platform dynamometer.

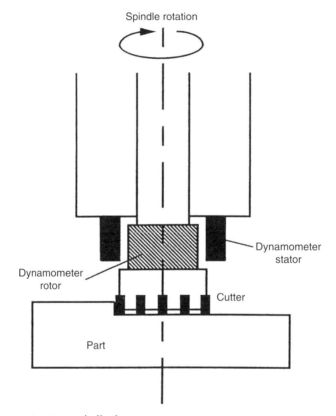

FIGURE 6.2 Rotor–stator type spindle dynamometer.

signals must be routed through a slip ring, radio transmitter, magnetic coil, or optical encoder to stationary amplifiers or recorders. An additional difficulty is that adapters are required to mount the dynamometer into the spindle nose and the tool into the dynamometer; if it is desired to use the dynamometer on a variety of machines with different spindle mounting methods and different toolholders, a large inventory of adapters must be maintained.

The frequency response of the measurement system is an important issue in testing at very high cutting speeds. Piezoelectric platform dynamometers typically have bandwidths between 800 and 1000 Hz. Filtering has been reported to increase effective dynamometer bandwidth [10–12], up to roughly 3500 Hz in a typical drilling case [13]. Special inserts with piezoelectric coatings have been designed to provide very high bandwidths (up to 40 MHz) for high-speed milling and acoustic emission studies [14, 15]. For accuracy, these inserts require careful dynamic calibration, since the signals they provide vary nonlinearly with the applied force. In tests at the highest cutting speeds, such as those produced by ballistic or gas gun test rigs, no dynamometer has a sufficiently high bandwidth, and forces must be inferred indirectly from the change in momentum estimated from workpiece velocity measurements [16–19].

The measurement of forces on production machinery presents special difficulties. Often it is not possible to shut machines down for extended periods to install instrumentation, or to use fixtures which significantly change the location of the tool point or workpiece surface. In these cases specially modified toolholders incorporating strain gages or piezoelectric force cells under the tool may be used to minimize installation time and changes in tool point position [20, 21]. In other cases, single or three-axis force transducers (e.g., piezoelectric cells) are installed in the base of fixtures, spindle carrier (at the interface of the carrier and spindle housing), turret housing behind the coupling, behind the outer race of the front spindle bearings, etc. The spindle bearings can be also instrumented with strain-gages to measure forces. Mean forces can be inferred from feed or drive motor current measurements [22, 23]. Current measurements provide little information on dynamic force variations; moreover, the calibration required to relate current levels to forces is often difficult to perform. Spindle motor power measurements are often used in machine tool monitoring.

Routine forces measurements are generally quite accurate. For commercial dynamometers, significant errors are likely to occur only if the cutting force is applied at a distance from the measuring elements; in this case a large torque acts on the dynamometer surface, and some load measuring elements may become overloaded. Dynamometer manufacturers provide technical documentation indicating acceptable force ranges and application distances for a particular instrument. In processes such as boring in which a large tool overhang is expected, specially instrumented tools may produce more accurate results than commercial platform dynamometers [24, 25]. The accuracy of specially built dynamometers is most often limited by calibration errors, especially for dynamic measurements at higher frequencies.

Force measurements from repeated tests under nominally identical conditions typically show variations of up to 10% around mean values [26]. These variations, however, are seldom due to instrument errors, but rather to variations in workpiece properties and tool cutting edge geometry or wear.

The chip thickness, which determines the shear angle (Section 6.8), is also sometimes measured in laboratory work. This parameter is more difficult to measure accurately than cutting forces. The average chip thickness is most commonly measured directly using a point micrometer or optical microscope. Depending on the material being machined, such direct measurements show a variability between 10 and 25% about mean values [26]. The chip thickness can also be inferred from a comparison of the length of the chip and the cutting distance. This method is effective in interrupted cutting tests, in which discrete sections of known length are cut, and in tests in which a long, continuous chip is produced and the cutting time is accurately known. In the latter case, if the chip is tightly curled or tangled, the ratio of lengths can be determined by weighing the chip [2]. Measurements using these methods usually show less scatter than direct measurements and are regarded as more reliable than micrometer measurements [2, 27]. Scatter in chip thickness measurements results from several sources, including the roughness of the free surface of the chip and variations in chip thickness resulting from shear-localized (type 4) chip formation or (for soft metals) melting of the chip near the cutting edge. In the last two cases, it is not as critical to measure the average chip thickness accurately, since the parameter has less physical significance.

6.3 FORCE COMPONENTS

A number of coordinate systems can be used to resolve the cutting force into directional components. Most analyses use coordinate systems in which one axis is parallel to either the cutting edge or the cutting velocity. For a simple oblique machining process, the resulting force components for mutually orthogonal coordinate directions for these two cases are shown in Figure 6.3. F_n and F_c are the components of force normal to the cutting edge and parallel to the cutting velocity respectively; F_p is the force component parallel to the cutting edge; F_z is the force component normal to the plane defined by F_n and F_p; and F_l is the force component normal to the plane defined by F_c and F_z. F_c, F_l, F_n, and F_p are related by the inclination angle λ through the equations

$$F_n = F_c \cos \lambda + F_l \sin \lambda \qquad (6.1)$$

$$F_p = -F_c \sin \lambda + F_l \cos \lambda \qquad (6.2)$$

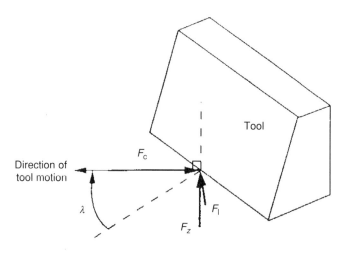

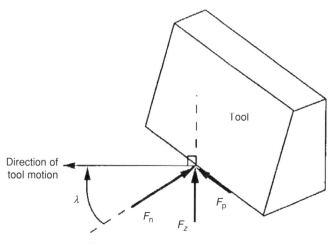

FIGURE 6.3 The F_c–F_l–F_z and F_n–F_p–F_z force systems in oblique cutting.

Mechanics of Cutting

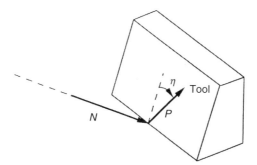

FIGURE 6.4 The forces N and P on the rake face of an oblique cutting tool.

$$F_c = F_n \cos \lambda - F_p \sin \lambda \tag{6.3}$$

$$F_1 = F_n \sin \lambda + F_p \cos \lambda \tag{6.4}$$

F_n and F_p are always positive. The positive senses of λ and F_1 should be defined so that both always have the same sign.

Some analyses use coordinate systems referred to the rake face of the tool. In this system (Figure 6.4), the most commonly defined force components are N and P, the force components normal and parallel to the rake face. N and P can be computed from the equations

$$N = F_n \cos \alpha - F_z \sin \alpha \tag{6.5}$$

$$P = \frac{F_n \sin \alpha + F_z \cos \alpha}{\cos \eta} \tag{6.6}$$

where α is the normal rake angle of the tool, measured in a plane normal to the cutting edge, and η is the chip flow angle, defined as the angle between the direction of chip flow and the normal to the cutting edge, measured in the plane of the rake face of the tool.

The chip flow angle can be measured directly using a microscope or a high-speed camera mounted near the tool [28], but this is rarely done. More commonly, the apparent chip flow angles due to cutting forces or the chip width, η_f or η_w, are measured. η_f is calculated from cutting forces using the relation

$$\tan \eta_f = \frac{F_p}{F_n \sin \alpha + F_z \cos \alpha} \tag{6.7}$$

η_w is determined from chip width measurements (Figure 6.5) using the formula

$$\cos \eta_w = \frac{b_c \cos \lambda}{b} \tag{6.8}$$

In most cases, η, η_f, and η_w are equal to within the accuracy of the measurements required to compute them. In the remainder of this chapter, we will assume equality of these parameters,

$$\eta = \eta_f = \eta_w \tag{6.9}$$

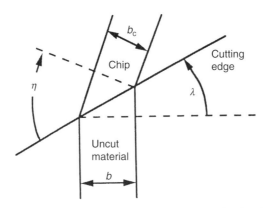

FIGURE 6.5 Relationship between width of cut b, width of chip b_c, inclination angle λ, and chip flow angle η in oblique cutting.

and use the notation η in equations. The assumption $\eta = \eta_f$ implies that the friction force P acts in the direction of chip flow. Equation (6.9) is inaccurate only when turning small diameter parts (i.e., those with diameter on the order of 2 cm), since in this case the curvature of the part influences the chip flow direction. In this case, the chip flows along an arc rather than a straight line, so that the concept of a chip flow direction loses meaning. Since chip width measurements are difficult to perform accurately, values of η determined from Equation (6.7) are usually more repeatable than those determined from Equation (6.8).

For a straight-edged oblique cutter, η is determined by the inclination angle λ:

$$\eta = \arctan[C_\eta \tan \lambda] \tag{6.10}$$

Experiments usually yield values of C_η between 0.7 and 1.2. C_η is often assumed to be 1.0; in this case, Equation (6.10) is referred to as "Stabler's rule" [29]. This assumption implies that there is no width change in the chip, and thus that the chip is formed by plain strain deformation. Deviations from Stabler's rule are usually smaller than the typical scatter in measurements. It is most accurate when λ or the rake angle α is small; when this condition is not met, the relation $C_\eta = \tan \alpha$ is more accurate when cutting steel (Figure 6.6) [30]. For a turning tool with a nose radius, η varies with the depth of cut and nose radius as well as λ and can be calculated using more complicated relations [31–34] as discussed in Section 8.6. Under these conditions, η is usually referred to as the effective lead angle.

F_c, F_z, and F_1 can be calculated from N and P using the relations

$$F_c = F_{nz} \cos \lambda + F_{pz} \sin \lambda \tag{6.11}$$

$$F_z = P \cos \eta \cos \alpha - N \sin \alpha \tag{6.12}$$

$$F_1 = F_{nz} \sin \lambda - F_{pz} \cos \lambda \tag{6.13}$$

where

$$F_{nz} = P \cos \eta \sin \alpha + N \cos \alpha \tag{6.14}$$

$$F_{pz} = P \sin \eta \tag{6.15}$$

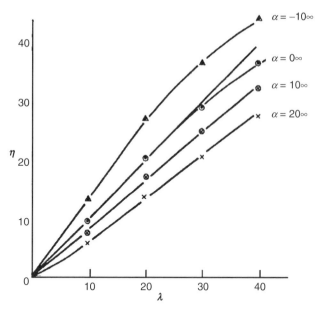

FIGURE 6.6 Relation between the chip flow angle η and inclination angle λ when cutting 1008 steel with tools with various rake angles. (After R.H. Brown and E.J.A. Armarego [30].)

In specific processes, force components are often related to the axes of motion of the machine tool. In turning and boring, the tangential (cutting), axial (feed), and radial force components F_t, F_a, and F_r shown in Figure 6.7 are used. F_t and F_a act in the cutting speed and feed directions, respectively. In milling the tangential (cutting), radial, and axial force components F_t, F_r, and F_a shown in Figure 6.8 are used. If a spindle-mounted dynamometer is used, the torque M on the cutter can also be measured. The axial force in turning is equivalent to the radial force in milling, and the radial force in turning is equivalent to the axial force in milling. In drilling, the thrust or axial force Th, torque M, and radial force F_r shown in Figure 6.9 are used. F_r rotates with time and is measured by measuring orthogonal components in directions fixed with respect to the machine tool.

The feed and radial forces are proportional to the tangential cutting force and the proportionality factor is affected by the depth of cut and lead angle. In rough turning with zero or small lead angle the ratio $F_t : F_a : F_r$ is 4:2:1. The ratio of the axial and radial forces is reduced as the lead angle increases. In addition, the axial and radial forces approach or even exceed the tangential force in finishing operations where the depth of cut is small. In peripheral (end)

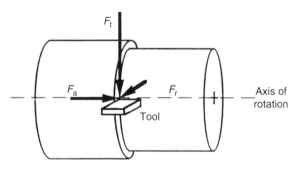

FIGURE 6.7 Forces on a turning tool.

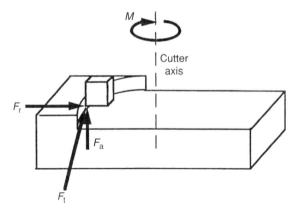

FIGURE 6.8 Forces on a cutting insert in face milling.

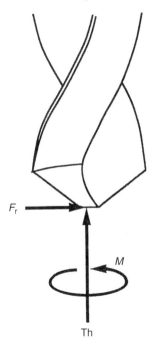

FIGURE 6.9 Forces in drilling.

milling operations, the radial force is approximately 30 to 50% of the tangential cutting force and the axial force is affected by the lead of the cutter and the edge corner geometry. In drilling, the total thrust Th is the sum of the chisel edge thrust and main cutting edge thrust. The chisel edge thrust depends on the length of the chisel edge as explained in Chapter 4 and Chapter 8. For small-diameter drills the thrust force is 1 to 1.5 times the tangential cutting force and for larger drills (especially spade or indexable drills) it is about one half.

6.4 EMPIRICAL FORCE MODELS

Cutting force estimates are often used to determine machine power requirements and bearing loads, and to design fixtures. In practice, forces are usually estimated using a variety of similar empirical equations [35–42].

For a given material, cutting forces depend most strongly on the feed rate and the width of cut. For a straight oblique cutting edge the forces depend linearly on the width of cut, provided this parameter is much larger than the feed per revolution. For a turning tool with a nose radius, which has curved cutting edges, the dependence on the corresponding parameter, the effective length of the cutting edge (or cutting edge engagement) is also almost linear. (Equations for the effective cutting edge engagement for various processes are given in Chapter 2.) Cutting forces are therefore often divided by the width of cut or effective edge length to determine the force per unit length of cutting edge. This parameter, which is important in determining if the tool will chip or break, increases most strongly with the feed rate. Increasing the rake angle generally reduces cutting forces; similarly, forces are usually lower for a tool with a formed-in chip breaker than for a flat-faced tool, since the effective rake angle is larger. The cutting forces increase significantly with increasing flank wear. Varying the cutting speed has a minor effect on cutting forces for most materials and cutting conditions. Speed effects are important, however, when cutting materials with temperature-sensitive material properties, such as titanium alloys and some steels, and when cutting over a wide speed range. Speed effects are also important in cutting soft metals at very low speeds, since a built-up edge may form and increase the effective rake angle, or when cutting at very high speeds when the momentum of the chip may become significant.

In orthogonal and oblique cutting with a straight cutting edge, the force per unit width of cut is usually assumed to vary exponentially with the feed rate and cutting speed, and to vary linearly or sinusoidally with the rake angle. For example, the forces per unit width of cut normal and parallel to the rake face of the tool can be calculated from

$$\frac{N}{b} = C_1 V^{a1} a^{b1} (1 - \sin \alpha)^{c1} \tag{6.16}$$

$$\frac{P}{b} = C_2 V^{a2} a^{b2} (1 - \sin \alpha)^{c2} \tag{6.17}$$

where N and P are the forces normal and parallel to the tool rake face, respectively, a is the uncut chip thickness, b is the width of cut, V is the cutting speed, α is the normal rake angle, and C_1, $a1$, $b1$, $c1$, C_2, $a2$, $b2$, and $c2$ are parameters which depend on the tool/workpiece material combination. Typical values of these parameters for some common work materials are given in Table 6.1. Other force components such as F_c, F_n, F_z, F_l, and F_p can be calculated

TABLE 6.1
Coefficients in Equations (6.16) and (6.17) When Cutting Common Materials with Tungsten Carbide Tools

Material	C1	a1	b1	c1	C2	a2	b2	c2
Gray cast iron (180 BHN)	1106.6	0.011	0.760	1.277	436.2	−0.042	0.595	0.065
Nodular cast iron (280 BHN)	1820.7	−0.092	0.649	0.849	1526.9	−0.091	0.400	−0.082
1018 steel (cold drawn)	2032.4	−0.080	0.730	1.425	889.8	0.000	0.641	0.958
10L45 steel (cold drawn)	1571.8	−0.101	0.682	0.853	567.4	−0.096	0.515	−0.491
319 aluminum (cast)	673.2	0.083	0.936	1.109	288.9	0.200	0.912	0.000
2024-T6 aluminum (cold drawn)	863.1	−0.018	0.800	1.007	566.2	0.000	0.688	0.000

Note: The values give N and P in N/mm² when V is given in m/min, a in mm, and α in degrees. They are valid for cutting speeds below 300 m/min, uncut chip thicknesses between 0.1 and 0.5 mm, and rake angles between −10° and 0° (iron and steel) or between 0° and 20° (aluminum alloys).

from estimates of N and P using the equations summarized in Section 6.3, or directly from equations similar to Equations (6.16) and (6.17), which are sometimes given in tooling catalogs or handbooks.

For tools with a nose radius, forces are often calculated by multiplying the area of material being cut by an empirical cutting pressure, which is in turn modeled using exponential equations. This approach is used in many simulation programs and is discussed in Section 8.6.

Exponential equations generally provide accurate force estimates. They tend to overestimate the force at low feed rates, leading to an overestimate of power requirements for machines used for finishing cuts, but the practical significance of this error is usually small. A useful feature of these equations is that they yield reasonable force estimates even when extrapolating beyond the range of cutting conditions used to generate baseline data. Such extrapolation should be ideally avoided but may be necessary in some cases to reduce testing requirements.

One limitation of the exponential models is that all variables appear independently, making it impossible to model interactions between variables. This limitation can be overcome by fitting polynomial equations which include cross-product terms. Such equations are easily fit to data from factorial experimental designs. The results often show statistically significant interactions between the feed rate and rake angle and, for some materials, between the cutting speed and all other inputs [13, 41, 42]. The gain in accuracy over exponential equations, however, is generally small except over narrow ranges of inputs (e.g., for low feed rates comparable to the chamfer on the tool edge). Equations including interaction terms also often yield unreasonable (or even negative) force estimates when extrapolated to conditions outside those used to gather the data.

Force components related to the cutting power, such as the tangential cutting force F_t in turning or the torque M in drilling, can also be estimated from tabulated values of the specific cutting power using the equations given in the next section. Values of the specific cutting power for common work materials can be obtained from machinability databases and tooling catalogs; values for some materials are given in Table 2.1. Estimates obtained in this way are not as accurate as those obtained from more complicated empirical relations, but are usually adequate for cutting power and bearing load estimates and can often be made without performing cutting tests.

6.5 SPECIFIC CUTTING POWER

The specific cutting power u_s, also called the unit cutting power, specific cutting energy, or unit cutting energy, is the power required to machine a unit volume of the work material. It is easily computed from measured cutting forces and is often used to compare the machinability of different materials, especially when comparative tool life data are unavailable.

The rotational speed is much larger than the feed speed in most cutting operations, so that only the torque or tangential force on the rotating element of the system must be considered in computing u_s. In *turning* and *boring*, the power consumption is given by the product of the cutting speed and tangential cutting force, VF_t. The volume of material removed per unit time is given by Equation (2.6), so that u_s is given by

$$u_s = \frac{F_t}{fd} \qquad (6.18)$$

In *drilling*, the power is determined by the product of the torque M and the rotational speed ω, and the volume of material removed per unit time is $(D^2 f\omega)/8$ where D is the drill diameter and f is the feed. u_s is thus given by

Mechanics of Cutting

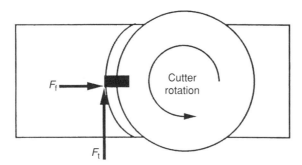

FIGURE 6.10 Forces on the tool in fly milling. The tangential force F_t, feed force F_f, and axial force F_a (directed out of the page) can be measured with a platform dynamometer when the tool is at full engagement.

$$u_s = \frac{8M}{D^2 f} \tag{6.19}$$

M can be measured directly using a drilling dynamometer. In *milling*, cutting forces and the cutting power vary continuously. Instantaneous values of u_s can be computed from measured cutting forces or torque. For a platform dynamometer, the tangential force F_t can be computed from measured forces if a single tooth cutter (or fly cutter) is used (Figure 6.10). The instantaneous specific cutting power in this case is given by

$$u_s = \frac{F_t}{fd} \tag{6.20}$$

For a spindle-mounted dynamometer, the torque on the cutter can be measured directly, and the instantaneous value of u_s is given by

$$u_s = \frac{2M}{Dfd} \tag{6.21}$$

Values for multitooth cutters can be calculated by summing results on individual inserts.

Typical values of u_s for various materials are listed in Table 2.1. u_s varies somewhat with cutting conditions. It usually increases as the feed decreases because the force component due to flank friction becomes increasingly significant. The specific energy is affected significantly by the cutting edge geometry (sharp, hone, or K-land as discussed in Chapter 4). u_s increases if the undeformed chip thickness, a, becomes smaller than the hone radius or K-land width due to increased plowing rather than shearing deformation. It also increases as the rake angle is decreased, since this increases the cutting force. Finally it normally decreases slightly as the cutting speed is increased. In using u_s to rate the machinability of various materials, similar cutting conditions should be used in tests on all materials.

In turning tests, u_s is sometimes divided into deformation and frictional energy components, u_d and u_f, which determine the proportions of the total energy expended in deforming the chip and in overcoming friction between the tool and the chip, respectively. If Stabler's rule, $\eta = \lambda$, is assumed, u_f and u_d are given by

$$u_f = \frac{P}{a_c b} \tag{6.22}$$

$$u_d = u_s - u_f \tag{6.23}$$

where P is the force parallel to the rake face of the tool (Section 6.3) and a_c, a, and b are the chip thickness, uncut chip thickness, and chip width, respectively u_f has the greater influence on tool life; an increase in u_f will usually increase temperatures along the tool–chip interface and lead to more rapid tool wear.

6.6 CHIP FORMATION AND PRIMARY PLASTIC DEFORMATION

The preceding sections have discussed the empirical study of cutting forces. Developing theoretical models based on solid or continuum mechanics requires consideration of deformation and frictional conditions in cutting.

Much of the plastic deformation in machining occurs in the formation of chips, which can be produced in a great variety of shapes and sizes. The first task in analyzing cutting processes is to identify the basic types of chips produced by distinct physical mechanisms. Four such basic types are shown in Figure 6.11, obtained by taking a cross section perpendicular to the cutting edge as illustrated in Figure 6.12. The first three types were defined by Ernst [43] and Merchant [44] in their early development of the theory of orthogonal cutting; the fourth was widely observed somewhat later as cutting speeds increased and new work materials were introduced. Type 1 chips are discontinuous chips of appreciable size. These are formed by a fracture mechanism when brittle materials such as beta brass are cut at low cutting speeds. Type 2 chips are continuous chips formed without a built-up edge on the tool. This is the most desirable type of chip. Type 3 chips are continuous chips formed with a built-up edge; these are usually encountered when machining soft, ductile metals, although many materials form a built-up edge at low cutting speeds. Type 4 chips are macroscopically continuous chips consisting of narrow bands of heavily deformed material alternating with larger regions of relatively undeformed material. These shear-localized (or catastrophic shear) chips can be formed when the yield strength of the work material decreases with temperature; under the proper conditions, rapidly heated material in a narrow band in front of the tool can become

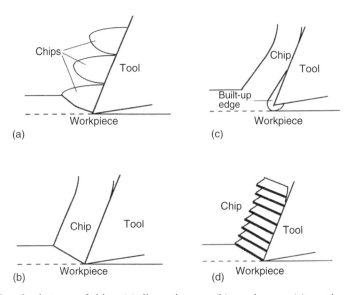

FIGURE 6.11 Four basic types of chips: (a) discontinuous, (b) continuous, (c) continuous with built-up edge (BUE), and (d) shear localized.

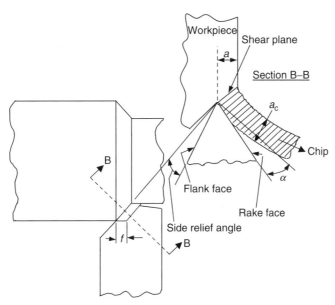

FIGURE 6.12 Cross section of chip formation perpendicular to cutting edge.

much weaker than the surrounding material, leading to localized deformation [45–47]. This type of chip is obtained when cutting hardened and stainless steels and titanium alloys at high cutting speeds [19, 48–53].

This system of classification is adequate for idealized analyses of cutting, but does not reflect some readily observed details of chip formation. Micrographs of etched and polished chips (Figure 6.13), especially specimens obtained from quick-stop devices, show that even macroscopically continuous chips are not uniform, as might be suggested from Figure 6.11. Most chips are smooth on the side formed by contact with the tool but exhibit a rough texture on the free surface. Grains in the chips are deformed in characteristic directions which may change through the thickness of the chip, particularly at low cutting speeds [54]. Deformation is also often concentrated in narrow bands. This is true of plastic deformation of metals in general, but suggests that type 1 and type 4 chips are formed by a similar mechanism, and that the macroscopic differences between these chips result from differences in the scale and degree of concentration.

Deformation conditions have been most widely studied and are best understood for continuous chip formation. In this type of chip formation, plastic deformation occurs in two regions: the primary deformation zone, which stretches from the tool tip to the free surface of the workpiece, and the secondary deformation zone at the tool–chip interface (Figure 6.14). This section describes deformation conditions in the primary deformation zone; the secondary deformation zone and tool–chip friction are discussed in Section 6.7.

The primary deformation zone can be studied experimentally using a number of methods. The most useful data comes from tests using "quick-stop" devices [55–64]. In these devices, the tool is fixed to the machine tool through some mechanism which permits it to be withdrawn from the workpiece very rapidly, abruptly interrupting the cut and leaving a partially formed chip on the workpiece. The chip sample can then be sectioned and examined; the boundaries of the deformation zones and the strains imparted to the chip can be identified by studying the workpiece grain structure or the deformation of a previously inscribed grid. The tool can never be withdrawn instantaneously, and given the small dimensions of the chip and deformation zone, even a minimal deceleration period can introduce errors. Also, the

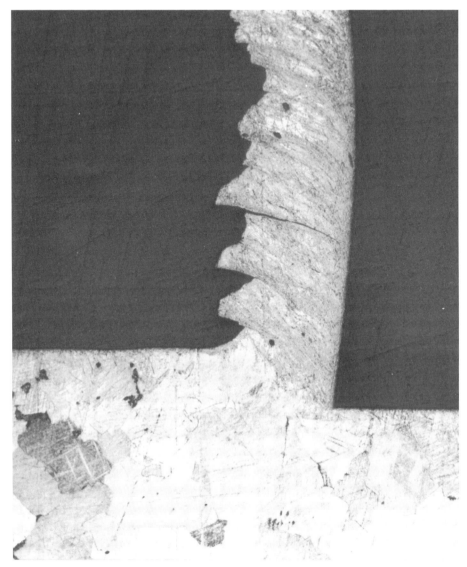

FIGURE 6.13 Typical metallographic section of a zinc chip, showing localized deformation and irregular free surface. (Courtesy of R. Stevenson.)

method provides only a picture of the deformation at a particular instant and gives no information on time-varying phenomena. This method, however, can be used at normal cutting speeds, and permits data to be obtained at the middle of the chip, rather than the edges.

The primary deformation zone can also be observed by taking high-speed motion pictures of the side of the workpiece [55, 65–70]. The earliest studies used very low cutting speeds (on the order of 1 cm/min). As better cameras have become available, results for normal cutting speeds have been reported [50, 69, 70]. The major drawback of this method is that the side of the cut is observed. This is an anomalous region where side spread may occur, so that conditions there may differ significantly from those prevailing through the bulk of the deformation zone. Similar methods combining still photography to freeze cutting action at an instant have also been proposed [71] but are subject to the same limitations.

Mechanics of Cutting

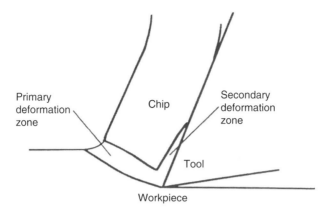

FIGURE 6.14 Deformation zones in cutting.

The stress distribution in the primary deformation zone can be studied using photoelastic methods [72]. These techniques can only be applied to transparent (e.g., plastic) work materials, and probably yield at best qualitative information on the deformation characteristics of ductile metals. Chip formation can also be observed in detailed machining studies using special devices inside a scanning electron microscope [73–76]; this method is limited to relatively soft metals and very low cutting speeds.

Overall the evidence from all methods is qualitatively consistent. The primary deformation zone is a narrow zone in which large strains are imparted to the chip. Most evidence indicates that the average thickness of the zone is on the order of one tenth of the thickness of the chip. At low speeds the zone is triangular in shape with the apex near the cutting edge, but as the cutting speed is increased this triangularity is reduced and the thickness of the zone becomes more constant. Most evidence also indicates that the curvature of the chip results directly from the primary deformation, rather than being produced by a subsequent process which produces residual stresses. It is found that the curvature of chips does not change as their outer layers are dissolved in an acid bath [77]; if residual stresses were responsible for the curvature, some change would be expected. Finally, it is often observed that the deformation zone extends below the tool point, imparting a residual stress to the machined surface.

The strains imparted in the deformation zone are among the largest observed in any deformation process [2, 64, 78–80]. The magnitude of the strain depends on how the strain is defined. In metal cutting, the geometric shear strain γ calculated using the platelet model of chip formation (Figure 6.15) is often used. In this model, the strain is given by [2]

$$\gamma = \frac{\Delta S}{\Delta y} = \tan(\phi - \alpha) + \cot \phi \qquad (6.24)$$

where α is the rake angle and ϕ is the shear angle, which depends on the rake and the cutting ratio r_c of the uncut chip thickness a to chip thickness a_c (see Section 6.8 below):

$$\tan \phi = \frac{r_c \cos \alpha}{1 - r_c \sin \alpha} \qquad (6.25)$$

$$r_c = \frac{a}{a_c} = \frac{\sin \phi}{\cos(\phi - \alpha)} \qquad (6.26)$$

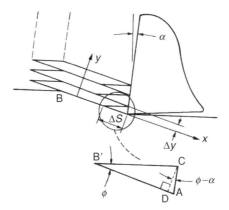

FIGURE 6.15 Platelet model for estimating shear strain in cutting. (After M.C. Shaw [2].)

Equation (6.24) can also be written as

$$\gamma = \frac{\cos \alpha}{\cos(\phi - \alpha) \sin \phi} \qquad (6.27)$$

Strains calculated using these formulas for cutting steel, brass, and aluminum are plotted in Figure 6.16 [27]. As can be seen, strains are on the order of 100% and are highest at low rake angles.

The strains are imparted in a narrow zone which is rapidly traversed, so that machining strain rates are also larger than those observed in any process except ballistic impact. The shear strain rate for the platelet model of chip formation (Figure 6.15) is given by [2]

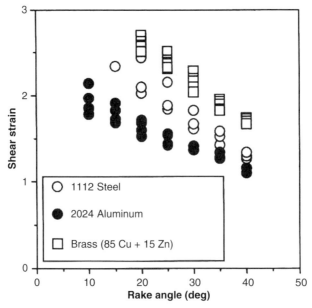

FIGURE 6.16 Shear strain when cutting steel, aluminum, and brass with tools with varying rake angles. (Data of D.M. Eggleston et al., [27].)

$$\dot{\gamma} = \frac{\cos \alpha}{\cos(\phi - \alpha)} \frac{V}{\Delta y} \tag{6.28}$$

where V is the cutting speed and Δy is the thickness of the platelets. When cutting steel at moderate cutting speeds, Δy is typically less than 25 µm. Strain rates measured in quick-stop tests when cutting steel at various cutting speeds are plotted in Figure 6.17 [55]. Although the strain rate varies considerably with the rake angle, rates on the order of 10^5sec^{-1} are observed at cutting speeds up to 250 m/min. The strain rate increases roughly linearly with cutting speeds, so that rates on the order of 10^6sec^{-1} are observed when machining at speeds above 1000 m/min.

Average primary deformation zone temperatures, estimated from infrared measurements [81], for cutting steel, aluminum, and brass at various cutting speeds are shown in Figure 6.18. Temperatures vary with the yield stress of the work material and are between 100 and 300°C for soft metals.

The shear flow stress of the work material in the primary deformation zone is usually larger than that measured in low-speed tension or torsion tests [82, 83]. Figure 6.19 illustrates this effect for free machining steel [84]. A number of unusual mechanisms, such as dependence of the flow stress on hydrostatic pressure [85, 86], have been proposed to account for the increase, but it now seems to be plausibly explained by strain and strain rate hardening. The strains in the primary deformation zone are sufficient to produce saturation strain hardening, and most metals also exhibit strain rate hardening, especially at strain rates over 10^4sec^{-1}. Flow stresses measured in low-speed cutting tests appear to be consistent with those measured in high-speed compression tests when strain, strain rate, and temperature effects are accounted for (Figure 6.20) [87].

The combination of strains, strain rates, and temperatures in the primary deformation zone, together with the frictional conditions to be discussed in the next section, cannot be reproduced in controlled material tests. As a result, the most difficult step in analyzing the mechanics of machining is describing the stress–strain behavior of the work material. Useful

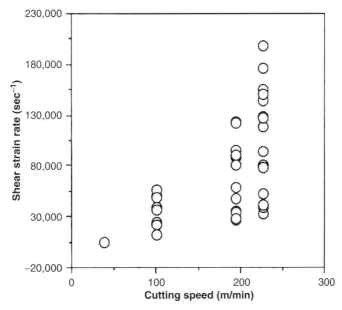

FIGURE 6.17 Shear strain rate in cutting when cutting steel at various cutting speeds. (Data of D. Kececioglu, *ASME Trans.* **80** (1958) 158–168.)

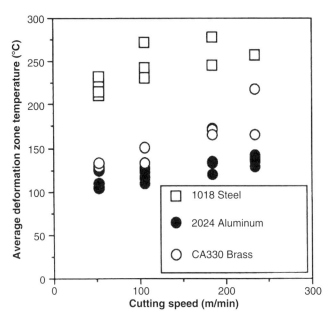

FIGURE 6.18 Average shear zone temperatures when cutting steel, brass, and aluminum at various cutting speeds. (Data of D.A. Stephenson [81].)

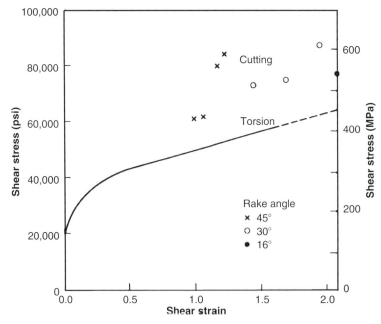

FIGURE 6.19 Shear stress of steel measured in cutting and torsion. (After M.C. Shaw and I. Finnie [84].)

analyses must be based on realistic constitutive assumptions, and tests must be defined to estimate parameters in constitutive models.

Machining tests have sometimes been proposed for characterizing material flow properties at high strains and strain rates [88–90]. Since the strains, strain rates, and temperatures are not uniform throughout the deformation zone, some model of chip formation must be

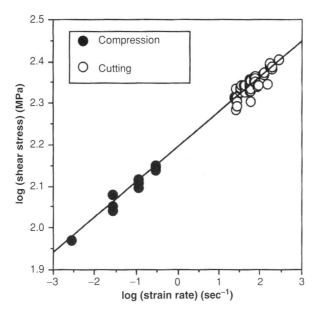

FIGURE 6.20 Shear stress of zinc in cutting and compression, corrected for strain and strain rate effects. (After R. Stevenson and D.A. Stephenson [87].)

assumed to interpret the results. These tests are therefore useful mainly for estimating parameters for particular models.

6.7 TOOL–CHIP FRICTION AND SECONDARY DEFORMATION

Friction between the tool and the chip in metal cutting influences primary deformation, built-up edge formation, cutting temperatures, and tool wear. An understanding of tool–chip friction is also necessary to develop accurate models for cutting forces and temperatures, since frictional stresses and heat fluxes are often used as boundary conditions.

The simplest way to characterize tool–chip friction is to define an effective friction coefficient, μ_e, as the ratio of the cutting force P parallel to the tool rake face to the force normal to the rake face, N:

$$\mu_e = \frac{P}{N} \qquad (6.29)$$

μ_e is sometimes converted to a friction angle, β, given by

$$\beta = \arctan(\mu_e) \qquad (6.30)$$

N and P can be estimated from cutting force measurements as discussed in Section 6.3. μ_e in cutting is usually larger than friction coefficients measured in conventional sliding friction tests; values above 1.0 are not uncommon. Increased friction in cutting results in part because the surface of the chip is newly formed and thus atomically clean. μ_e usually increases with the rake angle (Figure 6.21) [91] and also varies with the cutting speed. The friction coefficient often reaches a maximum over a narrow range of cutting speeds; in this range, the chip adheres strongly to the tool and may form a built-up edge (Figure 6.22) [53].

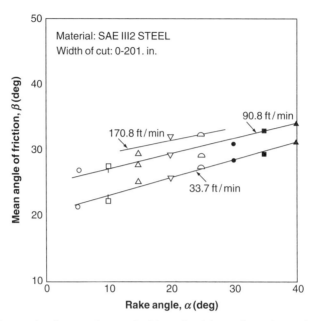

FIGURE 6.21 Friction angle when cutting steel with tools with varying rake angles. (After J.A. Bailey [91].)

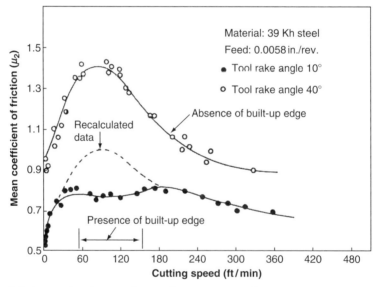

FIGURE 6.22 Friction angle when cutting steel at various cutting speeds, showing speed range over which built-up edge occurs. (After J.A. Bailey [91].)

The friction coefficient does not provide information on the distribution of stresses and strains across the tool–chip contact. Several experimental methods can be used to study the contact in more detail [92]. The most useful method employs special two-part tools (Figure 6.23) [93–97]. In this method the toolholder is modified so that the forces on the two parts can be measured independently. By varying the relative sizes of the parts and noting how the forces acting on them change, the gradient of the force distribution (the stress distribution)

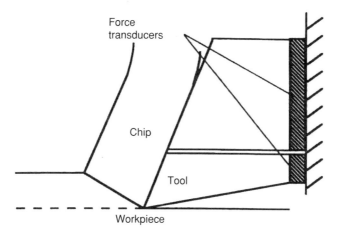

FIGURE 6.23 Composite tool for investigating the force distribution along the tool–chip interface.

can be measured, at least over parts of the interface. This method cannot be used to measure stresses at or very near the cutting edge because the parts of the tool cannot be made arbitrarily small.

Similar information can be obtained in tests with controlled contact length tools (Figure 6.24) [98–100]. In this method, contact stresses are inferred from differences in force components measured with tools with large recesses ground along the rake face to limit the tool–chip contact length to prescribed values. These results are not as useful as two-part tool results since limiting the contact length creates an artificial deformation geometry. The use of controlled contact tools greatly reduces the variability of force measurements, indicating that the significant scatter normally observed in repeated tests with conventional tools results from variations in tool–chip contact conditions.

Qualitative information on material behavior at the interface can be obtained from photoelastic studies [101–108] and from experiments in which soft metals such as lead are machined using transparent sapphire tools [109–112]. The normal stress distribution on the tool can also be inferred from the deformation of the cutting edge [113]. Information on frictional strains can be obtained by examining chip samples from quick-stop tests and from the high-speed motion picture studies discussed in the previous section. The total contact length, which is an important parameter in calculating cutting temperatures, can be measured by all of these methods. In addition, it can be estimated by examining the wear scar on the

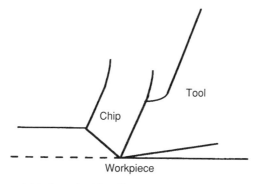

FIGURE 6.24 Tool with a restricted contact length.

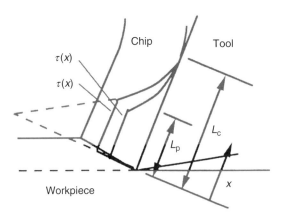

FIGURE 6.25 Tool–chip interface conditions.

tool [27, 114, 115] and by ultrasonic methods [116]. The contact length in orthogonal cutting can also be calculated from slip-line solutions [117] or more complicated numerical models [118]; however, cutting data are normally needed to estimate empirical parameters in the analytical results.

Most methods yield consistent results [63, 91, 92]. The experimental observations are summarized in Figure 6.25. The total contact length L_c depends most strongly on the chip thickness a_c. Some research [100, 105] indicates that the contact length is proportional to the product of the chip thickness and the effective friction coefficient:

$$L_c = C_f \mu_e a_c \qquad (6.31)$$

C_f varies between 1.0 and 1.5 for common test conditions. This equation can be simplified to

$$L_c = C_1 a_c \qquad (6.32)$$

where C_1 is a constant of proportionality. Reported values for C_1 range from 1.75 [119] to 2.0 [95] for ductile metals and approximately 1.5 [81] for brittle materials such as cast iron. Measured values of the ratio of the contact length to the chip thickness show considerable scatter (Figure 6.26), and Equations (6.31) and (6.32) should be regarded as rough approximations.

The tool–chip contact consists of two parts, a region of sticking or seizure friction near the cutting edge, in which the work material is deformed in shear by high frictional stresses, and a region of sliding friction away from the cutting edge, in which such deformation does not occur (Figure 6.25). Most researchers indicated that the two regions are of roughly equal length, and a half sticking, half sliding contact has been assumed in recent research on cutting temperatures [81] and acoustic emissions [120]:

$$\frac{L_p}{L_c} = 0.5 \qquad (6.33)$$

Experiments yield values of this ratio between 0.5 and 0.7, depending on the cutting conditions and tool angles.

The normal contract stress distribution $q(x)$ is found to be well approximated, at least over part of the contact, by a polynomial relationship. Based on photoelastic results, Zorev [105] proposed a single equation for the entire contact:

Mechanics of Cutting

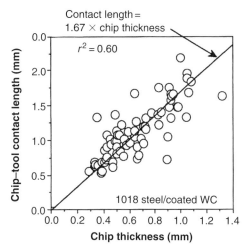

FIGURE 6.26 Relation between chip–tool contact length and chip thickness for 1018 steel workpieces cut at speeds up to 200 m/min. (Data of D.A. Stephenson [119].)

$$q(x) = Q_c \left[1 - \frac{x}{L_c}\right]^n \tag{6.34}$$

where Q_c and n are empirical constants. Zorev's results, however, were not accurate near the cutting edge due to secondary contact with the machined surface. Composite tool tests [93–97] usually suggest a two-part normal stress distribution in which the normal stress is constant in the plastic region and follows a polynomial variation in the sliding region:

$$q(x) = Q, \quad x \leq L_p \tag{6.35}$$

$$q(x) = Q_c \left[1 - \frac{x}{L_c}\right]^n, \quad x > L_p \tag{6.36}$$

As noted above, however, the two-part tool method is also inaccurate near the cutting edge, and stresses must be extrapolated over part of the sticking zone. From the data available, Equations (6.35) and (6.36) seem to describe the normal stress distribution more accurately than Equation (6.34) under most conditions.

Most data indicate that a two-part relation similar to Equations (6.35) and (6.36) is needed to describe the frictional stress distribution $t(x)$:

$$t(x) = k, \quad x = L_p \tag{6.37}$$

$$t(x) = \mu q(x), \quad x > L_p \tag{6.38}$$

In these relations μ is the true friction coefficient for the sliding region, which is generally larger than the effective coefficient μ_e, and k is the yield strength of the chip in shear, which is typically comparable to the tensile strength of the undeformed work material. These two equations can be approximated by [97]

$$t(x) = k\left[1 - \exp\left(\frac{-\mu q(x)}{k}\right)\right] \qquad (6.39)$$

The character of the contact also appears to be influenced by a number of factors which are difficult to control or to take into account theoretically, such as the presence of surface contaminants, the purity of the work material, and the cutting time [110]. Finally, it should be noted that cutting fluids often have little direct effect on tool–chip friction. Neat oils may penetrate part of the contact at low cutting speeds, reducing the sliding friction coefficient. Streams of cryogenically cooled fluid directed at the base of the chip may have a similar effect at higher speeds [121, 122]. The water-based fluids normally used at higher cutting speeds, however, function primarily as coolants and have little lubricating effect.

6.8 SHEAR PLANE AND SLIP-LINE THEORIES FOR CONTINUOUS CHIP FORMATION

The simplest theoretical models for cutting forces are based on shear plane and slip-line field analyses. In both of these approaches, the process is assumed to be steady in time and the tool is assumed to be infinitely sharp. The best-known analyses are applicable only to the two-dimensional case of orthogonal cutting (Figure 1.1) and assume that the work material is a rigid-perfectly plastic material.

Orthogonal cutting is a simple two-dimensional case of machining which can be approximated in planing, shaping, and end turning of a thin-walled tube. The shear plane model of orthogonal cutting (Figure 6.27) was first proposed by Piispanen [123, 124], Ernst [43, 125], and Merchant [44, 126, 127]. They assumed that the chip is formed by shear along a single plane inclined at an angle ϕ (the shear angle) with respect to the machined surface. This assumption is consistent with deformation patterns observed in many chip samples from quick-stop tests. ϕ can be calculated from the cutting ratio, r_c, using Equations (6.25) and (6.26). The geometric strain in the chip is given by Equation (6.27). The shear stress along the shear plane is assumed to be equal to the flow stress of the material in shear, k, and the resultant cutting force R is assumed to be directed at an angle β with respect to a normal to the rake face of the tool. β is the friction angle related to the effective friction coefficient μ_e through Equation (6.30).

It is also useful to define average sliding velocities along the shear plane, V_s, and along the rake face of the tool, V_c. Neglecting the effects of secondary shear at the tool–chip interface, V_s and V_c can be calculated from (Figure 6.28)

$$\frac{V}{\cos(\phi - \alpha)} = \frac{V_s}{\cos \alpha} = \frac{V_c}{\sin \phi} \qquad (6.40)$$

Conservation of mass also gives

$$Va = V_c a_c \qquad (6.41)$$

Equations (6.40) and (6.41) can be combined to derive Equation (6.26).

Under all these assumptions, an expression for the cutting and thrust forces, F_c and F_z, in orthogonal cutting can be derived geometrically:

$$F_c = \frac{kab \cos(\beta - \alpha)}{\sin \phi \cos(\phi + \beta - \alpha)} \qquad (6.42)$$

Mechanics of Cutting

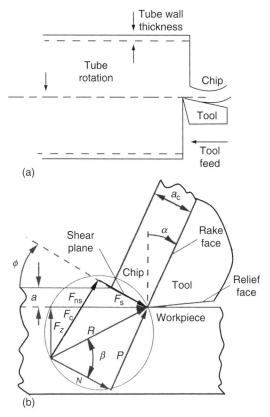

FIGURE 6.27 (a) End turning of a thin-walled tube to simulate orthogonal cutting. (b) Ernst and Merchant's shear plane theory of orthogonal cutting.

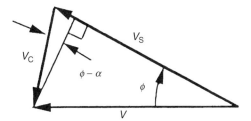

FIGURE 6.28 Velocity diagram for orthogonal cutting.

$$F_z = \frac{kab \sin(\beta - \alpha)}{\sin \phi \cos(\phi + \beta - \alpha)} \quad (6.43)$$

where F_z, the force normal to the cutting speed, is equivalent to the axial force F_a in conventional turning. These relations can be combined with Equations (6.18), (6.21), and (6.22) to derive alternate equations for the unit deformation and frictional power consumptions, u_d and u_f:

$$u_d = \frac{(F_c \cos \phi - F_z \sin \phi) \cos \alpha}{ab \cos(\phi - \alpha)} \quad (6.44)$$

$$u_\text{f} = \frac{(F_\text{c} \sin \alpha + F_z \cos \alpha) \cos \alpha}{ab \cos(\phi - \alpha)} \tag{6.45}$$

The sum of u_d and u_f can be proven to be equal to u_s, by expanding the above two equations, as expected and discussed in Section 6.5.

To apply these equations, the flow stress k and friction angle β must be specified. Both parameters are most accurately specified using cutting data. k and β can be calculated from measured cutting forces and chip thickness using the equations

$$k = \frac{(F_\text{c} \cos \phi - F_z \sin \phi) \sin \phi}{ab} \tag{6.46}$$

$$\beta = \arctan\left[\frac{F_z + F_\text{c} \tan \alpha}{F_\text{c} - F_z \tan \alpha}\right] \tag{6.47}$$

Typical values of k and β can be identified from a short series of tests. For ductile work materials such as low-carbon steels and aluminum alloys, k is usually found to be roughly equal to the low-speed yield stress, Y, which in turn is equal to roughly one third the penetration hardness, HB, in kg/mm^2. If tests cannot be performed, the assumptions $k = HB/3$ and $\beta = 45°$ can be used for rough calculations for these materials.

In addition to k and β, ϕ must be specified or computed to calculate forces from Equations (6.42) and (6.43). Merchant noted that the cutting power is given by $V \cdot F_\text{c}$, and assumed that ϕ would assume a value which minimized this power. Under this assumption, assuming k and β do not vary with ϕ, differentiating Equation (6.42) with respect to ϕ yields

$$\phi = \frac{\pi}{4} - \frac{\beta}{2} + \frac{\alpha}{2} \tag{6.48}$$

An identical result is reached if it is assumed that ϕ assumes a value which maximizes the shear stress on the shear plane [21].

The best-known slip-line field model was derived by Lee and Shaffer [128]. They proposed the slip-line field shown in Figure 6.29, in which all deformation occurs on a plane inclined at an angle ϕ with respect to the plane of cut and the material is stressed to the yield point in the triangular region ABC. From equilibrium considerations, the angle ACB can be shown to be equal to 45°, so that the shear angle is given by

$$\phi = \frac{\pi}{4} - \beta + \alpha \tag{6.49}$$

This equation yields $\phi = 0$, which is not possible, when $\beta = 45°$ and $\alpha = 0$; they proposed a second solution assuming built-up edge formation for this condition.

These analyses provide physical insight into the cutting process and, as discussed in the next chapter, have been the basis for successful analyses of steady-state cutting temperatures. They are also simple and can be readily checked against independent experimental evidence. A number of researchers have compared the predicted relationships between the angles ϕ, β, and α with measured data [27, 129–134]. In most cases, the agreement between measured and predicted values is poor for a broad data set, although both relations agree well with measurements for some materials. A typical comparison is shown in Figure 6.30 [27]. These solutions are therefore of value mainly because they provide physical insight and a convenient framework for interpreting experimental data, and not as quantitatively accurate methods for predicting cutting forces.

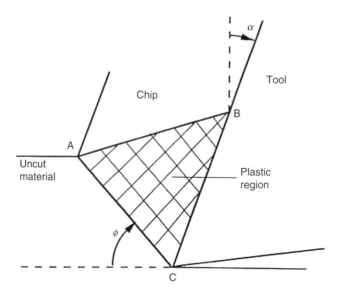

FIGURE 6.29 Lee and Shaffer's slip-line field solution for orthogonal cutting.

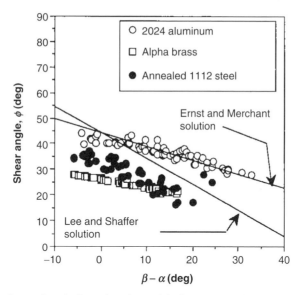

FIGURE 6.30 Comparison of typical cutting data with the Ernst and Merchant and Lee and Shaffer solutions for the relation between ϕ and $(\beta - \alpha)$. (After D.M. Eggleston et al., [27].)

A number of explanations for the poor agreement between these analyses and experimental data have been proposed. The minimum energy assumption used by Merchant is frequently questioned. As discussed in Section 6.11, however, this assumption is equivalent to the use of an upper bound theorem and is not inconsistent with the other assumptions. The most likely sources of the poor agreement are the material assumptions, namely that the work material is rigid-perfectly plastic and that friction on the rake face of the tool can be characterized by a simple friction coefficient.

Many later analyses have taken more general material assumptions into account. Both Piispanen and Merchant proposed shear plane solutions in which the material flow stress depends on the normal pressure on the shear plane. A large number of slip-line solutions have also been published. The best known is Palmer and Oxley's model for a work-hardening material [135] and various models assuming sticking friction [136–140]; many of these are applicable to controlled contact tools (Figure 6.24). Shaw [2] and Johnson and Mellor [141] have reviewed the slip-line solutions published before 1970. New shear plane and slip-line analyses are still reported [142–144]. It is difficult to judge the accuracy of the later analyses. Most are complicated and require inputs which cannot be estimated from well-defined procedures, which has prevented their wide application and comparison with independent experimental data.

6.9 SHEAR PLANE MODELS FOR OBLIQUE CUTTING

The analyses reviewed in the Section 6.8 are applicable only to orthogonal cutting. The shear plane model has been extended to three dimensions by several researchers [126, 145–147]. Slip-line analyses for oblique cutting, which yield similar results, have also been developed [148]. (Although most analyses yield similar numerical results, the equations differ considerably in form; also, in many cases different conventions are used for the positive direction of forces and the positive sense of the inclination angle λ, so that care should be taken in combining results from different analyses.) The angle definitions and shear angle solutions derived for orthogonal cutting can be used in the three-dimensional analyses provided all angles are measured in the plane normal to the cutting edge. To emphasize this point, many researchers use the subscript n to denote normal angles, that is, α_n for the normal rake angle and ϕ_n and β_n for the shear and friction angles defined by Equations (6.25) and (6.30). In the equations in this section, the subscript will be omitted with the understanding that all angles are measured in the normal plane.

Assumptions regarding the direction of the shear and frictional forces are also required to derive equations. Most commonly, it is assumed that the friction force acts in the direction of chip flow, and that the shear force acts in the direction of the shear strain. These additional assumptions probably contribute errors which are small compared to those associated with the assumptions already made in the orthogonal analyses. Under these assumptions, the cutting forces in oblique cutting are given by [148]

$$F_c = \frac{kab[\cos(\beta - \alpha)\cos\lambda + \sin\beta\sin\lambda\tan\eta]}{\sin\phi\cos(\phi + \beta - \alpha)} \quad (6.50)$$

$$F_z = \frac{kab\sin(\beta - \alpha)}{\sin\phi\cos(\phi + \beta - \alpha)} \quad (6.51)$$

$$F_l = \frac{kab[\cos(\beta - \alpha)\sin\lambda - \sin\beta\cos\lambda\tan\eta]}{\sin\phi\cos(\phi + \beta - \alpha)} \quad (6.52)$$

As would be expected, in the orthogonal case in which $\lambda = \eta = 0$, Equations (6.50) and (6.51) reduce to Equations (6.42) and (6.43), and Equation (6.52) yields $F_l = 0$. Once F_c, F_z, and F_l have been determined, F_n, F_p, N, and P can be calculated from the equations given in Section 6.3.

Estimates of η, β, and k are necessary to apply these equations. As in orthogonal cutting, reliable estimates are best made from a short series of cutting tests. η can be estimated from

Mechanics of Cutting

Equations (6.7)–(6.9). β can be determined from Equation (6.29) if P and N are determined from measured values of F_c, F_p, and F_z using Equations (6.5) and (6.6). k is most easily calculated from the deformation power per unit volume of material cut, u_d. Since it can be shown that u_d is equal to the product of k and the shear strain, γ, this gives

$$k = \frac{u_d}{\gamma} = \frac{(F_c/ab) - (P/a_c b)}{\gamma} \qquad (6.53)$$

where P can be calculated from Equation (6.6) and γ for oblique cutting is given by [2]

$$\gamma = \frac{\cot \phi + \tan(\phi - \alpha)}{\cos \eta_s} \qquad (6.54)$$

where η_s is the shear flow angle, given by

$$\eta_s = \arctan\left[\frac{\tan \lambda \, \cos(\phi - \alpha) - \tan \eta \, \sin \phi}{\cos \alpha}\right] \qquad (6.55)$$

If tests cannot be performed, the penetration hardness of the work material can be used to obtain a rough estimate of k as described in the previous section.

6.10 SHEAR ZONE MODELS

The most serious limitation of the shear plane theory is that it is based on restrictive material assumptions. The basic assumption that the deformation occurs on a single plane produces a velocity discontinuity along which the strain rate is infinite. Although geometric arguments can be used to compute an average effective strain rate, this assumption makes it difficult to include strain rate hardening, which is known to be a significant factor in high rate deformation processes, in the analysis. Also, since force but not moment equilibrium considerations are used to derive force relations, most shear plane models also assume that the shear stress on the shear plane is uniform, and that friction along the rake face of the tool can be characterized by a constant friction coefficient. As discussed in Section 6.7, the frictional assumption in particular is not consistent with experimental data.

More general material assumptions can be used if it is assumed that deformation occurs in a narrow band centered on the shear plane (Figure 6.31). Under this assumption the strain rate is finite. Moreover, moment equilibrium can also be included so that variations in stress along the shear plane and tool rake face can also be considered.

The best-known shear zone model of the cutting process is Oxley's theory [147, 149–153]. Oxley's analysis is complex and includes a numerical minimization scheme to determine the chip thickness and chip–tool contact length. A detailed description is given in his book [154]. Based on data from quick-stop tests such as those reported by Kececioglu [55], he assumes that the shear zone thickness is roughly one-tenth the shear zone length. Quick-stop data are also used to support assumptions about work material stream lines within the deformation zone. Based on these assumptions, the strain and strain rate at every point in the primary deformation zone can be calculated. Similar assumptions are used to compute the strain and strain rate distributions in the secondary deformation zone along the tool rake face. Stresses in both deformation zones are then calculated using an exponential constitutive equation,

$$\sigma = k\sqrt{3} = C\gamma^n \qquad (6.56)$$

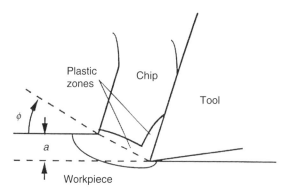

FIGURE 6.31 Shear zone model of Oxley.

where both C and n depend on the velocity modified temperature, T_{mod}, defined by [155]

$$T_{mod} = T\left[1 - \nu \frac{\dot{\gamma}}{\dot{\gamma}_0}\right] \qquad (6.57)$$

In this equation, T is the deformation zone temperature, $\dot{\gamma}$ is the strain rate, and ν and $\dot{\gamma}_0$ are constants. T can be calculated using procedures based on analytical or finite element temperature models, which will be discussed in the next chapter. Both compression and cutting tests are used to determine the variation of C and n with T_{mod}. Results have been presented mainly for hot rolled steel work materials; a typical example is shown in Figure 6.32. The velocity modified temperature was initially used to analyze low-speed, isothermal deformations, and its use in analyzing machining is sometimes questioned [150, 151]. However, Oxley has shown that the model generally yields accurate results for steel work materials.

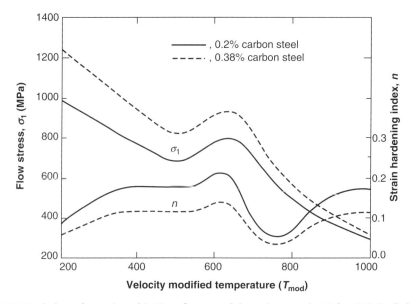

FIGURE 6.32 Variation of σ and n with T_{mod} for two plain carbon steels. (After P.L.B. Oxley [154].)

A similar analysis has been developed by Usui and coworkers [156–160]. As shown in Figure 6.33, Usui's original model [156–158] is a fully three-dimensional shear plane analysis which includes secondary cutting edge and nose radius effects. The chip thickness is predicted using a minimum energy approach, and the work material stress–strain response is modeled based on orthogonal cutting data taken under similar cutting conditions. The results are applicable to turning with a conventional single-point tool, plain milling, and groove cutting. Based on the computed forces and assumptions about the stress distributions in the contact zones, the rates of deformation and frictional heat generation are calculated and used as inputs to a finite-difference scheme to calculate the temperature distribution in the work material and tool near the cutting edge. The temperature and contact conditions are then used to estimate tool wear rates for carbide cutting tools. In later work, Usui and Shirakashi [159] have attempted to replace the orthogonal cutting data with finite element calculations; for these calculations, the material stress–strain response is modeled using results from high-speed, high-temperature tension tests.

Both Oxley's and Usui's models are complex and require specialized material test data and substantial computer code development for application. Because of these requirements they have not been widely applied outside of the research groups which developed them. Moreover, in the instances in which they have been compared to independent experimental data [161–163], small sets of measurements were used both to estimate model parameters and test calculated results, so that agreement between calculated and measured values is due in part to a curve-fitting effect. (The same concern arises when assessing the accuracy of finite element models.) These issues aside, Oxley's model is found to agree well with experimental data, particularly when modifications are made to include a larger range of material models (Figure 6.34) [163]. Both the Oxley and Usui models are more complete than earlier solutions in that they include temperature effects, and thus can be used to investigate phenomena such as tool wear [158–160] and shear-localized chip formation [164] which cannot be studied using simpler approaches. With the advent of increasingly powerful finite element models of machining, many based on commercially available solvers, it is unlikely that shear zone models will be widely implemented in the forms they were developed; however, the underlying physical algorithms and constitutive assumptions will be integrated into finite element models, as is already occurring with Usui's tool wear estimation algorithm [165] and Oxley's constitutive model [166].

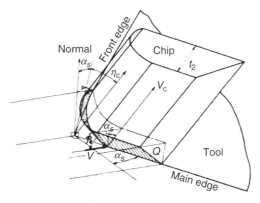

FIGURE 6.33 Usui's model of chip formation for a tool with a nose radius larger than the feed rate. (After E. Usui and A. Hirota [157].)

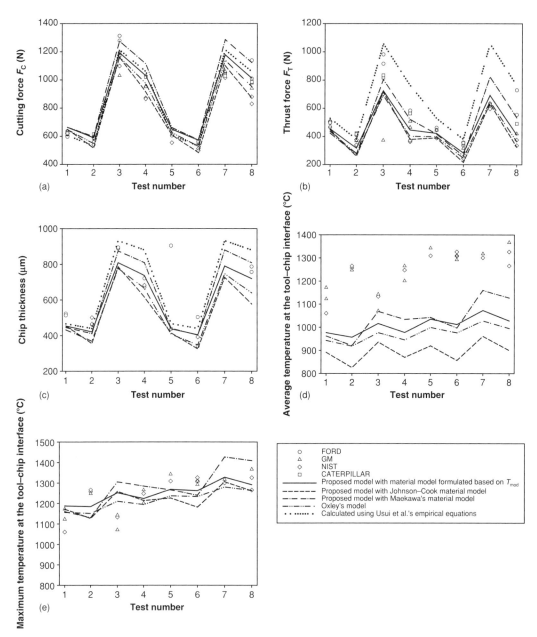

FIGURE 6.34 Comparison of the predictions of the Oxley's model with and without material modifications, Usui's model, and experimental data from four laboratories. (a) Cutting force, (b) thrust force, (c) chip thickness, (d) average temperature at the tool–chip interface, and (e) maximum temperature at the tool–chip interface. (After A.H. Abidi-Sedeh and V. Madhavan [163].)

6.11 MINIMUM WORK AND UNIQUENESS ASSUMPTIONS

Before describing finite element models of chip formation, two basic assumptions which have been widely made in analytical models should be considered: the assumption that the chip thickness will assume a value which minimizes the work required to form the chip, and the assumption that a unique deformation geometry exists for given cutting conditions and tool angles.

The chief difficulty in modeling the cutting process is that the size and shape of the chip are not determined by geometry and must be deduced from some physical condition. In analytical models the most commonly used condition has been the minimum energy criterion proposed by Merchant. Although this criterion has intuitive appeal, it has been widely questioned in the literature, generally on the grounds that it does not yield an accurate answer [167, 168]. Other conditions, such as conditions based on work hardening, have also been proposed [169], but seem to yield accurate results only for a limited range of materials [170].

The validity of the minimum energy criterion depends on the assumed constitutive model of the work material. Applying the minimum energy criterion is equivalent to seeking an upper bound solution and is therefore valid only when an equivalent limit theorem can be derived for the material model being considered. Limit theorems are usually derived from variational formulations in which the equations of motion, stress–strain equations, and boundary conditions are not solved explicitly. Instead, an approximate solution is sought by finding a stationary value of an appropriate functional of the stress or strain field. The functional must be chosen so that the Euler equations and natural boundary conditions obtained by setting its first variation equal to zero are the field equations and boundary conditions of the problem. In practice, appropriate functionals of the strain (or, equivalently, displacement or velocity) fields often have units of energy, and the stationary value can be shown to be a minimum, so that an approximate solution can be found by minimizing a scalar quantity with units of energy. This approach yields the minimum energy condition used by Merchant for a rigid-perfectly plastic material, but not for more general viscoplastic materials [171]. Similar conclusions can be drawn from variational principles used to formulate finite element models [172, 173]. This shows that the use of the minimum energy criterion in the simplest shear plane theories is valid and does not represent an independent physical assumption, and that the chief source of inaccuracy in these models is most likely the simple material model assumed.

A further question to be considered is whether the cutting forces and chip dimensions are uniquely determined by the current cutting conditions and tool angles, or whether they depend on initial conditions as well. Uniqueness is implicitly assumed in the shear plane theories, but was questioned at an early date by Hill [167], who developed an analysis to determine permissible ranges of the shear angle, rather than a unique value. Similar analyses have also been reported by Roth [174] and Dewhurst [139]. No experimental evidence for nonuniqueness, beyond the general variability of the process, has ever been presented. The only early experimental investigations of the question were those reported by Ota [175], whose results exhibited too much scatter to be conclusive, and Low [176, 177], who found that the chip thickness did not vary significantly with initial conditions. Recent experiments [178] using a CNC machining center, which permitted more accurate control of initial conditions, showed no significant variation of chip thickness or cutting forces with initial conditions when machining zinc and aluminum. The uniqueness assumption therefore appears to be well supported by the available experimental evidence.

6.12 FINITE ELEMENT MODELS

Due to their potential to accommodate more general geometric and material assumptions, there has long been interest in using finite element methods (FEM) to analyze machining. These methods hold promise not only for computing cutting forces, but also for quantities such as tool stresses and cutting temperature which are difficult to measure. This section focuses on general issues and force and stress modeling. Finite element modeling of temperatures, residual stresses, fixtures and machine tool structures are discussed in more detail in Chapter 7, Chapter 8 and Chapter 12.

The earliest finite element models were reported in the 1970s [179–182]. As computing power increased in the 1980s, improved models were developed [159, 172, 173, 183–189]. Particularly complete analyses were developed by Strenkowski and Caroll [184], Strenkowski and Moon [187], and Iwata et al. [172]; these were comprehensive steady state models with strain rate and temperature effects, which predicted cutting forces, chip dimensions, and the strain, strain rate, and temperature distributions in the cutting zone. Progress continued in the early 1990s, as numerical issues were addressed by researchers with backgrounds in structural analysis or metal forming analysis [161, 162, 165, 166, 190–197]. In recent years, there has been a great increase in finite element modeling of machining, driven by increased computing power and improved models, including commercially released and supported programs. Mackerle's bibliographies of papers on finite element analysis in machining [198, 199] list over 1000 items, 370 of which were published between 1996 and 2002.

Currently available models can be used to investigate or predict cutting forces, stress and temperature distributions, tool wear rates, chip breakage, and residual stresses. Although most are restricted to the two-dimensional case of orthogonal cutting, recent advancements in software technology have made three-dimensional simulations (such as drilling and milling) possible [200] (Figure 6.35).

In discussing finite element modeling of machining, it is helpful to concentrate on three broad areas:

- The formulation, which determines what problems the model will be best suited to, as well as what numerical issues will have to be addressed
- Material assumptions, specifically the workpiece constitutive model, tool–work friction model, thermal property assumptions, and procedures for estimating required parameters from cutting or noncutting tests
- Accuracy based on comparison with measured data

There are two basic formulations, Lagrangian and Eulerian, as illustrated in Figure 6.36 and Figure 6.37. In the Lagrangian formulation (Figure 6.36), the workpiece is fixed in space, and the tool is fed into it, imposing displacement boundary condition which drives plastic deformation. This approach is well suited to incipient and transient analyses, and to the study of such things as chip breaking and burr formation. It is not as well suited to studying steady-state machining, since calculations must be carried out over significant intervals; since small time steps must normally be used to ensure accuracy, this can create a large computational load even for the relatively simple case of orthogonal cutting. Determining a criterion for the separation of the chip from the workpiece was an issue in early Lagrangian analyses, but in recent work remeshing strategies combined with physically meaningful ductile fracture criteria have reduced concerns in this area. In Eulerian formulations (Figure 6.37), the tool is viewed as fixed in space and the work material is treated as a fluid flowing through a control volume in front of it. This eliminates the need to integrate from the incipient stage to obtain steady-state results, reducing the computational load. Fewer elements are also reportedly required for equivalent accuracy [184]. A disadvantage is that the dimensions of the chip must

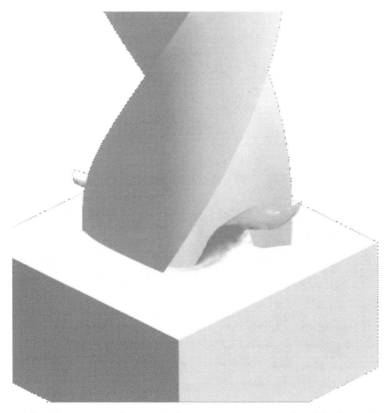

FIGURE 6.35 Finite element model of a drilling process to study the chip formation and torque. (Courtesy of WZL Aachen.)

be specified in advance; to produce a predictive model, an additional physical condition is required to determine the chip dimensions. Most commonly, a minimum energy principle or more general variational principle is used as discussed in the previous section. In this regard, the Eulerian models are similar to fluid-type variational models based on assumed chip flow lines [171]. Since computing power has steadily increased, almost all models developed since 1995 have employed Lagrangian formulations. A distinction in formulation can also be made between coupled and uncoupled analyses for models which include temperature effects; this issue will be discussed in Chapter 7.

Most current models assume combined sliding and sticking friction at the tool–chip interface, although they differ in methods of determining where sticking begins. The material constitutive response is usually modeled as elastic–plastic; although a great variety of functional forms are employed, almost all plasticity models account for strain hardening, strain rate hardening, and thermal softening. Iteration is required to include thermal softening in models in which the thermal and mechanical solutions are uncoupled. Many models assume constant thermal properties, but some do account for variations with temperature as discussed in Chapter 7. Determination of material parameters for specific models is often an issue. The common practice of quoting material constants from the literature is valid, but from a user's viewpoint it is preferable to have a defined test procedure to characterize new materials.

It is often difficult to assess the accuracy of models. Validation against a small set of force measurements is not convincing for models which have a large number of material parameters. Validation against field quantities, such as strain and temperature distributions, is often difficult since these quantities cannot be measured accurately. Accuracy is most easily

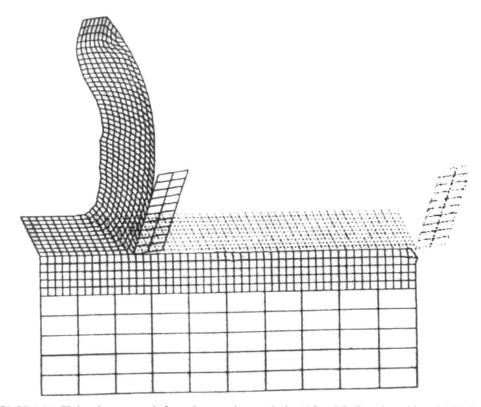

FIGURE 6.36 Finite element mesh for a Lagrangian analysis. (After J.S. Strenkowski and J.T. Caroll III [184].)

assessed if procedures for estimating material parameters are well-defined and large data sets of quantities which can be accurately and repeatably measured are available. Such quantities include forces, chip thickness, and the tool–chip interface temperature; quantities which are more difficult to measure repeatably but which still provide insight into accuracy include residual stresses, the tool–chip contact length, and chip-breaking characteristics.

As an example, we consider the most widely used of the currently available finite element models of machining, the model reported by Marusich and Ortiz [194] and marketed commercially as the AdvantEdge program [201]. It is a two-dimensional explicit Lagrangian model employing adaptive remeshing and subcycling between uncouple mechanical and thermal solutions. A two-stage rate hardening constitutive model is employed:

$$\left(1 + \frac{\dot{\varepsilon}^p}{\dot{\varepsilon}_0^p}\right) = \left(\frac{\bar{\sigma}}{g(\varepsilon^p)}\right)^{m_1} \quad \text{if } \dot{\varepsilon}^p \leq \dot{\varepsilon}_t^p \tag{6.58}$$

$$\left(1 + \frac{\dot{\varepsilon}^p}{\dot{\varepsilon}_0^p}\right)\left(1 + \frac{\dot{\varepsilon}_t^p}{\dot{\varepsilon}_0^p}\right)^{m_2/m_1 - 1} = \left(\frac{\bar{\sigma}}{g(\varepsilon^p)}\right)^{m_2} \quad \text{if } \dot{\varepsilon}^p > \dot{\varepsilon}_t^p \tag{6.59}$$

where is the effective Mieses stress, g is the flow stress, ε^p is the accumulated plastic strain, ε^p is the reference plastic strain rate, m_1 and m_2 are the low and high strain rate sensitivity exponents, respectively, and $\dot{\varepsilon}^p$ is the reference plastic strain rate separating the two regimes. A power hardening law with linear thermal softening is adopted for the flow stress:

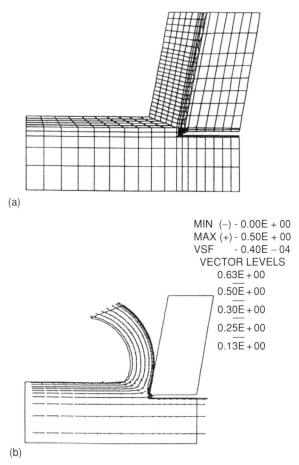

FIGURE 6.37 Finite element mesh (a) and velocity field (b) for an Eulerian analysis. (After J.S. Strenkowski and K.J. Moon [187].)

$$g(\varepsilon^p) = \sigma_0[1 - \alpha(T - T_0)]\left(1 + \frac{\varepsilon^p}{\varepsilon_0^p}\right)^{1/n} \quad (6.60)$$

where n is the hardening exponent, T is the current temperature, T_0 is a reference temperature, and σ_0 is the yield stress at T_0. The thermal and mechanical solutions are uncoupled. The temperature is treated as fixed in each stress calculation; temperatures are recomputed based on heat generation and used to update stress calculations iteratively until convergence is achieved.

A maximum plastic strain criterion is used for chip separation from the substrate; fracture occurs when the plastic strain reaches a critical value given by

$$\varepsilon_f^p \approx 2.48\, e^{-1.5p/\bar{\sigma}} \quad (6.61)$$

where p is the hydrostatic pressure. Combined sliding and sticking friction is assumed along the tool–chip interface. Sliding is assumed to occur until a maximum frictional stress is reached, after which sticking occurs and the stress is constant.

Material plasticity parameters are estimated from high-speed compression tests at different temperatures [201]. The sliding friction coefficient is determined from low-speed machining tests, or estimated from experience.

Marusich's initial paper did not describe experimental validation, but subsequent papers by AdvantEdge developers and users [202–206] have compared computations with measured values for a variety of steels and aluminum and titanium alloys. In general the predicted forces and chip characteristics from the most comprehensive finite element models are reasonably accurate; typical results are shown in Figure 6.38–Figure 6.41. Predictions of cutting temperatures, chip breaking, and residual stresses are less accurate but still generally reasonable as described in Chapter 7, Chapter 10, and Chapter 11.

6.13 DISCONTINUOUS CHIP FORMATION

The models described in Section 6.7–Section 6.11 are applicable to continuous chip formation. Although continuous chips are most commonly encountered in practice, many materials form discontinuous chips under some or all cutting conditions. The transition from continuous to discontinuous chip formation depends on the thermophysical properties and metallurgical state of the work material, as well as on the dynamics of the machine structure and cutting process. Many metals which have high hardness, hexagonal close-packed structure, and a low thermal conductivity (e.g., titanium alloys), form discontinuous chips. Other metals which form continuous chips under most conditions may form discontinuous chips at low

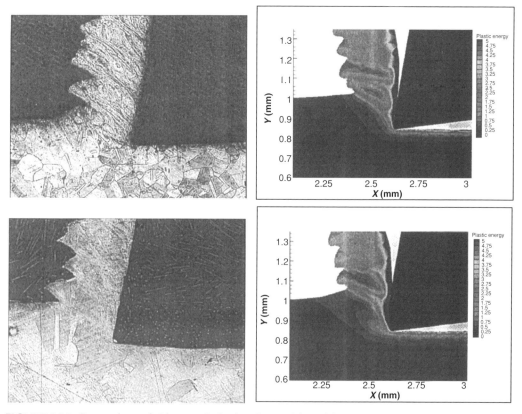

FIGURE 6.38 Comparison of chip morphologies observed in quick-stop tests and predicted by a finite element model for machining 316 stainless steel. (After M. Lundblad [205].)

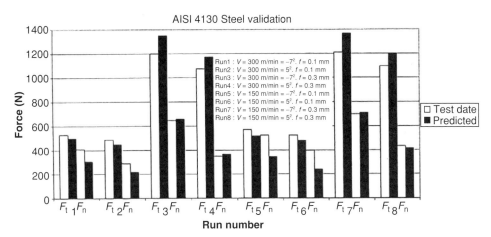

FIGURE 6.39 Comparison of measured cutting forces with forces predicted using a finite element model when cutting 4130 steel cut with coated carbide tools. (After S. Kalidas [204].)

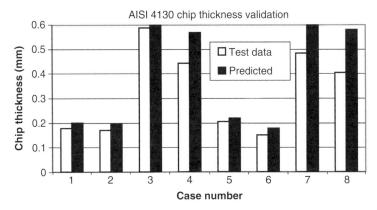

FIGURE 6.40 Comparison of measured and FEA-calculated chip thicknesses when cutting 4130 steel cut with coated carbide tools. (After T.D. Marusich et al., [203].)

cutting speed; for example, many low-carbon steels form discontinuous chips at cutting speeds less than 1 cm/sec^{-1}.

The formation of discontinuous chips was discussed qualitatively by several early researchers [124, 207, 208]. Although the mechanism of discontinuous chip formation varies somewhat for different work materials [209–211], it can be explained in broad terms with reference to Figure 6.42, which shows the stages of formation of discontinuous chips when cutting beta brass. The chip initially begins to form with a relatively high shear angle. As deformation continues, the chip slides up or adheres to the rake face. This increases the frictional force and causes the shear angle to decrease and the material to bulge. As the shear angle decreases the strain along the shear plane increases until a critical value is reached, producing a ductile shear fracture. The process then starts over. Although this description is inexact it emphasizes the importance of tool–chip friction and the ductility of the work material in discontinuous chip formation. When cutting steels, discontinuous chips are

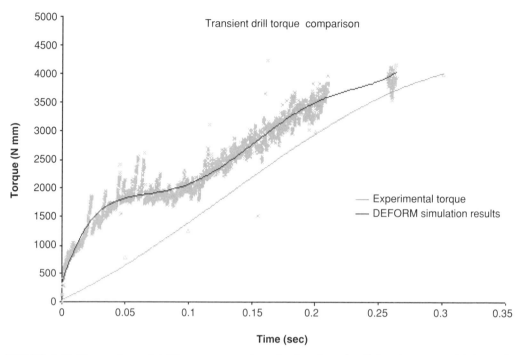

FIGURE 6.41 Comparison of drilling torque computed using the using finite element model shown in Figure 6.35 with experimental values. (Courtesy of WZL Aachen.)

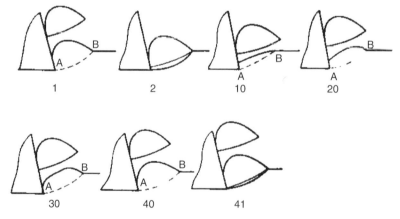

FIGURE 6.42 Formation of discontinuous chips when cutting beta brass at a speed of 1.25 cm/min. The numbers beneath successive stages are frame numbers from high-speed films; the tool feed rate is 0.0005 in./frame. (After N.H. Cook et al., [208].)

more likely to occur at low cutting speeds because tool–chip temperatures are low, increasing the strength of the work material and thus frictional stresses near the cutting edge. Similarly, the use of low or negative rake angles, which increases the effective friction coefficient, also promotes discontinuous chip formation. The importance of two additional factors is not clearly indicated by this description. Discontinuous chips are also more likely to be formed when the rigidity of the tool or part is low and when the work material contains inhomogeneities. Low system stiffness increases the elastic strain energy stored in the system, especially at

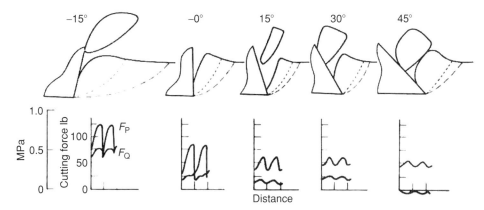

FIGURE 6.43 Variation of cutting forces with time for discontinuous chip formation when cutting beta brass with tools with varying rake angles. (After N.H. Cook et al., [208].)

high feed rates, and promotes crack propagation and fracture. Similarly, inhomogeneities in the work material produce stress concentrations which promote crack nucleation and propagation. Both of these factors are discussed in more detail by Shaw [2].

Discontinuous chip formation is inherently unsteady and produces periodic cutting forces. At low cutting speeds the amplitude of the force variations is large, especially for low or negative rake angles (Figure 6.43). At higher speeds, however, the amplitude of variation is smaller. When cutting brittle materials such as cast iron at conventional speeds, cutting force variations manifest themselves as apparent noise in measured signals, but individual peaks are not discernible unless the sampling rate is very high.

Analysis of the mechanics of discontinuous chip formation has long presented difficulties. Some work on predicting the initial and fracture shear angles from energy considerations has been reported [207, 212, 213], but the results have not agreed well with experimental data. Finite element models for discontinuous chip formation have been developed [214], and similar models for shear-localized chip formation [215] may also be adaptable to discontinuous chip formation. Although this work provides additional physical insight, results depend on a number of material parameters that are difficult to estimate without cutting tests, so that cutting forces for materials which form discontinuous chips must generally be modeled empirically.

6.14 BUILT-UP EDGE FORMATION

The built-up edge is an accumulation of heavily strained work material which collects on the cutting edge under proper conditions. It is an undesirable feature for several reasons. It reduces machining accuracy by changing the effective feed rate. It also reduces the quality of the machined surface because it periodically breaks off and reforms, introducing irregularities into the surface. The periodic breakage can also lead to chipping of the cutting edge. Finally, built-up edge is also associated with certain types of thermal cracking.

Built-up edge formation usually occurs at low cutting speeds. These speeds were much more typical of industrial practice earlier in the century, and much of the literature on the formation and avoidance of the built-up edge dates from this period. (Built-up edge formation is still a problem in low-speed processes such as drilling and end milling, especially for soft, ductile work material such as aluminum alloys.) The early work, much of which concentrated on steel work materials, has been reviewed in detail by Ernst and Martellotti [216] and Heginbotham and Gogia [217]. Nakayama et al. [218] has more recently clarified the relation of the built-up edge to cutting forces, temperatures, and surface finish. A number of researchers have described built-up edge formation for work materials other than steel [219–221], and Trent [222] has published a detailed discussion of the metallurgical aspects of the phenomenon.

Built-up edge formation is similar to discontinuous chip formation in that it is a time-varying process which depends heavily on tool–chip friction and the ductility of the work material. In fact, the built-up edge can be viewed as a partially formed discontinuous chip around which the undeformed work material flows. The time varying aspect of the phenomenon and the importance of tool–chip friction was demonstrated by Heginbotham and Gogia's quick-stop experimental studies using steel workpieces (Figure 6.44). They noted that the built-up edge starts as an embryonic structure to which successive layers adhere as cutting progresses, until it eventually attains a size and shape characteristic of the cutting conditions. The size and shape may vary considerably; Heginbotham and Gogia identified four distinct types of built-up edge for steel workpieces which occur over specific speed ranges. Ultimately the built-up edge breaks off and the process of formation repeats itself.

Physically, the built-up edge is comprised of heavily strained and hardened material. Trent [222] reports that the microhardness of the built-up edge can be more than twice as high as that of the surrounding chip, and that the strains involved in its formation are so high that they cannot be estimated because the grain structure of the material is no longer discernible. This

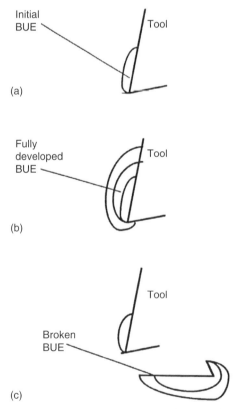

FIGURE 6.44 Formation of built-up edge (BUE) by addition of layers of work material. (a) Initial BUE. (b) Fully developed BUE composed of layers of work material. (c) Broken BUE and reformation of initial BUE.

observation further underscores the importance of the ductility of the work material in built-up edge formation, since less ductile work materials will fracture before such a structure can evolve.

As noted above, the built-up edge increases the effective rake angle of the tool and reduces cutting forces. A number of slip-line solutions have been published for cutting with a built-up edge [128, 217]; these are similar to those for conventional continuous chip formation, except that they include a roughly triangular region of dead material at the tool point which does not deform. Built-up edge formation has also been simulated using finite element models [223, 224].

Since the built-up edge is an undesirable feature, however, it is probably of greater interest to determine how it may be avoided. The standard method of reducing or eliminating built-up edge formation is to increase the cutting speed, which increases the tool–chip interface temperature and reduces the strength of the work material near the cutting edge. Other effective methods include applying a lubricant to reduce tool–chip friction, increasing the rake angle to reduce stresses at the tool point, and making the work material less ductile through cold work or the addition of inhomogeneities such as lead or sulfur particles in steel.

6.15 EXAMPLES

Example 6.1 In a turning operation on a mild steel tube with a 2.5 mm thickness, an end-cutting test was performed using a tool with a zero lead angle (which represents an orthogonal cutting test). The feed was 0.3 mm/rev, the rake angle of the tool was zero, and the cutting speed was 200 m/min. Several parameters were measured during this end-turning operation. The average chip thickness was measured with a point micrometer and found to be 0.7 mm. The tool–chip contact length (on the rake face) was estimated to be 0.5 mm. The cutting forces were measured with a dynamometer attached under the tool shank holder. The tangential cutting and axial (feed) forces were 900 and 600 N, respectively.

Calculate: (a) the mean angle of friction on the tool face, (b) the mean shear stress produced in cutting the workpiece, (c) the mean frictional stress at the tool face, and estimate (d) the estimated average strain and strain rate in the chip formation if the thickness of the shear zone is assumed to be 0.02 mm, and (e) the percentage of the total energy estimated to go into secondary deformation zone to overcome the friction at the tool–chip interface.

Solution:

(a) The mean friction coefficient on the rake face is given by Equation (6.30) using Figure 6.27 as

$$\mu = \tan(\beta) = \frac{P}{N} \Rightarrow \beta = \tan^{-1} \frac{P}{N}$$

but $\alpha = 0$, which implies $N = F_c$ and $P = F_z$ and gives

$$\beta = \tan^{-1}\left(\frac{600}{900}\right) = 33.7°$$

(b) The mean shear stress of the workpiece material during metal cutting (assuming uniform stress distribution on the shear plane) is

$$\tau_s = \frac{F_s}{A_s} = \frac{F_c \cos\phi - F_z \sin\phi}{(A_0/\sin\phi)} = \frac{(F_c \cos\phi - F_z \sin\phi)\sin\phi}{ab}$$

$$= \frac{(900 \cos 23.3° - 600 \sin 23.2°)\sin 23.2°}{(0.3)(2.5)} = 310 \text{ N/mm}^2$$

The cutting ratio from Equation (6.25) is $r_c = a/a_c = 0.3/0.7 = 0.429$. The mean shear angle between the direction of cutting speed and the shear plane is calculated from Equation (6.25)

$$\tan \phi = \frac{r_c \cos \alpha}{1 - r_c \sin \alpha} = \frac{0.429 \cos 0°}{1 - 0.429 \sin 0°} = 0.429 \Rightarrow \phi = 23.2°$$

(c) The mean frictional stress at the tool–chip interface is

$$\tau = \frac{P}{A_f} = \frac{R \sin \beta}{A_f} = \frac{\sqrt{F_c^2 + F_z^2} \sin \beta}{b l_f} = \frac{\sqrt{900^2 + 600^2} \sin 33.7°}{(2.5)(0.5)} = 480 \text{ N/mm}^2$$

(d) Shear strain, γ, at the shear zone is given by either Equation (6.24) or Equation (6.27)

$$\gamma = \cot \phi + \tan(\phi - \alpha) = \cot(23.2°) + \tan(23.2° - 0) = 2.76$$

or

$$\gamma = \frac{\cos \alpha}{\sin \phi \cos(\phi - \alpha)} = \frac{\cos 0°}{\sin(23.2) \cos(23.2 - 0)} = 2.76$$

also

$$u = \tau_s \cdot \gamma \Rightarrow \gamma = \frac{u_d}{\tau_s} = \frac{\text{Deformation power}}{\text{Shear stress along shear plane}}$$

or

$$\gamma = \frac{V_s}{V_n}$$

V_n is the velocity normal to shear plane as shown in Figure 6.28.

The shear strain is given by Equation (6.28) as

$$\dot{\gamma} = \frac{\gamma}{\Delta t} \quad \text{and} \quad \Delta t = \frac{\Delta y}{V_n} = \frac{\Delta y}{V \cdot \sin \phi} = \frac{0.02 \, (\text{mm}) \cdot 60 \, (\text{sec/min})}{200{,}000 \, (\text{mm/min}) \cdot \sin(23.2)} = 1.5 \cdot 10^{-5} \text{ sec}$$

where Δy is the thickness of the shear zone as illustrated in Figure 6.15

$$\dot{\gamma} = \frac{\gamma}{\Delta t} = \frac{2.76}{1.5 \times 10^{-5} \text{ sec}} = 1.84 \times 10^5 \text{ sec}^{-1}$$

(e) The fraction of friction energy is given by

$$\frac{u_f}{u_s} = \frac{PV}{F_c V} = \frac{P r_c}{F_c} = \frac{\sqrt{F_c^2 + F_z^2} \sin \beta r_c}{F_c} = \frac{\sqrt{900^2 + 600^2} \sin 33.7° \times 0.429}{900} = 0.29$$

Thus, the friction energy accounts for 29% of the total energy.

Example 6.2 Estimate the tangential (cutting) and feed forces for the same turning operation as in Example 6.1 if the rake angle is changed from 0° to –8°. The average chip thickness was measured to be 0.85 mm.

Solution: The mean shear angle should change because the rake angle of the tool was changed and it is calculated from Equation (6.24). The cutting ratio calculated from Equation (6.25) is $r_c = a/a_c = 0.3/0.85 = 0.353$. Therefore, the shear angle is

$$\tan \phi = \frac{r_c \cos \alpha}{1 - r_c \sin \alpha} = \frac{0.353 \cos(-8°)}{1 - 0.353 \sin(-8°)} = 0.368 \Rightarrow \phi = 20.2°$$

The shear angle is reduced from 23.2° to 20.2° as expected since the rake angle was increased. The cutting forces can be estimated using the mean shear strength of the work material calculated using the zero rake angle tool in Example 6.1. The mean friction angle can be estimated from Equation (6.30) or even better, by using the same mean friction angle calculated when cutting with zero rake angle in previous example. The cutting force can be estimated from the shear force. Therefore,

$$\tau_s = \frac{F_s}{A_s} = \frac{F_s}{(A_0/\sin \phi)} = \frac{F_s \sin \phi}{ab} \Rightarrow F_s = \frac{\tau_s ab}{\sin \phi}$$

$$= \frac{(310)(0.3)(2.5)}{\sin 20.2°} = 673 \text{ N}$$

The cutting and feed forces can be calculated from the force vector diagram in Figure 6.11 through the resultant force R

$$F_s = R \cos(\phi + \beta - \alpha) \Rightarrow R = \frac{F_s}{\cos(\phi + \beta - \alpha)}$$

$$= \frac{673}{\cos(20.2° + 26.6° + 8°)} = 1168 \text{ N}$$

Hence, the cutting and feed forces are equal to

$$F_c = R \cos(\beta - \alpha) \Rightarrow F_c = 1168 \cos(26.6° + 8°) = 961 \text{ N}$$

$$F_z = R \sin(\beta - \alpha) \Rightarrow F_z = 1168 \sin(26.6° + 8°) = 663 \text{ N}$$

6.16 PROBLEMS

Problem 6.1 In a turning operation on a cast iron tube with a lead angle of zero (which represents an orthogonal cutting test) the following conditions were noted: tube thickness = 3 mm; feed = 0.2 mm/rev; rake angle on the tool/insert body = −5°. The chip thickness was measured with a point micrometer to be 1.0 mm.

The cutting forces were measured with a dynamometer attached under the tool shank holder. The tangential (cutting) and feed forces were 800 and 400 N, respectively. Calculate the shear angle, the mean angle of friction on the tool face, and the mean shear strength of the workpiece material.

Problem 6.2 In an orthogonal cutting operation the rake angle is 8° and the coefficient of friction using a coated insert is 0.4. Determine the percentage change in chip thickness if an uncoated inset is used which results in double friction compared to coated insert.

Problem 6.3 An orthogonal cutting operation is being carried out using a 5 mm thick tube under the following conditions: feed = 0.1 mm/rev, chip thickness = 0.2 mm, cutting speed = 120 m/min, and rake angle = 10°. The forces were measured such as cutting force = 500 N and thrust force = 200 N. Calculate the percentage of the total energy that is dissipated in the shear plane during cutting.

Problem 6.4 In a turning operation of steel 1040 tube with a lead angle of zero (which represents an orthogonal cutting test) the following conditions were noted: tube thickness = 6 mm, feed = 0.2 mm/rev, and cutting speed 100 m/min. The selected tool has a rake angle of 10°. The chip thickness was measured with a point micrometer to be 0.4 mm. The specific energy of the material is 2300 N/mm².

Estimate the mean shear angle, the mean angle of friction on the tool face, and the mean shear stress of the workpiece material.

Show that the sum of the shearing power and friction power is equivalent to the total power during cutting.

Estimate the average strain and strain rate in the chip formation if the thickness of the shear zone is estimated to be 0.025 mm.

REFERENCES

1. J.T. Nicolson, Experiments with a lathe-tool dynamometer, *ASME Trans.* **25** (1904) 627–684
2. M.C. Shaw, *Metal Cutting Principles*, Oxford University Press, Oxford, 1984, Chapters 3, 7–9, and 12
3. G.F. Micheletti, B.F. von Turkovich, and S. Rossetto, Three force component piezo-electric dynamometer (ITM mark 2), *Int. J. Mach. Tool Des. Res.* **10** (1970) 305–315
4. G.H. Gautschi, Cutting forces in machining and their routine measurement with multi-component piezoelectric force transducers, *Proceedings of the 12th International Machine and Tool Design Research Conference*, Birmingham, UK, 1971, 113–120
5. J. Hoffmann and K.B. Pedersen, Design and calibration of dynamometer for CNC lathe, *Proc. NAMRC* **14** (1986) 183–188
6. K.J. Pedersen and J. Hoffmann, General analysis of dynamometers for metal cutting, *Proc. NAMRC* **14** (1986) 189–193
7. K. Nakazawa, Improvement of adaptive control of milling machine by non-contact cutting force detector, *Proceedings of the 16th International Machine and Tool Design Research Conference*, Birmingham, UK, 1975, 109–116
8. Y. Ikezaki, T. Takeuchi, and M. Sakamoto, Cutting force measurement of a rotating tool by means of optical data transmission, *CIRP Ann.* **33** (1984) 61–64
9. J. Tlusty, D.Y. Jang, and Y.S. Tarng, Measurement of the milling force over a wide frequency range, *Proc. NAMRC* **14** (1986) 273–280
10. W.A. Knight and M.M. Sadek, The correction for dynamic errors in machine tool dynamometers, *CIRP Ann.* **19** (1971) 237–245
11. Y.L. Chung and S.A. Spiewak, A model of high performance dynamometer, *ASME J. Eng. Ind.* **116** (1994) 279–288
12. N. Tounsi and A. Otho, Dynamic cutting force measuring, *Int. J. Mach. Tools Manuf.* **40** (2000) 1157–1170
13. D.N. Dilley, Accuracy, Vibration, and Stability in Drilling and Reaming, DSc Thesis, Washington Univeristy, St. Louis, MO, 2003
14. B. Bischoff, M. Hallen, T. Moser, T. Shi, D. Frohrib, and S. Ramalingam, Real time tool condition sensing. Part II. Fracture detection using a new transducer/failure identification system, in: M.K. Tse and D. Dornfeld, Eds., *Sensors for Manufacturing*, ASME, New York, 1987, 69–77

15. T. Shi, S. Ramalingam, D.A. Frohrib, and T. Moser, Tool fracture detection in the sub-millisecond time frame using machining inserts with integral sensors, *Proceedings of the USA–Japan Symposium on Flexible Automation*, ASME, New York, 1988, 1025–1033
16. G. Arndt, Ultra high speed machining: notes on metal cutting at speeds up to 7300 ft/s, *Proceedings of the 11th International Machine and Tool Design Research Conference*, Birmingham, UK, 1970, 533
17. J.P. Kottenstette and R.F. Recht, An ultra-high-machining facility, *Proc. NAMRC* **9** (1981) 318–325
18. J.P. Kottenstette and R.F. Recht, Ultra-high-speed machining experiments, *Proc. NAMRC* **10** (1982) 263–270
19. R. Komanduri, D.G. Flom, and M. Lee, Highlights of the DARPA Advanced Machining Research Program, *ASME J. Eng. Ind.* **107** (1985) 325–335
20. J.F. Pearson, W.D. Syniuts, and N.H. Cook, Development of an instrumented toolholder transducer, *ASME PED Vol. 2*, ASME, New York, 1980, 71–80
21. P.Y. Sun, Y.K. Chang, T.C. Wang, and P.T. Liu, A simple and practical piezo-electric shank type three-component dynamometer, *Int. J. Mach. Tool Des. Res.* **22** (1982) 111–124
22. Y. Altintas, Prediction of cutting forces and tool breakage in milling from feed drive current measurements, *ASME J. Eng. Ind.* **114** (1992) 386–392
23. Y.-H. Jeong and D.-W. Cho, Estimating cutting force from rotating and stationary feed motor currents on a milling machine, *Int. J. Mach. Tools Manuf.* **42** (2002) 1559–1566
24. R.A. Hallam and R.S. Allsopp, The design, development and testing of a prototype boring dynamometer, *Int. J. Mach. Tool Des. Res.* **2** (1962) 241–266
25. G.M. Zhang and S.G. Kapoor, Development of an instrumented boring bar transducer, *Proc. NAMRC* **14** (1986) 194–200
26. D.A. Stephenson, Material characterization for metal cutting force modeling, *ASME J. Eng. Mater. Technol.* **111** (1989) 210–219
27. D.M. Eggleston, R. Herzog, and E.G. Thomsen, Observations on the angle relationships in metal cutting, *ASME J. Eng. Ind.* **81** (1959) 263–279
28. I.S. Jawahir and C.A. van Luttervelt, Recent developments in chip control research and applications, *CIRP Ann.* **42** (1993) 659–693
29. G. Stabler, The fundamental geometry of cutting tools, *Proc. Inst. Mech. Eng.* **165** (1951) 14–21
30. R.H. Brown and E.J.A. Armarego, Oblique machining with a single cutting edge, *Int. J. Mach. Tool Des. Res.* **4** (1964) 9–25
31. L.V. Colwell, Predicting the angle of chip flow for single-point cutting tools, *ASME Trans.* **76** (1954) 199–204
32. K. Okushima and K. Minato, On the behavior of chip in steel cutting, *Bull. Jpn. Soc. Precision Eng.* **2** (1959) 58–64
33. H.T. Young, P. Mathew, and P.L.B. Oxley, Allowing for nose radius effects in predicting the chip flow direction and cutting forces in bar turning, *Proc. Inst. Mech. Eng.* **201C** (1987) 213–226
34. H.J. Fu, R.E. DeVor, and S.G. Kapoor, A mechanistic model for the prediction of the force system in face milling operations, *ASME J. Eng. Ind.* **106** (1988) 81–88
35. B.K. Srinivas, The Forces in Turning, SME Technical Paper MR82-947, 1982
36. K. Nakayama, M. Arai, and K. Takei, Semi-empirical equations for three components of resultant cutting force, *CIRP Ann.* **32** (1983) 33–35
37. C.A. Brown, A practical method for estimating machining forces from tool–chip contact area, *CIRP Ann.* **32** (1983) 91–93
38. G. Subramani, R. Suvada, S.G. Kapoor, and R.E. DeVor, A model for the force system for cylinder boring process, *Proc. NAMRC* **15** (1987) 439–446
39. J.W. Sutherland, G. Subramani, M.J. Kuhl, R.E. DeVor, and S.G. Kapoor, An investigation into the effect of tool and cut geometry on cutting force system prediction models, *Proc. NAMRC* **16** (1988) 264–272
40. I.A. Shareef, N. Wiemken, and J. Lis, Development of Mathematical Model for Prediction of Forces in Machining, SME Technical Paper MS90-262, 1990
41. D.A. Stephenson and J.S. Agapiou, Calculation of main cutting edge forces and torque for drills with arbitrary point geometries, *Int. J. Mach. Tools Manuf.* **32** (1992) 521–538

42. D.A. Stephenson and J.W. Matthews, Cutting forces when turning and milling cast iron with silicon nitride tools, *Trans. NAMRI/SME* **21** (1993) 223–230
43. H. Ernst, Physics of metal cutting, *Machining of Metals*, American Society for Metals, Metals Park, OH, 1938, 24.
44. M.E. Merchant, Mechanics of the metal cutting process. I. Orthogonal cutting and a type 2 chip, *J. Appl. Phys.* **16** (1945) 267–275
45. R.F. Recht, Catastrophic thermoplastic shear, *ASME J. Appl. Mech.* **31** (1964) 186–193
46. J.C. Lemaire and W.A. Backofen, Adiabatic instability in orthogonal cutting of steel, *Met. Trans.* **3** (1972) 477
47. J.Q. Xie, A.E. Bayoumi, and H.M. Zbib, A study on shear banding in chip formation of orthogonal machining, *Int. J. Mach. Tools Manuf.* **36** (1996) 835–847
48. R. Komanduri, T.A. Schroeder, J. Hazra, B.F. von Turkovich, and D.G. Flom, On the catastrophic shear instability in high-speed machining of ANSI 4340 steel, *ASME J. Eng. Ind.* **104** (1982) 121–131
49. R. Komanduri, Some clarification on the mechanics of chip formation when machining titanium alloys, *Wear* **76** (1982) 15
50. D. Lee, The effect of cutting speed on chip formation under orthogonal machining, *ASME J. Eng. Ind.* **107** (1985) 55–63
51. B.M. Manyindo and P.L.B. Oxley, Modelling the catastrophic shear type of chip when machining stainless steel, *Proc. Inst. Mech. Eng.* **200C** (1986) 349–358
52. T.J. Burns and M.A. Davies, A nonlinear dynamics model for chip segmentation in machining, *Phys. Rev. Lett.* **79** (1997) 447–450.
53. J. Barry and G. Byrne, The mechanism of chip formation in machining hardened steels, *ASME J. Manuf. Sci. Eng.* **124** (2002) 528–535
54. R. Stevenson, The morphology of machining chips formed during low speed quasi-orthogonal machining of CA360 brass and a model for their formation, *ASME J. Eng. Ind.* **114** (1992) 404–411
55. D. Kececioglu, Shear strain rate in metal cutting and its effects on shear flow stress, *ASME Trans.* **80** (1958) 158–168
56. W.F. Hastings, A new quick-stop device and grid technique for metal cutting research, *CIRP Ann.* **15** (1967) 109
57. J. Ellis, R. Kirk, and G. Barrow, The development of a quick-stop device for metal cutting research, *Int. J. Mach. Tool Des. Res.* **9** (1969) 321
58. P.K. Phillip, Study of the performance characteristics of an explosive quick-stop device for freezing cutting action, *Int. J. Mach. Tool Des. Res.* **11** (1971) 133
59. C. Spaans, A treatise on the streamlines and the stress, strain, and strain rate distributions, and on stability in the primary shear zone in metal cutting, *ASME J. Eng. Ind.* **94** (1972) 690–696
60. R.H. Brown and R. Komanduri, An investigation of the performance of a quick-stop device for metal cutting studies, *Proceedings of the 13th MTDR Conference*, 1973, 225–231
61. R.H. Brown, A double shear-pin quick stop device for rapid disengagement of a cutting tool, *Int. J. Mach. Tool Des. Res.* **16** (1976) 115–121
62. J.T. Black and C.R. James, The Hammer QSD — quick stop device for high speed machining and rubbing, *ASME J. Eng. Ind.* **103** (1981) 13–21
63. E.M. Trent, Metal cutting and the tribology of seizure: I. Seizure in metal cutting, *Wear* **126** (1988) 29–45
64. G. Boothroyd and W.A. Knight, *Fundamentals of Machining and Machine Tools*, 2nd Edition, Marcel Dekker, New York, 1989, Chapters 2 and 3
65. N.H. Cook and M.C. Shaw, A visual metal cutting study, *Mech. Eng.* **73** (1951) No. 11
66. W.B. Palmer and P.L.B. Oxley, Mechanics of orthogonal machining, *Proc. Inst. Mech. Eng.* **173** (1959) 623–654
67. A.I. Isayev and V.N. Gorbunova, A new filming method for investigating the process of plastic deformations in the zone of chip formation, *Res. Film* **3**:6 (1960) 349–356
68. T.H.C. Childs, A new visio-plasticity technique and a study of curly chip formation, *Int. J. Mech. Sci.* **18** (1971) 373–387

69. J.H.L. The, High speed films of the incipient cutting process in machining at conventional speeds, *ASME J. Eng. Ind.* **99** (1977) 262–268
70. V. Kalhori, M. Lundblad, and L.E. Lindgren, Numerical and experimental analysis of orthogonal metal cutting, *Manufacturing Science and Engineering*, ASME MED Vol. 6-2, ASME, New York, 1997
71. J.R. Crookall and D.B. Richardson, Use of photographed orthogonal grids and mechanical quick-stopping techniques in machining research, *Photography in Engineering*, The Institution of Mechanical Engineers, London, 1969, Paper 4
72. S. Ramalingam, A photoelastic study of stress distribution during orthogonal cutting. Part II. Photoplasticity observations, *ASME J. Eng. Ind.* **93** (1971) 538–544
73. A.C. Bell, S. Ramalingam, and J.T. Black, Dynamic metal cutting studies as performed on the SEM, *Proc. NAMRC* **1** (1973) 99–110
74. K. Iwata and K. Ueda, Crack nucleation and its propagation in discontinuous chip formation performed within a scanning electron microscope, *Proc. NAMRC* **3** (1975) 603–617
75. B.F. von Turkovich and M. Field, Survey on material behavior in machining, *CIRP Ann.* **30** (1981) 533–540
76. K. Iwata, K. Ueda, and K. Okuda, A study of mechanisms of burrs formation in cutting based on direct SEM observation, *J. Jpn. Soc. Precision Eng.* **48** (1982) 510–515
77. R.S. Hahn, Some observations on chip curl in metal cutting process under orthogonal cutting conditions, *ASME Trans.* **75** (1953) 538–544
78. J.K. Russel and R.H. Brown, Deformation during chip formation, *ASME J. Eng. Ind.* **87** (1965) 53–56
79. T.C. Hsu, An analysis of the plastic deformation due to orthogonal and oblique cutting, *J. Strain Anal.* **15** (1966) 375–379
80. S. Ramalingam, Deformation in orthogonal cutting, *ASME J. Eng. Ind.* **113** (1991) 121–128
81. D.A. Stephenson, Assessment of steady-state metal cutting temperature models based on simultaneous infrared and thermocouple data, *ASME J. Eng. Ind.* **113** (1991) 121–128
82. B.F. von Turkovich, Shear stress in metal cutting, *ASME J. Eng. Ind.* **92** (1970) 151–157
83. J.T. Black, Flow stress model in metal cutting, *ASME J. Eng. Ind.* **101** (1979) 403–415
84. M.C. Shaw and I. Finnie, The shear stress in metal cutting, *ASME Trans.* **77** (1955) 115–125
85. M.E. Merchant, Mechanics of the metal cutting process. II. Plasticity conditions in orthogonal cutting, *J. Appl. Phys.* **16** (1945) 318–324
86. S. Santhanam and M.C. Shaw, Flow characteristics for the complex stress state in metal cutting, *CIRP Ann.* **34** (1985) 109–111
87. R. Stevenson and D.A. Stephenson, The mechanical behavior of zinc during machining, *ASME J. Eng. Mater. Technol.* **117** (1995) 172–178
88. P.L.B. Oxley and M.G. Stevenson, Measuring stress/strain properties at very high strain rates using a machining test, *J. Inst. Metals* **95** (1967) 308
89. M.G. Stevenson and P.L.B. Oxley, An experimental investigation of the influence of strain-rate and temperature on the flow stress properties of a low carbon steel using a machining test, *Proc. Inst. Mech. Eng.* **185** (1970) 741
90. A.E. Bayoumi and M.N. Hamdan, Characterization of dynamic flow stress–strain properties through machining tests, *ASME Manuf. Rev.* **1** (1988) 130–135
91. J.A. Bailey, Friction in metal machining: mechanical aspects, *Wear* **31** (1975) 243–275
92. J.A. Arsecularatne, On tool–chip interface stress distributions, ploughing force and size effect in machining, *Int. J. Mach. Tools Manuf.* **37** (1997) 885–899
93. M.B. Gordon, Instrument for measuring the friction forces in metal cutting, *Mach. Tooling* **36**:7 (1965) 30–32
94. M.B. Gordon, The applicability of the binomial law to the process of friction in the cutting of metals, *Wear* **10** (1967) 274–290
95. S. Kato, K. Yamaguchi, and M. Yamada, Stress distribution at the interface between tool and chip in machining, *ASME J. Eng. Ind.* **94** (1972) 683–689
96. G. Barrow, W. Graham, T. Kurimoto, and Y.F. Leong, Determination of the rake face stress distribution in orthogonal machining, *Int. J. Mach. Tool Des. Res.* **22** (1982) 75–85

97. T.H.C. Childs and M.I. Mahdi, On the stress distribution between the chip and tool during metal turning, *CIRP Ann.* **38** (1989) 55–58
98. T.C. Hsu, A study of the normal and shear stresses on a cutting tool, *ASME J. Eng. Ind.* **88** (1966) 51–64
99. P.W. Wallace and G. Boothroyd, Tool forces and tool–chip friction in orthogonal machining, *J. Mech. Eng. Sci.* **6** (1964) 74–87
100. A. Bhattacharyya, On the friction process in metal cutting, *Proceedings of the 6th International Machine Tool Design and Research Conference*, 1963, 491–505
101. E.G. Coker and K.C. Chakko, An account of some experiments on the action of cutting tools, *Proc. Inst. Mech. Eng.* **117** (1922) 567–621
102. M. Okoshi and S. Fukui, Studies of cutting action by means of photoelasticity, *J. Soc. Precision Mech. Jpn.* **1** (1934) 598
103. L.C. Andreev, Photoelastic study of stresses in a cutting tool by means of cinematography, *Vestn. Mashinostr.* **38**:5 (1958) 54–57 (in Russian)
104. W.B. Rice, R. Salmon, and W.D. Syniuta, Photoelastic determination of cutting tool stresses, *Trans. Eng. Inst. Can.* **4** (1960) 20–23
105. N.N. Zorev, Interrelationship between shear process occurring along the tool face and on shear plane in metal cutting, *International Research in Production Engineering*, ASME, New York, 1963, 42–49
106. E. Usui and H. Takeyama, A photoelastic analysis of machining stresses, *ASME J. Eng. Ind.* **82** (1966) 303–308
107. W. Kattwinkel, Untersuchung an Schneiden Spannender Werkzeuge mit Hilfe der Spannungsoptik, *Ind. Anz.* **60** (1957) 29–36
108. H. Chaandrasekaran and D.V. Kapoor, Photoelastic analysis of tool–chip interface stresses, *ASME J. Eng. Ind.* **87** (1965) 495–502
109. E.D. Doyle, J.G. Horne, and D. Tabor, Frictional interactions between chip and rake face in continuous chip formation, *Proc. Roy. Soc. (London)* **A366** (1979) 173–183
110. P.K. Wright, J.G. Horne, and D. Tabor, Boundary conditions at the chip–tool interface in machining: comparisons between seizure and sliding friction, *Wear* **54** (1979) 371–390
111. P.K. Wright, Frictional interactions in machining: comparisons between transparent sapphire and steel cutting tools, *J. Met. Technol.* **8** (1981) 150–160
112. A. Bagchi and P.K. Wright, Stress analysis in machining with the use of sapphire tools, *Proc. Roy. Soc. (London)* **A409** (1987) 99–113
113. G.W. Rowe and A.B. Wilcox, A new method of determining the pressure distribution on a steel cutting tool, *J. Iron Steel Inst.* **209** (1971) 231–232
114. M.Y. Friedman and E. Lenz, Investigation of the tool–chip contact length in metal cutting, *Int. J. Mach. Tool Des. Res.* **10** (1970) 401–416
115. R.L. Woodward, Determination of plastic contact length between chip and tool in machining, *ASME J. Eng. Ind.* **99** (1977) 802–804
116. C. Spaans, A comparison of an ultrasonic method to determine the chip/tool contact length with some other methods, *CIRP Ann.* **19** (1971) 485–490
117. S. Ramalingam and P.V. Desai, Tool–Chip Contact Length in Orthogonal Machining, ASME Paper 80-WA/Prod-23, 1980
118. W.F. Hastings, P. Mathew, and P.L.B. Oxley, A machining theory for predicting chip geometry, cutting forces, etc. from work material properties and cutting conditions, *Proc. Roy. Soc. (London)* **A371** (1980) 569–587
119. D.A. Stephenson, Material characterization for metal cutting force modeling, *ASME J. Eng. Mater. Technol.* **111** (1989) 210–219
120. E. Kannatey-Asibu and D.A. Dornfeld, Quantitative relations for acoustic emissions in orthogonal cutting, *ASME J. Eng. Ind.* **103** (1981) 330–340
121. M. Mazurkiewicz, Z. Kubala, and J. Chow, Metal machining with high-pressure water-jet cooling assistance — a new possibility, *ASME J. Eng. Ind.* **111** (1989) 7–12
122. Increasing Machine Tool Productivity With High Pressure Cryogenic Coolant Flow, Manufacturing Technology Directorate, Wright Laboratory, Air Force Systems Command Report WL-TR-92-8014, May 1992

123. V. Piispanen, Lastunmuodostumisen Teoriaa, *Teknillinen Aikakauslehti* **27** (1937) 315–322 (in Finnish)
124. V. Piispanen, Theory of formation of metal chips, *J. Appl. Phys.* **19** (1948) 876–881
125. H. Ernst and M.E. Merchant, Chip formation, friction, and high quality machined surfaces, *Surface Treatment of Metals*, ASM, Cleveland, OH, 1941, 299–378
126. M.E. Merchant, Basic mechanics of the metal-cutting process, *ASME J. Appl. Mech.* **11** (1944) A168–A175
127. M.E. Merchant, Mechanics of the metal cutting process. II. Plasticity conditions in orthogonal cutting, *J. Appl. Phys.* **16** (1945) 318–324
128. E.H. Lee and B.W. Shaffer, The theory of plasticity applied to a problem of machining, *ASME J. Appl. Mech.* **18** (1951) 405–412
129. M.C. Shaw, N.H. Cook, and I. Finnie, Shear angle relationships in metal cutting, *ASME Trans.* **75** (1953) 273–288
130. J.H. Creveling, T.F. Jordan, and E.G. Thomsen, Some studies of angle relationships in metal cutting, *ASME Trans.* **79** (1957) 127–138
131. H.D. Pugh, Mechanics of the cutting process, *Proceedings of the IME Conference on Tech. Eng. Manufacture*, London, 1958, 237
132. S. Kobayashi and E.G. Thomsen, Metal cutting analysis — I. Re-evaluation and new method of presentation of theories, *ASME J. Eng. Ind.* **84** (1962) 63–70
133. E.J.A. Armarego, A note on the shear angle relation in orthogonal cutting, *Int. J. Mach. Tool Des. Res.* **6** (1966) 139–141
134. N. Ueda and T. Matsuo, An investigation of some shear angle theories, *CIRP Ann.* **35** (1986) 27–30
135. W.B. Palmer and P.L.B. Oxley, Mechanics of orthogonal machining, *Proc. Inst. Mech. Eng.* **173** (1959) 623–654
136. W. Johnson, Some slip-line fields for swaging or expanding, indenting, extruding and machining for tools with curved dies, *Int. J. Mech. Sci.* **4** (1962) 323–347
137. E. Usui and K. Hoshi, Slip-line fields in metal machining which involve centered fans, *International Research on Production Engineering*, ASME, New York, 1963, 61
138. H. Kudo, Some new slip line solutions for two-dimensional steady-state machining, *Int. J. Mech. Sci.* **7** (1965) 43–55
139. P. Dewhurst, On the non-uniqueness of the machining process, *Proc. Roy. Soc. (London)* **A360** (1978) 587–610
140. L. De Chiffre and T. Wanheim, Chip compression relationships in metal cutting, *Proc. NAMRC* **9** (1981) 231–234
141. W. Johnson and P.B. Mellor, *Engineering Plasticity*, Van Nostrand Reinhold, London, 1973, 467–493
142. F.C. Appl and S. Saleem, Prediction of shear angle using minimum energy principle and strain softening, *Trans. NAMRI/SME* **19** (1991) 113–120
143. T. Shi and S. Ramalingam, Slip line solution for orthogonal cutting with a chip breaker and flank wear, *Int. J. Mech. Sci.* **33** (1991) 689–704
144. A.G. Atkins, Modelling metal cutting using modern ductile fracture mechanics: quantitative explanations for some longstanding problems, *Int. J. Mech. Sci.* **45** (2003) 373–396
145. M.C. Shaw, N.H. Cook, and P.A. Smith, The mechanics of three-dimensional cutting operations, *ASME Trans.* **74** (1952) 1055–1064
146. N.N. Zorev, *Metal Cutting Mechanics*, Pergamon Press, Oxford, 1966, Chapter 6 (translated by H.S.H. Massey)
147. G.C. I Lin and P.L.B. Oxley, Mechanics of oblique machining: predicting chip geometry and cutting forces from work-material properties and cutting conditions, *Proc. Inst. Mech. Eng.* **186** (1972) 813–820
148. W.A. Morcos, A slip line field solution of the free continuous cutting problem in conditions of light friction at chip–tool interface, *ASME J. Eng. Ind.* **102** (1980) 310–314
149. R.G. Fenton and P.L.B. Oxley, Mechanics of orthogonal machining: allowing for the effects of strain rate and temperature on tool–chip friction, *Proc. Inst. Mech. Eng.* **178** (1969) 417–438

150. R.G. Fenton and P.L.B. Oxley, Mechanics of orthogonal machining: predicting chip geometry and cutting forces from work-material properties and cutting conditions, *Proc. Inst. Mech. Eng.* **184** (1970) 927–942
151. P.L.B. Oxley and W.F. Hastings, Minimum work as a possible criterion for determining the frictional conditions at the tool/chip interface in machining, *Philos. Trans.* **282** (1976) 565–584
152. W.F. Hastings, P. Mathew, and P.L.B. Oxley, A machining theory for predicting chip geometry, cutting forces etc. from work material properties and cutting conditions, *Proc. Roy. Soc. (London)* **A371** (1980) 569–587
153. P.L.B. Oxley, Machinability: a mechanics of machining approach, *On The Art of Cutting Metals — 75 Years Later*, ASME PED Vol. 7, ASME, New York, 1982, 37–83
154. P.L.B. Oxley, *The Mechanics of Machining*, Ellis Horwood, Chicester, UK, 1989
155. C.W. MacGregor and J.C. Fisher, Tension tests at constant true strain-rates, *ASME J. Appl. Mech.* **13** (1946) A11
156. E. Usui, A. Hirota, and M. Masuko, Analytical prediction of three dimensional cutting process. Part 1. Basic cutting model and energy approach, *ASME J. Eng. Ind.* **100** (1978) 222–228
157. E. Usui and A. Hirota, Analytical prediction of three dimensional cutting process. Part 2. Chip formation and cutting force with conventional single-point tool, *ASME J. Eng. Ind.* **100** (1978) 229–235
158. E. Usui, T. Shirakashi, and T. Kitagawa, Analytical prediction of three dimensional cutting process. Part 3. Cutting temperature and crater wear of carbide tool, *ASME J. Eng. Ind.* **100** (1978) 236–243
159. E. Usui and T. Shirakashi, Mechanics of machining — from 'descriptive' to 'predictive' theory, *On The Art of Cutting Metals — 75 Years Later*, ASME PED Vol. 7, ASME, New York, 1982, 13–35
160. E. Usui and T. Shirakashi, Analytical prediction of cutting tool wear, *Wear* **100** (1984) 129–151
161. M. Shatla, C. Kerk, and T. Altan, Process modeling in machining. Part I. Determination of flow stress data, *Int. J. Mach. Tools Manuf.* **41** (2001) 1511–1534
162. M. Shatla, C. Kerk, and T. Altan, Process modeling in machining. Part II. Applications of flow stress data to predict process variables, *Int. J. Mach. Tools Manuf.* **41** (2001) 1659–1680
163. A.H. Abidi-Sedeh and V. Madhavan, Effect of some modifications to Oxley's machining theory and the applicability of different material models, *Mach. Sci. Technol.* **6** (2002) 379–395
164. B.M. Maniyindo and P.L.B. Oxley, Modeling the catastrophic shear type of chip when machining stainless steel, *Proc. Inst. Mech. Eng.* **C200** (1986) 349–358
165. Y.-C. Yen, J. Soehner, H. Weude, J. Schmidt, and T. Altan, Estimation of tool wear of carbide tool in orthogonal cutting using FEM simulation, *Proceedings of the 5th CIRP International Workshop on Modeling of Machining Operations*, West Lafayette, IN, 2002, 149–160
166. K.W. Kim and H.-C. Sin, Development of a thermo-viscoplastic cutting model using finite element method, *Int. J. Mach. Tools Manuf.* **36** (1996) 379–397
167. R. Hill, The mechanics of machining: a new approach, *J. Mech. Phys. Solids* **3** (1954) 47–53
168. C. Rubenstein, A note concerning the inadmissibility of applying the minimum work criterion to metal cutting, *ASME J. Eng. Ind.* **105** (1983) 294–296
169. P.K. Wright, Predicting the shear plane angle in machining from work material strain-hardening characteristics, *ASME J. Eng. Ind.* **104** (1982) 285–292
170. A. Bagchi, Discussion on a previously published paper by P.K. Wright, *ASME J. Eng. Ind.* **105** (1983) 129–131
171. D.A. Stephenson and S.M. Wu, Computer models for the mechanics of three-dimensional cutting processes. I. Theory and numerical method, *ASME J. Eng. Ind.* **110** (1988) 32–37
172. K. Iwata, K. Osakada, and Y. Terasaka, Process modeling of orthogonal cutting by the rigid-plastic finite element method, *ASME J. Eng. Mater. Technol.* **106** (1984) 132–138
173. J.T. Carroll III and J.S. Strenkowski, Finite element models of orthogonal cutting with application to single point diamond turning, *Int. J. Mech. Sci.* **30** (1988) 899–920
174. R.N. Roth, The range of permissible shear angles in orthogonal machining allowing for variable hydrostatic stress on the shear plane and variable friction angle along the rake face, *Int. J. Mach. Tool Des. Res.* **15** (1975) 161–177

175. T. Ota, A. Shindo, and H. Fukuola, An investigation on the theories of orthogonal cutting, *Trans. Jpn. Soc. Mech. Engrs.* **24** (1958) 484–493 (in Japanese)
176. A.H. Low and P.T. Wilkinson, An Investigation of Non-Steady-State Cutting, Report 45, National Engineering Laboratory, Glasgow, 1962
177. A.H. Low, Effects of Initial Conditions in Metal Cutting, Report 65, National Engineering Laboratory, Glasgow, 1962
178. R. Stevenson and D.A. Stephenson, The effect of prior cutting conditions on the shear mechanics of orthogonal machining, *ASME J. Manuf. Sci. Eng.* **120** (1998) 13–20
179. K. Okushima and Y. Kakino, The residual stress produced by metal cutting, *CIRP Ann.* **10** (1971) 13–14
180. B.E. Klamecki, Incipient Chip Formation in Metal Cutting — A Three-Dimensional Finite-Element Analysis, PhD Thesis, University of Illinois, 1973.
181. A.O. Tay, M.G. Stevenson, and G. de Vahl Davis, Using the finite element method to determine temperature distributions in orthogonal machining, *Proc. Inst. Mech. Eng.* **188** (1974) 627–638
182. P.D. Muraka, G. Barrow, and S. Hinduja, Influence of the process variables on the temperature distribution in orthogonal machining using the finite element method, *Int. J. Mech. Sci.* **21** (1979) 445–456
183. R.T. Sedgwick, Numerical modeling of high-speed machining processes, *High Speed Machining*, ASME PED Vol. 12, ASME, New York, 1984, 141–155
184. J.S. Strenkowski and J.T. Caroll III, A finite element model of orthogonal metal cutting, *ASME J. Eng. Ind.* **107** (1985) 349–354
185. J.S. Strenkowski and G.L. Mitchum, An improved finite element model of orthogonal metal cutting, *Proc. NAMRC* **15** (1987) 506–509
186. T.H.C. Childs and K. Maekawa, A computer simulation approach towards the determination of optimum cutting conditions, *Strategies for Automation of Machining,* ASM, Materials Park, OH, 1987, 157–166
187. J.S. Strenkowski and K.J. Moon, Finite element prediction of chip geometry and tool/workpiece temperature distributions in orthogonal metal cutting, *ASME J. Eng. Ind.* **112** (1990) 313–318
188. K. Komvopoulos and S.A. Erpenbeck, Finite element modeling of orthogonal metal cutting, *ASME J. Eng. Ind.* **113** (1991) 253–267
189. T. Tyan and W.H. Yang, Analysis of orthogonal metal cutting processes, *Int. J. Numer. Methods Eng.* **34** (1992) 365–389
190. A.J. Shih and H.T.Y. Yang, Experimental and finite element predictions of residual stresses due to orthogonal metal cutting, *Int. J. Numer. Methods Eng.* **36** (1993) 1487–1507
191. Z.C. Lin and W.C. Pan, A thermoelastic–plastic large deformation model for orthogonal cutting with tool flank wear. Part I. Computational procedures, *Int. J. Mech. Sci.* **35** (1993) 829–840
192. Z.C. Lin and W.C. Pan, A thermoelastic–plastic large deformation model for orthogonal cutting with tool flank wear. Part II. Machining application, *Int. J. Mech. Sci.* **35** (1993) 841–850
193. G.S. Sekhon and J.L. Chenot, Numerical simulation of continuous chip formation during non-steady orthogonal cutting, *Eng. Comput.* **10** (1993) 31–48
194. T.D. Marusich and M. Ortiz, Modelling and simulation of high-speed machining, *Int. J. Numer. Methods Eng.* **38** (1995) 3675–3694
195. T. Ozel and T. Altan, Process simulation using finite element method — prediction of cutting forces, tool stresses, and temperatures in high-speed flat end milling process, *Int J. Mach. Tools Manuf.* **40** (2000) 713–738
196. T. Obikawa and E. Usui, Computational machining of titanium alloy — finite element modeling and a few results, *ASME J. Manuf. Sci. Eng.* **118** (1996) 208–215
197. Y.B. Guo and C.R. Liu, 3D FEA modeling of hard turning, *ASME J. Manuf. Sci. Eng.* **124** (2002) 189–199
198. J. Mackerle, Finite-element analysis and simulation of machining: a bibliography (1976–1996), *J. Mater. Process. Technol.* **86** (1999) 17–44
199. J. Mackerle, Finite element analysis and simulation of machining: an addendum, a bibliography (1966–2002), *Int. J. Mach. Tools Manuf.* **43** (2003) 103–114

200. C.E. Fischer and P. Chigurupati, Using computer simulation to understand and optimize high-speed machining, *High Speed Machining Conference*, SME, April 8–9, 2003
201. *Third Wave AdvantEdge Theoretical Manual*, Version 4.3, Third Wave Systems, Inc., Minneapolis, MN, 2002
202. T.D. Marusich and E. Askari, *Modeling Residual Stress and Workpiece Quality in Machined Surfaces*, Third Wave Systems, Inc., Minneapolis, 2002.
203. T.D. Marusich, C.J. Brand, and J.D. Thiele, A methodology for simulation of chip breakage in turning processes using an orthogonal finite element model, *Proceedings of the 5th CIRP International Workshop on Modeling Machining Operations*, West Lafayette, IN, 2002, 139–148
204. S. Kalidas, Cost effective tool selection and process development through process simulation, *Proceedings of the 2002 Third Wave AdvantEdge User's Conference*, Atlanta, GA, April 2002, Paper 2
205. M. Lundblad, Influence of cutting tool geometry on residual stress in the workpiece, *Proceedings of the 2002 Third Wave AdvantEdge User's Conference*, Atlanta, GA, April 2002, Paper 7
206. M. Russell, Modeling for machining, Third Wave applications at Sikorsky, *Proceedings of the 2002 Third Wave AdvantEdge User's Conference*, Atlanta, GA, April 2002, Paper 8
207. M. Field and M.E. Merchant, Mechanics of formation of the discontinuous chip in metal cutting, *ASME Trans.* **71** (1949) 421–430
208. N.H. Cook, I. Finnie, and M.C. Shaw, Discontinuous chip formation, *ASME Trans.* **76** (1954) 153–162
209. W.B. Palmer and M.S.M. Riad, Modes of cutting with discontinuous chips, *Proceedings of the 8th International MTDR Conference*, Manchester, 1967, 259–279
210. R. Komanduri and R.H. Brown, On the mechanics of chip segmentation in machining, *ASME J. Eng. Ind.* **108** (1968) 33–51
211. N. Ueda and T. Matsuo, An analysis of saw-toothed chip formation, *CIRP Ann.* **31** (1982) 81–84
212. T.H.C. Childs and G.W. Rowe, Physics in metal cutting, *Rep. Prog. Phys.* **36** (1973) 223–288
213. W.K. Luk and R.C. Brewer, An energy approach to the mechanics of discontinuous chip formation, *ASME J. Eng. Ind.* **86** (1964) 157–162
214. T. Obikawa, H. Sasahara, T. Shirakashi, and E. Usui, Application of computational machining method to discontinuous chip formation, *ASME J. Manuf. Sci. Eng.* **119** (1997) 667–674
215. J.Q. Xie, A.E. Bayoumi, and H.M. Zbib, FEA modeling and simulation of shear localized chip formation in metal cutting, *Int. J. Mach. Tools Manuf.* **38** (1998) 1067–1087
216. H. Ernst and M. Martelloti, The formation and function of the built-up edge, *Mech. Eng.* **57** (1935) 487
217. W.B. Heginbotham and S.L. Gogia, Metal cutting and the built-up nose, *Proc. Inst. Mech. Eng.* **175** (1961) 892–917
218. K. Nakayama, M.C. Shaw, and R.C. Brewer, Relationship between cutting forces, temperatures, built-up edge and surface finish, *CIRP Ann.* **14** (1966) 211–223
219. H. Takeyama and T. Ono, Basic investigation of built-up edge, *ASME J. Eng. Ind.* **90** (1968) 335–342
220. K. Iwata, J. Aihara, and K. Okushima, On the mechanism of built-up edge formation in cutting, *CIRP Ann.* **19** (1971) 323
221. H. Bao and M.G. Stevenson, An investigation of built-up edge formation in the machining of aluminum, *Int. J. Mach. Tool Des. Res.* **16** (1976) 165–178
222. E.M. Trent, Metal cutting and the tribology of seizure: II. Movement of work material over the tool in metal cutting, *Wear* **128** (1988) 47–64
223. E. Usui, K. Maekawa, and T. Shirakashi, Simulation analysis of built-up edge formation in machining low carbon steel, *Bull. Jpn. Soc. Precision Eng.* **15** (1981) 237–242
224. D.H. Howerton, J.S. Strenkowski, and J.A. Bailey, Prediction of built-up edge formation in orthogonal cutting of aluminum, *Trans. NAMRI/SME* **17** (1989) 95–102

7 Cutting Temperatures

7.1 INTRODUCTION

When metal is cut, energy is expended in deforming the chip and in overcoming friction between the tool and the workpiece. Almost all of this energy is converted to heat [1–4], producing high temperatures in the deformation zones and surrounding regions of the chip, tool, and workpiece (Figure 7.1).

Cutting temperatures are of interest because they affect machining performance. Temperatures in the primary deformation zone, where the bulk of the deformation involved in chip formation occurs, influence the mechanical properties of the work material and thus the cutting forces. For this reason, most of the more complete analyses of the mechanics of cutting use temperature-dependent constitutive models. Temperatures on the rake face of the tool have a strong influence on tool life. As temperatures in this area increase, the tool softens and either wears more rapidly through abrasion or deforms plastically itself. In some cases, constituents of the tool material diffuse into the chip or react chemically with the cutting fluid or chip, leading ultimately to tool failure. Since cutting temperatures increase with the cutting speed, temperature-activated tool wear mechanisms limit maximum cutting speeds for many tool–work material combinations. An understanding of temperatures in this region therefore provides insight into the requirements for tool materials and coatings. Finally, temperatures on the relief face of the tool affect the finish and metallurgical state of the machined surface. Moderate levels of these temperatures induce residual stresses in the machined surface due to differential thermal contraction, while high levels may leave a burned or hardened layer on the machined part.

This chapter describes the measurement of cutting temperatures, empirical observations on temperatures, analytical and numerical temperature models, and temperatures in interrupted cutting and drilling.

7.2 MEASUREMENT OF CUTTING TEMPERATURES

Cutting temperatures are more difficult to measure accurately than cutting forces. The cutting force is a vector completely characterized by three components, while the temperature is a scalar field which varies throughout the system and which cannot be uniquely described by values at a few points. For this reason, no simple analog to the cutting force dynamometer exists for measuring cutting temperatures; rather, a number of measurement techniques based on various physical principles have been developed. Particular methods generally yield only limited information on the complete temperature distribution.

7.2.1 THE TOOL–WORK THERMOCOUPLE METHOD AND RELATED TECHNIQUES

The most widely used method for measuring cutting temperatures is the tool–work thermocouple method (Figure 7.2) first developed in the 1920s [5–7]. This method uses the tool and

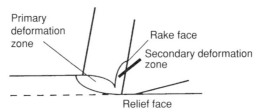

FIGURE 7.1 Areas of the cutting zone in which cutting temperatures are of practical interest.

workpiece as the elements of a thermocouple. The hot junction is the interface between the tool and the workpiece, and the cold junction is formed by the remote sections of the tool and workpiece, which must be connected electrically and held at a constant reference temperature. At least one leg of the circuit must be insulated from the machine tool, although in practice both legs are often insulated to eliminate noise from rotating elements of the system [8].

This method can only be used when both the tool and workpiece are electrical conductors, and thus cannot be used with many ceramic cutting tools. The thermoelectric power of the circuit is usually small and must be estimated by calibrating the circuit against a reference thermocouple. The calibration is critical to obtaining accurate results [1, 9–13]. Most calibration methods involve comparing the electromotive force (emf) produced by the tool and chip with that produced by a standard thermocouple when both are heated by a metal bath, welding torch, or induction coil (Figure 7.3). It is also often difficult to maintain the cold junction at a constant temperature; this is particularly true when small indexable tool inserts are used, since in this case a secondary hot junction may arise at the interface between the insert and the toolholder. This error can be minimized through a compensation circuit or by making electrical connections using wires made of materials which have a low thermoelectric power when coupled with the insert material [8, 9]. The requirement that one leg of the circuit be insulated from the machine tool can also create difficulties; the presence of the insulating material often reduces the stiffness of the machining system and leads to chatter in high-speed tests.

It is usually assumed that the tool–work thermocouple method measures an average interfacial temperature, although interpretation has sometimes been questioned [14]. In fact, the interpretation of tool–work thermocouple results depends on the thermoelectric

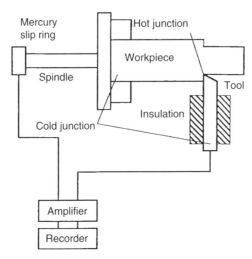

FIGURE 7.2 Tool–work thermocouple circuit for measuring cutting temperatures.

Cutting Temperatures

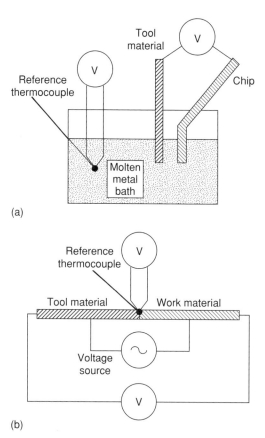

FIGURE 7.3 Methods of calibrating tool–work thermocouples. (a) Metal bath method described by Gottwein [6]. (b) Electric heating method described by Braiden [10].

characteristics of the materials involved [8]. For sharp tools, the quantity measured in the tool–work thermocouple method is the average thermal emf at the tool–chip interface. If the thermoelectric emf varies linearly with the temperature difference between the hot and cold junctions, this emf corresponds to the average interfacial temperature. For nonlinear temperature–emf relations, however, the temperature obtained by substituting the measured emf into the temperature–emf relation differs from the average interfacial temperature. For carbide tools and most common work materials, the difference between these temperatures is on the order of 5%, and it can be assumed that the tool–work thermocouple method measures the average interfacial temperatures. For material combinations with more nonlinear thermoelectric characteristics, however, this interpretation entails a more significant error.

Despite these difficulties, the tool–work thermocouple method has a number of advantages. The results are repeatable and correlate well with tool wear for carbide and high-speed steel tools. The measurements also show good time response, making the method suitable for measuring temperatures in thermally transient processes such as milling and for monitoring temperatures as an indication of tool wear [15]. The instrumentation required can also be built into the machine tool and operated reliably without constant readjustment. For these reasons, the tool–work thermocouple method is probably the best method for monitoring temperatures in routine machinability testing of common tool–work material combinations.

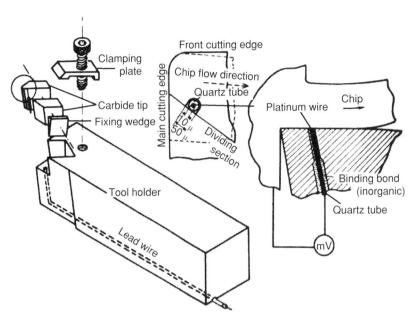

FIGURE 7.4 Insulated platinum wire embedded in a cutting insert to form a thermocouple with the chip. (After E. Usui et al., [23].)

Methods similar to the tool–work thermocouple technique include those in which insulated wires are embedded in the tool (Figure 7.4) or workpiece (Figure 7.5). In the first case, the hot junction is formed at the point of contact between the workpiece and the wire embedded in the tool; in the second case, the hot junction is formed when the tool cuts through the wire. In either case, the cold junction is formed in the same manner as in the tool–work thermocouple method. If the wire material is properly selected (e.g., a copper wire for a tungsten carbide tool [16]), the thermoelectric power of the circuit can be increased, improving the signal-to-noise ratio of the measurement. Also, the interpretation of the measurement is not an issue, since the temperature measured is clearly the temperature at the point where the hot junction is formed. If a number of tests are performed with wires at varying locations, the distribution of temperatures along the tool rake or relief face can be measured. These methods have been used most widely in drilling tests [16–21] but have also been applied in orthogonal cutting studies [22–24]. For drilling, the embedded wire can be replaced by an insulated, embedded foil, which allows mapping the temperature distribution across the drill lip in a single test [25]. The chief disadvantages of these methods are that they require careful calibration and tedious specimen preparation and data reduction. In the case of wires embedded in the workpiece, the time required for the tool and wire to come to thermal equilibrium also limits the maximum cutting speed that can be used.

7.2.2 Conventional Thermocouple Methods

Conventional thermocouples can be embedded in the tool [11, 16, 26–30] or workpiece [11, 31–34] to map temperature distributions. This approach has not been widely applied because of the extensive specimen preparation required. Since temperature gradients near the cutting zone are steep, its accuracy is limited by the placement accuracy of the thermocouples. The

FIGURE 7.5 Insulated copper wire embedded in a cast iron workpiece to form a thermocouple with a drill. (After J.S. Agapiou and D.A. Stephenson [16].)

resolution and accuracy of the measurements are also limited by the bead size of the thermocouple, by the difficulty of obtaining good thermal contact between the thermocouple bead and specimen, and by the fact that the temperature field is disturbed by the presence of the holes required to insert the thermocouples. In most cases, the embedded wire methods shown in Figure 7.4 and Figure 7.5 will yield more accurate results with less effort.

Conventional thermocouples can also be used to measure temperatures at points in the tool remote from the cutting zone (Figure 7.6). These remote measurements can then be used to back-calculate assumed temperature distributions along the cutting edge based on theoretical temperature fields [35–39]. (Full-field infrared cameras can also be used for this purpose [40–42].) The factors which limit the accuracy of such inverse methods are well known [43] and include the placement accuracy and temperature gradient effects of the embedded thermocouple methods. An added difficulty is the extrapolation effect which tends to magnify small errors in the remote measurements themselves. For these reasons, it is often difficult to obtain repeatable results with this approach. An advantage, however, is that the required thermocouples can be built into the toolholder, making the method attractive for routine measurements and process monitoring.

7.2.3 Metallurgical Methods

Metallic tool materials often undergo metallurgical transformations or hardness changes which can be correlated to temperature. This fact makes it possible to map temperature distributions in tools by sectioning the tools after cutting and performing metallographic or microscopic examinations. Appropriate procedures have been reported for both high-speed steel [44–49] and cemented carbide [50, 51] tools. These methods require postmortem measurements and are thus difficult to apply to routine testing. The limited results which have been

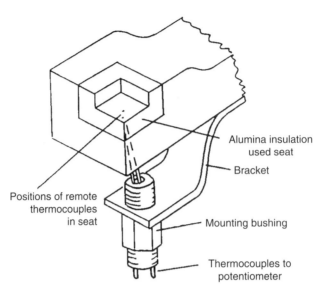

FIGURE 7.6 Toolholder with thermocouple welded to the seat to measure the temperature on the underside of a cutting insert. The alumina insulation is included to provide a controlled boundary condition and would not be used in routine testing. (After M.P. Groover and G.E. Kane [37].)

published, however, have shed valuable light on the nature of tool temperature distributions and the location of regions of maximum temperature (Figure 7.7).

In a related technique, split tools can be coated with powders which melt at specific temperatures to map isotherms within the tool [52]. This method can be used with cermet and ceramic tools, but does not appear to have been widely applied.

7.2.4 Infrared Methods

Cutting temperatures can also be estimated by measuring the infrared radiation emitted from the cutting zone. The best-known early studies were carried out by Schwerd [53], Reichenbach

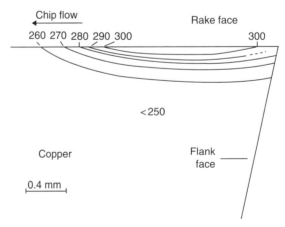

FIGURE 7.7 Temperature distribution in a high-speed steel tool used to cut copper, measured by a metallographic method. (After P.K. Wright [46].)

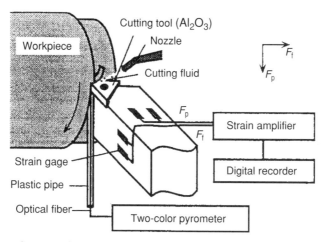

FIGURE 7.8 Fiber optic two-color pyrometer used to measure rake face temperatures. (After M. Al Huda et al., [66].)

[29], and Boothroyd [54, 55]. Reichenbach used a point sensor aligned with a small drilled hole to measure both shear plane and relief face temperatures. Since reliable point sensors have been available for some time, they have been applied by a number of researchers to measure rake and relief face temperatures using point sensors in both cutting and grinding [56–66] (Figure 7.8). Boothroyd took full-field infrared photographs of the cutting zone in low-speed experiments. Due to limited film sensitivity, he preheated his samples to a high temperature to generate a strong infrared signal. Later full-field infrared systems used electronic detectors rather than films, but still had limited spatial resolution and time response. They were used to map steady-state grinding temperatures [67, 68], but were not used until recently in cutting tests since the hot chip tended to dominate the signal, obscuring other features. The tehcnology has continued to advance, however, and new systems with much finer resolution and sensitivity are beginning to be used to routinely map temperature distributions in orthogonal and ballistic cutting operations [69–72] (Figure 7.9).

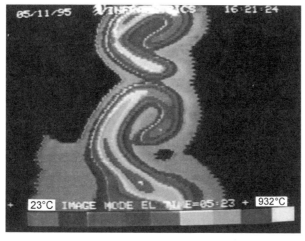

FIGURE 7.9 Full-field infrared thermal image of chip formation when machining 1045 steel at 170 m/min. (Courtesy of I.S. Jawahir, University of Kentucky.)

Infrared measurements are limited to exposed surfaces and cannot be used to directly measure temperatures in the interior of the chip. Because a finite amount of infrared energy is necessary for an accurate measurement, the spatial resolution of point measurements is often limited to an area on the order of 0.10 to 0.25 mm^2. As noted above, this limitation is becoming less serious as more sensitive detectors become available. As also noted above, the signal-to-noise ratio of measurements from many areas of the cutting zone may also be limited by the fact that the chip is normally much hotter than other areas and can dominate the infrared signal. This becomes less of an issue as spatial resolution improves. Most importantly, it is necessary to estimate the emissivity of the target to convert measured infrared intensities to temperatures. Estimating the emissivity is often difficult because it varies with both the temperature and the surface finish, so that assigning a value to specific points in the cutting zone is problematic without extensive calibration [70, 72]. If strict conditions on the variation of emissivities at different wavelengths are satisfied, the emissivity problem can be eliminated by using instruments which measure intensities at two infrared wavelengths simultaneously [62, 66].

Because of these limitations infrared measurements have traditionally been difficult to perform accurately and often do not produce repeatable results. The most accurate infrared methods appear to be those which measure intensities through predrilled holes, since in this case the emissivity over a narrow area can be accurately calibrated and the problem of the signal being overwhelmed by the infrared image of the chip can be eliminated.

7.2.5 OTHER METHODS

Temperature distributions can also be estimated by coating specimens with thermosensitive paints [73–76]. The limitations of this method are well documented [11]; it appears it is at best suitable for making qualitative comparisons. Similarly, the temperature of steel chips can be estimated in a crude fashion by noting their color, which gives an indication of the degree of temperature-dependent oxidation which they have undergone. This is also an inaccurate method suitable mainly for qualitative comparisons over limited ranges of cutting conditions.

7.2.6 SUMMARY

The average temperature along the tool rake face is most accurately measured using the tool–work thermocouple method, provided the materials involved permit its use. Tool–work thermocouple measurements are relatively easy to perform and yield repeatable results. The temperature distributions along the tool rake or relief faces can be measured accurately by cutting through an insulated wire, by metallographic methods, or by mounting an infrared sensor inside the workpiece and measuring the infrared intensity through a predrilled hole. These methods are much more tedious to perform, which makes them difficult to apply in broad studies over wide ranges of cutting conditions.

Temperatures within the deformation zones of the chip are much more difficult to measure. The only suitable methods are those using embedded thermocouples or full-field infrared sensors. These approaches can be tedious to apply and in the past have often not yielded repeatable results. Infrared methods are becoming more attractive as infrared sensors improve; the major difficulty in applying these methods is specifying the measurement target emissivity.

7.3 FACTORS AFFECTING CUTTING TEMPERATURES

The process parameter with the greatest influence on cutting temperatures is the cutting speed. Since cutting forces generally do not vary strongly with cutting speed, increasing the cutting speed increases the rate at which energy is dissipated through plastic deformation and friction, and thus the rate of heat generation in the cutting zone. Increasing the feed rate also increases heat generation and cutting temperatures. For moderate ranges of these variables, the cutting speed has a greater influence, and the tool–chip interface increases with the square root of the cutting speed but the third root of the feed [1]:

$$T_{int} \propto V^{0.5} a^{0.3} \tag{7.1}$$

Other parameters which affect the cutting force, such as the depth of cut and the rake angle, also influence cutting temperatures; changes in these parameters which increase the cutting force normally slightly increase cutting temperatures.

At very high cutting speeds, the tool–chip interface temperature does not vary as indicated in Equation (7.1), but rather approaches the melting temperature of the work material asymptotically. Kramer [77] reported that tool–chip interface temperatures approached the melting temperature of the chip when machining Inconel 718 at speeds up to 1200 m/min. Kottenstette [62] observed a leveling off of interfacial temperatures and evidence of chip melting when machining hardened 4340 steel at speeds greater than 3000 m/min. Chip melting also appears to occur when machining aluminum alloys with polycrystalline diamond tools at speeds over 700 m/min.

Material properties also strongly influence cutting temperatures. Cutting temperatures are higher for harder work materials because cutting forces and thus energy dissipation are increased. For materials of similar hardness, cutting temperatures increase with ductility, since more ductile materials can absorb more energy through plastic deformation. At a machining speed of roughly 300 m/min, the tool–chip interface temperature is usually roughly 400°C for an aluminum alloy with a Brinnel hardness of 100, 750°C for brittle gray cast iron with a Brinnel hardness of 200, and over 1000°C for ductile mild steel with a Brinnel hardness of 200.

Thermal properties of the work material which influence cutting temperatures include the thermal conductivity k and heat capacity ρc. Temperatures generally decrease as these parameters increase, since an increase implies that heat is more readily conducted away through the workpiece (higher k) or that the temperature increases more slowly for a given heat input (higher ρc). Increasing the thermal conductivity and heat capacity of the tool material also reduces temperatures, although the effect does not appear to be as marked as for the work material.

Another parameter which affects the thermal aspects of cutting is the ratio of the rate of material removal per unit depth of cut to the thermal diffusivity of the work material, $Va\rho c/k$, which is referred to as the thermal number in the metal cutting literature. In the general literature on heat transfer, it is defined as the Peclet number, Pe, which reflects the relative importance of mass transport and conduction. For low values of Pe, the speed of the tool with respect to the workpiece is relatively small compared to the thermal diffusivity of the material, and a comparatively large portion of the heat generated in the deformation zone conducts into the workpiece. For large values of Pe, on the other hand, almost all of the cutting remains in the chip and is transported out of the cutting zone. In conventional machining, Pe is generally on the order of 10^4 and the workpiece often heats up appreciably during cutting. In high-speed machining, Pe is more often on the order of 10^5, and the workpiece normally remains much cooler [40].

Finally, peak tool–chip interface temperatures are influenced by the tool nose radius and included angle [78]. Increasing the nose radius reduces the peak temperature by reducing the maximum uncut chip thickness and distributing frictional energy more evenly over the cutting edge. Reducing the included or wedge angle (by increasing the rake or relief angles) increases the peak temperature by reducing the area through which heat can diffuse from the cutting edge through the tool.

7.4 ANALYTICAL MODELS FOR STEADY-STATE TEMPERATURES

The difficulty of measuring cutting temperatures led to early research interest in analytical models for predicting temperatures. The best-known early analyses were simple solutions for plane heat sources based on the shear plane model of cutting mechanics (Section 6.8) [55, 79–83]. Although differing in details, these models were all based on the physical assumptions illustrated in Figure 7.10. The assumed work material enters at initial temperature θ_i and is heated by two plane heat sources of strength P_s and P_f, representing heating due to plastic deformation along the shear zone and frictional heating along the tool rake face. The material is assumed to emerge from the first source as a chip with a uniform temperature θ_s, and to undergo a further temperature rise θ_f in interacting with the second, frictional heat source. All models assume a steady state and neglect flank friction, which would introduce a third heat source.

In Loewen and Shaw's model [82], θ_s is given by

$$\theta_s = \frac{\Gamma_1 P_s}{\rho c a b V} + \theta_i = \frac{\Gamma_1 u_d}{\rho c} + \theta_i \qquad (7.2)$$

where P_s is the deformation power, which, for orthogonal cutting, is given by

$$P_s = V\left(F_c - P\frac{a_c}{a}\right) = F_s V_s \qquad (7.3)$$

and u_d is the deformation power given by Equation (6.23). Γ_1 is the proportion of the deformation energy entering the chip. F_c, P, L_c, L_p, a, and a_c are defined in Figure 6.3, Figure 6.4, Figure 6.25, and Figure 6.27; a is the undeformed chip thickness and b is the width of cut. Based on Jaeger's solution for a plane heat source sliding on a half space [84], Γ_1 can be estimated from the equation

$$\Gamma_1 = \frac{1}{1 + 1.328\sqrt{\frac{k\gamma}{\rho c V a}}} \qquad (7.4)$$

where k, ρ, and c are, respectively, the thermal conductivity, density, and specific heat of the work material, and γ is the strain in the chip defined by Equation (6.24).

Similarly, the frictional temperature rise θ_f is given by

$$\theta_f = \frac{0.377 \Gamma_2 P_f}{bk\sqrt{L_2}} \qquad (7.5)$$

where

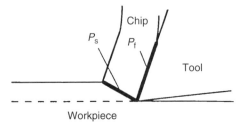

FIGURE 7.10 Two-source model of the cutting zone used in simple analyses of cutting temperatures.

$$L_2 = \frac{V_c L_c \rho c}{4k} \tag{7.6}$$

L_c is the tool–chip contact length, P_f is the frictional power, which, for orthogonal cutting, is given by

$$P_f = PV\frac{a}{a_c} \tag{7.7}$$

and Γ_2 is the proportion of frictional energy flowing into the chip. Again using Jaeger's friction slider solution, Γ_2 can be estimated by

$$\Gamma_2 = \frac{P_f(B/bk_t) + \theta_i - \theta_s}{P_f(B/bk_t) + 0.377(P_f/bk\sqrt{L_2})} \tag{7.8}$$

where

$$B = \frac{2}{\pi}\left[\sinh^{-1}(A_r) + A_r \sinh^{-1}\left(\frac{1}{A_r}\right) + \frac{1}{3}A_r^2\right] + \frac{2}{\pi}\left[\frac{1}{3A_r} - \frac{1}{3}\left(A_r + \frac{1}{A_r}\right)\sqrt{1 + A_r^2}\right] \tag{7.9}$$

k_t is the thermal conductivity of the tool material, and A_r is the aspect ratio of the contact area, given by

$$A_r = \begin{cases} b/2L_c & \text{(orthogonal tool)} \\ b/L_c & \text{(lathe tool with nose radius)} \end{cases} \tag{7.10}$$

The workpiece material properties k and ρc and the tool thermal conductivity k_t are assumed to vary with temperature in this model. To account for variations in material properties with temperatures, calculations are carried out iteratively from an initial guess temperature, with successive estimates of temperatures used to calculate new thermal properties until convergence is achieved.

For many metals, k varies approximately linearly with temperature, while ρc varies quadratically with temperature:

$$k(\theta) = k_0 + k_1\theta \tag{7.11}$$

$$\rho c(\theta) = \rho c_0 + \rho c_1 \theta + \rho c_2 \theta^2 \tag{7.12}$$

Values of the constants k_0, k_1, ρc_0, ρc_1, and ρc_2 for some common workpiece materials are listed in Table 7.1 [85]. Values of the tool thermal conductivity k_t for common tool materials are given in Table 7.2 [86–89].

The mean temperature rise of the chip surface along the tool face, θ_T, is equal to the sum of the mean shear plane temperature rise and the mean temperature rise due to friction in the tool–chip contact area:

$$\theta_T = \theta_s + \theta_f \tag{7.13}$$

Other simple analytical models for cutting temperatures use similar equations for calculating temperatures. The basic model of Trigger and Chao [80, 81] is equivalent to the Loewen–Shaw

TABLE 7.1
Values of Parameters in the Thermal Property Equations (7.11) and (7.12), for Common Workpiece Materials [85]

Material	k_0	k_1	ρc_0	ρc_1	ρc_2
1018 Steel	53.6	−0.027	3,124,025	6,480	−4.32
1070 Steel	49.3	−0.025	3,000,000	9,370	−7.0
2024 Al	112.5	0.256	2,437,866	1,515	−0.255
CA330 Brass	105.3	0.108	3,236,407	2,993	−3.74
Gray cast iron	49.0	−0.013	3,997,673	812.5	−4.27

Note: The values yield k in J/(sec m °C) and ρc in J/(m^3 °C) when θ is given in °C.

model. Boothroyd's model [55] uses Weiner's solution [90] for an inclined plane heat source on an infinite body to obtain a slightly different equation for Γ_1:

$$\Gamma_1 = \frac{\text{erf}\sqrt{Y_L}}{4Y_L} + [1 + Y_L]\text{erfc}\sqrt{Y_L} - \frac{\exp[-Y_L]}{\sqrt{\pi}}\left(\sqrt{Y_L} + \frac{1}{2\sqrt{Y_L}}\right) - \frac{\exp[-Y_L]}{\sqrt{\pi}}$$
$$\times \left(\sqrt{Y_L} + \frac{1}{2\sqrt{Y_L}}\right) \tag{7.14}$$

where

$$Y_L = \frac{V a \rho c \tan \phi}{4k} \tag{7.15}$$

In Boothroyd's model, the plane heat source on the rake is replaced by a distributed source, and the frictional temperature rise is calculated using a finite difference procedure. This is difficult to implement in practice, and other researchers have substituted alternative analytical schemes [88]. Other equations for the proportions of the heat flowing into the workpiece, chip, and tool have also been published [91–93] but do not appear to have been widely applied.

TABLE 7.2
Thermal Conductivities of Common Tool Materials at Approximately 100°C in J/(sec m °C) [86–89]

Material	k_t
M10 HSS	35.5
T1 HSS (tempered)	36.4
C2 Tungsten carbide	75
TiC cermet	22.5
Cold pressed Al_2O_3	15
Al_2O_3–TiC	18.3
Hot pressed silicon nitride	18.3

These models agree better with measured temperatures than the shear plane models agree with measured forces. One of the earliest experimental comparisons was carried out by Nakayama [94], who used measured temperatures on the face of a tube being cut to estimate the proportion of cutting heat entering the workpiece. He found that Hahn's partition function [79], which is an extension of Jaeger's solution accounting for oblique motion, agreed best with his data. Similar results have been reported by Boothroyd [55], who found that Weiner's solution was most accurate. More recently [85], temperatures calculated using four simple models were compared with tool–chip interface temperatures measured using the tool–work thermocouple method and shear zone temperatures estimated from infrared measurements. These results showed that the models based on Loewen and Shaw's analysis agreed well with measured values for materials which form a continuous chip, and that models based on Boothroyd's analysis overestimated measured temperatures. Typical results for low-carbon steel and aluminum workpieces are shown in Figure 7.11 and Figure 7.12. Boothroyd's model would yield better agreement if temperature-dependent thermal properties were used. All models overestimated temperatures when cutting cast iron, which form discontinuous chips. Leshock and Shin [13] found that temperatures predicted using the Loewen–Shaw model generally agreed with measured values for 4140 steel, although the model underestimated temperatures at low feeds, probably due to the neglect of flank friction.

A number of researchers have extended the basic models to apply to more general cutting conditions or to yield more complete information on temperature distributions. Chao and Trigger [95] extended their model to calculated relief face temperatures by adding a third heat source due to flank friction. Venuvinod and Lau [96] derived a model for temperatures in oblique cutting which reduces to the equivalent solutions of Chao and Trigger [80, 81] and Loewen and Shaw [82] for orthogonal cutting conditions. Average tool–chip interface temperatures calculated using this model agreed reasonably well with tool–work thermocouple measurements from tests on mild steel workpieces. Chao and Trigger [81], Wright et al. [97], and Venuvinod and Lau [96] have also extended the analysis to predict the temperature distribution along the rake face of the tool, rather than an average interfacial temperature. Wright's calculations compared well with temperature distributions measured by a metallurgical method in limited tests on mild steel workpieces. Grzesik and Nieslony [98] investigated more general heat partition models applicable to multilayered coated tools. The development of advanced finite element models as described in the next section has reduced interest in more general analytical models in recent years.

7.5 FINITE ELEMENT AND OTHER NUMERICAL MODELS

Finite element models for cutting temperatures were first investigated in the 1970s [99, 100]. The best-known early analysis was reported by Tay et al. [100]; this model formed the basis for much subsequent work [88, 101–104]. The chief advantage of these models was they predicted complete temperature distributions, rather than average temperatures over one or two surfaces. The input requirements, however, made them difficult to apply accurately to a broad set of cutting conditions. Tay's original model [100] and Muraka's model [101, 102] required measured cutting forces, chip properties, tool–chip contact lengths, and work material velocity distributions to predict temperatures; considerable experimental effort is required to generate this information. Subsequent researchers eliminated the need for measured velocity fields, so that temperature distributions could be based on cutting force and chip property inputs equivalent to those used in the simpler analytical models [103]. Since this information is predicted by the more complete models of the cutting mechanics, these analyses are suitable for use in coupled force and temperature models such as those reported by Oxley [105], Usui and Shirakashi [106, 107], and Strenkowski and Moon [108]. In these

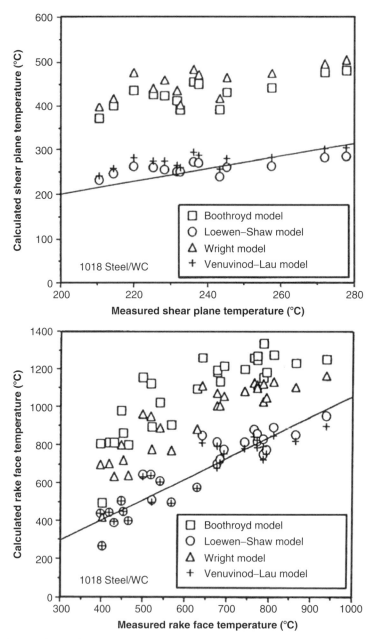

FIGURE 7.11 Comparison of measured average shear plane and rake face temperatures with values calculated using the models of Loewen and Shaw [82], Boothroyd [55] (as modified by Tay et al. [88]), Wright et al. [97], and Venuvinod and Lau [96]. The work material is 1018 steel cut at speeds up to 240 m/min. (After D.A. Stephenson [85].)

coupled models, initial predictions of forces and chip properties were used as inputs to temperature calculations, which in turn were used to modify work material flow characteristics; the two sets of calculations were carried out iteratively until convergence was achieved. This trend has been continued in the recent development of finite element models for machining, reviewed in

Cutting Temperatures

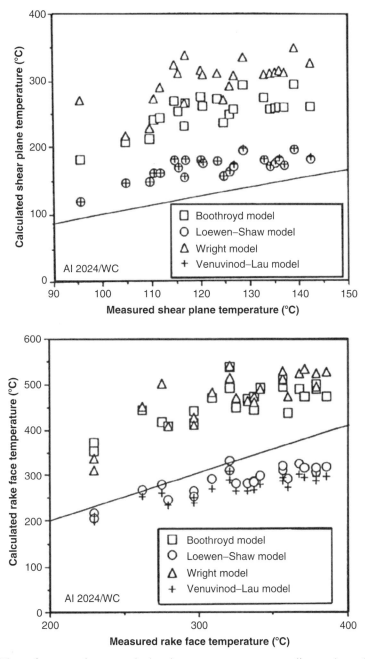

FIGURE 7.12 Plots of measured versus calculated temperatures corresponding to those shown in Figure 7.11 for 2024 aluminum workpieces. (After D.A. Stephenson [85].)

Section 6.12. The most complete current models are all coupled thermal and mechanical analyses [109–112].

Other numerical methods, such as the finite difference method [32, 113] and boundary element method [114, 115], have also been used to compute cutting temperatures. These methods often require less computing time than finite element calculations, but cannot be used with a broad range of tool geometries and field equations. A wide variety of specialized numerical methods for calculating temperature fields within tools have also been developed by researchers investigating inverse methods for monitoring temperatures using remote measurements [38, 39], the thermal contact resistance between the insert and tool holder [116], machining errors due to thermal expansion [117, 118], and temperatures in interrupted cutting [119–122].

Marusich and Ortiz's analysis [110, 123] is a representative example of current finite element models for machining temperatures. The governing thermal equation in this model is a weak form of the energy balance equation

$$\int_{B_t} \rho c \dot{T} \eta \, dV + \int_{\partial B_{tq}} h \eta \, dS = \int_{B_t} \mathbf{q} \cdot \nabla \eta \, dV + \int_{B_t} s \eta \, dV \quad (7.16)$$

In this equation, B_t is the boundary of a given deforming volume at time t (Neumann boundary), ρ is the density, c is the specific heat, T is the spatial temperature field, η is an admissible virtual temperature field, h is the outward heat flux through the surface, $\mathbf{q}$ is the heat flux, and s is the distributed heat source density. The rate of heat supply due to plastic deformation in chip formation is estimated as

$$s = \beta \dot{W}^p \quad (7.17)$$

where $\dot{W}^p$ is the plastic power per unit deformed volume (equivalent to u_d in Equation 6.23) and β is a coefficient of the order of 0.9. The rate of frictional heat generation on the rake face is given by

$$h = -\mathbf{t} \cdot \| \mathbf{v} \| \quad (7.18)$$

where $\mathbf{t}$ is the frictional traction and $\| \mathbf{v} \|$ is the jump in velocity across the tool–chip interface (i.e., the relative sliding speed). Based on infinite half-space solutions, the heat flux from Equation (7.18) is partitioned between the chip and tool using the formula

$$\frac{h_1}{h_2} = \frac{\sqrt{k_1 \rho_1 c_1}}{\sqrt{k_2 \rho_2 c_2}} \quad (7.19)$$

where h_1 and h_2 are the heat fluxes going to the chip and tool, respectively, k_1, ρ_1, and c_1 are the thermal properties of the chip, and k_2, ρ_2, and c_2 are the thermal properties of the tool.

The tool and the workpiece are assumed to obey Fourier's heat conduction law, which for material frame indifference is written in the form

$$\mathbf{q} = -\mathbf{D} \cdot \nabla \mathbf{T} \quad (7.20)$$

where $\mathbf{D}$ is the spatial conductivity tensor. For an isotropic lattice, $\mathbf{D}$ is related to the Cauchy–Green deformation tensor, $\mathbf{B}^e$, through

$$\mathbf{D} = -k \mathbf{B}^e \quad (7.21)$$

Inserting the finite element interpolation into Equation (7.16) results in the semidiscrete system of equations

Cutting Temperatures

$$\mathbf{C\dot{T} + KT = Q} \tag{7.22}$$

where **T** is the array of nodal temperatures, **C** is the heat capacity matrix, **K** is the conductivity matrix, and **Q** is the heat source array. The components of **C**, **K**, and **Q** are given by

$$C_{ab} = \int_{B_t} \rho c N_a N_b \, dV \tag{7.23}$$

$$K_{ab} = \int_{B_t} D_{ij} N_{a,i} N_{b,j} \, dV \tag{7.24}$$

$$Q_a = \int_{B_t} s N_a \, dV - \int_{\partial B_{tq}} h_\alpha N_a \, dV \tag{7.25}$$

In these equations, N_a and N_b are shape functions, $N_{a,i}$ and $N_{b,j}$ are normal contact forces, and h_α, $\alpha = 1, 2$ has the appropriate value for the chip or tool as in Equation (7.18). Integrating Equation (7.21) using a forward Euler algorithm yields

$$\mathbf{T}_{n+1} = \mathbf{T}_n + \Delta t \dot{\mathbf{T}}_n \tag{7.26}$$

$$\mathbf{CT}_n + \mathbf{K}_n \mathbf{T}_n = \mathbf{Q}_n \tag{7.27}$$

The use of the lumped heat capacity matrix, **C**, eliminates much equation solving and reduces the computational load.

It is difficult to assess the accuracy of finite element temperature models because the temperature distributions they predict cannot be easily measured. For some earlier models, calculations were compared with limited point temperature measurements from conventional thermocouples [99], tool–work thermocouple measurements [108], or partial tool temperature distributions measured using metallurgical techniques [103]. More recently, full-field infrared methods have become available for this purpose, and have been used to verify models to a limited extent [71]. Typical comparisons of temperatures calculated using finite element models with measurements are shown in Figure 7.13 and Figure 7.14. In both cases there is general qualitative agreement, but significant quantitative differences. This is to be expected given that all models, however general the geometries and boundary conditions they can accommodate, still employ significant simplifying assumptions for physical quantities not easily measured or specified. (One example is the use of Equation 7.19 to partition heat in Marusich and Ortiz's model.) Based on experience with earlier analytical models, steady-state finite element models which include temperature-varying thermal properties would be expected to be reasonably accurate for dry cutting. More validation and modeling experience is necessary to judge the quantitative accuracy of more general models. Unfortunately, many more researchers seem to be working on developing new models than on experimentally assessing and improving existing methods.

7.6 TEMPERATURES IN INTERRUPTED CUTTING

The temperature models discussed in Section 7.4 and Section 7.5 are in most cases applicable to steady-state processes such as orthogonal cutting and simple turning and boring operations. In most industrial processes the cutting speed or area of cut vary continuously, so that a thermal steady state is never reached. Examples include turning operations on complex parts using CNC machine tools and interrupted cutting processes such as milling. The

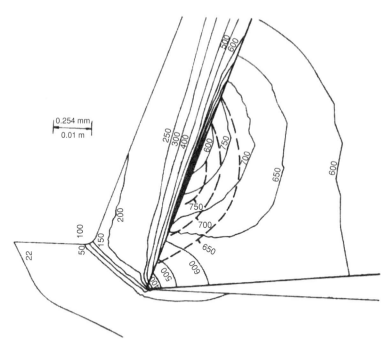

FIGURE 7.13 Temperature distributions in the workpiece, chip, and tool calculated using the finite element model developed by Stevenson et al. [103]. The broken lines in the tool show isotherms measured using a metallographic method. The work material is 12L14 steel cut at a speed of 106 m/min.

thermal aspects of interrupted cutting have been more widely studied than other non-steady-state processes [36, 69, 119–126] and are reviewed in this section. Models for transient temperatures in contour turning operations are reviewed in Section 8.2.

In the simplest case of interrupted cutting (Figure 7.15), the tool cuts through short lengths of metal interrupted by open regions of air. In this case the temperature response in

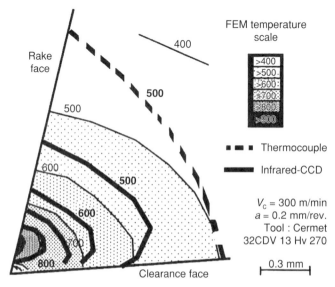

FIGURE 7.14 Comparison of tool temperature field calculated using a finite element model developed by M'Saoubi et al. [71] with full-field infrared temperature measurements.

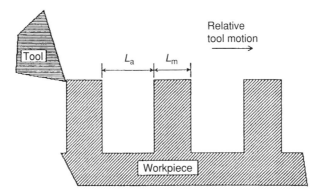

FIGURE 7.15 A simple interrupted cutting process in which the tool alternatively cuts through constant lengths of metal and air. (After D.A. Stephenson and A. Ali [69].)

the tool will clearly be periodic. In more general interrupted processes the lengths of metal and air cut may vary with time, as may the feed rate or depth of cut, and the temperature response is oscillatory but not necessarily periodic.

Cyclic variations in temperatures can lead to tool failure due to thermal fatigue. Thermal fatigue cracks develop in the tool as the surface layer expands and contracts in response to temperature variations, and eventually grow large enough to cause chipping of the cutting edge. This problem is commonly observed when milling steel with tungsten carbide cutters. The development of thermal cracks has been studied by Zorev and several other researchers [127–131]. As would be expected, these cracks grow more rapidly as peak temperatures during the heating cycle increase and as the difference between peak and low temperatures increases. Thus cracks are more likely to occur at high cutting speeds when cutting with a coolant. If these conditions cannot be avoided, thermal cracking can be reduced by changing the tool material grade to increase its thermal shock and fatigue resistance [132].

Peak temperatures are lower in interrupted cutting than in continuous cutting under the same conditions, leading to a reduction in chemical or diffusion wear. This type of wear is further reduced by the fact that peak temperatures are reached for short rather than extended periods. As illustrated in Figure 7.16 [69], peak temperatures depend more on the length of metal cut in continuous segments, rather than on the length of noncutting intervals. This indicates that the reduction in temperature is due primarily to interruption of the heating cycle during a transient phase, rather than from cooling between heating cycles. Physically, the temperature distribution at the tool–chip interface will not reach a steady state before the subsurface region of the tool near the cutting edge does; this requires a finite amount of energy, which cannot be supplied instantaneously, resulting in a period of transient (and lower) interfacial temperatures. If the ratio of the length of the transient period to the cutting speed increases with increasing cutting speed, it is possible that peak temperatures will decrease with cutting speed above a critical speed. Palmai [133] discussed this possibility and derived equations for calculating critical speeds from material properties. He also suggested that this phenomenon would explain the well-known results of Salomon [134, 135], who reported an increase in tool life with increasing cutting speed in very high-speed milling tests. This has not been observed in experiments at cutting speeds up to 300 m/min [69, 136]. Ming et al. [126] reported a decrease in cutting temperatures with increasing speed for end milling tests at speeds up to 700 m/min, but used an indirect temperature measurement method which would be sensitive to increased convective cooling at higher spindle speeds.

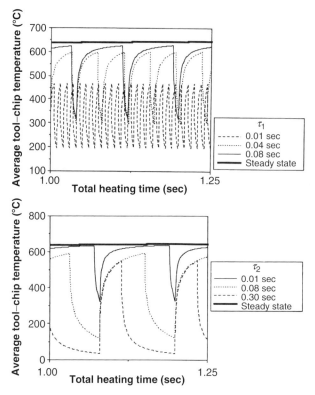

FIGURE 7.16 Variation in average interface temperatures in interrupted cutting when the length of the heating cycle (τ_1) and cooling cycle (τ_2) is varied. (After D.A. Stephenson and A. Ali [69].)

More controlled tests at higher cutting speeds would be worthwhile because verification of this phenomenon would have important practical implications.

7.7 TEMPERATURES IN DRILLING

Thermal conditions in drilling also differ significantly from those in orthogonal cutting, turning, and boring. The chip is formed at the bottom of the hole and remains in contact with the drill over a comparatively long distance, which increases tool temperatures. Temperatures in drilling are further increased by the fact that the drill point moves slowly into the portion of the work material being heated by chip formation; in turning, the work material approaching the cutting edge is generally cooler. Temperatures in drilling often do not reach a steady state, but rather increase with hole depth. For difficult-to-machine materials such as powder metals [137], this phenomenon limits the maximum hole depth which can be drilled without excessive tool wear. Increasing temperatures and the potential accumulation of hot chips at the bottom of the hole are serious problems in deep hole drilling, driving the use of horizonal setups and high pressure, through-tool coolant systems. Finally, the cutting speed varies with the radius of the drill, so that temperatures vary across the cutting edges. Speeds and temperatures are highest near the outer corner or margin of the drill, and temperature-activated margin wear often limits maximum spindle speeds.

Drilling temperatures have been studied both experimentally and theoretically for many years. The experimental work carried out before the mid-1960s has been reviewed by DeVries [20]. The most common method of measuring drill temperatures has been to embed

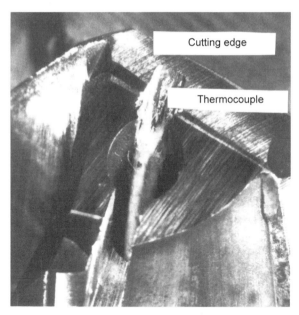

FIGURE 7.17 Thermocouple routed through the oil hole of a drill and welded near the cutting edge. (After J.S. Agapiou and D.A. Stephenson [16].)

thermocouples in the drill, often routing them through the oil holes of pressure-fed drills (Figure 7.17) [16, 137]. Tool–work and thin wire or foil thermocouple methods (see Figure 7.5) [17–21, 25], temperature-sensitive paints [76], and metallurgical methods [48, 49] have also been used. Typical signals from oil hole thermocouple are shown in Figure 7.18 [16]; they exhibit the characteristic increase in temperature with drilling depth. Differences in cutting temperatures between flutes on two-flute drills have also been observed experimentally [19], indicating that the cutting edges of the drill do not wear evenly and cut different amounts of material as drilling progresses; differences in edge temperatures also observed with asymmetric drills such as indexable drills, which do not have identical cutting edges.

The first analytical models for drill temperatures were reported in the 1960s. Early researchers [17, 138, 139] approximated the drill as a semi-infinite body and applied temperature analyses for orthogonal cutting to discrete segments of the cutting edge to compute the variation of temperature across the cutting edge. Later researchers treated the drill as finite and computed temperatures using finite difference [21, 140] or finite element [141] methods. In these analyses, the proportion of shear zone and frictional heat entering the drill was still computed using steady-state solutions such as Jaeger's solution for a friction slider on a half space. More recently, transient partition functions which reflect the fact that a larger percentage of this heat enters the tool in drilling than in other processes have been used. This approach has been used to develop a method of computing temperatures for drills with arbitrary point geometries based only on the drilling torque and material properties [16, 142, 143]. Despite the number of simplifying assumptions required in these analyses, their results have been found to agree well with limited thermocouple measurements (Figure 7.19).

As in other cutting processes, temperatures in drillling are most strongly affected by the spindle speed and the feed rate. Among geometric factors, the point angle has the largest influence. As the point angle is increased, the length of the drill cutting edge decreases, and temperatures increase as a roughly constant amount of heat diffuses into a smaller body. Increasing the helix angle, which reduces the drilling torque, does not seem to affect the drill

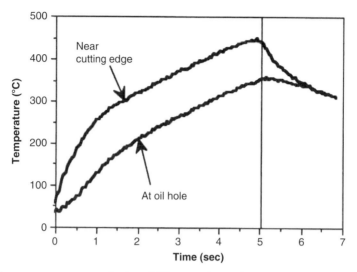

FIGURE 7.18 Temperatures measured when drilling cast iron with a carbide drill using thermocouples welded at the oil hole and near the cutting edge. The spindle speed is 1000 rpm and the feed rate is 0.457 mm/rev. (After J.S. Agapiou and D.A. Stephenson [16].)

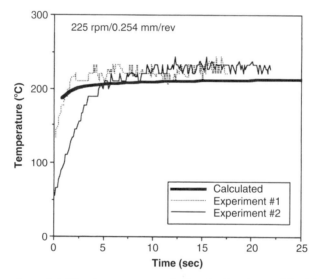

FIGURE 7.19 Comparison of drilling temperatures calculated using the model of Agapiou [15,123,124] with values measured by embedding a copper wire in the workpiece (Figure 7.5). (After J.S. Agapiou and D.A. Stephenson [16].)

temperature as strongly as would be expected since the bulk of the heat conducted into the drill comes from frictional contact rather than from shearing of the work material.

7.8 THERMAL EXPANSION

Thermal expansion can produce significant dimensional and form errors in precision machining processes. In many cases, errors are caused by the accumulation of hot chips on flat surfaces of the machine tool. A variety of methods for controlling these errors, including modifications of the machine design to eliminate flat surfaces, the use of coolants to ensure chip removal, and the use of constant-temperature fluid baths to control temperatures throughout the system, have been described [144]. Errors resulting from the conduction of cutting heat into the tool or workpiece are less common and have been less widely studied. The available work [117, 118, 145, 146] indicates that the expansion of the tool generally produces more significant errors than the expansion of the workpiece, largely because tool temperatures are usually higher than workpiece temperatures. The thermal expansion of the tool can be reduced by using brazed or clamped, rather than bonded, cutting tools [116], and by using toolholders cooled by cold fluids pumped through internal cooling passages [147]. The expansion of the workpiece produces more significant errors in precise holemaking processes such as cylinder boring. In these processes, thermal errors may be comparable to the mechanical errors produced by cutting forces [33, 148]. If adequate chip control can be ensured, thermal errors in boring can be reduced by increasing the cutting speed, since as the cutting speed increases the proportion of heat entering the workpiece decreases. Partly for this reason, carbide boring cutters have been replaced by ceramic or PCBN cutters designed for use at high cutting speeds in many automotive engine boring operations.

The thermal expansion of the workpiece has traditionally been a more serious problem in grinding than in conventional cutting operations. It has become a more significant issue recently in dry cutting operations. Hard turning, for example, is often carried out with ceramic tools which must be used dry. Since the cutting speed is relatively low and the specific cutting energy is high, this can result in a significant heat flux into the workpiece, and significant thermal distortions as shown in Figure 7.20 [149]. Dimensional and form errors due to this source can be minimized by machining part features with critical tolerances first, to avoid heating the part when cutting noncritical features, and by letting parts cool before gaging is performed in operations in which a compensation system is used to correct for tool wear. Thermal distortion of the workpiece is also a serious issue in dry drilling applications [150–154]. In these operations, the workpiece often heats up and expands ahead of the drill; when the drill is retracted, the workpiece contracts, reducing the drilled hole diameter. (In some cases, witness marks are visible in the hole, indicating that significant contraction occurs during drill retraction.) In high-precision operations, errors due to this source are commonly

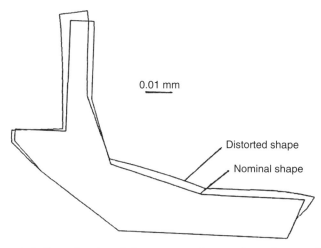

FIGURE 7.20 Computed dimensional and form errors due to thermal expansion when turning a hardened steel wheel spindle. (After D.A. Stephenson et al., [149].)

7.9 EXAMPLES

Example 7.1 Using Example 6.1 (in Chapter 6), estimate the maximum temperature at the chip–tool interface.

Solution: The maximum temperature at the chip–tool interface along the tool rake face is estimated using Equation (7.13). This requires the estimation of the temperature rise along the primary shear plane (Equation 7.2) and the frictional temperature rise (Equation 7.5). The deformation power along the shear plane in orthogonal cutting is calculated as

$$P = F_s V_s = (F_c \cos \phi - F_z \sin \phi) \frac{\cos \alpha}{\cos (\phi - \alpha)} V$$

The shear angle ϕ was calculated in Example 6.1 to be $23.2°$

$$P_s = (F_c \cos \phi - F_z \sin \phi) \frac{\cos \alpha}{\cos (\phi - \alpha)} V = \frac{(900 \cos 23.2° - 600 \sin 23.2°)(200 \text{ m/min})}{\cos (23.2°)}$$

$$P_s = 128{,}571 \text{ N m/min} = 2143 \text{ W} = 2143 \text{ J/sec}$$

The friction power at the chip–tool interface is equal to the total power minus the shear power (as discussed in the orthogonal case of a simplified problem where the friction power at the interface of the tool flank with the workpiece is neglected)

$$P_f = F_c V - P_s = (900 \text{ N})(200 \text{ m/min}) - 128{,}571 = 51{,}429 \text{ N m/sec} - 857 \text{ W}$$

The proportion of the primary shear energy entering the chip is Γ_1 (given by Equation 7.4). The workpiece material properties k and ρc and the tool thermal conductivity k_t vary with temperature as described in Equations (7.11) and (7.12). To account for variations in material properties with temperatures, calculations are carried out iteratively from an initial guess temperature, with successive estimates of temperatures used to calculate new thermal properties until convergence is achieved. This will be illustrated here. Using the properties of 1070 steel in Table 7.1 and Table 7.2 for the tool, Γ_1 at room temperature is

$$\Gamma_1 = \frac{1}{1 + 1.328 \sqrt{\frac{[48.55 \text{ J/(sec } m \text{°C)}](2.762)}{[3{,}274{,}800 \text{ J/(m}^3 \text{°C)}](3.333 \text{ m/sec})(0.0003 \text{ m})}}} = 0.867$$

The temperature rise along the primary shear plane is

$$\theta_s = \frac{\Gamma_1 P_s}{\rho c a b V} + \theta_i =$$

$$= \frac{(0.866)(2143 \text{ J/sec})}{[3{,}274{,}800 \text{ J/(m}^3 \text{°C)}](0.0003 \text{ m})(0.0025 \text{ m})(3.333 \text{ m/sec})} + 30°\text{C}$$

$$\theta_s = 257°\text{C}$$

Cutting Temperatures

The workpiece properties should be adjusted for an average temperature of the part at the primary shear zone since part of the heat generated along the shear plane is conducted into the workpiece adjacent to the shear zone where the next shear plane occurs. Therefore, by estimating the workpiece material properties k and ρc using Equations (7.11) and (7.12) at $(\theta_s + \theta_i)/2 = 144°C$, the temperature at the primary shear zone is estimated to be $\theta_s = 208°C$. One iteration is sufficient for adjusting the workpiece material properties.

The temperature rise of the chip at the interface with the tool is given by Equation (7.5). The constant L_2 is calculated using Equation (7.6)

$$L_2 = \frac{V_c L_c \rho c}{4k} = \frac{V a L_c \rho c}{4 a_c k} = \frac{(3.333 \text{ m/sec})(0.3 \text{ mm})(0.5 \text{ mm})[3{,}274{,}800 \text{ J}/(\text{m}^3 \,°\text{C})]}{4(0.7 \text{ mm})[48.55 \text{ J}/(\text{sec m}°\text{C})]}$$

$$L_2 = 12.05$$

The factors B and Γ_2 are calculated using Equations (7.9) and (7.8), respectively, so

$$B = 1.424 \quad \text{and} \quad \Gamma_2 = 0.841$$

The frictional temperature rise is given by Equation (7.5)

$$\theta_f = \frac{0.377 \Gamma_2 P_f}{bk \sqrt{L_2}} = \frac{0.377(0.9)(857 \text{ J/sec})}{(2.5 \text{ mm})[48.55 \text{ J}/(\text{sec m}°\text{C})](\sqrt{12.05})} = 645°C$$

Therefore, the maximum temperature of the chip at the chip–tool interface is estimated from Equation (7.13)

$$\theta_T = \theta_s + \theta_f = 238°C + 645°C = 883°C$$

The workpiece material properties should be adjusted for an average chip temperature. Considering the average chip temperature to be 450°C, the factors L_2, B, and Γ_2 are recalculated to be

$$L_2 = 27.22, \quad B = 1.424, \quad \Gamma_2 = 0.860$$

The maximum temperature of the chip at the chip–tool interface is estimated to be

$$\theta_T = \theta_s + \theta_f = 238°C + 560°C = 798°C$$

Example 7.2 Estimate the maximum temperature at the chip–tool interface if the cutting speed is reduced from 200 to 40 m/min in Example 7.1. The cutting and axial forces were increased about 4%. The chip thickness was measured and increased to 0.75 mm. The tool–chip contact length was 0.57 mm.

Solution: The deformation power along the shear plane is calculated in orthogonal cutting as in Example 7.1. The shear angle ϕ is calculated (as in Example 6.1) to be 21.8°

$$P_s = (F_c \cos \phi - F_z \sin \phi) \frac{\cos \alpha}{\cos(\phi - \alpha)} V = \frac{(936 \cos 21.8° - 624 \sin 21.8°)(40 \text{ m/min})}{\cos(21.8°)}$$

$$P_s = 27{,}456 \text{ N m/min} = 458 \text{ W} = 458 \text{ J/sec}$$

The friction power at the chip–tool interface is

$$P_f = F_c V - P_s = (936 \text{ N})(40 \text{ m/min}) - 27{,}456 = 166 \text{ W}$$

The proportion of the primary shear energy entering the chip is

$$\Gamma_1 = \frac{1}{1 + 1.328\sqrt{\dfrac{[45.7 \text{ J/(sec m }°\text{C)}](2.9)}{[4{,}200{,}449 \text{ J/(m}^3\,°\text{C)}](0.667 \text{ m/sec})(0.0003 \text{ m})}}} = 0.744$$

The workpiece properties are considered at the average temperature of 144°C (obtained in Example 7.1). The temperature rise along the primary shear plane is

$$\theta_s = \frac{(0.744)\,(458 \text{ J/sec})}{[4{,}200{,}449 \text{ J/(m}^3\,°\text{C)}]\,(0.0003 \text{ m})\,(0.0025 \text{ m})\,(0.667 \text{ m/sec})} + 30°\text{C}$$

$$\theta_s = 193°\text{C}$$

The temperature rise of the chip at the interface with the tool is estimated by Equation (7.5). The factors L_2, B, and Γ_2 are calculated using Equations (7.4), (7.9), and (7.8), respectively. The workpiece properties are considered at the average temperature of 300°C (based on the results in Example 7.1)

$$L_2 = 4.71, \quad B = 1.351, \quad \Gamma_2 = 0.690$$

The frictional temperature rise is given by Equation (7.5)

$$\theta_f = \frac{0.377\,\Gamma_2 P_f}{bk\sqrt{L_2}} = \frac{0.377(0.878)(166 \text{ J/sec})}{(2.5 \text{ mm})[41.8 \text{ J/(sec m }°\text{C)}](\sqrt{4.71})} = 191°\text{C}$$

Therefore, the maximum temperature of the chip at the chip–tool interface is estimated from Equation (7.13) as

$$\theta_T = \theta_s + \theta_f = 193°\text{C} + 191°\text{C} = 384°\text{C}$$

The comparison of the results in this example with those in Example 7.1 illustrates the importance of the frictional temperature rise along the rake face of the tool. The temperature at the tool–chip interface increased drastically with increase in cutting speed as compared to the temperature along the primary shear plane.

Example 7.3 Assume a 200 mm diameter steel tube is turned down to 195 mm. The inside diameter and the length of the tube are 180 and 350 mm, respectively. The cutting conditions are the same as in Example 7.2 (40 m/min cutting speed and 0.3 mm/rev feed). A sharp tool is used with very small nose radius. The lead angle of the tool is zero. Estimate the temperature effect on the physical dimensions of the workpiece for dry machining.

Solution: The greatest proportion of the heat generated in the shear zone is carried away by the chip, whereas a smaller portion is conducted back into the workpiece raising its temperature. Part of the heat flowing into the workpiece is removed with the chip generated in the following revolution. It becomes difficult to calculate the exact amount of heat carried away with the chips leaving the shear zone, as well as the temperature of the workpiece which is dependent on the cutting parameters (speed and feed) and the physical properties of the work material. Let us assume that this turning case is represented by the orthogonal case in Example 7.1 (since the radial force is small). The portion of the heat flowing in the workpiece is $(1 - \Gamma_1)$ that was estimated in Example 7.1 to be 0.242. The average temperature rise in the workpiece is given by the equation

$$\Delta\theta = \frac{(1-\Gamma_1)P_s t_m}{mc} = \frac{(1-0.744)(458 \text{ J/sec})(300 \text{ mm})/[(0.3 \text{ mm/rev})(64 \text{ rpm})]}{(\text{Workpiece volume})(7800 \text{ kg/m}^3)[502 \text{ J/(kg °C)}]} = 19°C$$

where t_m is the machining (cutting) time that the tool is in contact with the workpiece, m is the mass of the workpiece (or the mass of the machining section), and c is the specific heat of the workpiece material.

The workpiece size will expand with increasing temperature. The change in diameter (D) and length (L) is related to its change in temperature ($\Delta\theta$) by the following equation:

$$\Delta D = \Delta\theta D \alpha = (19°C)(180 \text{ mm})[0.0000117 \text{ mm/(°C mm)}] = 0.040 \text{ mm}$$

$$\Delta L = \Delta\theta L \alpha = (19°C)(300 \text{ mm})(0.0000117 \text{ mm/°C mm}) = 0.068 \text{ mm}$$

where α is the coefficient of thermal expansion for the workpiece material. The above equation assumes a uniform temperature distribution throughout the workpiece, which is not the case in metal cutting. Actual measurements have indicated the part to change temperature in a nonuniform manner. As the part temperature distribution is changing, not only size change occurs, but its geometry is distorted as well. Although the temperature uniformity assumption is generally incorrect, the above analysis is useful for quick and rough estimates of thermal growth. Finite element analysis can also be used to obtain better results.

7.10 PROBLEMS

Problem 7.1 The pump bore of an aluminum transmission case is rough bored at 400 m/min cutting speed and 0.2 mm/rev feed. The depth of cut is 1.5 mm. A diamond insert tool with zero nose radius, 8° rake angle, and zero lead angle (to minimize the radial force) is used. The diameter of the bore is 300 mm and the thickness of the wall at that section is 5 mm. The chip thickness was measured to be 0.4 mm. The tool–chip contact length was measured in the microscope to be 0.7 mm (which can be also estimated from Equation 6.31). Estimate the maximum tool–chip interface temperature and the possible bore expansion during boring.

REFERENCES

1. M.C. Shaw, *Metal Cutting Principles*, Oxford University Press, Oxford, 1984, Chapter 12
2. A.O. Schmidt, W.W. Gilbert, and O.W. Boston, A thermal balance method and mechanical investigation for evaluating machinability, *ASME Trans.* **67** (1945) 225–232
3. G.I. Epifanov and P.A. Rebinder, Energy balance of the metal cutting process, *Dokl. Akad. Nauka USSR* **66** (1949) 653–656 (in Russian; Henry Brutcher Translation No. 2394)
4. M.B. Bever, E.R. Marshall, and L.B. Ticknor, The energy stored in metal chips during orthogonal cutting, *J. Appl. Phys.* **24** (1953) 1176–1179

5. H. Shore, Thermoelectric measurement of cutting tool temperature, *J. Washington Acad. Sci.* **15** (1925) 85–88
6. K. Gottwein, Die Messung der Schneidentemperatur beim Abdrehen von Flusseisen, *Maschienenbau* **4** (1925) 1129–1135
7. E.G. Herbert, The measurement of cutting temperatures, *Proc. Inst. Mech. Eng.* **1** (1926) 289–329
8. D.A. Stephenson, Tool–work thermocouple temperature measurements — theory and implementation issues, *ASME J. Eng. Ind.* **115** (1993) 432–437
9. K.J. Trigger, R.K. Campbell, and B.T. Chao, A tool–work thermocouple compensation circuit, *ASME Trans.* **80** (1958) 302–306
10. P.M. Braiden, The calibration of tool/work thermocouples, *Proceedings of the 8th International MTDR Conference*, Birmingham, 1967, 653–666
11. B. Alvelid, Cutting temperature thermo-electrical measurements, *CIRP Ann.* **18** (1970) 547–354
12. G. Barrow, A review of experimental and theoretical techniques for assessing cutting temperatures, *CIRP Ann.* **22** (1973) 203–211
13. C.E. Leshock and Y.C. Shin, Investigation of cutting temperature in turning by a tool–work thermocouple technique, *ASME J. Manuf. Sci. Eng.* **119** (1997) 502–508
14. E.M. Trent, *Metal Cutting*, Butterworths, London, 1977, 58
15. N.F. Shillam, Machine Tool Control Systems, U.S. Patent Number 3,646,839, March 7, 1972
16. J.S. Agapiou and D.A. Stephenson, Analytical and experimental studies of drill temperatures, *ASME J. Eng. Ind.* **116** (1994) 54–60
17. M. Tsueda, Y. Hasegawa, and Y. Ishida, The study of cutting temperature in drilling (1) on the measuring method of cutting temperature, *Trans. JSME* **27** (1961) 1423–1430 (in Japanese)
18. S. Nishida, S. Ozaki, S. Nakayama, T. Shiraishi, and K. Nagura, Study on drilling II — lip temperature, *J. Mech. Lab. Japan* **8** (1962) 59–60 (in Japanese)
19. M. Tsueda, Y. Hasegawa, N. Nisina, and T. Hirai, Research on the cutting temperature of a drill point (2) the unequal temperature distribution along two cutting edges, *Trans. JSME* **28** (1962) 1076–1083 (in Japanese)
20. M.F. DeVries, Drill Temperature as a Drill Performance Criterion, ASTME Technical Paper MR68-193, 1968
21. K. Watanabe, K. Yokoyama, and R. Ichimiya, Thermal analyses of the drilling process, *Bull. Jpn. Soc. Precision Eng.* **11** (1977) 71–77
22. K.F. Meyer, Untersuchung an Keramischen Schneidstoffen, *Ind. Anz.* **11** (1962)
23. E. Usui, T. Shirakashi, and T. Kitagawa, Analytical prediction of three-dimensional cutting process. Part 3. Cutting temperature and crater wear of carbide tool, *ASME J. Eng. Ind.* **100** (1978) 236–243
24. M. Hirao, Determining temperature distribution on flank face of cutting tool, *J. Mater. Shaping Technol.* **6** (1989) 143–148
25. M. Bono and J. Ni, A method for measuring the temperature distribution along the cutting edges of a drill, *ASME J. Manuf. Sci. Eng.* **124** (2002) 921–923
26. D.L. Rall and W.H. Giedt, Heat transfer to, and temperature distribution in, a metal-cutting tool, *ASME Trans.* **78** (1956) 1507–1512
27. M.B. Hollander and J.E. Eglund, Thermocouple Technique Investigation of Temperature Distribution in the Workpiece During Metal Cutting, ASTME Research Report No. 7, 1957
28. M.C. Shaw, N.H. Cook, and P.A. Smith, Report on the Cooling Characteristics of Cutting Fluid, ASTME Research Report No. 19, 1958
29. G.S. Reichenbach, Experimental measurement of metal-cutting temperature distributions, *ASME Trans.* **80** (1958) 525–540
30. T.I. El-Wardany, E. Mohammed, and M.A. Elbestawi, Cutting temperature of ceramic tools in high speed machining of difficult-to-cut materials, *Int. J. Mach. Tools Manuf.* **36** (1996) 611–634
31. A.H. Quereshi and F. Koenigsberger, An investigation into the problem of measuring the temperature distribution on the rake face of a cutting tool, *CIRP Ann.* **14** (1966) 189–199
32. V.A. Ostafiev, A.A. Cherniavskaya, and V.A Sinopalnikov, Numerical calculation of non-steady-state temperature fields in oblique cutting, *CIRP Ann.* **32** (1983) 43–46

33. G. Subramani, M.C. Whitmore, S.G. Kapoor, and R.E. DeVor, Temperature distribution in a hollow cylindrical workpiece during machining: theoretical model and experimental results, *ASME J. Eng. Ind.* **113** (1991) 373–380
34. W. Grzesik, Experimental investigation of the cutting temperature when turning with coated indexable inserets, *Int. J. Mach. Tools Manuf.* **39** (1999) 355–369
35. M.P. Lipman, B.E. Nevis, and G.E. Kane, A remote sensor method for determining average tool–chip interface temperatures in metal cutting," *ASME J. Eng. Ind.* **89** (1967) 333–338
36. K.K. Wang, S.M. Wu, and K. Iwata, Temperature responses and experimental errors for multi-tooth milling cutters, *ASME J. Eng. Ind.* **90** (1968) 353–359
37. M.P. Groover and G.E. Kane, Continuous study in the determination of temperature in metal cutting using remote thermocouples, *ASME J. Eng. Ind.* **93** (1971) 603–608
38. D.W. Yen and P.K. Wright, A remote temperature sensing technique for estimating the cutting interface temperature distribution, *ASME J. Eng. Ind.* **108** (1986) 252–263
39. J.G. Chow and P.K. Wright, On-line estimating of tool/chip interface temperatures for a turning operation, *ASME J. Eng. Ind.* **110** (1988) 56–64
40. D.A. Stephenson, An inverse method for investigating deformation zone temperatures in metal cutting, *ASME J. Eng. Ind.* **113** (1991) 129–136
41. D. O'Sullian and M. Cotterell, Workpiece temperature measurement in machining, *Proc. Inst. Mech. Eng. J. Eng. Manuf.* **216B** (2002) 135–139
42. P. Kwon, T. Schiemann, and R. Kountanya, An inverse estimation scheme to measure steady-state tool–chip interface temperatures using an infrared camera, *Int. J. Mach. Tools Manuf.* **41** (2001) 1015–1030
43. J.V. Beck, B. Blackwell, and C.R. St. Clair, *Inverse Heat Conduction Ill-Posed Problems*, Wiley Interscience, New York, 1985, 13–36, 154–156
44. P.K. Wright and E.M. Trent, Metallurgical methods of determining temperature gradients in cutting tools, *J. Iron Steel Inst.* **211** (1973) 364–368
45. E.F. Smart and E.M. Trent, Temperature distribution in tools used for cutting iron, titanium and nickel, *Int. J. Prod. Res.* **13** (1975) 265–290
46. P.K. Wright, Correlation of tempering effects with temperature distribution in steel cutting tools, *ASME J. Eng. Ind.* **100** (1978) 131–136
47. B. Mills, D.W. Wakeman, and A. Aboukhashaba, A new technique for determining the temperature distribution in high speed steel cutting tools using scanning electron microscopy, *CIRP Ann.* **29** (1980) 73–77
48. B. Mills, T.D. Mottishaw, and A.J.W. Chisolm, The application of scanning electron microscopy to the study of temperatures and temperature distributions in M2 high speed steel twist drills, *CIRP Ann.* **30** (1981) 15–20
49. A. Thangaraj, P.K. Wright, and M. Nissle, New experiments on the temperature distribution in drilling, *ASME J. Eng. Ind.* **106** (1984) 242–247
50. P.A. Dearny and E. M Trent, Wear mechanisms of coated carbide tools, *Met. Technol.* **9** (1982) 60–75
51. P.A. Dearny, New technique for determining temperature distribution in cemented carbide cutting tool, *Met. Technol.* **10** (1983) 205–210
52. S. Kato, Y. Yamaguchi, Y. Watanabe, and Y Hiraiwa, Measurement of temperature distribution within tool using powders of constant melting point, *ASME J. Eng. Ind.* **98** (1976) 607–613
53. F. Schwerd, Ueber die Bestimmung des Temperaturfeldes beim Spanablauf, *Z. VDI* **77** (1933) 211–216
54. G. Boothroyd, Photographic technique for the determination of metal cutting temperatures, *Br. J. Appl. Phys.* **12** (1961) 238–242
55. G. Boothroyd, Temperatures in orthogonal metal cutting, *Proc. Inst. Mech. Eng.* **177** (1963) 789–802
56. E. Lenz, Die Temperaturmessung in der Kontaktzone Span–Werkzeug beim Drehvorgang, *CIRP Ann.* **13** (1966) 201–210
57. E. Lenz, Die Temperaturverteilung in der Kontaktzone Span–Werkzeug beim Drehen von Stahl mit Hartmetallwerkzeugen, *CIRP Ann.* **14** (1966) 137–144

58. O.D. Prins, The influence of wear on the temperature distribution at the rake face, *CIRP Ann.* **19** (1971) 579–584
59. B.T. Chao, H.L. Li, and K.J. Trigger, An experimental investigation of temperature distribution at tool-flank surface, *ASME J. Eng. Ind.* **83** (1961) 496–504
60. A.E. Focks, F.E. Westerman, P.E. Rentschler, J. Kemphaus, T.W. Shi, and M. Hoch, Heat flow patterns in superhard tools when cutting superalloys, *Proc. NAMRC* **13** (1985) 394–401
61. T. Ueda, A. Hosokawa, and A. Yamamoto, Measurement of grinding temperature using infrared radiation pyrometer with optical fiber, *ASME J. Eng. Ind.* **108** (1986) 247–251
62. J.P. Kottenstette, Measuring tool–chip interface temperatures, *ASME J. Eng. Ind.* **108** (1986) 101–104
63. J. Lin, S.-L. Lee, and C.-I. Wang, Estimation of cutting temperature in high speed machining, *ASME J. Eng. Mater. Technol.* **114** (1992) 289–296
64. E. Belotserkovsky, O. Bar-Or, and A. Katzir, Infrared fiberoptic temperature monitoring during machining procedures, *Meas. Sci. Technol.* **5** (1994) 451–453
65. P. Mueller-Hummel and M. Lahres, Infrared temperature measurement on diamond-coated tools during machining, *Diamond Relat. Mater.* **3** (1994) 765–769
66. M. Al Huda, K. Yamada, A. Hosokawa, and T. Ueda, Investigation of temperature at tool–chip interface in turning using two-color pyrometer, *ASME J. Manuf. Sci. Eng.* **124** (2002) 200–207
67. L. Kops and M.C. Shaw, Thermal radiation in surface grinding, *CIRP Ann.* **31** (1982) 211–214
68. L. Kops and M.C. Shaw, Application of infrared radiation measurements in grinding studies, *Proc. NAMRC* **14** (1983) 390–396
69. D.A. Stephenson and A. Ali, Tool temperatures in interrupted metal cutting, *ASME J. Eng. Ind.* **114** (1992) 127–136
70. L. Wang, K. Saito, and I.S. Jawahir, Infrared temperature measurement of curled chip formation in metal cutting, *Trans. NAMRI/SME* **24** (1996) 33–38
71. R. M'Saoubi, C. Le Calvez, B. Changeux, and J.L. Lebrun, Thermal and microstructural analysis of orthogonal cutting of a low alloyed carbon steel using an infrared-charge-coupled device camera technique, *Proc. Inst. Mech. Eng. J. Eng. Manuf.* **216B** (2002) 153–165
72. G. Sutter, L. Faure, A. Molinari, N. Ranc, and V. Pina, An experimental technique for the measurement of temperatures fields for the orthogonal cutting in high speed machining, *Int. J. Mach. Tools Manuf.* **43** (2003) 671–678
73. F. Penzig, Sichtbarmachen von Temperaturfeldern Durch Temperaturabhaengigen Farbenstriche, *Z. VDI* **83** (1939) 69
74. H. Schallbroch and M. Lang, Messung der Schnittemperatur Mittles Temperaturanzeigender Farbenstriche, *Z. VDI* **87** (1943) 15–19
75. U. Koch, Experimental and theoretical analysis of lathe tool temperature distribution in oblique cutting, *Proceedings of the 11th MTDR Conference*, Manchester, 1970, Vol. 1, 533–540
76. U. Koch and R. Levi, Some mechanical and thermal aspects of twist drill performance, *CIRP Ann.* **19** (1971) 247–254
77. B.M. Kramer, On tool materials for high speed machining, *ASME J. Eng. Ind.* **109** (1987) 87–91
78. A. Anagonye and D.A. Stephenson, Modeling cutting temperatures for turning inserts with various tool geometries and materials, *ASME J. Manuf. Sci. Eng.* **124** (2002) 544–552
79. R.S. Hahn, On the temperature developed at the shear plane in the metal cutting process, *Proceedings of the First U.S. National Conference on Applied Mechanics*, ASME, New York, 1951, 661–666
80. K.J. Trigger and B.T. Chao, An analytical evaluation of metal cutting temperatures, *ASME Trans.* **73** (1951) 57–68
81. B.T. Chao and K.J. Trigger, Temperature distribution at the tool–chip interface in metal cutting, *ASME Trans.* **77** (1955) 1107–1121
82. E.G. Loewen and M.C. Shaw, On the analysis of cutting-tool temperatures, *ASME Trans.* **76** (1954) 217–231
83. A.C. Rapier, A theoretical investigation of the temperature distribution in the metal cutting process, *Br. J. Appl. Phys.* **5** (1954) 400–405

84. J.C. Jaeger, Moving sources of heat and the temperature at sliding contacts, *Proc. Roy. Soc. NSW* **76** (1942) 203–224
85. D.A. Stephenson, Assessment of steady-state metal cutting temperature models based on simultaneous infrared and thermocouple data, *ASME J. Eng. Ind.* **113** (1991) 121–128
86. Y.S. Touloukian et al., *Thermophysical Properties of Matter*, IFI/Plenum, New York, 1970, Vol. 1, 1194–1196; Vol. 4, 1282–1284
87. G.A. Roberts and R.A. Cary, *Tool Steels*, 2nd Edition, ASM, Metals Park, OH, 1980, 709
88. A.O. Tay, M.G. Stevenson, G. de Vahl Davis, and P.L.B. Oxley, A numerical method for calculating temperature distributions in machining, from force and shear angle measurements, *Int. J. Mach. Tool Des. Res.* **16** (1976) 335–349
89. D. Bordui, Third generation silicon nitride, *Ceramic Cutting Tools and Applications*, SME, Dearborn, MI, 1989
90. J.H. Weiner, Shear-plane temperature distribution in orthogonal cutting, *ASME Trans.* **77** (1955) 1331–1341
91. G. Vieregge, Energieverteilung und Temperatur bei der Zerspannung, *Werkstatt und Betrieb* **86** (1953) 691–703
92. W.C. Leone, Distribution of shear-zone heat in metal cutting, *ASME Trans.* **76** (1954) 121–125
93. P.R. Dawson and S. Malkin, Inclined moving heat source model for calculating metal cutting temperatures, *ASME J. Eng. Ind.* **106** (1984) 179–186
94. K. Nakayama, Temperature rise in the workpiece during metal cutting, *Bull. Fac. Eng. Yokohama Natl. Univ.* **5** (1956) 1–10
95. B.T. Chao and K.J. Trigger, Temperature distribution at the tool–chip and tool–work interface in metal cutting, *ASME Trans.* **80** (1958) 311–320
96. P.K. Venuvinod and W.S. Lau, Estimation of rake temperatures in free oblique cutting, *Int. J. Mach. Tool Des. Res.* **26** (1986) 1–14
97. P.K. Wright, S.P. McCormick, and T.R. Miller, Effect of rake face design on cutting tool temperature distributions, *ASME J. Eng. Ind.* **102** (1980) 123–128
98. W. Grzesik and P. Nieslony, A computational approach to evaluate temperature and head partition in machining with multilayer coated tools, *Int. J. Mach. Tools Manuf.* **43** (2003) 1311–1317
99. W.M. Mansour, M.O.M. Osman, T.S. Sankar, and A. Mazzawi, Temperature field and crater wear in metal cutting using a quasi-finite element approach, *Int. J. Prod. Res.* **11** (1973) 59–68
100. A.O. Tay, M.G. Stevenson, and G. de Vahl Davis, Using the finite element method to determine temperature distributions in orthogonal machining, *Proc. Inst. Mech. Eng.* **188** (1974) 627–638
101. P.D. Muraka, Prediction of Temperatures in Orthogonal Machining Using the Finite Element Method, PhD Thesis, University of Manchester, 1977
102. P.D. Muraka, G. Barrow, and S. Hinduja, Influence of the process variables on the temperature distribution in orthogonal machining using the finite element method, *Int. J. Mech. Sci.* **21** (1979) 445–456
103. M.G. Stevenson, P.K. Wright, and J.G. Chow, Further developments in applying the finite element method to the calculation of temperature distributions in machining and comparisons with experiment, *ASME J. Eng. Ind.* **105** (1983) 149–154
104. T. Kagiwada and T. Kanauchi, Numerical analysis of cutting temperatures and flowing ratios of generated heat, *JSME Int. J. Series III* **31** (1988) 624–633
105. P.L.B. Oxley, *The Mechanics of Machining,* Ellis Horwood, Chicester, UK, 1989
106. E. Usui and T. Shirakashi, Mechanics of machining — from 'descriptive' to 'predictive' theory, *On The Art of Cutting Metals — 75 Years Later*, ASME PED Vol. 7, ASME, New York, 1982, 13–35
107. E. Usui and T. Shirakashi, Analytical prediction of cutting tool wear, *Wear* **100** (1984) 129–151
108. J.S. Strenkowski and K.J. Moon, Finite element prediction of chip geometry and tool/workpiece temperature distributions in orthogonal metal cutting, *ASME J. Eng. Ind.* **112** (1990) 313–318
109. K.W. Kim and H.-C. Sin, Development of a thermo-viscoplastic cutting model using finite element method, *Int. J. Mach. Tools Manuf.* **36** (1996) 379–397
110. T.D. Marusich and M. Ortiz, Modelling and simulation of high-speed machining, *Int. J. Numer. Methods Eng.* **38** (1995) 3675–3694

111. T. Ozel and T. Altan, Process simulation using finite element method — prediction of cutting forces, tool stresses, and temperatures in high-speed flat end milling process, *Int. J. Mach. Tools Manuf.* **40** (2000) 713–738
112. T. Obikawa and E. Usui, Computational machining of titanium alloy — finite element modeling and a few results, *ASME J. Manuf. Sci. Eng.* **118** (1996) 208–215
113. A.J.R. Smith and E.J.A. Armarego, Temperature prediction in orthogonal cutting with a finite difference approach, *CIRP Ann.* **30** (1981) 9–13
114. C.L. Chan and A. Chandra, A boundary element method analysis of the thermal aspects of metal cutting processes, *ASME J. Eng. Ind.* **113** (1991) 311–319
115. A. Chandra and C.L. Chan, Thermal aspects of machining: a BEM approach, *Int. J. Solids Struct.* **31** (1994) 1657–1693
116. S. Darwish and R. Davies, Investigation of the heat flow through bonded and brazed metal cutting tools, *Int. J. Mach. Tool Des. Res.* **29** (1989) 229–237
117. R. Ichimiya and K. Kawahara, Investigation of thermal expansion in machining operations (elongations of tool and workpiece in turning), *Bull. JSME* **14** (1971) 1363–1371
118. R. Ichimiya and Y. Usuzaka, Analysis of thermal expansion in face-cutting operations, *ASME J. Eng. Ind.* **96** (1974) 1222–1229
119. D.E. McFeron and B.T. Chao, Transient interface temperatures in plain peripheral milling, *ASME Trans.* **80** (1958) 321–329
120. K.K. Wang, K.C. Tsao, and S.M. Wu, Investigation of face-milling tool temperatures by simulation techniques, *ASME J. Eng. Ind.* **91** (1969) 772–780
121. H. Wu and J.E. Mayer, Jr., An analysis of thermal cracking of carbide tools in intermittent cutting, *ASME J. Eng. Ind.* **101** (1979) 159–164
122. R. Radulescu and S.G. Kapoor, An analytical model for prediction of tool temperature fields during continuous and interrupted cutting, *Materials Issues in Machining and The Physics of Machining Processes*, TMS, Materials Park, OH, 1992, 147–165
123. *Third Wave AdvantEdge Theoretical Manual*, Version 4.3, Third Wave Systems, Inc., Minneapolis, MN, 2002
124. T.-C. Jen, J.G. Gutierrez, and S. Eapen, Non-linear numerical analysis in interrupted cutting tool temperatures, *Numer. Heat Transfer Part A: Appl.* **39** (2001) 1–20.
125. I. Lazoglu and Y. Altintas, Prediction of tool and chip temperature in continuous and interrupted machining, *Int. J. Mach. Tools Manuf.* **42** (2002) 1011–1022
126. C. Ming, S. Fanghong, W. Haili, Y. Renwei, Q. Zhenghong, and Z. Shquiao, Experimental research on the dynamic characteristics of cutting temperature in the process of high speed milling, *J. Mater. Process. Technol.* **138** (2003) 468–471
127. N.N. Zorev, Machining steel with a carbide tool in interrupted heavy-cutting conditions, *Russ. Eng. J.* **43** (1963) 43–47
128. N.N. Zorev, Standzeit und Leistung der Hartmetall-Werkzeuge beim Unterbrochenen Zerspanen des Stahls mit Grossen Zerspanungsquerschnitten, *CIRP Ann.* **11** (1963) 201–210
129. T. Hoshi and K. Okushima, Optimum diameter and position of a fly cutter for milling 0.45 C steel, 195 BHN and 0.4 C steel, 167 BHN at light cuts, *ASME J. Eng. Ind.* **87** (1965) 442–446
130. P. M Braiden and D.S. Dugdale, Failure of carbide tools in intermittent cutting, *Materials for Metal Cutting*, ISI Special Publication 126, ISI, London, 1970, 30–34
131. A.J. Pekelharing, Cutting tool damage in interrupted cutting, *Wear* **62** (1978) 37–48
132. S.F. Wayne and S.T. Buljan, The role of thermal shock on tool life of selected ceramic cutting tool materials, *J. Am. Ceram. Soc.* **75** (1989) 754–760
133. Z. Palmai, Cutting temperature in intermittent cutting, *Int. J. Mach. Tools Manuf.* **27** (1987) 261–274
134. C. Salomon, Verfahren zur Bearbeitung von Metallen oder bei einter Bearbeitung durch Schneidende Werkzeuge sich aehnlich Verhaltenden Werkstoffen, German Patent 523594, 1931
135. R.I. King, *Handbook of High Speed Machining Technology*, Chapman and Hall, New York, 1985, 3–4
136. P. Lezanski and M.C. Shaw, Tool face temperatures in high speed milling, *ASME J. Eng. Ind.* **112** (1990) 132–135

137. J.S. Agapiou, G.W. Halldin, and M.F. DeVries, On the machinability of powder metallurgy austentitic stainless steels, *ASME J. Eng. Ind.* **110** (1988) 339–343
138. R.P. Hervey and N.H. Cook, Thermal Parameters in Drill Tool Life, ASME Paper 65-Prod-15, 1965
139. M.F. DeVries, U.K. Saxena, and S.M. Wu, Temperature distributions in drilling, *ASME J. Eng. Ind.* **90** (1968) 231–238
140. U.K. Saxena, M.F. DeVries, and S.M. Wu, Drill temperature distributions by numerical solutions, *ASME J. Eng. Ind.* **93** (1971) 1057–1065
141. K.J. Fuh, Computer Aided Design and Manufacture of Multifacet Drills, PhD Thesis, University of Wisconsin, 1987, Chapter 8
142. J.S. Agapiou and M.F. DeVries, On the determination of thermal phenomena during a drilling process. Part I. Analytical models of twist drill temperature distributions, *Int. J. Mach. Tools Manuf.* **30** (1990) 203–215
143. J.S. Agapiou and M.F. DeVries, On the determination of thermal phenomena during a drilling process. Part II. Comparison of experimental and analytical twist drill temperature distributions, *Int. J. Mach. Tools Manuf.* **30** (1990) 217–226
144. N. Ikawa, R.R. Donaldson, R. Komanduri, W. Koenig, P.A. McKeown, E.T. Moriwaki, and I.F. Stowers, Ultraprecision metal cutting — the past, the present and the future, *CIRP Ann.* **40** (1991) 587–594
145. R.G. Watts and E.R. McClure, Thermal Expansion of Workpiece during Turning, ASME Paper 68-WA/Prod-24, 1968
146. Y. Takeuti, S. Zaima, and N. Noda, Thermal-stress problems in industry. I. On thermoelastic distortion in machining metals, *J. Therm. Stresses* **1** (1978) 199–210
147. B. Krauskopf, Diamond turning: reflecting demands for precision, *Manuf. Eng.* **24**:5 (1984) 90–100
148. N.N. Kakade and J.G. Chow, Computer simulation of bore distortions for engine boring operation, *Collected Papers in Heat Transfer*, ASME HTD Vol. 123, 1989, 259–265
149. D.A. Stephenson, M.R. Barone, and G.F. Dargush, Thermal expansion of the workpiece in turning, *ASME J. Eng. Ind.* **117** (1995) 542–550
150. D.M. Haan, S.A. Batzer, W.W. Olson, and J.W. Sutherland, An experimental study of cutting fluid effects in drilling, *J. Mater. Process. Technol.* **71** (1997) 305–313
151. M.J. Bono and J. Ni, The effects of thermal distortions on the diameter and cylindricity of dry drilled holes, *Int. J. Mach. Tools Manuf.* **41** (2001) 2261–2270
152. M.J. Bono, Experimental and Analytical Issues in Drilling, PhD Thesis, Mechanical Engineering, University of Michigan, 2002
153. S. Kalidas, S.G. Kapoor, and R.E. DeVor, Influence of thermal effects on hole quality in dry drilling. Part 1. A thermal model of workpiece temperatures, *ASME J. Manuf. Sci. Eng.* **124** (2002) 258–266
154. S. Kalidas, S.G. Kapoor, and R.E. DeVor, Influence of thermal effects on hole quality in dry drilling. Part 2. Thermo-elastic effects on hole quality, *ASME J. Manuf. Sci. Eng.* **124** (2002) 267–274
155. T.C. Jen, G. Gutierrez, S. Eapen, G. Barber, H. Zhao, P.S. Szuba, J. Labataille, and J. Manjunathaiah, Investigation of heat pipe cooling in drilling applications. Part I. Preliminary numerical analysis and verification, *Int. J. Mach. Tools Manuf.* **42** (2002) 643–652

8 Machining Process Analysis

8.1 INTRODUCTION

Analysis methods have seen increasing use for machining process design and improvement in the automotive, aerospace, construction equipment, and cutting tool industries. Broadly, three types of analyses are performed: force, power, and cycle time analyses using kinematic simulations, structural analysis for clamping and fixturing using finite element methods, and detailed chip formation analyses using finite element models. This chapter describes kinematic simulations and structural finite element analyses. Finite element chip formation analyses, which are used especially in the cutting tool and aerospace industries, are described in Section 6.11, Section 7.5, and Section 10.7.

Kinematic simulations of machining processes are used to calculate cycle times and time histories of cutting forces and power. The inputs required include the part and tool geometries, tool paths, and measured cutting pressures for the tool–workpiece material of interest. The tool geometries and tool paths are preferably read directly from computer-aided design (CAD) and computer-aided manufacture (CAM) systems. Based on this information, the kinematic motions of the tool with respect to the workpiece as a function of time can be simulated, and the instantaneous area of material being cut (the interference between the tool and the workpiece) at any time can be computed from the tool path and part geometry. Forces are computed by multiplying the instantaneous area by the measured cutting pressures, and power is computed as the product of the cutting speed and the appropriate cutting force component. Accurate cycle time calculations require consideration of the transient response of the spindle and slides, times required for tool changes, pallet rotations, coolant purges, etc. These characteristics can be obtained from machine tool manufacturer's specifications, or measured if hardware is available. Computed force and power histories can be used to level forces and reduce cycle time with minimal testing, and as inputs to finite element analyses.

On a semantic note, many academic researchers call these types of programs mechanistic models [1, 2]. This is a statistician's term to distinguish a model based on a postulated physical mechanism (i.e., a *theoretical* model) from a purely empirical model [3]. In the case of kinematic simulations, the only physical mechanism postulated is that force equals pressure times area; since the pressure in question is measured, the underlying model is basically empirical. Engineers who apply these programs in industry often call them math-based analyses or simulations [4, 5]. Both these terms are also unsatisfactory, the first because it is vague and redundant, the second because most engineers understand simulation to mean discrete event simulation. Prudent engineers try to avoid calling these programs anything, lest they be drawn into a semantic argument.

Kinematic simulations for turning, boring, milling, and drilling are described in Section 8.2–Section 8.5. The use of baseline cutting force data to develop cutting pressure equations is described in Section 8.6. Typical applications are reviewed in Section 8.7.

Structural finite element analysis is used to estimate workpiece distortions due to clamping and machining. The objective of the analysis is to minimize such distortions for critical features, which may be accomplished by stiffening the part or fixture in directions of heavy loading, modifying the tool path or cutter geometry to direct forces in stiff or noncritical directions, choosing clamping and locating schemes which minimize clamping distortion and support compliant portions of the part, and minimizing clamping forces. If the required input information is readily available, finite element analysis permits a wider variety of options to be investigated more quickly and cheaply than through prototype part and fixture tests.

Technical issues in structural finite element analysis are discussed in Section 8.8, and typical application examples are given in Section 8.9.

8.2 TURNING

Turning (Figure 8.1) is one of the simplest cutting processes to simulate because the geometry and kinematic motions of the tool and workpiece are easily described. Moreover, the benefits of simulation are relatively easy to quantify when turning large volumes of parts on CNC lathes, since simulation can be used to reduce cycle times and thus the number of machines and capital investment required.

The earliest work on turning simulation was based on the shear plane cutting theory [6, 7]. As noted in the introduction, in more recent work [8–13] cutting forces are calculated by multiplying measured cutting pressures by the calculated uncut chip area. A common approach [8] is to relate cutting forces to the normal cutting pressure, K_n, the effective friction coefficient, K_f, and the effective lead angle, γ_{Le} (Figure 8.2). K_n and K_f are defined by

$$K_n = \frac{N}{A_c} \tag{8.1}$$

$$K_f = \frac{P}{N} \tag{8.2}$$

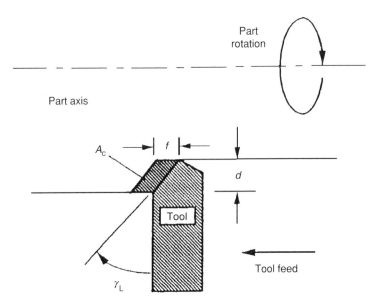

FIGURE 8.1 Uncut chip area A_c in turning.

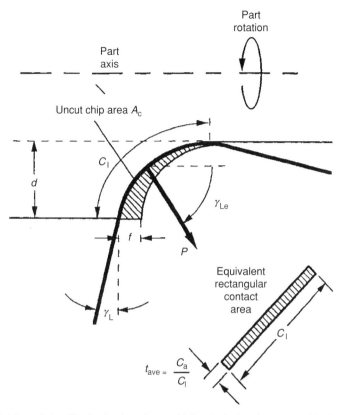

FIGURE 8.2 Definition of the effective lead angle γ_{Le}. (After D.A. Stephenson and P. Bandyopadhyay [4].)

where N and P are, respectively, the force components normal and parallel to the rake face of the tool (Figure 6.4) and A_c is the uncut chip area, which in most cases is adequately approximated by

$$A_c = fd \qquad (8.3)$$

where f is the feed per rotation and d is the depth of cut. This equation overestimates A_c for tools with a nose radius; for shallow depths of cut with such tools, more exact equations for A_c are available [11]. γ_{Le} determines the direction of the friction force and depends on the lead angle of the toolholder, the feed rate, depth of cut, and tool nose radius. The calculation of γ_{Le} and the estimation of K_n and K_f from baseline test data are discussed in Section 8.6. In terms of these parameters, the tangential, axial, and radial cutting forces (Section 6.3) are given by [14]

$$F_t = K_n A_c [\cos \alpha_b \cos \alpha + K_f (\cos \gamma_{Le} \sin \alpha + \sin \gamma_{Le} \sin \alpha_b)] \qquad (8.4)$$

$$F_a = K_n A_c [-\cos \alpha_b \sin \alpha + K_f (\cos \gamma_{Le} \cos \alpha)] \qquad (8.5)$$

$$F_r = K_n A_c [-\sin \alpha_b + K_f (\sin \gamma_{Le} \cos \alpha_b)] \qquad (8.6)$$

In these equations α_b is the back rake angle and α is the normal rake angle, which is related to the inclination angle, λ, and to the side rake, back rake, and lead angles, α_s, α_b, and γ_L (discussed in Chapter 4 and Chapter 6) through [15]

$$\lambda = \tan^{-1}[\tan \alpha_b \cos \gamma_L - \tan \alpha_s \sin \gamma_L] \qquad (8.7)$$

$$\alpha_n = \tan^{-1}[\cos \lambda (\tan \alpha_s \cos \gamma_L + \tan \alpha_b \sin \gamma_L)] \qquad (8.8)$$

These equations can be used to compute time histories of cutting forces in contour turning operations with varying speeds, feeds, or depths of cut. This is usually done by computing forces at fixed time steps based on the instantaneous cutting conditions at that time step, often determined from CAD/CAM data. Typical time-varying force plots are discussed in Section 8.7. This information can be used to identify part of the tool path in which forces are particularly high or low. The tangential force can also be multiplied by the spindle speed to determine the instantaneous cutting power, which can be compared to the machine tool's available power to identify portions of the cut over which stalling may occur. The calculations can then be used to modify tool paths, for example by adjusting the feed rate, to level the cutting forces and match the instantaneous power consumption to the machine's capability. In mass production applications for complex parts, this process can be used to significantly reduce cycle times.

The analysis in this section is applicable to quasi-steady processes and does not explicitly address tool wear, chip breaker, and vibration effects. The effect of tool wear on forces can be accounted for by multiplying K_n and K_f by correction factors [10]; for carbide tools, factors of 1.3 for K_n and 1.5 for K_f are typically used. The effect of chip breaking grooves in tools can be handled by modified or separately measured K_n and K_f values, or by defining an equivalent flat tool geometry based on groove parameters [13]. When deflections and vibrations cannot be neglected, a dynamic simulation which takes into account the compliances of the tool and part is required [11, 12]. In this case, the deflections are used to modify the calculated uncut chip area and thus the instantaneous force. The geometric component of the machined surface finish can also be calculated in dynamic simulations [12], although the wear component of the roughness is often more significant (Chapter 10). Finally, cutting temperatures can be calculated from simulated cutting forces using transient temperature models [16–18]. These more advanced simulations can, in principle, be used to investigate the effects of changes in process conditions on dynamic stability, accuracy, and tool life. They are not widely used at present, however, since methods of efficiently generating required inputs and unambiguously interpreting results have not been established.

8.3 BORING

There are two classes of boring operations, as shown in Figure 8.3: single-point boring, which is used as a finishing operation in cases in which the location of the bore axis is critical, and multitoothed boring, which is used as a roughing or semifinishing operation to remove metal more rapidly. Multitoothed boring is used especially in automotive engine manufacture.

Single-point boring is kinematically equivalent to turning, and the variation of static cutting forces for this case can be calculated using the equations similar to those given for turning in the previous section [9, 19]. These equations are adequate for estimating power requirements and bearing loads. The deflections of the boring bar and part, which determine bore accuracy in this case, are of particular importance in single-point boring. Depending on the rigidities of the various structural elements, either boring bar, part, or both must be modeled as compliant in this case. In Zhang and Kapoor's model [20], the part was treated as rigid while the bar was compliant. Stability limits for chatter predicted using this model were found to agree well with the results of turning experiments. Zhang and Kapoor have also published a random excitation model for calculating the surface finish which is applicable to finish boring [21, 22]. This model was found to agree well with data from boring tests on steel specimens.

Machining Process Analysis

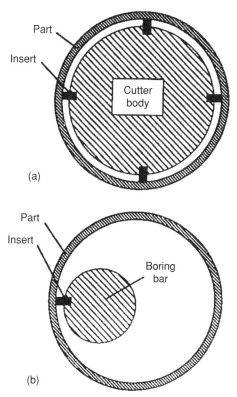

FIGURE 8.3 (a) Cylinder boring. (b) Single-point boring.

In multitoothed boring, the feed per tooth f_t can be substituted for the feed per revolution f in turning or single-point boring equations to calculate the uncut chip area and effective lead angle. With this substitution, these equations can be used to estimate power requirements and bearing loads in rough boring, although radial throw or runout between inserts has some effect on peak forces [23]. In practice, drilling simulations (as described in Section 8.5) are also used to calculate rough boring forces and power. When calculating deflections, the bar is usually treated as rigid [24–27] unless the bore being machined is interrupted, leading to asymmetric loading. However, as explained in Chapter 4 and Chapter 5, the structure of the boring bar and the toolholder-tool and toolholder–spindle interfaces would contribute significantly to part quality. The part deflection can be calculated either by using simulated cutting forces as inputs to a series of displacement solutions using a finite element model of the part [25, 27], or directly by multiplying the simulated forces and the compliance matrix of the finite element model [26]. Errors due to thermal distortion have also been simulated [26–28]. The heat entering the part is calculated from the cutting forces or power (as explained in Chapter 7), and used as an input for calculating the part temperature and thermal deflection distributions. Cylindricity errors calculated using Subramani's analysis [26, 28] have been compared to measurements from experiments on a cast iron engine block; the agreement between simulated and measured errors was reasonable (Figure 8.4). Simulation results indicate that the thermal and mechanical error components may be comparable. In many production applications, however, thermal errors are eliminated using flood coolants. In practice, it is found that distortions due to clamping and inaccuracies in the cast bore (especially those due to core shift in sandcast bores) are as significant as cutting deflections. The clamping distortions can often be modeled, but casting errors generally cannot. It may be possible to model the effect of casting inaccuracies by

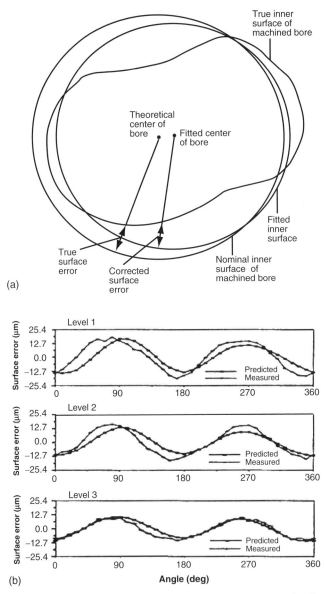

FIGURE 8.4 Comparison of simulated bore roundness errors with measured values for boring a cast iron cylinder. The feed rate and depth of cut are 0.254 mm/rev and 0.764 mm, respectively. Tools with a positive side rake angle, 5° lead angle, and 1.192 mm nose radius were used. (After G. Subramani et al., [26].)

running a series of analyses at the limits of the casting tolerance bands, but this does not appear to have been widely done. The machined surface finish is not modeled in cylinder boring since the bores are single-point finish bored, microsized, and honed in subsequent operations.

Simulations of turning and boring operations are useful to optimize the process parameters such as speed, feed, depth of cut, and number of passes with respect to dimensional and surface quality specifications. Improved boring process designs developed through simulation could conceivably reduce the number of boring passes, eliminate the need for some of these subsequent processes (i.e., microsizing or honing operations), and increase tool life in those which remain, but this does not appear to have been demonstrated in practice.

8.4 MILLING

Milling is more difficult to simulate than turning and boring because the kinematics are more complicated and because the cutting forces are periodic and excite harmonic vibrations. As discussed in Section 2.6, milling processes can be classified as face milling and peripheral (or end milling) operations. Because practical concerns differ, simulations are often applicable only to a particular class of processes.

8.4.1 Face Milling

In quasistatic analyses of face milling [14, 29–33], the variation of the uncut chip area with the cutter rotation is approximated by a sinusoidal function as discussed in Section 2.6, and forces are estimated by multiplying the uncut chip area by measured cutting pressures. Forces must be calculated for each insert in the cutter as a function of the cutter rotation angle; total forces are obtained by summing the forces on the inserts which are cutting. In Fu's well-known model [31], the tangential, radial, and axial forces acting on insert i when the cutter rotation angle is ϕ (Figure 8.5) are given by

$$\begin{Bmatrix} F_t(i,\phi) \\ F_r(i,\phi) \\ F_a(i,\phi) \end{Bmatrix} = K_t C_l(i,\phi) d_e(i,\phi) \begin{bmatrix} 1 + K_r \dfrac{\cos \eta_{Le} \tan \alpha_r}{\cos \alpha_a} \\ -\tan \alpha_r + K_r \dfrac{\cos \eta_{Le}}{\cos \alpha_a} \\ -\dfrac{\tan \alpha_a}{\cos \alpha_r} + K_r \dfrac{\sin \eta_{Le}}{\cos \alpha_a \cos \alpha_r} \end{bmatrix} \quad (8.9)$$

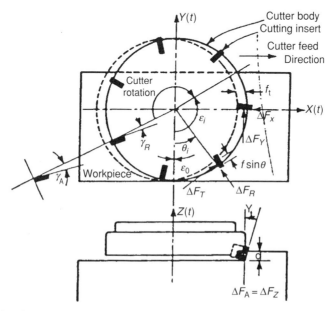

FIGURE 8.5 Cutting forces on an insert in face milling. (After H.J. Fu et al., [31].)

where K_t and K_r are empirical cutting pressures in the tangential and radial directions, γ_{Le} is the effective lead angle, α_a and α_r are, respectively, the axial and radial rake angles of the cutter, and η_{Le} is the effective chip flow angle. Methods for calculating K_t, K_r, and γ_{Le} are discussed in Section 8.6. $d_e(i, \phi)$ is the effective axial depth of cut, given by

$$d_e(i, \phi) = d - R \sin \eta_t + R \sin \eta_t \sin \theta_i(\phi) \tag{8.10}$$

where η_t is the spindle tilt angle. $\theta_i(\phi)$ is the angular position of insert i at cutter rotation angle ϕ, given by

$$\theta_i(\phi) = \phi + \theta_i(0) \tag{8.11}$$

where $\theta_i(0)$ is the initial spacing angle of insert i. For evenly spaced inserts,

$$\theta_i(0) = \frac{i-1}{n_t} 360° \text{ (evenly spaced inserts)} \tag{8.12}$$

For unevenly spaced inserts, $\theta_i(0)$ values must be specified as inputs. $C_1(i, \phi)$ is the uncut chip thickness for insert i at a rotational angle of ϕ, given approximately by

$$C_1(i, \phi) \approx f_t(i) \sin \theta_i(\phi) + e(i) - \varepsilon(i-1) \tag{8.13}$$

where f_t is the feed per tooth for insert i,

$$f_t(i) = f_r \frac{\theta(i, \phi) - \theta(i-1, \phi)}{360°} \tag{8.14}$$

and $\varepsilon(i)$ and $\varepsilon(i-1)$ are the radial runouts (or setup errors) at inserts i and $i-1$, which must be input. For large setup errors, Equation (8.13) becomes inaccurate near insert entry and exit, and a more complicated equation must be used to accurately estimate forces [31]. When the forces on individual inserts have been calculated, the total milling forces can be calculated by summing forces on all inserts cutting at a particular time:

$$\begin{Bmatrix} F_x(\phi) \\ F_y(\phi) \\ F_z(\phi) \end{Bmatrix} = \sum_1^{n_t} \delta(i, \phi) \begin{bmatrix} \cos \theta_i(\phi) & -\sin \theta_i(\phi) & 0 \\ \sin \theta_i(\phi) & -\cos \theta_i(\phi) & 0 \\ 0 & 0 & 1 \end{bmatrix} \begin{Bmatrix} F_t(i, \phi) \\ F_r(i, \phi) \\ F_a(i, \phi) \end{Bmatrix} \tag{8.15}$$

In this equation, $\delta(i, \phi)$ is a parameter equal to 1 for inserts which are cutting, and to 0 for inserts which are not. $\delta(i, \phi)$ must be determined from geometric constraints which depend on the part geometry; for complex parts a cross section of the CAD model is analyzed using a book-keeping algorithm to assign $\delta(i, \phi)$ values.

As in turning simulation, quasistatic models of face milling are used to compute time histories of milling forces and power. Typical examples are shown in Section 8.7. Simulated time histories are useful in estimating power requirements and in designing milling cutters to reduce forces in critical areas, especially for complex parts if the simulation is integrated into a CAD/CAM system so that the part geometry file can be read directly [34]. Forces predicted in quasistatic simulations can be used as inputs to analyses in which the workpiece and the spindle are modeled as compliant to calculate deflections [35]. Deflection calculations can be used to predict dimensional and flatness errors in rough milling. Quasistatic forces can also be used as inputs to dynamic analyses to calculate dynamic forces and stability limits [36–38].

Machining Process Analysis

For stable cutting conditions, static and dynamic forces do not differ significantly. Dynamic simulations are therefore most useful for predicting stability limits.

8.4.2 End Milling

Simulation can be used in end milling to optimize the cutting conditions, tool path, and the number of passes for deep cuts (i.e., die manufacturing) or thin wall structures. A number of quasistatic simulations of end milling forces have been reported [39–48]. Kline's well-known model [44, 45] is reviewed in this section. In this model, the cutting edge of each flute of the cutter is divided into N axial elements of height Δz (Figure 8.6). When spindle runout effects are included, the tangential and radial forces on the cutter at time t, $F_t(t)$ and $F_r(t)$, are given by

$$F_t(T) = \sum_{i=1}^{N} \left[\sum_{j=1}^{n_t} K_t t_{cij}(t) \Delta z \right] \tag{8.16}$$

$$F_r(t) = K_r F_t(t) \tag{8.17}$$

in these equations, n_t is the number of flutes on the cutter, K_t and K_r are empirical cutting pressures equivalent to those used in face milling models (Section 8.6), and $t_{cij}(t)$ is the uncut chip thickness for flute j at element i at time t, given by

$$t_{cij}(t) = f_t \sin \beta_{ij}(t) + d_{rij}(t) + n_j \tag{8.18}$$

$\beta_{ij}(t)$ is the angle of engagement of the ith element of flute j at time t, given by

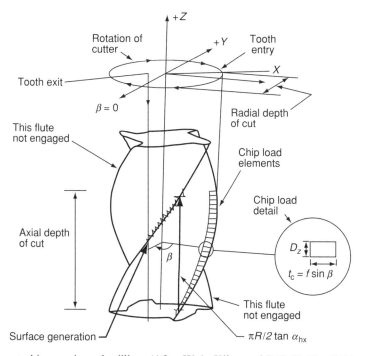

FIGURE 8.6 Uncut chip area in end milling. (After W.A. Kline and R.E. DeVor [44].)

$$\beta_{ij}(t) = -\theta(t) + (k-1)\gamma + z_i \frac{\tan \alpha_{hx}}{r_{ij}(t)} \qquad (8.19)$$

where γ is the angular spacing between flutes and α_{hx} is the helix angle of the cutter. $d_{rij}(t)$ is the change in radius from the current rotation to the previous rotation for flute j at element i, given by

$$d_{rij}(t) = r_{ij}(t) - r_{i-1,j}(t) \qquad (8.20)$$

where $r_{ij}(t)$ is the instantaneous radius for flute j at element i, given by

$$r_{ij}(t) = R + \rho \cos\left[\delta\xi - \xi + (j-1)\frac{2\pi}{n_t}\right] \qquad (8.21)$$

$$\delta\xi = z_i \frac{\tan \alpha_{hx}}{r_{ij}(t)} \qquad (8.22)$$

ξ is an angle of rotation of the cutter from a prescribed zero point (Figure 8.7) and z_i is the axial height of the ith element above the free end of the cutter,

$$z_i = \left(i - \frac{1}{2}\right)\Delta z \qquad (8.23)$$

where

$$\Delta z = \frac{D_z}{N} \qquad (8.24)$$

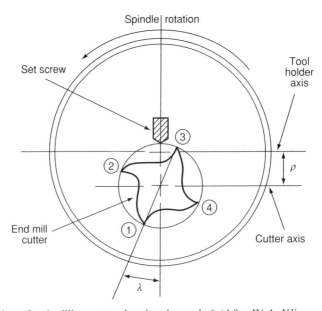

FIGURE 8.7 Top view of end milling cutter showing the angle ξ. (After W.A. Kline and R.E. DeVor [44].)

Machining Process Analysis

and D_z is the axial depth of cut. Somewhat similar formulation of the analytical modeling of end milling forces is discussed by Altintas [49].

Provided the runout can be accurately specified, quasi-steady forces calculated using this analysis agree well with measured values (Figure 8.8). End milling simulations can be adapted to calculate quasistatic forces in slab milling by restricting the region of contact between the cutter and the workpiece [34].

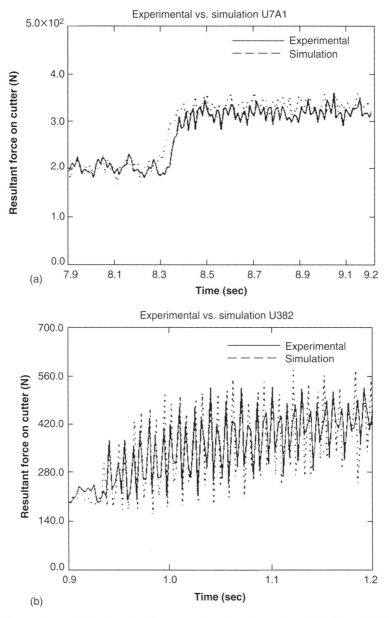

FIGURE 8.8 Comparison of simulated and measured cutting forces in end milling. (a) Milling a 1018 steel workpiece with a high-speed steel cutter with variable feed rate. (b) Milling a 2024 aluminum workpiece with a high-speed steel cutter with variable depth of cut. (After F.M. Kolartis and W.R. DeVries [50].)

As discussed by Smith and Tlusty [39], quasistatic force results can be used as initial inputs to dynamic simulations to predict dynamic forces [50], cutter and part deflections [50–53], and the machined surface characteristics [54, 55]. Stability of regenerative chatter vibrations and chatter stability lobes are discussed in Chapter 12. The stiffness of the cutter is lower in end milling than in face milling and most other processes, and in fact is often comparable to the part stiffness. Therefore, both the part and cutter are treated as flexible in the most detailed analyses [53]. Also, since deflections are often comparable to the stock removal and thus can significantly influence the instantaneous uncut chip area and cutting forces, the coupling between dynamic forces and deflections must also be taken into account to accurately simulate the process under general conditions.

End milling simulations have been applied primarily in airframe and aircraft engine manufacture [51]. In these applications thin, fin-like features are often produced in multiple passes on CNC milling machines. Simulations can be used to predict part deflections (Figure 8.9) and to modify tool paths to compensate for these deflections, leading to significant reductions in the number of passes required, especially in pocketing and cornering cuts.

8.4.3 Ball End Milling

The ball end milling process is similar to end milling, but cutting occurs at the curved end, rather than the straight periphery, of the cutter (Figure 8.10). A number of simulations of ball end milling have been reported [56–59]. These analyses are similar to straight end milling simulations, although a different geometric analysis is required to calculate the uncut chip area as explained in Example 2.6. Ball end milling is used primarily to machine three-dimensional contours on molds, dies, and similar parts. As in straight end milling, simulation can be used in principle to adjust tool paths to compensate for anticipated cutter deflections and to significantly reduce machining times. Simulated cutting forces in ball end milling compare well with measured values from tests on mild steel and titanium alloy parts [56, 58, 59]. Simulated surface errors have also been compared with measured values [57]; the agreement in this case is not as good because the reported models do not take the runout of the cutter into account.

8.5 DRILLING

Drilling differs from operations such as turning and face milling in that the cutting speed is low, the tool geometry is more complex, and high-speed steel tools are still common. The approach used in simulating other processes, in which forces are calculated by multiplying the

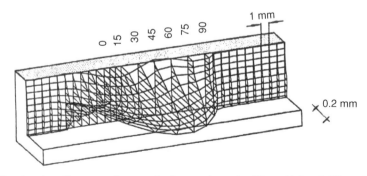

FIGURE 8.9 Simulated surface error in a pocketing cut in end milling. (After J. Tlusty [51].)

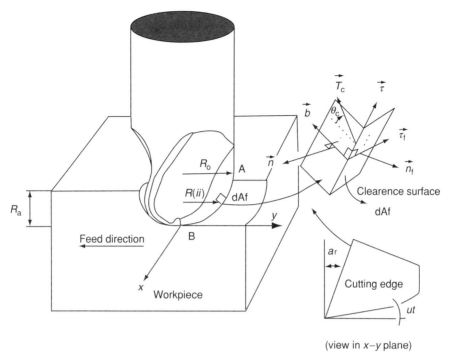

FIGURE 8.10 Uncut chip area in ball end milling. (After G. Yucesan and Y. Altintas [58].)

uncut chip area by measured cutting pressures, can be applied to drilling [60, 61] but is not well suited to the process. This is due in part to the influence of the drill point geometry on process performance; for a given spindle speed and feed rate, drills with different point configurations yield the same total uncut chip area but often produce significantly different forces and torques. In view of this fact, drilling forces can be more effectively simulated by dividing the drill's cutting edges into small segments which can be treated as oblique cutting edges, determining the forces on each element by oblique cutting measurements or calculations, and summing the results to calculate total loads. This approach does not require drilling tests whose results are applicable to specific point geometries, and thus permits investigation of the effect of varying the point geometry based on relatively few experiments. It also permits investigation of phenomena which involve only part of the cutting edge, such as drilling through cross holes or inclined exit surfaces.

As shown in Figure 8.11, a typical drill has three types of cutting edges. The main cutting edges or cutting lips account for most of the torque and power consumption, a significant portion of the axial thrust force, and the radial forces caused by cross holes, inclined exits, and point asymmetry. The central chisel edge generally produces much of the axial thrust force and affects the drill's centering accuracy and buckling behavior. The chisel edge is absent on a few point geometries such as inverted points and those found on indexable drills. The marginal cutting edges form the machined surface of the hole and may contribute to the torque. For simulation, the main cutting edges are of greatest interest, although the chisel edge must be considered to accurately predict the thrust force.

A number of methods of calculating main cutting edge forces based on oblique machining models have been reported [62–70]. In these analyses the main cutting edges are divided into N segments. The radial distance from the drill axis, the center of the ith segment, r_i, is

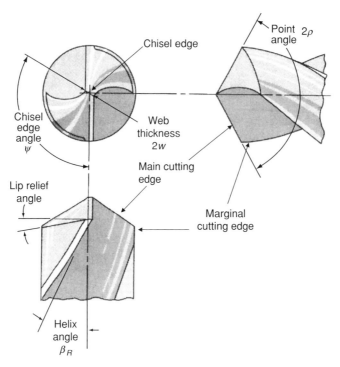

FIGURE 8.11 The point angle 2ρ, web thickness $2w$, chisel edge angle ψ, outer helix angle β_R, and cutting edges on a conventional drill.

$$r_i = r_0 + \left(i - \frac{1}{2}\right)\frac{R - r_0}{N} \qquad (8.25)$$

where r_0 is the radius from the drill axis to the chisel edge. For conventional drills with web thickness $2w$ and chisel edge angle ψ (Figure 8.11),

$$r_0 = \frac{w}{\cos(\psi - 90°)} \qquad (8.26)$$

The cutting speed V_i at segment i is

$$V_i = \Omega r_i \qquad (8.27)$$

The uncut chip thickness at a given radius, $t_{1j}(r_i)$, depends on the drill point angle and the number of flutes. It may vary from flute to flute at a given radius on drills with asymmetric point geometries, particularly if some flutes cut over only part of the total radius (e.g., indexable drills). For drills with symmetric point geometries,

$$t_{1j}(r_i) = \frac{f}{n_f}\sin[\rho(r_i)] \qquad (8.28)$$

where $\rho(r_i)$ is the point angle at radius r_i. The rake and inclination angles of the cutting edges also vary with radius. For a symmetric drill with straight cutting edges, the rake and inclination angles at radius r_i, $\alpha(r_i)$, and $\lambda(r_i)$, are given by [71]

$$\alpha(r_i) = \tan^{-1}\left[\frac{\tan[\beta(r_i)]\cos[\theta(r_i)]}{\sin[\rho(r_i)] - \cos[\rho(r_i)]\sin[\theta(r_i)]\tan\beta_R}\right] \quad (8.29)$$

$$\lambda(r_i) = \sin^{-1}\{\sin[\rho(r_i)]\sin[\theta(r_i)]\} \quad (8.30)$$

where $\theta(r_i)$ is the web angle at radius r_i, which for drills with straight cutting edges is given by

$$\theta(r_i) = \sin^{-1}\left(\frac{w}{r_i}\right) \quad (8.31)$$

$\beta(r_i)$ is the helix angle at radius r_i, given by

$$\beta(r_i) = \tan^{-1}\left[\frac{r}{R}\tan\beta_R\right] \quad (8.32)$$

β_R is the outer helix angle, which is specified in the drill design (Figure 8.11). For drills with asymmetric points or curved cutting edges, more complicated relations must be substituted for Equations (8.28) and (8.31) [68].

Once the geometric characteristics of the various elements have been calculated, the force components F_c, F_z, and F_l (Section 6.3) acting on each element can be calculated from relations of the form

$$F_{cji} = F_c[V(r_i), t_{1j}(r_i), \alpha(r_i), \lambda(r_i)]\Delta L(r_i) \quad (8.33)$$

$$F_{zji} = F_z[V(r_i), t_{1j}(r_i), \alpha(r_i), \lambda(r_i)]\Delta L(r_i) \quad (8.34)$$

$$F_{lji} = F_l[V(r_i), t_{1j}(r_i), \alpha(r_i), \lambda(r_i)]\Delta L(r_i) \quad (8.35)$$

where

$$\Delta L(r_i) = \frac{R - r_0}{N\sin[\rho(r_i)]} \quad (8.36)$$

The functional forms of the relations can be determined using an oblique cutting theory [63–65] or empirically from end turning test data [67–69]. Alternatively, empirical equations for the forces normal and parallel to the rake face, N and P, can be used and converted to the components F_c, F_z, and F_l using Equations (6.11)–(6.15) [70].

When the force components acting on each element have been determined, the drilling torque M and main cutting edge (mce) contribution to the thrust force $F_{t\,mce}$ can be calculated from

$$M = \sum_{j=1}^{n_f}\sum_{i=1}^{N} dM_{ij} \quad (8.37)$$

$$F_{t\,mce} = \sum_{j=1}^{n_f}\sum_{i=1}^{N} dF_{tij} \quad (8.38)$$

where (Figure 8.12)

$$dM_{ij} = r_i F_{cij} \quad (8.39)$$

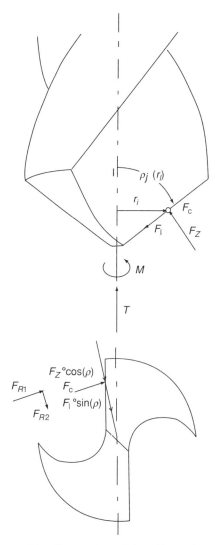

FIGURE 8.12 Relation between cutting forces at a point on the main cutting edge and drilling thrust, torque, and radial forces. (After D.A. Stephenson and J.S. Agapiou [68].)

$$dF_{tij} = F_{zij} \sin[\rho_j(r_i)] - F_{lij} \cos[\rho_j(r_i)] \qquad (8.40)$$

If the oblique cutting relations, Equations (8.33)–(8.35), are accurate, torques calculated using Equation (8.37) generally agree well with measured values (Figure 8.13). Thrust values calculated using Equation (8.38) agree reasonably well with measured values for drills without a chisel edge and for drilling with a pilot hole with a diameter larger than the drill's web thickness; in the more general case, however, the main cutting edges may account for less than half the total thrust, so that Equation (8.38) significantly underestimates measured values (Figure 8.14). Accurate thrust contributions in this case can be obtained by adding an estimate the chisel edge contribution to the thrust, F_{tce}, to the main cutting edge contribution:

$$F_t = F_{t\,mce} + F_{tce} \qquad (8.41)$$

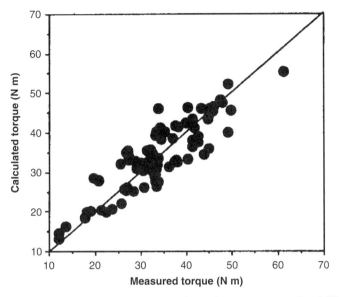

FIGURE 8.13 Comparison of simulated and measured steady-state torques for drilling cast iron with tungsten carbide drills with varying point geometries. (After D.A. Stephenson and J.S. Agapiou [68].)

A number of analyses for calculating F_{tce} have been reported [64, 72–75]. In these analyses, the chisel edge cutting action is usually modeled as some combination of an orthogonal cut and a wedge indentation. The results are applicable to restricted ranges of conditions and

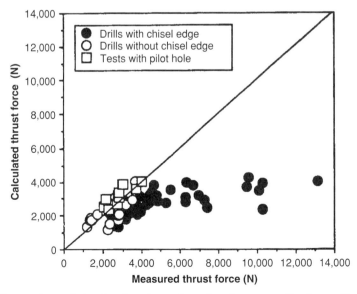

FIGURE 8.14 Comparison of simulated main cutting edge thrust contributions with the total thrust measured in cast iron drilling tests. The simulation underestimates the thrust in cases in which the chisel edge thrust is significant. (After D.A. Stephenson and J.S. Agapiou [68].)

require tests to estimate empirical constants. More broadly applicable results can be obtained by fitting empirical equations to chisel edge thrust contributions measured in tests with and without pilot holes. F_{tce} is well approximated by a relation of the form

$$F_{tce} = C_{ce} f^a L_{ce}^b H^c \qquad (8.42)$$

where C_{ce}, a, b, and c are empirical constants, L_{ce} is the chisel edge length, which is usually equal to two times r_0, and H is the Brinell hardness of the work material. Separate equations are required for each type of chisel edge geometry (straight, bevel, crown, etc.). Typical values for drilling gray cast iron are given in Table 8.1.

For drills with asymmetric points, the drill experiences a net radial force normal to the drill axis. The components of this force due to a particular flute, F_{R1j} and F_{R2j} (Figure 8.12), can be calculated from the relations

$$F_{R1j} = \sum_{i=1}^{N} dF_{R1ji} \qquad (8.43)$$

$$F_{R2j} = \sum_{i=1}^{N} dF_{R2ji} \qquad (8.44)$$

where

$$dF_{R1ji} = F_{cji} \qquad (8.45)$$

$$dF_{R2ji} = F_{zji} \cos[\rho(r_i)] + F_{1ji} \sin[\rho(r_i)] \qquad (8.46)$$

The net radial force rotates with the drill, so that the component in a direction X fixed with respect to the workpiece, $F_X(t)$, varies sinusoidally with time:

$$F_X(t) = A \cos(\Omega t) \qquad (8.47)$$

The relation between the amplitude A and F_{R1j} and F_{R2j} depends on the number of flutes on the drill. For two-fluted drills,

TABLE 8.1
Estimates of Parameters in Equation (8.42) for Drilling Pearlitic Gray Cast Iron with Tungsten Carbide Drills

	Drill Point Geometry	
Parameter	Conventional/Split	Helical
C_{ce}	121.5	32.1
a	0.97	1.00
b	1.10	0.42
c	0.54	0.82

Note: The values predict Th_{ce} in N when C_e is given in mm, f in mm/rev, and H in kg/mm². Computed values are valid for two-fluted drills; for three-fluted drills, the estimate should be multiplied by 1.5.

Machining Process Analysis

$$A = \left[(F_{R11} - F_{R12})^2 + (F_{R21} - F_{R22})^2 \right]^{1/2} \qquad (8.48)$$

More complicated relations are required for three- and four-fluted drills [68]. Radial forces calculated using these relations agree reasonably well with measured values [68].

Drilling torque and force simulations can be used to estimate drill loads and power requirements as functions of the point geometry, feed rate, and spindle speed. Torque and force simulations are especially useful for complicated tools which produce holes with multiple diameters, such as combined drilling/counterboring tools or tapered reamers (Figure 8.15), or when the workpiece exhibits special features which produce unbalanced loads, such as cross holes, irregular cast holes, or inclined exit surfaces (Figure 8.16). To investigate these applications, calculations must be carried out over a series of time steps as the drill advances, with the limits on the summations in Equations (8.37), (8.38), (8.43), and (8.44) being adjusted based on geometric considerations to reflect which portions of the cutting edges are actually cutting. In this case, simulated results can be used to determine bearing load and power requirements for various tool designs and feed-speed cycles, with the ultimate goal of reducing

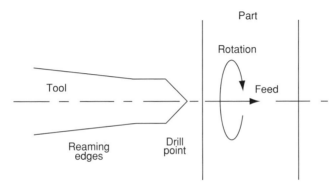

FIGURE 8.15 Drilling a tapered hole in a thin part with a combined drilling and reaming tool. (After D.A. Stephenson and P. Bandyopadhyay [4].)

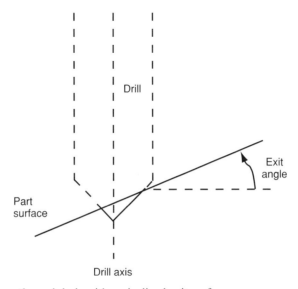

FIGURE 8.16 Drilling a through hole with an inclined exit surface.

the cycle time or number of passes required. Force and torque simulations can also be used as inputs to drill temperature models (Section 7.7) to estimate the drill tip temperature as a function of hole depth.

An important issue in precision and deep-hole drilling is the bending and wandering of the drill tip, which can lead to hole straightness errors and reduced drill life. Predicting these errors accurately is difficult because the drill interacts with and is supported by the hole wall. Simulations which can provide more than a relative comparison of expected errors for different process designs can probably be developed only by combining models of the rigidity of freely vibrating drills [69, 76–81] with experimental data on the effective stiffness of the supporting wall.

8.6 FORCE EQUATIONS AND BASELINE DATA

All the process simulations described in this chapter use baseline cutting force data to predict steady-state cutting forces. The collection of baseline data is often the critical step in applying simulation in practice. The accuracy of the baseline data determines the accuracy of all simulated outputs. Moreover, the effort required to generate these data determines whether the use of simulation in a particular application will reduce development time and testing requirements.

The baseline data required for each tool/workpiece material combination consist of measured cutting forces for a range of cutting conditions, which can be used to fit cutting force or pressure equations as functions of controllable inputs over the test range. The input conditions varied usually include the cutting speed, feed rate, and depth of cut, and in some cases may also include tool angles, the tool nose radius, the edge preparation, and the chip-breaker geometry.

In turning, boring, and face milling, in which indexable inserts are commonly used, forces are usually characterized by two cutting pressures and an effective lead angle (Figure 8.2). The advantage of this approach is that variations in the direction of the resultant cutting force for tools with different nose radii are accounted for by the effective lead angle. This reduces baseline testing requirements since the nose radius does not have to be included as an independent variable in test plans. Some analyses, such as those described for milling in Section 8.4, use the tangential and radial cutting pressures, K_t and K_r, defined by

$$K_t = \frac{F_t}{A_c} \quad (8.49)$$

$$K_r = \frac{F_r}{A_c} \quad (8.50)$$

In these equations, F_t and F_r are the tangential and radial cutting forces (Section 6.3) and A_c is the uncut chip area (Section 8.2). K_t is proportional to the unit cutting power u_s used to estimate the cutting power (Section 2.2 and Section 6.5). Other analyses, such as those described for turning in Section 8.2, use the normal and frictional cutting pressures, K_n and K_f, defined in Equations (8.1) and (8.2). K_f is the effective friction coefficient (Section 6.7).

Several equations for calculating the tangential, axial, and radial cutting forces from various combinations of cutting pressures have been reported. Equation (8.9) (Section 8.4) relates K_t and K_r to these cutting forces for face milling; these equations can be used for turning or boring by substituting $A_c = d_e(i,\phi) \cdot C_1(i,\phi)$ in the leading term. K_n and K_f are related to the tangential, axial, and radial force components by Equations (8.4)–(8.8) in Section 8.2.

γ_{Le}, the effective lead angle, determines the direction of the friction force and is similar to the chip flow angle used in more basic analyses [82–84]. For relatively deep cuts in which the ratio of the depth of cut to the tool nose radius is large (i.e., greater than 5), the effective lead angle is approximately equal to the lead angle. For other cases, a numerical procedure is required to calculate γ_{Le}. Most procedures differ in assumptions about the distribution of frictional pressure along the cutting edge. Fu's method [31] assumes the frictional pressure is uniformly distributed; Subramani's method [24], which is more accurate, assumes the pressure distribution is proportional to the chip thickness distribution. A detailed description of this method is given by Kuhl [8]. A curve fit to results computed using Subramani's method yields the following relation for the effective lead angle [4]:

$$\tan \gamma_{Le} = 0.5053 \tan \gamma_L + 1.0473 \frac{f}{r_n} + 0.4654 \frac{r_n}{d} \qquad (8.51)$$

In this equation, r_n is the tool nose radius, f is the feed per revolution, and d is the depth of cut, all given in millimeters (mm). This relation is valid for $-5° < \gamma_L < 45°$, 0.5 mm $< d <$ 5.0 mm, 0.1 mm $< f <$ 1.0 mm, 0.4 mm $< r_n <$ 4.4 mm, and $r_n > f$. For conditions outside this range, the exact procedure should be used.

The cutting pressures, K_n, K_f, K_t, and K_r, vary with the uncut chip thickness, tool angles, and cutting speed. The uncut chip thickness has the strongest effect; generally pressures increase at low chip thicknesses due to the increased importance of flank friction. K_n and K_f are usually calculated from empirical equations of the general form [4]

$$K_n = C_n t_{ave}^{a_n} V^{b_n} (1 - \sin \alpha)^{c_n} \qquad (8.52)$$

$$K_f = C_f t_{ave}^{a_f} V^{b_f} (1 - \sin \alpha)^{c_f} \qquad (8.53)$$

where C_n, a_n, b_n, c_n, C_f, a_f, b_f, and c_f are empirical constants and t_{ave} is the average uncut chip thickness computed assuming an equivalent rectangular contact (Figure 8.2). Values of these parameters for five tool–work material combinations are given in Table 8.2. In the literature, the rake angle dependence is often neglected, and the cutting speed dependence is sometimes neglected, so that K_n and K_f are given by Equations (8.52) and (8.53), respectively, with $c_n = c_f = 0$ or with $b_n = b_f = c_n = c_f = 0$. In these cases, separate equations are required for each cutter geometry. A number of similar cutting pressure equations have been reported [1, 9, 14, 39, 85–87].

When the depth of cut is large compared to the tool nose radius, t_{ave} is approximately equal to $f \cos \gamma_L$, where γ_L is the lead angle (see Equation 2.3). In other cases t_{ave} must be calculated using numerical methods used for the effective lead angle. A curve fit to numerical results yields the equation [4]

$$t_{ave} = 0.5334 \left(\frac{r_n}{f}\right)^{0.0921} \left(\frac{r_n}{d}\right)^{-0.3827} (f \cos \gamma_L)^{1.0317} \qquad (8.54)$$

As for the effective lead angle, this relation is valid for $5° < \gamma_L < 45°$, 0.5 mm $< d <$ 5.0 mm, 0.1 mm $< f <$ 1.0 mm, 0.4 mm $< r_n <$ 4.4 mm, and $r_n > f$.

For processes such as drilling and end milling, which are often carried out using solid tools, the cutting edges of the tool can be divided into small segments which can be approximated as straight-edged oblique cutting tools (Figure 1.1). In these cases, cutting pressures are sometimes not used; rather, empirical equations for specific cutting force components such as Equations (8.33)–(8.35) are used [56, 68, 69]. Often these equations are fit to data from end

TABLE 8.2
Parameter Estimates for Five Common Tool–Work Material Combinations [4]

Work material	Gray cast iron	Nodular iron	390 Al	356 Al	1018 Steel
Hardness (BHN)	170	270	110	73	163
Tool material	Silicon nitride	Coated WC	PCD	PCD	Coated WC
C_n	1227	1730	470	356	2119
a_n	−0.338	−0.336	−0.243	−0.475	−0.231
b_n	−0.121	−0.089	0.060	−0.040	−0.080
c_n	1.190	1.183	1.065		1.149
C_f	0.405	0.304	0.303	0.453	0.453
a_f	−0.363	−0.258	−0.306	−0.368	−0.095
b_f	0	0	0.025	−0.085	−0.090
c_f	−2.126	−1.392	−1.810		−0.233

turning tests on thin-walled tubes (Figure 8.17). For a given tool–workpiece material combination, forces in end turning depend on the cutting speed, feed rate, cutting edge width, and tool rake and inclination angles. Polynomial equations or exponential equations of the type described in Section 6.4 can be used; the exponential equations generally yield more accurate results, especially when extrapolating to cutting conditions outside the test range. If equations are fit to the normal and frictional cutting forces, N and P, they can be related to cutting pressures. In this case cutting pressure data from turning or face milling tests can be used for material combinations for which no end turning data are available [4] as discussed below.

Cutting forces calculated from baseline cutting pressure or force equations and the effective lead angle agree reasonably well with experimental data [4]. Typical errors for the

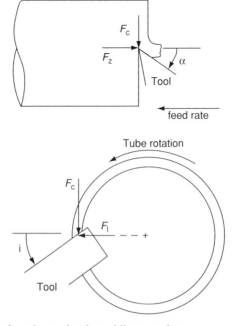

FIGURE 8.17 End turning of a tube to simulate oblique cutting.

tangential, axial, and radial forces in turning are 5, 10, and 10%, respectively. The accuracy is best for roughing cuts and decreases as the feed and depth of cut are reduced; this is especially true for the radial component of the force. The accuracy of baseline cutting force or pressure equations carries over into process simulations in which they are used to compute forces, especially those components which determine cutting power and depend primarily on the tangential forces. Typical examples for drilling and milling processes are shown in Figure 8.18 and Figure 8.19.

For new materials, collecting baseline data involve measuring forces in a series of tests in which the cutting speed, feed rate, depth of cut, and (in many cases) the tool geometry are varied. Generally full or half fraction factorial experiments using three or more levels of feed and at least two levels of the cutting speed and depth of cut are required; this results in test plans between 25 and 50 tests per insert type or tool geometry. On CNC machines, testing effort can be reduced by writing programs to perform a series of short cuts at various combinations of the cutting speed and feed rate; the entire test plan can then be completed quickly by running the program for each depth of cut and tool geometry. All test

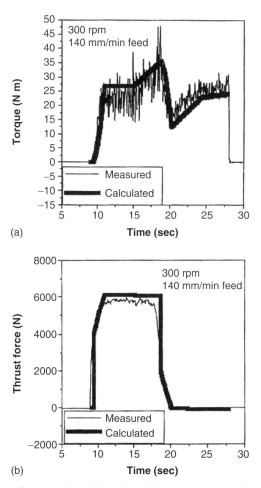

FIGURE 8.18 Comparison of measured and simulated forces and torque for the combined drilling and reaming process shown in Figure 8.15. The spindle speed is 300 rpm and the feed rate is 140 mm/min. (After D.A. Stephenson and P. Bandyopadhyay [4].)

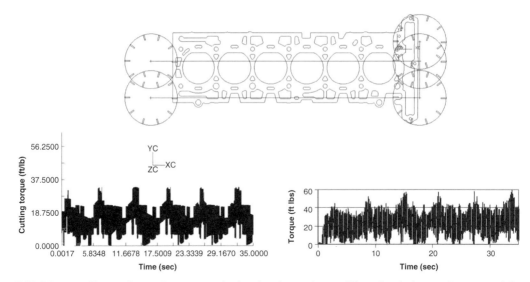

FIGURE 8.19 Comparison of torques calculated using a face milling simulation and measured for cutting the joint face of an aluminum engine head. (After D.A. Stephenson [90].) The calculated torque history plot is shown on the left, below the part geometry and cutter path; the measured torque history is shown on the right. The average calculated and measured torques are 31.6 and 33.2 N m, respectively.

conditions should be repeated at least once; if measured forces vary by more than roughly 10% between repeated runs, it is likely that there is an error in the program or that the tool is wearing rapidly or chipping due to excessive cutting speeds or feeds.

Gathering baseline data for several material combinations can involve a significant testing effort and, as noted at the beginning of this section, can reduce or even eliminate the benefit of applying simulations. At present, baseline testing requirements make simulation most suitable for mass production applications involving relatively few material combinations, and less attractive for batch production operations involving many materials. Some research on reducing testing requirements has been reported [4, 67, 85, 86]; this has focused on process-independent testing methods which produce data useful for simulating more than one process. This work suggests that turning, boring, and face milling processes can be simulated from a common set of turning or milling data [4, 85], or even from orthogonal cutting data similar to that used for drilling simulation [86].

8.7 PROCESS SIMULATION APPLICATION EXAMPLES

In mass production applications, machining process simulation can be used both when purchasing new equipment and when processing or reprocessing parts for existing equipment. For new equipment, simulation can be used to define machine horsepower and bearing requirements, and to create tool paths which make full use of available power during roughing cuts and which do not contain excessive noncutting time. For complex parts, processing steps can be moved between operations to balance work content. Accurate process simulations can be used to evaluate a wider range of potential machine types, process sequences, and tool paths than would be possible through physical testing, and to select the option which minimizes investment. In processing a new part on existing equipment, simulation is used to tailor the process to the machinery; in this case, the machine's horsepower and transient response characteristics are fixed constraints rather than options to be selected. It is also frequently necessary to analyze an existing process to accommodate a part change (which often involves adding machined stock) or volume increase without additional investment. Reasonable objectives of this type can often be met by increasing feeds during roughing cuts, increasing cutting speeds if new tool grades are available, using rapid traverse through noncutting portions of tool paths, etc. Simulation in current processes helps to increase productivity, part quality, and reduce tooling and production cost.

Machining process simulation is also used to compute time histories for input to finite element analyses as discussed below. Force and torque calculations are also useful in designing special drills and milling cutters. Finally, simulations can be of benefit for visualization and training. Since production machining equipment has heavy guarding, waste time during tool motion is often not directly observable, but can be detected through graphic simulation. Engineers with relatively little experience can also use simulations to learn CNC code, the effects of various tool angles, and similar material without cutting test parts.

Machining simulation is not currently useful for directly solving tool life and chip control problems. There are numerous models which purport to predict tool life as a function of cutting speed, feed, etc. for various materials. In most cases, these models assume tool life ends when a defined level of flank wear occurs. In practice, however, tool life depends on part tolerances. A tool will be removed from service when a part's dimension, surface finish, or edge condition is out of specification; this will occur at various levels of flank wear depending on the part geometry and part tolerances. It might be possible to tailor a tool life model to a narrow group of operations in which the end of tool life is consistently known; this would require extensive experimentation, however, and it is not clear what additional information the simulation would provide. Simulation can, however, provide qualitative insight into tool life issues, since measures which reduce cutting forces generally increase tool life. Similarly, chip control problems are best investigated by increasing or interrupting feeds, changing tool angles, and testing inserts with various chip breaking patterns; the physical complexity of the chipbreaking process, and the fact that such problems often arise from variability in the incoming material, make it difficult to add value through simulation.

Figure 8.20–Figure 8.24 show typical applications of process simulations [5, 88–90]. In Figure 8.20, two potential methods of milling and engine block front face are investigated. The results show that the larger cutter gives better flatness error at a constant cycle time since the cutter path is shorter and a lower feed rate can be used. This result is obvious once the analysis is performed but was not obvious beforehand. Routine analytical evaluations of this kind can be done more quickly than cutting tests, and provide more physical insight into the process.

Figure 8.21 shows an analysis of a multitool turning operation on a transmission reverse drum assembly, which consists of a mild steel housing containing a welded hardened bearing race. The operation was exhibiting excessive tool cost and downtime. Investigation of the initial operation (Figure 8.21a) showed that CBN inserts were being used on all materials, not just the hardened race, and that speeds and feeds were incorrect. In addition, insert shapes were not standardized, increasing tool room inventory. Analysis was used to design a new operation (Figure 8.21b) with standardized inserts in which CBN tooling was used only for the hardened race. Speed was reduced, and feed rates were adjusted (generally increased) to level forces and reduce cycle time. The new operation reduced tool costs by a factor of 8 with a slightly lower cycle time. Downtime was also reduced since less frequent tool changes were required.

Figure 8.22 shows a torque history plot for a step drill designed to replace two single-diameter drills. The single-diameter drills broke frequently due to material buildup and chip clogging. Since a tool change was eliminated, the step drill could be run at a lower feed rate to eliminate these problems. Analysis was used to determine the maximum feed at which target thrust and torque values would not be exceeded. When implemented the step drill eliminated three hours of downtime due to tool breakage per week.

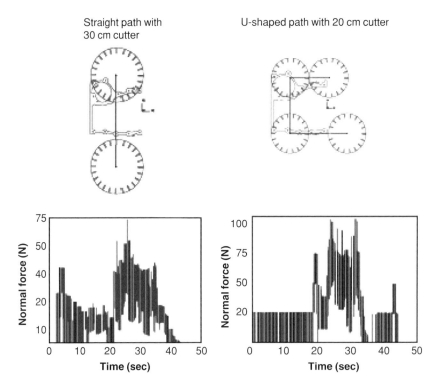

FIGURE 8.20 Analysis of two options for milling an engine block front face. (After D.A. Stephenson [88].) For a fixed cycled time, case (a) yields 30% better flatness accuracy than case (b), since the shorter tool path permits use of a lower feed rate, resulting in lower forces normal to the cut surface.

Figure 8.23 shows an engine head casting cubing application. Simulation was used to reprocess this operation to accommodate a volume increase without adding an additional machining center. Cutters were commonly removed because parts were out of dimension, due in part to axial deflection of the milling spindle as tool wear increased. A number of new cutters with different insert and pocket geometries were investigated analytically, and a new cutter design which yielded the same axial force at a higher feed rate (and lower cycle time) was identified. Axial forces also increased more slowly with insert wear for this design, leading to a reduction in tooling costs as well as cycle time in production.

Figure 8.24 shows force simulations from a boring operation on an engine hydraulic-lash adjuster. This component was made of 1013 steel and measured 27 mm long by 12 mm in diameter. The bore depth was 24 mm, and the diameter varied from 8 to 9.5 mm. An 8 mm boring bar with a triangular coated carbide insert was initially used, but exhibited an unacceptable failure rate. Simulation showed that the initial tool path resulted in load spikes on the insert. Feed rates were adjusted to level the forces and eliminate the load spikes while maintaining cycle time. Peak loads on the insert were reduced by a factor of 2, greatly reducing tool breakage.

8.8 FINITE ELEMENT ANALYSIS FOR CLAMPING, FIXTURING, AND WORKPIECE DISTORTION APPLICATIONS

Structural finite element analysis is used to estimate workpiece distortions due to both clamping and machining forces.

Machining Process Analysis

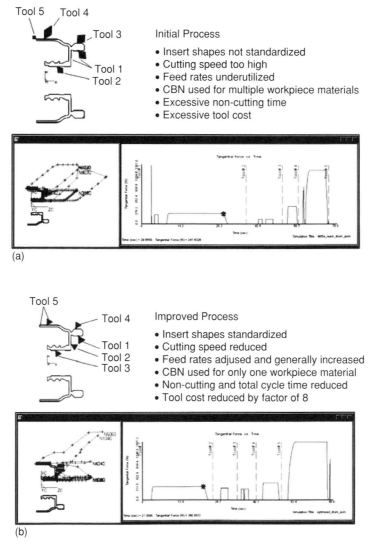

FIGURE 8.21 Analysis to optimize a turning operation on a transmission reaction drum assembly. (After D.A. Stephenson [90].) (a) Initial process. (b) Improved process.

Clamping distortions are more easily calculated. The inputs required are the clamping and locating points and the clamping forces. Finite element models of both the part and fixture structure are generally required; attempts to replace the fixture with equivalent boundary conditions, such as springs or displacement constraints, save computing time but generally yield inaccurate results. The part finite element model used for structural design is usually adequate, although some mesh refinement near the clamping points may be needed. If a fixture model is not available, one must be created; second-order tetragonal elements yield the most accurate results. The support and clamping elements in contact with the part are modeled. The interfaces between the part and fixture should be modeled using contact elements with friction for optimum accuracy. A static analysis in which the clamping loads are applied at the clamping points yields a distortion prediction. The major unknowns are usually the friction coefficients at the contact points; these can be determined experimentally

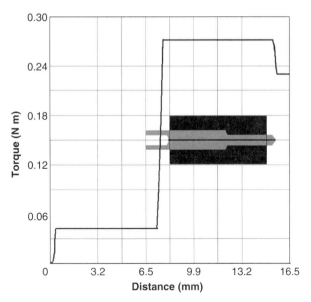

FIGURE 8.22 Torque history plot for a step drill designed to replace two single-diameter drills. Since a tool change is eliminated, the step drill can run at a lower feed rate without increasing the cycle time. (After D.A. Stephenson [88].)

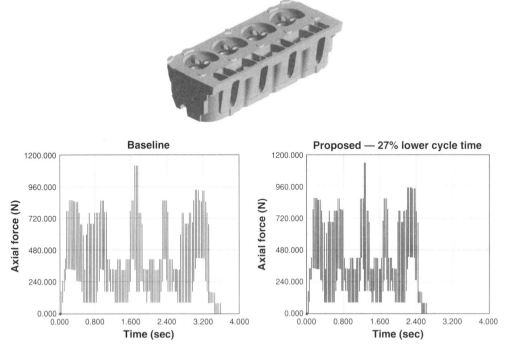

FIGURE 8.23 Axial force plots for baseline and proposed tool paths and milling cutters for an engine head cubing operation. (After D.A. White [89].)

if the solution is sensitive to these variables. In overconstrained clamping schemes (i.e., for four-point locating schemes on planes) locator dimensional variations are also significant. This type of analysis is most often used for thin-walled, compliant parts; in this case it is rarely necessary to model additional elements of the machine tool structure.

Machining Process Analysis

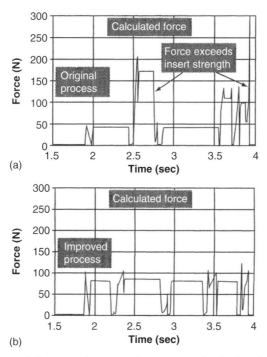

FIGURE 8.24 Analysis to optimize a turning operation on an engine hydraulic-lash adjuster. (After Ref [5].) (a) Initial process. (b) Improved process.

When modeling distortions induced by machining forces, more elements of the system must generally be considered. As shown in Figure 8.25, machining forces act between the tool and the part, and may cause deflections of two broad structural assemblies: the tool, toolholder, and machine tool structure on one side, and the part and fixture on the other. To compute deflections, cutting force histories must be estimated, often using the kinematic simulations described in the preceding sections, and applied to structural finite element models of both assemblies [91]. In some operations, however, the compliance of one element of the system (tool/toolholder/machine structure or part/fixture) may be much larger than the other, so that the other element can be treated as rigid. As noted below, sequential or iterative analyses may be required in applications in which machining significantly changes the structural compliance of the part or in which cutting force and deflections are coupled.

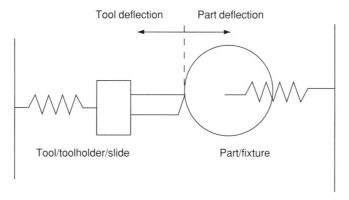

FIGURE 8.25 Components of dimensional errors in machining due to deflections.

In most turning and cylinder boring simulations, the tool can be assumed to be rigid and the part compliant [11, 25–27]. In this case the compliance of the part is usually determined from a finite element model. Deflections are then calculated from the formula

$$\mathbf{U} \cong \mathbf{K}^{-1}\mathbf{F} \tag{8.55}$$

where $\mathbf{U}$ is the nodal displacement vector, $\mathbf{K}^{-1}$ is the compliance matrix of the model, and $\mathbf{F}$ is the nodal force vector. Simulated cutting forces can be used to construct the nodal force vector at any given time. Two approaches, illustrated in Figure 8.26, can then be used to calculate deflections. In the first, the simulated force vector as used as an input to a finite element analysis [25, 27]. As the tool moves with respect to the workpiece, new force vectors are calculated at regular intervals and used as inputs to further analyses. In the second approach, the compliance matrix of the part is extracted from an initial analysis and stored [26]; the simulation program can then calculate deflections from Equation (8.55) at each time step without the need for further finite element analyses. The second approach requires much less computing time, since only one finite element analysis is needed, and should be used whenever the compliance of the part does not change significantly as a result of cutting. The first approach is required, however, in cases in which machining removes enough material to reduce the part's stiffness. In this case, the finite element model must be updated in addition to the force vector at each time step.

In face milling, the workpiece and fixture are often more rigid than the tooling and machine structure. In this case, the deflections are calculated by multiplying simulated forces by the compliance of the tooling. The compliance of the tooling is usually determined by the axial or radial compliance of the spindle, which can be measured or calculated from a finite element model. The variation of the spindle compliance with spindle speed must be taken into account in variable speed applications. In face milling thin-walled parts, the tooling is more rigid than the part and fixture, and analyses similar to those described for turning and boring are required.

In end milling, either the tool or workpiece may be treated as compliant. In the first case, the tool is usually modeled as a simple spring–mass–dashpot system characterized by an effective mass, stiffness, and damping coefficients estimated from experimental data. These models are similar to models for face milling with a radially compliant spindle. In the case in which the tool is treated as rigid, the compliance of the part is usually estimated from a finite element model.

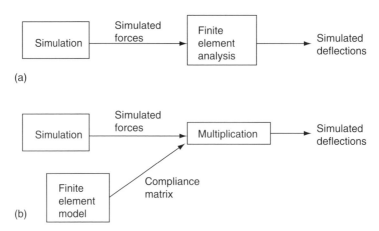

FIGURE 8.26 (a) Simulated forces used as inputs to a series of finite element analyses to calculate deflection. (b) Multiplication of simulated forces with a stored compliance matrix to calculate deflections.

Deflections are calculated from finite element analyses using simulated forces as inputs. Some work has been reported on modeling both the tooling and workpiece as compliant and accounting for the dependence of cutting forces on deflections [52]. In this case the compliances are modeled as in simpler analyses and iterative methods are used to compute forces.

Verifying the accuracy of simulation models for deflections is difficult because deflections are difficult to measure accurately, and because the stiffness and damping characteristics of the system often must be determined experimentally and thus may effectively "calibrate" the model. Most models have only been compared with post-process data on part quality, often for very simple parts. This method assumes that all deviation of the machined surface from specified dimensions is due to deflection, although in reality setup and machine motion errors also contribute to these deviations. Bearing these considerations in mind, simulated deflections in boring and end milling generally agree well with post-process part variations (Figure 8.4 and Figure 8.27).

Simulation models for deflections are most useful for designing tooling and selecting process conditions to maximize part accuracy. In the case of compliant parts, they can also be used to ensure that fixtures are supporting the part adequately over areas where high cutting forces are expected. Deflection models are not as easy to implement as static force simulations, since additional inputs to characterize the stiffness of the machining system are required. The results are also not as general, and in fact are valid only for a single machine.

Simulations which predict deflections can also be used to analyze the dynamic response of the machining system under given conditions, and specifically to determine if the system will be stable dynamically. This subject is discussed in Chapter 12 and in review papers [39, 51].

8.9 FINITE ELEMENT APPLICATION EXAMPLES

Very few production applications of structural finite element analyses have been reported in the literature [88, 90, 92, 93]. In most of these, only clamping distortions have been considered, and analysis results have been used primarily for qualitative comparisons, that is, for ranking a group of design options or fixturing concepts from best to worst. Such qualitative rankings are quite reliable provided no gross modeling errors were made. It is more difficult to use the results for quantitative assessments, for example to determine if a given operation will produce a part within tolerance [92]. Finite element studies generally underestimate distortions when compared to measurements, partly because finite element models are generally stiffer than real parts, and partly because elements of the system which are compliant in reality are assumed to be rigid in the analysis. As a general guideline predicted distortions or stresses should be multiplied by a safety factor when making quantitative assessments; 1.5 is typically an adequate factor for routine analyses for which proper element types and boundary conditions are known.

Figure 8.28–Figure 8.33 show examples of finite element analyses applied to machining fixturing problems. Figure 8.28 shows a clamping study for a torque converter housing. The original fixture design called for clamping on a large bore in the center of the part. This bore has little supporting structure, however, and clamping distortions in this area can lead to large flatness deviations due to the asymmetric geometry of the part. Finite element analysis calculations prior to prototype build showed that much lower flatness deviations would result from clamping on a smaller side bore with more supporting structure.

Figure 8.29 shows a sensitivity study for a channel plate milling fixture which had exhibited periodic flatness variations. The results showed that it is critical to ensure that the center clamp was properly seated and engaged during each cycle. Analysis can be of great benefit in eliminating potential error sources when troubleshooting complex problems; in this particular example, the error in question occurred once or twice per day, so that considerable

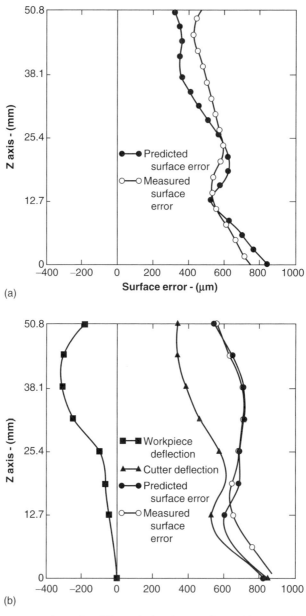

FIGURE 8.27 Surface errors in end milling of aluminum with a high-speed steel cutter: (a) rigid workpiece; (b) compliant workpiece. (After W.A. Kline et al., [43].)

engineering attention and continuous gaging would have been required to solve the problem without analysis.

Figure 8.30 shows an analytical study of the effect of varying clamping pressure on milled deck face flatness for an engine block. The station in question had both front and rear clamps actuated by independent hydraulic cylinders. Since the front of the block is structurally weaker than the rear, reduced distortion (and improved flatness) can be achieved by using unbalanced

Machining Process Analysis

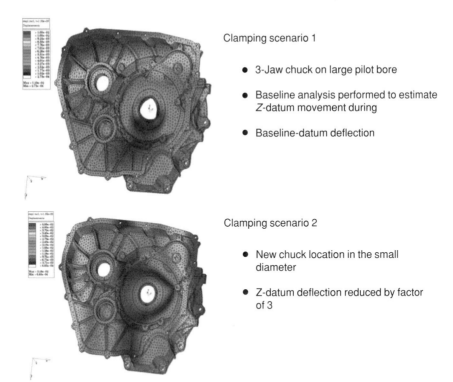

FIGURE 8.28 Finite element clamping study for a transmission converter housing. (After D.A. Stephenson [90].)

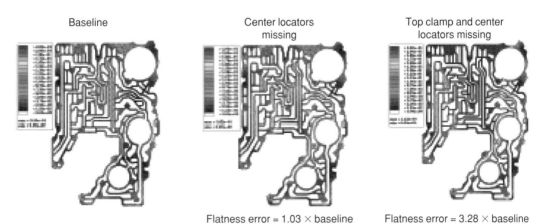

FIGURE 8.29 Analysis to determine the sensitivity of part flatness to locating and fixturing errors when finish milling a channel plate. (After D.A. Stephenson [88].)

clamping forces, with the highest force applied by the rear clamps. For the best combination of pressures, the flatness was reduced to less than 10% of the baseline level for equal loading.

Figure 8.31 and Figure 8.32 show an analytical study of the deckface flatness for a cylinder head due to (1) seat and guide pressing and (2) part clamping at the milling station. The objective in this analysis is whether a single-pass milling operation can be used to finish the deckface or a two-pass milling (semifinish and finish passes) is required. The overall deckface flatness prediction of 38 μm due to seat and guide interference fit was within 5 to 10% of the measured values as

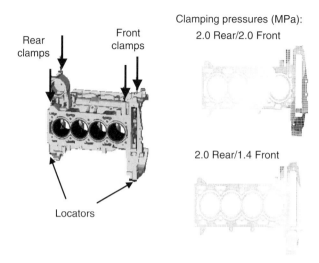

FIGURE 8.30 Analytical study of varying clamping pressures to control flatness in engine block deck milling. (After D.A. Stephenson [88].)

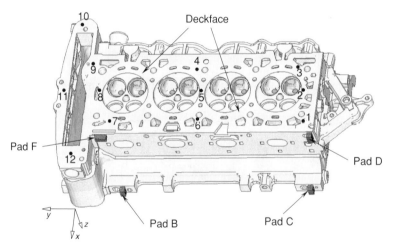

FIGURE 8.31 Twelve points on an engine head deckface to estimate distortion (supporting pads also shown). (After D.A. Stephenson et al., [93].)

explained in Ref. [89]. The deckface distortion due to clamping forces using several cases of locating clearance in the 4–2–1 locating scheme was evaluated to be in the range of 0.010 to 0.050 mm depending on the relative height error among the four support pads. The static deckface distortion due to the cutting forces during milling was analyzed and the results were about 10 to 15 µm. This analysis can be used to decide: (1) if a single-pass milling operation can be used to finish the deckface because the flatness is determined by the sum of the clamping and cutting distortions, and (2) if the single-pass milling should be performed before or after the pressing of the seats and guides by comparing the required deckface flatness part quality to the 38 µm distortion.

Figure 8.33 shows a study of locating schemes for an engine block. For the operation in question the part is to be located on its rear face and clamped from the top; the cylinder head banks are to be milled, and the shape and location of the locating pads which minimized the bank face flatness deviation must be determined. Finite element calculations show that using a round central pad and two square pads greatly reduces the flatness deviation when

Machining Process Analysis

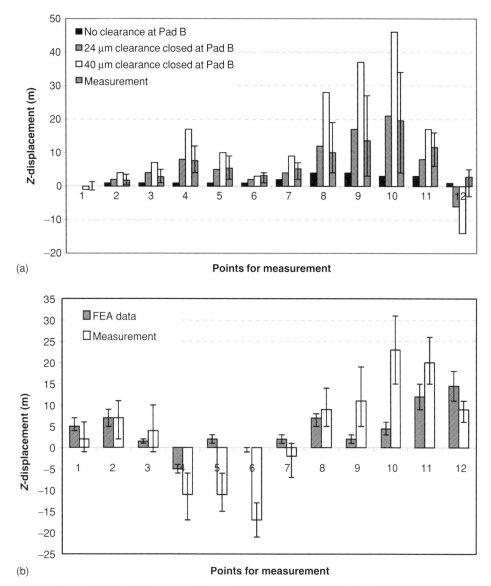

FIGURE 8.32 (a) Comparison of deckface displacements due to seat and guide pressing computed using finite element analysis with measurements. (b) Comparison of deckface displacements due to clamping with different initial clearances. (After D.A. Stephenson et al., [93].)

compared to various combinations of square pads, and that the best diameter for the round pad is 150 mm. Qualitative analytical comparisons of this type done prior to prototype build can improve accuracy while reducing testing costs and lead time.

8.10 EXAMPLES

Example 8.1 Calculate the forces acting on an end milling cutter with diameter of 25 mm and two straight flutes performing an up-milling process. The axial and radial depths of cuts are 10 and 4 mm, respectively. The feed per tooth (chip load) is 0.1 mm, and the spindle speed is 8000 rpm.

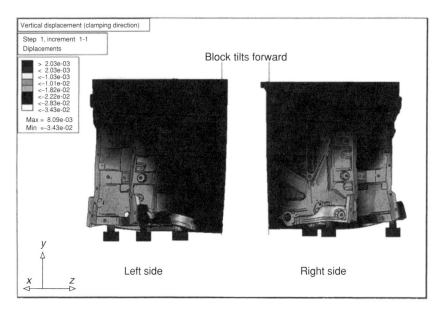

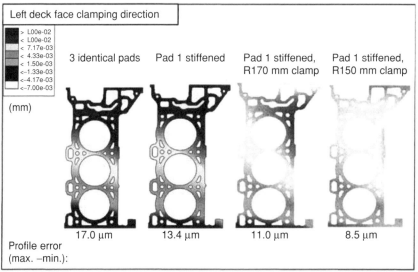

FIGURE 8.33 Analytical study to determine the best geometry for locating pads in an engine block deck milling operation. (After D.A. Stephenson [90].)

Solution: The end milling operation is illustrated in Figure 2.11 and a cross section of up-milling with a two-fluted cutter is shown in Figure 8.34a. The uncut chip thickness changes from zero to maximum as discussed in milling Section 2.2 and illustrated in Figure 2.12 and Figure 2.14. The resultant force on the cutter can be decomposed into three forces, feed (horizontal) force F_f, normal force F_n (in the direction of the width of cut), and axial force F_a as illustrated in Figure 6.8 and Section 8.4. The resultant force on each cutting edge can be decomposed into three forces, the tangential force (tangent to the tool perimeter in the direction of the peripheral cutting velocity) F_t, the radial force F_r (along the radius of the cutter), and the axial force (normal to the other two forces along the axis of the tool). The tangential force is the main cutting force. The other two forces are estimated from the

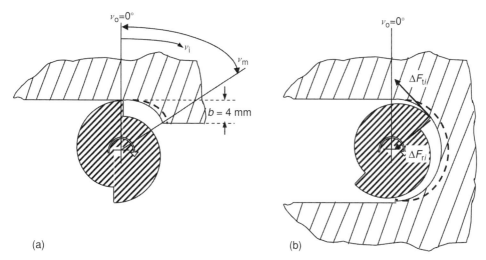

FIGURE 8.34 (a) Cross section of up-milling with a two-fluted cutter. (b) End mill fully immersed in the workpiece.

tangential force as $F_r = 0.3 F_t$ and $F_a = F_r \tan \lambda$, where λ is the helix of the cutter. If the helix angle is 45°, the axial force is equal to the radial force. The undeformed chip thickness is given by Equation (2.21) or $a_i = f \sin \nu_i$. The tangential force as a function of the angle of rotation relative to the workpiece is

$$F_{ti} = u_s d_a f \sin \nu_i \tag{8.56}$$

where u_s is the specific cutting energy of the workpiece material (see Chapter 2 and Chapter 6). If the cutter is performing up-milling (Figure 8.34a), as in the case of peripheral milling in Figure 2.14, the tooth engages over the arc of the width of cut. The engagement angle is determined by Equations (2.21)–(2.24). So $\nu_0 < \nu_i < \nu_m$, where in the above case $\nu_0 = 0°$ and $\nu_m = 47°$.

The tangential and radial forces acting on the cutting edges can be decomposed into components in the feed direction, F_x, and normal to feed direction, F_y, in the workpiece coordinate system through the engagement angle:

$$F_x = \sum_{i=1}^{n_t} (F_{ti} \cos \nu_i + F_{ri} \sin \nu_i)$$
$$= u_s d_a f \sum_{i=1}^{n_t} (\sin \nu_i \cos \nu_i + 0.3 \sin^2 \nu_i) \tag{8.57}$$

$$F_y = \sum_{i=1}^{n_t} (F_{ti} \sin \nu_i - F_{ri} \cos \nu_i)$$
$$= u_s d_a f \sum_{i=1}^{n_t} (\sin^2 \nu_i - 0.3 \sin \nu_i \cos \nu_i) \tag{8.58}$$

where n_t is the number of teeth engaged in the workpiece at an instant, d_a is the axial depth of cut (equivalent to d in Chapter 2), and the radial depth of cut is b in Figure 2.11 for end milling. The maximum forces are obtained when the maximum number of teeth is engaged in

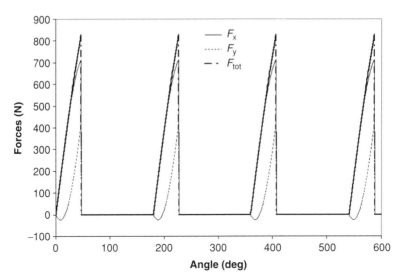

FIGURE 8.35 Forces on the cutter in Example 8.1.

the workpiece. One rotation of the cutter is divided into K intervals that define the angular integration. The forces are periodic as shown in Figure 8.35. The resultant force (F_{tot}) is also given in the graph. Every 360° the teeth pass through the same location with respect to the engagement angle ν. An abrupt drop of these forces will occur if all cutting edges are disengaged from the cut because the radial width of cut is small or the number of edges around the cutter is small (i.e., two-flute end mill). A small computer program can be generated to calculate the F_x and F_y force components using Equations (8.57) and (8.58). These forces can be calculated at small increments of the angle of cutter rotation ν_i. The angle of engagement for 4 mm width of cut is $0° < \nu_i < 47°$, and the forces are plotted in Figure 8.35. The forces are maximum at the exit of cut because the chip thickness starts with zero and gradually increases to maximum in up-milling (Figure 8.34).

Example 8.2 Calculate the forces acting on a slot milling operation for the same tool and conditions used as in Example 8.1. Estimate the cutting torque and power required by the spindle.

Solution: The analysis of a slot milling operation (with the end mill fully immersed in the workpiece as shown in Figure 8.34b) is similar to that in partial immersion in previous example. Since the cutter diameter is fully engaged in the workpiece, $0 < \nu_i < 180°$ and the forces are given by Equations (8.57) and (8.58). The forces acting on the teeth are estimated and shown in Figure 8.36. The forces are sinusoidal because the chip thickness starts with zero and gradually increases to a maximum at 90° rotation and gradually reduces to zero at 180°. The total force is zero only at the zero chip load location occurring at $\nu_i = 0°$.

The instantaneous cutting torque and power on the spindle are

$$M = \frac{D}{2} \sum_{i=1}^{n_t} F_{ti} \quad \text{and} \quad P = V \sum_{i=1}^{n_t} F_{ti} \qquad (8.59)$$

where F_t is the tangential force and D is the cutter diameter. The torque and power have the same trend as the tangential force given by Equation (8.57) and illustrated in Figure 8.37. The average spindle power obtained from the instantaneous graph or Equation (2.29) is 7.2 kW.

Machining Process Analysis

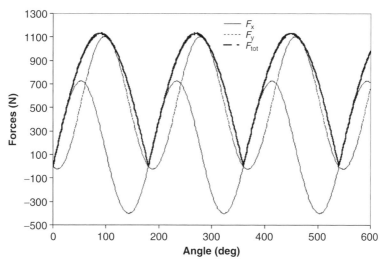

FIGURE 8.36 Forces on the teeth in Example 8.2.

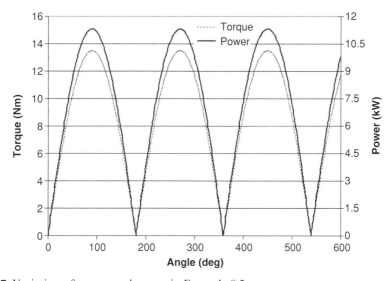

FIGURE 8.37 Variation of torque and power in Example 8.2.

Example 8.3 Calculate the forces acting on an end milling cutter with diameter of 25 mm and two helical smooth edges performing an up-milling process. The axial and radial depths of cut are 10 and 9 mm, respectively. The feed per tooth (chip load) is 0.1 mm, and the spindle speed is 8000 rpm.

Solution: A typical end milling cutter with helical flutes is shown in Figure 2.11. The helix provides a gradual engagement and disengagement of the cutting edge with the radial depth of cut contact. An unwound cutting edge has a straight line contact along the axial depth of cut inclined at the helix angle λ. Therefore, any point (element j) on the cutting edge above the free end of the cutter will lag by an angle ψ (as illustrated in Figure 8.38):

$$\psi_j = \frac{2z_j \tan \lambda}{D} \qquad (8.60)$$

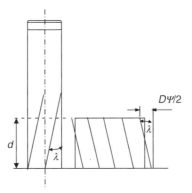

FIGURE 8.38 Definition of the angle ψ in Example 8.3.

where D is the diameter of the cutter and z_j is the distance of element j along the cutting edge from its free end. The cutting edge is divided into N elements (N is the number of axial integration steps). Therefore, the immersion angle of any point (element) along the cutting edge at distance z above the free end of the cutter will have an immersion angle of $(\nu - \psi)$. The chip thickness will be different for each point along the cutting edge according to the immersion angle (as illustrated in Figure 8.6), compared to a constant chip thickness with straight flute end mills discussed in Example 8.1. The analysis of each point along the cutting edge is similar to that discussed in Example 8.1 with an immersion angle of $(\nu - \psi)$.

The cutting edge of each flute of the cutter is divided into N axial elements of height Δz (Figure 8.6). The tangential and radial forces on the cutter at time t, $F_t(t)$, and $F_r(t)$, are given by Equations (8.16) and (8.17), respectively (see Section 8.4). The F_x and F_y forces are calculated using Equations (8.57) and (8.58) by summing the elemental forces along each of the flutes:

$$F_x = \sum_{j=1}^{N} \left[\sum_{i=1}^{n_t} F_{ti} \cos(\nu_i - \psi_i) - F_{ri} \sin(\nu_i - \psi_i) \right] \qquad (8.61)$$

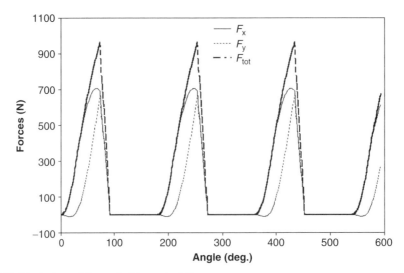

FIGURE 8.39 Variation of forces in Example 8.3.

where $(\nu_i - \psi_i)$ is the angle of the cutter rotation for each element in relation to the zero reference point ($\nu_i = 0°$ in Figure 8.34). This means that while the cutter rotates an angle ν_i, each element j along the cutting edge will rotate an angle $(\nu_i - \psi_i)$. The variation of the forces for a helical end mill is shown in Figure 8.39. The ratio F_x/F_y is smaller for a helical tool compared to a straight fluted one. The forces increase at a slower rate with a helical tool even though the resultant force is higher for a helical tool because the length of contact is longer as illustrated by the contact angle in the horizontal axis.

8.11 PROBLEMS

Problem 8.1 Calculate the instantaneous forces acting on the end milling cutter in Example 8.1 when performing a down-milling process. Explain the significant difference in the cutting force response between the down- and the up-milling cases discussed in Example 8.1.

Problem 8.2 Calculate the instantaneous forces acting on the end milling cutter in Example 8.1 assuming the end mill has four flutes. Compare the cutting force response of the four-flute cutter with that of two-flute cutter in Example 8.1.

REFERENCES

1. K.F. Ehman, S.G. Kapoor, R.E. DeVor, and I. Lazoglu, Machining process modeling: a review, *ASME J. Manuf. Sci. Eng.* **119** (1997) 655–663
2. E.J.A. Armarego, J.A. Arsecularatne, P. Mathew, and S. Verezub, A CIRP survey on available predictive performance models of machining operations, *Proceedings of the 52nd Annual CIRP Assembly*, San Sebastian, Spain, August 2002, 61–76
3. G.E.P. Box, W.G. Hunter, and J.S. Hunter, *Statistics for Experimenters*, Wiley, New York, 1978, 293–294
4. D.A. Stephenson and P. Bandyopadhyay, Process independent force characterization for machining simulation, *ASME J. Eng. Mater. Technol.* **119** (1997) 86–94
5. Delphi Corporation, Simulated action, *Cutting Tool Eng.* **56**:3 (2004) 69–73
6. W.R. DeVries and M.S. Evans, Computer graphics simulation of metal cutting, *CIRP Ann.* **33** (1984) 81–88
7. C.J. Li and W.R. DeVries, The effect of shear plane length models on stability analysis in machining simulation, *Proc. NAMRC* **16** (1988) 195–201
8. M.J. Kuhl, The Prediction of Cutting Forces and Surface Accuracy for the Turning Process, MS Thesis, Mechanical Engineering, University of Illinois, 1987
9. J.W. Sutherland, G. Subramani, M.J. Kuhl, R.E. DeVor, and S.G. Kapoor, An investigation into the effect of tool and cut geometry on cutting force system prediction models, *Proc. NAMRC* **16** (1988) 264–272
10. D.A. Gustafson, The Effect of Tool Geometry and Wear on the Cutting Force System in Turning, MS Thesis, Mechanical Engineering, University of Illinois, 1990
11. W.J. Endres, J.W. Sutherland, R.E. DeVor, and S.G. Kapoor, A dynamic model of the cutting force system in the turning process, *Monitoring and Control for Manufacturing Processes*, ASME PED Vol. 44, ASME, New York, 1990, 193–212
12. G.M. Zhang, S. Yerramareddy, S.M. Lee, and S.C.-Y. Lu, Simulation of intermittent turning processes, *ASME J. Eng. Ind.* **113** (1991) 458–466
13. G. Parakkal, R. Zhu, S.G. Kapoor, and R.E. Devor, Modeling of turning process cutting forces for grooved tools, *Int. J. Mach. Tools Manuf.* **42** (2002) 179–191
14. F.M. Gu, S.G. Kapoor, R.E. DeVor, and P. Bandyopadhyay, An approach to on-line cutter runout estimation in face milling, *Trans. NAMRI/SME* **19** (1991) 240–247
15. N.H. Cook, *Manufacturing Analysis*, Addison-Wesley, Reading, MA, 1966, 73

16. D.A. Stephenson, T.-C. Jen, and A.S. Lavine, Cutting tool temperatures in contour turning: transient analysis and experimental verification, *ASME J. Manuf. Sci. Eng.* **119** (1997) 494–501
17. T.-C. Jen and A. Anagonye, An improved transient model of tool temperatures in metal cutting, *ASME J. Manuf. Sci. Eng.* **123** (2001) 30–37
18. T.-C. Jen, S. Eapen, and G. Gutierrez, Nonlinear numerical analysis in transient cutting tool temperatures, *ASME J. Manuf. Sci. Eng.* **125** (2003) 48–56
19. F. Atabey, I. Lazoglu, and Y. Altintas, Mechanics of boring processes — Part I, *Int. J. Mach. Tools Manuf.* **43** (2003) 463–476
20. G.M. Zhang and S.G. Kapoor, Dynamic modeling and analysis of the boring machining system, *ASME J. Eng. Ind.* **109** (1987) 219–226
21. G.M. Zhang and S.G. Kapoor, Dynamic generation of machined surfaces. Part 1. Description of a random excitation system, *ASME J. Eng. Ind.* **113** (1991) 137–144
22. G.M. Zhang and S.G. Kapoor, Dynamic generation of machined surfaces. Part 2. Construction of surface topography, *ASME J. Eng. Ind.* **113** (1991) 145–153
23. F. Atabey, I. Lazoglu, and Y. Altintas, Mechanics of boring processes — Part II — Multi-insert boring heads, *Int. J. Mach. Tools Manuf.* **43** (2003) 477–484
24. G. Subramani, R. Suvada, S.G. Kapoor, and R.E. DeVor, A model for the prediction of force system for cylinder boring process, *Proc. NAMRC* **15** (1987) 439–446
25. W.E. Sneed, Chip removal simulation to predict part error and vibration, *Proceedings of the ASME Computers in Engineering Conference*, Vol. 2, 1987, 447–455
26. G. Subramani, S.G. Kapoor, and R.E. DeVor, A model for the prediction of bore cylindricity during machining, *ASME J. Eng. Ind.* **115** (1993) 15–22
27. N.N. Kakade and J.G. Chow, Finite element analysis of engine bore distortions during boring operation, *ASME J. Eng. Ind.* **115** (1993) 379–384
28. G. Subramani, M.C. Whitmore, S.G. Kapoor, and R.E. DeVor, Temperature distribution in a hollow cylindrical workpiece during machining: theoretical model and experimental results, *ASME J. Eng. Ind.* **113** (1991) 373–380
29. P.C. Subbarao, R.E. DeVor, S.G. Kapoor, and H.J. Fu, Analysis of cutting forces in face milling of high silicon casting aluminum alloys, *Proc. NAMRC* **10** (1982) 289–296
30. Z. Ruzhong and K.K. Wang, Modeling of cutting force pulsations in face milling, *CIRP Ann.* **32** (1983) 21–26
31. H.J. Fu, R.E. DeVor, and S.G. Kapoor, A mechanistic model for the prediction of the force system in face milling operation, *ASME J. Eng. Ind.* **106** (1984) 81–88
32. F.M. Gu, S.G. Kapoor, R.E. DeVor, and P. Bandyopadhyay, A cutting force model for face milling with a step cutter, *Trans. NAMRI/SME* **20** (1992) 361–367
33. A.D. Spence and Y. Altintas, A solid modeller based milling process simulation and planning system, *ASME J. Eng. Ind.* **116** (1994) 61–69
34. S. Jain and D.C.H. Yang, A systematic analysis of the milling operation, *Computer-Aided Design and Manufacture of Cutting and Forming Tools*, ASME PED Vol. 40, ASME, New York, 1989, 55–63
35. H. Schulz, Optimization of precision machining by simulation of the cutting process, *CIRP Ann.* **42** (1993) 55–58
36. S.J. Lee and S.G. Kapoor, Cutting process dynamics simulation for machine tool structure design, *ASME J. Eng. Ind.* **108** (1986) 68–74
37. H.S. Kim and K.F. Ehmann, A cutting force model for face milling operations, *Int. J. Mach. Tools Manuf.* **33** (1993) 651–673
38. M.A. Elbestawi, F. Ismail, and K.M. Yuen, Surface topography characterization in finish milling, *Int. J. Mach. Tools Manuf.* **34** (1994) 245–255
39. S. Smith and J. Tlusty, An overview of modeling and simulation of the milling process, *ASME J. Eng. Ind.* **113** (1991) 169–175
40. F. Koenigsberger and A.J.P. Sabberwal, An investigation into the cutting force pulsations during the milling process, *Int. J. MTDR* **1** (1961) 15
41. J. Tlusty and P. MacNeil, Dynamics of the cutting forces in end milling, *CIRP Ann.* **24** (1975) 248
42. R.E. DeVor, W.A. Kline, and W.J. Zdeblick, A mechanistic model for the force system in end milling with application to machining airframe structures, *Proc. NAMRC* **8** (1980) 297–303

43. W.A. Kline, R.E. DeVor, and J.R. Lindberg, The prediction of cutting forces in end milling with application to cornering cuts, *Int. J. MTDR* **22** (1982) 7–22
44. W.A. Kline and R.E. DeVor, The effect of runout on cutting geometry and forces in end milling, *Int. J. MTDR* **23** (1983) 123–140
45. W.A. Kline, R.E. DeVor, and I. Shareef, The prediction of surface accuracy in end milling, *ASME J. Eng. Ind.* **104** (1982) 272–278
46. J.H. Ko, W.-S. Yun, D.-W. Cho, and K.F. Ehmann, Development of a virtual machining system. Part 1. Approximation of the size effect for cutting force prediction, *Int. J. Mach. Tools Manuf.* **42** (2002) 1595–1605
47. W.-S. Yun, J.H. Ko, D.-W. Cho, and K.F. Ehmann, Development of a virtual machining system. Part 2. Prediction and analysis of machined surface error, *Int. J. Mach. Tools Manuf.* **42** (2002) 1607–1615
48. W.-S. Yun, J.H. Ko, H.U. Lee, D.-W. Cho, and K.F. Ehmann, Development of a virtual machining system. Part 3. Cutting process simulation in transient cuts, *Int. J. Mach. Tools Manuf.* **42** (2002) 1617–1626
49. Y. Altintas, *Manufacturing Automation — Metal Cutting Mechanisms, Machine Tool Vibrations, and CNC Design*, Cambridge University Press, Cambridge, UK, 2000
50. F.M. Kolartis and W.R. DeVries, A mechanistic dynamic model of end milling for process controller simulation, *ASME J. Eng. Ind.* **113** (1991) 176–183
51. J. Tlusty, Effect of end milling deflections on accuracy, *Handbook of High Speed Machining Technology*, Chapman and Hall, New York, 1985, 140–153
52. D. Montgomery and Y. Altintas, Dynamic peripheral milling of flexible structures, *ASME J. Eng. Ind.* **114** (1992) 137–145
53. D. Montgomery and Y. Altintas, Mechanism of cutting force and surface generation in dynamic milling, *ASME J. Eng. Ind.* **113** (1991) 160–168
54. F. Ismail, M.A. Elbestawi, R. Du, and K. Urbasik, Generation of milled surface including tool dynamics and wear, *ASME J. Eng. Ind.* **115** (1993) 245–252
55. M.A. Elbestawi, F. Ismail, and K.M. Yuen, Surface topography characterization in finish milling, *Int. J. Mach. Tools Manuf.* **34** (1994) 245–255
56. M.Y. Yang and H.D. Park, The prediction of cutting force in ball-end milling, *Int. J. Mach. Tools Manuf.* **31** (1991) 45–54
57. E.M. Lim, H.Y. Feng, and C.H. Menq, The prediction of dimensional errors for machining sculptured surfaces using ball-end milling, *Manufacturing Science and Engineering*, ASME PED Vol. 64, ASME, New York, 1993, 149–156
58. G. Yucesan and Y. Altintas, Mechanics of ball end milling process, *Manufacturing Science and Engineering*, ASME PED Vol. 64, ASME, New York, 1993, 543–551
59. C.G. Sim and M.Y. Yang, The prediction of the cutting force in ball end milling with a flexible cutter, *Int. J. Mach. Tools Manuf.* **33** (1993) 267–284
60. V. Chandrasekharan, S.G. Kapoor, and R.E. DeVor, A mechanistic approach to predicting the cutting forces in drilling: with application to fiber-reinforced composite materials, *ASME J. Eng. Ind.* **117** (1995) 559–570
61. V. Chandrasekharan, S.G. Kapoor, and R.E. DeVor, A mechanistic model to predict the cutting force system for arbitrary drill point geometry, *ASME J. Manuf. Sci. Eng.* **120** (1998) 563–570
62. E.J.A. Armarego and C.Y. Chen, Drilling with flat rake face and conventional twist drills — 1. Theoretical investigation, *Int. J. MTDR* **12** (1972) 17–35
63. E.J.A. Armarego and S. Wiriyacosol, Thrust and torque prediction in drilling from a cutting mechanics approach, *CIRP Ann.* **28** (1979) 87–91
64. A.R. Watson, Drilling model for cutting lip and chisel edge and comparison of experimental and predicted results. Parts I–IV, *Int. J. MTDR* **25** (1985) 347–404
65. D.A. Stephenson and S.M. Wu, Computer models for the mechanics of three-dimensional cutting processes. Part II. Results for oblique end turning and drilling, *ASME J. Eng. Ind.* **110** (1988) 38–43
66. K.J. Fuh, Computer-Aided Design and Manufacture of Multifacet Drills, PhD Thesis, Mechanical Engineering, University of Wisconsin, 1987, Chapter 6

67. D.A. Stephenson, Material characterization for metal cutting force modeling, *ASME J. Eng. Mater. Technol.* **111** (1989) 210–219
68. D.A. Stephenson and J.S. Agapiou, Calculation of main cutting edge forces and torque for drills with arbitrary point geometries, *Int. J. Mach. Tools Manuf.* **32** (1992) 521–538
69. D.M. Rincon, Coupled Force and Vibration Modeling of Drills with Complex Cross Sectional Geometries, PhD Thesis, Mechanical Engineering, University of Michigan, 1994, Chapter 3
70. H.T. Huang, C.I. Weng, and C.K. Chen, Prediction of thrust and torque for multifacet drills (MFD), *ASME J. Eng. Ind.* **116** (1994) 1–7
71. D.F. Galloway, Some experiments on the influence of various factors on drill performance, *ASME Trans.* **79** (1957) 191–231
72. R.A. Williams, A study of the basic mechanics of the chisel edge of twist drills, *Int. J. Prod. Res.* **8** (1970) 325–343
73. R.A. Williams, A study of the drilling process, *ASME J. Eng. Ind.* **96** (1974) 1207–1215
74. S. Bera and A. Bhattacharyya, On the determination of the torque and thrust during drilling of ductile materials, *Proceedings of the 8th International MTDR Conference*, Birmingham, UK, 1967, 879–892
75. C.A. Mauch and L.K. Lauderbaugh, Modeling the drilling process — an analytical model to predict thrust force and torque, *Sensors and Controls in Manufacturing*, ASME PED Vol. 20, ASME, New York, 1990
76. E. Magrab and D.E. Glisin, Buckling loads and natural frequencies of twist drills, *ASME J. Eng. Ind.* **196** (1984) 196–204
77. O. Tekinalp and A.G. Ulsoy, Modeling and finite element analysis of drill bit vibrations, *ASME J. Vibr. Acoust. Stress Reliabil. Des.* **111** (1989) 148–155
78. O. Tekinalp and A.G. Ulsoy, Effects of geometric and process parameters on drill transverse vibrations, *ASME J. Eng. Ind.* **112** (1990) 189–194
79. M. McColl and R. Ledbetter, CAD applied to twist drills, *Proc. Inst. Mech. Eng.* **207B** (1993) 251–256
80. J.Z. Yan, V. Jaganathan, and R. Du, A new dynamic model for drilling and reaming processes, *Int. J. Mach. Tools Manuf.* **42** (2002) 299–311
81. P.V. Bayly, M.T. Lamar, and S.G. Calvert, Low-frequency regenerative vibration and the formation of lobed holes in drilling, *ASME J. Manuf. Sci. Eng.* **124** (2002) 275–285
82. L.V. Colwell, Predicting the angle of chip flow for single point cutting tools, *ASME Trans.* **76** (1954) 199–204
83. H.T. Young, P. Mathew, and P.L.B. Oxley, Allowing for nose radius effects in predicting the chip flow direction and cutting forces in bar turning, *Proc. Inst. Mech. Eng.* **C201** (1987) 213–226
84. J. Wang, Development of a chip flow model for turning operations, *Int. J. Mach. Tools Manuf.* **41** (2001) 1265–1274
85. D.A. Stephenson and J.W. Matthews, Cutting forces when turning and milling cast iron with silicon nitride tools, *Trans. NAMRI/SME* **21** (1993) 223–230
86. E. Budak and Y. Altintas, Prediction of milling force coefficients from orthogonal cutting data, *Manufacturing Science and Engineering*, ASME PED Vol. 64, ASME, New York, 1993, 453–459
87. S. Jayaram, S.G. Kapoor, and R.E. DeVor, Estimation of the specific cutting pressures for mechanistic cutting force models, *Int. J. Mach. Tools Manuf.* **41** (2001) 265–281
88. D.A. Stephenson, Casting and Machining Process Analysis at GM Powertrain, SAE Technical Paper 2002-01-0622, 2002
89. D.R. White, Machining applications at GM Powertrain, *Proceedings of the 2002 Third Wave AdvantEdge User's Conference*, Atlanta, GA, April 2002, Paper 6
90. D.A. Stephenson, Machining process analysis at General Motors Powertrain, *Proceedings of the 2004 Third Wave AdvantEdge User's Conference*, Gaithersburg, MD, March 2004, Paper 2
91. D. T.-Y. Huang and J.-J. Lee, On obtaining machine tool siffness by CAE techniques, *Int. J. Mach. Tools Manuf.* **41** (2001) 1141–1163
92. J.S. Agapiou, E. Steinhilper, F. Gu, and P. Bandyopadhyay, Modeling machining errors on a transfer line to predict quality, *SME J. Manuf. Processes* **5** (2003) 1–12
93. J.Q. Xie, J.S. Agapiou, D.A. Stephenson, and P.M. Hilber, Machining quality analysis of an engine cylinder head using finite element methods, *SME J. Manuf. Processes* **5** (2003) 170–184

9 Tool Wear and Tool Life

9.1 INTRODUCTION

Tool life is often the most important practical consideration in selecting cutting tools and cutting conditions. Tools which wear or otherwise fail slowly have comparatively long service lives, resulting in reduced production costs and more consistent dimensional and surface finish capability. For these reasons, tool life is the most common criterion used to rate cutting tool performance and the machinability of materials.

An understanding of tool life requires an understanding of the ways in which tools fail. Broadly, tool failure may result from wear, plastic deformation, or fracture. Tool wear may be classified by the region of the tool affected or by the physical mechanisms which produce it. The dominant type of wear in either case depends largely on the tool material. The common types of wear and wear measurement methods are discussed in Section 9.2 and Section 9.3; tool wear mechanisms and material aspects of wear are discussed in Section 9.4 and Section 9.5. Tools deform plastically or fracture when they are unable to support the loads generated during chip formation. The physical mechanisms leading to plastic deformation are similar to those resulting in wear and are also discussed in Section 9.4 and Section 9.5. Common fracture mechanisms are discussed in Section 9.9. The general discussions in these sections are applicable primarily to continuous turning operations. Special wear and fracture concerns in drilling and in milling are discussed in Section 9.10 and Section 9.11.

A primary goal of metal cutting research has been to develop methods of predicting tool life from a consideration of tool failure mechanisms. One of the objectives in this chapter is to provide a qualitative understanding of the basic physics of tool wear so that the gradual wear of a tool can be described quantitatively. Tool life testing and methods for predicting tool life are discussed in Section 9.6–Section 9.8. In considering these topics, it is important to bear in mind the distinction between tool wear and tool life. It is often possible to predict tool wear rates based on test data or physical considerations, but this does not translate into a prediction of tool life in a general sense, since tool life depends strongly on part requirements. In practice, tools are removed from service when they no longer produce an acceptable part. This may occur when the part's dimensional accuracy, form accuracy, or surface finish are out of tolerance, when an unacceptable burr or other edge condition is produced, or when there is an unacceptable probability of gross failure due to an increase in cutting forces or power. Tools used under the same conditions in different operations may have quite different usable lives depending on critical tolerances or requirements. Because of this fact, methods of predicting tool life are useful primarily for comparative purposes, for example, in ranking expected levels of tool life for different work materials, tool materials, or cutting conditions; realistically, they cannot be expected to yield an accurate estimate of tool life in a given application unless prior application data for similar parts are available.

Because of the practical importance of tool life, the subject may be discussed from a variety of viewpoints in a general survey of machining theory and practice. The purpose of this chapter is to clarify the physical mechanisms which lead to tool failure and to identify general strategies for reducing failure rates and increasing tool life. The discussion is meant to complement related discussions on tool life concerns when selecting tool materials (Chapter 4) and when rating machinability (Chapter 11).

9.2 TYPES OF TOOL WEAR

The principal types of tool wear, classified according to the regions of the tool they affect, are shown in Figure 9.1 [1–12]. Wear occurs on both the rake and relief faces of the tool due to several mechanisms as discussed in Section 9.3. Wear on the relief face is called *flank wear* (Figure 9.1a) and results in the formation of a wear land (Figure 9.2 and Figure 9.3). Rubbing of the wear land against the machined surface damages the surface and produces large flank forces which increase deflections and reduce dimensional accuracy. As discussed in the next section, flank wear most commonly results from abrasion of the cutting edge. The extent of flank wear is characterized by the average or maximum land width (Figure 9.2). The flank wear rate changes with time as shown in Figure 9.4. After an initial wearing in period corresponding to the initial rounding of the cutting edge, flank wear increases slowly at a steady rate until a critical land width is reached, after which wear accelerates and becomes severe. Flank wear progress can be monitored in production by examining the tool or (more commonly) by tracking the change in size of the tool or machined part. The flank wear land is generally of uniform width, with thicker sections occurring near the ends. Flank wear can be minimized by increasing the abrasion and deformation resistance of the tool material, and by the use of hard coatings on the tool.

Rake face or *crater wear* (Figure 9.1b) produces a wear crater on the tool face (Figure 9.2 and Figure 9.5). Moderate crater wear usually does not limit tool life; in fact, crater formation increases the effective rake angle of the tool and thus may reduce cutting forces. However, excessive crater wear weakens the cutting edge and can lead to deformation or fracture of the tool, and should be avoided because it shortens tool life and makes resharpening the tool difficult. The extent of crater wear is characterized by the crater depth KT. Crater wear also varies with time in a manner similar to flank wear. As discussed in Section 9.4, severe crater wear usually results from temperature-activated diffusion or chemical wear mechanisms. Crater wear can be minimized by increasing the chemical stability of the tool material or by decreasing the tool's chemical solubility in the chip; this can be done by applying coatings as discussed in Chapter 3. Reducing the cutting speed is also effective in controlling crater wear.

Tools used in rough turning often develop *notch wear* (Figure 9.1c and Figure 9.6) on the tool face, especially at the point of contact between the tool and the unmachined part surface or free edge of the chip. Depth of cut notching usually results from abrasion [13] and is especially common when cutting parts with a hard surface layer or scale, or work-hardening materials which produce an abrasive chip (e.g., stainless steels and nickel-based alloys). Notch wear may also result from oxidation if a coolant is used, or by chemical reactions or corrosion at the interface between the tool and the atmosphere [7]. Severe notch wear makes resharpening the tool difficult and can lead to tool fracture, especially with ceramic tools. Notch wear can be reduced by increasing the lead angle, which increases the area of contact between the tool and part surface, by varying the depth of cut in multipass operations, and by increasing the hot hardness and deformation resistance of the tool material.

Nose radius wear (Figure 9.1d and Figure 9.3) occurs on the nose radius of the tool, on the trailing edge near the end of the relief face. It resembles a combined form of flank and notch wear, and results primarily from abrasion and corrosion or oxidation [14]. Severe nose radius wear degrades the machined surface finish.

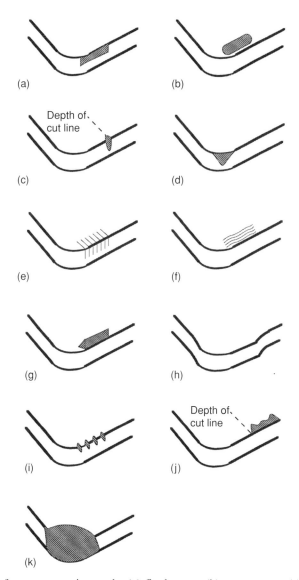

FIGURE 9.1 Types of wear on cutting tools: (a) flank wear; (b) crater wear; (c) notch wear; (d) nose radius wear; (e) comb (thermal) cracks; (f) parallel (mechanical) cracks; (g) built-up edge; (h) gross plastic deformation; (i) edge chipping or frittering; (j) chip hammering; (k) gross fracture.

Thermal and mechanical cracking (Figure 9.1e,f and Figure 9.7) usually results from cyclic loading of the tool in interrupted cutting or when machining materials which generate high tool-chip temperatures. Two types of cracks may occur: cracks perpendicular to the cutting edge, which usually result from cyclic thermal loads, especially when a coolant is used, and cracks parallel to the cutting edge, which usually result from cyclic mechanical loads. Crack formation leads to rapid tool fracture or chipping. Tool failure due to cracking in interrupted cutting is discussed in more detail Section 9.11.

Edge buildup (Figure 9.1g and Figure 9.8) most often occurs when cutting soft metals, such as aluminum alloys, at low cutting speeds. It results when metal adheres strongly to the cutting edge, building up and projecting forward from it. As discussed in Section 9.10, buildup is also a serious problem in drilling operations; it may occur at the outer corner of the spiral point drills

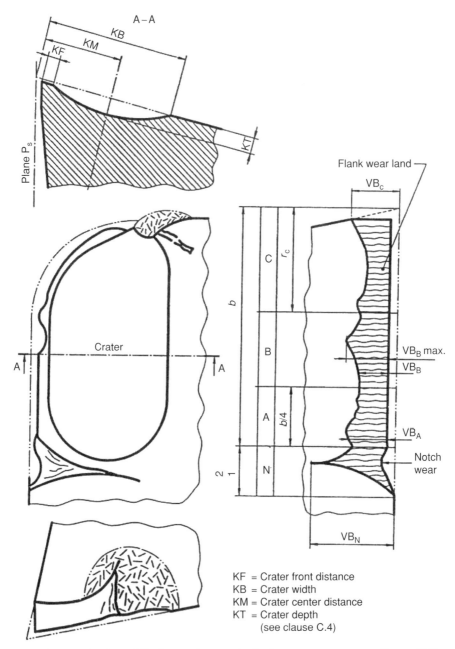

FIGURE 9.2 Characterization of flank wear land and rake face wear crater according to ISO standard 3685. (After Ref [9].)

because the chip becomes thin at this point. Built-up edge (BUE) formation is undesirable because it changes the effective depth of cut (or hole diameter) and because it is often unstable, leading to poor surface finish and tool chipping. The mechanics of BUE formation is described in Section 6.14. BUE formation can be minimized by using a more positive rake angle, tools with smooth surface finishes (<5 to 10 μm), coolant with increased lubricity, higher-pressure coolants directed on the rake face, and higher cutting speeds.

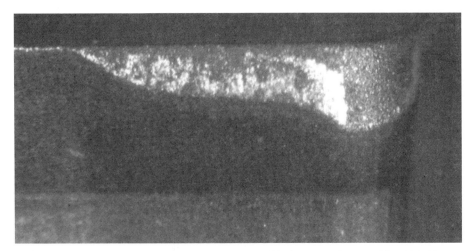

FIGURE 9.3 Severe flank and nose radius wear on a carbide insert used to machine 390 Al.

Plastic deformation (Figure 9.1h) of the cutting edge occurs when the tool is unable to support the cutting pressure over the area of contact between the tool and chip. Cutting edge deformation usually occurs at high feed rates, which produce high cutting edge loads, or at higher cutting speeds, since the hardness of the tool decreases with increasing cutting speed and temperature. Excessive cutting edge deformation results in a loss of dimensional accuracy, poor surface finish, and severe flank wear or tool fracture.

Edge chipping or *frittering* (Figure 9.1i) occurs when cutting with brittle tool materials, especially ceramics and polycrystalline compacts, or when cutting work materials which include hard or abrasive particles, such as metal matrix composites or aluminum–silicon alloys. Vibration due to excessive cutting forces or low system stiffness can also lead to chipping. Chipping results in poor surface finish and increased flank wear and may lead to tool breakage. As discussed in Section 9.9, chipping can often be controlled by changing the tool edge preparation or by increasing the fracture strength of the tool material.

Chip hammering (Figure 9.1j) occurs when cutting materials that form a tough or abrasive chip (e.g., stainless steels, nicked-based alloys) with ceramic tools. It occurs when the chip curls back and strikes the tool face away from the cutting edge. It leads to chipping or pitting of the

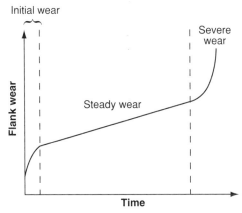

FIGURE 9.4 Variation of the flank wear rate with cutting time, showing the initial wear, steady wear, and severe wear periods.

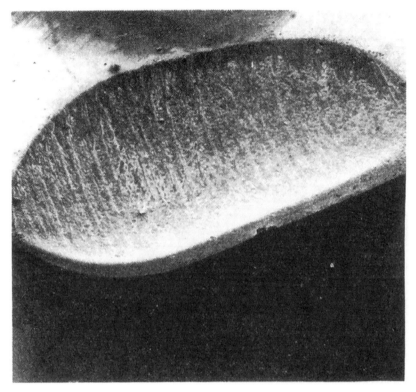

FIGURE 9.5 Crater wear on a cemented carbide tool used to machine plain carbon steel. (After A.T. Santhanam et al., [2].)

tool surfaces and in extreme cases to tool fracture. Chip hammering results from improper chip control and can be eliminated by changing the lead angle, depth of cut, feed rate, or tool nose radius to alter the chip flow direction.

Tool fracture or breakage (Figure 9.1k) results in the catastrophic loss of the cutting edge and a substantial portion of the tool. The causes of fracture are discussed in Section 9.9. General strategies for eliminating fracture include reducing cutting forces, using stronger or more rigid tooling setups, and using tools with increased fracture toughness.

9.3 MEASUREMENT OF TOOL WEAR

Flank and crater wear are the most important and thus the most widely measured forms of tool wear. Flank wear is most commonly used for tool wear monitoring since it occurs in virtually all machining operations.

Tool wear is most commonly measured by examining the wear scar on the tool using a microscope or stylus tracing instrument. Toolmaker's microscopes with micrometer stages, often fitted with computer control and imaging systems, are widely used (Figure 9.3 and Figure 9.5–Figure 9.8). Computer-controlled systems are particularly convenient for rapid and accurate studies; a disadvantage, however, is that they normally provide only two-dimensional information. Stylus instruments are similar to profilometers (Chapter 10), but may have a longer traverse length and are often equipped with ground diamond styluses to reduce errors due to stylus radius effects [10]. These require greater time and skill to operate but can provide a three-dimensional map of the wear scar.

FIGURE 9.6 Depth of cut notch on a cemented carbide tool used to machine a nickel alloy. (After A.T. Santhanam et al., [2].)

Since the initially sharp cutting edge becomes rounded and the original edge is destroyed, it is sometimes necessary to establish a datum line from which flank wear progress can be measured. Video imaging procedures provide a means of comparing sharp and worn edges which establish and preserve a datum line. Similarly, photographs of the cutting edge at intervals during the tool's life can be used to record flank wear progress. Other techniques, such as gradually lapping down the tool from the top edge [15] and optical contour mapping, can also be used [15, 16].

On the tool flank, the average and maximum values of the flank wear land width, VB_{ave} and VB_{max}, are normally measured (Figure 9.2) [9]. (If significant notch wear is present, the notch depth is measured separately, and VB_{max} is defined as the maximum width of the central section of the wear land.) The volume of material worn off the flank, v_w, can be calculated approximately from the average land width, VB_{ave}, using the equation [17]

$$v_w = \frac{VB_{ave}^2 b \tan \theta}{2} \tag{9.1}$$

where θ is the relief angle and b is the width of cut. As discussed in Section 9.6, most standard tool life tests use flank wear criteria to define the end of tool life.

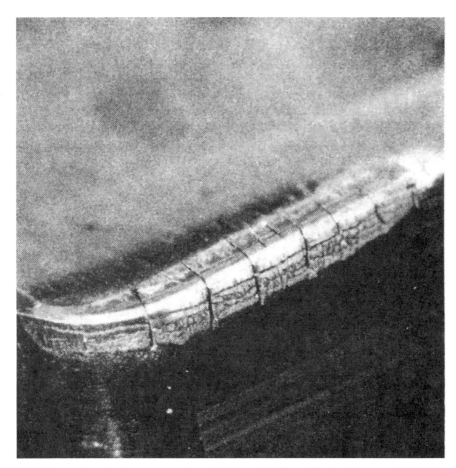

FIGURE 9.7 Thermal cracks in a cemented carbide insert. (After A.T. Santhanam et al., [2].)

On the rake face, the crater width KB, crater depth KT, and crater land width KF are most commonly measured (Figure 9.2). The total volume of the wear crater, v_{cr}, can be calculated approximately from the equation [17]

$$v_{cr} = \frac{2b(\mathrm{KB} - \mathrm{KF})\mathrm{KT}}{3} \quad (9.2)$$

The reduction of the volume of the cutting tool is sometimes used to indicate the degree of tool wear. Methods used for measuring the wear volume include: (1) geometric determination of the flank and crater wear volumes using Equations (9.1) and (9.2); (2) weighing the tool [18, 19]; and (3) radiotracer methods [20] as described below.

The specific wear rate η can be calculated by assuming that the volume of tool material worn away is proportional to the area of the contact surface, A, and the length of sliding, L_s:

$$\eta = \frac{1}{A} \frac{dv}{dL_s} \quad (9.3)$$

The specific wear rate for flank wear, η_f, is [21, 22]

Tool Wear and Tool Life

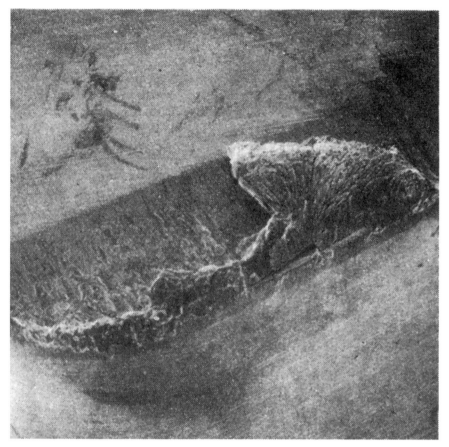

FIGURE 9.8 Edge buildup on a cemented carbide tool used to machine a nickel alloy at a low cutting speed. (After A.T. Santhanam et al., [2].)

$$\eta_f = \sin\theta \frac{d\,VB}{dL_s} \tag{9.4}$$

where θ is the relief angle. The specific wear rate for crater wear, η_{cr}, is

$$\eta_{cr} = \frac{2}{3}\frac{d\,KT}{dL_c} \tag{9.5}$$

where L_c is the tool–chip contact length.

The crater wear grows both in depth (KT) and width (KB − KF). As crater wear progresses, the width of the land between the crater and cutting edge, KF, and the location of the maximum depth, KM, changes. As KF decreases and the crater boundary merges into the cutting edge, the effective rake angle α_{eff} approaches

$$\alpha_{eff} = \alpha_0 + \arctan\left(\frac{KT}{KM}\right) \tag{9.6}$$

where α_0 is the nominal rake angle. As KT decreases, the edge weakens and catastrophic failure becomes more likely.

Microscope and stylus measurements are time-consuming and cannot be used for tool wear monitoring. A great variety of online tool wear sensing methods based on optical, pneumatic, electric, displacement, and force measurements have been proposed. A detailed discussion of these methods is beyond the scope of this book and can be found in review articles [23–26]. Of the available methods, those based on force and power measurements seem to be the most practical. The axial and radial forces in turning are much more sensitive to flank wear than the tangential force, so that the ratio of the axial or radial force to the tangential force is often strongly correlated to tool flank wear.

Beginning in the 1950s, radiotracer methods were developed for measuring instantaneous wear rates [20, 27–30]. In these methods, the tool was irradiated so that some of its constituents became radioactive (e.g., the carbon in WC tools). A Geiger counter was then placed next to the chip stream to monitor how much material was being removed from the tool as a function of time. These methods had excellent resolution and provided early convincing evidence of diffusion wear of tools. This method is difficult and expensive, however, and is not widely used today.

9.4 TOOL WEAR MECHANISMS

The physical mechanisms which produce the various types of wear described in Section 9.2 depend on the materials involved, tool geometry, and the cutting conditions, especially the cutting speed.

Adhesive and abrasive wear are the most significant types of wear at lower cutting speeds. *Adhesive* or *attritional wear* [3, 7, 8] occurs when small particles of the tool adhere or weld to the chip due to friction and are removed from the tool surface. It occurs primarily on the rake face of the tool and contributes to the formation of a wear crater. Adhesive wear rates are usually low, so that this form of wear is not normally practically significant. However, significant adhesive wear may accompany BUE formation, since the BUE is also caused by adhesion, and can result in chipping of the tool. As noted in the previous section, this occurs primarily when cutting soft work materials at low speeds, and in drilling. Methods for reducing BUE formation are discussed in Section 6.14.

Abrasive wear occurs when hard particles abrade and remove material from the tool. The abrasive particles may be contained in the chip, as with adhering sand in sand-cast parts, iron carbides from foundry chill in cast iron, martensite, austenite, and other hard phases in steels, free silicon particles in aluminum–silicon alloys, and fibers in metal-matrix composites [31, 32]. They may also result from the chip form or from a chemical reaction between the chips and cutting fluid, as with powder metal steels (which form a powdery chip) or cast irons alloyed with chromium. Abrasion occurs primarily on the flank surface of the tool. Abrasive wear by hard particles entrained in the cutting fluid is sometimes called *erosive wear*. Abrasive wear is usually the primary cause of flank wear, notch wear, and nose radius wear, and as such is often the form of wear which controls tool life, especially at low to medium cutting speeds. Abrasive wear also affects crater wear [33].

Both adhesive and abrasive wear can be described quantitatively by an equation of the form [34]

$$v = \frac{k_w N L_s}{H} \quad (9.7)$$

where v is the volume of material worn away, k_w is the wear coefficient, N is the force normal to the sliding interface, L_s is the distance slid, and H is the penetration hardness of the tool.

This equation shows that an effective method of controlling wear is to increase the hardness H of the tool. This can be done directly, by using tools made of harder materials, or indirectly, by coating the tool with a hard surface layer. Equation (9.7) also indicates that reducing cutting forces, which reduces N, should also reduce wear due to these mechanisms. This is most easily done by increasing the tool rake angle, although this may also reduce the cutting edge strength and lead to tool deformation or chipping.

As the cutting speed is increased, adhesive and abrasive wear rates increase for two reasons. First, the distance slid in a given time, L_s, also increases with the cutting speed. Second, increasing the cutting speed increases cutting temperatures. The hardness of the tool material typically drops with increasing temperature. This phenomenon, known as *thermal softening*, not only leads to increased abrasive wear, but may also result in plastic deformation of the cutting edge. Typical hardness–temperature relations for cemented carbides and ceramics are shown in Figure 9.9. Tools made of materials other than high-speed steel are composed of hard particles held together by a metallic or ceramic binder; the binder, rather than the hard particles, controls thermal softening behavior. This form of wear can be reduced by reducing the binder content or altering the binder composition which increases the tool's hot hardness, or by reducing the cutting speed to reduce cutting temperatures.

As the cutting speed is increased further, temperature-activated wear mechanisms become predominant. These include diffusion, oxidation, and chemical wear. These forms of wear depend on the chemical compatibility of the tool and workpiece materials; the cutting speeds at which these forms of wear become significant depend in addition on the tool–chip interface temperature and the melting temperature of the chip. Generally, these forms of wear can be reduced by machining at lower speeds to reduce cutting temperatures, or by coating the tool with a hard layer of chemically inert material to act as a buffer between the tool substrate and the chip, cutting fluid, and atmosphere [14].

In *diffusion* or *solution wear*, a constituent of the tool material diffuses into or forms a solid solution with the chip material. This weakens the tool surface and results in a wear crater on the rake face of the tool. Severe cratering ultimately leads to tool failure due to

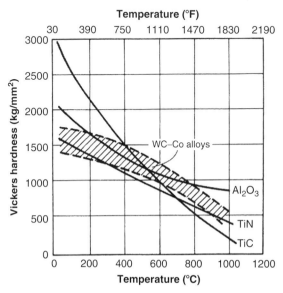

FIGURE 9.9 Variation of penetration hardness with temperature for cemented carbide and ceramic work materials. (After A.T. Santhanam et al., [2].)

breakage. The diffusion wear rate depends primarily on the solubility of the tool material in the work material and the contact time between the tool and chip at elevated temperatures, and increases exponentially as the cutting temperature increases. Diffusion wear can be reduced by changing tool materials to a less soluble grade.

Oxidation occurs when constituents of the tool (especially the binder) react with atmospheric oxygen. It most often occurs near the free surface of the part, where the hot portion of the tool in and around the tool–chip contact region is exposed to the atmosphere. Oxidation often results in severe depth of cut notch formation and can be recognized by the fact that the tool material is typically discolored in the region near the notch. Oxidation of wear debris or particles of the work material may also result in the production of hard oxide particles which increase abrasive wear. Oxidation does not occur with aluminum oxide based ceramic tools.

Chemical wear or *corrosion*, caused by chemical reactions between constituents of the tool and the workpiece or cutting fluid, produces both flank and crater wear, with flank wear dominating as the cutting speed is increased. Chemical wear scars are smooth compared to wear scars produced by other mechanisms and may appear to be deliberately ground into the tool. This type of wear is commonly observed when machining highly reactive materials such as titanium alloys [35]. Chemical wear may also result from reactions with additives [e.g., free sulfur or chlorinated extreme pressure (EP) additives] in the cutting fluid. (EP additives, in fact, are used to reduce adhesive wear by producing controlled chemical wear [14].) The surface layer of the tool is changed to the reaction product, which is typically soft and wears rapidly by abrasion. Changing the tool material (or coating) or the additives in the cutting fluid will often reduce this type of wear.

As noted above and in Chapter 4, coating the tool with a thin layer of hard, chemically inert material can reduce abrasive, diffusion, oxidation, and chemical wear and permit the tool to be used at higher cutting speeds. The coating itself generally wears by abrasion, but may also fail by *spalling* due to mechanical or thermal fatigue. Advances in coating application methods have reduced the incidence of spalling in recent years. When the coating is worn away, other forms of wear attack the substrate of the tool; since coated tools are used at higher cutting speeds than uncoated tools, tool failure usually follows rapidly. Coating failure can be prevented by changing the coating materials or thickness, by machining dry, and by changing the exit angle in interrupted cuts.

Methods for diagnosing and reducing various forms of tool wear are summarized in Table 9.1. Reducing the cutting speed is effective in reducing many types of wear, but since this reduces the metal removal rate, it should be considered only as a last resort in most operations. It should be noted, however, that reducing the cutting speed has no impact on productivity in many transfer machining operations, specifically those in which the cutting time is substantially lower than the cycle time of the line. In transfer machine operations, the cycle time should be determined by optimizing the speed and feed in critical operations; in the other operations, the cutting speed should be reduced to use all of the available cutting time and increase tool life.

9.5 TOOL WEAR — MATERIAL CONSIDERATIONS

The types and mechanisms of tool wear were discussed from a general viewpoint in Section 9.2 and Section 9.4. The dominant form of wear in a given application depends on a number of factors, the most important of which are the tool material, work material, and cutting speed. Not all types or mechanisms of wear are observed with all tool materials, and chemical, oxidation, and diffusion wear are particularly sensitive to the materials involved. In this section, the common types of wear encountered when machining with tools of specific materials are briefly discussed. The discussion focuses in particular on how various failure modes limit the range of application of tools made of specific materials. A detailed description

TABLE 9.1
Mechanisms, Characteristics, and Countermeasures for Common Types of Tool Wear and Failure

Type of Wear	Mechanism	Characteristics	Countermeasures
Flank wear	Abrasion	Even wear scar	Use harder tool material
			Use coated tool
			Filter cutting fluid
			Clean parts
			Refine part microstructure
			Reduce feed rate (*)
	Thermal softening		Reduce speed (*)
	Cutting above center height	Poor finish	Check insert height
	Edge deformation	Edge deformed	See below under "Edge deformation"
	Feed too low	Poor finish	Increase feed
Crater wear	Diffusion	Rapid wear rate	Reduce cutting speed (*)
			Improve cooling ability of coolant
			Increase coolant volume and pressure
			Direct coolant toward chip–tool interface
	Chemical wear	Smooth wear scar	Change tool or coating material or coolant
	Feed too low	Poor finish	Increase feed rate
Notch wear	Abrasion	Occurs at part free surface	Vary depth of cut
			Use harder tool
			Increase lead angle
	Oxidation	Discoloration	Change coolant
			Reduce speed (*)
			Change tool material or coating
Nose radius wear	Abrasion	Rough, uneven scar	Reduce feed (*)
			Use harder tool
			Increase nose radius
Edge cracking	Thermal fatigue	Cracks normal to edge	Reduce cutting speed (*)
			Machine dry
			Use tougher tool
	Mechanical fatigue	Cracks parallel to edge	Reduce feed (*)
			Use tougher tool
Built-up edge	Adhesion	Poor surface finish	Increase cutting speed
			Increase rake angle
			Increase lubricity of coolant
Edge deformation	Overload	Occurs rapidly	Reduce feed (*)
	Thermal softening		Use harder tool
			Reduce speed (*)
			Increase coolant supply
			Use low-friction coating
Edge chipping	Abrasion (hard spots)		Inspect work material
			Use tougher tool
			Use stronger edge preparation
			Reduce feed (*)
	Improper chip breaker		Increase chip-breaker land width
	Vibration	Noise, chatter marks	Increase system stiffness

continues

TABLE 9.1
Mechanisms, Characteristics, and Countermeasures for Common Types of Tool Wear and Failure — Continued

Type of Wear	Mechanism	Characteristics	Countermeasures
			Use tougher tool
			Use stronger edge preparation
			Reduce depth of cut (*)
	Intermittent coolant		Increase coolant supply or cut dry
	Improper seating		Check seat condition
	Adhesion	Built-up edge	See above under "Built-up edge"
Chip hammering	Improper chip flow	Damage away from edge	Change chip flow direction (change lead angle or tool nose radius)
Gross fracture	Overload	Occurs rapidly	Use tougher tool
			Increase nose radius
			Use stronger edge preparation
			Reduce feed (*)
			Reduce depth of cut (*)
	Vibration	Noise, chatter marks	Increase system stiffness
			Use tougher tool
			Decrease nose radius
			Use stronger edge preparation
			Reduce depth of cut (*)

Note: This table presents general information which is most directly applicable to continuous turning and boring. Countermeasures marked with an asterisk (*) reduce the metal removal rate and should be used only as a last resort in general-purpose machining.

of all possible failure mechanisms, especially those due to chemical wear, is beyond the scope of this book but can be found in many of the references cited.

High-speed steel (HSS) tools most commonly fail by abrasion, often aggravated by thermal softening, by plastic deformation, and by adhesion and edge buildup [36]. The hot hardness of HSS decreases rapidly at temperatures above roughly 540°C [37]; at cutting speeds which produce temperatures in excess of this level, rapid abrasive wear and plastic deformation occur. This limits the usable speed for HSS tools to roughly 35 m/min when machining soft steels. Adding cobalt to HSS increases the hot hardness and permits their use at somewhat higher cutting speeds (up to 50 m/min). When machining aluminum alloys and other nonferrous metals, which generally melt at temperatures below 600°C, thermal softening does not usually limit tool life. For these materials, abrasion due to hard particles in the work material (e.g., adhering sand in sand-cast parts and SiC fibers in metal-matrix composites), BUE formation, or burring limit tool life. BUE formation is a particularly serious problem for HSS tools because the major constituent of the tool material, iron, is a metal with a comparatively high chemical affinity with, and thus tendency to adhere to, common work materials. Coating HSS tools with TiN or other thin ceramic layers can increase their resistance to abrasion and BUE formation, permitting their use at higher speeds. The tempering temperature of a HSS tool also has a major influence on its wear resistance and performance [38]. HSS tools generally fail by thermal softening before temperatures are reached at which diffusion or chemical wear become significant. Also, tools made of HSS generally do not chip because of the material's high fracture toughness.

Sintered tungsten carbide (WC) tools may fail by abrasion, edge chipping, plastic deformation, and diffusion, oxidation, and chemical wear. Abrasion and plastic deformation accelerate at cutting temperatures above roughly 700°C, the temperature range in which the hot hardness of most grades begins to decrease rapidly (Figure 9.9) [2]. (The hot hardness of specific carbide grades depends on the carbide grain size and binder content, and micrograin grades in particular are effective at higher temperatures.) For low-carbon steels and cast irons, temperatures of this magnitude occur at cutting speeds above roughly 80 m/min [15, 39]. Chipping most commonly results from hard inclusions in the work material or from tool vibrations caused by inadequate system stiffness.

Diffusion wear occurs when cutting ferrous work materials with uncoated WC tools when the cutting temperature exceeds 750°C [8], which for low-carbon steels usually occurs at cutting speeds above 150 m/min. Steel has a high affinity for carbon, which diffuses out of the hard WC particles and into the chip stream. Diffusion can be inhibited by adding a small amount of TiC or TaC to the WC, since this lowers the solubility of the carbide phase in iron [7, 40]. Steel cutting carbide grades therefore commonly include such additives. Diffusion may also be observed when cutting cast irons with WC tools, but at higher cutting speeds (roughly 200 m/min) because iron produces shorter chips and thus lower tool–chip interface temperatures.

Chemical wear may result from reactions with the work material or the cutting fluid. Fluids containing high concentrations of free sulfur additives are particularly likely to produce chemical wear; in this case, the cobalt binder in the tool reacts with the sulfur to form cobalt sulfide, a soft salt which is rapidly removed by abrasion [1]. Cutting fluids containing chemically combined rather than free sulfur should be used to eliminate this mechanism of wear. When machining titanium alloys, which are highly reactive, the carbon in both WC and polycrystalline diamond (PCD) tools may react with the chip to form a titanium carbide interlayer, which promotes rapid diffusion wear [35].

Oxidation wear also usually attacks the cobalt binder of WC tools [1]. Atmospheric oxygen penetrates the tool–chip contact primarily at the free surface of the workpiece, producing a wear notch. In the interior of the cut, oxidation may result from air entrained in the cutting fluid, since most methods of applying the cutting fluid result in aeration of the fluid. Once oxygen combines with the binder, the WC particles are removed rapidly from the weakened tool matrix. Oxidation wear can be distinguished from other forms of chemical wear by discoloration of the tool material near the wear scar. Oxidation is generally insignificant at temperatures below 700°C [14].

Coated WC tools exhibit a two-stage wear process [2, 41–43]. While the coating is intact, abrasive wear due to hard particles in the workpiece predominates. The wear rate is lower than for uncoated tools because thermal softening is less pronounced. Eventually the coating fails either through excessive abrasive wear or through spalling, which is particularly common in interrupted processes such as milling. Since coated tools are used at higher cutting speeds than uncoated tools, rapid cratering of the substrate due to diffusion or chemical wear usually follows coating failure.

Coating life depends on the coating material, thickness, and method of deposition. The two most common classes of coating materials are TiN-based (gold) coatings and Al_2O_3-based (black) coatings. (As discussed in detail in Chapter 4, other coating materials are also used, and multiphase coatings consisting of layers of various materials are common.) Al_2O_3-based coatings are harder (Figure 9.9) and more chemically inert at high temperatures and provide better abrasion and cratering resistance; they are preferred when cutting cast and nodular irons. Gold coatings, however, reduce friction and thus cutting temperatures and flank wear at higher cutting speeds. They provide longer tool life when cutting steels and cast iron at high cutting speeds (Figure 9.10) [2]. Gold coatings are also more commonly used when cutting aluminum alloys because they are less likely to produce a BUE.

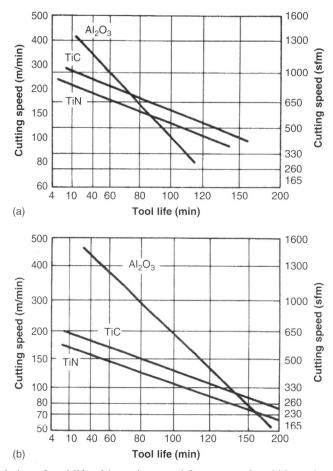

FIGURE 9.10 Variation of tool life with cutting speed for cemented carbide tools coated with various materials when cutting 1045 steel (a) and gray cast iron (b). Al_2O_3 coatings provide superior abrasion resistance and yield the longest tool lives at low cutting speeds, but TiN and TiC coatings reduce tool–chip friction, cutting temperatures, and crater wear and provide increased tool life at higher cutting speeds. (After A.T. Santhanam et al., [2].)

The coating thickness must be controlled within a narrow range to maximize tool life. If the coating is too thin, it will wear rapidly and fail due to abrasion. If the coating is too thick, however, differences in the thermal expansion coefficient between the coating and substrate will lead to large thermoelastic stresses in the coating and cause coating failure due to spalling. The optimum coating thickness is usually on the order of 0.010 mm [2].

More details on coating materials and methods are given in Chapter 4.

Cermet tools [44–48] typically fail by mechanisms similar to those for WC tools, that is, by abrasion, plastic deformation, edge chipping (often accompanied by BUE), diffusion and chemical wear, and depth of cut notching due to oxidation. The discussions of abrasion, plastic deformation, and diffusion, chemical, and oxidation wear given above for WC tools are applicable for cermet tools as well. However, cermet tools are often manufactured with nickel rather than cobalt binders, so that the reactions involved in chemical and oxidation wear differ from those encountered with WC tools.

As discussed in Chapter 4, TiC-, TiN- or TiCN-based cermets in both hard and tough grades have been developed. Generally, these materials have hot hardness characteristics comparable to WC, are chemically stable at higher temperatures, and have a lower fracture toughness and thermal shock resistance. Because of their greater chemical stability, they are used at higher cutting speeds than WC tools, and under appropriate conditions provide better tool life at comparable cutting speeds. The use of higher cutting speeds also reduces BUE formation, which can lead to edge chipping and limit tool life for hard cermets at lower cutting speeds. Because of their lower fracture toughness, edge chipping is a common failure mechanism for cermet tools; it is controlled by avoiding the use of positive rake angles and by using honed or chamfered edge preparations. Fracture concerns also limit allowable depth of cuts and prevent the use of many cermet grades in rough machining, interrupted cutting, and in the machining of hard steels. Cracking due to thermal shock also occurs when cermets are used with a cutting fluid; to avoid failures due to this mechanism, cermet tools should be used dry or with copious flood coolant.

Alumina-based (or oxide) ceramic tools fail by abrasive notch and flank wear, mechanical cracking, plastic deformation, edge chipping, gross fracture, thermal shock, and diffusion or chemical wear [49–54].

Alumina tools have a high hardness at temperatures up to 1000°C and thus exhibit little abrasive wear when used to cut cast iron, nickel-based superalloys, and aluminum–silicon alloys at speeds up to 300 m/min [50]. Alumina tools have found limited success in the high speed machining of steels due to their poor thermal shock resistance and fracture strength. Polycrystalline cubic boron nitride (PCBN) tool are often a better choice for these applications.

Chipping and fracture failures are common due to the brittleness of these materials. These may result from hard inclusions in the work material, high cutting forces encountered when machining at low speeds or with heavy depths of cut, vibration, improper exit conditions in interrupted cutting, or thermal shock, especially in interrupted cutting applications. Because of their low fracture resistance, alumina tools are often used only for finish or semifinish cuts at high cutting speeds. Fracture due to vibration can be reduced by limiting the depth of cut and by using extremely rigid tooling and fixturing setups. Chipping and fracture can be controlled by using honed or chamfered edge preparations (see Chapter 4). The fracture toughness and thermal shock resistance of alumina tools can be increased by adding ZrO_2, TiC, TiN, or SiC whiskers to the tool [47, 48], and these grades are commonly recommended for interrupted cutting. As discussed below, however, these additives reduce the tool's chemical stability. Finally, since even toughened grades of alumina have limited thermal shock resistance, alumina ceramics are best suited for dry, continuous cutting. The use of coolants invariably leads to fracture due to thermal cracking, and thermal cracking is also observed in dry interrupted cutting [50, 51]. As in the case of excessive notch or flank wear, the best course of action under these conditions is often to switch to PCBN tooling.

Alumina is chemically stable at temperatures up to 1200°C [50]. Diffusion and chemical wear are most often observed when machining steels, which produce temperatures in excess of this level. Chemical stability is a particular problem for toughened grades containing TiC, TiN, and SiC whiskers [48]. The titanium additives are soluble in iron at high temperatures and produce a glassy phase which wears rapidly. Similarly, SiC whiskers also react chemically with iron at high temperatures, leaving holes in the tool matrix which contribute to rapid wear or fracture of the tools. As a result, plain rather than toughened grades are best suited for continuous cuts on steel.

TiC- and TiN-based ceramics fail by mechanisms similar to those observed for alumina-based ceramics. The major difference is that these materials have a high chemical affinity for iron, so that diffusion and chemical wear are more serious problems when cutting ferrous work materials [50].

Silicon nitride tools [50, 51, 55] have a lower hot hardness but higher fracture toughness than alumina-based ceramic tools. They are well suited for machining cast iron in both continuous and interrupted operations.

In continuous cutting applications, Si_3N_4 tools fail primarily by abrasive wear. Abrasion may result from hard inclusions in the work material, especially adhering sand or carbide inclusions in sand castings, or by thermal softening. Thermal softening occurs at temperatures above 900°C with hot pressed Si_3N_4 tooling because hot pressed tools contain a glassy binder phase, which begins to melt at this temperature [51]. This limitation makes Si_3N_4 tooling unsuitable for machining steels and nickel alloys under most conditions. The use of thin CVD coatings on Si_3N_4 substrates can increase chemical and abrasive wear resistance and shows some promise for overcoming this limitation.

In interrupted cutting, Si_3N_4 tooling often fails by edge chipping or fracture. Due to its higher fracture toughness, chipping is usually less of a problem than for alumina tools [56] and can often be controlled by using hones and (especially) chamfered cutting edge preparations.

Si_3N_4 tooling is not used to machine aluminum alloys due to the high solubility of silicon in aluminum. Similarly, it is not used to machine titanium alloys because titanium reacts with nitrogen. As with alumina tooling, Si_3N_4 tooling has poor thermal shock resistance and should therefore be used dry.

Polycrystalline cubic boron nitride tools are hard and chemically stable at high temperatures and have excellent thermal shock resistance [57, 58]. They most commonly fail by abrasion, edge chipping, chemical wear, and thermal shock.

Because of the high hardness of PCBN, abrasive wear rates are low. For conventional cast irons, the wear rate may be imperceptible, particularly for tools with large nose radii. This makes PCBN especially attractive for precision operations such as engine cylinder boring. Although these operations can also be performed using silicon nitride tooling, the use of PCBN can lead to significantly higher tool life.

Abrasive wear rates are higher when machining hardened iron or steel components, chilled irons, cobalt- and nickel-based superalloys, and powder metals [57, 59, 60]. Abrasion generally produces flank rather than notch wear in these applications. The wear rate is often controlled by the binder phase which softens thermally. Abrasive wear rates may be reduced by using harder grades with ceramic rather than metallic binders, and by applying copious flood coolants in continuous cutting operations, especially finishing operations. Chemical wear also attacks the binder phase in PCBN tools; it can also be reduced by using ceramic rather than metallic binders in most cases, or by switching to alumina tools in continuous cuts.

Edge chipping is common in interrupted cutting. Chipping may also result in both continuous and interrupted cutting from vibrations caused by inadequate system stiffness, especially when cutting hardened irons or steels. Sharp cutting edges and positive insert geometries should be avoided in these applications; negative rake tools with chamfered edge preparations will provide better tool life. Thermal shock also occurs in interrupted cutting and results in thermal cracking of the tool, especially if a coolant is applied. Thermal shock can be reduced by cutting dry or by increasing the coolant supply.

As with ceramic tools, a variety of additional problems may be encountered when machining soft materials or when cutting at low speeds with PCBN tools. These include edge buildup, excessive flank wear, cratering or spalling, and poor surface finish. The measures for counteracting these problems, summarized in Table 9.1, also generally work for PCBN tools, but the most economic solution may be to use higher cutting speeds and lower feed rates, or (for soft materials) to switch to carbide or cermet tooling.

Polycrystalline diamond tools are the hardest tools available and provide excellent abrasive wear resistance. They are particularly well suited for machining nonferrous materials such as aluminum–silicon and magnesium alloys. Since these materials melt at low temperatures and produce low cutting pressures, tool lives on the order of months can be achieved, and maximum cutting speeds are limited by spindle capabilities rather than tool material limitations.

PCD tooling is not as well suited for machining ferrous alloys due to the high chemical affinity of iron for carbon. At higher cutting speeds carbon in diamond tools diffuses into the chip, resulting in edge weakening and fracture. This limits the maximum speed at which PCD tools can be used for these materials to the range typical of WC tooling. As noted above, PCD tools are also not well suited for cutting titanium alloys because of the high chemical affinity of titanium to carbon [35].

PCD tools may also fail by edge chipping when cutting harder materials due to their limited fracture toughness. Chipping problems are aggravated by the fact that positive rake angles and sharp edge preparations must normally be used to prevent edge buildup. Toughened PCD grades and diamond-coated WC grades suitable for machining ferrous alloys at low cutting speeds have been developed, but to date it has often been more economical to use WC or cermet tooling in these applications.

9.6 TOOL LIFE TESTING

A number of standardized tool life tests have been developed to help rank the performance of cutting tool materials or the machinability of workpiece materials. These include the ISO standard tests for single-point turning [9], face milling [61], and end milling [62], equivalent tests defined by national standard organizations, the ASTM bar turning test [63], and the Volvo end milling test [64, 65]. Various nonstandard tests such as drilling tests are also used for these purposes.

The standard tests strictly define the tool and workpiece geometries, cutting conditions, machine tool characteristics, and tool life criteria needed to construct repeatable tool life curves. They typically use a maximum average flank wear criterion to define tool life; in the ISO turning test, for example, tool life is assumed to be over when the average flank wear reaches 0.5 mm under roughing conditions (VB = 0.5 mm in Figure 9.2). The flank wear criterion is used because flank wear is normally the most desirable form of wear (given that some form of wear is unavoidable); the flank wear rate is relatively low and repeatable, so that tools which fail due to flank wear have comparatively long and consistent lives. (Other failure criteria based on crater wear, notch wear, surface finish, edge condition, or dimensional variations may also be used for tool–work material combinations for which flank wear is not the critical failure mode.) The use of a flank wear criterion, however, requires periodic measurement of the wear land width during testing. Since these measurements must be made offline with a microscope, such measurements contribute significantly to the time required to perform the tests. Constructing tool life curves at two cutting speeds using the ISO turning test, for example, often requires roughly 40 h of machine time.

The standard tests are applicable mainly to steel and iron work materials cut with HSS or WC tools. The standard material used as a baseline in most tests is 1045 steel. These tests were developed by steel and insert manufacturers, and are useful primarily for ranking the machining performance of various WC or HSS tool grades, or the machinability of various steel and iron alloys. The standard tests are generally not used for softer work materials such as aluminum alloys, or for coated carbide, ceramic, or polycrystalline tool materials. In the case of softer work materials, the flank wear rate may be quite low, and tool failures may more often result from edge buildup and microchipping, surface finish degradation, or burr formation; also, relative machinability rankings may be of less interest since these materials are relatively easy to machine. The tests are not used for advanced tool materials cutting nonferrous work materials because a prohibitive amount of material would have to be machined to produce the required level of flank wear.

Specialized tests such as drilling tests are often used on materials such as powder metals or special steel alloys which are difficult to produce in large quantities in bar form [66, 67]. Drilling tests are attractive for these applications because they require relatively small amounts of material. An added advantage is that direct wear measurements of the tool are often not required. In most cases drills are tested to failure, and the number of holes required to produce failure is the primary test output; the drill thrust force, torque, hole surface finish, or hole diameter may be monitored, however, as an indication of the progress of drill wear through the tests. Since direct tool wear measurements are not required, these tests require substantially less machine time (e.g., 8 to 12 h of machine time, rather than 40) than the standard ISO tests.

Many accelerated wear tests have also been proposed. In these tests, the cutting speed is increased significantly beyond the general application range for the tool–work material combination to produce rapid tool failures. One well-known test of this type is the face turning test [17] in which a turning tool is used to cut the face of a cylinder rotating at a constant spindle speed. The tool starts cutting at the center and is fed outward. The cutting speed increases with the radial distance from the center, until failure occurs; the radius being cut at failure is used as a relative measure of machinability. The disadvantage of this and similar tests is that tool failure often results from thermal softening or diffusion or chemical wear, rather than from flank wear as is usually desired in practice.

In conducting tool life tests, it is essential to use consistent tool and work materials. When possible, a single batch of cutting tools should be used for all tests to minimize variations between tools due to differences in microstructure, heat treatment, and grinding geometry. This requirement becomes increasingly important as the complexity of the tool increases; it is most critical when testing drills and end mills. It is also desirable to conduct tool life tests under dry cutting conditions, as this eliminates variation due to coolant concentration and condition. The scatter inherent in tool life testing can be illustrated by considering a recent round robin test in which the tool life was measured by ten laboratories using steel from a single heat, tools from a single batch, and identical toolholders, but different machines [65]. The scatter in these data was statistically significant, and the longest tool life measured was more than twice as long as the shortest tool life. The scatter was much greater at the lowest cutting speed used in the test, since at the low speed several tool wear mechanisms were active. Most tool life test results converge at higher cutting speeds since the tool life becomes short and temperature-activated wear mechanisms dominate.

It should be emphasized that both the standard and specialized tests are useful largely for comparison purposes, that is, for ranking the machinability of a group of work materials or the cutting performance of a group of tool grades. (Machinability ratings derived from standard tool life tests may also be useful for cost estimating in job shop production.) They do not generally provide an accurate indication of the tool life to be expected in most production operations. As noted in the introduction, tool life in a production operation depends on the part tolerances, and is effected by the part geometry (and rigidity), equipment condition, toolholding and fixturing method in addition to cutting conditions. Also, dominate wear mechanisms in production may vary significantly from the flank wear observed when cutting dry with uncoated tools. Realistic estimates of tool life in production operations can only be obtained from pilot test results or experience.

9.7 TOOL LIFE EQUATIONS

Since tool life has a strong economic impact in production operations, the development of quantitative methods for predicting tool life has long been a goal of metal cutting research. As noted in Section 9.1, tool life depends as much on part requirements as on the tool material and cutting conditions, making it difficult to develop general methods of predicting tool life. A number of simple tool life equations at least qualitatively applicable to many machining operations, however, have been developed.

The most widely used tool life equation is the Taylor tool life equation [68], which relates the tool life T in minutes to the cutting speed V through an empirical tool life constant, C_t:

$$VT^n = C_t \tag{9.8}$$

When T is specified in minutes, C_t is the cutting speed which yields 1-min tool life. The exponent n determines the slope of the tool life curve (Figure 9.11) and depends primarily on the tool material; typical values are 0.1 to 0.17 for HSS tools, 0.2 to 0.25 for uncoated WC tools, 0.3 for TiC- or TiN-coated WC tools, 0.4 for Al_2O_3-coated WC tools, and 0.4 to 0.6 for solid ceramic tools [5, 69–71]. The constant C_t varies widely with the tool material, work material, and tool geometry; is typically on the order of 100 m/min for rough machining of low-carbon steels. Extensive tables of n and C_t values for common work and tool materials as a function of cutting conditions are given by Kronenberg [71].

The basic Taylor equation reflects the dominant influence of the cutting speed on tool life, but does not account for the smaller but significant effects of the feed rate and the depth of cut. For this reason, a modified version of Taylor's equation, called the extended Taylor equation [70], is often used:

$$VT^n f^a d^b = K_t \tag{9.9}$$

For HSS tools, typical values of n, a, and b are 0.17, 0.77, and 0.37, respectively, when T is given in minutes, V in ft/min, and d and f in inches [72]. K_t varies considerably with the rake angle of the tool, but is typically approximately 500 for mild steels and 200 for cast iron [70, 71]. Values of K_t, n, a, and b for specific tool grades and common work materials are sometimes tabulated in tooling catalogs. In general, $n < a < b$ because cutting speed has a larger influence on tool life than feed and depth of cut, assuming that tool life is limited by thermal damage and abrasion as expected for a properly designed cutting process. However, in cases where chipping and fracture failures are predominant, the exponents a and b are smaller than the exponent n.

The extended Taylor equation treats the influences of the feed rate and depth of cut independently. An alternative equation proposed by Colding [73] combined the influence of these parameters by assuming tool life depends on the effective chip thickness, ECT:

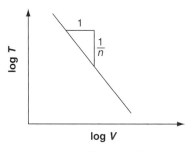

FIGURE 9.11 A Taylor tool life curve, showing linear variation between tool life and cutting speed when plotted on a log–log scale. The slope of the curve is negative and equal in magnitude to the inverse of the exponent n in the Taylor tool life equation.

$$T = AV^m \text{ECT}^p \tag{9.10}$$

ECT is equivalent to the average chip thickness computed in simulation programs (Chapter 8) and depends on the feed, depth of cut, lead angle, and tool nose radius. Equation (9.10) involves fewer experimental constants than the extended Taylor equation and yields more broadly applicable results since tool nose radius and lead angle effects are accounted for in the ECT calculation.

Many similar relations have been proposed [6]. Simple tool life equations are useful mainly for comparative purposes. They are useful for ranking the general machining performance of insert grades or similar alloys as discussed in the previous section. Taylor tool life tests performed according to defined standards [9] are also useful in ensuring results reported by different laboratories are consistent; this is an important concern in large machinability testing programs involving several independent laboratories [65]. Tool life predictions based on these relations, however, are generally not quantitatively reliable for the reasons discussed in connection with standard tool life tests in the previous section. The results are most accurate quantitatively for general purpose, low-speed machining using conventional machine tools, in which part dimensional tolerances are typically on the order of 0.1 mm. The results are least accurate for high-speed machining, in which tool failure often results from crater wear; in precision machining, in which dimensional tolerances are much smaller; and in mass production, in which tool lives on the order of hours rather than minutes are desirable.

First- and second-order mathematical models for tool life can also be fit to experimental data [71, 74–77]. Unlike the empirical equations discussed above, which are intended to be general, the first- and second-order mathematical models are narrowly applicable to the range of conditions used to generate the test data. Using the cutting speed, feed rate, and depth of cut as independent variables, a first-order mathematical model has the form

$$\ln T = b_0 + b_1 \ln V + b_2 \ln f + b_3 \ln d \tag{9.11}$$

while a second-order model has the form [75–77]

$$\begin{aligned}\ln T = {} & b_0 + b_1 \ln V + b_2 \ln f + b_3 \ln d \\ & + b_{11}(\ln V)^2 + b_{22}(\ln f)^2 + b_{33}(\ln d)^2 \\ & + b_{12} \ln V \ln f + b_{13} \ln V \ln d + b_{23} \ln f \ln d\end{aligned} \tag{9.12}$$

Equations of this type are used to select optimum economic cutting conditions for a narrow range of applications involving a single tool–work material combination; in contrast to the empirical equations described above, they can be used for complex cutting operations in addition to simple operations such as single-point turning.

The tool life equations described above treat tool life as a deterministic property. In reality, tool life exhibits scatter and is probabilistic in nature (Figure 9.12). Sources of tool life variability in practice include: (a) variation in the workpiece material hardness, microstructure, composition, and surface characteristics; (b) variation in the cutting tool material, geometry, and grinding methodology; (c) variations in coolant concentration and condition; and (d) vibrations due to the compliance of the machine tool structure, workpiece, fixture, toolholder, and tool body. Various probabilistic tool life models based on response surface methods and similar techniques have been proposed [78–91], but most are difficult or inefficient to apply in practice. For example, one common approach is to study tool reliability, or the probability of failure-free operation for a specified time in a specified manufacturing environment. In cases in which tools fail randomly, the reliability is described by either an

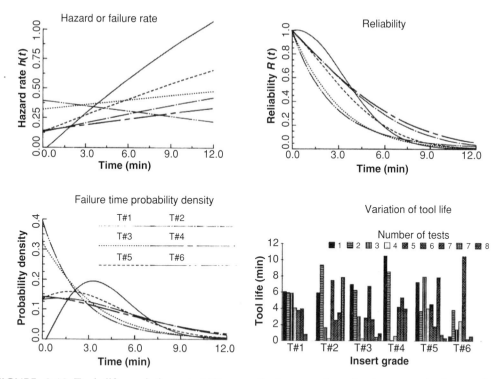

FIGURE 9.12 Tool life variation at 0.75 mm flank wear for alumina ceramic inserts from different manufacturers when turning 52100 steel at 600 m/min cutting speed, 0.5 mm/rev feed, and 2.54 mm depth of cut, together with corresponding failure rate, reliability, and probability distribution curves.

exponential or Weibull distribution [81, 86, 92]. This requires large number of samples, between 50 and 100, which are generally not available because tools are often removed at regular safe intervals. An additional difficulty with statistical reliability testing is the cost and time required to generate an operational profile and uncertainty when creating the operational profile.

9.8 PREDICTION OF TOOL WEAR RATES

A number of researchers have developed models for predicting the rate of tool wear due to specific mechanisms [11, 22, 93–100].

The best-known analysis was developed by Usui and coworkers [94] based on the shear zone model of cutting mechanics discussed in Section 6.10. This model predicts the wear rates due to adhesive, abrasive, and (indirectly) diffusion wear for a carbide tool when cutting steel. Based on earlier work [95, 96] and assumptions about the thermal softening behavior of carbide tools, they derive a differential equation relating the incremental volume of material worn away at a given point of the tool, dv, and the incremental distance slid, dL_s:

$$dv = C_1 q \exp\left[-\frac{C_2}{\theta}\right] dL_s \qquad (9.13)$$

where q is the normal stress at the point in question, θ is the interfacial temperature, and C_1 and C_2 are empirical constants. C_1 is proportional to the tool hardness.

Based on composite tool measurements like those discussed in Section 6.7, Usui assumed that the stress on the rake face of the tool was given by an additional exponential equation:

$$q(x) = \exp\left\{D_1\left[1 - \frac{x}{L_c}\right]\right\} \quad (9.14)$$

where x is the distance from the cutting edge, L_c is the tool–chip contact length, and D_1 is an empirical constant. He assumed that the stress on the flank face of the tool was constant, and measured temperatures on both the rake and flank faces using the embedded wire method described in Section 7.2.

Under all these assumptions, the constants C_1, C_2, and D_1 can be estimated from a series of orthogonal cutting tests, and the total volume of material worn away as a function of time can be calculated by substituting Equation (9.14) into Equation (9.13) and integrating with respect to x and time. Usui presents data showing that the functional form of Equation (9.13) is reasonable, and that the equation generally yields accurate results when cutting low-carbon steels with an uncoated carbide tool (Figure 9.13 and Figure 9.14).

This model is useful because it illustrates the importance of the tool hardness (through C_1) and interfacial temperature on the wear rate. It is difficult to apply in practice because of the

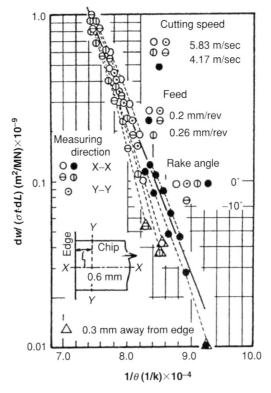

FIGURE 9.13 Variation of normalized crater wear rate with cutting temperature when cutting 1020 steel with a P20 carbide tool. The dashed lines represent theoretical predictions. (After E. Usui et al., [94].)

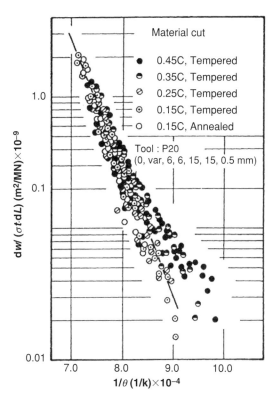

FIGURE 9.14 Variation of normalized crater wear rate with cutting temperature when cutting various low-carbon steels with a P20 carbide tool. (After E. Usui et al., [94].)

extensive tests required to estimate temperature distributions and empirical constants. In principle, testing requirements could be reduced by estimating the stress and temperature distributions using finite element analyses; preliminary research in this direction has been reported [101].

Models have also been derived to calculate wear rates due to diffusion [97–99]. The best known is Kramer and Suh's model [97], which is applicable to WC tools used to cut steel. For this material combination, they assume the crater wear rate is governed by the equation

$$\frac{dW}{dt} = K\left[-D\frac{dc}{dy} + CV_y\right]_{y=0} \qquad (9.15)$$

where K is the ratio of the molar volumes of the tool material and the workpiece material, D is the diffusivity of the slowest diffusing tool constituent into the work material, c is the concentration of the tool material in the chip, C is the equilibrium concentration or solubility of the tool material in the work material, and V_y is the bulk velocity of the work material normal to the rake face (Figure 9.15). This equation identifies the thermodynamic properties of the tool–work material combination which control the diffusion wear rate. It is difficult to apply because the equilibrium concentration C is difficult to specify [100]. It has been used qualitatively to rank the cutting performance of various tool coatings [35].

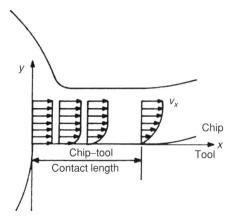

FIGURE 9.15 Definition of variables used in Equation (9.15) (After B.M. Kramer and N.P. Suh [97].)

9.9 TOOL FRACTURE AND EDGE CHIPPING

Tool fracture occurs when the tool is unable to support the cutting force over the tool–chip contact area. When fracture occurs near the cutting edge and results in the loss of only a small part of the tool, it is referred to as edge chipping or frittering. Chipped tools produce poor surface finishes but are often still usable. Fracture which occurs away from the cutting edge and results in the loss of a substantial portion of the tool is referred to as gross fracture or breakage; when this occurs, the tool is unusable and must be changed.

Since fracture results when the tool is loaded beyond its structural limit, it can be prevented either by reducing the cutting load or by increasing the structural strength of the tool. Since reducing the cutting load usually requires reducing the metal removal rate and thus the productivity of the process, increasing the structural strength of the tool should be attempted first. The structural strength of the tool can be increased by increasing the tool nose radius, using negative rather than positive rake angles, using a thicker insert, and increasing the lead angle; these measures spread the load over a larger volume of the tool material and reduce maximum tool stresses. Switching to a tougher or more fracture resistant tool grade may also be effective. The property of the tool which controls fracture toughness is the transverse rupture strength, which varies with the chemical composition and binder content. Generally, increasing the binder content of the tool increases the fracture toughness (Figure 9.16). Adding strengthening agents such as whiskers or titanium phases to ceramic tools also increases fracture strength. As discussed in Section 9.4, however, tools with increased binder contents or strengthening agents may have reduced resistance to abrasive and chemical wear. Edge chipping can be often controlled by using honed or chamfered, rather than sharp edge preparations. The optimum chamfer length varies with the tool material but is usually comparable to the feed rate; recommendations for chamfer lengths and angles are often given in tooling catalogs.

If strengthening methods are ineffective, load reduction measures must be used. In determining which measures to use, it is useful to distinguish between initial fractures, which occur at the onset of cutting, and regular or intermittent fractures, which occur either predictably or unpredictably after the tool has cut for some time.

Initial fracture usually results from overloading or from the use of cracked or damaged tools. When tools consistently break at the onset of cutting, the cutting force is too large for

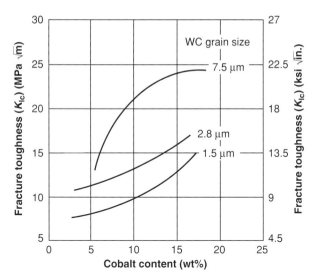

FIGURE 9.16 Variation of fracture toughness with binder content for cemented carbide tools with various grain sizes. (After A.T. Santhanam et al., [2].)

the edge configuration. This type of fracture can be prevented by reducing the feed rate or depth of cut, or by increasing the cutting speed. When initial fractures occur unpredictably, they may result from tool damage, and particularly from cracks or nicks in the tool resulting from improper tool handling. This is particularly common for brittle ceramic or cermet tools and can be prevented by handling tools carefully before use and by avoiding excessive tool clamping pressures.

Regular fractures occur at predictable periods and usually result from tool wear in continuous cutting or from fatigue in interrupted cutting. In continuous cutting, progressive flank wear increases stresses in the tools, while crater and notch wear reduce cutting edge strength. In interrupted cutting, thermal or mechanical fatigue can result in cracks in the cutting edge which ultimately lead to fracture. The methods for reducing wear rates and crack formation discussed in Section 9.2 should reduce the occurrence of regular fractures.

Intermittent fractures occur at irregular intervals and may result from hard inclusions in the work material or from intermittent vibration or chatter problems. If hard inclusions such as carbide inclusions in steel or adhering sand in sand castings are responsible, the best course of action is to inspect incoming workpieces and discard or repair those containing inclusions. If this is not practical or effective and such fractures cannot be tolerated, the feed rate or depth of cut should be reduced or a tougher tool grade should be used. Vibration problems are usually easily recognized because they create excessive noise or leave characteristic marks on the machined surface; they are most commonly eliminated by increasing the stiffness of the tool, toolholder, or part fixture, by changing the cutting speed, or by reducing the depth of cut. Machine tool vibrations are discussed in detail in Chapter 12.

Tool fracture is generally of the brittle type and results from excessive tensile stresses in the tool. A number of finite element analyses for determining the stress distribution within the tool for given edge and boundary conditions have been reported [51, 102]. The inputs to these analyses are the normal and frictional stresses at the tool–chip interface, the tool geometry, and the transverse rupture strength of the tool; the outputs are the tensile stress distribution in the tool and, ultimately, the maximum cutting loads which can be sustained without exceeding the maximum allowable tensile stress for a given tool geometry (Figure 9.17). With

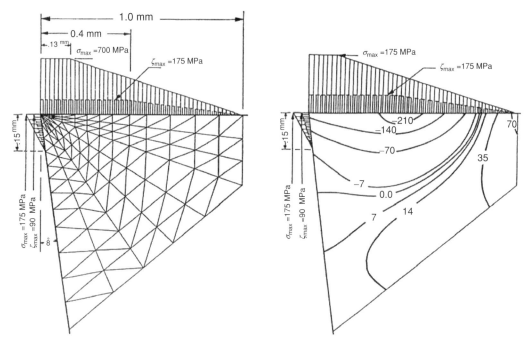

FIGURE 9.17 Finite element model of the tip region of a cutting tool, and contours of maximum principal stresses for a given edge loading. (After J. Tlusty and Z. Masood [102].)

further development, these analyses may permit the prediction of the maximum allowable feed rate for a given tool geometry, and may be useful in determining optimum edge conditions (particularly chamfer lengths and angles) for given edge loadings. Since the transverse rupture strength of the tool depends on temperature, combined analyses which determine both the stress and temperature distributions in the tool will be required to achieve this goal.

9.10 DRILL WEAR AND BREAKAGE

Characteristic wear patterns for conventional twist drills are shown in Figure 9.18 and Figure 9.19. Four types of wear are commonly observed: chisel edge wear, lip wear, margin wear, and crater wear [103–106].

Chisel edge wear occurs at the central chisel edge of the drill and results from abrasion or plastic deformation. This form of wear may double or triple the thrust force on the drill and affects the drill's centering accuracy.

Lip or *flank wear* occurs along the relief face of the drill's cutting lips and results from abrasion, plastic deformation, insufficient flank relief caused by improper point grinding, or excessive dwelling at the bottom of blind holes. Lip wear increases the thrust force, power consumption, and maximum lip temperature, which in turn leads to increased thermal softening and further wear. Lip wear also increases the size of the burr produced when drilling through holes. When lip wear becomes excessive, the drill may cease to cut and fail by chatter or breakage.

Margin or *land wear* occurs at the outer corner of the cutting lip or on the land which contacts the drilled surface. Margin wear results from abrasion, thermal softening, or

Tool Wear and Tool Life

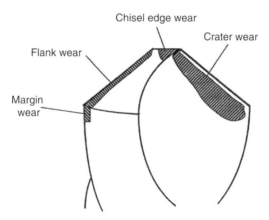

FIGURE 9.18 View of drill point showing regions where flank, nose radius, margin, and crater wear occur. The drill is rotating clockwise when viewed from the top.

FIGURE 9.19 Photomicrograph of a drill point showing chisel edge and flank wear.

diffusion wear. Excessive margin wear results in poor hole size control and surface finish. Generally, margin wear produces an undersized hole unless it is accompanied by BUE formation or centering errors, in which case it produces an oversized hole.

Crater wear occurs on the flute surface and results from thermal softening or diffusion wear. Moderate crater wear is usually not of concern, but excessive cratering weakens the drill's cutting edge and can lead to edge deformation, chipping, or breakage.

At low to medium cutting speeds, abrasive wear is usually dominant and most wear occurs at the outer corner near the margins. At higher cutting speeds, in addition to abrasive wear, workpiece deposits may form on the drill lips and the margins, especially when drilling aluminum. Generally, the tool life decreases as the speed is increased when drilling abrasive materials. The feed rate should be maintained above a minimum level to avoid rubbing along the drill lips.

When wear is caused primarily by abrasives in the work material, the number of holes to failure can be estimated from an abrasive wear model. In general, the number of holes to failure, N, depends directly on the feed f and is inversely proportional to the cutting speed V:

$$N = C_1 \frac{\text{VB} \cdot f}{V^a} \quad (9.16)$$

The time to failure T should be independent of the feed and inversely proportional to the cutting speed:

$$T = C_2 \frac{\text{VB}}{V^{a+1}} \quad (9.17)$$

In these equations, C_1, C_2, and a are empirical constants [107]. VB is the width of flank wear.

Methods for reducing various types of drill wear are summarized in Table 9.2. Generally, chisel edge wear results from excessive force and can be reduced by reducing the feed, while margin and crater wear result from excessive heat and can be reduced by reducing the spindle speed. Cutting temperatures at a given cutting speed can be reduced by using flood coolants, or by using coolant-fed drills with compressed air or pressurized coolants provided the drilling machine is equipped to use such drills. Flank wear due to abrasion can be reduced by increasing the feed, provided this does not reduce accuracy or cause chip clogging. Increasing the feed reduces the distance slid per unit time or hole depth, increasing the number of holes which can be drilled to failure; this initially surprising effect is reflected in Equation (9.16) and is observed in tests [108]. When drilling through holes, wear is increased due to torsional vibrations and feed surging at breakthrough [105]; reducing the feed rate prior to breakthrough can increase drill life on CNC machines. For blind holes, wear increases with the dwell time at the bottom of the hole as the feed is reversed. The dwell time cannot be adjusted on many machines, but on CNC machines equipped with high acceleration/deceleration feed slides, excessive dwelling should be avoided.

In many cases, it is more economical to change the drill material rather than reduce the spindle speed or feed rate. Increasing the drill's hot hardness will reduce most forms of wear and is especially effective in reducing margin wear. A significant increase in drill life can thus usually be obtained by using an HSS–Co or solid WC drill instead of an HSS drill. Coating drills with TiN reduces both margin and crater wear. Because margin wear is reduced, coated drills produce holes of constant diameter throughout their lives.

Drill breakage and chipping may result from overloading, hard spots in the work material, chip packing, misalignment, spindle runout, vibration, hole interruption, or improper point grinding [109]. Measures for eliminating breakage due to these causes are also summarized in Table 9.2.

Overloading usually produces an immediate failure and can be eliminated by reducing the feed rate. Similarly, as discussed in Section 9.9, breakage due to hard spots in the work material can be reduced by reducing the feed rate if hard spots cannot be detected by improved inspection of incoming work.

Chip packing occurs when the chips become lodged in the flute, so that further chip evacuation is impossible. It occurs especially when drilling deep holes. Chip packing can be prevented by changing the flute profile (e.g., to a parabolic flute), by changing (generally increasing) the helix angle, by decreasing the feed rate, by changing the orientation of the spindle from vertical to horizontal, and on special machines by drilling from the underside

TABLE 9.2
Mechanisms, Characteristics, and Countermeasures for Common Types of Drill Wear and Failure

Type of Wear	Mechanism	Characteristics	Countermeasures
Flank wear	Abrasion	Even wear scar	Use harder drill
			Filter fluid
			Increase feed
	Thermal softening		Reduce speed (*)
Margin wear	Thermal softening	Hole undersized	Use harder drill
			Reduce speed (*)
			Use corner radius or spiral point
Crater wear	Diffusion	Wear crater present	Reduce speed (*)
			Use harder drill
			Use coated drill
			Increase helix angle
Chisel edge wear	Abrasion		Reduce feed
			Use harder drill
			Thin web
Corner chipping	Interrupted hole	Occurs at breakthrough	Use corner radius
			Reduce feed at exit (*)
			Use three- or four-fluted drill
	Vibration	Margin damaged	Increase system rigidity
			Use bushing
			Reduce feed (*)
Drill breakage	Overload	Occurs immediately	Reduce feed (*)
			Reduce helix angle
	Misalignment	Margin damaged	Check alignment
			Use bushing
	Centering error	Occurs near entry	Use self-centering point
			Spotface or centerpunch
			Reduce feed at entry (*)
			Pilot drill (*)
	Chip packing	Flutes clogged	Peck drill
			Increase flute space
			Use parabolic flute
			Increase helix angle
			Increase coolant supply filter coolant
	Vibration	Noise, margin damaged	Increase system rigidity
			Use tougher drill
	Feed surging	Occurs at exit	Reduce feed at exit (*)
	Flank or heel rubbing	Noise, heel damaged	Regrind point to increase clearance
Hole oversize	Edge buildup		Increase spindle speed
			Use corner radius or spiral point
			Use coated drill
	Poor alignment	Margin worn	Check alignment
			Use bushing

Note: Countermeasures marked with an asterisk (*) reduce the metal removal rate and should be used only as a last resort in general purpose machining.

rather than the top of the part. On properly equipped machines, using high pressure through the spindle coolant is also effective. Finally, on CNC machines, the use of a pecking cycle in which the drill is periodically withdrawn from the bottom of the hole can also eliminate chip packing, although this increases the cycle time and may increase flank wear.

If the spindle is misaligned or runs out, the drill will often exhibit markings due to impact or rubbing on the margin. In this case the runout and alignment of the spindle should be checked and adjusted as necessary. Spindle runout is especially critical when using solid WC drills; for these drills, the runout should not exceed 0.001 in.

Excessive vibration can usually be detected by the noise it produces or by marks on the machined surface. As discussed in Chapter 12, vibration can be eliminated by increasing the stiffness of the toolholder, spindle, or part, or by reducing the spindle speed. On CNC machines, reducing the feed rate at breakthrough can reduce vibration when drilling through holes.

Hole interruption due to inclined exits or intersecting holes induces vibration and can lead to chipping, breakage, straightness errors, and burr formation. These effects can be minimized by using three- or four-fluted drills rather than two-fluted drills [110], by grinding a corner break or corner radius on the drill point, by using a guide bushing, and by reducing the feed rate at breakthrough on CNC machines.

If a drill point is ground improperly and has inadequate clearance, the heel of the flank will drag, producing noise and often drill breakage. Used tools will have worn spots on the heel in this case. This problem, which is becoming more rare as CNC drill grinders become more common, can be eliminated by correcting the grinding procedure to provide adequate clearance.

Drills are sometimes broken in practice by running the spindle backwards or by indexing the fixture before retracting the drill from the part. These problems, which result from programming errors, are easily corrected.

9.11 THERMAL CRACKING AND TOOL FRACTURE IN MILLING

In milling, tools are subjected to cyclic thermal and mechanical loads and may fail by mechanisms not observed in continuous cutting. Two common failure mechanisms unique to milling are thermal cracking and exit failure.

The cyclic variations in temperature in milling induce cyclic thermal stresses as the surface layer of the tool expands and contracts. This can lead to the formation of thermal fatigue cracks near the cutting edge. In most cases, such cracks are perpendicular to the cutting edge and begin forming at the outer corner of the tool, spreading inward as cutting progresses (Figure 9.7). The growth of these cracks eventually leads to edge chipping or tool breakage. This problem is especially common when milling steel with WC tools [111–117]. Thermal cracks form and grow more rapidly as peak temperatures during the cutting cycle increase and as the difference between peak and low temperatures increases. As discussed in Section 7.6, peak temperatures during the heating cycle increase as the cutting speed and the length of the heating cycle (i.e., the cutter immersion) increase. The difference between the peak and low temperatures increases when a cutting fluid is applied.

Thermal cracking can be reduced by reducing the cutting speed (or spindle rpm) or by using a tool material grade with a higher thermal shock resistance [117]. In applications in which a coolant is applied, adjusting the coolant volume can also reduce crack formation. An intermittent coolant supply or insufficient coolant volume promote crack formation; if a steady, copious volume of coolant cannot be supplied, tool life can often be increased by switching to dry cutting.

When using coated WC tooling, the coating may fail by spalling due to a similar thermal fatigue mechanism. This type of coating failure can be prevented by changing to a more shock

Tool Wear and Tool Life

resistant coating grade or by eliminating the coolant in wet operations. Coolants should be avoided with many coating grades; for example, coolant should not be used when milling cast iron with aluminum oxide coated tools. Tooling catalogs normally provide coolant application guidelines for specific grades.

Edge chipping is also common in milling. Chipping may occur when the tool first contacts the part (entry failure) or, more commonly, when it exits the part (exit failure). Entry and exit failures have been most widely studied for face milling steel with WC tooling, but fracture mechanisms and appropriate countermeasures are applicable to other material combinations.

Entry failures most commonly occur when the outer corner of the insert strikes the part first (Figure 9.20) [115]. This is more likely to occur when the cutter rake angles are positive rather than negative. Entry failure is therefore most easily prevented by switching from positive to negative rake cutters. When entry failure occurs with negative rake cutters, it can be prevented by using a cutter with a lower lead angle [115], or by reducing the cutter diameter or moving the cutter center toward the part to increase the entry angle.

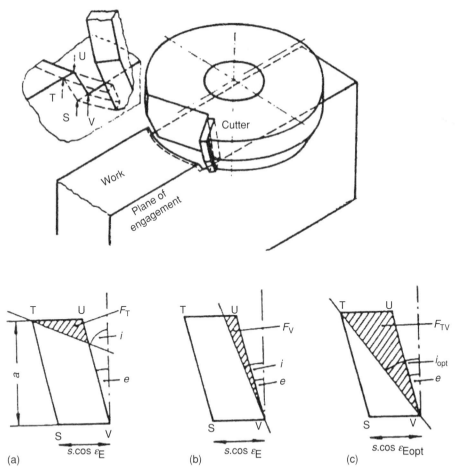

FIGURE 9.20 Initial contact between the workpiece and insert in face milling, described by defining the parallelogram STUV, where S is the outer corner of the insert. The insert should strike the workpiece at point U first to minimize the chance of breakage. When possible, the lead and entry angles should also be adjusted to produce the maximum possible engagement (c) before points T and V enter the workpiece. (After H. Opitz and H. Beckhaus [115].)

Exit failures usually occur when an exit burr or foot is formed as discussed in Chapter 11. As shown in Figure 9.21, the interaction between the burr and the tool as the tool exits can induce large stresses near the cutting edge. The measures for reducing burr formation —

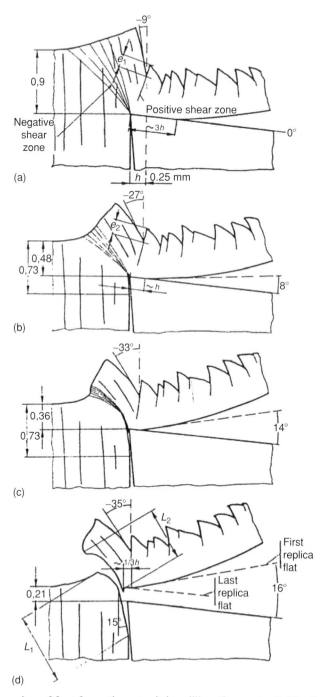

FIGURE 9.21 Progression of foot formation at exit in milling (from a to d). The formation of the foot induces a large tensile stress on the tool face which can lead to fracture. (After A.J. Pekelharing [116].)

discussed in Chapter 11 — are effective in preventing exit failures. Exit failures can also be prevented by using a larger diameter cutter to increase the exit angle, and by switching from negative to positive rake cutters to increase edge strength.

Finally, the methods discussed in Section 9.9 for preventing fractures in general, such as the use of chamfered or honed edge preparations, tougher tool grades, and reduced feeds or depths of cut, are also effective in preventing both entry and exit failures.

9.12 TOOL WEAR MONITORING

Tool wear is one important factor contributing to the variation of the cutting forces and surface finish. Worn tools are often changed on a statistical basis at a rate dictated by the shortest life expectancy for multi-tool operations. In such cases, significant useful life of the tools may be wasted and system productivity may be reduced. Alternatively, an in-process tool wear monitoring and control approach may be used to predict in-process tool wear throughout the life of specific tools.

Tool wear can be monitored by either direct or indirect measurements. Direct methods involve the measurement of the wear and evaluating the volumetric loss from the tool due to tool wear using radioactive, optical, laser scan micrometer, or electrical resistance sensors [118]. Indirect measurements of tool wear are made by relating tool wear to other cutting parameters such as part quality, surface roughness, cutting force or torque, motor power (or current), acoustic emission, or vibration [119–121].

Tool monitoring can reduce tool damage and downtime, and improve part quality. It offers the potential to characterize, in real time, the efficiency of the metal removal process. Over the past three decades, a substantial amount of research has been conducted on machining process monitoring and control [122–127]. Research studies and several production applications have shown that there is a good correlation between many cutting parameters and flank wear [128, 129]. All monitoring systems use sophisticated signal processing techniques or wear estimation. Monitoring system performance is highly dependent on the quality of the transducer data available and the signal processing technique. In some cases, more than one sensor are used to improve the reliability of the monitoring system [130].

Several tool monitoring systems have been developed to protect the machine tool from potential damage due to collision, missing tools, tool breakage, or tool wear as illustrated in Figure 9.22. Spindle power monitors are now common on contemporary machine tools. Sensitivity can be a problem, especially with small diameter tools, because of the inertia and friction present in the system. Strain gages mounted in the spindle bearings and piezoelectric load cells mounted under the tool are more sensitive to cutting forces. Wireless sensor technology provides much higher flexibility for measuring forces, vibration, etc. because they can be installed in the rotary component of the spindle.

9.13 EXAMPLES

Example 9.1 Estimate the parameters for Taylor's tool life equations (Equation 9.8) for a new grade carbide indexable insert. An alloy steel bar is being machined using a turning operation with the carbide insert to evaluate the tool life in accordance with the ISO 3685 STD. The cutting conditions used in this test are: 2.54 mm depth of cut and 0.18 mm/rev feed.

Solution: Several turning tests were performed under extensive controllable cutting conditions to minimize variations and scatter in tool life. Several inserts were evaluated at four different cutting speeds : 90, 120, 150, and 180 m/min. Two inserts from the same production batch were tested at 90 m/min, four inserts were tested at 120 m/min, six inserts were tested at

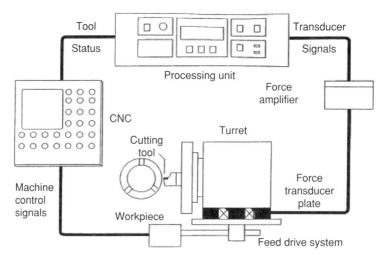

FIGURE 9.22 Tool wear monitoring system.

150 m/min, and eight inserts were tested at 180 m/min. More inserts were tested at higher speeds because the tool life was shorter and a larger variation was expected due to more rapid tool wear. The flank wear was measured at different time intervals during turning. The average tool wear data obtained from the repeated tests are summarized in Table 9.3 and plotted in Figure 9.23.

The flank wear limit based on the response in Figure 9.23 and previous experience was assumed to be 0.5. The tool life in minutes corresponding to this wear limit was obtained by taking the intercepts of the wear curves with this limit as shown in Figure 9.23. The tool life results are given in Table 9.4 and plotted in Figure 9.24. The tool life data are plotted on a log–log scale in Figure 9.25. The data represent a straight line on the log–log scale since

$$V \cdot T^n = C_t \Rightarrow \log V + n \log T = \log C_t \Rightarrow \log V = \log C_t - n \log T$$

TABLE 9.3
Experimental Tool Wear Data from Turning Operation

Tool Test Number	Cutting Speed (m/min)	Cutting Time (min)	Flank Wear (mm)	Tool Test Number	Cutting Speed (m/min)	Cutting Time (min)	Flank Wear (mm)
1	180	1	0.14	3	120	12	0.37
1	180	2	0.35	3	120	15	0.47
1	180	3	0.49	3	120	20	0.71
1	180	4	0.64	4	90	6	0.11
2	150	2	0.18	4	90	12	0.21
2	150	4	0.3	4	90	25	0.27
2	150	6	0.46	4	90	35	0.31
2	150	8	0.66	4	90	40	0.37
3	120	2	0.11	4	90	45	0.42
3	120	4	0.16	4	90	50	0.5
3	120	6	0.23	4	90	55	0.6
3	120	8	0.29	4	90	60	0.78

Tool Wear and Tool Life

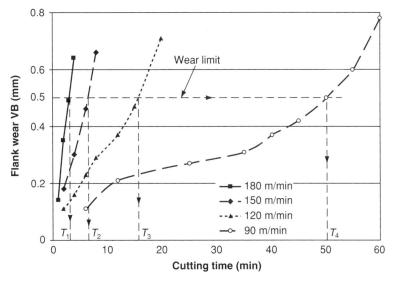

FIGURE 9.23 Graph of tool wear data for Example 9.1, showing method of intercepts for determining tool life for a given cutting speed and wear limit.

The slope of the straight line, n, is given by

$$n = \frac{\log V_1 - \log V_4}{\log T_4 - \log T_1} = \frac{\log 90 - \log 150}{\log 3 - \log 50} = 0.25$$

While the intercept of the line with the vertical axis at time 1 gives the constant C_t since

$$\log V = \log C_t - n \log T \Rightarrow \log V = \log C_t - n \log(1)$$
$$\Rightarrow \log V = \log C_t - n(0) \Rightarrow C_t = V \quad \text{for } T = 1$$

The value of constant C_t can be also obtained from the data by

$$VT^n = C_t \Rightarrow C_t = VT_1^n = VT_2^n = VT_3^n = VT_4^n$$
$$\Rightarrow C_t = (180)(3)^{0.25} = 237$$

Hence, Taylor's tool life equation for this case is

$$V \cdot T^{0.25} = 237$$

TABLE 9.4
Tool Life at Different Speeds

Test	Cutting Speed (m/min)	Tool Life (min)
4	90	50
3	120	16
2	150	6
1	180	3

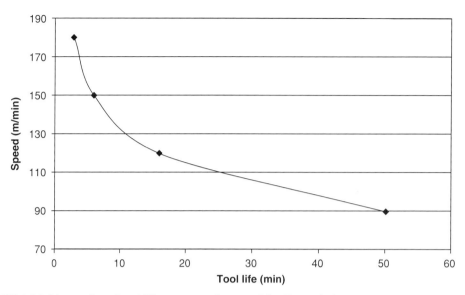

FIGURE 9.24 Linear plot of tool life versus cutting speed for Example 9.1.

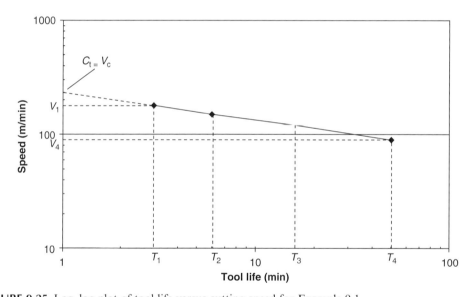

FIGURE 9.25 Log–log plot of tool life versus cutting speed for Example 9.1.

Example 9.2 The following table of tool life values was obtained from several tool life tests carried out by drilling steel with carbide drills. The drill diameter is 8 mm and the hole depth 25 mm. The mean of the tool life corresponding to different speeds and feeds are given in the following table:

	V (m/min)	f (mm/rev)	T (min)
1.	70	0.15	120
2.	70	0.25	105
3.	140	0.15	24

Tool Wear and Tool Life

Compute the parameters of the extended Taylor tool life equation, Equation (9.9), for these data.

Solution: The extended Taylor tool life equation is

$$VT^n f^a = K_t$$

Let us tabulate: $V_1 = V_2 = 70$ and $V_3 = 140$; $f_1 = f_3 = 0.15$ and $f_2 = 0.25$. The following equality is obtained using cases 2 and 3 in the above data:

$$\frac{V_3}{V_1} = \frac{T_1^n}{T_3^n} \quad \text{or} \quad \frac{140}{70} = \left(\frac{120}{24}\right)^n \Rightarrow n = \frac{\ln 2}{\ln 5} = 0.43$$

Likewise,

$$\frac{f_1^a}{f_2^a} = \frac{T_2^n}{T_1^n} \quad \text{or} \quad \left(\frac{0.15}{0.25}\right)^a = \left(\frac{100}{120}\right)^{0.43} \Rightarrow a = 0.43 \frac{\ln 0.83}{\ln 0.5} = 0.11$$

$$K_t = VT^n f^a = 70 \times 120^{0.43} \times 0.15^{0.11} = 445$$

Hence, the tool life equation is $VT^{0.43} f^{0.11} = 445$.

This equation may be expressed in two different forms in a rectangular coordinate system as shown in Figure 9.26. These graphs illustrate the considerations for the choice of cutting speeds and feeds as explained in more detail in Chapter 13. Both graphs illustrate that the selection of the cutting speed is more important than the feed. An increase in cutting speed of 100% (from 70 to 140 m/min) can decrease the tool life by as much as 80% as shown in Figure 9.26a. However, an increase in feed of 67% (from 0.15 to 0.25 mm/rev) decreases the tool life by only 13%. The effect of the combination of speed and feed on tool life is shown in Figure 9.26b. For example, if the cutting speed of 110 m/min is selected with a requirement of 30 min tool life, the feed selected should be below 0.6 mm/rev. If the speed is increased to 130 m/min, the feed must be reduced to a maximum of 0.13 mm/rev for the same tool life of 30 min. This figure illustrates that the effect of speed on tool life is more significant than the effect of feed, although both affect the material removal rate.

Example 9.3 Estimate the improvement in tool life if the clearance (relief) angle of a tool is increased from 5° to 11°. Assume that the tool is considered worn at a specified amount of flank wear VB_e.

Solution: As explained in Chapter 4, the cutting edge geometry (as shown in Figure 9.27) and strength are affected by both the rake angle (α) and primary clearance angle (θ). The tool performance improves by increasing the rake angle (assuming the workpiece material and process allow) and the clearance angle assuming that the cutting edge does not fracture as discussed in Chapter 4. The width of flank wear is related to the depth of flank wear (defined as DB in the feed direction as illustrated in Figure 5.27). The relationship of the width of flank wear to the tool geometry is

$$\frac{VB}{DB} = \cot\theta - \tan\alpha$$

Figure 9.28 shows the relationship between the width of flank wear land VB and the clearance angle for zero and 8° rake tools. Assuming that the flank wear characteristics stay the same for different clearance angles, the width of flank wear is proportional to the $\cot\theta$ (see above

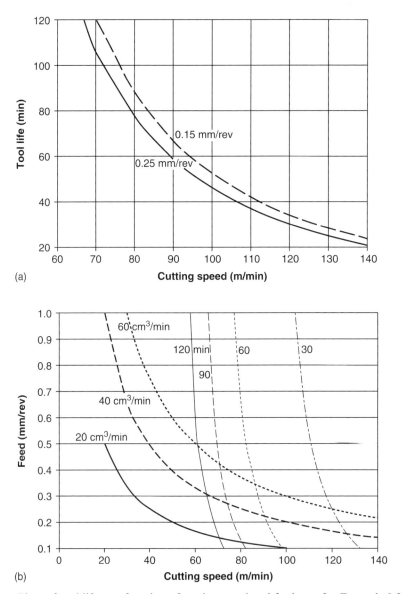

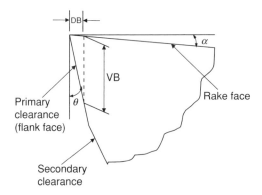

FIGURE 9.26 Plots of tool life as a function of cutting speed and feed rate for Example 9.2.

FIGURE 9.27 Definition of the rake angle (α) and primary clearance angle (θ).

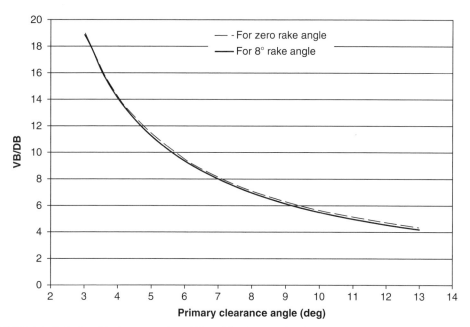

FIGURE 9.28 Relationship between the width of flank wear land VB and the clearance angle for zero and 8° rake tools for the data in Example 9.3.

equation). Therefore, the width of the flank wear is reduced significantly by increasing the primary clearance angle. The rake angle has no influence on the width of flank wear as shown in Figure 9.28. However, large clearance and rake angles will weaken the cutting edge significantly and the reliability of the tool will be reduced as explained in Chapter 4. VB/DB decreases with increasing clearance angle, which results in longer tool life even though the cutting tool edge length DB is reduced significantly. DB is important in turning or boring operations and can be compensated for if a relation between DB to tool life is known.

9.14 PROBLEMS

Problem 9.1 The parameters of the Taylor tool life equation for a carbide tool used in steel machining were estimated to be $n = 0.4$ and $C_t = 450$. Calculate the percentage increase or reduction in tool life if the cutting speed is (a) reduced by 20% and (b) increased by 30%.

Problem 9.2 Determine the constants of Taylor tool life equation, n and C_t, for the three tool materials shown in Figure 9.29.

Problem 9.3 Tool life tests were conducted with a carbide tool at 3 mm depth of cut and 0.5 mm feed using three different workpiece materials. The tool life was evaluated at several speeds as summarized in Table 9.5.

1. Plot the Taylor tool life curves in logarithmic scale.
2. Determine the Taylor tool life equation ($VT^n = C_t$) for each workpiece material.
3. What are some of the shortcomings of Taylor's equation: $VT^n = C_t$?
4. Referring to the 1020 CRS data in Table 9.5

   (a) find the cutting speed (V) for 60 min of tool life,
   (b) what would the tool life be if the cutting speed (V) is increased 20%?

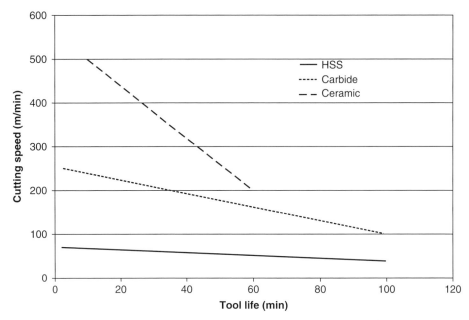

FIGURE 9.29 Tool life as a function of cutting speed for three tool materials (for Problem 9.2).

TABLE 9.5
Tool Life Data for a Carbide Cutting Tool With Three Different Workpiece Materials

Workpiece Material	Cutting Speed (m/min)	Tool Life (min)
(A) Gray 30 cast iron	30	350
	45	115
	60	53
	75	29
	90	17.5
	105	11.5
(B) 1020 CRS	105	900
	120	520
	135	330
	150	200
	165	145
	180	100
(C) Pearlitic malleable (BHN 180)	60	580
	75	270
	90	150
	105	80
	120	50
	135	35
	150	24
	180	13

REFERENCES

1. C.F. Barth, Turning, *Handbook of High Speed Machining Technology*, Chapman and Hall, New York, 1985, 173–196
2. A.T. Santhanam, P. Tierney, and J.L. Hunt, Cemented carbides, *Metals Handbook*, 10th Edition, Vol. 2, ASM, Materials Park, OH, 1990, 950–977
3. P.K. Wright and A. Bagchi, Wear mechanisms which dominate tool life in machining, *J. Appl. Metalworking* **1** (1981) 15–23
4. G.T. Smith, *Advanced Machining: The Handbook of Cutting Technology*, IFS/Springer Verlag, London, 1989, 186–190, 227–258
5. M.C. Shaw, *Metal Cutting Principles*, Oxford University Press, Oxford, 1984, Chapter 11
6. E.J.A. Armarego and R.H. Brown, *The Machining of Metals*, Prentice-Hall, New York, 1969, Chapter 7
7. E.M. Trent, *Metal Cutting*, Butterworths, London, 1977, Chapter 6
8. N.H. Cook, Tool wear and tool life, *ASME J. Eng. Ind.* **95** (1973) 931–938
9. Tool Life Testing with Single-Point Turning Tools, ISO Standard 3685:1993(E), 1993
10. M.R. Abou-Zeid and S.E.D.M.K. Oweis, Cutting Tool Wear, SME Technical Paper TE81-956, 1981
11. I. Ham, Fundamentals of Tool Wear, ASTME Technical Paper MR68-617, 1968
12. V.C. Venkatesh and M. Satchithanandam, A discussion on tool life criteria and total failure causes, *CIRP Ann.* **29** (1980) 19–22
13. M. Lee, J. Horne, and D. Tabor, *Proceedings of the International Conference on Wear of Materials*, ASME, New York, 1979, 460–469
14. T. Kurimoto and G. Barrow, The influence of aqueous fluids on the wear characteristics and life of carbide cutting tools, *CIRP Ann.* **31** (1982) 19–23
15. H. Takeyama and R. Murata, Basic investigation of tool wear, *ASME J. Eng. Ind.* **85** (1963) 33–38
16. S.M. Wu and R.N. Meyer, Optical contour mapping of cutting tool crater wear, *Int. J. MTDR* **6** (1966) 153–170
17. N.H. Cook, *Manufacturing Analysis*, Addison-Wesley, Englewood Cliffs, NJ, 1966, 60–61, 69
18. S. Popov, *Nucleonics* **19**:76 (1961)
19. P.F. McGoldrick and M.A.M. Hijazi, The use of a weighing method to determine a tool wear algorithm for end-milling, *Proceedings of the 20th International MTDR Conference*, 1979, 345–349
20. S.N. Mukherjee and S.K. Basu, Multiple regression analysis in evaluation of tool wear, *Int. J. MTDR* **7** (1967) 15–21
21. R.A. Etheridge and T.C. Hsu, The specific wear rate in cutting tools and its application to the assessment of machinability, *CIRP Ann.* **17** (1970) 107–117
22. A. Ber and M.Y. Friedman, On the mechanism of flank wear in carbide tools, *CIRP Ann.* **15** (1967) 211–216
23. G.F. Michelletti, W. Koenig, and H.R. Victor, In process tool wear sensors for cutting operations, *CIRP Ann.* 25 (1976) 483–496
24. N.H. Cook, Tool wear sensors, *Wear* **62** (1980) 49–57
25. J.I. El Gomayel and K.D. Bregger, On-line tool wear sensing for turning operations, *ASME J. Eng. Ind.* **108** (1986) 44–47
26. N. Constantindes and S. Bennet, An investigation of methods for the on-line estimation of tool wear, *Int. J. Mach. Tools Manuf.* **27** (1987) 225–237
27. M.E. Mercanht, H. Ernst, and E.J. Krabacher, Radioactive cutting tools for rapid tool life testing, *ASME Trans.* **75** (1953) 549–559
28. B. Colding and L.G. Erwall, Wear studies of irradiated carbide cutting tools, *Nucleonics* **11** (1953) 46–49
29. N.H. Cook and A.B. Lang, Criticism of radioactive tool life testing, *ASME J. Eng. Ind.* **85** (1963) 381–387
30. G. Lume and B. Anderson, A study of the wear processes of cemented carbide cutting tools by a radioactive tracer technique, *Int. J. MTDR* **10** (1970) 79–93
31. H. Chen and A.T. Alpas, Wear of aluminum matrix composites reinforced with nickel-coated carbon fibres, *Wear* **192** (1996) 186–198

32. F. Bergman and S. Jacobson, Tool wear mechanisms in intermittent cutting of metal matrix composites, *Wear* **179** (1994) 89–93
33. W. Kim and P. Kwon, Understanding the mechanisms of crater wear, *Trans. NAMRI/SME* **29** (2001) 383–390
34. E. Rabinowicz, *Friction and Wear of Materials*, Wiley, New York, 1965, Chapters 6 and 7
35. B.M. Kramer, On tool materials for high speed machining, *ASME J. Eng. Ind.* **109** (1987) 87–91
36. P.K. Wright and E.M. Trent, Metallurgical appraisal of wear mechanisms and processes on high-speed-steel cutting tools, *Met. Technol.* **January** (1974) 1323
37. T.J. Drozda and C. Wick, *Tool and Manufacturing Engineers Handbook*, Vol. 1: Machining, SME, Dearborn, MI, 1988, Chapter 1
38. B.F. Von Turkovich and W.E. Henderer, On the tool life of high speed steel tools, *CIRP Ann.* **27** (1978) 35–38
39. D.A. Stephenson, Assessment of steady-state metal cutting temperature models based on simultaneous infrared and thermocouple measurements, *ASME J. Eng. Ind.* **113** (1991) 121–128
40. E.M. Trent, Some factors affecting wear on cemented carbide tools, *Proc. Inst. Mech. Eng.* **166** (1952) 64–73
41. P.A. Dearnley and E.M. Trent, Wear mechanisms of coated carbide tools, *Met. Technol.* **9** (1982) 60–75
42. P.A. Dearnley, Rake and flank wear mechanisms of coated cemented carbides, *Surf. Eng.* **1** (1985) 43–58
43. A.T. Santhanam and G.P. Grab, Innovations in coated carbide cutting tools, *Tool Materials for High-Speed Machining*, ASM, Materials Park, OH, 1987, 67–76
44. W.W. Gruss, Cermet cutting tools: turning with small tools made of cermet grade, *Ceramic Cutting Tools and Applications*, SME, Dearborn, MI, 1989, paper 4
45. R.D. Gantt, Finish boring steel bearing liners with cermets, *Ceramic Cutting Tools and Applications*, SME, Dearborn, MI, 1989, paper 6
46. K. Kitajima, I. Matsune, Y. Tanaka, and K. Kishimoto, Cutting Performance of Solid Cermet End Mill, SME Technical Paper TE89-164, 1989
47. C. Wick, Cermet cutting tools, *Manuf. Eng.* **99**:6 (1987) 35–40
48. H.K. Toenshoff, H.G. Wobker, and C. Cassel, Wear characteristics of cermet cutting tools, *CIRP Ann.* **43** (1994) 89–92
49. W.W. Gruss, Turning steel with ceramic cutting tools, *Tool Materials for High Speed Machining*, ASM, Material Park, OH, 1987, 105–115
50. S. Lo Casto, E. Lo Valvo, and V.F. Ruisi, Wear mechanisms of ceramic tools, *Wear* **160** (1993) 227–235
51. H.K. Toenshoff and S. Batsch, Wear mechanisms of ceramic cutting tools, *Ceram. Bull.* **67** (1988) 1020–1025
52. S.T. Buljan and S.F. Wayne, Wear and design of ceramic cutting tool materials, *Wear* **133** (1989) 309–321
53. M. Masuda, T. Sato, et al., Cutting performance and wear mechanism of alumina-based ceramic tools when machining austempered ductile iron, *Wear* **174** (1994) 147–153
54. G.K.L. Goh, L.C. Lim, et al., Transitions in wear mechanisms of alumina cutting tools, *Wear* **201** (1996) 199–208
55. J. Vleugels, P. Jacobs, et al., Machining of steel with Sialon ceramics: influence of ceramic and workpiece composition on tool wear, *Wear* **189** (1995) 32–44
56. H.K. Tonschoff and B. Denkena, Tribological aspects of interrupted cutting with ceramic tool materials, *Trans. NAMRI/SME* **19** (1991) 191–198
57. C. Wick, Machining with PCBN tools, *Manuf. Eng.* **101**:1 (1988) 73–78
58. K. Shintani and Y. Fujimura, Effective use of CBN tool in fine cutting (continuous and intermittent turning), *Strategies for Automation of Machining*, ASM, Materials Park, OH, 1987, 117–126
59. G. Dawson and T.R. Kurfess, *Wear Trends of PCBN Cutting Tools in Hard Turning*, White Paper, Hardinge Inc., 2001
60. K. Liu, X.P. Li, et al., CBN tool wear in ductile cutting of tungsten carbide, *Wear* **255** (2003) 1344–1351

61. Tool Life Testing in Milling — Part 1: Face Milling, ISO Standard 8688-1, 1989
62. Tool Life Testing in Milling — Part 2: End Milling, ISO Standard 8688-2, 1989
63. Standard Method for Evaluating Machining Performance of Ferrous Metals Using and Automatic Screw/Bar Machine, ASTM Standard E618-81, 1981
64. The Volvo Standard Machinability Test, Standard 1018.712, Volvo Laboratory for Manufacturing Research, Trollhattan, Sweden, 1989
65. A.J. DeArdo, C.I. Garcia, R.M. Laible, and U. Eriksson, A better way to assess machinability, *Am. Machinist* **137**:5 (1993) 33–35
66. J.J. Fulmer and J.M. Blanton, Enhanced Machinability of P/M Parts Through Microstructure Control, SAE Technical Paper 940357, 1994
67. Y. Matsushima, M. Nakamura, H. Takeshita, S. Akiba, and M. Katsuta, Improvement of drilling machinability of microalloyed steel, *Kobelco Technol. Rev.* **17** (1994) 38–43
68. F.W. Taylor, On the art of cutting metals, *ASME Trans.* **28** (1907) 31–35
69. T. Floyd, High efficiency turning, *Cutting Tool Eng.* **March** (1993) 55–58
70. N.E. Woldman and R.C. Gibbons, *Machinability and Machining of Metals*, McGraw-Hill, New York, 1951, 47–53
71. M. Kronenberg, *Machining Science and Application*, Pergamon Press, Oxford, 1966, 170–204
72. O.W. Boston, *Metal Processing*, Wiley, New York, 1941
73. B.N. Colding, A three-dimensional, tool-life equation — machining economics, *ASME J. Eng. Ind.* **81** (1959) 239–250
74. H.J.J. Kals and J.A.W. Hijink, A computer aid in the optimization of turning conditions in multicut operations, *CIRP Ann.* **27** (1978) 465–469
75. S. Rossetto and A. Zompi, Stochastic tool life model, *ASME J. Eng. Ind.* **103** (1981) 126–130
76. S. Ramalingam, Tool Life Distribution, ASME Paper 77-WA/PROD-40, 1977
77. W.J. Zdeblick and R.E. DeVor, Open loop adaptive control methodology, *Proceedings of the 7th NAMRC*, SME, Dearborn, MI, 1979, 300–306
78. S.M. Wu, Tool life testing by response surface methodology. Parts I and II, *ASME J. Eng. Ind.* **86** (1964) 105–116
79. R. Vilenchich, K. Strobele, and R. Venter, Tool life testing by response surface methodology coupled with a random strategy approach, *Proceedings of the 13th International MTDR Conference*, 1972, 261–266
80. E. Kuljanic and T.F.R. Rijeka, Random strategy method for determining tool life equations, *CIRP Ann.* **29** (1980) 351–356
81. S. Ramalingam and J.D. Watson, Tool-life distributions. Part 1. Single-injury tool-life model; Part 2. Multiple-injury tool-life model, *ASME J. Eng. Ind.* **99** (1977) 523–531
82. S. Ramalingam, Y.I. Peng, and J.D. Watson, Tool-Life Distributions. Part 3. Mechanism of Single Injury Tool Failure and Tool Life Distribution in Interrupted Cutting, ASME Paper 77-WA/PROD-39, 1977
83. S. Ramalingam and J.D. Watson, Tool-Life Distributions. Part 4. Minor Phases in Work Material and Multiple-Injury Tool Failure, ASME Paper 77-WA/PROD-40, 1977
84. S.M. Pandit, Data Dependent Systems Approach to Stochastic Tool Life and Reliability, ASME Paper 77-WA/PROD-33, 1977
85. K. Hitomi, N. Nakamura, and S. Inoue, Reliability Analysis of Cutting Tools, ASME Paper 78-WA/PROD-9, 1978
86. B.F. Von Turkovich and W.E. Henderer, On the tool life of high speed steel tools, *CIRP Ann.* **27** (1978) 35–38
87. S.M. Pandit and C.H. Kahng, Reliability and life distribution of ceramic tools by data dependent systems, *CIRP Ann.* **27** (1978) 23–27
88. S. Rossetto and R. Levi, Fracture and wear as factors affecting stochastic tool-life models and machining economics, *ASME J. Eng. Ind.* **99** (1977) 281–287
89. A.I. Daschenko and V.N. Redin, Control of cutting tool replacement by durability distributions, *Int. J. Adv. Manuf. Technol.* **3** (1988) 36–60
90. J.M. Pan, W.J. Kolarik, and B.K. Lambert, Mathematical model to predict system reliability to tooling for automated machining systems, *Int. J. Prod. Res.* **24** (1986) 493–505

91. S. Rossetto and A. Zompi, A stochastic tool-life model, *ASME J. Eng. Ind.* **103** (1981) 126–130
92. L.M. Leemis, Lifetime distribution identities, *IEEE Trans. Reliabil.* **R35** (1986) 170–174
93. J. Taylor, Tool wear, life and surface finish, *Proceedings of the International Production Engineering Research Conference*, ASME, New York, 1963, 130–136
94. E. Usui, T. Shirakashi, and T. Kitegawa, Analytical prediction of three dimensional cutting process. Part 3. Cutting temperature and crater wear of carbide tool, *ASME J. Eng. Ind.* **100** (1978) 236–243
95. M.C. Shaw and S.O. Dirke, On the wear of cutting tools, *Microtechnic* **10** (1956) 187
96. K.J. Trigger and B.T. Chao, The mechanism of crater wear of cemented carbide tools, *ASME Trans.* **78** (1956) 1119
97. B.M. Kramer and N.P. Suh, Tool wear by solution: a quantitative understanding, *ASME J. Eng. Ind.* **102** (1980) 303–309
98. E. Kannatey-Asibu, Jr., A transport-diffusion equation in metal cutting and its application to analysis of the rate of flank wear, *ASME J. Eng. Ind.* **107** (1985) 81–89
99. V.C. Venkatesh, On a diffusion wear model for high speed steel tools, *ASME J. Eng. Ind.* **100** (1978) 436–441
100. B.M. Kramer, An Analytical Approach to Tool Wear Prediction, PhD Thesis, Mechanical Engineering, Massachusetts Institute of Technology, Cambridge, MA, 1979
101. Y.-C. Yen, J. Soehner, H. Weude, J. Schmidt, and T. Altan, Estimation of tool wear of carbide tool in orthogonal cutting using FEM simulation, *Proceedings of the 5th CIRP International Workshop on Modeling of Machining Operations*, West Lafayette, IN, 2002, 149–160
102. J. Tlusty and Z. Masood, Chipping and breakage of carbide tools, *ASME J. Eng. Ind.* **100** (1978) 403–412
103. D.F. Galloway, Some experiments on the influence of various factors on drill performance, *ASME Trans.* **79** (1957) 191–231
104. S. Sonderberg, O. Vingsbo, and M. Nissle, Performance and failure of high speed steel drills related to wear, *Wear of Materials — 1981*, ASME, New York, 1981, 456–467
105. K. Subramanian and N.H. Cook, Sensing of drill wear and prediction of drill life, *ASME J. Eng. Ind.* **99** (1977) 295–301
106. W.E. Henderer, Relationship between alloy composition and tool life of high speed steel twist drills, *ASME J. Eng. Mater. Technol.* **114** (1992) 459–464
107. J. Alverio, J.S. Agapiou, and C.H. Shen, High speed drilling of 390 aluminum, *Trans. NAMRI/SME* **18** (1990) 209–215
108. R.T. Coelho, S. Yamada, D.K. Aspinwall, and M.L.H. Wise, The application of polycrystalline diamond (PCD) tool materials when drilling and reaming aluminum based alloys including MMC, *Int. J. Mach. Tools Manuf.* **35** (1995) 761–774
109. G.T.E. Valenite Corporation, Pocket Drilling Guide, 1987
110. J.S. Agapiou, An evaluation of advanced drill body and point geometries in drilling cast iron, *Trans. NAMRI/SME* **19** (1991) 79–89
111. N.N. Zorev, Machining steel with a carbide tool in interrupted heavy-cutting conditions, *Russ. Eng. J.* **43** (1963) 43–47
112. N.N. Zorev, Standzeit und Leistung der Hartmetall-Werkzeuge beim Unterbrochenen Zerspanen des Stahls mit Grossen Zer-spanungsquerschnitten, *CIRP Ann.* **11** (1963) 201–210
113. T. Hoshi and K. Okushima, Optimum diameter and position of a fly cutter for milling 0.45 C steel, 195 BHN and 0.4 C steel, 167 BHN at light cuts, *ASME J. Eng. Ind.* **87** (1965) 442–446
114. P. M Braiden and D.S. Dugdale, Failure of carbide tools in intermittent cutting, *Materials for Metal Cutting*, ISI Special Publication 126, ISI, London, 1970, 30–34
115. H. Opitz and H. Beckhaus, Influence of initial contact on tool life when face milling high strength materials, *CIRP Ann.* **18** (1970) 257–264
116. A.J. Pekelharing, Cutting tool damage in interrupted cutting, *Wear* **62** (1978) 37–48
117. S.F. Wayne and S.T. Buljan, The role of thermal shock on tool life of selected ceramic cutting tool materials, *J. Am. Ceram. Soc.* **75** (1989) 754–760
118. T. Matsumura and E. Usui, Self-adaptive tool wear monitoring system in milling process, *Trans. NAMRI/SME* **29** (2001) 375–382

119. S.M. Pandit and S. Kashou, A data dependent systems strategy of on-line tool wear sensing, *ASME J. Eng. Ind.* **104** (1982) 217–223
120. P.K. Ramakrishna Rao, P. Prasad, et al., On-line wear monitoring of single point cutting tool using vibration techniques, *NDE Science & Technology, Proceedings of the 14th World Conference on Non-Destructive Testing*, New Delhi, December 8–13, 1996, 1151–1156
121. L. Dan and J. Mathew, Tool wear and failure monitoring techniques for turning — a review, *Int. J. Mach. Tools Manuf.* **30** (1990) 579–598
122. J.J. Park and A.G. Ulsoy, On-line flank wear estimation using an adaptive observer and computer vision. Part 1. Theory, *ASME J. Eng. Ind.* **115** (1993) 31–36
123. J.J. Park and A.G. Ulsoy, On-line flank wear estimation using an adaptive observer and computer vision. Part 2. Experiment, *ASME J. Eng. Ind.* **115** (1993) 37–43
124. R.G. Landers, A.G. Ulsoy, and R. Furness, Monitoring and control of machining operations, in: O. Nwokah, Ed., *Mechanical Systems Design Handbook*, CRC Press, Boca Raton, FL, 2001
125. L. Wang, M.G. Mehrabi, and E.K. Asibu, Hidden Markov model-based tool wear monitoring in turning, *ASME J. Manuf. Sci. Eng.* **124** (2002) 651–658
126. L. Wang, M.G. Mehrabi, and E.K. Asibu, Tool wear monitoring in reconfigurable machining systems through wavelet analysis, *Trans. NAMRI/SME* **29** (2001) 399–406
127. M.C. Lu and E. Kannatey-Asibu, Analysis of sound signal generation due to flank wear in turning, *International ME Congress & Exposition*, Orlando, FL, 2000
128. P.K. Ramakrishna Rao, P. Prasad, et al., Acoustic emission technique as a means for monitoring single point cutting tool wear, *NDE Science & Technology, Proceedings of the 14th World Conference on Non-Destructive Testing*, New Delhi, December 8–13, 1996, 2513–2518
129. G.F. Michelleti, Tool wear monitoring through acoustic emission, *CIRP Ann.* **38** (1989) 99–102
130. C. Scheffer and P.S. Heyns, Wear monitoring in turning operations using vibration and strain measurements, *Mech. Systems Signal Process.* **15** (2001) 1185–1202

10 Surface Finish and Integrity

10.1 INTRODUCTION

Many parts are machined to produce specific surface characteristics because they have features such as bearing, locking, or gasketing surfaces which require a consistent surface finish. In many applications, especially finishing operations, the surface finish requirement restricts the range of tool geometries and feed rates which can be used. Moreover, since the machined surface finish becomes rougher and less consistent as the tool wears, stringent finish requirements may also limit tool life and thus strongly influence machining productivity and tooling costs.

There are two components or features to a machined surface finish. The first is the ideal or geometric finish, which is the finish that would result from the geometry and kinematic motions of the tool. The ideal finish can be calculated from the feed rate per tooth, the tool nose radius, and the tool lead angle. It is the predominant component of the finish in operations in which tool wear and cutting forces are low, for example, when machining aluminum alloys with diamond tools. The second component is the natural finish, which results from tool wear, vibration, machine motion errors, and work material effects such as inhomogeneity, built-up edge (BUE) formation, and rupture at low cutting speeds. Unlike the ideal finish, the natural finish is difficult to predict in general. It is often the predominant component of the finish when machining steels and other hard materials with carbide tooling, or when machining inhomogeneous materials such as cast iron or powder metals.

Engineering surfaces manufactured by different processes have different topographies. For example, a milled surface is spatially inhomogeneous, a ground surface may have pits and troughs, and a honed surface has cross-hatched grooves. Such surface features often cannot be modeled analytically; therefore, statistical parameters have traditionally been used to describe their influence on part function.

The first part of this chapter describes methods of measuring and predicting the machined surface finish. The measurement of surface finish is discussed in Section 10.2; the discussion concentrates on stylus-type measurements, which are most common in practice. The surface finish in turning, boring, milling, drilling, reaming, and grinding operations is discussed in Section 10.3–Section 10.6.

These sections consider only the texture of the machined surface. Of equal importance in many applications is the surface integrity, which can be defined broadly as the metallurgical and mechanical state of the machined surface. Surface integrity can be assessed using microhardness measurements or microstructural analyses which reveal microcracks, phase transformations, melted and redeposited layers, and similar features. Residual stresses in machined surfaces are discussed in Section 10.7. White layer formation in steels and surface burn in grinding are considered in Section 10.8 and Section 10.9. Surface integrity is influenced by the thermal effects in cutting (Chapter 7) and is best understood for ferrous work materials.

10.2 MEASUREMENT OF SURFACE FINISH

10.2.1 Stylus Measurements

The finish of machined surfaces is most commonly measured with a stylus-type profile meter or profilometer, an instrument similar to a phonograph which amplifies the vertical motion of a stylus as it is drawn across the surface [1–3]. The output of the profilometer is a two-dimensional profile of the traced surface segment, amplified in the directions both normal and along the surface to accentuate surface contours and irregularities (Figure 10.1).

On a gross scale, the surface profile of a nominally smooth surface gives an indication of the surface's shape, waviness, and roughness (Figure 10.2). The *shape* of the surface is the macroscopic surface contour. Shape errors may result from errors in the machine tool guideways or machined part, distortions due to clamping forces or subsequent heat treatment,

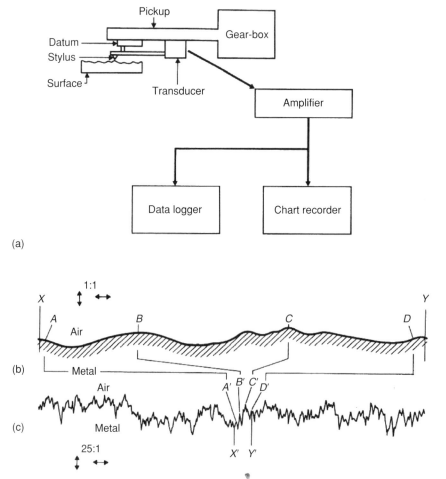

FIGURE 10.1 Schematic illustration of a profilometer (a), and typical unamplified (b) and amplified (c) surface profile traces. (After T.R. Thomas [2].)

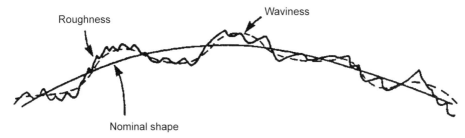

FIGURE 10.2 The shape, waviness, and roughness of a surface. (After D.J. Whitestone [1].)

and tool wear. For nominally flat surfaces, the shape is referred to as the slope or lay of the surface. *Waviness* refers to variations in the surface with relatively long wavelengths or, equivalently, lower freqencies. Waviness may result from clamping errors, errors in the tool or cutter geometry, or vibrations of the tool, or workpiece. As discussed in Section 10.4, the spindle tilt in face milling operations also produces a waviness or shape error. *Roughness* is the term for surface profile variations with wavelengths shorter than those characteristic of waviness. Roughness has a geometric component dependent on the feed rate, tool nose radius, tool lead angle and cutting speed, as well as a natural component resulting from tool wear, inhomogeneities in the work material, higher-frequency vibrations of the machining system, and damage to the surface caused by chip contact. In measurement practice, roughness, waviness, and lay are generally distinguished by cutoff wavelengths.

Many parameters have been proposed to characterize surface roughness [1, 2, 4–9]. The commonly used parameters, together with standard measurement methods, are defined in national and international standards, which are broadly equivalent but differ somewhat in detail. The American national standard is ASME B46.1-2002 [10]; a series of ISO and DIN standards cover the same subjects, the most relevant in the current context being ISO 4288:1996, ISO 4287:1997, and DIN 4768:1990 [11–13]. These standards define the parameters most often used for inspection and tolerancing of machined surfaces: the arithmetic average roughness R_a, maximum peak height R_p, maximum valley depth R_v, peak-to-valley height R_t, ten-point height R_z, and bearing ratio t_p.

The parameters R_a, R_v, R_p, and R_t are all defined with respect to a centerline of a filtered stylus trace (Figure 10.3). Filtering is performed to remove the slope and waviness components of the trace. Once filtering is performed, the centerline is determined as the mean line of

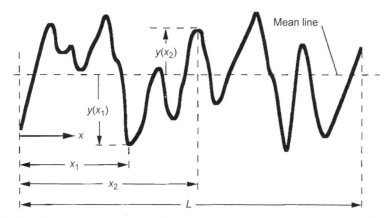

FIGURE 10.3 Definition of parameters used to compute the roughness measures R_a, R_p, R_v, and R_t.

the surface profile. Profile deviations from the centerline are designated as y. The average roughness R_a is defined as the average absolute deviation of the workpiece from the centerline:

$$R_a = \frac{1}{L} \int_0^L |y(x)| \, dx \tag{10.1}$$

R_p is the maximum deviation of a peak above the centerline encountered within the sampling length (Figure 10.3):

$$R_p = \max y(x), \quad 0 < x < L \tag{10.2}$$

where x is the distance along the trace and L is the sampling length. Similarly, R_v, the maximum depth of valley below the centerline, is defined as

$$R_v = |\min y(x)|, \quad 0 < x < L \tag{10.2a}$$

R_t, the maximum peak-to-valley deviation, is equal to

$$R_t = R_p + R_v \tag{10.3}$$

The ten-point height R_z (Figure 10.4) is measured using the unfiltered profile, and is defined as the difference between the average values of the five highest peaks Y_{pi} and the five lowest valleys Y_{vi}:

$$R_z = \frac{1}{5} \sum_{i=1}^{5} (y_{pi} - y_{vi}) \tag{10.4}$$

The bearing ratio t_p (Figure 10.5) is a function of the depth p below the highest peak and is defined as the ratio of the total length of the profile below the depth p to the total trace length L:

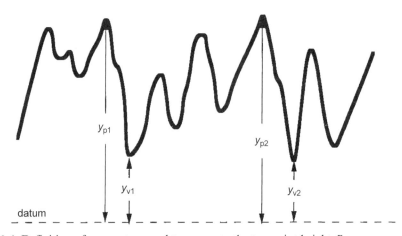

FIGURE 10.4 Definition of parameters used to compute the ten-point height R_z.

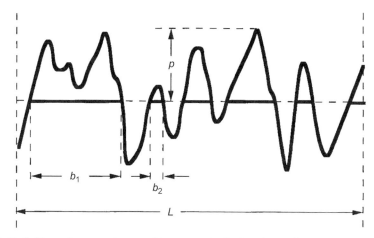

FIGURE 10.5 Definition of parameters used to compute the bearing ratio t_p.

$$t_p = \frac{\sum_{i=1}^{N} b_i}{L} \qquad (10.5)$$

where N is the number of segments over which the profile height exceeds p over the entire trace length.

R_a is the most commonly specified roughness parameter and is well suited for monitoring the consistency of a machining process. R_a increases with tool wear and is also sensitive to changes in the condition of the coolant or work material. R_z is similar to R_a but is more sensitive to the presence of high peaks or deep valleys and gives better sense of deviations from the mean. R_t, R_p, R_v, and similar parameters are sensitive to the presence of high peaks and deep scratches which would affect oil loss on sliding or sealing surfaces [8]. R_t is a good indicator of potential crack propagation, which is more likely when there are many high peaks and low valleys. The bearing area curve provides information on how the surface profile affects the tribology of bearing surfaces. It indicates how well two surfaces might mate for sealing or wear (e.g., meshing gears and cylinder bores). The related parameter P_c (representing the number of peaks above a given height in a given length or area) has been used as an indicator for the performance of surfaces designed for plating, painting, forming, lubricant retention.

Other surface parameters have been used to characterize certain functional surfaces. The motif parameters [14, 15] have been used in the French automotive industry to describe significant components of the surface roughness and waviness not adequately characterized by more traditional methods; these parameters correlate well with the functions of specific mechanical parts. In addition, the DIN R_k parameters [16, 17], originally developed for the assessment of honed and plateau honed surfaces of cylinder bores, have been adapted to characterize general bearing area curves. Five parameters are used: R_k, R_{pk}, R_{vk}, M_{r1}, and M_{r2}. R_k is the core roughness depth, R_{pk} is the reduced peak height and describes the top portion of the surface that wears away first, and R_{vk} is the valley depth which indicates the oil retaining capability of deep troughs.

Surface profiles can also be analyzed using normal distribution statistics such as standard deviation, skew, and kurtosis. In this analysis, the standard deviation provides a measure of the variability of the surface, and the skew and kurtosis are useful in characterizing surfaces for bearing and locking applications [1, 3]. Skew can distinguish between asymmetrical profiles of the same R_a and R_t. Negative skew indicates a predominance of valleys, while positive skew indicates a predominance of peaks. Plateau honed surfaces generally have negative skew, representing a plateau honing; ground surfaces typically have zero skew, while turned, EDMed, or bead blasted surfaces typically have positive skew.

Parameter variation in surface roughness characterization may be caused by the surface topography or by the measurement and data processing methods [9]. Variations due to the latter cause may result from improper selection of instrument factors such as the stylus geometry and sampling interval and length. The stylus radius has a significant influence on the measurement of surface geometry, especially for smooth surfaces. For ground surfaces, the stylus diameter is typically 0.01 mm. A stylus with a diameter of 0.005 mm may be used for finer finishes, while a 0.02 mm diameter stylus is used for rougher surfaces generated by turning, milling, and other operations. There are two types of stylus systems: skidded and skidless. The former has a skid or foot with a much larger diameter which rides on the surface at the same time as the stylus and defines the tracing datum for the stylus measurement. The stylus of a skidless instrument is free to move up and down with respect to the instrumentation datum of the tracing unit, which provides better accuracy. The stylus force (preload) should be between 50 and 200 mg to avoid surface damage and maintain measurement consistency.

The sampling or cutoff length has a strong influence on measurement results. It determines the longest wavelength included in the profile analysis, and distinguishes roughness irregularities from waviness and form error irregularities. The sampling length is controlled by the profile filter, a low-pass filter that removes the longer wavelength components of the surface profile. Standard cutoff values are 0.08, 0.25, 0.8, 2.5, and 8 mm. 2RC phase corrected Gaussian filters are normally used. When a filtering limit for short wavelength (high frequency) profile components is not used in the instrument, a limit is determined by the stylus geometry, the electrical amplifier characteristics, or the sampling interval. The evaluation length is the length over which the trace and the roughness parameters are calculated; it is usually smaller than the traversing length so that the stylus starting and stopping intervals are not included in the analysis. The evaluation length should be at least five times the sampling length for statistical purposes.

Practical engineering surfaces normally contain nonstationary or special features such as spikes, pits, and troughs. As a result, the statistical parameters used to characterize surface roughness vary with the number of samples, correlation between samples, and the number, type and location of surface features measured in the samples [9]. Measurement errors may result from both instabilities in the tracing speed and the kinematic characteristics of the measurement system in relation to the tracing speed [18, 19]. The measurement datum and data processing datum could contribute to parameter variations; the significance of these errors is surface dependent [20, 21]. Varying the sampling interval within a proper range does not have much effect on amplitude-related parameters, but does affect spacing, slope and curvature parameters significantly [22, 23].

10.2.2 Other Methods

The main limitations of profilometer methods for characterizing surfaces are that profilometer measurements are two-dimensional and do not provide information on the three-dimensional characteristics of the surface, and that the measurements are time-consuming.

A number of alternative methods for characterizing surface topography have developed in response to these limitations. In order to characterize surfaces in three dimensions, special scanning or tracking profilometers, which measure a series of parallel traces, have been developed (Figure 10.6) [24]. Optical techniques have also been developed (Figure 10.7) [25]. Some are three-dimensional methods which require only noncontact measurements.

Surface Finish and Integrity 557

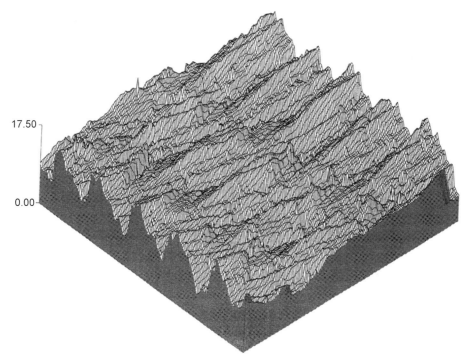

FIGURE 10.6 Three-dimensional profile of a typical end milled surface, measured using a traversing profilometer. (After K.J. Stout et al., [24].)

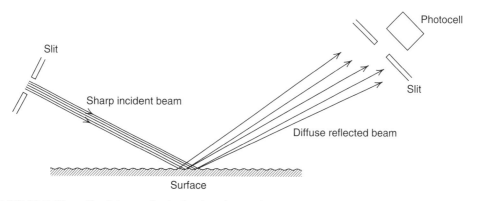

FIGURE 10.7 The reflectivity method of estimating surface roughness. (After E. Rabinowicz [25].)

For example, optical interferometry systems using a basic interferometer, phase-shifting interferometry (PSI) and vertical scanning interference (VSI) microscopy have been developed. The phase measuring mode makes it possible to measure smooth surfaces with a vertical resolution as low as 0.1 nm. The vertical scanning mode provides a measurement range up to 500 mm. Other alternatives using capacitive and pneumatic transducers are described in Ref. [1].

10.3 SURFACE FINISH IN TURNING AND BORING

A turned or bored surface usually shows a uniform roughness distribution without significant waviness (Figure 10.8). In this case, the surface finish is adequately characterized for most purposes by the average roughness parameters R_t and R_a.

As discussed in Section 10.1, there are two components to the machined surface finish: the ideal or geometric finish, and the natural finish. Since turning and boring are generally single-point operations, the geometric roughness is easily calculated from the tool angles and feed. For a sharp-nosed tool (Figure 10.9), the geometric peak-to-valley roughness R_{tg} is given by [26]

$$R_{tg} = \frac{f}{\cot \kappa_{re} + \cot \kappa'_{re}} \quad (10.6)$$

where f is the feed and κ_{re} and κ'_{re} are defined in Figure 10.9. Since the contact region between the tool and part is triangular, the average geometric roughness R_{ag} is equal to one fourth the peak-to-valley roughness R_{tg}:

$$R_{ag} = \frac{R_{tg}}{4} = \frac{f}{4(\cot \kappa_{re} + \cot \kappa'_{re})} \quad (10.7)$$

For a tool with a nose (or corner) radius (Figure 10.9), the depth of cut is usually smaller than the nose radius, especially in finish turning and boring. In this case, the geometric roughness is

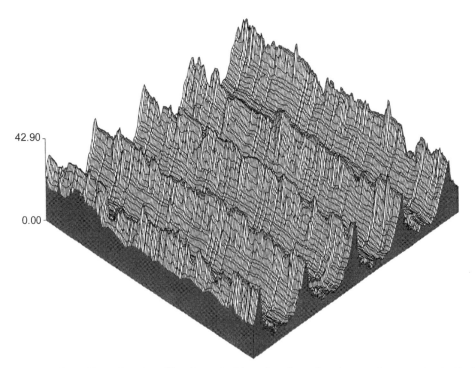

FIGURE 10.8 Three-dimensional profile of a typical bored surface, showing regular grooving in the feed direction. (After K.J. Stout et al., [24].)

Surface Finish and Integrity

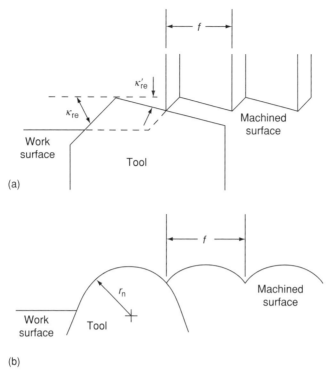

FIGURE 10.9 (a) Schematic illustrations for determining the geometric component of roughness when turning with a sharp-nosed tool and (b) a tool with a nose radius.

independent of the tool angles κ_{re} and κ'_{re}, and is determined by the feed per revolution f and nose radius r_n [3, 26]:

$$R_{tg} = \frac{f^2}{8r_n} \tag{10.8}$$

$$R_{ag} = \frac{0.0321 f^2}{r_n} \tag{10.9}$$

These equations provide a lower bound for the roughness obtained in practice and indicate that smoother surfaces can be generated by using a smaller feed, larger tool nose radius, and larger tool lead angle. In practice, all three methods are used.

Wiper nose radius inserts (Figure 10.10) are effective in improving surface finish in turning and facing operations. The wiper geometry consists of a crown with a large radius adjacent to the standard corner radius. It reduces the effective cutting edge angle and increases the effective lead angle, and cuts the peak of the feed mark. It can be used for both positive and negative rake inserts in both roughing and finishing as explained in Chapter 4. As shown in Figure 10.11, wiper inserts can reduce average roughness by a factor of 2 or more; the degree of improvement increases with the feed rate. *Trigon and 80° rhombic inserts* produce a perfect corner radius with no need for adjustment of the tool offset. *Square, triangular, and 55° rhombic inserts* do not produce a perfect corner radius, which may affect workpiece dimensions.

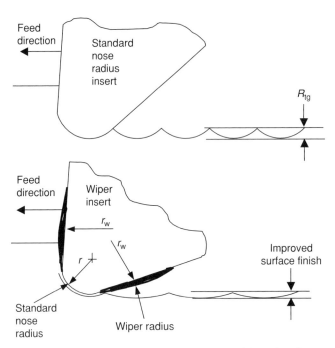

FIGURE 10.10 Illustration of how the surface finish is improved with a wiper insert versus the standard nose radius insert.

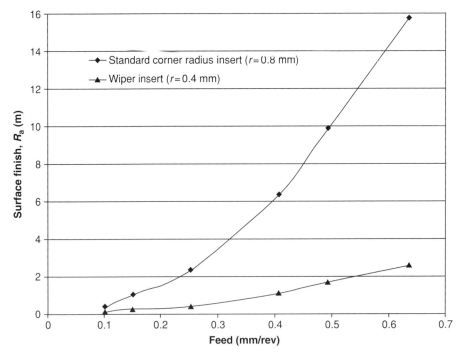

FIGURE 10.11 Comparison of the surface finish generated by a wiper insert versus the same type and size insert without wiper. The standard insert has 0.8 mm nose radius compared to 0.4 mm for the wiper insert.

The natural component of the roughness results from tool wear, errors in the machine motion, inconsistencies in the work material, discontinuous chip formation, and machining system vibrations. This component of roughness is difficult to predict. When the work material forms continuous chips, and tool wear and system vibrations are not excessive, the natural and geometric components of roughness are generally of equal magnitude, the actual roughness R_a is usually less than twice the geometric roughness. The natural component of the roughness becomes smaller as the cutting speed is increased, so that the actual roughness approaches the geometric roughness (Figure 10.12). This is particularly true for soft materials prone to BUE formation; lower speeds promote BUE, which tears and galls the machined surface. The ratio of roughness is also reduced by increasing the rake angle and applying cutting fluid (coolant) properly. Proper coolant use can affect surface texture by reducing both BUE and tool wear. The machined surface finish can also be affected by the surface finish of the cutting edge. Polishing cutting edges generally improves surface finish; a polished uncoated insert can result in a 10 to 20% improvement in surface finish. When using carbide tooling, a smoother finish is usually obtained with coated rather than uncoated inserts, assuming that the finish of the coated inserts is smoother and more uniform than that of uncoated inserts. Ceramic inserts can perform as well as coated carbides when they are new, but the surface finish deteriorates much more drastically with worn ceramics than worn carbide inserts due to differences in the wear patterns of these materials.

The surface roughness of hard-turned steel has been found to be comparable to that of ground surfaces [27–30]. The tool, toolholder, and machine requirements for machining hardened materials are somewhat different than those for softer materials as discussed in Chapter 2 and Chapter 11. Generally, the equations used for predicting roughness in finish turning cannot be extrapolated to predict the surface roughness in hard turning. The relationship between surface roughness and feed is not monotonic in hard turning, and the overall variation of the roughness is much smaller than for ordinary steels [30].

In summary, the surface roughness produced in turning and boring depends on the feed rate, tool geometry, tool wear, and work and tool material characteristics. The surface roughness produced by these operations can be reduced by:

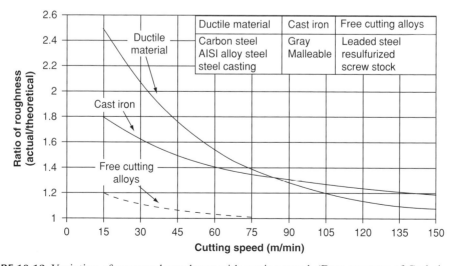

FIGURE 10.12 Variation of measured roughness with cutting speed. (Data courtesy of Carboloy, Inc.)

1. increasing the tool nose radius, provided chatter does not result;
2. decreasing the end cutting edge angle;
3. decreasing the feed rate, although this reduces the production rate;
4. decreasing the depth of cut;
5. increasing the tool lead angle (for tools with no nose radius), provided chatter does not result;
6. using sharp uncoated rather than coated tool grades (for carbide tooling), provided excessive tool wear does not result;
7. increasing the cutting speed, provided chatter and excessive tool wear do not result;
8. increasing the rake angle;
9. using the proper cutting fluid, fluid pressure and volume, and application method;
10. improving workpiece material by adding free machining additives or increasing the hardness.

10.4 SURFACE FINISH IN MILLING

The discussion of surface finish in turning and boring in the previous section is qualitatively applicable to milling. The finish in milling is also affected by a number of factors not present in turning; these factors result mainly from differences in tooling and process kinematics.

Face milling is carried out using multitoothed cutters. Due to setup or grinding errors, the cutting edges all cut at slightly different feed rates and depths of cut, especially with indexable cutters. Vibration due to the interrupted nature of the process and changes in cutter position caused by spindle and cutter runout produce further variations in the effective feed rate and depth of cut of each cutting edge. The effective feed rate also varies with the angle of the cutting edge from the feed direction. As a result of all these effects, the surface finish in milling is less uniform than that in turning and boring, in which the machined surface is formed by a single cutting edge cutting at a comparatively constant feed rate and depth of cut (Figure 10.13). The finish has a spatial variation which is not reflected in simple roughness parameters such as R_t and R_a. Measurements of R_t or R_a at specified points on the workpiece surface can be used to monitor the process for consistency; for more general applications, three-dimensional methods of assessing the surface finish should be considered.

When face milling with radiused inserts, the surface finish depends on the insert radius and on the effective feed rate. The geometric component of the average peak-to-valley cusp height, R_{tg}, and average roughness, R_{ag}, measured in the feed direction can be calculated approximately using the formulas

$$R_{tg} = \frac{f_{eff}^2}{8r_n} \tag{10.10}$$

$$R_{ag} = \frac{0.0321 f_{eff}^2}{r_n} \tag{10.11}$$

where r_n is the nose radius of the inserts and f_{eff} is the effective feed per tooth, given by

$$f_{eff} = f_t \cos\left(\frac{z}{R}\right) \tag{10.12}$$

where f_t is the feed per tooth, R is the cutter radius, and z is the distance between the measurement trace and the centerline of the cutter (Figure 10.14). These equations are

Surface Finish and Integrity

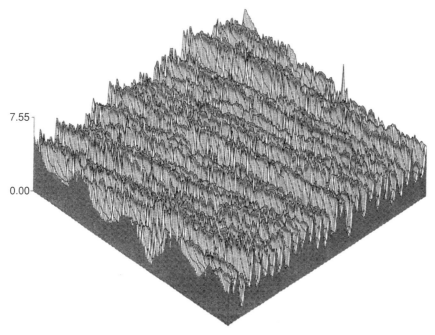

FIGURE 10.13 Three-dimensional profile of a typical fly milled surface, showing curvature of grooves in the feed direction. For face milling with multitoothed cutters, the groove height would also vary periodically. (After K.J. Stout et al., [25].)

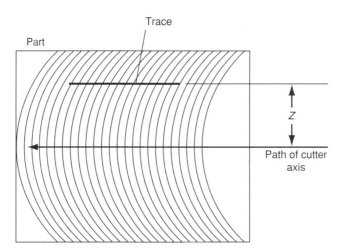

FIGURE 10.14 Schematic illustration for determining the effective feed rate for traces parallel to the feed direction.

exact for a single-toothed cutter (fly mill) with no spindle runout; for multitoothed cutters, the degree of approximation involved depends on cutter setup errors and runout.

In finish milling, fine finishes are often produced using cutters with corner chamfered inserts or wiper inserts (Figure 10.15) [3]. When using corner chamfer inserts, the chamfer should be parallel to the machined surface (perpendicular to the cutter axis), and the feed per tooth should be less than the chamfer length (see Chapter 4). Under these conditions, the

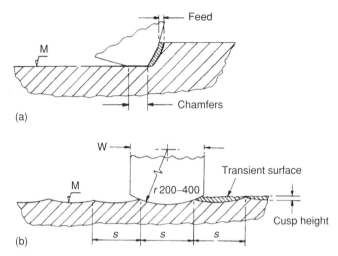

FIGURE 10.15 Milling with corner chamfer (a) and wiper (b) inserts. (After G.T. Smith [3].)

finish is generated by the cutting edge which cuts deepest. As the cutter grinding or setup accuracy improves, all cutting edges tend to cut at the same level, producing an overlapping effect which improves the machined surface finish. Wiper inserts are crowned inserts mounted behind the regular cutting inserts on a cutter. When properly applied, they improve surface roughness and waviness by deforming (or smearing) the roughness cusps produced by the regular inserts. Wipers usually have a crown radius between 200 and 400 mm. As rules of thumb, wipers should cut at a depth between 0.05 and 0.1 mm; the feed rate should be roughly half (or at least 0.75 mm less than) the wiper length. When large cutters with many wipers are used, the resulting finish depends significantly on cutter setup face runout errors. Wiper inserts can produce finer finishes than corner chamfer inserts. They are used primarily on materials such as cast iron or brass which produce short chips; when used with steel, they can impede chip flow, generating large tangential and axial forces which can lead to chatter.

In face milling applications on dedicated or transfer equipment, the spindle is tilted slightly in the direction of feed to provide relief or "dish" behind the cut (Figure 10.16). Spindle tilt produces a concave machined surface and results in a flatness error. The depth of the concavity d_f can be calculated from the Kirchner–Schulz formula [3]

$$d_f = \tan\theta \left[\frac{D_e}{2} - \left(\frac{D_e^2}{4} - \frac{e^2}{4} \right)^{1/2} \right] \quad (10.13)$$

where D_e is the effective diameter of the cutter, e is the width of the workpiece, and θ is the spindle tilt angle.

In peripheral milling (slab or end milling), the final finish is also produced by multiple cutting edges and is generally less uniform than in turning. The geometric component of the average peak-to-valley roughness height, R_{tg}, can be calculated approximately from Equation (10.10) with the feed per tooth f_t substituted for the effective feed f_{eff} [31]. More exact equations which take into account the trochoidal nature of the tool motion with respect to the workpiece are available [32, 33], but the increase in accuracy is usually small compared to the natural roughness component. Reducing spindle runout and cutter grinding or setup errors reduces the average roughness. Reducing these errors causes all cutting edges to cut at

Surface Finish and Integrity

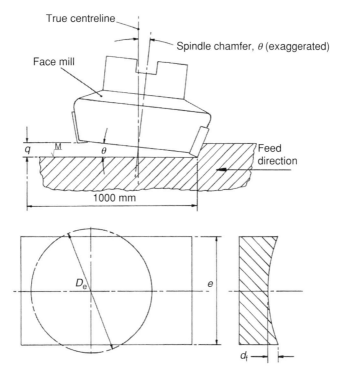

FIGURE 10.16 Waviness error produced by spindle tilt or chamfer in face milling. (After G.T. Smith [3].)

more uniform depth. The radial cutting force will contribute to the shape of the surface (a static form error perpendicular to the surface) because end mills are generally the most flexible part in the machine tool system due to low aspect ratio. The form error on the surface becomes more complex for helical end mills than straight flutes [34–38]. The tool static deflection is affected by the number of teeth cutting simultaneously. For example, if there is only one tooth in contact with the workpiece at any time, the surface form error (shape) is small in end milling (see Chapter 2 and Chapter 4) for a straight flute cutter because the uncut chip thickness becomes zero when the cutting edge passes a line perpendicular to the machine surface.

10.5 SURFACE FINISH IN DRILLING AND REAMING

The machined surface in drilling is produced by a combined cutting and rubbing action; the marginal cutting edges near the drill point initially cut the surface, and the margins and lands rub and burnish the surface as the drill enters the hole [3]. The component of the finish due to cutting depends primarily on the feed rate per revolution, while the rubbing component depends on the margin design, land width, and the plastic response (hardness and ductility) of the work material. Relatively little can be said quantitatively about the finish in drilling; the process is generally regarded as a roughing operation, so that the drilled surface roughness is not normally monitored as an indication of process consistency and has not been the subject of much research. Holes requiring fine finishes are normally finished by boring, reaming, burnishing, or honing. (Gun drilling can also produce fine surface finishes since the guide pads burnish the hole wall.) Likewise, drills with double or triple margins (or "G" drills) will produce significantly better surface finish (30 to 60% improvements) than standard two-flute drills. Finally, in cases in which indexable drills can be used, wiper inserts can be incorporated to improve hole surface finish.

BUE formation, which is common in drilling, degrades the hole surface finish and increases the hole size error. Methods of controlling BUE formation are discussed in Section 6.14; increasing the cutting speed or using through-the-tool coolant (when practical) are often the most effective control measures.

The finish produced by boring was discussed in Section 10.3. Like drilling, reaming produces a surface by a combined cutting and rubbing action. The reamed surface finish depends most strongly on the reamer's chamfer geometry, land width, and relief geometry. A double-chamfer geometry will produce a smoother finish than a standard 45° entry chamfer. The finish also improves with an increase in the land width and a decrease in the land relief angle. However, the use of double chamfers, wide lands, and low land relief angles increase the reamer's susceptibility to chatter and BUE formation. As with gun drills, gun reamers or single flute reamers with carbide or diamond pads produce very smooth surface finishes. Reamer manufacturers provide technical data for selecting optimum values of these design parameters for finish reamers as a function of the reamer diameter and work material.

In many reaming operations, especially those involving tapered reamers, the reamed surface roughness can be further reduced by allowing the reamer to dwell from 1 to 3 sec at the bottom of the stroke before retracting the tool. This approach, however, can reduce tool life and increase the likelihood of BUE and white layer formation.

10.6 SURFACE FINISH IN GRINDING

Grinding is often used to produce parts with fine surface finishes and tight tolerances. The surface finish and tolerance are related in grinding, since tolerances comparable to the average roughness are often specified. When the wheel is dressed frequently so that the wheel wear is not a significant variable, the ground surface finish depends primarily on the grinding conditions, wheel type, and wheel dressing method.

A number of equations for the ideal or geometric roughness in grinding have been reported; typical results are summarized by Malkin [39]. The actual roughness obtained, however, is normally much larger than the ideal roughness. In practice, the roughness can be predicted more reliably from empirical equations.

Empirical roughness data in cylindrical plunge grinding are usually well characterized by an equation of the form [39]

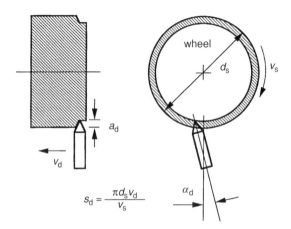

FIGURE 10.17 Single-point diamond dressing.

$$R_a = R_1 \left[\frac{v_w a}{v_s}\right]^x \quad (10.14)$$

where v_w is the workpiece velocity, v_s is the wheel velocity, a is the depth of cut, and R_1 and x are empirical coefficients for a given wheel type. The exponent x is usually between 0.15 and 0.6. The depth of cut a in Equation (10.14) is the depth of cut during the spark-out phase of the grinding process, which is smaller than the nominal stock removal. Increasing the spark-out time generally improves the ground surface finish; if complete spark-out occurs, so that the depth of cut diminished to zero before the wheel is retracted, the final roughness is reduced by about one half. The quantity in brackets in Equation (10.14) is the average uncut chip thickness.

The finish in straight surface grinding is found empirically to depend only on the speed ratio v_w/v_s; the roughness increases with this ratio, and to a first approximation is independent of depth of cut. In creep-feed grinding, the empirical roughness is generally proportional to $(v_w/v_a)\sqrt{a}$.

Dressing parameters strongly influence the ground surface finish since they determine the topography of the wheel surface. For single-point dressing (Figure 10.17), the wheel topography depends on the dressing lead s_d and dressing depth a_d; if these parameters are known, Equation (10.14) for the roughness in cylindrical plunge grinding can be refined by replacing the coefficient R_1 with a different empirical coefficient R_2 multiplied by a term including these factors [40]:

$$R_a = R_2 s_c^{1/2} a_d^{1/4} \left[\frac{v_w a}{v_s}\right]^x \quad (10.15)$$

For *rotary diamond dressing*, the wheel roughness is found to depend on the angle δ at which the diamonds initially cut the wheel surface; in this case, Equation (10.14) can be refined to include this parameter [41]:

$$R_a = R_3 \delta^{1/3} \left[\frac{v_w a}{v_s}\right]^x \quad (10.16)$$

where R_3 is an empirical coefficient.

The above equations are generally valid for a newly dressed wheel, although their accuracy is sensitive to the dressing conditions and geometry. However, as the wheel wears, they do not predict the surface finish accurately. Wheel breakdown is very difficult to predict because it is affected by several parameters which are difficult to quantify. Roughness values from these equations provide lower bounds for the surface roughness; they are most useful in predicting the effect of varying grinding conditions on the roughness when the wheel type and work material are held constant.

The condition of the dressing tool also influences the ground surface finish; the optimum combination of the dressing lead, dressing depth and dressing tool sharpness influence the surface finish. A finer finish is obtained when the dressing tool is sharp rather than dull. The nature of the dressing cycle also has an influence. In single-point dressing, for example, dressing should be carried out in a single pass, with the tool being withdrawn from the wheel prior to retraction; if the tool is allowed to cut on both the feed and retraction strokes, interfering threads will be cut on the wheel which will result in a nonuniform finish on the ground workpiece [42].

The properties of the wheel which have the strongest influence on the ground surface finish are the wheel grit size, grit spacing, effective diameter, and hardness. Smoother surface finishes are usually obtained with fine-grained wheels; as the wheel grit size increases, the effective spacing of cutting edges decreases, so that roughness peaks are more closely spaced and thus shorter. For similar reasons, the ground roughness also decreases with increasing grain density. With other factors held constant, the average chip thickness and ground roughness both decrease with increasing effective wheel diameter. Finally, smoother finish is usually obtained as the wheel hardness is increased. For example, wheels with hard shellac bonds are used to produce smooth finish on hardened workpieces [26]. As discussed in Chapter 4, the effective hardness of the wheel depends to some extent on the grinding conditions as well as the wheel structure.

Finally, it should be noted that the ground surface finish deteriorates markedly if chatter occurs. The methods of controlling chatter in general cutting processes discussed in Chapter 12 are broadly applicable to grinding. Since the specific cutting energy and contact forces are high in grinding, system stiffness is a particular concern, and the condition of the main spindle bearings and related elements should be closely monitored [42].

10.7 RESIDUAL STRESSES IN MACHINED SURFACES

Machined surfaces often exhibit residual stresses, which are induced both by differential plastic deformation and by surface thermal gradients [26, 43, 44]. Stresses due to plastic deformation are obviously mechanically induced, but those due to thermal gradients may reflect phase transformations or chemical reactions. These stresses increase with tool wear, since both deformation forces and tool–workpiece frictional heating increase. For a sharp tool, significant residual stresses typically do not occur at depths much greater than 50 μm below the surface; for worn tools, however, significant stresses may occur at 5 or 10 times this depth.

Residual stresses are undesirable for two reasons. First, residual stresses give rise to residual strains, which can cause significant distortions in thin-walled workpieces. This has long been an issue in airframe structure and turbine blade machining. (A related phenomenon occurs when machining thin-walled diccastings; in this case machining often results in deformation due to removal of a cast residual stress layer.) Second, tensile surface residual stresses reduce the fatigue resistance of the surface [45]. This is a particular issue for parts subject to cyclic loading, such as bearing races or cam lobes. A standard method for countering both these effects is to remove most material in initial machining passes, leaving a small amount of stock to remove in a final pass to clean up any distortion and significant surface stresses. This approach adds cost but is often effective.

In most cases, plastic deformation results in compressive residual stress, while thermal gradients in the absence of phase transformations result in tensile residual stresses. This is illustrated in a simplified fashion in Figure 10.18 [46]. Plastic deformation of the surface generally produces an initial tensile stress behind the tool, which results in a compressive stress after elastic recovery due to overstraining. The workpiece surface layer is heated by plastic deformation and tool–workpiece friction and is therefore at a higher temperature than the substrate of the workpiece. When the part cools the surface layer, which had expanded at a high temperature, contracts more than the substrate. The differential contraction results in a residual tensile stress in the surface. When surface temperatures are particularly high, such as when cutting hard materials with high specific energies or when cutting using a worn tool, the slope of the temperature gradient into the workpiece and peak residual stresses increase. The situation is more complex if phase transformations or chemical reactions occur in the surface layer. It should be noted that applying a flood coolant can eliminate high workpiece surface temperatures and thus tensile residual stresses in the machined surface; this is a standard method for combating residual stress, especially in grinding.

Residual stresses and microhardness distributions in machined surfaces have been documented since the early 1950s, leading to the development of models for determining the depth

Surface Finish and Integrity

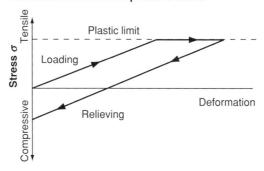

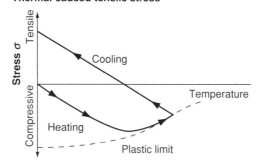

FIGURE 10.18 Simplified schematic illustration of the formation of residual stresses in machined surface layers. Plastic deformation tends to give rise to compressive stress, while thermal gradients tend to give rise to tensil stresses. (After M. Reinhardt et al., [46].)

of the hardened layer [47]. More recently, finite element methods of machining have been used to predict residual stresses from computed stress and temperature distributions [48–56]. The accuracy of these predictions depends on the accuracy of constitutive models and the stress and temperature. In the few reported studies in which residual stresses computed using commercially released finite element programs have been compared to x-ray diffraction measurements, the agreement has been inconsistent [54, 55, 57].

Residual stress formation can be reduced by reducing plastic deformation, frictional heating, and temperature gradients at the tool–workpiece interface; this can be done by applying flood coolants and avoiding the use of dull tools as discussed above, and by reducing the cutting speed or increasing the relief angle on the tool. Residual stress formation is a particular problem in grinding due to the high cutting speed, high specific energy, and blunt relief geometry of the cutting edges characteristic of the process. Residual stresses in grinding are most effectively controlled by reducing the grinding wheel speed, increasing the work speed, or applying more effective flood coolant.

10.8 WHITE LAYER FORMATION

White layer formation occurs when cutting ferrous materials, especially steels [3, 39, 58]. The white layer is a surface layer which has undergone microstructural alterations caused by excessive surface temperatures and air hardening. It is resistant to standard etchings, so that it appears white under an optical microscope (or featureless in a scanning electron microscope). The white layer has the same chemical composition as the substrate, but due to its different microstructure it has different mechanical properties, and most significantly increased hardness. In grinding, the white layer is reported to be composed of a hard layer of untempered martensite with a softer layer of tempered martensite often forming beneath it [59], although a variety of austenitic and other microstructures appear to occur in other cases [60–62]. Regardless of its microstructure, the white layer acts as a stress riser which significantly reduces the fatigue life of the part. White layer formation should therefore be avoided if possible, and any white layer detected on the part should be removed by additional machining or grinding.

White layers occur in processes in which high surface temperatures are generated in the absence of sufficient cooling by cutting fluids, specifically in grinding [39, 42, 61, 63–65], hard turning [58, 61, 62], drilling [3, 66], and electrodischarge machining [58]. In drilling, white layer formation results from the heating produced by both the cutting action of the point and the rubbing of the margin [3]. Although much of the affected layer may be removed by the drill as it penetrates the workpiece (Figure 10.19), some sections near the hole entry and at significant depths may remain following completion of the operation. The white layer in drilling not only reduces the fatigue resistance of the workpiece but also interferes with subsequent operations such as reaming, threading, and honing. White layer formation in drilling can be controlled by reducing the feed rate or cutting speed, avoiding the use of worn drills, and improving cooling through the use of through-the-tool coolant. In grinding, white layer formation results from frictional heating of the workpiece surface by the abrasive grains. It can be controlled through the same measures used to reduce residual stress formation; the measures discussed in the next section to reduce surface burning are also effective. In hard turning, white layer formation increases with tool wear, and is combated by reducing tool wear rates or changing tools more frequently.

10.9 SURFACE BURN IN GRINDING

Surface burning concerns often limit the maximum wheel speed or stock removal in grinding operations. Surface burning has been studied in detail mainly for carbon and alloy steel workpieces [39, 42, 64, 65]. Burning is accompanied by metallurgical and chemical phenomena such as oxidation (which produced the characteristic burn marks on the surface), tempering, residual stresses, and phase transformations; apart from esthetic concerns, burning should be avoided because these phenomena lead to a reduction in the workpiece fatigue life. When no oxide layer is visible (e.g., when it is removed during spark-out), the occurrence of burning can be detected through microhardness measurements (Figure 10.20) or ferrographic analysis of the grinding swarf [64]. Burning is obviously a thermal problem that can be controlled by reducing the workpiece surface temperature [39]. Thermal analyses indicate that burning occurs when a critical temperature, which depends on the specific grinding energy, is exceeded. The critical specific energy for burning, u^*, is found to depend almost linearly on a factor computed from the grinding conditions (Figure 10.21):

$$u^* \propto d_e^{1/4} a^{-3/4} v_w^{-1/2} \tag{10.17}$$

where d_e is the effective wheel diameter, a is the depth of cut, and v_w is the workpiece speed. Altering grinding conditions to increase the value of this factor increases cooling and the critical specific energy for burning. The critical specific energy is larger in creep-feed grinding than in other types of grinding because the stock removal is comparatively large. Burning can also be reduced by monitoring the wheel condition to ensure that the wheel remains sharp

Surface Finish and Integrity

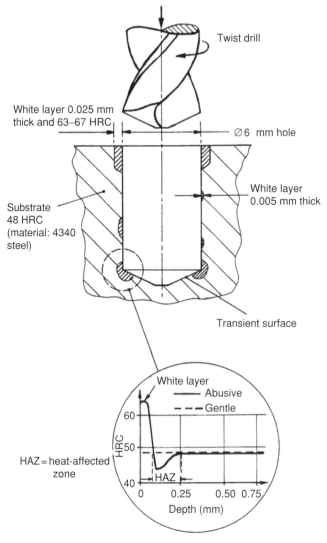

FIGURE 10.19 White layer formation when drilling steel. (After G.T. Smith [3].)

(which reduces frictional heating); force measurement systems may be used to monitor wheel sharpness. Burning can also be reduced by switching from water-based to oil-based grinding fluids. Oil-based fluids reduce heat buildup because they have better lubricity [42]. Burn limits and residual stress predictions can also be computed using grinding simulation programs and used to suitably adjust infeed cycles, the number of passes, and wheel characteristics to maximize productivity [67].

10.10 EXAMPLES

Example 10.1 A turning insert with a sharp corner is used in a finishing operation. The lead angle of the tool (or side-cutting-edge angle, SCEA) is 20°. The minor cutting edge angle (ECEA) is 10°. What maximum feed is allowed if the surface finish requirement is 3 μm?

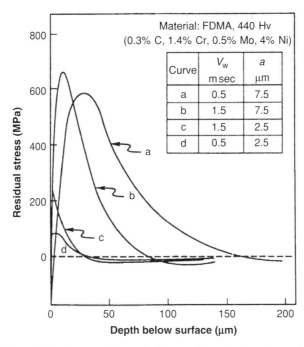

FIGURE 10.20 Variation of hardness with workpiece surface depth when grinding steel with and without surface burning. (After S. Malkin [39].)

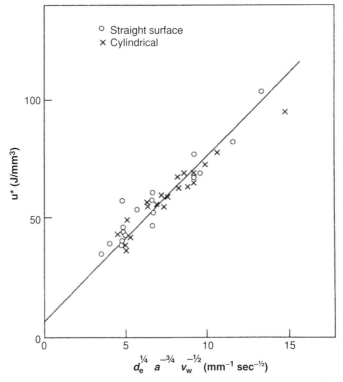

FIGURE 10.21 Variation of critical specific energy for surface burning u^* with $d_e^{1/4} a^{-3/4} v_w^{-1/2}$. (After S. Malkin [39].)

Solution: Equation (10.7) gives the average geometric roughness of a surface in turning with a sharp-nosed tool.

$$f = 4R_a[\cot \kappa_{re} + \cot \kappa'_{re}]$$
$$= 4(0.003)[\cot(90° - 30°) + \cot(10°)]$$
$$= 0.075 \text{ mm}$$

The feed was calculated by assuming that the geometric roughness generated by the insert is equal to the surface finish requirement. This neglects the natural part of the surface finish as discussed in Section 10.3. The maximum feed should be less than the ideal roughness calculated above. Therefore, if the natural part of the finish is assumed to be 40% of the geometric (based on previous experience for the particular cutting conditions, workpiece, coolant, and machine tool), the maximum feed should be:

$$f_{max} = 0.075/1.4 = 0.054 \text{ mm}$$

Example 10.2 What corner radius should be used on an insert to give an arithmetic mean surface roughness of 1 μm? The feed selected for this turning operation is 0.1 mm/rev. What should be the speed range selection if the workpiece material is aluminum?

Solution:

(a) The average geometric roughness of a surface in turning with a nose radius tool is given by Equation (10.9)

$$r_n = \frac{0.0321 f^2}{R_{ag}} = \frac{0.0321(0.1)^2}{0.001} = 0.321 \text{ mm}$$

Therefore, the standard corner radius size 1 (corresponding to 0.4 mm) is selected. It is understood that the surface finish with the 0.4 mm radius should be higher than 1 μm since the natural part of the roughness is not included in the above calculation. If it is assumed that the natural part of the roughness is, in the worse case, equal to the geometric, the size 2 nose radius (corresponding to 0.8 mm) should be selected so that the average geometric roughness is significantly lower than the upper limit of 1 μm. The surface roughness can be also maintained to 1 μm by either reducing the feed or using the same insert geometry with a wiper. However, reducing the feed will result in lower productivity, while the wiper design will result in lower tool life.

(b) The higher speed range of the machine should be used to turn this aluminum workpiece in order to minimize BUE formation and the natural part of the surface roughness.

10.11 PROBLEMS

Problem 10.1 A 250 mm diameter milling cutter using 16 square inserts with a nose radius of 0.8 mm is used to finish a 180 mm wide surface. The centerline of the cutter is offset by 20 mm from the centerline of the part. The feed per tooth is 0.1 mm. Determine the average geometric roughness of the surface along the centerline of the part and 20 mm away from the sides of the part.

Problem 10.2 A 30 mm diameter indexable drill uses a special trigon insert with 0.4 mm nose radius at the outer corner. The feed is 0.2 mm/rev. Determine the average geometric roughness of the surface in the hole. How can the surface finish of the hole be improved using such an indexable drill?

REFERENCES

1. D.J. Whitestone, *Handbook of Surface Metrology*, Institute of Physics Publishing, Philadelphia, PA, 1994, 5–33
2. T.R. Thomas, Stylus instruments, in: T.R. Thomas, Ed., *Rough Surfaces*, Longman Group, Harlow, UK, 1982, Chapter 2
3. G.T. Smith, *Advanced Machining*, Springer-Verlag, New York, 1989, 198–223
4. J. Peters, P. Vanherck, and M. Sastrodinoto, Assessment of surface topography analysis techniques, *CIRP Ann.* **28** (1979) 539–554
5. E.G. Thwaite, Measurement and control of surface finish in manufacture, *Precision Eng.* **3** (1981) 97–104
6. T.R. Thomas, Characterization of surface roughness, *Precision Eng.* **6**:4 (1984) 207–217
7. G.H. Schaffer, The many faces of surface texture, *Am. Machinist Automat. Manuf.* **June** (1988) 61–68
8. P.M. Noaker, The well-honed competitive edge, *Manuf. Eng.* **115**:3 (1995) 53–59
9. W.P. Dong, P.J. Sullivan, and K.J. Stout, Comprehensive study of parameters for characterizing three-dimensional surface topography I: Some inherent properties of parameter variation, *Wear* **159** (1992) 161–171
10. ASME B46.1–2002, Surface Texture (Surface Roughness, Waviness, and Lay) (Revision of ASME B46.1–1995)
11. ISO 4288:1996, Geometrical Product Specifications (GPS) — Surface Texture: Profile Method — Rules and Procedures for the Assessment of Surface Texture (replaces ISO 4288:1985)
12. ISO 4287:1997, Geometrical Product Specifications (GPS) — Surface Texture: Profile Method — Terms, Definitions and Surface Texture Parameters (replaces ISO 4287-1:1984)
13. DIN 4768:1990, Determination of Roughness Parameters R_a, R_z, R_{max} by Means of Stylus Instruments; Terms, Measuring Conditions
14. J. Boulanger, The MOTIFS method: an interesting complement to ISO parameters for some functional problems, *Int. J. Mach. Tools Manuf.* **32** (1992) 203–209
15. ISO 12085: 1996, Geometrical Product Specifications (GPS) — Surface Texture: Profile Method — Motif Parameters
16. U. Schneider, A. Steckroth, N. Rau, and G. Hubner, An approach to the evaluation of surface profiles by separating them into functionally different parts, *Surf. Topogr.* **1** (1988) 71–83
17. R.A. Aronson, The secrets of surface analysis, *Manuf. Eng.* **October** (1995) 57–64
18. J.I. McCool, Assessing the effect of stylus tip radius and flight on surface topography measurements, *ASME J. Tribol.* **106** (1986) 203–211
19. A. Konczakowski and M. Shiraishi, Sampling error in a/d conversion of the surface profile signal, *Precision Eng.* **4** (1982) 49–60
20. E.J. Davis and K.J. Stout, Stylus measurement techniques: a contribution to the problem of parameter variation, *Wear* **83** (1982) 49–60
21. H. Dagnall, *Exploring Surface Texture*, Rank Taylor Hobson, Leicester, UK, 1986.
22. G.T. Smith, Industrial Metrology, Springer Verlag, London, 2002, 3–67.
23. T.G. King, Rms skew and Kurtosis of surface profile height distributions: some aspects of sample variation, *Precision Eng.* **2** (1980) 207–215
24. K.J. Stout, E.J. Davis, and P.J. Sullivan, *Atlas of Machined Surfaces*, Chapman and Hall, London, 1990
25. E. Rabinowicz, *Friction and Wear of Materials*, Wiley, New York, 1965, 38–43
26. G. Boothroyd and W.A. Knight, *Fundamentals of Machining and Machine Tools*, 2nd Edition, Marcel Dekker, New York, 1989, 166–173
27. T. Ekstedt, The challenge of hard turning, *Carbide Tool J.* **19**:5 (1987) 21–24
28. W. Konig, R. Komanduri, and H.K. Toenshoff, Machining of hard materials, *CIRP Ann.* **33** (1984) 417–427

29. T.I. El-Wardany and M.A. Elbestawi, Performance of ceramic tools in hard turning, *Proceedings of the 6th International Conference of PEDAC*, Alexandria, Egypt, 1992
30. T.I. El-Wardany, M.A. Elbestawi, M.H. Attia, and E. Mohamed, Surface finish in turning of hardened steel, *Engineered Surfaces*, ASME PED-Vol. 62, ASME, New York, 1992, 141–159
31. M.C. Shaw, *Metal Cutting Principles*, Oxford University Press, Oxford, 1984, 487–519
32. E.J.A. Armarego and R.H. Brown, *The Machining of Metals*, Prentice-Hall, Englewood Cliffs, NJ, 1969, 187–188
33. N.H. Cook, *Manufacturing Analysis*, Addison-Wesley, Reading, MA, 1966, 79–80
34. S. Smith and J. Tlusty, An overview of modeling and simulation of the milling process, *ASME J. Eng. Ind.* **113** (1991) 169–175
35. W.A. Kline, et al., The prediction of surface accuracy in end milling, *ASME J. Eng. Ind.* **104** (1982) 272–278
36. E. Budak and Y. Altintas, Flexible milling force model for improved surface error predictions, *Proceedings of the 1992 Engineering System Design and Analysis Conference*, Istanbul, ASME PD-Vol. 47–1, ASME, New York, 1992, 89–94.
37. J.W. Sutherland and R.E. DeVor, An improved method for cutting force and surface error prediction in flexible end milling systems, *ASME J. Eng. Ind.* **108** (1986) 269–279
38. Y. Altintas, *Manufacturing Automation, Metal Cutting Mechanics, Machine Tool Vibrations, and CNC Design*, Cambridge University Press, Cambridge, UK, 2000, 70–72
39. S. Malkin, *Grinding Technology: Theory and Applications of Machining with Abrasives*, Ellis Horwood, Chichester, UK, 1989, Chapters 4, 6, and 7
40. S. Malkin and T. Murray, Comparison of single point and rotary dressing of grinding wheels, *Proc. NAMRC* **5** (1977) 278
41. T. Murray and S. Malkin, Effects of rotary dressing on grinding wheel performance, *ASME J. Eng. Ind.* **100** (1978) 297
42. R.S. Hahn, Trouble-shooting surface finish/integrity grinding problem, *Proceedings of the Milton C. Shaw Grinding Symposium*, ASME PED-Vol. 16, ASME, New York, 1985, 409–414
43. E.K. Henriksen, Residual stresses in machined surfaces, *ASME Trans.* **73** (1951) 69–76
44. C.R. Liu and M.M. Barash, Variables governing patterns of mechanical residual stress in a machined surface, *ASME J. Eng. Ind.* **104** (1982) 257–264
45. R.A. Roggie and J.J. Wert, Influence of surface finish and strain hardening on near surface residual stress and the friction and wear behavior of A2.D2 and CPM–10V tool steels, *Proceedings of the International Conference on Wear of Materials*, Vol. 8, Orlando, FL, 1991, 497–501
46. M. Reinhardt, L. Markworth, and S. Knodt, Modeling of difficult to cut materials, *Proceedings of the 2002 Third Wave AdvantEdge User's Conference*, Atlanta, GA, April 2002, Paper 5
47. Y. Matsumoto, M.M. Barash, and C.R. Liu, Residual stress in the machined surface of hardened steel, *High Cutting Speed*, SME, Dearborn, MI, 1986, 193–204
48. J. Mackerle, Finite-element analysis and simulation of machining: a bibliography (1976–1996), *J. Mater. Process. Technol.* **86** (1999) 17–44
49. K. Okushima and Y. Kakino, The residual stress produced by metal cutting, *CIRP Ann.* **10** (1971) 13–14
50. C. Wiesner, Residual stresses after orthogonal machining of AISI 304: numerical calcuation of the thermal component and comparison with experiments, *Metall. Trans. A* **23** (1992) 989–996
51. T. Shirakashi, T. Obikawa, H. Sasahara, and T. Wada, Analytical prediction of the characteristics within machined surface layer (1st report): the analysis of the residual stress distribution, *J. Jpn. Soc. Precision Eng.* **59** (1993) 1695–1700
52. A.J. Shih and H.T.Y. Yang, Experimental and finite element predictions of residual stresses due to orthogonal metal cutting, *Int. J. Numer. Methods Eng.* **36** (1993) 1487–1507
53. R. Liu and Y.B. Guo, Finite element analysis of the effect of sequential cuts and tool–chip friction on resdiual stresses in a machined layer, *Int. J. Mech. Sci.* **42** (2002) 1069–1086
54. M. Lundblad, Influence of cutting tool geometry on residual stress in the workpiece, *Proceedings of the 2002 Third Wave AdvantEdge User's Conference*, Atlanta, GA, April 2002, Paper 7
55. M. Russel, Modeling for machining, *Proceedings of the 2002 Third Wave AdvantEdge User's Conference*, Atlanta, GA, April 2002, Paper 10

56. C. Shet and X. Deng, Residual stresses and strains in orthogonal metal cutting, *Int. J. Mach. Tools Manuf.* **43** (2003) 573–587
57. D.A. Stephenson, Machining process analysis at General Motors Powertrain, *Proceedings of the 2004 Third Wave AdvantEdge User's Conference*, Gaithersburg, MD, March 2004, Paper 2
58. Y.K. Chou and C.J. Evans, White layers and the thermal modeling of hard turned surfaces, *Int. J. Mach. Tools Manuf.* **39** (1999) 1863–1881
59. M.C. Shaw and A. Vyas, Heat affected zones in grinding steel, *CIRP Ann.* **43** (1994) 279–282
60. S. Kompella S.P. Moylan, and S. Chandrasekar, Mechanical properties of thin surface layers affected by material removal processes, *Surf. Coat. Technol.* **146/147** (2001) 384–390
61. H. Toeshoff, H.-G. Wobker, and D. Brandt, Hard turning: influences on the workpiece's properties, *Trans. NAMRI/SME* **23** (1995) 215–220
62. A.M. Abrao and D.K. Aspinwall, The surface integrity of turned and ground hardened bearing steel, *Wear* **196** (1996) 279–284
63. R.S. Hahn, Grinding, *Handbook of High Speed Machining Technology*, Chapman and Hall, New York, 1985, 348–354
64. D. Cantillo, S. Calabrese, W.R. DeVries, and J.A. Tichey, Thermal considerations and ferrographic analysis in grinding, *Grinding Fundamentals and Applications*, ASME PED-Vol. 39, ASME, New York, 1989, 323–333
65. R. Snoeys, M. Maris, and J.R. Peters, Thermally induced damage in grinding, *CIRP Ann.* **27** (1978) 571–581
66. B.J. Griffiths, White layer formations at machined surfaces and their relationship to white layer formation at worn surfaces, *ASME J. Tribol.* **107** (1985) 165–171
67. G. Xiao, R. Stevenson, I.M. Hanna, and S.A. Huckerp, Modeling of residual stress in grinding of nodular cast iron, *ASME J. Manuf. Sci. Eng.* **124** (2002) 833–839

11 Machinability of Materials

11.1 INTRODUCTION

"Machinability" can refer to either the ease or difficulty of machining a material, as when discussing machinability ratings [1–10], or to the body of knowledge and practice accumulated on the machining of a particular material, as in machinability handbooks or databases [2, 6, 11].

When used in the first sense, machinability is often regarded as a material property and assigned a numerical rating. As shown schematically in Figure 11.1, however, the ease or difficulty of machining a particular part is affected by a variety of factors. These include the chemical composition, mechanical, and thermophysical properties of the work material, the rigidities of the machine tool, part, and fixture, the tool material, and the cutting speeds and feeds. Machinability is thus really a property of a machining system operating under a given set of conditions.

Although quantitative machinability ratings should not be misconstrued as material properties, they are often useful for comparison or ranking purposes. For example, they may be used to rank a list of material choices for a given part from most to least machinable in a general sense. In these cases the primary consideration is usually the overall machining cost, which depends most strongly on the tool life which can be achieved under the assumed production conditions. Machinability ratings are also useful for general cost estimating calculations. In job shop production, for example, the cost of machining a part made of an unfamiliar material can be estimated by comparing its machinability rating with that of a material which the shop has more experience with.

As noted above, the term "machinability" also refers to the accepted machining practices for a given material [2, 6, 11]. Machinability data collected in handbooks or computer databases consist of recommended cutting speeds, feed rates, and depths of cut for specific work materials. Separate values are normally given for different operations and tool material grades. Machinability data are generally gathered from production experience and summarizes machining conditions which yield acceptable tool life and part quality under common operating conditions. The recommended speeds and feeds represent only initial starting conditions which should be modified as needed to optimize machining performance in a given application. Since machining practice changes over time due to new developments in tool materials and machine tool capabilities, machinability data must be updated frequently to remain current. Extensive current machinability data are available from laboratories specializing in machinability testing and from cutting tool manufacturers in the form of insert grade advisers [12, 13]. Common machinability criteria, tests, and indices are discussed in Section 11.2. Chip control and burr formation, which have a significant impact on the machinability of soft, ductile alloys, are discussed in Section 11.3 and Section 11.4. General machining practices and machinability concerns for selected common engineering materials are described in Section 11.5. This material is intended to supplement the discussion on tool material applications in Chapter 4 and Chapter 9.

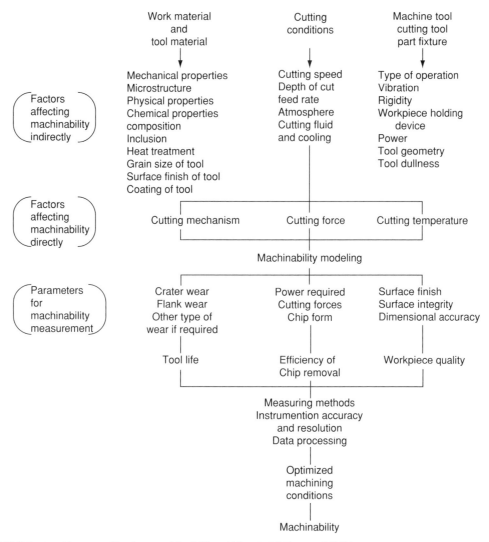

FIGURE 11.1 Factors affecting machinability. (After J. ElGomayel [10].)

11.2 MACHINABILITY CRITERIA, TESTS, AND INDICES

The most commonly used criteria for assessing machinability are [2, 6–8]:

1. *Tool life or tool wear rates.* This is the most meaningful and common machinability criterion for ferrous materials [14, 15]. Tool wear affects both the quality and cost of the machined part. Machinability is said to increase when tool wear rates decrease (or tool life increases) under the machining conditions of interest. Ratings based on wear rates are generally applicable to a restricted range of cutting conditions; when conditions change, for example, when the cutting speed is substantially increased or decreased, the dominant tool wear mechanism and tool wear rate may change. This fact is particularly relevant when ranking the machinability of a group of materials under different cutting conditions.

2. *Chip form and tendency to burr.* Materials which produce short chips which are easily managed and disposed of are more machinable than those which produce long, unbroken chips or small, powder-like chips. The chip form is particularly important for applications such as drilling for which chip breaking and disposal concerns may limit production rates. Ductile materials which form long, unbroken chips also have a greater tendency to form burrs, especially as the tool wears. Chip form and burr behavior are often used to assess the machinability of soft, ductile alloys, especially aluminum alloys, since these materials are normally machined with diamond tools and do not generate a great deal of wear. Chip control issues and burr formation are discussed in the next two sections.

Less commonly used machinability criteria include the following [8, 16]:

3. *Achievable surface finish.* Generally, machinability increases as the surface finish achievable under a given set of cutting conditions improves. As discussed in Chapter 10, the average roughness is the most common parameter used to assess surface quality in machining tests. This criterion is most useful in ranking different classes of materials, for example, conventional versus powder metal steels. This criterion is not applicable in roughing cuts.
4. *Achievable tolerance.* Machinability increases as the tolerance which can be achieved under a given set of cutting conditions decreases. As with the surface finish criterion, this criterion is most useful when comparing different classes of materials. It is more commonly used to assess the machinability of woods or plastics than metals; for metals, it is most commonly used in medium and high production applications. Tolerance criteria are also often used to define the end of tool life in tests involving form tools (e.g., the ASTM bar turning test [17]).
5. *Functional or surface integrity.* Materials which are subject to metallurgical damage such as residual stresses or hard layers are less machinable than those which are not.
6. *Cutting forces or power consumption.* Machinability increases as cutting forces and power consumption decrease for the cutting conditions of interest. Lower cutting forces generally imply lower tool wear rates, better dimensional accuracy (due to decreased deflections), and increased machine tool life (due to reduced loads on bearings and ways). Cutting force or power consumption measurements require much less material and machine time than tool life tests, and are particularly suitable for screening the machinability properties of a large group of materials to identify a few promising candidates for tool life tests. Cutting force measurements from short series of tests, however, have limited resolution and detect only gross differences in machinability.
7. *Cutting temperature.* Cutting temperature measurements such as tool-work thermocouple measurements (Chapter 7) can be used to compare the machinability of materials under a given set of cutting conditions. Machinability increases as cutting temperatures decrease because many tool wear mechanisms are temperature activated; the average temperature is therefore generally correlated with tool life, and thus with machinability. Some tool materials such as ceramics are comparatively insensitive to tool temperatures, so that small increases in temperature do not affect or may even improve their performance. Like cutting force measurements, these measurements require less time and material than tool life tests and are particularly well suited for screening purposes, but have limited resolution.
8. *Mechanical properties.* Mechanical properties of work material, such as the hardness, yield strength, ductility, and strain hardening exponent, have occasionally been correlated to machinability. The correlation exists within a group or class of metals. For

example, steels in the same class generally show negative correlation between machinability and both hardness and yeild stress. Generally, low values of hardness and yield stress are favorable. The same is true with respect to hardness within the general group of cast irons: a hardness effect is not as obvious for aluminum alloys, however. Hardness tests are used to evaluate work material uniformity between lots to ensure consistent machinability. Machinability within some classes of materials, such as heat resistant alloys and austenitic stainless steels, decreases with increasing work-hardening rate. A significant increase in hardness will occur in a thin layer of the machined surface for these materials. The depth of the work-hardened layer and the level of hardness should be kept to minimum especially if the feed per edge is closed to the depth of the work-hardened layer. In such cases, the most positive and sharp cutting edge should be selected to reduce the cutting energy.

Tool life is the most widely used machinability criterion. Therefore, as noted in Section 11.1, tool life tests are often used to assess machinability. Standardized tool life tests are described in Section 9.6. Traditionally, variants of the ISO turning test [18] or ASTM bar turning test [17] have most commonly been used for machinability assessment. These are very general tests which do not uniquely define all cutting parameters and test procedures. They do, however, provide an excellent framework for developing simpler test procedures optimized for specific applications, such as the simplified test strategy shown in Figure 11.2. Since these tests are time consuming and require a large supply of uniform materials, there has long been interest in developing test methods which require less machine time and material. The most commonly used tests of this type are the Volvo end milling test [19, 20] and various drilling tests [21–24], which are also described in Section 9.6. Less common tests, which are described in detail by Smith [5], include: face and taper turning tests, in which the cutting speed increases linearly with time due to the workpiece geometry (with the maximum speed achieved before tool failure used as the measure of machinability); tests in which the cutting speed is increased in steps, rather than continuously (permitting the use of standard bar specimens); and various nonmachining tests in which combinations of physical or microstructural properties thought to be related to tool wear rates are measured. The limited evidence available indicates most machining tests, whether based on turning, drilling, or milling, yield equivalent ranking results for high-speed steel (HSS) and uncoated carbide tools [25].

In mass production applications in which transfer machines and other serial production systems are used, tools are often changed according to a regular schedule (e.g., after one or two shifts of production). In this case small increases in tool life have less impact on production costs than the maximum production rate which can be achieved while maintaining some minimum tool life. For these operations, accelerated tool wear tests at various metal removal rates are sometimes used to assess machinability. Tests of this type have not been standardized because the required minimum tool life varies considerably from application to application.

One factor which should be considered in machinability testing is the consistency of the work material. As shown in Figure 11.1, machinability is affected by composition, hardness, microstructure, and inclusion morphology, all of which are influenced by the fabrication history of the material. If a uniform lot of work material of sufficient size for testing is not available, different grades with different fabrication histories may be used in the test, introducing confounding effects into the results. The size of the test workpiece should be standardized when possible to eliminate variation due to specimen temperature rise caused by heat entering the workpiece; this is especially important when drilling tests are used to assess machinability. Test conditions should be selected so that chatter does not occur. The

Machinability of Materials

Test purpose	Machinability evaluation for a family of work materials. Evaluation of other parameters (i.e., cutting fluids, tool materials, coatings, etc.).
Machine tools	Must be capable of maintaining constant cutting speed by variable spindle rpm. Must have adequate horsepower to maintain the cutting speed under the cutting load both new and worn tools. CNC lathes of suitable size and power are considered as the ideal test bed.
Work material	Must be obtained from one supplier and from the same batch or heat. All test pieces must have the same diameter and length.
Toolholder	Use one repeatable holder if possible. Otherwise, the same style from the same supplier must be used.
Tool material	Use an ANSI insert style, all from the same supplier, the same batch of material, and the same batch of manufacturing.
Cutting fluid	"Dry" — do not use coolant if possible. "Wet" — specify the coolant type, concentration, flow rate, pressure, temperature, and the way it is supplied at the cutting zone (i.e., nozzle location, nozzle size, filtration system).
Test procedure	Specify the cutting conditions. Specify the machining and data collection procedure.
Measuring equipment	Specify the type of instruments. Specify the procedure used for the measurements.
Cutting conditions	Select the speeds and feeds based on the objectives of the tests. Slightly higher values than those recommended (i.e., *Machinability Data Handbook*) would accelerate the testing time.
Measuring parameters	Tool wear/tool life/cutting forces/cutting power/cutting temperature/surface finish/chip formation/dimensional accuracy

FIGURE 11.2 Suggested strategy for machinability testing.

characteristics of the material should be uniform through the core. If the fabrication procedure, composition, or heat treatment produce a material with a serious change in microstructure from outer to inner diameters, a more suitable sample size and machining process should be selected.

To this point, we have discussed machinability from a comparative (ranking) perspective. For cost estimating purposes, there has long been interest in developing absolute or quantitative measures for machinability [1]. The most commonly used measure is the machinability rating or index, I_m, defined by [13]

$$I_m = 100 \times \frac{(V_{60})_{mat}}{(V_{60})_{ref}} \qquad (11.1)$$

where $(V_{60})_{mat}$ is the cutting speed at which the material being rated yields a 60-min tool life for a specified feed rate, depth of cut, tool material, and tool geometry, and $(V_{60})_{ref}$ is the cutting speed which yields a 60-min tool life for a reference material with a defined machinability rating of 100 under the same conditions. This is the machinability index which is most commonly used in tables of material properties included in engineering handbooks. In the United States, SAE B1113 sulfurized free-machining steel has traditionally been used as the reference material, although in recent years SAE 1045, 1018 [26], or 1212 [7] have been used. The index is also less commonly calculated using 20- or 90-min tool life speeds; all three definitions yield equivalent results for common work materials cut with HSS or uncoated carbide tools.

Numerous other machinability indices, often computed from more complicated equations which include metallurgical or chemical content parameters, have been proposed [1, 27]. These parameters were more meaningful in an absolute sense when the range of cutting materials available was restricted to HSS and uncoated WC. The development of coated carbide, ceramic, PCBN, and PCD tools has greatly reduced the absolute accuracy of such indices, so that they are now useful primarily for comparative or ranking purposes.

11.3 CHIP CONTROL

Chip control is an important issue in many machining operations, particularly turning and drilling operations involving ductile work materials such as low-carbon steels and aluminum alloys. Chip control usually involves two tasks: breaking of chips to avoid the formation of long, continuous chips which can become entangled in the machinery, and removal of chips from the cutting zone to prevent damage to the machined surface and errors due to thermal distortion. Because a very large spectrum of chip forms are encountered in practice, no general theory of chip control is available. Rather, a number of strategies applicable to particular operations or materials, usually based on test results, have been developed. This section summarizes the more common approaches and is intended to provide physical insight into chip-breaking and control issues. More detailed descriptions of specific strategies are given in review articles on the subject [28–32].

In turning, chip breaking is usually accomplished by directing the chip toward an obstacle to produce a bending stress sufficient to fracture it as shown in Figure 11.3. If the chip is modeled as a beam which can sustain a maximum bending strain of ε_b, the chip will break when [28]

$$h_c \left[\frac{1}{R_L} - \frac{1}{R_0} \right] \geq \varepsilon_b \qquad (11.2)$$

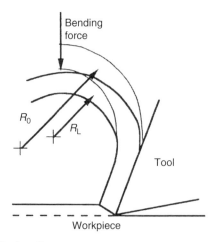

FIGURE 11.3 Chip breaking by bending.

where h_c is the distance between the neutral bending plane in the chip and the chip surface (usually roughly half the chip thickness [33]), and R_0 and R_L are the natural and obstructed radii of chip curl, respectively. The quantitative accuracy of this relation is difficult to assess because an appropriate fracture strain ε_b is difficult to estimate and the inherent radius of curl R_0 varies considerably in experiments; however, it does reflect qualitative dependencies which can be exploited to promote chip breaking.

Equation (11.2) indicates that chip breaking is promoted when ε_b is reduced and when the distance to the neutral plane, h_c, is increased. ε_b can be reduced by imparting prior cold work to the part or by adding alloying elements which reduce ductility. For steel, ε_b can sometimes be reduced by increasing the cutting speed, since this increases the chip temperature and may produce a brittle blue oxide layer on the chip surface. h_c can be increased by increasing the feed rate and thus the chip thickness. These options are not always feasible in practice, but the underlying physical concepts are in accord with shop experience, since chip breaking is normally more of a problem in finishing operations on relatively ductile (e.g., stamped) parts than in roughing cuts on cold-worked (e.g., forged) parts.

Equation (11.2) also indicates that a continuous chip will not break if it is permitted to curl naturally, that is, if $R_L = R_0$. The chip must therefore be guided toward an obstruction to produce a bending strain. This can be accomplished by properly directing chip flow. In turning, the chip flow direction depends on the radius of the workpiece, the lead and back rake angles of the tool, and the ratio of the depth of cut to the tool nose radius [34–39]. In turning, as shown in Figure 11.4, common strategies include directing the chip flow parallel to

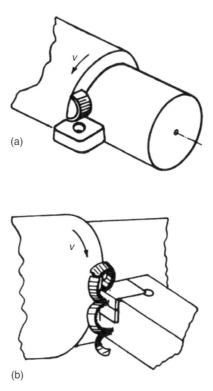

FIGURE 11.4 Breaking of the chip against the workpiece surface (a) or toolholder (b). (After S. Kaldor et al., [45].)

the workpiece axis, so that the chip breaks against the workpiece surface, or imparting side flow to the chip so that it breaks against the toolholder. The first type of breakage can be promoted by using a small lead angle and tool nose radius; the second type is accomplished by using larger lead and back rake angles and a large tool nose radius.

More effective chip breaking in turning can be ensured by clamping an obstruction-type chip breaker to the insert (Figure 11.5) or by using inserts with formed-in chip breakers. Formed-in chip breakers come in a great variety of forms but may be classified as groove- or pattern-type (Figure 11.6).

Obstruction-type breakers work by imparting curvature to the chip to produce a bending stress [40–45]. The placement of the obstruction is important in effective operations; it must be moved closer to the cutting edge as the feed is reduced. The advantage of obstruction chip breakers is that they can be adjusted to operate effectively over a wide range of feed rates. The main disadvantage is that they require careful setup and adjustment.

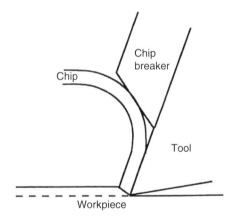

FIGURE 11.5 Obstruction chip breaker.

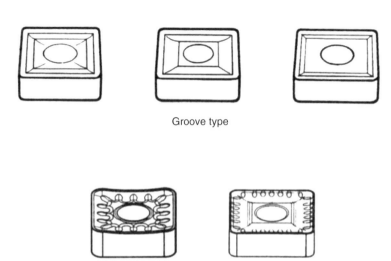

FIGURE 11.6 Typical chip-breaking patterns formed into inserts.

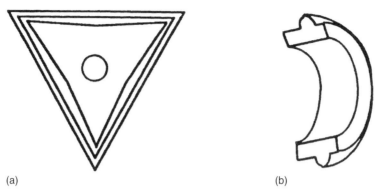

FIGURE 11.7 (a) Insert with formed-in chip-breaking groove. (b) Chip with thickened section produced by grooved insert. (After S. Kaldor et al., [45].)

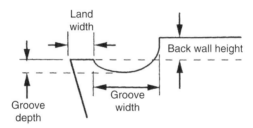

FIGURE 11.8 Parameters characterizing chip-breaking groove.

Groove-type chip breakers have various functions. The groove may guide the chip toward an obstruction and produce a thick section in the chip which promotes breaking (Figure 11.7). More commonly, the groove imparts a bending stress by decreasing the chip's radius of curvature. The dimensions and position of the groove must be selected properly for effective operation. As shown in Figure 11.8, the geometry of the groove is characterized by the land width, groove width, groove depth, back wall height, and groove geometry. The land width is the most critical parameter and must be adjusted based on the feed rate. Smaller land widths are used in finish cutting to ensure that the chip flows into the groove; larger land widths are used for roughing cuts to increase the strength of the cutting edge. The groove depth and back wall height are not as critical as the land width but should also be increased for roughing cuts to increase the bending stress.

The main advantages of groove-type chip breakers are that they require no setup and that they increase the effective rake angle and reduce the tool–chip contact length. They can therefore reduce cutting forces and increase tool life when compared to flat-faced, negative rake inserts. The main disadvantage is that they are effective over relatively narrow ranges of feeds and cannot be adjusted. Double-groove designs are sometimes used to increase the effective range of operation in semifinish and finish machining; this approach is not used in rough machining because it reduces edge strength.

Integral pattern-type chip breakers come in a variety of forms which consist primarily irregularly-shaped depressions and raised dots. They function by a number of mechanisms and may serve to guide chip flow, impart bending stresses, and induce stress concentrations. The number of possible patterns is infinite, and reliable inserts are available for many common operations. As with groove-type chip breakers, the main advantage of pattern-type chip breakers is that they require no setup, and the main disadvantage is that they are effective over a limited range of feed rates.

The range of applicability of particular chip-breaker designs cannot be accurately predicted by analysis. In practice, therefore, recommendations of insert manufacturers must be relied upon in selecting a design for a particular application. Insert manufacturers provide two types of charts to guide insert selection: charts of recommended inserts for particular combinations of the feed and depth of cut (Figure 11.9), and charts of the effective chip-breaking range for a particular insert design (Figure 11.10). These charts are based on limited tests, usually involving only one work material, tool nose radius, or cutting speed. They provide useful initial insert designs. In high-volume operations, additional tests should be conducted to accurately establish chip-breaking limits for the work material and cutting conditions involved.

Chip breaking and removal are also important in drilling, since the allowable penetration rate is often limited by chip-breaking characteristics. Chip breaking can be promoted by modifying the flute profile or drill point [46]. Common modifications, shown in Figure 11.11, include special flute profiles which compress the chip to increase bending stresses, grooves in the flute or flank profile to split and produce thick sections in the chip, and stepped cutting edges to split the chip. Chip breaking in drilling can also be promoted by using multifacet point geometries with cusps which split the chip, and by using indexable drills with chip-breaking inserts.

Active chip-breaking strategies can also be used for chip control. The most common strategy is to vary the feed rate to periodically interrupt the cut or to produce thick sections in the chip which will break more readily. On numerically controlled machines, the feed made be increased or halted at prescribed points to control the chip length [47, 48]. A cam-driven, hydraulic, or piezoelectric system may also be added to make the tool vibrate in the feed direction at an amplitude comparable to the feeds [49, 50]. This latter strategy is particularly

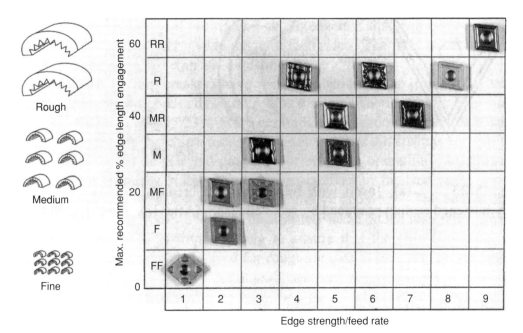

FIGURE 11.9 Chart of chip-breaking patters recommended for cutting steel at various feed rates and depths of cut.

Machinability of Materials

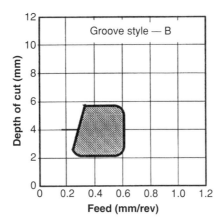

FIGURE 11.10 Typical chip-breaking chart for a particular insert. Separate charts are required for each work material and cutting speed.

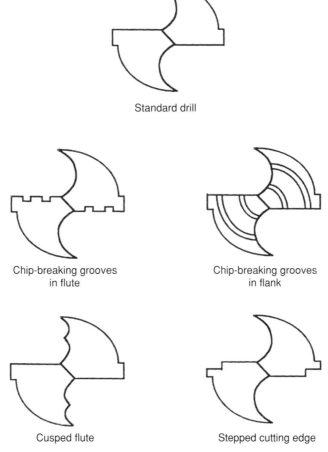

FIGURE 11.11 Drill point and flute modifications to improve chip breaking.

common in the drilling of small holes, where axial vibration of the drill is sometimes used to break chips [51]. The chief limitations of these methods are that they reduce the stability of the machining process and produce an irregular surface finish. High-pressure (6,000 to 40,000 psi) jets of water-based or cryogenic cutting fluids, directed at the base of the tool–chip interface, have been used to break chips when cutting difficult to machine materials such as titanium alloys or stainless steels [52–56].

Even if chips are broken, they may cause difficulties if they are not cleared from the machine tool or machined part. Chips are generally hot when they emerge from the cutting zone. If they collect on the machine tool or within the machined part, they can produce thermal distortions which will lead to dimensional or form errors in the machined part. This mechanism can be a source of significant errors in precision machining processes such as the boring of engine cylinders. The common method of evacuating chips is to use a stream of coolant to carry chips into a chip conveyor system. Some care must be taken to avoid damaging tools if they have poor thermal shock resistance and are normally used dry; this is the case for some ceramic tools (e.g., silicon nitride tools) and for oxide-coated carbide tools. In these cases compressed air can be used instead of a water-based coolant. Chip removal can also be improved by feeding pressurized coolant through the spindle for a rotary tool or through the toolholder for a turning insert; this is only an option on machines equipped with high-pressure, high-volume coolant systems.

11.4 BURR FORMATION AND CONTROL

A phenomenon similar to the formation of chips is the formation of burrs at the end of a cut, especially in milling and drilling. In most cases, burrs are partially formed chips left at points along the workpiece too weak to support the forces involved in complete chip formation. Burrs are undesirable because they present a hazard in handling machined parts and can interfere with subsequent assembly operations. Burrs are also associated with certain types of tool wear such as notch formation in turning and exit fracture in milling [57].

The most general studies of burr formation have been carried out by Gillespie et al. [58–60], who identified three principal types of burrs. The first is the Poisson or compression burr (Figure 11.12), which results from the workpiece material's tendency to bulge in the direction parallel to the cutting edge when compressed by flank forces on the tool. For large flank forces, such as those typical of worn tools, burrs of considerable size can be produced by this mechanism. The second and most common type of burr is the rollover burr (Figure 11.13), formed when a partially formed chip bends in the direction of the cutting velocity at the end of the cut. The third is the tear or breakout burr (Figure 11.14), formed when work material ruptures rather than deforms at the end of the cut. Gillespie and Blotter [58] analyzed the

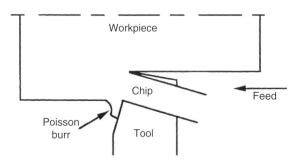

FIGURE 11.12 Poisson burr.

Machinability of Materials

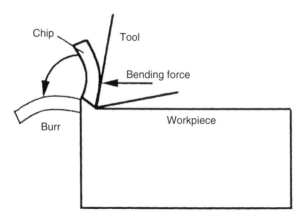

FIGURE 11.13 Rollover burr formed by bending.

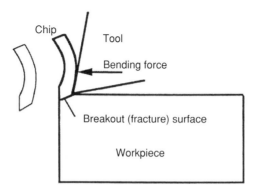

FIGURE 11.14 Workpiece tearing or breakout.

formation of Poisson and rollover burrs and derived equations for their approximate size under given cutting conditions. The quantitative accuracy of these equations is difficult to assess due to the difficulty of specifying some inputs, but they are useful in identifying measures which can minimize burr size. An alternative but similar classification scheme has been proposed by Nakayama and Arai [61].

The mechanical property that determines a given work material's tendency to burr is ductility, which is quantified by the percent elongation or strain to fracture (equivalent to the fracture strain ε_b, defined in the previous section). Considering basic burr formation mechanisms, it is clear that materials which can deform to large strains before fracturing are much more likely to form compression or rollover burrs, while materials which fracture at low strains will often break out before they burr. Therefore, burr formation is a particular problem for highly ductile materials, including pure metals (e.g., pure Cu, SHG Mg, and Al-00 aluminum), some soft low-carbon steels (e.g., 1008 and 1010), and highly ductile aluminum alloys (such as A356). For some of these materials, burring and chip-breaking characteristics are the major factors determining tool life and machinability.

More specific studies of burr formation in milling and drilling have also been reported. In milling the most widely studied burr is the foot or negative shear burr, which is a special type of tearout burr associated with chipping of the cutting edges of milling tools [62–66]. As with more common rollover burrs they are formed by deformation in the direction of the cutting velocity, but the deformation and rupture are produced by shearing rather than bending. Tool failures due to foot formation can be eliminated by changing the disengage angle of the

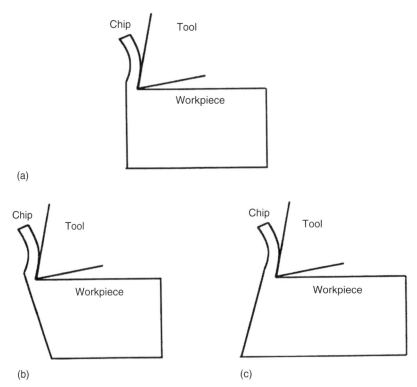

FIGURE 11.15 (a) Normal, (b) acute, and (c) oblique exit surfaces. The bending and shearing strength of the workpiece is greatest for the oblique surface and smallest for the acute surface.

milling cutter so that the force is not directed perpendicular to the exit surface [67]. Changes in tool geometry, such as the use of a larger axial rake angle and positive rather than negative radial rake angles when possible, can also be effective. Finally, burr size can be reduced by inclining the exit surface of the workpiece with respect to the machined surface to increase its bending strength (Figure 11.15), although this is seldom an option in existing operations.

Exit burrs in drilling are generally of the rollover type. They typically have an irregular appearance, but for highly ductile materials may deform into a cap or crown structure (Figure 11.16) [68]. As shown in Figure 11.17, the most common type forms when the annulus of work material cut away as the drill exits become too weak to support the thrust force. Changes in drill point geometry can reduce the size of these burrs [67, 69]. One effective method is to use a spiral-pointed drill (Figure 11.18) rather than a conventional conical drill. This drill not only increases the bending strength of the annulus throughout the exit phase, but also reduces the thrust force near the end of exit and directs a larger proportion of it in the direction parallel to the exit surface. The size of these burrs can sometimes also be reduced by using inverted or fishtail point drills, which exit at the outer margin rather than the center, or by imparting axial vibrations to the drill [51].

Burrs are a more serious problem for worn cutting tools, since cutting forces usually increase with tool wear. Measures which increase tool life will therefore also often reduce burr size. Since the strength of the workpiece will always decrease to zero near the end of the cut, however, burrs will inevitably be formed in many machining operations, and deburring operations will often be required before machined parts can be assembled. Manual deburring is common in small lot production. In mass production, wire brush, vibratory, centrifugal, and

Machinability of Materials 591

FIGURE 11.16 Cap (left) and crown (right) burrs formed in through drilling of 304L stainless steel. (After J. Kim et al., [68].)

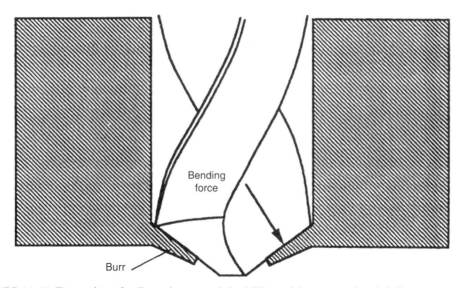

FIGURE 11.17 Formation of rollover burr at exit in drilling with a conventional drill.

water jet methods are often used. Many robotic and abrasive methods have been developed, but their application has been limited due to cost and environmental concerns. Detailed descriptions of specific deburring methods are given in deburring handbooks [60, 61].

11.5 MACHINABILITY OF ENGINEERING MATERIALS

The machinability of any alloy is primarily influenced by its base metal; aluminum alloys are generally highly machinable, ferrous alloys exhibit moderate machinability, and high-temperature alloys generally have poor machinability. The microstructure of the metal also plays an important role in machinability. Some important microstructural parameters include the grain size, the number and size of inclusions, and the type(s) of metallic structures present. In general, hard structures and fine grains result in lower tool life. Machinability improves for softer structures and coarse grains. Small hard inclusions in the matrix promote abrasive tool

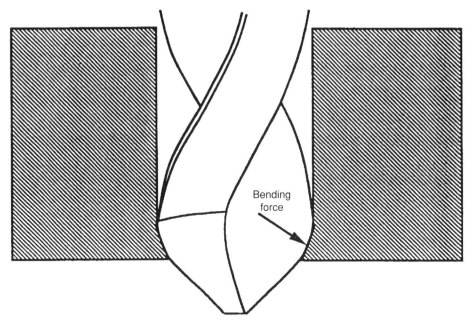

FIGURE 11.18 Formation of exit burr in drilling with a spiral-pointed drill. The burr size is smaller than for a conventional drill since the bending force is smaller and the bending strength of the workpiece at exit is larger.

wear. Soft inclusions are often beneficial to tool life. Materials which work-harden rapidly have lower machinability than those that do not. The chemical composition of a metal is very important and has a complex effect on machinability within the metal family. The primary manufacturing process of a material or component will also affect its machinability.

The remainder of this section discusses specific factors affecting the machinability of common classes of ferrous and nonferrous alloys.

11.5.1 Magnesium Alloys

Magnesium alloys are the most machinable of the common structural metals in terms of tool life, cutting forces, power consumption, and surface finish. Because magnesium alloys have a hexagonal structure and low ductility, cutting forces and power consumption are typically 50 to 80% lower than for aluminum alloys under comparable conditions [2, 70–73]. Magnesium alloys also have a low melting point (typically less than 650°C), so that cutting temperatures are low [74]. Due to their brittle nature, they form short, segmented chips, which restrict the length of the contact between the tool and chip. Finally, magnesium has a high thermal conductivity and does not alloy with steel. For all these reasons, tool wear rates due to thermal softening and diffusion are low for WC and HSS tooling. Most information indicates tool life when machining magnesium alloys dry is roughly five times longer than when machining aluminum wet under otherwise comparable conditions [70, 72]. Tool lives on the order of hundreds of hours, or dozens of production shifts in mass production, were achievable in the early 1960s [75]. More recently, PCD tools have been used to machine cast magnesium parts [76], resulting in tool lives approaching one million parts in some milling and drilling operations [70]. Magnesium alloys acquire fine surface finishes under normal cutting conditions; average roughnesses down to 0.1 μm can be obtained by turning or milling at high or low speeds, with or without cutting fluids [71], so that grinding and polishing operations are often unnecessary.

There are two chief concerns in machining magnesium: the risk of fire and built-up edge (BUE) formation.

Magnesium burns when heated to its melting temperature. In machining magnesium, fires are most likely to occur when thin chips or fines with a high surface area-to-volume ratio are produced and allowed to accumulate. The source of ignition may be frictional heating caused by a tool which is dull, broken, improperly ground, or allowed to dwell at the end of a cut, or sparks generated externally to the machining process. To minimize the risk of fire, the following practices should be observed [70–72]:

1. Sharp tools with the largest possible relief angles should be used.
2. Heavy feed rates, which produce thick chips, should be used when possible.
3. Allowing tools to dwell against the part at the end of the cut should be avoided.
4. Chips should be collected and disposed of frequently.
5. Appropriate coolants should be used when fine feeds are necessary.

Since magnesium chips react with water to form magnesium hydroxide and free hydrogen gas, water-based coolants have traditionally been avoided. The accepted practice was to machine dry when possible and to use a mineral oil coolant when necessary [77, 78]. Dry machining magnesium parts in high volumes, however, presented long-term housekeeping problems, especially when processes such as drilling or tapping which produced fine chips were used. Airborne fines could collect over time in areas such as space heaters and electric motor vents, leading to unexpected fires [77]. In the last 20 years, water-based coolants which can be used with magnesium without excessive hydrogen generation have been developed. These coolants are now used in production in some installations and reportedly increase tool life and reduce the risk of fire when compared to dry machining. Grinding or polishing of magnesium should be avoided when possible but, when necessary, should be carried out using dedicated machinery equipped with a wet-type dust collection system [70].

BUE formation is observed when machining cast magnesium–aluminum alloys dry with HSS or carbide tools. The metallurgical factors controlling BUE formation are discussed in Ref. [73]. BUE formation can be reduced or eliminated by applying a mineral oil coolant or by switching to PCD tooling. Flank buildup due to adhesion may also occur at cutting speeds over 600 m/min, especially when cutting dry; this may result in increased cutting forces, chatter, and poor surface finish [79].

Extensive tables of recommended speeds, feeds, depths of cut, and tool geometries for various operations are available from magnesium suppliers [70, 80]. Generally, turning and boring are carried out at speeds between 700 and 1700 m/min, feed rates greater than 0.25 mm/rev, and depths of cut up to 12 mm using tools with positive (up to 20°) rake angles, large relief angles, and honed edges. Face milling may be carried out at speeds up to 3000 m/min, feed rates between 0.05 and 0.5 mm/tooth, and depths of cut up to 12 mm. Finer feeds and higher speeds are used in finishing. The milling cutters used have positive cutting geometries and roughly one third fewer teeth than cutters of comparable diameters used to machining other metals.

11.5.2 Aluminum Alloys

Aluminum alloys are also among the most machinable of the common metals [2, 81]. Cutting forces are generally low, and because aluminum is a good heat conductor and most alloys melt at temperatures between 500 and 600°C, cutting temperatures and tool wear rates are also low. Tool life approaching one million parts can be obtained in some mass production operations using PCD tooling. When cut under proper conditions with sharp tools, aluminum alloys acquire fine finishes through turning, boring, and milling, minimizing the necessity for grinding and polishing operations.

Aluminum is commonly machined with HSS, carbide, and PCD tooling. Silicon nitride-based ceramic tools are not used with aluminum because of the high solubility of silicon in aluminum. The major machinability concerns with aluminum alloys include tool life, chip and burr characteristics, edge build-up, and surface finish. Tool life is a concern especially with alloys containing hard inclusions such as aluminum oxide, silicon carbide, or free silicon (e.g., the hypereutectic aluminum–silicon alloys discussed below).

Two major classes of commonly machined aluminum alloys are cast alloys, used in automotive powertrain and component manufacture, and wrought or cold-worked alloys, used especially in structural applications (e.g. airframe manufacture).

The most commonly machined cast aluminum alloys by volume are cast aluminum–silicon alloys, which are used extensively in automotive applications. From a machining viewpoint it is common to distinguish between eutectic alloys, containing 6 to 10% silicon, and hypereutectic alloys containing generally 17 to 23% silicon [2, 82, 83]. In the eutectic alloys (e.g., 319, 356, 380, 383, and most special piston alloys), silicon is encountered only in a eutectic phase. In the hypereutectic alloys (e.g., 390), the silicon content exceeds the eutectic limit, so that particles of free silicon are contained in the matrix. From a tool life viewpoint, most eutectic alloys, when properly tempered, present few difficulties, so that long tool life can be achieved at relatively high cutting speeds. Speeds up to 450 m/min can be used when turning with carbide tools [2], and speeds as high as 5000 m/min can be achieved in some milling applications with PCD tooling. Of the common eutectic alloys, the most easily machined are 319 and 380. The presence of hard inclusions limits achievable cutting speeds and tool life when machining hypereutectic alloys. The free silicon particles are much harder than the surrounding matrix, and produce accelerated abrasive wear with conventional tools and wear and chipping with PCD tools. Cutting speeds are typically limited to roughly 100 m/min when turning with uncoated carbide tools [2], and to 1000 m/min when milling with PCD tools. The carbide grades used for cast iron are suitable for machining aluminum alloys. In drilling, wear becomes excessive at speeds above 200 to 400 m/min with carbide drills, although diamond-coated drills may be used at speeds up to 600 m/min [83, 84]. Sand casting produces a coarser microstructure with larger silicon particles than permanent mold casting, and thus yields poorer tool life and lower allowable cutting speeds; die casting yields the finest microstructure [83–85].

Iron, which is present in all casting alloys, also forms hard inclusions which have a negative impact on tool life and machinability. Iron is deliberately added to die casting alloys to overcome die soldering; the addition of up to 0.2% manganese to these alloys refines the iron structure and improves tool life [81]. Many machinability concerns for cast alloys containing other alloying elements, such as magnesium, zinc, or copper, are identical to those for the corresponding wrought alloys discussed below.

356 Aluminum, a corrosion-resistant alloy, is sometimes used in place of 319 or 380 aluminum in marine and high-performance engine applications. It contains less copper and approximately 1% more silicon than these alloys, and consequently wears tools more rapidly by abrasion. Most indications are the wear rate increases by 50 to 100% compared to 319 Al. Wear is a particular issue in tapping and milling operations. Roll taps rather than cut taps are generally used to improve tap life. Since the material is highly ductile, burr formation often limits tool life in milling, even at modest insert wear levels [86]. Chip control problems are also experienced in drilling operations. Proper tempering (to T6) can improve machinability, adding strontium to the alloy is also reported to be effective. Generally, this material must be machined at lower speeds and higher feeds than other hypoeutectic alloys to achieve adequate tool life.

Wrought or cold-worked alloys are classified according to primary alloying elements and temper. A four-digit classification system is used in which the first digit indicates the primary alloying element. An excellent summary of the grade contents and machinability concerns was given by Chamberlain [81]; the discussion in the following paragraphs is taken largely from this source.

Grades designated 1xxx are commercially pure grades without significant alloy content. Grades designated 3xxx have manganese as the principal alloying element. Grades 1xxx and 3xxx are soft grades which cannot be heat treated and which have similar machining characteristics. They generally present few tool life problems but are subject to chip control, BUE, and surface finish and integrity problems as discussed below.

Grades designated 2xxx have copper as the primary alloying element, although other elements (e.g., magnesium) may be specified. Copper forms hard particles in the matrix which increase tool wear; proper tempering should be used to ensure that the copper is taken into solution. Alloys graded as 2xxx generally have the best machinability of the wrought aluminum alloys. Alloy 2011, which contains bismuth and lead, and 2024 are common free-machining grades which yield excellent tool life and chip characteristics in turning and drilling applications. Grades designated 5xxx have magnesium as the principal alloying element. These are soft alloys which generally are almost as easily machined as the 2xxx grades. Grades designated 7xxx have zinc as the principal alloying element but may also contain specified levels of copper, magnesium, chromium, and zirconium. These are also soft grades which are generally easily machined, although they produce continuous chips which may present disposal problems at high cutting speeds.

Silicon and magnesium are the principal alloying elements in grades designated 6xxx. These grades are difficult to machine because the magnesium and silicon are tied up in hard Mg_2Si particles. Alloy 6262 is the easiest material in this series to machine because it contains bismuth added specifically to improve machinability. A commonly encountered grade in this series is 6061, used for welding and corrosion resistance applications; it is a relatively poor heat conductor and must be cut at speeds lower than those used for 2024 to avoid melting of the chip and consequent surface finish and dimensional accuracy problems.

Grades designated 4xxx have silicon as the principal alloying element. These grades, which are used primarily in forgings and for welding and brazing, are generally the most difficult to machine because they contain free silicon. Grades designated 8xxx include lithium alloys and some miscellaneous grades with typically yield intermediate tool life.

As noted above, aside from tool life, the major concerns in machining aluminum are chip control and surface finish and integrity. Chip control is a particular concern in drilling and in turning at high cutting speeds. Drills with high helix angles and open (parabolic) flute designs are best suited for aluminum alloys because they minimize chip obstruction and packing. Polishing flutes in the direction of chip flow is also effective in preventing these problems, as is the use of through-the-tool coolant. In turning, chip control problems and BUE occur for alloys which form continuous chips, such as the 7xxx series of wrought alloys. Strategies for controlling continuous chips in this case are discussed in Section 11.3. Other alloys, notably most cast alloys and wrought alloys of the 2xxx series, generally form short, segmented chips which are more easily disposed of; even for these alloys, however, flood coolant is usually required for chip control at higher cutting speeds.

Surface finish problems occur especially when machining soft alloys such as 319 and the 1xxx and 3xxx series wrought alloys. For these grades, BUE may form on the cutting edge, and the material itself may tear rather than form a chip, producing an uneven or gouged machined surface. These problems are more serious for carbide than for PCD tools. The most common approach for minimizing these problems is to increase the cutting speed, provided this does not result in chip control difficulties; other effective strategies include using flood coolant and sharp, uncoated tools with high positive rake angles and polished rake faces. TiN coatings, especially for drills, are also often found to be effective in reducing friction and edge buildup when machining cast alloys.

11.5.3 Metal Matrix Composites

Metal matrix composites (MMCs) are metals reinforced with particles, fibers, or whiskers of harder materials. In most cases, the metal matrix is an aluminum alloy, although steel, magnesium, and titanium MMCs are also available [87]. Common reinforcing materials include boron, silicon carbide (SiC), aluminum oxide, graphite, tungsten, and various proprietary fibers such as KAOWOOL or SAFFIL. In general, MMCs are more difficult to machine than the unreinforced base material due to their increased abrasiveness. In the case of whisker- and fiber-reinforced composites, delamination of the fibers from the matrix during machining may present additional concerns. Composites reinforced with chopped or short fibers are less abrasive and more easily machined than those reinforced with longer fibers.

Aluminum MMCs reinforced with boron and SiC are machined using practices similar to those used for hypereutectic cast aluminum–silicon alloys (e.g., 390 aluminum), which were discussed in the previous section [87]. Due to the abrasive nature of these materials, PCD [88, 89], diamond tipped [90–92], or CVD diamond-coated tools [93] are normally required. Typically these tools provide close tolerances and excellent surface finish and integrity when conventional aluminum machining practices are followed with slight modifications. The tool life, however, is reduced due to increased abrasive wear and tool chipping. BUE formation is also a concern with certain base matrix materials (e.g., 6061 aluminum). Aluminum MMCs reinforced with proprietary fibers other than SiC can also generally be turned and milled using standard aluminum machining practices with good results. In most cases, longer tool life is achieved with these materials than with SiC reinforced MMCs because the proprietary fibers are less abrasive than SiC.

Aluminum MMCs reinforced with alumina are more difficult to machine than SiC-reinforced materials. Relatively little data on machining of these materials have been published [87]. In some cases carbide rather than PCD tooling may provide better tool life due to its increased toughness and reduced tendency to chip.

Holemaking operations such as drilling, reaming, and tapping are typically the most critical cutting operations with composite materials. Feed is the most significant parameter affecting drill life and tool forces. An increase in the feed will result in an increase in tool forces and significant improvement in tool life. Feeds must generally be higher than 0.10 mm/rev to achieve acceptable tool life. High-pressure/high-volume coolant should be used to eject SiC particles generated during hole production.

Delamination of fiber-reinforced materials is a particular concern in drilling, especially for through holes. Although the tendency to delaminate is to a large extent a material property depending on the strength of the bond between the matrix and reinforcing fiber, delamination can be inhibited by using sharp drills with high helix angles. For more abrasive materials this greatly limits the tool life which can be achieved with conventional HSS drills.

Magnesium MMCs are machined using practices similar to those used for aluminum MMCs. Titanium MMCs are more difficult to machine than aluminum MMCs, and are often machined using lasers or other nontraditional methods, rather than the conventional chip-forming metal cutting processes.

Steel MMCs are typically composed of roughly 25 to 45% ultra-hard titanium carbide grains by volume dispersed in a high-carbon–high-chromium tool steel matrix. They are more difficult to machine than the base steel materials. Steel MMC parts are most readily processed by machining to the desired shape in the annealed state (at 44 to 50 R_c hardness) using cutting tools and conditions similar to those used for the base material with similar hardness. Following this, the parts are heat treated and tempered to 65 to 70 R_c; the critical dimensions are then ground.

11.5.4 COPPER ALLOYS

Copper and its alloys, the brasses and bronzes, are in most cases easily machinable, although notable exceptions exist. They typically have melting points lower than 1050°C; temperatures of this magnitude are sufficient to cause softening of many types of tools, but are rarely achieved at the tool point due to the high thermal conductivity of these materials and the fact that they are frequently cut at lower cutting speeds. Brasses, for example, are often machined in wire or bar form on automatic screw machines (e.g., in the manufacture of electrical components or hydraulic line fittings); the small stock diameter limits the maximum attainable cutting speed. They are typically machined using HSS or carbide tooling. PCD tooling could, in principle, be used to machine many copper alloys, but does not appear to have been widely used in practice.

Pure copper and brasses are normally machined in wrought or drawn bar form. Pure copper is a difficult material to machine because it is highly ductile. It produces high cutting forces, especially at low cutting speeds, which may lead to excessive deflection, vibration, and breakage of tools, especially drills [2]. Its high ductility leads to difficulties in breaking chips and to excessive burr formation. It also yields a poor machined finish at low cutting speeds. Pure copper should be cut at relatively high cutting speeds (greater than 200 m/min) using sharp tools; deburring is often required following machining. The machinability of copper improves with increasing cold work, since this reduces ductility. Alloying electrical grades with sulfur or tellurium to form free cutting grades also improves chip-breaking characteristics and machinability.

Alloying copper with zinc and to form brass greatly improves its machinability. Tool forces are reduced for 70/30 (single phase or α) brass and especially 60/40 (two-phase or α–β) brass [2]. α-Brass yields discontinuous chips and a poor surface finish at low cutting speeds; other types of brass (e.g., red and cartridge brass) produce continuous chips which are easier to break than those obtained with pure copper but which still may present disposal problems. Chip breaking is improved by the addition of lead, which precipitates into soft inclusions which provide lubrication and weak points in the chip. Leaded free-machining brasses, generally containing 2 to 3% lead by weight, are regarded as easily machinable materials. They are used especially in turning at cutting speeds up to 300 m/min, although much higher speeds can be achieved with larger diameter stock using carbide tooling.

Cast copper alloys are commonly used for fittings, bearings, and bushings. From a machinability viewpoint, these alloys can be divided into three groups [94]. The first group includes single-phase alloys rich in copper but containing lead. As with leaded free-machining brass, these are easily machined because they contain lead particles which facilitate chip breaking. The second group includes grades with two or more phases, in which the secondary phases are harder or more brittle than the matrix. Materials in this group include many aluminum bronzes, silicon bronzes, and high-tin bronzes. The brittle phases improve chip breakability, so that these materials generally yield short, broken chips. Nonetheless, the hard phases also increase abrasive tool wear, so that the machinability of these grades is classified as moderate. The third group includes high-strength manganese bronzes and aluminum bronzes with high iron or nickel contents. These materials are difficult to machine because they produce continuous chips and contain hard phases which increase tool wear.

11.5.5 Cast Iron

Iron is usually machined in cast form and is used extensively in the manufacture of internal combustion engines and machine tool structures. Three basic types of cast iron are in wide use for these applications: gray iron, malleable iron, and nodular or ductile iron. Two other forms, white iron and compacted graphite iron, are also encountered but are less common.

Gray cast iron [95–99] often consists primarily of pearlite (typically 85% of the matrix) and ferrite, although primarily ferritic grades also exist and most grades contain some steadite or austenite. It contains between 3 and 5% graphite in flake or lamellar form. Flake graphite reduces ductility, facilitates chip breaking, and acts as a natural internal lubricant. Gray iron therefore produces comparatively small cutting forces for its hardness and forms short, easily broken chips. Its limited ductility also reduces the tool–chip contact length and limits maximum cutting temperatures. For these reasons, it is usually classified as an easily machinable material. It is used extensively for engine blocks and heads and other automotive components because of its good machinability, low cost, and high internal damping.

Gray iron castings are often used in mass production applications. For a fixed iron chemistry, tool life in these applications correlates strongly with the casting hardness, which depends on the cooling rate and metallurgical state. Casting hardness increases as the cooling rate is increased. Generally machining processes in such applications are designed to provide acceptable tool life for a range of casting hardness values. If shakeout times are reduced, resulting in more rapid cooling, castings with hardness values exceeding the specified limit may be produced. When this is the case, the foundry practice must be altered to reduce the incoming hardness or the machining processes must be redesigned to provide acceptable tool life with harder iron. Inspection of incoming castings to ensure that their hardness is within an acceptable range can help reduce intermittent tool life problems caused by occasional hard castings. Annealing castings prior to machining can also significantly increase tool life and machinability.

Other machinability problems encountered with gray iron can often be traced to one of two causes: hard inclusions formed during solidification, and sand adhering to or entrained in the cast layer. These phenomena typically result from poorly designed or controlled casting process and are best eliminated through changes in foundry practices. Hard inclusions include iron carbide phases and martensite inclusions. Both are most commonly caused by excessively rapid cooling or foundry chill, which is a particular problem at the corners or in thin sections of castings, and by segregation during solidification. Hard inclusions increase abrasive wear rates and may also contribute to tool chipping or breakage. They are best eliminated at the foundry by redesigning the casting process so that all sections of the casting cool at the same rate; this can be accomplished by altering gating, changing shakeout practices, redesigning castings to eliminate thin sections, and (less commonly) by adding alloying elements to the iron. Adhering sand or dross in the cast layer can be eliminated by properly cleaning castings, generally by shot blasting. Eliminating sand entrained in the cast layer requires changing the binder in the sand. A less common source of machining problems is foundry swell or casting distortion; the amount of stock to be removed in roughing cuts may be increased beyond process capabilities in the case of excessively distorted castings. As with other defects, casting swell or distortion should be solved at the foundry by altering core assembly methods and by careful inspection of finished castings.

Alloying elements have an impact on the machinability of gray iron, although their influence is usually less pronounced than that of the casting hardness. Adding copper or tin to the iron usually refines the pearlite structure and reduces carbide formation, resulting in an increase in machinability. Adding chromium or nickel to castings to increase their abrasive wear resistance (e.g., in the cylinders of engine blocks) increases casting hardness and abrasive tool wear. Adding phosphorus in concentrations about 0.15% results in steadite formation, which increases abrasive wear rates and reduces machinability.

Gray iron is a notoriously dirty material to machine. Machining iron at high speeds in mass production applications can result in a buildup of dust, composed primarily of graphite, in the machining system. The graphite dust has a negative impact on machine life, and in some cases the dust may contain particles of alloying elements such as chromium which react with the cutting fluid to form abrasive oxides which accelerate spindle bearing and way wear. Frequent cleaning of machinery and the use of flood coolants (when the tool material permits) to reduce dust dispersal are effective in reducing machine failures due to this cause. The machinability of pearlitic gray irons generally improves as the pearlite structure becomes finer, although the alloying elements added to refine pearlite may counteract this effect. Ferritic gray irons are generally softer and more easily machined than pearlitic irons, although they may be subject to BUE formation. Free ferrite may also react with the binder in PCBN tooling, greatly reducing tool life; this problem is not commonly observed or well understood. Steadite and austentite phases increase hardness and result in increased abrasive wear and a reduction in machinability.

Gray iron can be machined using coated or uncoated carbide, alumina and silicon nitride ceramics, or PCBN tooling [100–102]. It can also be finished turned using cermets. Coated carbides can be used for turning and milling at speeds up to 150 m/min. Flood coolants are normally used with carbide tooling to reduce dust dispersal. Tool life on the order of 1000 or 2000 parts can be achieved with carbide tooling. Silicon nitride tooling is used in turning, milling, and boring at speeds between 800 and 1300 m/min. Silicon nitride tools are used without coolants to prevent tool failures due to thermal shock. Chamfered edge preparations are normally used to prevent tool chipping, especially in interrupted operations. Tool life between 2,000 and 10,000 parts can be achieved with silicon nitride tooling. Depths of cut should be greater than the thickness of the cast layer; when this is not possible, alumina-based ceramic tooling should be used for enhanced wear resistance. PCBN tooling is used especially in milling and finish boring at speeds similar to those for silicon nitride. Coolants are often used to prevent dust dispersal and thermal expansion of the workpiece but is not required for acceptable tool life. Tool life of over 10,000 parts (often up to 50,000 parts) can be achieved with PCBN tooling. Negative rake cutting geometries are used for turning, boring, and milling operations with all these tool materials. Gray iron can be drilled using HSS, HSS–Co, or carbide drills. For HSS drills, the cutting speed is usually kept below 25 m/min to provide adequate tool life (>1000 parts). Micrograin carbide drills with through-the-tool coolant can be used to improve drill life (often to greater than 30,000 parts at a cutting speed of 80 m/min) and drilled hole quality [103–105].

Malleable iron, sometimes called ARMA steel, is more ductile than gray iron and generally harder to machine. It produces a longer chip–tool contact length and higher cutting temperatures. A frequently quoted general rule is that malleable iron yields 25% higher tool life than free cutting steel under comparable conditions [99]. Traditionally, malleable iron has been machined using coated carbide tools; a black oxide coating is normally preferred because it increases abrasive wear resistance. Silicon nitride tooling has traditionally not been used for malleable iron because cutting temperatures exceed the melting temperature of the glassy-phase binders in these tools. Recently developed coated silicon nitride grades, however, are reportedly suitable for machining malleable iron [106]. Relatively little information on the machinability of malleable iron is available in the literature, although tooling catalogs provide some information on suitable speeds, feeds, and depths of cuts for specific tool grades. Much of the information discussed below for ductile iron can be applied in a general sense to malleable irons. Pearlitic or decarburized structures at or near the surface of ferritic malleable iron castings reduce machinability.

In *ductile* or *nodular iron*, the graphite has a spherical or nodular, rather than flake, structure. As the name implies, nodular irons have significant ductility (with elongations between 5 and 15%). They are used in applications requiring fatigue resistance, for example, for crankshafts, camshafts, bearing caps, and clutch housings.

Due to their increased ductility, nodular irons produce longer tool–chip contact lengths and higher cutting temperatures than gray iron. The nodular structure of the graphite also inhibits its dispersal at the tool–chip interface and thus its lubricating effect. As a result they are typically more difficult to machine than gray irons; in fact, nodular iron is comparable to cast steel from a machinability viewpoint. The machinability of nodular irons depends on their microstructure, alloy content, hardness, and ductility.

Nodular irons may have a primarily pearlitic, ferritic, or austentitic structure. Pearlitic structures are common and generally result in intermediate hardness and ductility. Irons with a primarily ferritic structure are softer and more ductile than pearlitic grades and provide roughly equivalent machinability. In pearlitic grades, the graphite nodules are usually encased in ferrite. Increasing the thickness of the ferrite layer around the nodule appears to reduce machinability in general, perhaps by impeding graphite dispersal and lubrication. Grades containing steadite and (particularly) austenite are harder and more abrasive than other grades and are the most difficult to machine.

The effect of alloying elements on the machinability of nodular iron is similar to that for gray irons [95]. Adding copper or tin reduces hardness and refines the pearlite structure and results in an increase in machinability. Adding phosphorous promotes steadite formation and reduces machinability. Adding chromium, nickel, and manganese increases abrasive tool wear and reduces machinability. The amount of inoculant added has a significant effect on tool life [107].

Increasing either hardness or ductility reduces machinability. The influence of ductility appears to be stronger than that of hardness within the normally encountered ranges of these variables. For example, a grade with a Brinnel hardness of 280 and an elongation of 6% typically yields a longer tool life than a grade with a Brinnel hardness of 270 and elongation of 10%.

Nodular irons are normally turned or milled with coated carbide or PCBN tooling. Silicon nitride tooling is not used because cutting temperatures exceed the melting temperature of the glassy binder phases in these tools, resulting in rapid tool wear. In contrast to gray iron, positive rake tooling may be used with some nodular irons to reduce cutting forces and chatter. Black oxide coatings are preferred because they increase abrasive wear resistance. Tool life generally increases with the application of a flood coolant provided this does not result in spalling of the coating due to thermal shock. The allowable cutting speeds in turning, boring, milling, and drilling with carbide tools are typically roughly half those quoted above for gray iron. The allowable speeds for PCBN tools, which are used especially with harder grades, are higher, but much lower than those used to cut gray iron with silicon nitride tools [101, 102]; recommendations for specific tool grades are available from tooling manufacturers.

Unlike gray irons, nodular irons are subject to flank buildup and burr formation [108–110]. The strategies discussed in Section 11.4 are effective in reducing burr formation. Flank buildup can be reduced or eliminated by decreasing the feed rate, using positive rake angles, increasing clearance angles, and by using a flood coolant.

Foundry defects such as chill, adhering sand, and swell, affect the machinability of nodular irons in the same manner as gray iron as discussed above.

Compacted graphite iron has a structure intermediate between gray and ductile iron and is used especially for diesel engine blocks and bearing caps, where it provides increased strength. Most data indicate its machinability is also intermediate between gray and nodular iron

[99, 111]. Current production practices are closer to those for nodular than gray iron. This is in part because complex castings nominally made of compacted graphite iron may exhibit a range of microstuctures, including nodular structures in some sections. Few foundries are capable of producing compacted graphite castings of consistent quality.

White irons and corrosion-resistant *high-silicon gray irons* are the most difficult cast irons to machine because they are brittle and abrasive [99]. Alloyed white irons containing nickel are generally ground to finished dimensions, but can be turned and bored if rigid tooling setups and ceramic tools are used. Electrodischarge and electrochemical machining methods should also be considered for these materials. High-silicon irons have high hardness and are also normally ground to finish dimensions. Adding carbon or phosphorous reportedly improves their machinability, but degrades other mechanical properties which may be critical in applications for which they are considered.

11.5.6 Carbon and Low Alloy Steels

Steels vary greatly in chemical content and microstructure. The next two sections discuss the machinability of carbon, low alloy, and stainless steels, concentrating primarily on wrought or bar products. The factors affecting the machinability of these materials are well understood in a general sense. Cast, hardened, and high alloy steels are not discussed in detail. These materials are harder than carbon and low alloy steel and in general are much more difficult to machine. An indication of the tool life which can be expected for these materials as compared to low alloy steels is provided by tables of machinability indices [26, 112]. Recommendations for machining specific alloys can be obtained from specialized studies [113], review articles [114], or from steel or tooling manufacturers who have experience with the material in question.

The machinability of steel depends on hardness, chemistry, microstructure, mechanical state, and work-hardening characteristics. As with most materials, machinability decreases with increasing hardness. Increasing the hardness increases cutting forces and temperatures and stresses at the tool point; for HSS and sintered carbide tools, excessive temperatures and stresses lead to plastic collapse of the cutting edge. Graphs of limiting speeds and feed rates for preventing plastic collapse at specified hardness levels are available [2] (Figure 11.19). The mechanical state of the material, particularly the degree of prior cold work, influences cutting forces and chip-breaking characteristics. Annealed steels produce lower cutting forces and temperatures, but due to increased ductility may produce chips which are more difficult to break; cold-worked materials produce higher forces but more easily broken chips. Alloys which work-harden rapidly produce more rapid tool wear, especially notch wear, than those which harden more slowly. As discussed below, chemistry and microstructure influence the distribution of hard particles within the matrix and thus abrasive tool wear rates. Macro inclusions (>150 μm) negatively affect machinability and could result in the sudden tool failure. Some of the undesirable inclusions are Al_2O_3 and Ca because they are hard and abrasive; FeO and MnO have a similar but less severe effect.

Carbon steels may be classified as low carbon (containing less than 0.3% carbon; AISI grades 1005 to 1029), medium carbon (containing between 0.3 and 0.6% carbon; AISI grades 1030 to 1059), and high carbon (containing over 0.6% carbon; AISI grades 1060 to 1095) [112]. As-rolled low-carbon steels consist primarily of ferrite, although other phases, particularly pearlite, are present. The pearlite content (and, generally, the hardness) increases with increasing carbon content. As-rolled medium-carbon steels have a pearlite–ferrite structure, and are predominately pearlitic when the carbon content exceeds 0.4%. As-rolled high-carbon steels have a pearlitic matrix, with cementite, the hardest component of steel, predominating for steels with carbon content greater than 0.8%. In general, as-rolled carbon steels become harder and less machinable as the carbon content increases. Heat treating carbon steels may

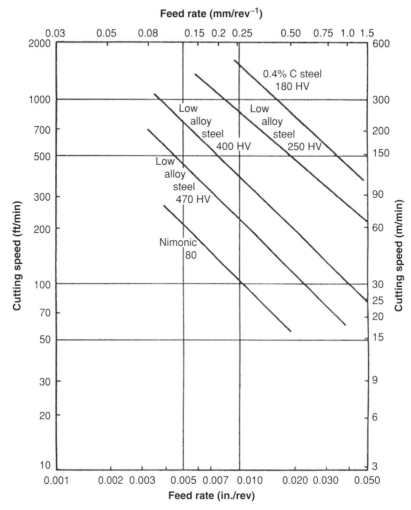

FIGURE 11.19 Conditions of deformation of cemented carbide tools when cutting steels of different hardness. (After E.M. Trent [2].)

result in the formation of bainite or tempered martensite phases. Although these phases are harder than pearlite or ferrite, they serve to help break chips, so that tempering generally increases machinability.

Carbon steels are most commonly turned, bored, and milled with coated carbide tooling. As discussed in Chapter 3, steel-cutting carbide grades contain TiC to reduce crater wear; TiN-based (gold) coatings are also often used to reduce cratering due to diffusion. Negative rake tools with honed edge preparations are most commonly used, and turning inserts usually have molded-in chip breakers. Carbon steels can also be finish turned with cermets, turned and bored with alumina-based ceramic tooling, and turned, bored, and milled with PCBN tooling. Drilling, end milling, and tapping of carbon steels may be performed using HSS, HSS–Co, or solid carbide tooling; rotary tools are often coated with TiN-based coatings to reduce cratering. Most carbon steel machining operations are carried out using a flood coolant to control tool and workpiece temperatures.

Low-carbon steels generally have the highest machinability of the carbon steels. They are typically turned at speeds of roughly 200 m/min and feed rates of roughly 0.15 mm/rev.

Milling is carried out at somewhat higher speeds and lower feed rates. Drilling is carried out at a variety of speeds which are generally below 20 m/min for a HSS drill. The feed rate often has a stronger influence on maximum penetration rates than the speed because it influences drill breakage. In the as-rolled or annealed conditions, machinability is generally best for steels with carbon contents between 0.15 and 0.25% [112]. At lower carbon levels, the material is more ductile and adheres to the tool, increasing the difficulty of breaking chips and resulting in BUE formation, which leads to poor surface finish. Very low carbon grades such as AISI 1010 or 1008 are notoriously susceptible to these difficulties in the annealed condition. Moreover, since these materials are often manufactured from scrap metal when obtained in bar form, they may show significant variation over time in mass production operations unless the consistency of incoming material is ensured through careful inspection. Cold working decreases ductility and usually increases machinability; it is particularly effective in eliminating chip-breaking problems and BUE formation.

Medium-carbon steels are harder than low-carbon steels and yield higher cutting forces under equivalent conditions. They also contain higher proportions of harder phases, particularly cementite, which accelerates abrasive tool wear. As a result they are machined at lower cutting speeds than low-carbon steels; speeds should be progressively decreased as the carbon content is increased [112]. Annealing and normalizing these steels coarsens the pearlite structure and improves machinability. Medium-carbon steels are less subject to chip-breaking difficulties and BUE formation than low-carbon steels.

High-carbon steels contain a greater proportion of hard phases (e.g., cementite) than medium-carbon steels and therefore produce higher cutting forces, temperatures, and abrasive tool wear rates. As a result, they are machined at lower cutting speeds and feed rates; as with medium-carbon steels, the allowable cutting speed for acceptable tool life decreases with increasing carbon content. As noted above, grades containing more than 0.8% carbon have excess cementite in the matrix in the as-rolled or air-cooled condition and are particularly difficult to machine. Annealing tends to consolidate the cementite into larger particles and typically improves machinability [112]. A spheroidized microstructure also improves the machining characteristics of higher carbon steels. Assembling carbides into spheroids improves the cutting action and reduces tool wear. Coarse spheroidite or pearlite structures provide better machinability than fine structures. High-carbon steels tend to air harden, so that surface integrity problems such as residual stress and white layer formation are a particular concern if the coolant volume is insufficient or inconsistent. High-carbon steels typically present few chip control problems and are not prone to BUE formation. Hardened high-carbon steels contain abrasive martensite and bainite phases and are the most difficult of the plain carbon steels to machine [115]. In addition, they are prone to white layer formation under certain conditions [116–118] as discussed in Chapter 10. When hardening is required, steels should be rough machined prior to hardening, and hard turned or ground to their final dimension following hardening.

Adding alloying elements to steel generally increases hardness in the as-rolled and annealed conditions, resulting in decreased machinability [112]. Many alloying elements are added to increase strength or wear resistance, and combine with carbon to form very hard, abrasive carbides, which reduce tool life; examples include chromium, nickel, and manganese [2]. Not surprisingly, the impact of alloying elements on machinability depends on the alloy content; low or lean alloy steels machine much more like carbon steels than corresponding high alloy grades [119]. A spheroidized structure in alloy steels results in improved overall machining performance because it reduces hardness and arranges the hard carbide phase into spheroids, which reduces the abrasive action of the carbides. Tempered alloy steels produce better surface finishes than annealed steels because the tempered structure reduces or eliminates BUE formation. As noted above, machinability tables can be used to assess the tool life to be expected from specific high alloy grades as compared to plain carbon grades. Formulas for computing an equivalent carbon content from the alloy content can also be used for this purpose [120]. A detailed discussion of the machinability of specific alloy families, related to their metallurgy, is given by Finn [112].

There are alloying elements, however, which are added to steel specifically to increase machinability at conventional and high speeds [121]. These include lead, sulfur, manganese sulfide, phosphorous, calcium, bismuth, selenium, and tellurium [2, 113, 122]. Typically these additives, which are often used in combination, result in insoluble inclusions in the matrix. In addition to the content, the size, shape, and distribution of the inclusions affect machinability. These inclusions cause the metal matrix to deform more easily and facilitate crack propagation, resulting in reduced cutting forces, enhanced chip breakability, and improved surface finish. The resulting grades are designated variously as free machining, free cutting, or enhanced machining steels.

Sulfurized free-machining steels have a high sulfur content. The sulfur is often added in the form of manganese sulfide, MnS, a solid lubricant which forms inclusions in the matrix [2, 112]. During cutting, the MnS coats and lubricates the rake face of the tool, reducing friction, tool–chip temperatures, and tool wear rates. The inclusions also enhance chip breaking. Depending on the amount of MnS, other mechanical properties of the steel such as corrosion resistance, ductility, toughness, formability, and weldability may be negatively impacted. A wide variety of sulfurized grades which trade off one or more of these properties for enhanced machinability are available.

Leaded free-machining steels contain lead, which also forms inclusions in the steel matrix which serve to lubricate the rake face of the tool and to break chips. Most leaded grades also include other free-machining additives such as MnS and phosphorous; as a result, they generally provide better machinability than nonleaded sulfurized grades. Leaded steels have become less common as concerns about occupational exposure to lead have grown. This has led to interest in the development of nonleaded free-machining steels (e.g., grades containing bismuth). In mass production applications in which a number of machines are connected to a single recirculating cooling system, strict care should be taken to change the coolant at regular intervals if leaded steels are machined. If this is not done, lead may build up to unacceptable levels in the coolant sump.

Calcium is sometimes added to deoxidizing agents in the final stages of steel production. Calcium deoxidation reduces abrasive tool wear and improves machinability, especially for high-carbon grades [112]. As with other enhanced machining grades, the increase in machinability results from the formation of soft inclusions in the matrix. The chemical composition of the inclusions depends on the other deoxidizing elements used.

Microalloyed steels [23, 123–125] contain small amounts of hard elements such as vanadium and tungsten. They are most often intended for use in an as-forged condition in applications which normally require forged and heat-treated (quenched and tempered) steels, for example, for high-strength crankshafts. The aim of microalloying is to increase strength and wear resistance in the as-forged condition significantly, so that subsequent quenching and tempering is unnecessary. Most test results indicate that as-forged microalloyed steels are significantly more machinable than forged, quenched, and tempered carbon grades. In these tests, the hardness of the microalloyed steel is generally significantly lower than that of the quenched and tempered steels.

11.5.7 Stainless Steels

Stainless steels contain a high proportion of chromium, generally in excess of 11%. They are generally difficult to machine due to their high tensile strength, high ductility, high work hardening rate, low thermal conductivity, and abrasive character. This combination of properties often results in high cutting forces, temperatures, and tool wear rates, as well as a susceptibility to notch wear, chip-breaking difficulties, BUE formation, and poor machined surface finish [2, 126]. Alloying elements can be added to reduce some of these difficulties, however, resulting in free cutting grades with comparatively good machinability.

Stainless steels are usually classified into four categories depending on their primary constituent of the matrix: ferritic, martensitic, austenitic, and duplex (combined ferritic/austenitic). They may also be classified based on their heat treatment (precipitation-hardenable versus non-precipitation-hardenable alloys) or machining characteristics (free-machining versus non-free-machining grades).

Ferritic stainless steels are alloyed primarily with chromium, although molybdenum, titanium, or niobium may be added to some grades to improve corrosion resistance or as-welded properties. Ferritic alloys are generally more machinable than other alloys [126, 127]. They are used in an annealed or cold-worked condition but are not heat treated, so that their hardness is comparatively low. Their machinability generally decreases with increasing chromium content.

In addition to chromium, martensitic alloys may contain carbon, molybdenum, and nickel to increase strength. The machinability of martensitic stainless steels is influenced by hardness, carbon content, nickel content, and metallurgical structure [127]. As with most materials, increasing hardness typically reduces tool life and machinability. Increasing the carbon content increases the proportion of abrasive chromium carbides in the matrix and reduces tool life and machinability. Increasing the nickel content increases the annealed hardness and also reduces machinability. The metallurgical factor which has the strongest influence on machinability is the proportion of free ferrite in the matrix; generally machinability increases with free ferrite content.

Austenitic stainless steels contain nitrogen, carbon, and nickel or manganese in addition to chromium. They exhibit high strength, ductility, and toughness, and are typically more difficult to machine than ferritic or martensitic stainless steels. Specific difficulties encountered when machining austenitic stainless steels include high wear rates due to high cutting forces and temperatures, BUE formation, chip control problems, poor surface integrity (hardened machined surfaces), and a tendency to chatter. Poor tool life is related to the annealed hardness, which increases with increasing nitrogen content. Increasing the carbon content increases the work-hardening rate and also decreases machinability. Abrasive carbon/nitrogen compounds may form in the matrix and reduce tool life; these can be controlled by adding titanium or niobium. As with other stainless steels, hardness increases and machinability decreases with increasing nickel content. Imparting moderate cold work to the material typically increases machinability by reducing the tendency for BUE formation and improving the machined surface finish and integrity.

Duplex alloys have a chemistry similar to austenitic stainless steels but are generally more difficult to machine due to their high annealed strength. Machining duplex alloys can be particularly challenging because no standard enhanced machining grades are available.

Enhanced or free-machining ferritic, martensitic, and austenitic stainless steels contain alloying elements intended to improve machinability. Common machinability additives include sulfur, selenium, tellurium, lead, bismuth, and phosphorous. As with carbon and low alloy steels, these additives are effective because they form compounds which have low solubility in the matrix and precipitate as inclusions which serve to lubricate the tool–chip interface and to break chips. Sulfur is the most commonly used additive, followed by selenium. Machinability can also be improved by varying the deoxidizing agents used in steelmaking to control oxide inclusions in the matrix.

General guidelines for machining stainless steels include [126]:

- use lower cutting speeds and metal removal rates than for carbon steels;
- use rigid tooling and fixturing to avoid chatter;
- maintain feed above a minimum level to avoid poor surface integrity;
- use sharp tools with a fine finish to avoid BUE formation;
- use proper cutting fluids with sufficient flow rates for heat removal.

Detailed guidelines for machining various alloy families with HSS and carbide tools are given in Ref. [126]. Stainless steels may also be machined with ceramic and PCBN tooling; guidelines for the application of specific tool grades can be obtained from tooling manufacturers.

11.5.8 Powder Metal Materials

Powder metal (P/M) parts are near-net shaped parts produced by sintering metal powders under pressure. They generally require little machining, although grinding, drilling, and threading operations are often necessary to achieve tight tolerances and produce features such as transverse or threaded holes. P/M parts may be made of many materials such as aluminum, iron, nickel, and copper. Machining practices for these materials are similar to those of cast or wrought materials with similar chemical compositions, but are affected by unique features of P/M materials including porosity, variations in density and thermal conductivity, and increased abrasiveness. This section concentrates primarily on the machinability of P/M structural steels. The influence of properties such as porosity on machinability are similar for sintered metals made of other materials.

Generally, increasing the density of P/M materials increases machinability. Optimum machinability is often achieved for matcrials with dcnsitics bctwccn 90 and 95% of thc theoretical maximum [128]. In this case, the machinability of P/M parts is comparable to that of corresponding cast or wrought parts. Machinability increases with density because porosity decreases. Increased porosity has a negative effect on machinability for a number of reasons [128, 129]. Porosity causes discontinuous contact between the tool and the workpiece, resulting in an interrupted cutting action which increases dynamic and localized stresses on the tool. As a result, the tool wears more rapidly and may chip because it is subject to shock loading and increased vibration. Porosity also affects thermal properties. Increasing porosity generally reduces thermal conductivity, resulting in higher cutting temperatures and increased chemical (thermally activated) wear. Materials with densities greater than 95% of the theoretical maximum are often produced by forging; in this case increased hardness may offset the effect of reduced porosity and result in decreased machinability.

The increased cutting temperatures caused by high porosity may also accelerate abrasive wear. Abrasive wear rates are generally higher even for low-porosity P/M materials than for corresponding cast or wrought materials because P/M materials form a powdery chip which may include free particles of hard materials. As temperatures increase, there is a tendency toward increased oxidation of the machining debris, resulting in a higher concentration of abrasive oxide particles.

Optimum machinability can also be obtained by machining materials in the presintered condition [128]. Presintering is performed to volatilize and burn off lubricants used in pressing, prior to heating parts under pressure to amalgamate the metal powder. Presintered parts are easier to machine than sintered parts because they are typically softer and less abrasive. However, machining in the presintered condition reduces achievable accuracy (tolerance) and finish characteristics.

Finally, the machinability of P/M parts can be increased through use of free-machining additives [128, 130]. Additives are effective for P/M materials for essentially the same reasons they are effective in cast irons and wrought steels; they result in the formation of inclusions in the material matrix which serve to lubricate the tool–chip contact and reduce cutting friction and temperature. Additives also generally reduce cutting forces, reduce BUE formation, improve surface finish, and increase chip breakability. For P/M steels, manganese sulfide is the most common and effective additive, followed by sulfur and lead. Bismuth, graphite, tellurium, and selenium are also usually effective. Adding copper is also effective when it results in a significant increase in thermal conductivity. Iron–copper P/M steels are used extensively in automotive applications for bearing caps, connecting rods, and valve seats and guides; care should be taken in these applications in reducing copper content to reduce costs, as this may result in greatly reduced tool life.

11.5.9 Titanium Alloys

Titanium alloys have a high strength-to-weight ratio and are used in many aerospace applications. Pure titanium undergoes a metallurgical transformation at about 830°C, changing from a hexagonal close-packed structure (α-phase) to a body-centered cubic structure (β-phase) [131]. Adding alloying elements can significantly change the transformation temperature, so that the beta phase is stabilized and can be retained at room temperature. There are four main groups of titanium alloys: unalloyed titanium, which sometimes contains small amounts of oxygen or iron for increased strength; α alloys, which contain α-phase stabilizers such as aluminum, oxygen, nitrogen, and carbon; α–β alloys, which contain both α- and β-phases and which are alloyed with both α- and β-stabilizers; and β alloys, which are alloyed with β-stabilizers such as molybdenum, vanadium, niobium, copper, and silicon. Most alloys used in aerospace applications, for example, Ti–6Al–4V, are α–β alloys. Generally, increasing the β-phase content increases strength and decreases machinability, although in some cases increasing the β-phase content improves chip formation and results in an increase in machinability.

Titanium alloys are regarded as difficult to machine for the following reasons [131, 132]:

- they maintain high strength at high temperatures, increasing cutting forces and tool stresses;
- they produce thin chips, which increases cutting temperatures and stresses at the tool cutting edge;
- they have comparatively low thermal conductivities, which further increases cutting temperatures;
- they have a high chemical reactivity with almost all tool materials at elevated temperatures;
- due to their thermomechanical properties, they often produce discontinuous or shear-localized chips;
- they have a low modulus of elasticity, which can lead to excessive deflection of the workpiece and to chatter;
- they are susceptible to surface damage during machining, and yield poor machined surface finishes under many conditions;
- they may ignite during machining due to the high cutting temperatures often generated.

Because of all these factors, tool life and allowable machining rates are lower for titanium alloys than for most other metals. In particular, the choice of tool materials is limited due to high chemical reactivities. In most cases, titanium alloys are machined using HSS or uncoated carbide tools. High-cobalt HSS (HSS–Co grades) typically yield better tool life than straight

HSS grades [131, 132]. Among straight grades, highly alloyed materials such as T5, T15, M33, and the M40 typically perform better than general purpose grades such as M1, M2, M7, and M10 [132]. For carbide tooling, straight grades such as C2 (ISO K20) perform better than the steel-cutting grades such as C8 (ISO grade PO1). The steel cutting grades, which are alloyed with TiC and TaC, wear more rapidly by diffusion; possible metallurgical reasons for this are discussed by Trent [2]. Coated carbide tools also wear more rapidly than uncoated grades [133]. Among advanced materials, ceramics such as sialon and alumina-based materials are not suitable for machining titanium due to rapid chemical and abrasive wear [133]. PCBN and PCD tooling reportedly provide acceptable performance [131, 132], but are seldom used due to their higher cost and the low production rates dictated by the chip-formation characteristics of the material.

As noted above, titanium alloys produce serrated or shear-localized chips under most cutting conditions [131]. As discussed in Section 6.6, shear-localized chip formation results from the high strength, low thermal diffusivity, and temperature-softening behavior of the material. This type of chip formation can lead to large variations in cutting forces and induce chatter. The poor thermal properties of titanium may also lead to poor surface integrity in the machined surface due to the occurrence of steep temperature gradients and differential cooling of the surface layer. Titanium also adheres strongly to common tool materials, resulting in high friction and a tendency for surface damage due to deformation and BUE formation.

As a result of these material constraints, the following general guidelines should be followed in machining titanium alloys [132]:

- low cutting speeds should be used to limit cutting temperatures;
- high feed rates should be maintained to avoid surface damage;
- high coolant volumes should be maintained to reduce temperatures and clear chips;
- sharp tools with positive rake angles and proper clearance should be used to avoid BUE formation;
- dwelling of the tool against the workpiece should be avoided to reduce surface damage;
- rigid tooling and fixturing setups should be used to avoid excessive workpiece deflections and chatter.

Turning should be carried out using carbide tooling when possible [132]. Negative rake tools are used for roughing with carbide tools, while positive rakes are used for finish turning and with HSS tooling.

Milling titanium alloys is more difficult than turning or boring because chips may adhere to the tool during noncutting periods in interrupted cutting, resulting in tool chipping or breakage [134]. Milling operations are often carried out at lower cutting speeds than turning or boring, and HSS rather than carbide tooling is often preferred. Climb milling is preferred to conventional milling when possible to reduce tool chipping [132]. When face milling large surfaces, the spindle should be tilted so that the trailing cutting edges do not rub against the machined surface.

Drilling should be carried out using sharp drills with high point angles [135] or spiral point geometries [132]. Flood or (preferably) through-the-tool coolant should be used to dissipate heat. Dwelling of the drill in the bottom of the hole should be avoided; positive feed mechanisms are useful in avoiding dwell. Better tool life is obtained with carbide-tipped than with HSS drills; when HSS drills are used, chromium or oxide coatings are often effective in reducing galling at the margins.

More detailed recommendations on tool geometries for turning, milling, and drilling, as well as recommended practices for reaming, tapping, grinding, and nontraditional machining processes, are discussed in Ref. [132].

11.5.10 Nickel Alloys

Nickel alloys are generally difficult to machine. They have high strength and ductility and work-harden rapidly [2, 136, 137]. Because of this combination of properties, they are subject to many of the machining difficulties encountered with austenitic stainless steels.

The work-hardening characteristics of nickel are the source of many machining difficulties [136]. Since they work-harden rapidly, they produce high tool stresses and temperatures, which can accelerate tool wear. The work-hardening rate is highest for annealed or hot-worked materials; one common method of improving machinability is to impart cold work to the material prior to machining when possible. The effect of work hardening can also be reduced by using sharp tools with positive rake angles. High feed rates and depths of cut are also recommended to reduce damage to the machined surface caused by excessive frictional heating. The high forces generated during machining can also lead to distortion of the part and to chatter. To prevent distortion, it is often advisable to rough parts to near the finished dimension, stress relieve them, and then finish machine. This is particularly true for the Group D alloys discussed below. Microstructural features also have some influence on machinability. As with cast iron and free-machining steels, the presence of graphite or sulfide phases in the matrix improves machinability by reducing friction and improving chip breaking. On the other hand, the presence of hard phases such as carbides, nitrides, oxides, and silicates increases abrasive tool wear and reduces tool life.

From a machining viewpoint, nickel alloys may be classified into five groups [136]: Group A, consisting of alloys containing more than 95% nickel (e.g., commercially pure nickel grades); Group B, consisting of most nickel–copper alloys (e.g., Monel 400 and Nilo alloy 48); Group C, consisting of solid solution nickel–chromium and nickel–chromium–iron alloys (e.g., Inconel 600 and Incoloy 800); Group D, consisting of age-hardenable alloys (e.g., Incoloy 925 and Inconel 718); and Group E, consisting of the free-machining Monel R-45 alloy. Group A alloys have moderate strength and are subject to BUE formation and chip control problems in the annealed or hot-worked state. When possible, they should be cold worked prior to machining. Group B alloys have higher strength and lower ductility than Group A alloys. They should be machined in the cold-worked condition for optimum results. Group C alloys are subject to hard phase formation and should be cold drawn and stress relieved prior to machining. Group D alloys have high strength and hardness and are the most difficult to machine. They produce high cutting forces and are subject to distortion as discussed above. They should be machined in the unaged condition; material which has been solution annealed and quenched or rapidly air cooled has the lowest hardness and best machinability. The Group E alloy is the most machinable of the nickel alloys and is suitable for high production applications in automatic bar machines.

Nickel alloys are most commonly machined using HSS and carbide tooling [138]. Because they have low chemical reactivity with nickel, alumina-based ceramics and PCBN tools are also suitable [139]. Straight alumina ceramics may fracture in interrupted cutting operations; research suggests that whisker-reinforced ceramics may be suitable for these operations [140]. As with steel, the high cutting temperatures generated makes silicon nitride-based ceramic tools unsuitable for machining nickel alloys. Widespread use of ceramic tooling has been inhibited by the fact that many nickel alloys form shear-localized chips at high cutting speeds [141, 142], which effectively limits the maximum cutting speeds which can be employed (often to below 30 m/min).

Carbide tools are most commonly used in turning operations. Positive rake angles, large nose radii, and molded-in chip breakers are used to prevent BUE formation, tool fracture, and chip control problems. Surface finish is influenced by the feed and depth of cut; an increase in depth of cut improves the surface finish generated with coated carbide tools and leads to deterioration in the finish when uncoated tools are used [143, 144]. Milling is often carried out using HSS cutters because carbide cutting edges tend to chip or fracture in interrupted operations. Rigid tool and fixturing setups are required to maintain accuracy and prevent chatter. General purpose drilling can be carried out using HSS drills [138]. As with titanium alloys, positive feed equipment should be used to prevent dwelling of the drill at the bottom of the hole. Crankshaft drills and carbide-bladed spade drills provide better performance in deep hole applications. Specific recommendations on tool geometries and cutting conditions for these and other processes are summarized in Ref. [136].

11.5.11 Depleted Uranium Alloys

Depleted uranium alloys are byproducts of enrichment processes in which the ^{235}U is extracted from natural uranium. They are machined on a large scale at high production rates to produce weapons components, armor, calorimeter plates, ballistic penetrators, radiation shielding, gyroscope rotors, flywheels, sinker bars, and aircraft counterweights [145, 146]. Three classes of alloys can be distinguished: dilute alloys, containing less than 0.4 wt% of alloying elements; lean alloys, containing between 0.4 and 4 wt% of alloying elements; and stainless alloys, containing more than 4 wt% of alloying elements. Common alloying elements include titanium, which increases hardness, and niobium, which increases corrosion resistance. Metallurgically, these alloys usually consist of an orthorhombic α uranium matrix with carbides and oxides as principal inclusions [145]. Complex tetragonal (β) and body-centered cubic (γ) crystal structures also occur [146].

Although depleted uranium alloys are in many ways free cutting, they exhibit a number of material properties which create machining difficulties. These include:

- high ductility, adhesiveness, and tendency to gall;
- abrasiveness;
- high work hardening;
- low modulus;
- reactivity with tools and coolants;
- pyrophoricity;
- anisotropic thermal expansion;
- shape memory;
- toxicity and radioactivity.

The negative consequences of some of these properties have been discussed with reference to other materials. High ductility, adhesiveness, and tendency to gall limit cutting speeds and can lead to BUE, chip breaking, and burring problems as discussed for pure copper, 356 Al, and 1008 and 1010 steel. High abrasiveness and work hardening can increase abrasive and notch tool wear as discussed for stainless steels and nickel alloys. Low modulus and high reactivity can limit tool material choices and achievable tolerances and promote chatter as discussed for titanium alloys. The countermeasures adopted to address these difficulties for other materials are broadly applicable to machining depleted uranium. The problems more specific to depleted uranium alloys are pyrophoricity, anisotropic thermal expansion, shape memory, and toxicity and radioactivity.

Pyrophoricity is the tendency to catch fire. Uranium alloys oxidize rapidly when exposed to air, and may generate self-sustaining fires when machined into thin chips at low feeds. (A similar problem occurs with magnesium alloys, but a spark or external heat source is required to produce a fire in the magnesium case.) To combat this problem, chips are normally collected in a pan and submerged beneath coolant to reduce temperatures. Water-based coolants are preferred. Mineral oil coolants can also be used, but in this case fires must be extinguished before temperatures exceeding the flash point of the oil are generated. In either case the chip pan should be emptied frequently so that large masses of chips do not accumulate. Finish machined unranium components are also often coated after fabrication to limit oxidation.

Due to its crystal structure, unalloyed uranium expands in two directions and contracts in the third when heated. Uranium alloys also exhibit shape memory, a tendency to return to their original shape after a temperature change. Because of this property, it is also advisable to avoid working the material in one direction, which can impart a preferred orientation to the crystal structure. Both the anisotropic thermal expansion and shape memory properties dictate the use of sharp tools and controlled temperature coolant to minimize distortion due to temperature effects. In precision applications, stock may be left for a light finish pass after roughing; the finish pass is often performed several days after roughing to allow the material to stabilize, and often also after heating to the β transition temperature followed by rapid cooling.

Depleted uranium is a toxic and radioactive material which can present serious exposure hazards (referred to in the nuclear industry by the stoic euphemism "health physics"). In the United States, detailed safety protocols are provided by the regulating agencies which license its use.

Depleted uranium can be machined on standard CNC machine tools. Turning, facing, single-point threading, and milling should be performed with sharp tools to limit heat generation and cold working. Carbide tooling is generally used. Drilling may be performed with solid carbide drills coated with a sulfur paste lubricant. Spiral taps with variable rakes and oxide coatings are used for internal threading. Grinding should be avoided but can be performed with silicon carbide wheels and water-based coolants when necessary. Recommended cutting speeds and feeds for these operations are summarized in Ref. [145]. Tolerances of ± 0.05 mm are reportedly achievable in turning and milling [146].

REFERENCES

1. E.J.A. Armarego and R.H. Brown, *The Machining of Metals*, Prentice-Hall, New York, 1969, 246–253
2. E.M. Trent, *Metal Cutting*, Butterworths, London, 1977, 139–180
3. M.C. Shaw, *Metal Cutting Principles*, 3rd Edition, Oxford University Press, Oxford, 1984, 6–7
4. G. Boothroyd and W.A. Knight, *Fundamentals of Machining and Machine Tools*, 2nd Edition, Marcel Dekker, New York, 1989, 148–151
5. G.T. Smith, *Advanced Machining*, Springer-Verlag, New York, 1989, 167–198
6. C.F. Barth, Turning, in: R.I. King, Ed., *Handbook of High Speed Machining Technology*, Chapman and Hall, New York, 1985, 174–179
7. T.J. Drozda and C. Wick, Eds., *Tool and Manufacturing Engineers Handbook*, 4th Edition, SME, Dearborn, MI, 1983, Vol. 1, 40–59
8. N.H. Cook, What is machinability? *Influence of Metallurgy on Machinability*, ASM, Metals Park, OH, 1975, 1–10
9. C. Zimmerman, S.P. Boppama, and K. Katbi, Machinability test methods, *Metal Handbook*, Vol. 16: Machining, 9th Edition, ASM, Materials Park, OH, 1989, 639–647
10. J. ElGomayel, Fundamentals of the Chip Removal Process, SME Technical Paper MR77-256, 1977
11. Metcut Research Associates, *Machinability Data Handbook*, Cincinnati, OH, 1982
12. P. Balakrishnan and M.F. DeVries, A review of computerized machinability data base systems, *Proc. NAMRC* **10** (1982) 348–356
13. L.-Z. Lin and R. Sandstrom, Evaluation of machinability data, *J. Testing Eval.* **22** (1994) 204–211

14. M.P. Groover, A Survey on the Machinability of Metals, SME Technical Paper MR76-269, 1976
15. G.F. Micheletti, Work on machinability in the co-operative group C of CIRP and outside this group, *CIRP Ann.* **18** (1970) 13–30
16. D.A. Stephenson, Tool–work thermocouple temperature measurements — theory and implementation issues, *ASME J. Eng. Ind.* **115** (1993) 432–437
17. Standard Method for Evaluating Machining Performance of Ferrous Metals Using an Automatic Screw/Bar Machine, ASTM Standard E618–81, 1981
18. Tool Life Testing with Single-Point Turning Tools, ISO Standard 3685:1993(E), 1993
19. The Volvo Standard Machinability Test, Volvo Laboratory for Manufacturing Research, Trollhattan, Sweden, Standard 1018.712, 1989
20. A.J. DeArdo, C.I. Garcia, R.M. Laible, and U. Eriksson, A better way to assess machinability, *Am. Machinist* **137**:5 (1993) 33–35
21. G. Lorenz, Comparative drill performance tests, *Machinability Testing and Utilization of Machining Data*, ASM, Metals Park, OH, 1979, 147–163
22. J.J. Fulmer and J.M. Blanton, Enhanced Machinability of P/M Parts Through Microstructure Control, SAE Technical Paper 940357, 1994
23. Y. Matsushima, M. Nakamura, H. Takeshita, S. Akiba, and M. Katsuta, Improvement of drilling machinability of microalloyed steel, *Kobelco Technol. Rev.* **17** (1994) 38–43
24. J.S. Agapiou, G.W. Halldin, and M.F. DeVries, On the machinability of powder metallurgy austentitic stainless steels, *ASME J. Eng. Ind.* **110** (1988) 339–343
25. V.C. Venkatesh and V. Narayanan, Machinability correlation among turning milling and drilling processes, *CIRP Ann.* **35** (1986) 59–62
26. Bar Products Group, American Iron and Steel Institute, *Steel Bar Product Guidelines*, Iron and Steel Society, Warrendale, PA, 1994, 164–166
27. A. Henkin and J. Datsko, Influence of physical properties on machinability, *ASME J. Eng. Ind.* **85** (1963) 321
28. K. Nakayama, Chip control in metal cutting, *Bull. Jpn. Soc. Precision Eng.* **18**:2 (1984) 97–103
29. C. Spaans, A Systematic Approach to Three-Dimensional Chip Curl, Chip Breaking and Chip Control, SME Technical Paper MR70-241, 1970
30. W. Kluft, W. Konig, C.A. Van Luttervelt, K. Nakayama, and A.J. Pekelharing, Present knowledge of chip control, *CIRP Ann.* **28**:2 (1979) 441–455
31. I.S. Jawahir, A survey and future predictions for the use of chip breaking in unmanned systems, *Int. J. Adv. Manuf. Technol.* **3**:4 (1988) 87–104
32. I.S. Jawahir and C.A. Van Luttervelt, Recent developments in chip control research and applications, *CIRP Ann.* **42** (1993) 659–693
33. P.D. Liu, R.S. Hu, H.T. Zhang, and X.S. Wu, A study on chip curling and breaking, *Proeedings of the 29th Matador Conference*, 1992, 507–512
34. G. Stabler, The fundamental geometry of cutting tools, *Proc. Inst. Mech. Eng.* **165** (1951) 14–21
35. L.V. Colwell, Predicting the angle of chip flow for single-point cutting tools, *ASME Trans.* **76** (1954) 199–204
36. K. Okushima and K. Minato, On the behavior of chip in steel cutting, *Bull. Jpn. Soc. Precision Eng.* **2** (1959) 58–64
37. R.H. Brown and E.J.A. Armarego, Oblique machining with a single cutting edge, *Int. J. Mach. Tool Des. Res.* **4** (1964) 9–25
38. H.T. Young, P. Mathew, and P.L.B. Oxley, Allowing for nose radius effects in predicting the chip flow direction and cutting forces in bar turning, *Proc. Inst. Mech. Eng.* **201C** (1987) 213–226
39. H.J. Fu, R.E. DeVor, and S.G. Kapoor, A mechanistic model for the prediction of the force system in face milling operations, *ASME J. Eng. Ind.* **106** (1988) 81–88
40. K. Nakayama, A study on chip breaker, *Bull. JSME* **5** (1962) 142–150
41. T.L. Subramanian and A. Bhattacharyya, Mechanics of chip breakers, *Int. J. Prod. Res.* **4** (1965) 37–49
42. A.R. Trim and G. Boothroyd, Action of obstruction type chip former, *Int. J. Prod. Res.* **6** (1968) 227
43. C. Spaans and P.F.H.J. Van Geel, Break mechanisms in cutting with a chip breaker, *CIRP Ann.* **19** (1970) 87–92

44. B. Worthington, The operation and performance of a groove type chip breaker, *Int. J. Prod. Res.* **14** (1976) 529–558
45. S. Kaldor, A. Ber, and E. Lenz, On the mechanism of chip breaking, *ASME J. Eng. Ind.* **101** (1979) 241–249
46. M. Ogawa and K. Nakayama, Effects of chip splitting nicks in drilling, *CIRP Ann.* **34** (1985) 101
47. I.I. Shilin and K.G. Sadolevskaya, Chip breaking by interrupted feed, *Mach. Tooling* **36**:3 (1965) 30–32
48. H. Takayama, H. Sekiguchi, and K. Takada, One solution for chip hazard in turning — study on automatic programming for numerically controlled machines (1st report), *J. Jpn. Soc. Precision Eng.* **36** (1970) 150–156
49. P.F. Ostwald, Dynamic Chip Breaking: Can It Overcome the Surface Finish Problem? ASTME Paper MR67-228, 1967
50. F. Rasch, Hydraulic chip breaking, *CIRP Ann.* **30** (1981) 333–335
51. H. Takayama and S. Kato, Burrless drilling by means of ultrasonic vibration, *CIRP Ann.* **40** (1991) 83–86
52. M. Mazurkiewicz, Z. Kubala, and J. Chow, Metal machining with high-pressure water-jet cooling assistance — a new possibility, *ASME J. Eng. Ind.* **111** (1989) 7–12
53. Increasing Machine Tool Productivity With High Pressure Cryogenic Coolant Flow, Manufacturing Technology Directorate, Wright Laboratory, Air Force Systems Command Report WL-TR-92-8014, May 1992
54. A.G. Ringler, High velocity coolant distribution system for improved chip control and disposal, *Strategies for Automation of Machining: Materials and Processes*, ASM, Metals Park, OH, 1987, 147–155
55. J.D. Christopher, The Influence of High Pressure Cryogenic Coolant on Tool Life and Productivity in Turning, SME Technical Paper MR90-249, 1990
56. R.R. Lindeke, F.C. Schoenig, A.K. Khan, and J. Haddad, Machining of α–β titanium with ultra-high pressure through the insert lubrication/cooling, *Trans. NAMRI/SME* **19** (1991) 154–161
57. M. Arai and K. Nakayama, Boundary notch on cutting tool caused by burr and its suppression, *J. Jpn. Soc. Precision Eng.* **52** (1986) 864–866
58. L.K. Gillespie and P.T. Blotter, The formation and properties of machining burrs, *ASME J. Eng. Ind.* **98** (1976) 66–74
59. L.K. Gillespie, *Deburring Capabilities and Limitations*, SME, Dearborn, MI, 1976, Chapters 3–5
60. L.K. Gillespie and R.E. King, Eds., *Robotic Deburring Handbook*, SME, Dearborn, MI, 1987
61. K. Nakayama and M. Arai, Burr formation in metal cutting, *CIRP Ann.* **36** (1987) 33–36
62. A.J. Pekelharing, Exit failure in interrupted cutting, *CIRP Ann.* **27** (1978) 5–10
63. S. Nakamura and A. Yamamoto, Influence of disengage angle upon initial fracture of tool edge, *Bull. Jpn. Soc. Precision Eng.* **19** (1985) 169–174
64. T.C. Ramaraj, S. Santhanam, and M.C. Shaw, Tool fracture at the end of a cut: I. Foot formation, *ASME J. Eng. Ind.* **110** (1989) 333–338
65. S.L. Ko and D.A. Dornfeld, A study on burr formation mechanism, *Symposium on Robotics*, ASME DSC Vol. 11, ASME, New York, 1988, 271–282
66. S.L. Ko and D.A. Dornfeld, Analysis and modelling of burr formation and breakout in metal, *Mechanics of Deburring and Surface Finishing Processes*, ASME PED Vol. 38, ASME, New York, 1989, 79–91
67. H. Ernst and W.A. Haggarty, The spiral point drill — a new concept in drill point geometry, *ASME Trans.* **80** (1958) 1059–1072
68. J. Kim, S. Min, and D.A. Dornfeld, Optimization and control of drilling burr formation of ANSI 304L and AISI 4118 based on drilling burr control charts, *Int. J. Mach. Tools Manuf.* **41** (2001) 923–936
69. S. Zaima, A. Yuki, and S. Kamo, Drilling of aluminum plates with special type point drill, *J. Jpn. Inst. Light Met.* **18** (1986) 269–276, 307–313
70. Machining Magnesium, Dow Chemical Company, Midland, MI, 1982, Form No. 141-480-82
71. *Metals Handbook*, Vol. 2: Properties and Selection: Nonferrous Alloys and Pure Metals, 9th Edition, ASM, Metals Park, OH, 1979, 549–551

72. H.J. Morales, Magnesium, Machinability and Safety, SAE Technical Paper 800418, 1980
73. M. Videm, R.S. Hansen, N. Tomac, and K. Tonnesen, Metallurgical Considerations for Machining Magnesium Alloys, SAE Technical Paper 940409, 1994
74. K.T. Kurihara, H. Kato, and T. Tozava, Cutting temperature of magnesium alloys, *ASME J. Eng. Ind.* **103** (1981) 254–260
75. O. Hoehne and D. Korff, The Design and Production of Light Metal Castings at Volkswagen, SAE Technical Paper 800B, 1964
76. N. Tomac, K. Tonnessen, and F.O. Rasch, PCD tools in machining magnesium alloys, *Eur. Mach.* **May/June** (1991) 12–16
77. A. Spicer, J. Kasi, C. Billups, and J. Pajec, Machining Magnesium with Water Based Coolants, SAE Technical Paper 910415, 1991
78. K. Tonnessen, N. Tomac, and F.O. Rasch, Machining magnesium alloys with use of oil–water emulsions, *8th International Colloquium, Tribology 2000*, Esslingen, January 1992, 18.7-1–18.7-9.
79. N. Tomac and K. Tonnessen, Formation of flank build-up in cutting magnesium alloys, *CIRP Ann.* **41** (1991) 79–82
80. Machining Magnesium, Hydro Magnesium (now Norsk Hydro), Oslo, Norway, 1988, Booklet No. BA01186
81. B. Chamberlain, Machinability of aluminum alloys, *Metals Handbook*, Vol. 2: Properties and Selection: Nonferrous Alloys and Pure Metals, 9th Edition, ASM, Metals Park, OH, 1979, 187–190
82. R.E. DeVor and J.C. Miller, Machinability of high silicon cast aluminum alloy with carbide and diamond cutting tools, *Proc. NAMRC* **9** (1981) 296–304
83. J. Alverio, J.S. Agapiou, and C.-H. Shen, High speed drilling of 390 aluminum, *Trans. NAMRI/SME* **18** (1990) 209–215
84. W. Koenig and D. Erinski, Machining and machinability of aluminum cast alloys, *CIRP Ann.* **32** (1983) 535–540
85. J.L. Jorstad, Influence of aluminum casting alloy metallurgical factors on machinability, *Modern Casting* **December** (1980) 47–51
86. S.D. Jones and R.J. Furness, An experimental study of burr formation for face milling 356 aluminum, *Trans. NAMRI/SME* **25** (1997) 183–188
87. H.E. Chandler, Machining of metal-matrix composites and honeycomb structures, *Metals Handbook*, Vol. 16: Machining, 9th Edition, ASM, Materials Park, OH, 1989, 893–901
88. C. Lane, Machining characteristics of particulate-reinforced aluminum, *Fabrication of Particulate-Reinforced Metal Composites*, ASM, Materials Park, OH, 1990, 195–201
89. G.A. Chadwick and P.J. Heath, Machining metal matrix composites, *Met. Mater.*, **February** (1990) 73–76
90. W.S. Ricci, S.E. Swider, and T.J. Moores, Mechanisms of tool wear when diamond machining composites, *Procedings of the 1987 Eastern Manufacturing Technology Conference*, Springfield, MA, 1987, 3-190–3-214.
91. C. Lane, Drilling and tapping of SiC particle-reinforced aluminum, *A Systems Approach to Machining*, ASM, Materials Park, OH, 1993
92. J. Bunting, Drilling advanced composites with diamond veined drills, *A Systems Approach to Machining*, ASM, Materials Park, OH, 1993, 99–104.
93. C. Lane and M. Finn, Observations on using CVD diamond in milling MMCs, *Materials Issues in Machining and the Physics of Machining Processes*, TMS, Warrendale, PA, 1992, 39–51
94. Copper, *Metals Handbook*, Vol. 2: Properties and Selection: Nonferrous Alloys and Pure Metals, 9th Edition, ASM, Metals Park, OH, 1979, 388–390
95. Machining of cast irons, *Metals Handbook*, Vol. 16: Machining, 9th Edition, ASM, Materials Park, OH, 1989, 648–665
96. Gray iron, *Metal Handbook*, Vol. 1: Properties and Selection: Irons and Steels, 9th Edition, ASM, Materials Park, OH, 1978, 23–53
97. R.T. Wimber, Machinability of pearlitic gray cast iron, *Proc. NAMRC* **9** (1981) 290–295
98. T.H. Wickenden, Data on machinability and wear of cast iron, *SAE Trans.* **23** (1925) 181–189
99. J.R. Davis, Ed., *ASM Specialty Handbook: Cast Irons*, ASM, Materials Park, OH, 1996, 244–254

100. D. Bordui, Third generation silicon nitride, *Ceramic Cutting Tools and Applications*, SME, Dearborn, MI, 1989
101. C. Wick, Machining with PCBN tools, *Manuf. Eng.* **101**:1 (1988) 73–78
102. M. Goto, T. Nakai, and S. Nakatani, Cutting Performance of PCBN for Cast Iron, SAE Technical Paper MR91-176, 1991
103. J.S. Agapiou, Design characteristics of new types of drill and evaluation of their performance drilling cast iron — I. Drills with four major cutting edges, *Int. J. Mach. Tools Manuf.* **33** (1993) 321–341.
104. J.S. Agapiou, Design characteristics of new types of drill and evaluation of their performance drilling cast iron — II. Drills with three major cutting edges, *Int. J. Mach. Tools Manuf.* **33** (1993) 343–365.
105. J.S. Agapiou, An evaluation of advanced drill body and point geometries in drilling cast iron, *Trans. NAMRI/SME* **19** (1991) 79–89
106. Cerasiv GmbH, Anwendungstechnik: SPK-Schneidkeramik, SPK-Cermet, WURBON, SPK Tools, Florence, KY, 1995
107. Clean Iron Production and Machining Technology Project, Report, University of Alabama at Birmingham, July 1996
108. K.J. Trigger, L.B. Zylstra, and B.T. Chao, Tool forces and tool–chip adhesion in the machining of nodular cast iron, *ASME Trans.* **74** (1952) 1017–1027
109. I. Ham, K. Hitomi, and G.L. Thuering, Machinability of nodular cast irons. Part I. Tool forces and flank adhesion, *ASME J. Eng. Ind.* **83** (1961) 142–154
110. K. Hitomi and G.L. Theuring, Machinability of nodular cast irons. Part II. Effect of cutting conditions on flank adhesion, *ASME J. Eng. Ind.* **84** (1962) 282–288
111. D. Sahm, Werkzeugverschleißmessungen an 23 Stationen einer Transferstraße für Motorblöcke bei der Umstellung auf den neuen Werkstoff GGV, Abschlußbericht, Angebot Nr. 5387 Teil 4–5, PTW, TH-Darmstadt, 1996
112. M.E. Finn, Machining of carbon and alloy steels, *Metals Handbook*, Vol. 16: Machining, 9th Edition, ASM, Materials Park, OH, 1989, 666–680
113. A.M. Abrao, D.K. Aspinwal, and M.L.H. Wise, Tool Life and Workpiece Surface Integrity Evaluations when Machining Hardened AISI H13 and AISI E52100 Steels with Conventional Ceramics and PCBN Tool Materials, SME Technical Paper MR950159, 1995
114. Machining of tool steels, *Metals Handbook*, Vol. 16: Machining, 9th Edition, ASM, Materials Park, OH, 1989, 708–732
115. S.V. Subramanian, H.O. Gekonde, and J. Gao, Microstructural engineering for hard turning, *Proceedings of the 40th MWSP Conference*, ISS, 1998
116. S. Akcan, S. Shah, et al., Characteristics of white layers formed in steel by machining, ASME MED Vol. 10, 1999, 789–795
117. Y. Chou and C.J. Evans, Process effects on white layer formation in hard turning, *Trans. NAMRI/SME* **26** (1998) 117–122
118. Y.B. Guo and J. Sahni, A comparative study of hard turned and cylindrically ground white layers, *Int. J. Mach. Tools Manuf.*, 44 (2004) 135-145
119. D.W. Murray, Machining performance of steels, *Iron Steel Engr.* **April** (1967) 123–128
120. M.E. Finn, G.A Beaudoin, R.M. Nishizaki, S.V. Subramanian, and D.A.R. Kay, Influence of steel matrix on machinability of high carbon steels in automatic machining, *Strategies for Automation of Machining: Materials and Processes*, ASM, Materials Park, OH, 1987, 43
121. S.V. Subramanian, H.O. Gekonde, et al., Inclusion engineering of steels for high speed machining, *CIM Bull.* **91** (1998) 107–115
122. S.V. Subramanian, D.A.R. Kay, and J. Junpu, Inclusion engineering for the improved machinability of medium carbon steels, *Proceedings of the Symposium on Inclusions and Their Influence on Material Behavior*, Chicago, IL, 1988 (ASM paper No. 8821-003)
123. J.H. Hoffman and R.J. Turonek, High Performance Forged Steel Crankshafts — Cost Reduction Opportunities, SAE Technical Paper 910139, 1991
124. A.R. Chambers and D. Whittaker, Machining characteristics of mircoalloyed forging steels, *Met. Technol.* **11** (1984) 323–333

125. V. Ollilainen, T. Lahti, H. Pontinen, and E. Heiskala, Machinability comparison with substituting microalloyed forging steel for quenched and tempered steels, *Fundamentals of Microalloying Forging Steels*, TMS, Warrendale, PA, 1987, 461–474

126. T. Kosa and R.P. Ney, Sr., Machining of stainless steels, *Metals Handbook*, Vol. 16: Machining, 9th Edition, ASM, Materials Park, OH, 1989, 681–707

127. D.M. Blott, Machining wrought and cast stainless steels, *Handbook of Stainless Steels*, McGraw-Hill, New York, 1977, Section 24, 2–30

128. H.E. Chandler, Machining of powder metallurgy materials, *Metals Handbook*, Vol. 16: Machining, 9th Edition, ASM, Materials Park, OH, 1989, 879–892

129. J.S. Agapiou, G.W. Halldin, and M.F. DeVries, Effect of porosity on the machinability of P/M 304L stainless steel, *Int. J. Powder Metall.* **25** (1989) 127–139

130. J.S. Agapiou and M.F. DeVries, Machinability of powder metallurgy materials, *Int. J. Powder Metall.* **24** (1988) 47–57

131. A.R. Machado and J. Wallbank, Machining of titanium and its alloys — a review, *Proc. Inst. Mech. Eng.* **B204** (1990) 53–59

132. H.E. Chandler, Machining of reactive metals, *Metals Handbook*, Vol. 16: Machining, 9th Edition, ASM, Materials Park, OH, 1989, 844–857

133. P.A. Dearnley and A.N. Grearson, Evaluation of principal wear mechanisms of cemented carbides and ceramics used for machining titanium alloy IMI 318, *Mater. Sci. Technol.* **2** (1986) 47–58

134. Basic Design Guide — RMI Titanium, RMI Co., Niles, OH

135. H. Barish, Quality drills contribute to successful titanium tooling, *Cutting Tool Eng.*, **February** (1988) 38–41

136. R.W. Breitzig, Machining nickel alloys, *Metals Handbook*, Vol. 16: Machining, 9th Edition, ASM, Materials Park, OH, 1989, 835–843

137. K. Uehare, High-speed machining Inconel 718 with ceramic tools, *CIRP Ann.* **42** (1993) 103–106

138. E.O. Ezugwu and C.J. Lai, Failure modes and wear mechanisms of M35 high-speed steel drills when machining Inconel 901, *J. Mater. Process. Technol.* **49** (1995) 295–312

139. B.M. Kramer and P.D. Hartung, Theoretical considerations in the machining of nickel-based alloys, *Cutting Tool Materials*, ASM, Metals Park, OH, 1981, 57–74

140. A.D. Johnson, A.R. Thangaraj, and K.J. Weinmann, The influence of mechanical properties on the performance of Al_2O_3–SiC_w ceramics in the machining of Inconel 718, *Materials Issues in Machining and the Physics of Machining Processes*, TMS, Warrendale, PA, 1992, 187–202

141. P.K. Wright and J.G. Chow, Deformation characteristics of nickel alloys during machining, *ASME J. Eng. Mater. Technol.* **108** (1986) 85–93

142. R. Komanduri and T.A. Schroeder, On shear instability in machining nickel-based superalloy, *ASME J. Eng. Ind.* **108** (1986) 93–100

143. I.A. Choudhury and M.A. El-Baradie, Machinability assessment of Inconel 718 by factorial design of experiment coupled with response surface methodology, *J. Mater. Process. Technol.* **55** (1999) 30–39

144. M. Alauddin, M.A. El-Baradie, and M.S. Hashmi, Optimization of surface finish in end milling Inconel 718, *J. Mater. Process. Technol.* **56** (1996) 54–65

145. J.A. Aris, Machining of uranium and uranium alloys, *Metals Handbook*, Vol. 16: Machining, 9th Edition, ASM, Materials Park, OH, 1989, 874–878

146. *User's Guide to MSC Materials: Depleted Uranium*, Manufacturing Sciences Corporation, Oak Ridge, TNMetal Cutting Theory and Practice

12 Machining Dynamics

12.1 INTRODUCTION

As discussed in previous chapters, the machine tool, cutting tool, part, and fixture form a complex system consisting of several coupled structural elements. During cutting, a substantial amount of energy is dissipated through plastic deformation and friction. Some of this energy is transmitted to the structural elements of the system, inducing vibrations (relative motion between the tool and workpiece). These vibrations should be minimized because they degrade machining accuracy and the machined surface texture; moreover, under unfavorable conditions they may become unstable, leading to chatter, which can cause accelerated tool wear and breakage, accelerated machine tool wear, and damage to the machine tool and part. Vibration is a particularly serious problem in fine finishing operations such as grinding, and in processes such as boring which employ compliant tooling [1]. Unstable vibrations are a major factor limiting production rates in many operations, especially at high cutting speeds.

A number of analytical and theoretical methods have been developed to study the dynamic behavior of the machining system. The study of machining dynamics has two basic objectives: (1) to identify rules for designing stable machine tools, and (2) to identify rules for choosing dynamically stable cutting conditions. The dynamic analysis of the machining system is complicated by the physical complexity of machine tool systems and the cutting process, the difficulty of estimating dynamic properties of joints between structural components, and the fact that the system is time varying since components move relative to each other during the process.

The design of machine tools, cutting tools, and tool holders was discussed in Chapter 3–Chapter 5. Engineers have traditionally relied on the inherent rigidity of the machining system to control deflections, which cause dimensional errors in the machined part [2–4]. This rigidity and the internal damping of the system are also the primary design features which control vibration amplitudes and stability. Although machine structural elements are designed for rigidity, they all exhibit structural vibration modes which may be excited during cutting operations [5–7]. Moreover, chatter may also result from vibration of the part or fixture, which often are more compliant than the machine structure.

This chapter describes analytical and experimental methods of investigating machining dynamics to address the two objectives stated above. Topics covered include analysis methods, dynamics of lumped parameter systems, types of machine tool vibrations, forced vibrations, self-excited vibrations, experimental methods for stability analysis, chatter detection, and passive and active methods for chatter suppression and control.

12.2 VIBRATION ANALYSIS METHODS

Several methods can be used to analyze the structural dynamics of machine tool systems [8–15]. These include kinematic simulations, lumped mass and massless beam methods, the receptance method with beams having distributed mass, and finite element analysis (FEA).

There are several simulation programs for the large-scale transient analysis of controlled mechanical systems such as machine tool systems. These programs can be used to perform kinematic and dynamic analyses of a machine structure. The kinematic simulation tools not only provide position, velocity, and acceleration information at each time step for massless mechanisms, but also all internal reaction forces at joints and constraints. They can also be used to perform dynamic analyses involving the mass properties of the moving bodies and forces acting on the bodies. They also handle flexible bodies with nonlinear components and they can efficiently perform static structural optimization based on the modal potential energy. They can be very useful for the static analysis of machine tools during design optimization because they can simulate the static structural deflections of several machine configurations and different load cases in a short time frame.

The structural elements of a machine tool system are continuous with distributed mass. For many analysis purposes, however, the various components can be treated as a collection of discrete or lumped masses connected by springs and dampers. This idealization is the basis for many experimental methods for analyzing the dynamics and stability of machine tools. The dynamics of lumped mass systems is described in detail in Section 12.3.

Among continuous or distributed mass methods, finite element methods are the most practically useful approaches for analyzing machining systems because they permit use of the most realistic assumptions and because they can be used not only for dynamic analysis, but also for static and thermal analysis using the same model.

Two approaches can be used to apply the finite element method to predict the dynamic behavior of the machining system. In the first approach, a single large model of the whole system is employed. In the second approach, the overall system is subdivided into smaller subsystems (machine components) which are analyzed separately; the results of all the subsystems are linked together in an overall system structural response program using a generalized building block approach to predict the structural characteristics of the whole system. The latter method provides more flexibility because individual components of the machining system can be changed and analyzed individually, but requires more model building time.

Several element types have been used to model machine tools in the past, including plate, shell, beam, and solid elements. In more recent practice, automeshers using either tetragonal or cubic (hex) elements have been increasingly used. These greatly reduce model building time, although some manual refinement of the mesh is often still necessary. First-order elements are most commonly used, although in some cases (e.g., bending loads on cantilever-like structures) coarser meshes of second-order elements yield better results with less computing time.

Ideally, the machine tool structure's static and dynamic behavior should be evaluated through FEA before hardware is built. At this stage, several design variations can be compared and structural parameters can be optimized. Unfortunately, machine tool stiffness and damping ratios generally cannot be predicted, but must be measured, and a number of other nonlinear structural effects are difficult to account for in the design stage. In practice, therefore, FEA provides reasonable static stiffness assessments for major components, but the dynamic behavior of the entire system can only be modeled after hardware is built and system properties such as the dynamic stiffness, damping percentages, natural frequencies, and mode shapes are determined experimentally.

12.3 VIBRATION OF DISCRETE (LUMPED MASS) SYSTEMS

Machine tools are composed of several components and therefore can be considered multimass vibrators. Although the structural elements of a machine tool are geometrically complex and have continuously distributed masses, it is possible in some instances to simulate the behavior of a machine tool structure under the influence of dynamic loading by considering

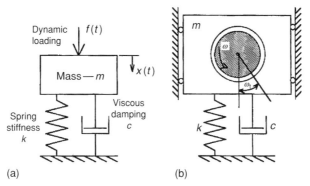

FIGURE 12.1 Free-body diagram for forced vibrations with harmonic excitation of an SDOF mass–spring–viscous damper system. (a) Mass subjected to forced vibration. (b) Unbalanced motor running at constant velocity.

several discrete masses connected together. Although practical systems such as machine tools are multiple degree of freedom (MDOF) and have some degree of nonlinearity, they can be represented as superposition of single degree of freedom (SDOF) spring–mass–dashpot vibrators for many purposes. For example, the mechanical model in Figure 12.1a can represent a cutting process in which m is the mass of the tool and the damping coefficient c and spring stiffness k can be determined through modal analysis of the machine tool. x is the displacement of the center of the tool relative to the workpiece. The structure is assumed to be flexible in the X-direction only. This reduces the model to SDOF. This SDOF model is appropriate, for example, when a thin-walled workpiece is the most flexible part of the structure and is compliant primarily in one direction.

The dynamic performance (or stiffness) of a linear mechanical system of this type can be described by a set of transfer functions or by the resonance frequencies and their associated displacements at the different points, which are called associated modes of vibration.

Several methods have been proposed to examine the dynamic behavior and system stability, including: (1) the s-plane approach, (2) the frequency (j) plane approach, and (3) the time-domain approach [16–18].

The s-plane approach describes the response and performance of a system in terms of the complex frequency variable s. It uses the Laplace transformation to transform the differential equations representing the system motions into algebraic equations expressed in terms of the complex variable s. The transfer function of a linear system, representing the input–output relationship, is then obtained from the algebraic equations. The root locus method can then be used to evaluate the stability of the system by examining the poles and zeros of the transfer function on the s-plane.

The frequency response of a system is defined as the steady-state response of the system to a sinusoidal input signal. The output of a linear system to such an input is also sinusoidal with the same frequency in the steady state; it differs from the input waveform only in the amplitude and phase angle. This method is easily applied to experimentally determine the frequency response and the transfer function of a system. The transfer function in the frequency response method can be also obtained from the transfer function in the s-plane by replacing s with j, where $j = \sqrt{-1}$. The disadvantage of the frequency response method for analysis and design is the indirect link between the frequency domain and the time domain. The frequency domain approach is also limited in applicability to linear time-invariant systems; it is particularly limited when considering multivariable control systems due to the emphasis on the input–output relationship of transfer functions. The frequency response method analyzes a system based on the amplitude and phase equations and curves, which is an advantage for the analysis and design of machining systems.

In the time-domain method, the system response is described by a set of state variables. The state variables are those variables that determine the future behavior of a system when the present state of the system and the excitation signals are known [17–21]. The time-domain approach can be used not only for linear systems but for nonlinear, time-varying, and multivariate systems.

In this section, we review the analysis of lumped mass oscillators based on the frequency domain approach, the approach most commonly applied in experimental modal analysis. The dynamic behavior of SDOF discrete systems will be discussed first. This will provide the basis for a discussion of machine component interactions or MDOF systems.

12.3.1 Single Degree of Freedom Systems

The analysis of an SDOF system provides basic physical insight into the dynamic behavior of vibrating systems in general. The commonly analyzed system of this type, shown in Figure 12.1a, consists of a mass connected to ground by a spring and viscous damper in parallel. The system is excited by a periodic force and vibrates at the forcing frequency. The frequency response method is often used to analyze such a system, generally under the assumption that the driving forcing function is sinusoidal. Physically, a sinusoidal forcing function may result from imbalance in a motor running at constant angular velocity as shown in Figure 12.1b; the four rubber isolators on which the motor legs are mounted represent the spring of the system across which the vibration occurs. The equation of motion for the SDOF system shown in Figure 12.1 with mass m, spring stiffness k, and viscous damping is

$$m\ddot{x}(t) + c\dot{x}(t) + kx(t) = f(t) \tag{12.1}$$

where c is the coefficient of viscous damping, or damping constant, which has dimensions of force per unit velocity (N s/m or kg/s). $x(t)$, $\dot{x}(t)$, and $\ddot{x}(t)$ are, respectively, the displacement, velocity, and acceleration of the externally applied load(s). The natural angular frequency of free undamped oscillations, ω_n (rad/sec), and the damping ratio, ζ, the basic dynamic characteristics of an SDOF system, are

$$\omega_n = \sqrt{\frac{k}{m}} \quad \text{and} \quad \zeta = \frac{c}{c_c} = \frac{c}{2\sqrt{km}} \tag{12.2}$$

The damping ratio ζ is the ratio of the actual damping c to the critical damping c_c, which is the smallest value of c for which the free damped motion is nonoscillatory, that is, the level of damping which would just prevent vibration. The damping ratio is very small for machine tools; typically for machine tools $\zeta \leq 0.05$, which means that the actual damping is 5% of the amount which would just prevent vibration. The damped natural angular frequency and the resonance angular frequency of the system, ω_{dn} and ω_r, are related to the undamped natural frequency by

$$\omega_{dn} = \omega_n\sqrt{1-\zeta^2} \tag{12.3}$$

$$\omega_r = \omega_n\sqrt{1-2\zeta^2} \tag{12.4}$$

The resonance frequency is the frequency at which resonance occurs and the compliance increases. For machine tool systems, the differences between the frequencies ω_n, ω_{dn}, and ω_r is usually not significant because the damping ratio is small.

All mechanical systems exhibit damping, as evidenced by the fact that the amplitude of free vibrations diminishes with time. The effect of damping on the system response can be ignored when the damping ratio is small and the system is not excited at a frequency near resonance. Accurately modeling damping becomes most critical when the system is excited near resonance. Since the effect of damping is not generally known before the machine tool is analyzed, an attempt should be made to model damping until experiments prove it to be insignificant.

Damping forces arise from several sources, including friction between lubricated sliding surfaces, air or fluid resistance, electric damping, internal friction due to the imperfect elasticity of real materials, internal strain, looseness of joints, and other complex causes [6]. There are several forms of damping including viscous, hysteretic, and dry frictional damping. Each assumed model of damping represents different idealization of the actual damping and results in a different predicted response. For example, hysteretic and viscous damping limit resonant amplitudes, while dry frictional damping does not; viscous damping affects the frequency of the resonant peak, whereas hysteretic and dry frictional damping do not. The differences in the predicted response diminish as the difference between the forcing frequency and the natural frequency increases. Viscous damping is the simplest case to analyze mathematically because it is proportional to velocity. Therefore, the resisting forces of complex systems, such as machine tools, are often replaced by equivalent viscous dampers to simplify mathematical analysis. The equivalent damping coefficient is chosen to dissipate the same amount of energy per cycle as the actual resisting force. It is important to note that a large part of the energy dissipation occurs at the toolholder–spindle interface.

It is often convenient to characterize a linear system by its response to a specific sinusoidal input force given by

$$f(t) = F e^{j\omega t} \tag{12.5}$$

where F is the forcing amplitude, $j = \sqrt{-1}$, t is time, and ω is the exciting frequency in rad/sec. The circular frequency ω is related to the frequency f in hertz (Hz) through

$$f = \frac{\omega}{2\pi} \tag{12.6}$$

Modal analysis is performed using the Fourier transform $X(\omega)$ of the displacement $x(t)$:

$$X(\omega) = \int_{-\infty}^{\infty} x(t) e^{-j\omega t} \, dt \tag{12.7}$$

The Fourier transforms of the time derivative of a function can be determined by multiplying the Fourier transform of the function by $j\omega$:

$$\int_{-\infty}^{\infty} \dot{x}(t) e^{-j\omega t} \, dt = j\omega X(\omega)$$
$$\int_{-\infty}^{\infty} \ddot{x}(t) e^{-j\omega t} \, dt = -\omega^2 X(\omega) \tag{12.8}$$

Taking the Fourier transform of both sides of Equation (12.1) therefore yields

$$(-\omega^2 m + j\omega c + k) X(\omega) = F(\omega) \tag{12.9}$$

where the forcing function $f(t)$ is given by Equation (12.5) and $F(\omega)$ is the Fourier transform of $f(t)$.

The steady-state response of this system, which is present as long as the forcing function is active, is given by

$$X(\omega) = G(\omega)F(\omega) \qquad (12.10)$$

where

$$G(\omega) = \frac{X(\omega)}{F(\omega)} = \frac{1}{-\omega^2 m + j\omega c + k}$$

$$= \frac{1}{k}\frac{1}{1 - r^2 + 2j\zeta r} \qquad (12.11)$$

$$r = \frac{\omega}{\omega_n} \qquad (12.12)$$

$G(\omega)$, the *frequency response function* (FRF) of the system, is the ratio of the complex amplitude of the displacement (which is a harmonic motion with frequency ω) to the magnitude F of the forcing function. In other words, it is the amplitude of vibration produced by a unit force at the frequency ω. The factor $1/k$ is the static flexibility or compliance of the system (the deflection for a unit force). The frequency response function is also called the magnification factor or *transfer function* (TF) in metal cutting, although in the general literature on vibration the transfer function is usually defined in terms of Laplace transforms (*s*-plane method).

The frequency response function is a complex quantity and contains both real and imaginary parts, given by

$$\mathrm{Re}[G(\omega)] = \frac{k - m\omega^2}{(k - m\omega^2)^2 + (c\omega)^2} = \frac{1}{k}\frac{1 - r^2}{(1 - r^2)^2 + (2\zeta r)^2}$$
$$\mathrm{Im}[G(\omega)] = \frac{-c\omega}{(k - m\omega^2)^2 + (c\omega)^2} = \frac{1}{k}\frac{-2\zeta r}{(1 - r^2)^2 + (2\zeta r)^2} \qquad (12.13)$$

The real part represents the *mobility* of the system, while the imaginary part represents the *inertance*. The magnitude of the FRF is given by

$$|G(\omega)| = \sqrt{\frac{|X(\omega)|}{|F(\omega)|}}$$

$$= \frac{1}{k\sqrt{(1 - r^2)^2 + (2\zeta r)^2}} \qquad (12.14)$$

which represents the *dynamic compliance* of the system.

The fact that the imaginary part is negative indicates that the cosine component of the forced amplitude must lag the sine component. Therefore, the displacement X lags the disturbing force F by the phase angle $\phi(\omega)$ given by

$$\phi(\omega) = \tan^{-1}\left\{\frac{\mathrm{Im}[G(\omega)]}{\mathrm{Re}[G(\omega)]}\right\} = \tan^{-1}\left[\frac{2\zeta r}{1 - r^2}\right] \qquad (12.15)$$

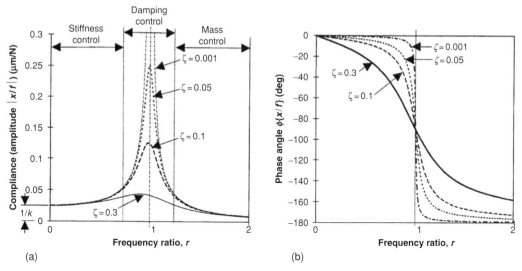

FIGURE 12.2 Response curves (the Bode plot, consisting of two graphs) for the system in Figure 12.1 using several damping ratios. (a) Compliance/amplitude (of FRF) versus frequency. (b) Phase versus frequency.

Plots of the magnitude $|G(\omega)|$ and phase angle $\phi(\omega)$ as functions of ω together define the system frequency response and are known as *Bode plots* (Figure 12.2) presented in polar coordinates. The phase ranges from 0° to 180° and the response lags the input by 90° at resonance.

The compliance and phase of the FRF are plotted as functions of the frequency ratio r in Figure 12.2. Resonance occurs when the response frequency equals the natural frequency of the system ($\omega = \omega_r \approx \omega_n$ for $\zeta = 1$). The amplitude of the response at resonance is $1/2k\zeta$, and is limited only by the amount of damping in the system. The static stiffness k can be determined from the compliance versus frequency curve and is equal to the reciprocal of the compliance at zero frequency (Figure 12.3). However, the dynamic compliance and stiffness differ from the static values. Low-frequency exciting forces cause displacements determined by the static compliance. The drive point mobility becomes the mobility of the spring at very low frequencies. The amplitude of the displacement (or the dynamic compliance) increases with increasing excitation frequency up to the resonance frequency. Any further increase in the excitation frequency will then result in a decrease in dynamic compliance. The mobility of the mass dominates at frequencies much larger than the resonance frequency. The compliance in the mass-controlled region at high frequencies approaches $1/\omega^2$. The stiffness and mass mobilities balance each other at frequencies near the resonance, so that only the mobility of the damper is left. Therefore, the system is said to be damping controlled at frequencies near the resonance frequency.

There are several different force–response relationships, summarized in Table 12.1, that are often used in vibration analysis [18]. The *impedance* and mobility are often of interest. The impedance of the SDOF system in Figure 12.1 is

$$\frac{F}{v} = c + j\left(m\omega - \frac{k}{\omega}\right) \tag{12.16}$$

where v is the velocity. The real part of the impedance is called the mechanical resistance. The imaginary part is called the mechanical reactance; $m\omega$ is the mechanical inertance term, while

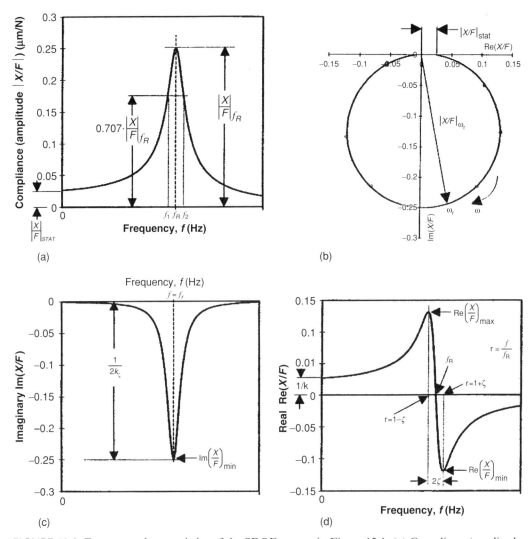

FIGURE 12.3 Frequency characteristics of the SDOF system in Figure 12.1. (a) Compliance/amplitude (of FRF) versus frequency. (b) Real part versus imaginary part (Nyquist plot). (c) Imaginary part versus frequency. (d) Real part (of FRF) versus frequency.

TABLE 12.1
Different Types of Frequency Response [135]

FRF	Representation
Displacement/force	Receptance
Force/displacement	Dynamic stiffness
Velocity/force	Mobility
Force/velocity	Impedance
Acceleration/force	Inertance
Force/acceleration	Apparent mass

ω/k is the mechanical compliance term. The system's response is damping controlled if the mechanical resistance term is dominant (Figure 12.2); the system's response is stiffness controlled if the mechanical compliance term is dominant. If the mechanical inertance term is dominant, the system's response is mass controlled. The FRF can be presented as displacement (shown in Figure 12.2), velocity, or acceleration. Acceleration is currently the accepted method of measuring modal response as will be discussed later. When $\omega = \omega_n$, the frequency response is approximately equal to the asymptote shown in Figure 12.2 and it is called the static stiffness line with slope of 0, 1, and 2 for displacement, velocity, and acceleration responses, respectively.

The real and imaginary parts of the FRF can be plotted as functions of frequency to identify the system as shown in Figure 12.3c and d. At resonance, the real part of the transfer function is zero, and the imaginary part of the transfer function reaches a maximum. The imaginary component approaches zero at frequencies away from the resonance frequency. The damping ratio can be determined from the real curve, and the amplitude can be determined from the imaginary curve.

A third method of representing the FRF is to plot the real part versus the imaginary part to obtain a *Nyquist plot* (or a vector response plot). In this representation, also called the operative receptance locus, harmonic response locus, or harmonic receptance, the magnitude is plotted against the phase angle in polar form. Values of magnitude and phase angle at various frequencies represent points along the locus of the polar plot. This permits the characteristics of vibration to be combined in a single polar curve as shown in Figure 12.3b. The compliance or degree of flexibility is represented by the polar radius, while the phase is given by the angle between the polar vector and the positive real axis. This representation can be used for phase plane analysis, a graphical method for determining system stability [22–25]. For an SDOF system, the polar plot is almost circular when hysteretic damping is used; the diameter of the circle increases with decreasing damping ratio.

The effect of the damping ratio on the system response is shown in Figure 12.2, in which the magnitude of the frequency response (dynamic compliance) and the phase angle are plotted for various values of the damping ratio. Two characteristics are clear: (a) the compliance decreases almost proportionally with increasing damping for all frequencies; and (b) the maximum compliance at higher damping values (>0.2) occurs at a frequency lower than the resonant frequency. The phase angle is small for small values of the frequency ratio, and approaches 180° asymptotically for very large frequency ratios. This means that the amplitude of vibration is in phase with the exciting force for $r \leq 1$ and out of phase for $r \geq 1$. At resonance the phase angle is $-90°$ for all values of viscous damping.

There are several methods for determining the damping ratio [18, 26, 27] from measured data. It is easily determined using the half-power method with the compliance–frequency curve (Figure 12.3) using the formula

$$\zeta \approx \frac{\Delta f}{2 f_R}; \quad \Delta f = f_2 - f_1 \tag{12.17}$$

f_1 and f_2 represent the half power points which can be measured from the plot. This method can be used for every resonance increase in an MDOF system. The damping ratio for an SDOF system can also be calculated from the formula

$$\zeta = \frac{|G(0)|}{2|G(\omega_n)|} \tag{12.18}$$

Finally, the damping ratio can be estimated from the decay curve of the system as discussed in Ref. [27].

12.3.2 MULTIPLE DEGREE OF FREEDOM SYSTEMS

In general, the dynamic response of a cutting tool, machine tool, or workpiece–fixture structural system cannot be described adequately by an SDOF model as the response usually includes time variations of the displacement shape as well as amplitude. For some analysis purposes, machine tool structural elements can be idealized as discrete lumped masses connected by springs and dampers or a distributed parameter system. In this idealization, the system is treated as an MDOF system, which is a natural extension of the SDOF system discussed above.

MDOF systems have one natural frequency for each degree of freedom (or mass). Each natural frequency has a corresponding characteristic deformation pattern or mode shape. The free vibration of the system can be described as a linear combination of vibrations in the individual modes.

The real and imaginary parts of the FRF of an MDOF system are shown in Figure 12.4. The polar plot for a system with multiple natural frequencies has multiple loops, one for each natural frequency in the frequency response curve. Polar plots for three milling machines in three directions are shown in Figure 12.5; they clearly show the structural characteristics for each machine in the selected directions. Polar curves are often used to identify the parameters of MDOF systems (i.e., the appropriate values for masses and spring and damper constants) as will be discussed below.

The equations of motion for an MDOF system in matrix form are

$$[M]\{\ddot{x}\} + [C]\{\dot{x}\} + x[K]\{x\} = \{f(t)\} \tag{12.19}$$

where $[M]$, $[C]$, and $[K]$ are, respectively, the mass, damping, and stiffness matrices for the system. The elements m_{ij}, c_{ij}, and k_{ij} in the $[M]$, $[C]$, and $[K]$ matrices represent the inertial, damping, and spring forces, respectively, acting at position i due to a unit acceleration of mass, velocity, and displacement, respectively, at position j, holding all other masses fixed.

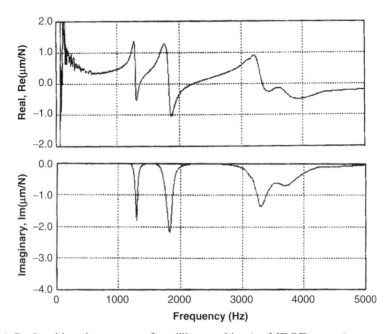

FIGURE 12.4 Real and imaginary parts of a milling machine (an MDOF system).

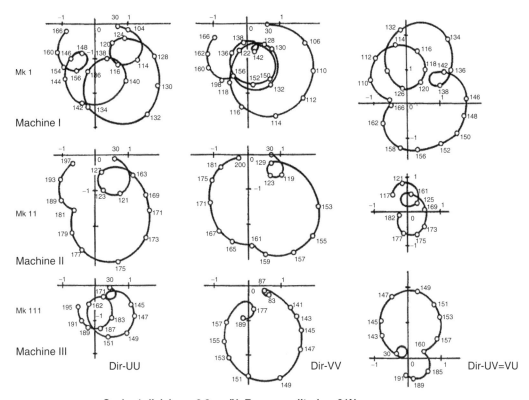

Scale: 1 division = 0.2 μm/N, Force amplitude = 31N

FIGURE 12.5 Polar curve of the flexibility of three different milling machines in three directions. (After M.M. Sadek and S.A. Tobias [121].)

The equations of motion in the above system are generally coupled but can be uncoupled through coordinate transformations so that each equation in the matrix system can be solved separately. This is accomplished using a modal matrix [ψ] as the transformation matrix:

$$\{x\} = [\psi]\{q\} \tag{12.20}$$

In this equation, $\{x\}$ is the displacement vector in the local coordinates and $\{q\}$ is the displacement vector in the principal (orthogonal) coordinates. A unique displacement vector called the mode shape exists for each distinct natural frequency and damping ratio. For a system with N degrees of freedom, [ψ] is formed by placing the eigenvectors $\{\psi_i\}$ for each mode i in the columns of an $N \times N$ matrix:

$$[\psi] = [\{\psi_1\}\{\psi_2\}\{\psi_3\} \cdots \{\psi_N\}] \tag{12.21}$$

As discussed below, the eigenvector for a mode i represents the relative amplitude of vibration of the various masses, or mode shape, for vibrations at the natural frequency of the mode ω_{ni}.

The eigenvalues and eigenvectors for the system described by Equation (12.19) are determined by solving the corresponding undamped free vibration equations,

$$[M]\{\ddot{x}\} + [K]\{x\} = \{0\} \tag{12.22}$$

Values of a scalar λ which solves the equation

$$([K] - \lambda[M])\{x\} = \{0\} \tag{12.23}$$

are sought; for an N-DOF system, there are N such values, determined by solving the associated characteristic equation,

$$|[K] - \lambda[M]| = 0 \tag{12.24}$$

The resulting N eigenvalues λ_i are the squares of the natural frequencies of the system for each mode:

$$\lambda_i = \omega_{ni}^2 \tag{12.25}$$

For this homogeneous, undamped case, the eigenvalues are real. An eigenvector or mode shape $\{\psi_i\}$, called the normal mode vector, is associated with every undamped natural frequency ω_{ni}. The eigenvectors $\{\psi_i\}$ are determined by substituting the corresponding eigenvalue λ_i into Equation (12.22).

By substituting Equation (12.20) into Equation (12.19) and multiplying by the transpose of the modal matrix, Equation (12.19) can be transformed to

$$[M_q]\{\ddot{q}\} + [C_q]\{\dot{q}\} + [K_q]\{x\} = [\psi]^T\{f\} \tag{12.26}$$

where $[M_q]$, $[C_q]$, and $[K_q]$ are, respectively, the mass, damping, and stiffness matrices in principal coordinates (discussed below). The principal modes of vibration are orthogonal in the sense that, for two modes i and j with eigenvectors $\{\psi_i\}$ and $\{\psi_j\}$ and eigenvalues $\lambda_i \neq \lambda_j$, the following relationships hold:

$$\{\psi_i\}^T[M]\{\psi_j\} = \{\psi_j\}^T[M]\{\psi_i\} = 0 \tag{12.27}$$

$$\{\psi_i\}^T[C]\{\psi_j\} = \{\psi_j\}^T[C]\{\psi_i\} = 0 \tag{12.28}$$

and

$$\{\psi_i\}^T[K]\{\psi_j\} = \{\psi_j\}^T[K]\{\psi_i\} = 0 \tag{12.29}$$

If these conditions are true, the matrices $[M_q]$, $[C_q]$, and $[K_q]$ are diagonal matrices and are referred to as the principal mass, damping, and stiffness matrices, respectively. They provide the mass, damping, and stiffness coefficients in normal coordinates. The mass and stiffness matrices can be diagonalized easily through the following transformations:

$$[\psi]^T[M][\psi] = [M_q] \quad \text{and} \quad [\psi]^T[K][\psi] = [K_q] \tag{12.30}$$

However, the damping matrix cannot generally be diagonalized simultaneously with the mass and stiffness matrices. When damping is present, the modes have phase relationships that complicate the analysis. The eigenvalues are real and negative or complex with negative real parts [28]. The complex eigenvalues occur as conjugate pairs and therefore their eigenvectors also consist of complex conjugate pairs. The treatment of damping is discussed in more detail below.

The eigenvalues and eigenvectors for free vibration of an undamped MDOF system are determined by solving Equation (12.22). The forced vibration of an undamped MDOF system can be analyzed based on Equation (12.26). The steady-state solution for the elements q_i in the displacement vector $\{q\}$ for harmonic excitation is

$$\{q_i\} = \frac{\{\psi_i\}^T\{f\}}{k_{qi} - m_{qi}\lambda_i} \tag{12.31}$$

The displacements in the local coordinates are given by Equation (12.20).

Damping of the machine tool must be considered when its frequency response is needed at or near a natural frequency. The solution of Equation (12.26) can be simplified by assuming that the damping matrix is linearly related to the mass and stiffness matrices:

$$[C] = a[M] + b[K] \tag{12.32}$$

where a and b are constants. This idealization is often used and is referred to as the *proportional damping* assumption because the damping matrix is proportional to the mass and stiffness matrices. This assumption allows the equation of motion, Equation (12.19) or (12.26), to be uncoupled by the same transformation used for the undamped system [29]; therefore, the modal analysis method used above for an undamped system can be used to determine the eigenvalues and eigenvectors of an MDOF system with proportional damping. In this case, the principal damping matrix is diagonalized by the transformation

$$[C_q] = [\psi]^T(a[M] + b[K])[\psi] \\ = a[M_q] + b[K_q] \tag{12.33}$$

The equations of motion are given by Equation (12.26); the solutions are

$$\{q_i\} = \frac{\{\psi_i\}^T\{f\}}{k_{qi} - m_{qi}\omega_i^2 + jc_{qi}\omega} \quad \forall i = 1, 2, \ldots, N \tag{12.34}$$

The mode shapes of a system with proportional damping are the same as for an undamped system since they have the same modal matrix. Physically, this means that, when the system vibrates at a particular mode shape, every point reaches the maximum position at the same time. However, different modes have different phase relations relative to the exciting force for forced vibrations.

Caughey [30] showed that the proportionally condition, Equation (12.32), is sufficient but not necessary for a damped system to have principal modes. The required condition for obtaining principal modes in a damped system is that the transformation adapted above to diagonalize the damping matrix must also uncouple the equations of motion, Equation (12.26). This requirement is less restrictive than Equation (12.32).

A lightly damped system (specifically with damping ratios ζ_i less than 0.2 for all modes) can be analyzed assuming that the equations of motion are uncoupled by the modal matrix determined for the system without damping [28]. Under this assumption, the matrix $[\psi]$ is assumed to be orthogonal with respect to not only $[M]$ and $[K]$, but also $[C]$, that is, that Equation (12.28) holds true. It implies that any off-diagonal terms resulting from the operation

$$[C_q] = [\psi]^T[C][\psi] \tag{12.35}$$

are small and can be neglected. This assumption, called *modal damping*, has great practical application since the damping ratios for machine tools generally satisfy the light damping condition. It should be emphasized that this concept is based on the normal coordinates for the undamped system, and that damping ratios must be specified in those coordinates.

Under this assumption, the resultant response of an MDOF system when an exciting force $\{f\}$ is applied in different points is given in terms of the N normal modes by equation

$$\{x\} = \sum_{i=1}^{N} \{\psi_i\}\{q_i\} \qquad (12.36)$$

Also under this assumption, the cross frequency response function G_{ij} for the displacement $\{x_i\}$ of point i excited by a force $\{F_j\}$ at a point j is given by

$$G_{ij}(\omega) = \frac{X_i(\omega)}{F_j(\omega)} = \sum_{p=1}^{N} \frac{S_{ijp}}{1 - r_p^2 + 2j\zeta r_p} \qquad (12.37)$$

where X_i and F_j are the amplitude of the corresponding displacement and force. If $i=j$ the direct frequency response is obtained. S_{ijp} is the cross ($i \neq j$) or direct ($i=j$) modal flexibility.

The case of arbitrary viscous damping, in which the above proportionality condition is not satisfied and the off-diagonal terms in matrix $[C_q]$ in Equation (12.35) cannot be neglected, can be analyzed by the transformation method developed by Duncan and coworkers [31] and later used by Foss [32]. Such systems are generally highly damped and have eigenvalues with significant imaginary parts. In this case, the real modal matrix $[\psi]$ cannot uncouple the equations of motion since the damping matrix does not satisfy the orthogonality condition. The Duncan–Foss approach transforms the N second-order equations of motion, Equation (12.19), into $2N$ uncoupled first-order equations. This leads to $2N$ eigenvalues which are real and negative in the case of critically damped systems and complex with negative real parts in the case of underdamped systems (such as machine tools). This transformation is described in Refs. [31–33].

In metal cutting, the number of degrees of freedom and natural frequencies of interest from a practical viewpoint is generally limited. The machining system or any part of the system can often be simplified by reducing it to a few major modes of vibration; however, even then the determination of the compliance and damping characteristics for the system and its components is not simple because the system varies as the machine axes (i.e., column, table, spindle, workpiece, etc.) move. In addition, the metal cutting process itself influences the system vibration when the tool is in contact with the workpiece. In general, it is important to determine the natural frequencies of the machine tool or its major structural elements so that forced vibrations at resonant frequencies can be avoided. The frequencies of the disturbing forces generated during conventional cutting are typically limited and therefore only a limited number of modes with relatively low natural frequencies must be considered.

12.4 TYPES OF MACHINE TOOL VIBRATION

Machine tools are subject to three basic types of vibration: free, forced, and self-excited vibrations [34].

Free or *natural* vibrations (Figure 12.6a) occur when the stable system is displaced from its equilibrium position by shock; in this case, the system will vibrate and return to its original position in a manner dictated by its structural characteristics. Since machine tools are designed for high stiffness, this type of vibration seldom causes practical problems, and will not be considered further.

A *forced* vibration (Figure 12.6b) occurs when a dynamic exciting force is applied to the structure. Such forces are commonly induced by one of the following three sources:

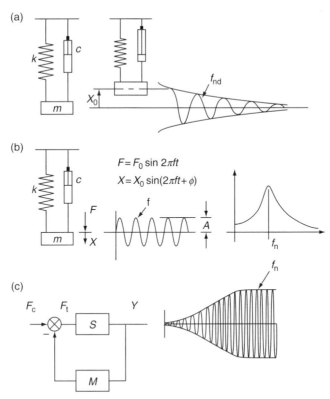

FIGURE 12.6 Schematics of various vibrations: (a) free; (b) forced; (c) unstable self-excited. (Courtesy of D3 Vibrations, Inc.)

(1) alternating cutting forces such as those induced by (1a) inhomogeneities in the workpiece material (i.e., hard spots, cast surfaces, etc.), (1b) built-up edge (which forms and breaks off periodically), (1c) cutting forces periodically varying due to changes in the chip cross section, and (1d) force variations in interrupted cutting (i.e., in turning a nonround or slotted part, milling, and broaching); (2) internal source of vibrations, such as (2a) disturbances in the workpiece and cutting tool drives (caused by worn components, i.e., bearing faults, defects in gears, and instability of the spindle or slides), (2b) out of balance forces (rotating unbalanced members, i.e., masses in the spindle or transmission), (2c) dynamic loads generated by the acceleration/deceleration or reversal of motion of massive moving components; and (3) external disturbances transmitted by the machine foundation [35]. Force vibration is discussed in more detail in Section 12.5.

Self-excited vibration or *chatter* (Figure 12.6c) is induced by variations in the cutting forces (caused by changes in the cutting velocity or chip cross section), stick-slip dry friction, built-up edge, metallurgical variations in the workpiece material, and regenerative effects. This is indicated by the control loop schematics in Figure 12.6c and Figure 12.7. This type of vibration is referred to as chatter and is the least desirable type of vibration because the structure enters an unstable vibration condition.

Chatter is a complex phenomenon which depends on the design and configuration of both the machine and tooling structures, on workpiece and cutting tool materials, and on machining regimes. The stiffness of the tool, spindle, workpiece, and fixture are important factors. The cutting stiffness of the workpiece material is also an important factor; for example, steels have a greater tendency than aluminum to cause chatter. Cutting conditions, such as depth of

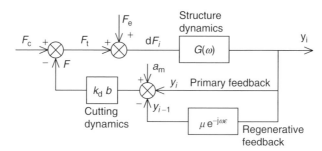

FIGURE 12.7 Block diagram representation of the machining process with noise input.

cut, width of cut, and cutting speed, greatly affect the onset of chatter. The chatter resistance of a machine tool is often expressed in terms of the maximum allowable width of cut b_{lim}. Forced vibrations are often easily detectable during the development stage or final inspection of a machine tool and can be reduced or eliminated. However, chatter occurrence may not be easily detected during the runoff stage unless the machine tool is thoroughly tested [36, 37]. In addition, because it is a complex and typically nonlinear phenomenon, chatter may occur only under certain seemingly random conditions and may come and go sporadically. The elimination of chatter in a particular machining process can be a laborious exercise and frequently can be accomplished only by reducing the production rate in conventional spindles.

Chatter occurs because the damping of the machine tool system is not sufficient to absorb the portion of the cutting energy transmitted to the system. The practical significance of the chatter depends on the type of operation, whether it is a finishing or roughing operation, the surface finish requirements, tool wear characteristics, the allowable acoustic noise, and on its propagation to surrounding equipment. Chatter is never desirable because it can lead to accelerated machine tool wear, reduced part quality, and catastrophic tool failure. Chatter becomes more significant as cutting speeds increase since the exciting forces approach the natural frequencies of the system. It is often difficult to overcome chatter, but progress can be made through the proper selection of cutting conditions, improved design of the machine tool structure and spindle, and improved vibration isolation.

Chatter suppression typically refers to increasing the machining process stability. A related term, noise attenuation, is usually understood to mean improving the workpiece surface finish. In practice, the stability problem (chatter suppression) and the surface finish problem (noise attenuation) are coupled because the occurrence of vibration and especially chatter reduces machined surface quality.

Two approaches may be taken to solve chatter problems. The first is to choose or change cutting conditions such as the cutting speed, feed, depth of cut, tool geometry, coolant, etc., to optimize the metal removal rate while operating in a stable regime. This is the test cuts approach (that detects and corrects). The second is to analyze the dynamic characteristics of the machining system to determine the stable operating range, and to suggest improvements to the system design which can extend this range. The second approach is often called the stability chart method (prediction and avoidance) and will be explained in Section 12.6 and Section 12.7. Active chatter suppression is discussed in Section 12.9.

12.5 FORCED VIBRATION

The cutting process generates forced vibrations through transient cutting forces, especially in interrupted cutting. The forcing frequency is usually the spindle rotational frequency and harmonics, or the tooth impact frequency and harmonics. In side or face milling, the

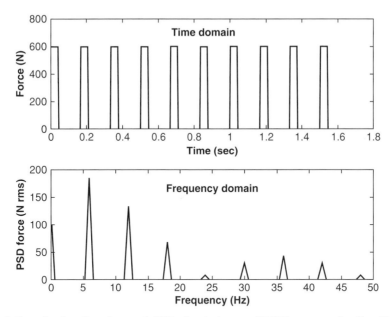

FIGURE 12.8 Step forcing function and PSD simulating an SDOF system of a flymill single insert rotating at 360 rpm during in a 1/2 slot. (Courtesy of D3 Vibrations, Inc.)

frequency of the forced vibration equals the product of the tool/spindle rotational frequency and the number of teeth on the tool (or tooth passing frequency). The forcing frequency is easily changed by adjusting the spindle rpm or the number of teeth on the tool. Forced vibrations decay when the excitation is removed.

Figure 12.8 shows a flymilling operation with a single insert at 360 rpm. The top plot is a simplified time-domain forcing function showing the force turning on and off as the tooth enters and exits the cut. The bottom plot shows the frequency-domain forcing function for that time-domain force. The tooth passing frequency or fundamental forcing frequency (FFF) and harmonics are present because of this step-function. The FFF equals 6 Hz (1 tooth × 360 rpm ÷ 60 sec/min) and the harmonics are integer multiples of the FFF (12, 18, 24, etc.).

Forced vibration was analyzed mathematically in Section 12.3 for single and multiple DOF systems. The amplitude of forced vibrations depends on the amplitude of the exciting force and on dynamic stiffness of the machine tool, cutting tool, and the workpiece, which are often an order of magnitude lower than the corresponding static values. The FRF of the part used for Figure 12.8 is given in Figure 12.9 and it shows the compliance or flexibility of the part at every frequency. For a linear system the displacement of the part at each frequency can be determined from the FFF and FRF:

$$\text{Displacement (mm)} = \text{FFF (N)} \times \text{FRF (mm/N)} \tag{12.38}$$

The displacement amplitude is shown in Figure 12.10, where the displacement is greatest at 12 Hz, which is the first harmonic of the forcing frequency. Even though the force is less at 12 Hz, the displacement is larger because the dynamic flexibility is near the resonant frequency. To characterize the ability to resist this kind of vibration, an index of machine tool dynamic stiffness is used:

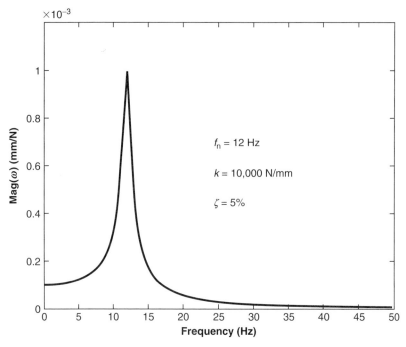

FIGURE 12.9 Frequency response function (FRF) of an SDOF part used for Figure 12.8. (Courtesy of D3 Vibrations, Inc.)

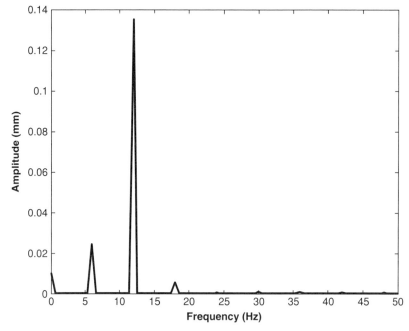

FIGURE 12.10 Estimated SDOF part displacement amplitude as a function of FFF (Figure 12.8) and FRF (Figure 12.9). (Courtesy of D3 Vibrations, Inc.)

$$C_d = \frac{F}{A_{max}} \tag{12.39}$$

where F is the disturbing force and A_{max} is the resonant amplitude of vibration.

Forced vibrations can be produced by all machining operations, but are especially important in finish and fine machining operations in which surface waviness is unacceptable. Forced vibrations have their greatest practical impact when the exciting frequency is near one of the system's natural frequencies. For instance, if the rotational speed of a single-point boring bar (rev/sec) is very close to or equal to a natural frequency of the bar, the bored surface will exhibit excessive waviness due to the vibration of the bar. The effect of the number of teeth on a milling cutter is shown in Figure 12.11; the frequency of the torque variations increases while the peak torque decreases as the number of teeth on the cutter increases, assuming that all cutters run at the same speed and feed rate.

Forced vibrations due to periodic forces of defective machine components (such as unbalanced motors) typically become significant only when they excite a system resonance. In this case, the problem can generally be eliminated by changing the machine structure to change the resonant frequency.

Forced vibrations can lead to instability when machining at high speeds if the spindle speed is large enough for the tooth passing frequency to approach a dominant natural frequency of the system. For example, if the measured dominant natural frequency for a motorized spindle shaft is 500m Hz, the spindle rpm corresponding to this natural frequency is 30,000 rpm. Similarly, if the measured dominant natural frequency for a long end mill with six teeth is 400 Hz, the critical spindle rpm N_{cr} producing the same tooth passing frequency is 4000 rpm based on the following equation:

$$N_{cr} = \frac{60 f_n}{n_t} = \frac{30 \omega_n}{\pi n_t} \tag{12.40}$$

where n_t is the number of teeth on the cutter.

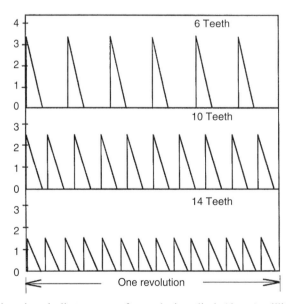

FIGURE 12.11 Variations in spindle torque or forces during climb (down) milling with a zero axial and radial rake cutter.

12.6 SELF-EXCITED VIBRATIONS (CHATTER)

Self-excited or self-induced vibrations occur because the dynamic cutting process forms a closed-loop system [34, 38]. Disturbances in the system (i.e., vibrations which affect cutting forces) are fed back into the system and over time, under appropriate conditions, may result in instability. Self-excited vibrations do not result directly from external forces, but draw energy from the cutting process itself. The regenerative effect has become the most commonly accepted explanation for machine tool chatter [7, 39–43].

The characteristic features of self-excited vibrations are: (a) the amplitude increases with time, until a stable limiting value is attained; (b) the frequency of the vibration is equal to a natural frequency or critical frequency of the system; and (c) the energy supporting the vibration is obtained from a steady internal source.

Self-excited vibrations can be distinguished from forced vibrations by the fact that they disappear when cutting stops; forced vibrations exist and often persist whether or not the tool is engaged. The forced vibration caused by an interrupted cut can be distinguished from self-excited vibration by performing tests at two different cutting speeds. If the frequency of vibration changes and is equal to some multiple of the spindle speed in both tests, the vibration is forced; if the frequency of vibration does not change significantly, the vibration is self-excited.

Self-excited vibrations arise spontaneously and may grow in amplitude until some nonlinear effect provides a limit. Their occurrence is determined by the stability of the machine–tool–workpiece system. The cutting process is considered unstable if growing vibrations are generated; then the cutting tool may either oscillate with increasing amplitude or monotonically recede from the equilibrium position until nonlinear or limiting restraints appear. On the other hand, the cutting process is considered stable if the excited cutting tool approaches the equilibrium position in either a damped oscillatory fashion or asymptotically.

Self-excited vibrations in cutting result from variations in the chip thickness, depth of cut, and cutting speed, which result in cutting force variations. The system is considered statically unstable if displacement-dependent forces are causing chatter. A system excited by velocity-dependent forces is said to be dynamically unstable.

Chip formation under unstable conditions (due to part or fixture deflection, or tool/spindle deflection) results in variation of the shear angle and cutting forces. Force variations lead to excessive vibration if the rigidity of the system is low. Force variations can lead to machine vibration, which in turn can cause additional force fluctuations by inducing variations in the uncut chip thickness. The variation of the uncut chip thickness due to vibrations during the previous pass (in turning and single-point boring) or previous tooth (in milling, reaming, and multitooth boring) cause additional force fluctuations. When the dynamic cutting force is out of phase with the instantaneous relative movement between the tool and workpiece, this leads to the development of self-excited vibration. This type of instability is called *regenerative chatter* because the vibration reproduces itself in subsequent revolutions through the generation of the waviness. It is the most practically significant form of self-excited vibration and is the main type discussed in the remainder of this section.

A second type of self-excited vibration, *nonregenerative chatter*, occurs without undulation. There are other mechanisms that could lead to dynamic instability, such as the dependence of the cutting force on the cutting velocity. These types of chatter are relatively uncommon and have not been widely studied; one type of nonregenerative chatter, mode coupling, is discussed briefly at the end of this section.

12.6.1 REGENERATIVE CHATTER; PREDICTION OF STABILITY CHARTS (LOBES)

The dynamic cutting process and the machine tool structure represent a three-dimensional MDOF system as shown in Figure 12.12a. The dynamic machining process can be represented as a closed looped system by the block diagram shown in Figure 12.12b. Dynamic fluctuations of the cutting force and tool position relative to the workpiece occur in all machining processes because workpiece and tool are not infinitely stiff. This relative motion leaves an undulation on the machined surface with amplitude y_i. Oscillations on the surface left by the tool are removed by the succeeding tooth (in milling, reaming, and multitooth boring) or by the tool during the following revolution of the workpiece (in turning and single-point boring), resulting in a subsequent oscillation of amplitude y_{i-1}. The tooth cutting a wavy surface experiences a variable force that causes additional tool vibration. If the phase relation between the cutting force and surface oscillations is unfavorable, this can lead to vibrations of increasing amplitude, or regenerative chatter.

Variations in the uncut chip cross section (shown in Figure 12.13) occur as a result of deflections of the tool and the workpiece. The contact length between the tool and workpiece

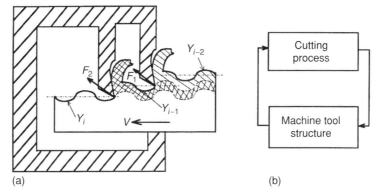

FIGURE 12.12 The dynamic machining process. (a) Closed-loop diagram of the structure, tool, and cutting process. (b) Diagram for deriving stability limits. (After J. Tlusty [38].)

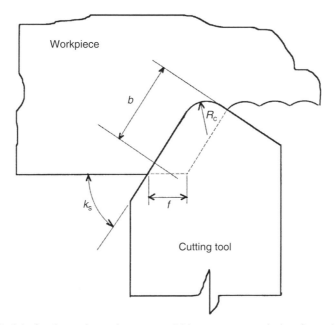

FIGURE 12.13 Model of a dynamic cutting system (chip geometry variation for a single-point cutting tool).

is a function of the doc, lead angle, and tool corner radius (Figure 12.13). The length of the contact for the cutting edge is

$$b = \frac{d - r_n + r_n \cos(k_s)}{\sin(k_s)} + \frac{\pi r_n k_s}{180} \quad (12.41)$$

The uncut chip thickness is given by Equation (2.2). Therefore, any movement of the tool and the workpiece will result in a corresponding change in the width and thickness of the chip (db and da, respectively). The changes in the chip cross-sectional area lead to proportional variation on the cutting force [d$F = f($d$a)$].

The conditions for the limit of stability of the machine tool structure and cutting process system can be explained with reference to Figure 12.12. Several assumption and simplifications are made: (a) the cutting process takes place in one plane; (b) the machine tool structure is represented by an SDOF mechanical system; (c) the system is linear; (d) the direction of the variable component of the cutting force is constant and lies in the same plane as the cutting velocity; and (e) the variable component of the cutting force depends only on vibration in the direction normal to the cut surface Y.

The direction of principal vibration X, shown in Figure 12.14, forms an angle α with the normal to the machined surface Y. The direction of the cutting force F is inclined by an angle β from the normal Y. The average cutting speed is V and the width of cut is b. The chip thickness variation due to the surface wave Y_{i-1} produced by subsequent cuts depends on the phase lag ε with the surface wave Y_i left by the previous revolution. The number of waves between cuts is

$$n_p + \frac{\varepsilon}{2\pi} = \frac{f}{N} \quad (12.42)$$

at spindle speed N (in rev/min) and the chatter frequency f (Hz). n_p is the largest possible integer (number of whole waves) such that $\varepsilon/2\pi < 1$, and ε is the phase of inner modulation y_i to the outer modulation y_{i-1}. In other words, the number of cycles of vibration (waves) between subsequent revolutions is an integer plus a fraction. When the vibration frequency is a whole number multiple of the rotational speed ($\varepsilon = 0°$), the vibration allows the tool to follow

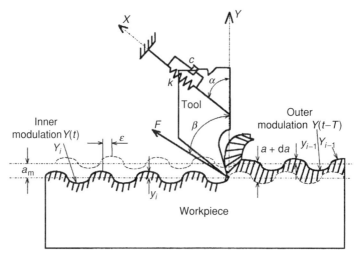

FIGURE 12.14 Diagram of regenerative chatter due to the cutting of an undulated surface.

the previous wavy surface and self-excited vibration is not present. The maximum chip thickness variation occurs at $\varepsilon = 180°$.

Many researchers have studied machining stability and several theories [3, 4, 38, 44] have been proposed to explain self-excited chatter. The stability of chatter vibrations was initially analyzed by Tobias and Fishwick [45], Tlusty and Polacek [46], and Merritt [47] using linear theory. Several nonlinearities such as multiple regeneration, interrupted cut, process damping, and large vibrations resulting in zero chip thickness are neglected in linear stability analyses. In this book, the well-known and typical theories of Tlusty and Tobias are briefly summarized.

12.6.2 Tlusty's Theory

Tlusty proposed a simple analysis that assumes that the dynamic cutting force is proportional to the undeformed chip thickness [4, 44, 47]. The vibration of the tool in the direction normal to the cut surface during the ith cut is

$$y_i = Y_i \sin \omega t = X_i \cos \alpha \sin \omega t \tag{12.43}$$

and the mean chip thickness variation is

$$\begin{aligned} a &= a_m + da = a_m + y_{i-1} - y_i \\ &= a_m + (x_{i-1} - x_i) \cos \alpha \end{aligned} \tag{12.44}$$

where da is the variable component of the chip thickness and y_{i-1} is the amplitude of the surface undulation. The magnitude of the dynamic force variation depends on the relative motion between the cutting tool edge and the workpiece surface and on the angle between the cutting force and the direction of principal vibration. The force on any tooth in a cutter is proportional to the chip thickness. Therefore, the variable force component or regenerative force is

$$\begin{aligned} dF_i &= k_d b da = k_d b (y_{i-1} - y_i) \\ &= k_d b (x_{i-1} - x_i) \cos \alpha \end{aligned} \tag{12.45}$$

where k_d is the specific dynamic cutting stiffness or specific force, which is assumed to be a material constant. k_d can be determined from the specific energy of the workpiece material as discussed in Chapter 6. The force depends not only on the feed per tooth and on the deflection of the cutter, but also on the surface which was left by previous teeth. The assumption that the cutting force is changing in phase with y is not really true and Equation (12.45) represents only an approximation of the real physical system. The generated vibration either grows or diminishes depending on several parameters (i.e., k_d, b, N). The variable force excites the vibration of the machine tool in the ith cut, represented by an SDOF system (Figure 12.15). The amplitude of vibration during the ith cut is given by Equations (12.9) and (12.10):

$$x_i = dF_i \cos(\alpha - \beta) \frac{1}{k} \frac{1}{1 - r^2 + 2j\zeta r} \tag{12.46}$$

The relation between the amplitudes for the ith and $(i-1)$th cuts can be determined by substituting dF_i from Equation (12.45) into Equation (12.46):

$$\frac{y_i}{y_{i-1}} = \frac{x_i}{x_{i-1}} = \frac{G(\omega)}{G(\omega) + (1/k_d b)} \tag{12.47}$$

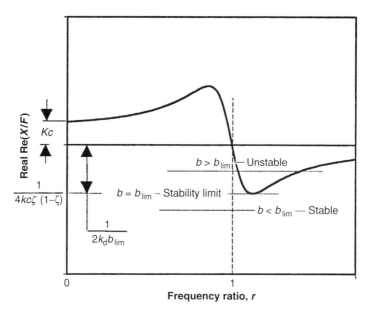

FIGURE 12.15 Schematic representation of the stability limit for an SDOF system.

where

$$G(\omega) = \frac{Y(\omega)}{F(\omega)} = \frac{Y(\omega)}{F_x(\omega)} u = u G_d(\omega) \tag{12.48}$$

$$u = \cos(\alpha) \cos(\alpha - \beta) \tag{12.49}$$

and

$$G_d(\omega) = \frac{1}{k} \frac{1}{1 - r^2 + 2j\zeta r} \tag{12.50}$$

where the transfer function (TF) $G(\omega)$ represents the response in the direction Y to a force acting in the direction of the cutting force. $G_d(\omega)$ is the direct transfer function defined in the direction X and u is the directional orientation factor.

Based on regenerative chatter theory [48], the state of the dynamic cutting process is described by

$$\left| \frac{y_i}{y_{i-1}} \right| = \begin{cases} > 1 & \text{unstable} \\ = 1 & \text{at stability limit} \\ < 1 & \text{stable} \end{cases} \tag{12.51}$$

Regenerative instability will occur when the vibratory motion increases with time, in which case the magnitude of Equation (12.47) is greater than 1 for some frequencies. The borderline of stability is obtained when the magnitude of the expression is equal to unity. Therefore, equating the magnitude of Equation (12.47) to 1,

$$\text{Im}\{G(\omega)\} = 0 \quad \text{and} \quad \text{Re}\{G(\omega)\} = \frac{-1}{2k_d b} \tag{12.52}$$

This equation is the simplest form of the condition for the limit of stability. The real part of the frequency response function is obtained from Equation (12.13):

$$\text{Re}\{G(\omega)\} = \frac{u}{k}\frac{1-r^2}{(1-r^2)^2 + (2\zeta r)^2} \tag{12.53}$$

with extreme values of

$$\text{Re}\{G(\omega)\}_{\min} = -\frac{u}{k}\frac{1}{4\zeta(1+\zeta)} = \frac{-1}{4k_c\zeta(1+\zeta)} \tag{12.54}$$

for $r = \sqrt{1+2\zeta}$ and $\zeta \ll 1$ as shown in Figure 12.15. The maximum chip width for stable cutting (or in other words, the limit of stability) can be calculated from Equations (12.47) and (12.52) assuming that the dynamic characteristics of the system (the cutting stiffness) are known. The cutting stiffness k_d is defined as the increment in the magnitude of the cutting force per unit increment of depth of cut at unit chip width.

The above analysis considers a phase shift between two subsequent cuts. The dynamic portion of the instantaneous chip thickness is

$$y_{i-1} - y_i = \mu y_i(t-\varepsilon) - y_i(t) + r(t) \tag{12.55}$$

where ε represents the time between the production of an undulated surface and the renewed cutting of that wavy surface (i.e., one revolution in turning as shown in Figure 12.14). $r(t)$ represents external disturbances which affect the chip thickness. μ is the overlap factor which lies in the interval [0, 1] (i.e., for plunge cutting $\mu = 1$ and for thread cutting $\mu = 0$). The amplitude of vibration of the $(i-1)$th cut is

$$y_{i-1} = \mu y_i e^{-j\omega\varepsilon} \tag{12.56}$$

The stability condition can be redefined by substituting Equation (12.56) into Equation (12.47) to yield

$$G(\omega)(1 - \mu e^{-j\omega\varepsilon}) = \frac{-1}{k_d b} \tag{12.57}$$

There is full overlap between two consequent cuts in many operations ($\mu = 1$), which means that the full undulation produced during one revolution is removed in the next.

The process of self-excitation and regenerative chatter can be represented by the block diagram of the closed loop system shown in Figure 12.7. In this model, the external disturbances which affect the chip thickness $r(t)$ are not directly accounted for but two new force inputs are considered, $F_c(t)$ and $F_e(t)$, both of which represent noise affecting the process. $F_c(t)$ represents the cutting noise while $F_e(t)$ represents noise from external sources. The cutting noise results from material inhomogeneity, workpiece out-of-roundness, and the force variations that occur during the chip formation process. The external noise results from rough bearings, spindle unbalance, floor vibrations, and similar sources. The force input representing the external noise $F_e(t)$ is not transmitted by the cutting tool. This force acts directly on the workpiece–machine structure and must induce motion of the structure before a tool force results. Analysis of this loop using feedback control theory also yields Equation (12.57). Therefore,

$$[(\text{Re}\{G(\omega)\} + \text{Im}\{G(\omega)\})(\cos\varepsilon - j\sin\varepsilon - 1)]k_d b = 1$$
$$\text{Re}\{G(\omega)\}(\cos\varepsilon - 1) + \text{Im}\{G(\omega)\}\sin\varepsilon$$
$$+ j([\text{Im}\{G(\omega)\}(\cos\varepsilon - 1)] - \text{Re}\{G(\omega)\}\sin\varepsilon) = \frac{1}{k_d b} \quad (12.58)$$

The imaginary part is zero, hence

$$\text{Im}\{G(\omega)\}(\cos\varepsilon - 1) - \text{Re}\{G(\omega)\}\sin\varepsilon = 0$$
$$\sin\phi(\cos\varepsilon - 1) - \cos\phi\sin\varepsilon = 0$$
$$\sin(\phi - \varepsilon) = \sin\phi \quad (12.59)$$
$$\Rightarrow \phi = \frac{\pi + \varepsilon}{2}$$

The real part is equal to 1, hence

$$[\text{Re}\{G(\omega)\}(\cos\varepsilon - 1) + \text{Im}\{G(\omega)\}\sin\varepsilon]k_d b = 1$$
$$b_{cr} = \frac{1}{k_d[\text{Re}\{G(\omega)\}(\cos\varepsilon - 1) + \text{Im}\{G(\omega)\}\sin\varepsilon]} \quad (12.60)$$

and since

$$\tan\phi = \frac{\text{Im}\{G(\omega)\}}{\text{Re}\{G(\omega)\}} = \frac{\sin\varepsilon}{\cos\varepsilon - 1} \quad (12.61)$$

this yields

$$b_{cr} = \frac{-1}{2k_d \, \text{Re}\{G(\omega)\}} \quad (12.62)$$

This stability criterion was proposed by Merritt [47], who gave Tlusty and Polacek [46] credit for prior publication. The limit of stability occurs at the critical width of cut; the minimum borderline of stability is given by

$$(b_{cr})_{min} = \frac{-1}{2k_d(\text{Re}\{G(\omega)\})_{min}}$$
$$(\text{Re}\{G(\omega)\})_{min} = |(\text{Re}\{G(\omega)\}_{neg\,max}| \quad (12.63)$$

b_{cr} is a limiting case, below which no chatter will occur and above which chatter may occur. The above solution is valid only for negative values of $\text{Re}\{G(\omega)\}$ since b_{cr} is a physical quantity. This solution represents an orthogonal cutting case of an SDOF system that represents a turning process with zero lead angle well. For a tool with a lead angle κ and a cutting force at an angle σ (from a line normal to the machined surface), the stability lobes move up resulting in higher stability. The limit of stability is given by the oriented transfer function

$$\text{Re}\{G_1(\omega)\} = \sin^2\kappa \, \cos\sigma \, \text{Re}\{G(\omega)\} \quad (12.64)$$

The FRF is generally either that of the workpiece or the tool (the weakest of the two), typically the boring tool or spindle. The milling process is better treated in time-domain simulation, but can be analyzed as a turning process by fixing the tool at a given position. Equations (12.63) and (12.64) can be used for a simplified SDOF milling process with n_{tc} teeth in the cut, in which case

$$(b_{cr})_{min,\ milling} = \frac{(b_{cr})_{total}}{n_{tc}} = \frac{(b_{cr})_{min}}{n_{tc}} \qquad (12.65)$$

and $(b_{cr})_{total}$ is given by Equation (12.63). In the case of end milling, $\kappa = 90°$ and in the case of face milling $\kappa = 0°$. The direction of the cutting force σ is zero in milling.

In many cases, it is assumed that the structural dynamics of the machine tool can be represented by an SDOF system based on a second-order dynamic model as discussed above. Under this assumption $Re\{G(\omega)\}_{min}$ is given by Equation (12.52) and b_{lim} is given by

$$b_{lim} = \frac{2k_c\zeta(1+\zeta)}{k_d} \simeq \frac{2k_c\zeta}{k_d} \qquad (12.66)$$

b_{lim} can also be calculated from Equation (12.62) because the $Re\{G(\omega)\}$ can be measured in Figure 12.3. If the value $\zeta = 0.05$ is adopted as a nominal value for machine tool structures [49], the critical value of the gain ratio is

$$b\frac{k_d}{k_c} = 0.105 \qquad (12.67)$$

As a result, for the second-order structural dynamic model with $\zeta = 0.05$, the stability of the overall system will be improved if the critical gain ratio is greater than 0.105. This theory should be applied only when the modes of the machine tool structure are well separated, as is usually the case in drilling and face milling and sometimes the case in turning.

An MDOF system can be analyzed using a similar approach because the real part of the TF of a complex system is simply a sum of the TFs of the individual modes involved [44]. Therefore,

$$G(\omega) = u_1 G_{d1}(\omega) + u_2 G_{d2}(\omega) + \cdots = \sum_{i=1}^{N} u_i G_{di}(\omega) \qquad (12.68)$$

An example of this method is given in Figure 12.16 for three and four DOF systems.

For example, milling cutter can be assumed to have two orthogonal degrees of freedom as shown in Figure 12.17. Two modes of vibration are considered, one in the feed direction X and the other in the normal direction Y. The cutter is assumed to have n_{tc} teeth in the cut (in contact with the workpiece at any instant) and a zero helix angle. More than one tooth cuts at a time, and the orientation of the force F changes. The cutting forces excite the structure, causing deflections in the two perpendicular directions in the plane of the cut. These deflections are carried to rotating tooth number (j) in the radial or chip thickness direction by its projection onto v_j, $v_j = X \sin(v_j) + Y \cos(v_j)$; where the angle v_j is the instantaneous angular immersion of tooth j. The critical axial depth of cut is

$$b_{cr} = \frac{-1}{2k_d n_{tc}(u_x\ Re\{G_x(\omega)\} + u_y\ Re\{G_y(\omega)\})} \qquad (12.69)$$

where the directional orientation factors for X and Y, respectively, are

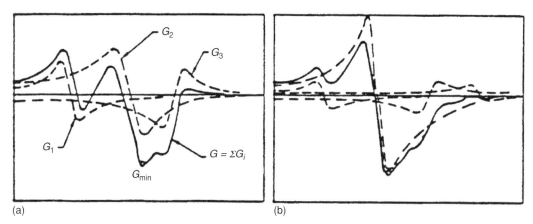

FIGURE 12.16 The real part of oriented transfer functions generated from various combinations of modes (a) for a 3DOF and (b) a 4DOF system. (After F. Koenigsberger and J. Tlusty [99].)

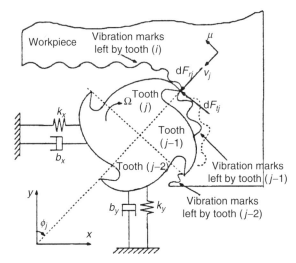

FIGURE 12.17 Dynamic model of milling with two degrees of freedom. (After Y. Antilas and E. Budak [50].)

$$u_x = \cos\alpha \; \sin\nu = \sin(\beta + \nu)\sin\nu$$
$$u_y = \cos(\beta + \nu)\cos\nu \qquad (12.70)$$

More details on the analysis of milling processes can be found in Refs. [50–53].

The procedure for determining the value of $b_{\lim}$ for a given oriented transfer function for every possible chatter frequency series corresponding to different spindle speeds N is as follows:

1. Select a frequency f.
2. Determine the values of $\text{Re}\{G(\omega)\}$, $\text{Im}\{G(\omega)\}$, and ν.
3. Calculate the value of ε.
4. Determine the speed N from Equation (12.42) for several values of n_p ($n_p = 1, 2, 3, \ldots$). For milling the frequency f is divided by the number of teeth in the milling cutter n_t in Equation (12.42).

5. Calculate $b_{\lim}$ from Equation (12.62) or (12.69).
6. Rearrange and plot the pairs (b_{cr}, N) in ascending order of N.

A simple way to graph the lobes is to use the solution of the SDOF cutting process and plot $b_{cr}(\omega)$ versus the spindle rpm $N(\omega)$ using the equation

$$N(\omega) = \frac{60}{n_t} \frac{\omega}{-\mathrm{atan}\left(-\mathrm{Im}\{G(\omega)\}/[b(\omega)k_d \, \mathrm{Mag}\{G(\omega)\} + \mathrm{Re}\{G(\omega)\}]\right) \pm 2\pi n_p} \quad (12.71)$$

This is the solution of the SDOF system based on the stability condition given by Equation (12.57). A typical stability chart for a machine tool is shown in Figure 12.18. Three borderlines of stability are shown; these are called lobed, tangent, and asymptotic borderlines [37]. The lobed borderline of stability is the exact borderline obtained through the above procedure. The tangent borderline is a hyperbola touching all the unstable ranges of the lobed borderline. The asymptotic borderline is determined using $b = (b_{cr})_{\min} = b_{\lim}$ calculated from Equation (12.63). The asymptotic borderline of stability defines the maximum width of cut which results in stable cutting at all spindle speeds. However, a larger width of cut can be used when the proper spindle speed is selected (between lobes). The lobed borderline indicates that the system exhibits conditional stability above one specific speed and below another specific speed. In addition it indicates that the danger of instability at fairly low speeds is reduced due to process damping [54, 55]. The stable speed ranges (lying between the unstable ranges of speed) become wider at higher speeds. The lobed borderline of stability is therefore of practical use primarily at high cutting speeds. It is complex to apply in practice because it requires a different stability chart for each possible position of the movable elements of the machine tool. Commercially available software allows measurements of the frequency response of the cutter–toolholder system without cutting and without running the spindle. Such software converts the frequency response data to useable dynamically optimized cutting parameters using stability lobes [34, 56]. In addition, a safety zone along the lobes can be plotted to compensate for the variation and uncertainty of some of the factors in the stability limit including the directional orientation factors.

The effects of the individual cutting parameters on stability lobes are as follows: (1) The limit of stability is controlled by the upper limit of chip width b in turning and by the axial depth of cut in milling. (2) The axial depth of cut in milling is affected by the number of teeth

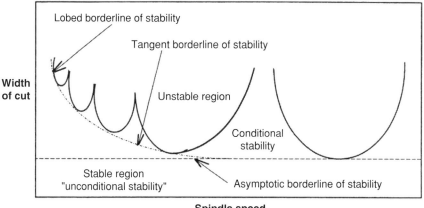

FIGURE 12.18 Typical stability chart for a machine tool.

n_{tc} in the cut (or the radial immersion for the cutter discussed in Chapter 2). (3) The axial depth of cut is proportional to the dynamic stiffness (or product of stiffness and damping) of the machine and workpiece structure and inversely proportional to the specific force. This means that the axial depth of cut decreases with increasing workpiece material hardness since k_d increases. (4) The limit of stability in milling is determined by the limit of the MRR affected by the product of the axial and radial depths of cut. (5) The spindle rpm, N, and the number of teeth in the cutter affect the stability lobes and the selection of b_{lim}. (6) The feed per tooth has a small effect on the limit of stability but does affect the amplitude of vibration in unstable regions.

12.6.3 Shear Plane Method

An earlier, less convenient method of stability analysis was the shear plane method developed by Tobias and Fishwick [45, 57, 58]. This quasiempirical method was intended to develop stability charts based on a mathematical representation of the machining process with the necessary constants obtained experimentally. The dynamic cutting force expression in this theory is

$$dF = K_1 \cdot da + K_2 \cdot df_r + K_3 \cdot dN \tag{12.72}$$

where da, df_r, and dN are, respectively, changes of the chip thickness, feed rate, and rotational (spindle) speed from one state to another which cause variations in the cutting force components. The coefficients K_1, K_2, and K_3 are defined by

$$K_1 = \left(\frac{\partial F}{\partial a}\right)_{df_r = dN = 0} ; \quad K_2 = \left(\frac{\partial F}{\partial f_r}\right)_{da = dN = 0} ; \quad K_3 = \left(\frac{\partial F}{\partial N}\right)_{da = df_r = 0} \tag{12.73}$$

Assuming that the influence of the vibration $x(t)$ on the cutting speed is negligibly small, the above expression can be simplified to

$$dF = K_1 \cdot da + K_2 \cdot df_r \tag{12.74}$$

Under regenerative conditions, the dynamic force is

$$dF = K_1 \cdot [x(t) - \mu \cdot x(t - \tau)] + K_2 \cdot \frac{dx}{dt} \tag{12.75}$$

where μ is the overlap factor indicating the degree of overlap between subsequent cuts (i.e., for plunge cutting $\mu = 1$ and for thread cutting $\mu = 0$). K_1 is the chip thickness coefficient and K_2 is the penetration coefficient. τ is the characteristic of regenerative chatter and reflects the time between subsequent revolutions or teeth ($\tau = 1/N \cdot n_t$). Considering the weakest mode of the machine structure, the equivalent vibratory system is acted on by the force element dF and the equation of motion is

$$m \cdot \ddot{x}(t) + c \cdot \dot{x}(t) + k \cdot x(t) = -dF \tag{12.76}$$

Substituting Equation (12.75) into Equation (12.76) and assuming the solution to be of the form $x(t) = A \cos(\omega t)$, the following differential equation is obtained:

$$\ddot{x}(t) + \omega_n^2 \left(\frac{1}{Q\omega_n} + \frac{K_1}{k}F_2 + \frac{K*}{kN}\right)\dot{x}(t) + \omega_n^2\left(1 + \frac{K_1}{k}F_1\right)x(t) = 0 \tag{12.77}$$

where $K^* = K_2 N$ is a constant. For the critical state of stability the total damping is zero, and from Equation (12.77)

$$\left(\frac{1}{Q\omega_n} + \frac{K_1}{k} F_2 + \frac{K^*}{kN} \right) = 0 \qquad (12.78)$$

and

$$\omega^2 = \omega_n^2 \cdot \left(1 + \frac{K_1}{k} \cdot F_1 \right) \qquad (12.79)$$

with

$$\omega_n^2 = \frac{k}{m}, \quad Q = \frac{k}{c \cdot \omega_n} = \frac{1}{2\zeta} \qquad (12.80)$$

and the coefficients

$$F_1 = 1 - \mu \cdot \cos\frac{\omega}{N}, \quad F_2 = \frac{\mu}{\omega} \cdot \sin\frac{\omega}{N} \qquad (12.81)$$

Equations (12.78) and (12.79) define the stability conditions for the particular mode of vibration; using these equations, the relations between Q and N and ω and N can be solved. The stability chart is generated using Q as the ordinate and the parameter $n_t N/f_n$ as the abscissa. This chart defines stable and unstable regions similar to those defined in Figure 12.18 depending on the magnitude and sign of the penetration rate coefficient K^* [57]. The asymptotic borderline of stability is obtained when $K^* = 0$; the tangent and lobed borderline of stability are generated when $K^* > 0$; finally, when $K^* < 0$, the low unstable speed ranges in the stability chart are displaced downward and the risk of instability at low speeds increases. The parameters K_1 and K_2 are functions of the material, tool geometry, feed, chip width, and similar factors. These coefficients can be measured through dynamic experiments [3]. In general, one or more modes of vibration can become unstable, so that stability charts (Figure 12.18) should be generated for each mode in practice, however, only a single mode of vibration is unstable in many operations [3], so that stability analysis can be carried out using a single stability chart. One of the fundamental problems in this approach is the prediction of the cutting forces while cutting. It should be also noted that a number of other different models have been proposed with varying degrees of success [58, 59].

12.6.4 OTHER METHODS

Regenerative vibrations may also be represented by a graph in the complex plane as shown in Figure 12.19. In this figure, curve I represents the dynamic receptance $G(\omega)$ of the machine itself, while the straight line II represents that of the cutting process $H(\omega)$. If these two lines are in contact or intersect, instability will occur, otherwise the system is stable.

The cutting process can also be described by the equation

$$H(\omega) = \frac{1}{2 \cdot K} \qquad (12.82)$$

where K is the cutting force coefficient. The larger the coefficient K, the smaller the $H(\omega)$, meaning that the two lines I and II in Figure 12.19 are more likely to contact or intersect. The

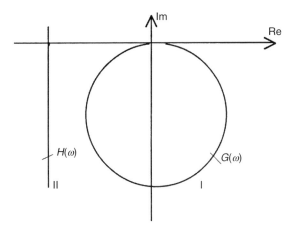

FIGURE 12.19 Operative receptance for an SDOF structure.

machine tool can be also characterized by the allowable K_{lim}; this parameter is difficult to measure but can be calculated from

$$K_{lim} = b_{lim} \cdot K_f \qquad (12.83)$$

where K_f is the cutting force coefficient per unit width of cut under specific working conditions. Hence, K_{lim} can be determined by varying the cutting width b and keeping K_f constant by adopting fixed values for the other cutting conditions.

Another method for predicting the limiting width of cut b_{lim} is the *incremental stiffness method* using Equation (12.62) [60]. This method uses an incremental cutting stiffness k_i instead of the dynamic cutting coefficient k_d in Equation (12.62). The two force components under the assumption of orthogonal cutting conditions, the thrust and the main cutting forces, are measured at a given cutting depth for a series of cutting speeds and for two feed rates; the cutting stiffness k_i is then computed using a method described in Ref. [60].

The above analytical methods for predicting stability lobes are based on simplified linear models which assume an average tooth position, force direction, and number of teeth in the cut, and which do not account for process damping. They cannot be used to predict limiting levels of chatter or to model nonlinear springs, backlash, and similar real phenomena. These methods are applicable mainly to operations such as turning which resemble a basic orthogonal cutting operation in which the direction of the cutting forces and chip thickness do not vary with time. For rotating tooling applications such as milling, in which the chip thickness, cutting forces, and direction of excitation vary and are intermittent, a more realistic and more detailed method for generating stability charts is *time domain simulation* (TDS) [31, 46, 49, 51–53, 58–67]. Even though TDS is a computationally intensive method, it is attractive in many cases because it can include many of the complex characteristics of real machining systems.

As an example, a milling cutter can be assumed to have two orthogonal degrees of freedom with one or more masses, springs, and dashpots along each direction as shown in Figure 12.17. The lumped parameters representing the modal stiffness, mass, and damping can be extracted from the transfer functions in two orthogonal directions. Velocity-dependent process damping and basic process nonlinearities arising when the tool exits the cut due to excessive vibration can be considered using TDS with this idealization. For example, Montgomery and Altintas [53] considered process damping and stiffness in a time domain simulation model, and predicted the surface finish produced by chatter in milling.

TDS can be used for SDOF or MDOF systems with closed-loop behavior described by a second-order system equivalent to Equation (12.1). In this method the cutter is advanced in small steps; 360 steps per natural period are required to match TDS results to frequency-domain solutions [62]. For each increment, the forces on each tooth are calculated and summed vectorialy. The acceleration produced by the force is used to calculate displacements in both directions; the resulting differential difference equation is numerically integrated. This approach accounts for change in chip thickness throughout the cut including the entry and exit transients. The limiting width of cut is determined by running simulations at various widths of cut for each spindle speed of interest. Convergence on b_{lim} is achieved by considering increments and decrements of the axial depth of cut, based on the stability conclusion at each axial depth of cut considered, while gradually reducing the magnitude of the step changes. The peak-to-peak (PTP) amplitude of the displacement and force are calculated for each simulation and plotted versus the spindle speed for various axial depths of cut as shown in Figure 12.20 [51, 54]. For each simulation, only the PTP amplitude is recorded for force and displacement.

Figure 12.20 was constructed for high-speed milling with a 9.5 mm diameter two-flute carbide end mill, which is the same system used to compute the results shown in Figure 12.4. The individual lines correspond to axial depths of cut from 0.5 to 7 mm in steps of 0.5 mm (14 axial depths of cut). The spindle speeds varied from 25,000 to 45,000 in 500 rpm steps (40 speeds), so that the information in the figure is a composite of 560 simulation runs. The graph plots the force in the X-direction versus spindle speed. The stable zones are above 40,000 rpm and between 25,000 and 29,000 rpm. For axial depths of cut below 4 mm, all cuts were stable. The depth of cut lines which have unstable speed ranges are marked. In the unstable regions, the PTP forces may become quite large. TDS can provide limited information about the surface roughness generated by considering the vibration and the trajectory of the teeth along the cutting path that the tool center follows.

In general, the identification of critical chatter frequencies in milling is not a trivial task either experimentally or theoretically. Milling force or acceleration power spectra typically exhibit several peaks [68]. Some result from tooth pass excitation, while others are due to

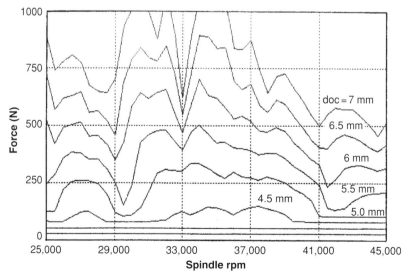

FIGURE 12.20 Peak-to-peak diagram for a two-flute, 9.5 mm diameter carbide end mill, illustrating an algorithm for optimizing the material removal rate.

regenerative effects or the natural frequency of the tool. In the case of milling, therefore, there are multiple chatter frequencies compared to the single well-defined chatter frequency typical of an unstable turning process. The analysis of the milling process is more complex because the direction and level of the cutting force changes with tool rotation and because the cut is interrupted. Tooth pass excitation results in a parametric excitation of the system, and the governing equation of motion is a delay-differential equation (DDE). Stability properties have been studied through analysis of these DDEs [41, 69–86] and through frequency analysis of chatter signals [66, 69, 87]. Some chatter frequencies are related to unstable periodic motions about stable stationary cutting states. In this case, subcritical (Hopf) bifurcation occurs, as demonstrated experimentally by Shi and Tobias [88] and analytically by Stépán and Kalmár-Nagy [89]. Similarly, periodic doubling bifurcation is also a typical mode of stability loss in milling processes [76–78, 86]. The nonlinear analysis of Stépán and Szalai [90] showed that this period doubling bifurcation is also subcritical.

Low radial immersion milling has become an increasingly important process due to advances in machining centers, which have enabled rapid adoption of high-speed milling, and due to increased near-net shape manufacturing of aerospace components. Increasingly, there has been a shift within the aerospace community to replace assembly-intensive sheet metal buildups with monolithic aluminum structures. To further eliminate the waste of monolithic component production, many part features are now being cast, formed, or forged near-net shape. This has driven the need to better understand stability boundary and the corresponding surface location error in interrupted cutting processes as the radial depth of cut or radial immersion is varied (Figure 12.21). The relationship between the direction of tool rotation and feed defines two types of partial immersion milling operations: up-milling and down-milling (Figure 12.22). Both operations can produce the same result, but their dynamic and stability properties are not the same. Partial immersion milling operations are characterized by the number N of teeth and the radial immersion ratio RDOC/D, where RDOC is the radial depth of cut and D the diameter of the tool (see Chapter 2). Several researchers have shown the stability properties for low radial engagements differ from those observed when cutting a full slot [76–78, 82, 86, 91]. For instance, Davies et al. examined interrupted cutting processes for an infinitesimal cutting period to formulate a dynamic map [76, 92]. Stability predictions were made from the map transition matrix eigenvalues. Stépán, Insperger, and Balachandran have also developed an approach to formulate dynamic maps for stability prediction using a method called semidiscretization [68, 77, 83]. Bayly, Halley, and Mann

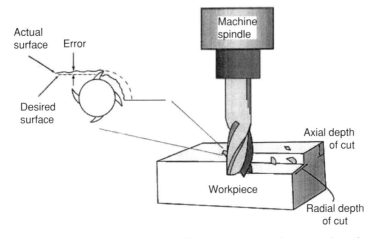

FIGURE 12.21 Schematic diagram showing the milling process is an interrupted cutting process.

Machining Dynamics 651

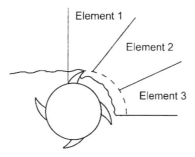

FIGURE 12.22 The temporal finite element approach divides the cutting motion into elements. (Courtesy of B. Mann, University of Florida.)

have presented an approach called temporal finite element analysis to examine interrupted cutting in turning and milling [82, 93–97]. This method forms an approximate analytical solution by dividing the time in the cut into a finite number of elements (Figure 12.22). The approximate solution for the cutting motion is then matched with the exact solution for noncutting time periods to obtain a discrete linear map. The formulated dynamic map is then used to determine stability and the steady-state surface location error (Figure 12.21). Figure 12.23 shows the effect of changing the radial immersion on the stability boundaries for three different radial depths of cut. One interesting result from the three plots is that a larger axial depth can be obtained for low radial immersion cutting processes.

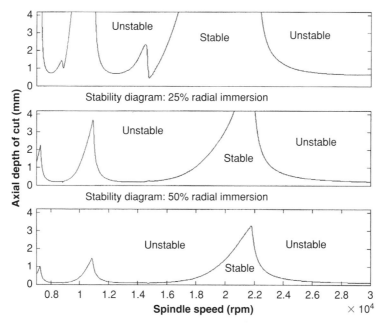

FIGURE 12.23 Example of low radial immersion stability lobes for three different radial depths of cut. These plots show it is possible to obtain a larger axial depth of cut in finishing operations. (Courtesy of B. Mann, University of Florida.)

12.6.5 NONREGENERATIVE CHATTER; MODE COUPLING

One type of nonregenerative chatter occurs when the tool vibrates relative to the workpiece in at least two directions in the plane of cut as shown in Figure 12.24. The machining system is modeled by a 2DOF mass–spring system, with orthogonal axes of major flexibilities and a common mass. The characteristics of the vibration are such that the tool follows a closed elliptical path relative to the workpiece as indicated by the arrows in the contour in front of the tool. During the periodic motion of the tool from portion 1 to 2 along the contour the resultant cutting force is in the opposite direction of the tool motion and energy is dissipated from the system. However, during the other half of the of the contour, from portion 2 to 1, where the motion and cutting force act in the same direction, energy is supplied in the system, which increases the vibratory energy of the tool. The force F tends to be larger over the lower portion of the elliptical contour than over the upper portion because it is located deeper in the cut; therefore, the input energy is larger than the total energy losses per cycle, which leads to increasing vibration amplitudes, thus producing negative effective damping. This type of instability is usually referred to as *mode coupling* [44].

The mode coupling principle may be explained by a simple form of a 2DOF system as shown in Figure 12.24. The system is assumed to be linear as long as the tool does not leave the cut. The width of cut for the threshold of stability depends directly on the difference between the two principal stiffness values, and chatter tends to occur when the two principal stiffnesses are close in magnitude. This type of nonregenerative chatter may occur in thread turning and shaping, operations during which the tool does not cut into the surface generated by the previous revolution of the workpiece. In addition, the regenerative mode may be hindered if the phase shift between two subsequent passes or teeth is constrained to zero ($\varepsilon = 0$), or even when milling with special tools such as those with alternating helix [98]. Nonregenerative chatter of this type cannot exist for an SDOF system [44]. The equations of motion of the system in directions X_1 and X_2 (as shown in Figure 12.24) are given by Equation (12.45). This type of vibration can be characterized by Equation (12.57) as explained in Refs. [99, 100].

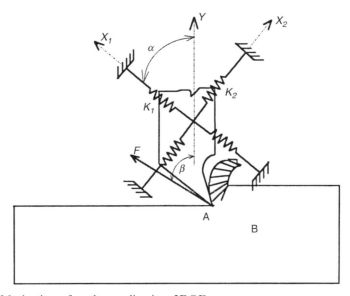

FIGURE 12.24 Mechanism of mode coupling in a 2DOF system.

Machining Dynamics

Mode coupling can be studied using the analysis for regenerative chatter with $\varepsilon = 0°$. The limiting width of the chip for nonregenerative instability is approximately twice that for regenerative chatter [44], so that

$$(b_{\lim})_n \approx 2(b_{\lim})_r \qquad (12.84)$$

Therefore, this type of instability will not occur for conditions where successive passes of the cutting tool overlap each other. The theory of mode coupling has been applied in boring and turning, operations in which two modes of the system tend to be close to each other in frequency.

12.7 CHATTER PREDICTION

As noted above, chatter is undesirable because it limits production rates and leads to poor machined surface finish and accelerated machine and cutting tool wear. Chatter detection and prediction are therefore of great practical importance in machine tool monitoring and dynamic performance testing applications [101]. Several methods have been proposed for chatter detection and/or prediction. Some of the parameters used for this purpose include (a) the cross-correlation coefficient between the dynamic thrust force and cutting forces [102], (b) the static deflection of a workpiece [103], (c) the RMS value of the vibration signal [104], (d) the peak of the power spectrum of the vibration signal [105], and (e) other parameters determined from spectral analysis of the force or vibration signal [54, 106].

Chatter prediction actually means the prediction of the stability limits of the machining process. Obviously this information is important when determining optimal cutting conditions for programming automated machines. There are two prevailing techniques for this purpose: *limit chip analysis* and the *stability chart method*.

Limit chip tests are conducted under conditions that have been standardized to some extent. The tests yield the limiting axial depth of cut at which chatter occurs. This method is the basis of ASME standard B5.54 for chatter limit evaluation of machining centers [36]. Care should be used in interpreting results of these tests because the critical depth of cut is influenced not only by the dynamic characteristics of the machine tool (as assumed in the standard), but also by the workpiece, fixture, toolholder, cutting tool characteristics, and cutting conditions. This method is therefore primarily useful for ranking the dynamic performance of various machines, rather than for predicting stability limits in specific applications.

Stability chart methods are based on a number of analysis techniques such as those discussed above in the regenerative chatter section. The most common approaches make use of experimentally measured transfer functions of the cutting process and the machine tool [48]. The limit of stability is computed using classical control theory. In a similar approach, Weck et al. [107] proposed using the measured stability lobes (shown in Figure 12.18) to select chatter free spindle speeds. Since stability lobes are narrow at low cutting speeds, this approach is effective primarily in the higher cutting speed ranges. In another similar approach, the tooth passing frequency is equated to the chatter frequency, which is close to the natural frequency of the structure [108]. This method minimizes the phase between the inner and outer modulations and increase the chatter-free depth of cut. Below the limit of stability shown in Figure 12.18, the force and sound spectra are dominated by the tooth frequency, runout, harmonics, and noise. Above the limit of stability, a prominent chatter frequency appears; the spindle speed is selected so that the tooth passing frequency does not excite the structure at this frequency. This approach requires different stability charts for different cutting positions because the conditions under which chatter occurs vary with the cutting position. Commercial analysis systems are available for measuring the FRF of a system, generating a stability, and determining optimum process conditions [34].

12.7.1 Experimental Machine Tool Vibration Analysis

The first task in analyzing machine tool vibration problems is to determine what type of vibration is present. Methods of distinguishing between forced and self-excited vibrations were discussed in Section 12.5. The machine structure has an infinite number of natural modes of vibration. The direction of relative motion between the cutting edge and the workpiece is determined by the principal vibration directions for each natural mode of vibration. The machine frame can be idealized as a sum of several elementary systems representing each natural frequency of the frame assuming that the natural frequencies are well separated (i.e., that no nearly equal pairs of natural frequencies are present). Each elementary system consists of an equivalent mass m, an equivalent linear spring with constant k, and an equivalent viscous damper with constant c. The stability of the cutting process is determined by the stability behavior of all active modes. It is generally necessary to consider only the least stable modes (the modes with the lowest natural frequencies) in most applications.

Due to the complexity of the phenomena involved, experimental methods are generally used in practice once machines have been built. Experimental analysis is used for two major purposes: (a) to measure the modes of vibration and mode shapes; and (b) to obtain a mathematical model of the structure. Two types of tests are used for vibration measurement. In the first type (i), one parameter such as a force component, acceleration, or similar signal is measured during machine operation either while the spindle is rotating or while the cutting tool is engaged in the workpiece. In the second type (ii), the response output and an input are measured simultaneously. Type (i) tests are generally used to define a mathematical model of a process or for online machine control. These tests address only the second measurement objective (b) discussed above; a response measurement alone cannot be used to determine whether a particularly large response magnitude is due to strong excitation or to structural resonance. Type (ii) tests are commonly performed because the measurement of the input and output signals simultaneously can be used to completely define the system vibration characteristics. These tests can be used to address either of the measurement objectives (a) and (b) discussed above. In these tests, the structure or cutting tool is excited using an input signal which is often stronger than the disturbances expected in normal service.

In order to determine the characteristics of the system from the measured response, it is necessary for the response to have the same frequency response characteristics (poles) as the system, or to correct the measured response to remove the poles and zeros of the excitation [109]. In general, the measured response contains the poles of the system under study and of the input. Therefore, if the input force is not measured, it is not possible, without some prior knowledge about the input, to determine if the poles of the response are truly system characteristics. The poles of the response in the frequency range of interest are system characteristics if no poles or zeros exist in the force spectrum in this range. The input to the system should be measured to avoid this potential problem.

The signal analysis techniques commonly used to quantify an experimentally measured signal are summarized in Figure 12.25. *Signal magnitude analysis* is used primarily for monitoring and provides information about a single measured signal. The other two techniques are used for both single and dual signal analysis.

The relationship between the input and output in frequency domain is described by the transfer function (TF). The TF is an important dynamic characteristic of structures as discussed in previous sections. The measurement of the TF is also required for machine tool structure identification. The natural frequencies, damping ratios, and mode shapes can be determined from the TF.

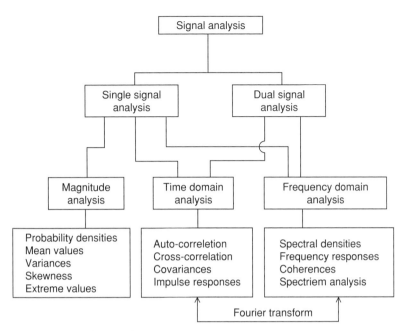

FIGURE 12.25 Signal analysis techniques.

12.7.2 MEASUREMENT OF TRANSFER FUNCTIONS

The problem of determining the TF of the cutting process or its inverse (generally called the dynamic cutting coefficient) is a central problem in machining dynamics theory and practice. The experimental methods developed to measure the TF of the cutting process can be broadly classified into two main groups: static methods and dynamic methods.

Static methods are based on measurements of static force components. Their analytical formulations were discussed in the previous section as the shear plane method and the incremental stiffness method. In static methods, the TF of the cutting process is determined indirectly because some system model is assumed in the analysis.

Dynamic methods measure the dynamic parameters directly and are subdivided into four classes: stiffness methods, dynamometer methods, frequency-response functions, and time series methods.

Stiffness methods measure the cutting process stiffness and damping based on the pulse response of the test rig measured both while cutting and while idling. The cutting process characteristics are calculated from differences in the measurements taken under both conditions [63, 64]. This approach is accurate because the cutting process adds stiffness and damping to the machining system in addition to that available from the machine tool and toolholder structures.

Dynamometer methods allow measurement of the tool vibration and cutting forces during chatter conditions. In these methods, the cutting tool–toolholder system is excited using a vibration exciter while the toolholder is mounted on a dynamometer. The two main concerns with this method are: (1) the dynamic response of the dynamometer must be accounted for by using either compensating circuits [65] or a lightweight dynamometer; (2) it is difficult to account for outer modulation and to combine it precisely in phase with the inner modulation signal measured nearly simultaneously on the tool [110–114].

In recent years, computerized *modal analysis* systems have become increasingly common [115–120]. These systems use the frequency-response function method to identify the natural frequencies, damping ratios, and mode shapes of the structure associated with specific modes of vibration. The methods used are collectively referred to as experimental modal analysis. The vibration measurements can be made while the machine tool, the cutting tool in the spindle, or the workpiece in the fixture are excited by known disturbances, which as noted above may be stronger than those expected in normal service. This method is known as the *excitation test method* because the frequency-response functions are measured using excitation at single or multiple points [108, 120]. It is important to include sufficient points in the test to completely describe all modes of interest. If the excitation points are not chosen carefully or if enough response points are not measured, a particular mode may not be adequately represented. Frequency responses can be measured independently using single point excitation or simultaneously with multiple point excitations.

The measured vibration data of a machine tool (the TF data) are subjected to a range of curve-fitting procedures in an attempt to find the mathematical model which provides the closest description of the observed response. After the modal frequencies are found, the machine may be excited at each such frequency to identify each mode shape. Based on this procedure, the weak mode of the machine can often be identified; this provides direction for improving the system design.

Single-point excitation is commonly used in modal analysis. The main purpose of the excitation test method is to exert and exciting force which simulates the cutting force between the workpiece and tool at the same time the response is measured. This test should performed under conditions as close to the practical working conditions as possible.

The measurement process involves three equally important components: (1) the correct support and excitation of the structure; (2) the correct measurement of the input force and displacement response; and (3) proper signal processing, which depends on the type of test used.

The structure is generally excited either by connecting it to a vibration generator or shaker, or by using some form of transient input [120]. The source signal can take several forms: harmonic sinusoidal (from an oscillator), random (from a noise generator), periodic (from a signal generator such as sinewave generator), or transient (from a pulse generating device or by applying an impact with a hammer). The magnitudes of the applied exciting force and the responses of the structure can be analyzed using either a spectrum (Fourier) analyzer or a frequency-response analyzer. Responses measured may be accelerations, velocities, or displacements.

Types of devices used to excite the structure include mechanical exciters (out-of-balance rotating masses), electromagnetic exciters (often based on a moving coil in magnetic field) and electrohydraulic exciters, electrodynamic exciters, and impulse hammers [120, 121]. The manner in which the exciter is attached to the structure should be chosen to ensure that the excitation and response act in the same direction.

The impulse response or hammer or impact test is a relatively simple means of exciting the structure. In this test, a small sensor is mounted at a location of interest and the structure is struck with an instrumented hammer. Both the input and response signals are measured. The hammer can be fitted with various tips and heads which serve to extend the input frequency and force level ranges. The magnitude of the impact is determined primarily by the mass of the hammer head.

The time domain input and response signals are usually transformed into FRFs using a fast Fourier transform (FFT) analyzer [122]. All modal parameters are estimated from these FRFs, which in principle should provide an accurate representation of the actual structural response [116, 117, 123, 124]. FFT analysis requires an analog to digital data conversion. The signals must be preprocessed (filtered) and digitized in a manner which prevents frequency aliasing; for this purpose, a sampling rate at least twice as large as the highest significant frequency component in the signal should be used. For unknown structures and exploratory studies, it is often advisable to tape the signals using an analog tape recorder with a high bandwidth, so that the signals can be replayed and resampled at higher sampling rates if required. Spectrum analysis can be also performed to determine the frequency characteristics of the signal.

Modal parameters are extracted using a two-step process. In the first step, the TF is estimated using an FFT spectrum analyzer and the global parameters (natural frequencies and damping ratios) are identified. There are three common methods for parameters estimation: the simple quadrature peak pick, the resonant frequency detector fit, and the multifunction, MDOF curve fit [109, 120]. The magnitude of the TF can be determined by curve fitting based on the natural frequencies and damping ratios at each resonance. If the modes are not heavily coupled, an SDOF curve fit is acceptable for each mode. In many circumstances, however, modes are heavily coupled; under this condition, an MDOF curve fit must be used. The damping ratio is determined based on the methods discussed in Section 12.3.

In the second step the mode shapes are determined. The measurement of modal coefficients may be accomplished using one of two methods: quadrature peak pick or least-squares circle fit. Three main categories of system model can be identified: the spatial model (of mass, stiffness and damping properties), the modal model (consisting of the natural frequencies and mode shapes), and the response model (consisting of a set of frequency response functions). The completion of the second step yields a minimum parameter linear mathematical model of the particular machine tool structure. The block diagram of a typical experimental setup is shown in Figure 12.26.

Chatter and the cutting process transfer function can also be modeled using a *time series approach*. Various approaches [125–130] can be used to model machine tool chatter as a self-excited random vibration based on time series methods. The time series, obtained by sampling the continuous vibration signal at uniform intervals, is a realistic modeling technique which

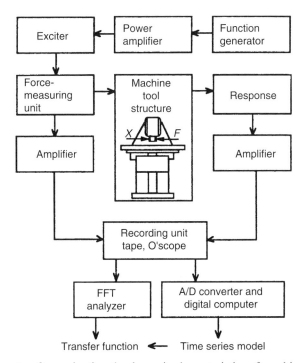

FIGURE 12.26 Basic setup for evaluating the dynamic characteristics of machine tools.

takes into account the unknown factors in chatter and its random nature. In the time series approach, the random vibration systems is excited by white noise when sampled at uniform intervals. The underlying structure is extracted first in the time domain by a parametric modeling procedure and subsequently transformed into frequency domain. The system is represented by linear stochastic systems and is represented by a discrete autoregressive moving average (ARMA) model of order (n, m), defined as

$$x_t - \phi_1 \cdot x_{t-1} - \phi_2 \cdot x_{t-2} - \cdots - \phi_n \cdot x_{t-n}$$
$$= a_t - \theta_1 \cdot a_{t-1} - \theta_2 \cdot a_{t-2} - \cdots - \theta_m \cdot a_{t-m} \qquad (12.85)$$

where x_t is the system's response series, a_t is a discrete white noise series, ϕ_i are the autoregressive parameters, and θ_i are the moving average parameters. The estimated model will best approximate the structure in terms of the mean and covariance because it satisfies the conditions

$$E[a_t] = 0 \quad \text{and} \quad E[a_t \cdot a_{t-k}] = \delta_k \cdot \sigma^2 \qquad (12.86)$$

where δ_k is the Kronecker delta function and E is the expected value operator. The strategy used to determine the adequate orders of the model n and m is discussed in Refs. [131, 132].

Once the model is known, the modal natural frequencies and damping ratios can be determined by solving the characteristic equation [126]

$$\lambda^n - \sum_{j=1}^{n} \phi_j \cdot \lambda^{n-j} = 0 \qquad (12.87)$$

for the n roots λ. The ARMA model allows the definition of the transfer function in the general form

$$T(B) = \frac{1 - \sum_{k=1}^{n-1} \theta_k \cdot B^k}{1 - \sum_{k=1}^{n} \phi_k \cdot B^k} \qquad (12.88)$$

where B is the backshift operator, that is, $x_{t-k} = B^k \cdot x_t$.

12.8 VIBRATION CONTROL

The dynamic behavior of a machining system can be improved by reducing the intensity of the sources of vibration for the machine tool, toolholder, and cutting tool. Several sources, primarily stiffness and damping, have a significant impact on forced and self-excited vibrations. As explained in Chapter 3 stiffness affects the accuracy of machine tools by reducing structural deformations due to cutting forces, while damping accelerates the decay of transient vibrations. The stability of the cutting process against vibration and chatter can be improved by several approaches:

1. optimizing the design of the machine tool using both analytical and experimental methods to provide maximum static and dynamic stiffness;
2. selecting the best toolholder device for the particular tool and application;
3. selecting the proper bearing types, configurations, and installation geometry to provide maximum stiffness and damping;

4. isolating the system from vibration forces and using active or passive dynamic absorbers;
5. increasing the effective structural damping and using tuned vibration dampers;
6. selecting optimum cutting conditions, especially the spindle speed;
7. selecting special cutting tool geometries;
8. increasing the part stiffness;
9. for force vibration, decrease cutting force, increase part stiffness, change tooth passing frequency away from resonance frequency of structure (see Section 12.5);
10. for self-excited vibration, decrease do and number of teeth in cut or change tooth passing frequency to match resonance frequency.

12.8.1 Stiffness Improvement

The static stiffness, the ratio of the deflection to the applied static force at the point of application, can be measured for all three coordinate axes of the machine tool. The main contributors to deformation between tool and workpiece are the contact deformations in movable and stationary joints between components of the machine structure and fixture, the toolholder–spindle interface, and the tool–toolholder interface. Individual components of the structure are generally designed for high stiffness and make a comparatively insignificant contribution to the deflection.

The stiffness of the structure is determined primarily by the stiffness of the most flexible component in the loading path. This component should be reinforced to enhance stiffness. Therefore, the contribution of each machine component to the overall deflection should be estimated analytically or experimentally [48].

The effective static stiffness of a machine tool may vary within wide limits as discussed in Chapter 3. The overall stiffness can be improved by placing the tool and workpiece near the main column, by using rigid tools, toolholders, and clamps, by using rigid supports and clamps in the fixture, and by securely clamping all machine parts which do not move with respect to each other. Stiff foundations or well-damped mountings are required.

The bearing design (size, type, distance between bearings, and shaft type as explained in Chapter 3) has a strong influence on the static and dynamic behavior of machine tools. For example, a solid shaft increases the spindle stiffness, but its substantially greater mass reduces the spindle natural frequency, which is undesirable for high-speed spindles.

Although a decrease in rigidity in a machine tool is generally undesirable, it may be tolerated when it leads to a desirable shift in natural frequencies (especially in high speed machining) or is accompanied by a large increase in damping or by a beneficial change in the ratio of stiffnesses along two orthogonal axes, which can result in improved nonregenerative chatter stability [3, 48]. However, reducing the tool–workpiece compliance is not always possible in practice, and other approaches, discussed next, should be examined before productivity is sacrificed by reducing the depth of cut to ensure stable operation.

12.8.2 Isolation

Vibration isolation is the reduction of vibration transmission from one structure to another via some elastic device; it is an important and common component of vibration control. Vibration isolation materials, such as rubber compression pads, metal springs, and inertia blocks, may be used [6, 133]. Rubber is useful in both shear and compression; it is generally used to prevent transmission of vibrations in the 5 to 50 Hz frequency range. Metal springs are used for low frequencies (>1.5 Hz). Inertia blocks add substantial mass to a system, reducing the mounted natural frequency of the system and unwanted rocking motions, and minimizing alignment errors through an increase in inherent stiffness.

12.8.3 DAMPING AND DYNAMIC ABSORPTION

The overall damping capacity of a structure depends on the damping capacity of its individual components and more significantly on the damping associated with joints between components (i.e., slides and bolted joints [35]). The typical contributions of various components to the overall damping capacity of a machine tool are as follows: mechanical joints (log decrement 0.15), steel welded frames (structural damping 0.001), cast iron frames (0.004), material damping in polymer–concrete (0.02), and granite (0.015) [6, 35, 134].

The overall damping of machine tools varies, but the log decrement is usually between 0.15 and 0.3. While structural damping is significantly higher for frame components made of polymer–;concrete compositions or granite (as discussed in Chapter 3), the overall damping does not change significantly since the damping of even these materials is small compared to damping due to joints. A significant damping increase can be achieved by filling internal cavities of the frame components with special materials (i.e. replicated internal viscous dampers [134]) as discussed in Chapter 3. Resonant structural vibrations can be reduced by applying a dynamic absorber or layers of damping material on the surfaces of the structure [133].

The effect of increasing the damping ratio from 0.02 to 0.2 on the stability of a lathe having a resonant frequency of 60 Hz is shown in Figure 12.27 [135]; the asymptotic borderline of stability is increased from 0.04 to 0.48 mm, which corresponds to an increase of a factor of 12 in the effective cutting stiffness.

A dynamic absorber is an alternative form of vibration control. It consists of a secondary mass attached to the primary vibrating component via a spring which can be either damped or undamped. This secondary mass oscillates out of phase with the main mass and applies an

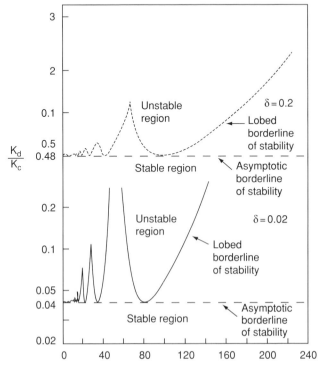

FIGURE 12.27 Effect of the damping ratio on the stability chart for a typical SDOF structure with a natural frequency of 60 Hz. (After C.C. Ling et al., [135].)

inertial force (via the spring) which opposes the main mass. For maximum effectiveness, the natural frequency of the vibration absorber is tuned to match the frequency of the exciting force. Auxiliary mass dampers can be used on machine columns, spindles, and rams.

Tuned dynamic vibration absorbers have been used with considerable success in milling and boring applications [35, 136]. A tunable tool provides a controlled means of adjusting the dynamic characteristics of the tool in a particular frequency range. A tuning system minimizes the broadband dynamic response of the tool without requiring cutting tests or trial and error tuning. A very common vibration damper used in boring bars consists of an inertial weight or a spring-mounted lead slug fitted into a hole bored into the end of boring bar; bars so equipped are often called antivibration boring bars. The weight helps damp bar motion and prevents chatter. The chatter resistance of boring bars can also be increased by using different materials in the bar structure as discussed in Chapter 4 [137]. Impact dampers can be also installed in the toolholder, spindle, or ram to absorb vibration energy.

Active control of structures can be also used to suppress vibration and chatter in machining. Actively controlled dynamic absorbers use sensors and force actuators together in a closed-loop control system to alter the dynamic characteristics of a structure so that it possesses greater damping and stiffness characteristics [138, 139]. Accurate system identification (both in terms of sufficient model order and parameter accuracy), hybrid high-speed control, and proper power force actuators are essential elements of any successful structural or tool vibration control system. These methods are discussed in more detail in the next section.

12.8.4 Tool Design

Reductions of both forced and self-excited vibrations for multiple cutting edges tool structures (such as milling cutters, end mills, and reamers) have been achieved by employing unequally spaced cutting edges (nonuniform tooth pitch), variable axial rake, variable helix cutting edges (alternating helix on adjacent flutes), lip height error, or serrated and undulated edges as discussed in Chapter 4 [44, 140–142]. These cutters are often called white nose cutters.

All these tool design variations increase stability by reducing the effective tooth passing frequency and disturbing the regeneration of surface waviness (the phase between the inner and outer modulations). Their effects are most pronounced over particular ranges of cutting speeds which depend on the cutter geometry (L/D ratio, lead angle, helix angle, etc.), the configuration of the particular part feature to be machined, and the cutter path (up-milling versus down-milling, end mill diameter versus diameter of circular interpolation, etc.). Alternating the helix increases pitch variations along the axial depth of cut, improving both the dynamic performance and operative speed range of the cutter. An optimal distribution of spacings or helix angle variations between the teeth can be found, but the nonuniform pitch or helix cutters can be used within a limited speed range constrained by the dominant structural mode. Theoretical analysis of stability for milling with such special cutters is described in Refs. [44, 47].

Sharp tools are more likely to chatter than slightly blunted tools. Therefore, a lightly honed cutting edge can be used to avoid chatter. Negative rakes and small clearance angles minimize chatter occurrence. In general, it is important to use the smallest possible tool nose radius which gives acceptable tool life, because a small nose radius can alleviate the regenerative effect as explained in Chapter 4 [143]. Finally, combination tools which generate complex surfaces should incorporate design features which inhibit chatter. For example, combined reaming, countering, and chamfering tools require a cylindrical land not only along the reaming and counterboring margin sections, but also a narrow land along the chamfer section to prevent chatter on the chamfered surface which could occur with a conventional sharp chamfering design.

12.8.5 Variation of Process Parameters

Cutting conditions, especially the cutting speed, directly affect chatter generation. At a given spindle speed, the widths and depths of cut are usually limited by the chatter threshold. As indicated in the stability lobe diagram (Figure 12.18), a small increase or decrease in speed may stabilize the cutting process. Small changes in the cutting speed are particularly effective in increasing stability in milling operations.

Automatic regulation of the spindle speed for stable cutting can be used in CNC machine tools. If it is known that there is a gap in the stability lobe diagram, then that frequency is used while increasing the depth of cut until the maximum machine/spindle power is reached [34, 107, 108, 144, 145]. This approach is currently only applicable to uniform pitch cutters. Chatter suppression systems employing this principle are discussed in the next section. Case studies indicate that the use of spindle speed variations provides more flexibility than the use of variable pitch cutters [146].

12.9 ACTIVE VIBRATION CONTROL

Active machine tool control systems use feedback from sensors to regulate the tool position and cutting conditions (speed and feed) to prevent or suppress chatter. Several quantities can be measured for this purpose, including the relative displacement between the tool and the workpiece, the cutting force, the tool acceleration, and the sound spectrum of the process.

The first active chatter control system was built by Comstock et al. [147], who used the relative displacement between tool and workpiece, measured by an inductive transducer, as the control variable. A block diagram of the control scheme is shown in Figure 12.28. The noise signals F_e and F_c act on the machine structure to produce motion y_m. y_i is the displacement of workpiece relative to the cutting tool. The term *chase control* is used for this control scheme because a variation in the relative tool and workpiece position gives rise to an error which causes the controller to supply a corrective motion y_t to the cutting tool; in effect, the tool attempts to catch or "chase" the workpiece and cancel the error. It is assumed that the displacement transducer measures the tool to workpiece centerline distance rather than the tool to workpiece surface distance. All except gross surface irregularities can be eliminated through use of a proximiting gage with a sensing head whose dimensions are large compared to the spacing of the irregularities. Signals corresponding to gross irregularities and workpiece out of roundness can be eliminated through use of a high-pass filter; a high-pass filter with a corner frequency of approximately 10 Hz will substantially attenuate out of roundness signals for a wide range of practical spindle speeds. By removing both surface ripple and out of

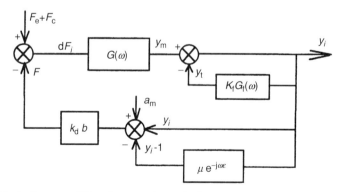

FIGURE 12.28 Block diagram of the chase mode control system.

roundness signals from the control signal, the controller will respond only to workpiece motion. The stability criterion for the controlled system is given by Equation (12.57). Comparison of this stability criterion to that of the uncontrolled process indicates that the increase in stability achieved is almost proportional to the increase of the controller gain, as would expected for an ideal controller. The case of a nonideal controller is discussed in Ref. [49].

The basic advantage of chase mode control is that it is capable of attenuating noise and improving stability at the same time. However, the instrumentation required may present problems. A proximity gage, used to measure tool to workpiece relative motion, may interfere with the cutting system (chips and tooling) or become contaminated by cutting fluids and chips. Additionally, a proximity gage actually measures the relative distance with respect to the workpiece surface rather than the actual tool to workpiece centerline distance; signal components due to surface irregularities, unless eliminated by filtering, will cause the controlled tool to follow the irregular surface rather than maintain a constant position with respect to the workpiece centerline. In general, the lower the order of the controller, the more effective filtering it will provide. In the case of a first-order controller, the corner frequency needs to be no higher than the frequency of the machine structural mode responding to the noise.

A second active control system was built by Nachtigal and Cook [148]. The block diagram of this feed-forward control scheme, which is called *predictive control*, is shown in Figure 12.29. The cutting noise F_c is the monitored parameter. The principal advantage of a predictive mode control system is the relative simplicity of the instrumentation, since cutting force gaging is not physically dependent on the geometry of the workpiece. However, there are two basic disadvantages to this scheme: (a) it is not effective for controlling the external noise force F_e, and (b) in order to reduce the effects of the cutting noise F_c, the controller must accurately simulate the dynamics of the machine–workpiece structure. This is difficult because the structure will generally have several prominent vibration modes and, moreover, its dynamic characteristics change as a function of the position of the moving components of the machine tool. A predictive control scheme is particularly sensitive to changes in the dynamic characteristics of the machine tool because it is not based on a feedback loop.

The chase control scheme was reported to increase stability by a factor of 35, while the predictive control scheme was reported to increase it by a factor of 15. In the chase control scheme, stability basically depends on the dynamic order of the controller and, at least theoretically, a factor of 1000 increase is easily attainable. In contrast, the predictive control scheme cannot achieve an increase in stability of more than a factor 20 even if an adaptive controller is used. Another advantage of the chase control scheme over the predictive control scheme is its effectiveness in attenuating both F_c and F_e noise effects.

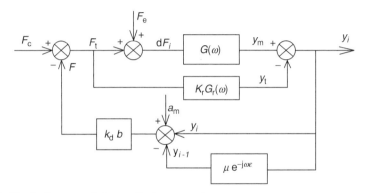

FIGURE 12.29 Block diagram of the predictive control system.

An advantage of feed-forward (predictive) control schemes is that the tool motion begins essentially at the same time as the workpiece motion. Thus, theoretically, one should be able to reduce relative motion to zero. However, since it is an open loop method, the tool moves without prior knowledge of workpiece motion, introducing an additional source of uncertainty. A disadvantage of chase control schemes is that they require high controller gain in order to perform well, otherwise the control signal always lags the noise signal; the use of a high gain factor may cause stability problems in the control loop itself. The stability condition requires that $K_r G_r(\omega) < G(\omega)$.

Second generation active machine tool control systems were developed by Nachtigal and coworkers [149–151]. The primary problem they confronted in all cases was the sensitivity of the controller to machine dynamic variations. In one attempt to overcome this limitation, the wideband active control scheme [150] was designed; it was actually the same as Nachtigal's original system but incorporated an additional lead compensator. The maximum increases in stability for this control scheme were less than those reported for the original system due to the effects of the lead function. Another attempt was based on the fact that changing dynamics can be efficiently accounted for using an adaptive control scheme [151]. This system adaptively varied the controller natural frequency to maintain a predetermined phase relationship between the cutting force and tool servo responses. This feature made the adaptive control scheme capable of maintaining a greater machining rate even with relatively large variations in the machine tool natural frequency. However, the increase in stability was not as large as in Nachtigal's original system, although it was greater than that of the wideband controller scheme.

Third generation active control schemes were multivariable control schemes designed by Mitchell and Harrison [152]. They used the best features of each type of control scheme, and, simultaneously, eliminated some of their drawbacks. The predictive controller was an open-loop, feed-forward system, whereas the chase controller was a closed-loop, feedback system; the multivariable controller was basically a closed-loop, feed-forward system. A block diagram of the control system, which employs a hardware observer, is shown in Figure 12.30. Observer theory was applied so that the active control scheme could effectively reduce chatter as well as attenuate noise effects. The state estimator or observer can be effective because it

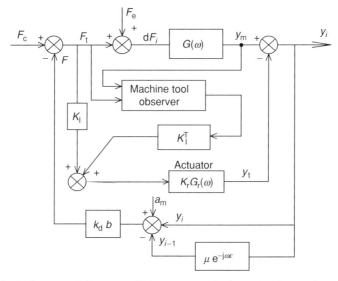

FIGURE 12.30 Block diagram of the controlled system employing a hardware observer.

is not easy to determine the machine tool dynamic characteristics, which depend on the positions of movable elements, and because it is difficult to measure the necessary states of the rotating workpiece, such as workpiece center line position, velocity, or acceleration. An observer is actually a closed-loop system itself which uses the same input signal as the observed system and compares its output with the observed system output. The observer tends to force the tool to follow the workpiece motion.

The actuator–observer control scheme takes advantage of the predictive properties of feed forward systems. Additionally, since it is a closed-loop system, it tends to be insensitive to parameter variations. This scheme requires an actuator faster than the mode it has to control. The predictive type control requires an actuator of about the same speed as the controllable mode, whereas the chase controller can be slower than the mode it is controlling.

A number of other techniques have been also proposed with varying degrees of success. One common approach employs active suppression of chatter by programmed variation of the spindle speed [153–159]. Jemielniak and Widota [160] claimed, based on a simplified analysis of dynamic orthogonal cutting in which the time variation of the dynamic system within each spindle revolution was neglected, that the speed must be continuously varied in order to prevent the dynamic cutting process from locking itself into the most favorable phase for chatter. The continuously varying spindle speed damps the chatter vibration amplitudes, which in turn varies the phase shift between the inner and outer modulations on the cut surface [156]. Speed variations during machining disturb the regeneration of waviness responsible for chatter. Sinusoidal speed variation was found to be the most favorable form of oscillation for the spindle drive system for improving chatter resistance in milling [157]. Methods employing speed oscillation require a high-performance, high-torque delivery spindle drive system which can deliver speed oscillations with a wide range of amplitudes and frequencies over a short time interval.

Altintas and Chan [146] developed an in-process chatter detection system which used the measured sound spectrum as the control variable and suppressed chatter through spindle speed variation. A remotely positioned microphone was used to measure the sound pressure during end milling. The in-process chatter detection and avoidance algorithm is outlined in Figure 12.31. The machine operator inputs the chatter threshold factor for the sound spectrum, the nominal speed, and the amplitude and frequency of speed variation. The sound signals are sampled at a frequency at least five times higher than the possible chatter frequency. The search is carried out within a possible chatter frequency range, and precautions

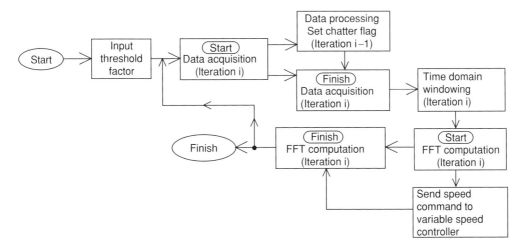

FIGURE 12.31 Chatter avoidance algorithm. (After Y. Altintas and P.K. Chan [146].)

are taken to avoid the tooth passing frequency harmonics. In addition, the average magnitude in a specified low-frequency (chatter free) range is computed. Chatter is assumed to be present when the maximum magnitude exceeds the low-frequency spectrum average by a factor greater than the threshold factor. If chatter is detected, speed variation signals are sent to the spindle speed controller; otherwise a constant speed signal is sent. The spindle speed is varied in a sinusoidal manner to disrupt the regenerative effect of chatter [158]. The difficulty of effectively varying the spindle frequency at high frequencies limits this technique's effectiveness.

In contrast to systems which vary the spindle speed during machining, Weck et al. [107] proposed the use of stability lobes (as shown in Figure 12.18) for selecting chatter free spindle speeds. Chatter was detected by comparing the maximum spindle torque with an allowable threshold. The spindle speed was varied slowly until one of the stable lobes was penetrated. Since stability lobes are narrow at low cutting speeds, this approach is probably effective only at higher cutting speeds.

A similar approach equates the tooth passing frequency to the chatter frequency, which is close to the natural frequency of the structure [108]. This approach is used in the CRAC or Harmonizer® system developed by Tlusty and coworkers [161]. This system selects spindle speeds for which favorable phasing prevents chatter [54, 144, 145, 162–164] and allows the axial depth of cut to be increased to maximize the metal removal rate. The Harmonizer system operates in such a way that if a frequency is detected in an audio spectrum which is not the tooth passing frequency or a harmonic of the tooth passing frequency, then a new spindle speed is selected which would make the tooth passing frequency equal to the detected chatter frequency. Furthermore, the system is interfaced with the machine controller which commands a spindle speed change. A few trial speeds may be required before an acceptable speed is found.

The principle of the Harmonizer approach is illustrated in Figure 12.32 and Figure 12.33. The cut using the speed and depth of cut indicated by "a" is unstable because it falls inside the

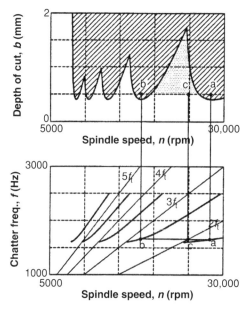

FIGURE 12.32 Stability lobe diagram and chatter frequency diagram showing the basic operation of the CRAC system. (After W.R. Winfough and S. Smith [54].)

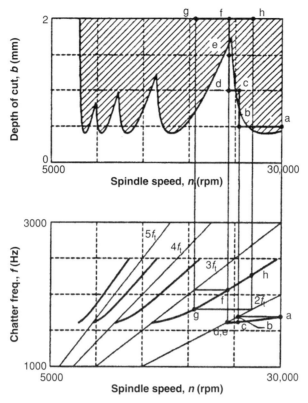

FIGURE 12.33 Stability lobe diagram and chatter frequency diagram showing an algorithm for optimizing the material removal rate. (After W.R. Winfough and S. Smith [54].)

stability lobes shown in Figure 12.32. The corresponding chatter frequency which would be detected is also marked "a" in the chatter frequency diagram. The Harmonizer system regulates the spindle speed and proportionally the feed (to keep the chip load constant) so that the second harmonic of the tooth passing frequency is equal to the detected chatter frequency; this corresponds to a horizontal move from a to c in the figure. Similarly, if the cut "b" was initially selected, the Harmonizer system will move the process to the cut marked c. This is the first function of the Harmonizer system. The second function is to maximize the MRR by increasing the axial depth of cut in the stable zone as indicated in Figure 12.33. The two plots completely describe the spindle speed–chatter frequency relationship typical for most tools. Therefore, the Harmonizer system commands a speed change to the cut marked b and then commands an increase in the axial depth of cut until the cut is unstable as indicated by the cut marked c. At this point another spindle speed change is used to move to the cut marked d, which is stable. The depth of cut is further increased until chatter occurs for the cut is indicated by f. Stable spindle speeds are searched at either sides if f (to g and h) but are not found as expected. Therefore, the conditions of the last stable cut (marked e) are recorded. The MRR at the conditions of cut e (obtained by reducing the spindle speed by 20% while increasing the depth of cut by 20%) is greater than that using conditions of cut a or b. There are several obstacles to overcome in order to ensure effective and reliable audio detection. Audio detection has been found very effective when a frequency-based approach is used instead of a typical RMS or other time-based approaches [34, 39].

A final group of chatter suppression schemes employs the *forecasting control* (FC) strategy. In this strategy, the machining process is considered to be unknown, and the onset of chatter is predicted by fitting autoregressive models to suitable forms of the acceleration signal. The models are then used as a basis for forecasting corrective signals [147]; when the process approaches instability, appropriate corrective actions can be taken before the problems actually occur. In this control strategy, the dynamic characteristics of the machining process are determined online. Prior knowledge of the system is helpful but not required. The most significant advantage of using online identification of the process dynamics is that the estimated model can vary with time, so that the time varying characteristics of the machining process are not ignored under this scheme. An ARMA or AR stochastic model is generally used to represent the system [126]. This model is adequate, in a statistical sense, and is justified physically by the fact that, as long as the process is stable, it can be efficiently described by a linear differential equation with constant coefficients. In this case the corresponding discrete-time system is a difference equation in the form of a proper ARMA model [131]. After the process dynamics have been determined some quantity which is important for the stability of the process is computed. The values of this quantity create a new time history; a new ARMA or AR model is used to describe the updated time series. Using the forecasting idea, as explained in Ref. [131], the values of this series are forecast and tendencies toward instability are detected. If such tendencies are present, corrective actions are taken to ensure stability.

A block diagram representation of the FC strategy is shown in Figure 12.34. It consists of five main functional blocks: the machining process, process identification, forecasting and parameter updating, decision and modification, and drive blocks. The decision and modification block decides if there is any need to issue new speed and feed commands and, if corrective actions are necessary, what their magnitudes should be. The FC strategy has been applied successfully in turning operations [165, 166]. A nonlinear mathematical model for chatter has been also proposed in Ref. [167].

As an example, the FC strategy was applied to a vibration signal measured on the tailstock of the lathe measured using an accelerometer. In one case, an AR(2) model was used as a useful approximation of the machining dynamics [166]. The acceleration signal filtered using a band pass filter. The quality used to assess the stability of the machining process was the damping ratio ζ, computed from the parameters of the AR(2) model. The order of the model and the initial values of the autoregressive parameters were determined offline. The parameters of the AR(2) model were not constant and were updated using a

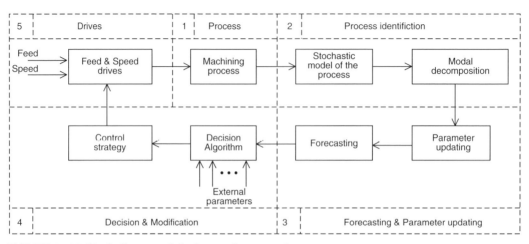

FIGURE 12.34 Block diagram of the forecasting control strategy.

recursive procedure known as Durbin's method. In this way, different values of the damping ratio were computed during each cycle and a new time history, consisting of these values, was created. An AR(2) stochastic model was also used to represent this time history:

$$\zeta_t - \phi_1 \cdot \zeta_{t-1} - \phi_2 \cdot \zeta_{t-2} = a_t \qquad (12.89)$$

The autoregressive parameters were updated every time a new value of the damping ratio was known. For this purpose, a sequential updating procedure was used. The FC strategy used the one-step ahead forecast given by

$$\hat{\zeta}_t = \phi_{t,1} \cdot \zeta_t + \phi_{t,2} \cdot \zeta_{t-1} \qquad (12.90)$$

As long as these forecasts had probability limits not including the minimum allowable value of the damping ratio, no corrective action was taken. However, if the probability limits included the value proper corrective actions were taken to alter the speed or feed to ensure stability.

In another case, an AR(6) stochastic model was used to represent the machining process dynamics. The quantity used to determine the stability of the system was not the damping ratio, since this would be very time-consuming with a higher-order model. Instead, the maximum value of the power spectral density (PSD) of the process [165] was used. The use of this parameter was justified by the fact that the PSD generally exhibits a peak as chatter increases. The maximum values of the PSD formed a new time series which was modeled by an AR(2) stochastic model:

$$s_m - \phi_1 \cdot s_{m-1} - \phi_2 \cdot s_{m-2} = a_m \qquad (12.91)$$

Then, according to the FC strategy, a two-step ahead forecast was computed:

$$\hat{s}_{m,t}(2) = \phi_{1,t} \cdot \hat{s}_{m,t}(1) + \phi_{2,t} \cdot s_{m,t} \qquad (12.92)$$

where

$$\hat{s}_{m,t}(1) = \phi_{1,t} \cdot \hat{s}_{m,t} + \phi_{2,t} \cdot s_{m,t-1} \qquad (12.93)$$

The calculated $\hat{s}_{m,t}(2)$ is then compared with some critical value S_{mc} to decide whether chatter is likely to develop or not. This critical value is calculated from the actual data during each cutting process by multiplying the average of the peak values of the PSD by a gain factor (i.e., $S_{mc} = GS_{ma}$, where $G = 1.5$ typically). Corrective action is necessary when $\hat{s}_{m,t}(2)$ is greater than S_{mc}. Further details of this scheme are given in Ref. [165].

12.10 EXAMPLES

Example 12.1 A face milling operation is to be designed with maximum spindle speed of 5000 rpm. Cutters with diameter 50 to 100 mm can be used and the number of inserts varies between four and eight (i.e., the 50 mm has five inserts, 75 mm has five or six inserts, and the 100 mm has seven to eight inserts). It is assumed that self-excited chatter is not present. As mentioned in Section 12.5, the forcing frequency of a linear forced vibration in a machining process can be changed by the operator through the machining parameters, which can be used to improve the design of a process. The characteristics of the part were measured (using the modal or "tap" test as explained in Section 12.7) and the frequency-response function (FRF) is given in Figure 12.35. Assume that the cutting tool is much stiffer than the part.

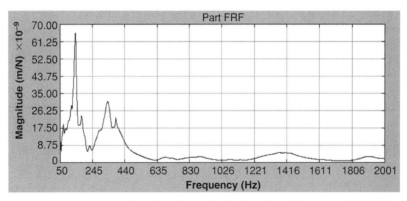

FIGURE 12.35 Frequency response function of part in X-direction. (Courtesy of D3 Vibrations, Inc.)

A. Select the rpm that will place the fundamental forcing frequency (FFF) at 245 Hz.
B. Determine the harmonics of the FFF.
C. For quality reasons, would you recommend using the maximum spindle rpm or the rpm found in part A for the various cutters with different number of teeth?

Solution:

A. The spindle rpm can be selected using Equation (12.40) so that the FFF is 245 Hz. The FFF of 245 Hz was selected to lie between the two peaks (modes) at $f_1 = 140$ Hz and $f_2 = 340$ Hz in the FRF to avoid excessive vibration. The corresponding rpm for the different number of teeth in the cutter is given in Table 12.2.

B. The first five harmonics of the FFF, obtained by multiplying the FFF by 2, 3, 4, 5, and 6, are given in Table 12.3.

C. The tooth passing frequencies at 5000 rpm for the various number of teeth, computed from Equation (12.40), are given in Table 12.4. Running the milling cutters with four and five teeth at this speed is not recommended because the corresponding tooth passing frequencies are near the second natural frequency of 340 Hz as shown in Figure 12.35. Running the cutters with six, seven, or eight teeth at this speed is recommended because the tooth passing frequencies are far from the second natural frequency.

TABLE 12.2
Tooling rpm for Cutters With Different Number of Teeth at an FFF of 245 Hz

Number of Teeth	rpm for 245 Hz
4	3675
5	2940
6	2450
7	2100
8	1838

TABLE 12.3
First Six Harmonics (in Hz) of the FFF in Example 12.1

1st harmonic	490
2nd harmonic	735
3rd harmonic	980
4th harmonic	1225
5th harmonic	1470
6th harmonic	1715

TABLE 12.4
FFF for Tools with Different Number of Teeth in Example 12.1

Number of Teeth	FFF at 5000 rpm	Recommend
4	333	No
5	417	No
6	500	Yes
7	583	Yes
8	667	Yes

Example 12.2 The characteristics of a part were measured using a modal test; and the FRF is given in Figure 12.36. A combination tool consisting of a two-flute twist drill and a spotfacer was used in a machining center. A spindle speed which is a multiple of 500 rpm between 4,000 and 10,000 rpm must be selected. Assume self-excited vibration is not present.

A. Select the rpm range where the FFF, first, and second harmonic fall at frequencies at which the compliance is below 375×10^{-9} m/N on the FRF.
B. If the force magnitude at the FFF is 100 N, at first harmonic is 40 N, and at the second harmonic is 15 N, estimate the displacement at each frequency and sum them for the following spindle speeds: 4000, 6000, and 8000 rpm.

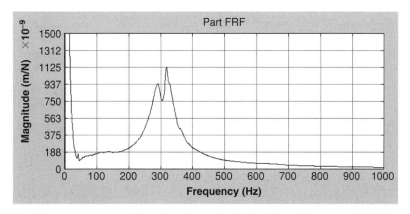

FIGURE 12.36 Frequency response function of part in X-direction. (Courtesy of D3 Vibrations, Inc.)

Solution:

A. The tooth passing frequencies or FFF, calculated from Equation (12.40) for a range of spindle rpms at 500 rpm increments, are given in Table 12.5. The corresponding first and second harmonic frequencies for the FFF are also given in the table. The objective is to evaluate and reject the cutting speeds generating FFF and harmonics frequencies within the bell curve of the natural frequency with a magnitude greater than 375×10^{-9} m/N in the FRF shown in Figure 12.36. The frequencies to avoid are between 240 and 375 Hz. Any rpm between 6000 and 7000 rpm will reduce the amount of forced vibration. Therefore, a spindle speed between 6000 and 7000 rpm should be used.

B. The displacement at the three cutting speeds is estimated for each frequency (FFF, first harmonic, second harmonic) as illustrated in Table 12.6 for the three spindle speeds of 4000, 6000, and 8000 rpm. For example, the three tooth passing frequencies at 4000 rpm are FFF = 133 Hz, first harmonic = 267 Hz, and second harmonic = 400 Hz, and the corresponding compliances are estimated from Figure 12.36 and given in the table. The displacements are calculated based on the applied force and the compliance at a particular frequency: 0.00019 mm/N × 100 N = 0.019 mm, 0.00055 mm/N × 40 N = 0.022 mm, and 0.0002 mm/N × 15 N = 0.003 mm, respectively, at FFF, first harmonic, and second harmonic. The total displacement of the cutter at 4000 rpm is the sum of the three above displacements, or 0.044 mm. The displacement at 6000 rpm is the lowest, 0.033 mm, as expected from the analysis in part A. Since the phasing of the frequencies is not considered, the sum value may be different than estimated; however, this can still be a useful qualitative analysis. A time domain plot for the 6000 rpm displacement is shown in Figure 12.37 for the FFF, first harmonic, second harmonic, and total.

Example 12.3 The FRF for 17 points in the Z-direction on the pump face of a transmission valve body structure was measured using a modal "tap" test. Compliance in the Z-direction controls the flatness of a critical surface on this part. The FRF is given in Figure 12.38. Select

TABLE 12.5
Calculation of the FFF and the 1st and 2nd Harmonics of the Tooling System in Example 12.2

Spindle rpm	FFF (Hz)	1st Harmonic (Hz)	2nd Harmonic (Hz)
4,000	133	267	400
4,500	150	300	450
5,000	167	333	500
5,500	183	367	550
6,000	200	400	600
6,500	217	433	650
7,000	233	467	700
7,500	250	500	750
8,000	267	533	800
8,500	283	567	850
9,000	300	600	900
9,500	317	633	950
10,000	333	667	1,000

Machining Dynamics 673

TABLE 12.6
Calculation of the Displacement FFF at Various rpms for the Tooling System in Figure 12.36

	Frequency (Hz)							Estimated Total Displacement (μm)
	133	200	267	400	533	600	800	
Compliance (m/N × 10⁻⁹)	190	230	560	230	90	40	20	
Displacement (μm) at 4000 rpm	19		22	3				44
Displacement (μm) at 6000 rpm		23		9		1		33
Displacement (μm) at 8000 rpm			56		4		0	60

a spindle speed (rpm) for a 5-tooth milling cutter so the FFF and first two harmonics do not fall on a resonant peak. The wall thickness of the pump face is narrow, so that only one tooth is generally engaged (i.e., this is a low immersion cut).

Solution: The high compliance frequencies are 585, 707, 1022, 1420, 1595, 2325, and 2968 Hz as shown in Figure 12.38. Harmonics are caused by the tooth impacts, and are found at integer multiples of the FFF. For example, the 585 Hz FFF for the surface has harmonics at 1170, 1755 Hz, etc. as shown in Figure 12.39. When trying to reduce forced vibrations, it is most important to have FFF and lower harmonics away from high-response areas. The

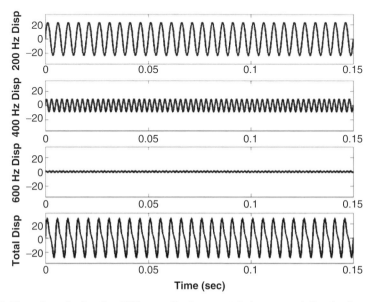

FIGURE 12.37 Time domain for the 6000 rpm displacement (micrometers) for the frequency response function of the part in Figure 12.36. (Courtesy of D3 Vibrations, Inc.)

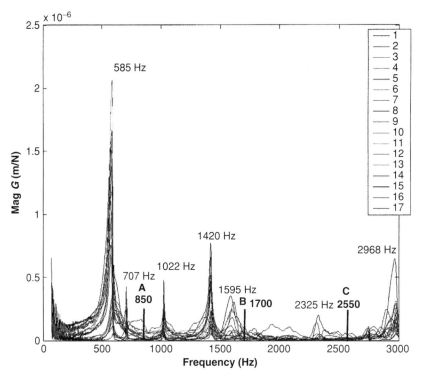

FIGURE 12.38 Frequency response function of the pump face of a transmission valve body.

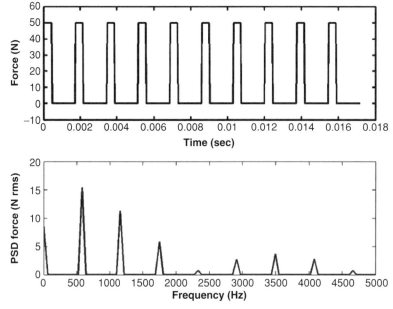

FIGURE 12.39 Simplified fundamental forcing frequency of the face milling cutter force at 7000 rpm with five teeth used for the pump face of a transmission valve body.

spindle speeds for the 5-tooth cutter having tooth passing frequencies at or near the high compliant frequencies of the surface can be identified using Equation (12.40). For example, 7000 rpm will result in an FFF of 583 Hz for the tool (7000 rpm × 5/60 = 583 Hz). If the FFF is 850 Hz, the first and second harmonics are 1700 and 2550 Hz, respectively. These three frequencies are plotted on the FRF plot in Figure 12.38 (illustrated by the marks A, B, and C, respectively). This figure shows that the force vibration will be avoided at these frequencies because they do not fall near FRF peaks. The spindle speed corresponding to these frequencies is 10,200 rpm.

Example 12.4 A face milling cutter with 30 inserts running at 1997 rpm introduces vibration into a machine tool. A microphone was used to measure time- and frequency-domain signals as shown in Figure 12.40. How can the problem be corrected?

Solution: The audio signal in the figure indicates that the third harmonic near 4000 Hz must be near a resonant frequency, leading to excessive forced vibration. The displacement at these frequencies is much higher than at the FFF, first harmonic, and second harmonic, although there is a higher force at 1000, 2000, and 3000 Hz. Note that the harmonic at 4000 Hz is four times the tooth impact (or FFF) of 998.5 Hz. Under normal vibration the conditions, the higher harmonic would be a small fraction in amplitude compared to the tooth impact frequency. Hence, it most likely this milling cut exhibits resonance at 3994 Hz. This indicates that the cutting speed (rpm) should be changed so that no harmonic falls near 4000 Hz. It is suggested to increase or reduce the spindle speed by between 5 and 20% [34] as long as instability does not develop at the new spindle speed. Therefore, if the spindle speed is decreased by 10% to 1800 rpm, the resonance vibration at this mode will be eliminated. However, as discussed in Section 12.5, it is sometimes difficult to identify the best speed if the system has several modes close to each other, since a spindle change may move another nearby harmonic into resonance.

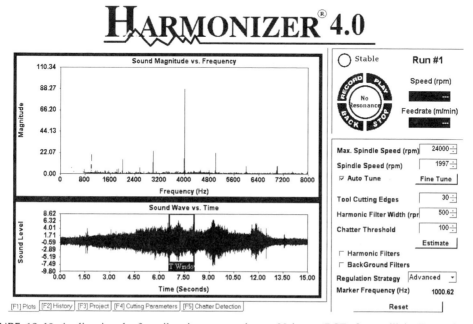

FIGURE 12.40 Audio signal of a vibration case using a 30-insert PCD face mill in Example 12.4. (Harmonizer, Courtesy of Manufacturing Laboratories, Inc.)

Example 12.5 A full slot is milled in aluminum 7075 using a two-flute end mill. The diameter of the end mill is 20 mm and its length extending from the toolholder is 140 mm ($L/D = 7$). The maximum spindle speed of the CNC machine is 12,000 rpm. A stability chart for the tool mounted in the spindle is shown in Figure 12.41 for a full slot.

  A. Select the rpm that provides the highest stable axial depth of cut (ADOC). What is this maximum depth of cut?
  B. What rpm would you select if the engineering process required you to have a 500 rpm cushion on either side of your rpm, and a depth of cut a minimum of 0.5 mm below the stability line?
  C. If the spindle speed could be increased to 15,000 rpm maximum, could improvements be made in the process? What improvements?
  D. What would happen if the tool had three flutes?

Solution: This example shows how stability charts can be used to design a milling process. The stability chat of the cutter for full-immersion slotting is shown in Figure 12.41. It was measured using a modal test on the end mill clamped in the spindle through a toolholder. The modal parameters of the end mill in the spindle converted to the displacement domain were identified using modal analysis software and are given in Table 12.7. The parameters are identical in both *x*- and *y*-directions because the FRFs in these directions are similar. This stability chart provides the minimum stable depth of cut as a function of spindle rpm. The shaded section inside the lobes is the unstable region. Note that there are hardly any stable pockets at speeds below 4000 rpm.

A. From the chart, the minimum stable depth of cut is approximately 1.2 mm. The most stable spindle speed below 12,000 rpm is approximately 9,750 rpm, where the stable depth of cut reaches a maximum of approximately 5 mm. This maximum depth of cut will result in the maximum material removal rate without chatter.

B. The same stability pocket should be used, but the stable ADOC must be reduced so that the required conditions of a 500 rpm cushion on either side of the selected rpm, and an ADOC a minimum of 0.5 mm below the stability line are met. These conditions are met using a spindle speed of 9250 rpm and an ADOC of 2.5 mm.

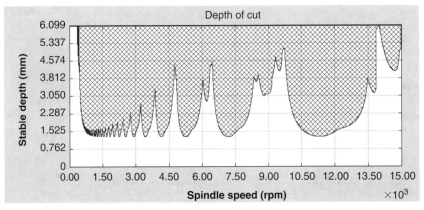

FIGURE 12.41 Stability chart for a two-flute ($L/D = 7$) end milling a full slot in aluminum part in Example 12.5. (Courtesy of D3 Vibrations, Inc.)

TABLE 12.7
Identified Modal Parameters of the End Mill With the Spindle in Either X- or y-Direction

Mode	f (Hz)	K (N/m)	ζ
1	424	1.26×10^7	0.0205
2	492	2.36×10^7	0.0268
3	641	2.38×10^7	0.0233
4	882	8.33×10^7	0.0188

C. The ideal stability pocket for spindle speed lies at about 14,000 rpm, with a stable ADOC of 6.1 mm. A 15,000 rpm maximum spindle speed would permit use of this more robust region, and provide a throughput improvement because both the speed and ADOC are higher than those discussed in part B. Selecting 2.5 mm ADOC at 15,000 rpm provides a 62% productivity gain and should yield similar quality; moreover, the process would be more robust because the width of the pocket between lobes is wider.

D. Increasing the number of flutes moves the stability lobes down and left in Figure 12.42. This is seen by comparing the stability chart for the three-flute cutter, shown in Figure 12.42, to that of the two-flute cutter in Figure 12.41. The achievable ADOC is not as high for the three-flute end milling tool in this instance. There is a large robust region between 9,000 and 14,000 rpm; however, the ADOC is not as great as the two-flute. In this case, even if feed/tooth is held constant, the two-flute cutter would yield a higher MRR.

Example 12.6 A full slot is milled in aluminum 7075 using a two-flute end mill. The diameter of the end mill is 9.52 mm and its length extending from the toolholder is 105 mm ($L/D = 11$). The stability chart for the tool mounted in the spindle is shown in the top graph in Figure 12.43. The bottom graph in the figure shows the estimated chatter frequency at the chosen rpm. This helps distinguish between self-excited vibration (chatter) and the forced vibrations discussed in Example 12.1 and Example 12.2.

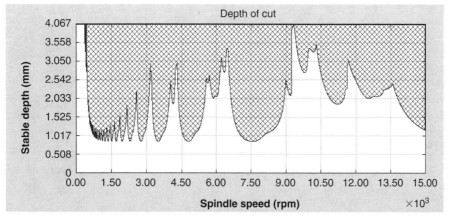

FIGURE 12.42 Stability chart for a three-flute ($L/D = 7$) end milling a full slot in aluminum part in Example 12.5. (Courtesy of D3 Vibrations, Inc.)

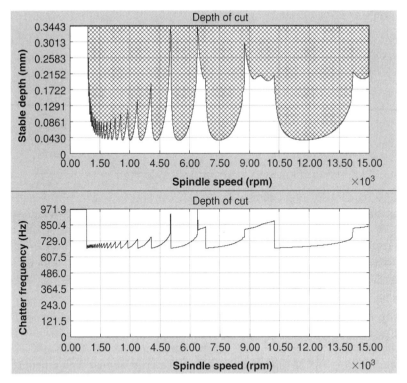

FIGURE 12.43 Stability chart and the corresponding chatter frequency charts for a two-flute ($L/D =$ 11) end milling a full slot in aluminum workpiece in Example 12.6. (Courtesy of D3 Vibrations, Inc.)

A. At what frequency will the tool chatter if it is rotating at 7500 rpm with an ADOC of 0.12 mm?
B. What is the fundamental forcing (tooth passing) frequency (FFF)?
C. What is the ratio of chatter frequency to the FFF?

Solution: The chatter frequency diagram is generated from the solution of an SDOF system using Equation (12.57), which has three unknowns: (b, ω, ε); ω is the tooth pass frequency, which is related to the spindle rpm [$\omega = 60/(n_t \cdot \text{rpm})$]. In Figure 12.43, b is plotted versus rpm in the top graph, and the chatter frequency is plotted versus rpm in the bottom graph.

A. The spindle speed is 7500 rpm and the ADOC is 0.12 mm. Since these cutting conditions lie in the shaded area in the lobe diagram, it is clear that the tool will chatter. The chatter frequency is estimated from the chatter frequency graph to be approximately 670 Hz. In addition, vibration analysis was performed in this case using an audio analysis (as explained in Section 12.9) and yielded a chatter frequency of 662 Hz while cutting as shown in Figure 12.44. This measurement agrees with the stability chart estimate.

B. The FFF of the two-flute cutter at 7500 rpm, calculated from Equation (12.40) ($f_n = 2 \times 7500/60$), is 250 Hz.

C. The ratio of chatter frequency to the FFF is $670/250 = 2.68$. This ratio is a noninteger multiple of the FFF, which indicates that the vibration is self-excited. A forced vibration would be an integer multiple (i.e., the FFF or one of its harmonics).

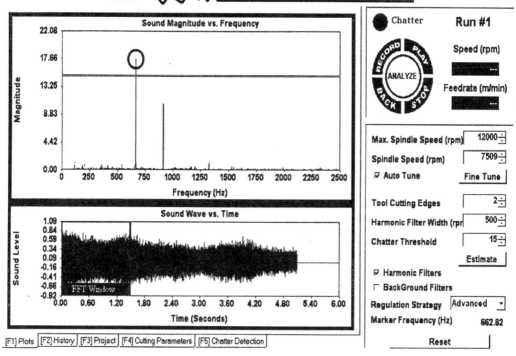

FIGURE 12.44 Audio signal of a two-flute ($L/D = 11$) end milling of aluminum at 7500 rpm and 0.125 mm axial depth of cut showing a chatter frequency of 662 Hz in Example 12.6. (Courtesy of Manufacturing Laboratories, Inc.)

Example 12.7 The part and cutting process described in Example 12.3 is used in this example to illustrate how a knowledge of the material removal rate (MRR) based on the stability lobe diagram can be used to select cutting conditions. The MRR chart given in Figure 12.45 was generated based on the stability chart in Figure 12.41 using Equation (2.29) and assumes constant feed per tooth of 0.125 mm/tooth for the 20 mm diameter cutter. The stable ADOC at each spindle speed in the MRR calculation was obtained from Figure 12.41.

A. Determine the maximum MRR if the following maximum speeds are allowed: (a) 10,000 rpm, (b) 12,000 rpm, and (c) 15,000 rpm.
B. Determine the time required to remove 1000 cm³ of material for the three conditions above.
C. Compute the time improvement for a speed decrease from 12,000 to 10,000 rpm based on the stability chart. Note that the ADOC is about 1.5 mm from Figure 12.41.
D. If the process only required a 1.0 mm ADOC, what stable rpm is the most productive given a 12,000 rpm maximum spindle speed?

Solution:

A. The maximum MRR for a given speed is estimated from the MRR stability chart by finding the rpm intercept with the stability boundary. The maximum MRRs for 10,000,

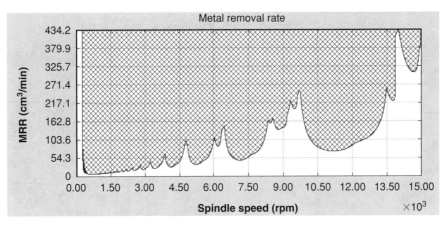

FIGURE 12.45 Material removal rate (MRR) stability chart for the 20 mm diameter two-flute ($L/D = 7$) end milling of aluminum at 0.125 mm/tooth based on the stability chart in Figure 12.41 in Example 12.7. (Courtesy of D3 Vibrations, Inc.)

12,000, and 15,000 rpm are 110, 90, and 380 cm³/min, respectively. Note that a higher spindle speed is not always better, as the maximum MRR is not obtained at the maximum spindle rpm in this case.

B. The time requires to remove 1000 cm³ of material is calculated as $t_m = \text{Vol/MRR}$. The machining times for the 10,000, 12,000, and 15,000 rpm are 9, 11.1, and 2.6 min, respectively.

C. The MRR at 12,000 rpm is approximately 90 cm³/min at the stability boundary; therefore, it takes 11.1 min to manufacture the part. By choosing a lower speed of 10,000 rpm where ADOC is larger and MRR is 110 cm³/min (greater than that at the higher speed of 12,000 rpm), the cycle time reduces by 2.1 min or a 19% gain.

D. A spindle speed of 12,000 rpm is the most productive since it is the machine maximum and a 1.0 mm ADOC is below the stability limit at 12,000 rpm.

Example 12.8 A face milling operation is performed on a cast aluminum workpiece with $K_c = 2.8 \times 10^8$ N/m². A 125 mm diameter milling cutter with PCD inserts is used. The cutting parameters are: 4 mm ADOC, 7000 rpm maximum spindle speed, 40 KW maximum spindle power, 0.1 to 0.3 mm/tooth suggested feed range for the end mill, and 8 sec maximum allowable cycle time. The surface to be milled is 600 mm long and 100 mm wide. The feed distance for the cutter is 800 mm. The dynamic characteristics of both the part and the tooling were evaluated and the face mill was found to be stiffer than part in this case. The part dynamic characteristics are $f_n = 763$ Hz, $k_d = 2.2 \times 10^7$ N/m, and $\zeta = 0.01$ based on an SDOF system.

A. Analyze the dynamics of the system by plotting the transfer function and the stability chart using the dynamic characteristics of the weakest component of the machine–tool–workpiece system.
B. Determine the following cutting parameters: rpm, number of teeth for the face milling cutter, feed, and actual cycle time. (Note: The 125 mm face milling cutter is available with 6, 8, 10, and 12 inserts.)

Solution: When determining stability for high-speed applications, only the rpm, number of teeth in the cutter, number of teeth engaged in the workpiece, and ADOC and radial depth of cut (RDOC) are significant factors in regenerative chatter. The product of the ADOC and RDOC affects stability.

A. The workpiece is represented by a single degree of freedom system since a single natural frequency was measured in the modal test. The system dynamic characteristics are $f_n = 763$ Hz, $k_d = 2.2 \times 10^7$ N/m, and $\zeta = 0.01$ or 1.0% as given above. The real and imaginary parts of the transfer function and corresponding stability diagram are plotted on top in Figure 12.46. The real and the imaginary parts of the transfer function were generated using Equation (12.13). The stability lobes we generated for different numbers of teeth in the

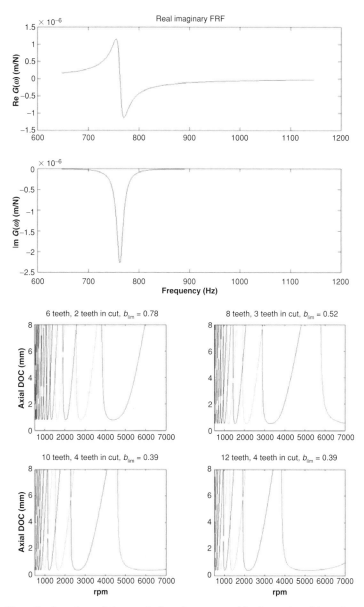

FIGURE 12.46 Transfer function of the workpiece (represented by its real and imaginary parts) and the stability charts a cutter with different number of teeth.

125 mm face milling cutter using Equations (12.62) and (12.69) and the procedure discussed in Section 12.6. Since b_{lim} is not high enough on any of the tooth designs, the chosen rpm is important. The procedure requires the calculation of the spindle rpm from Equation (12.71) for each stability lobe $n_p = 1, 2, 3, \ldots$ for a range of frequencies around the dominant mode in the transfer function ($f_n = 763$ Hz).

B. The maximum robust rpms for each tooth design were chosen to lie within the middle of the wider stability pocket between lobes and are summarized in Table 12.8. Since the workpiece is aluminum and diamond inserts are used in the face milling cutter, the rpm is not a limiting factor in this case. The torque (= force × cutter radius), and power (= force × velocity) could be limiting factors. Assuming middle range for the feed of 0.2 mm/tooth, the cutting force is estimated either from simulation (as explained in Chapter 8) or from the cutting power (as explained in Chapter 2). Cutting force results based on simulation are provided in Table 12.9 for the four teeth designs. The force per cutting edge is the same for all cutters since the geometry of the cutters (insert type and geometry, axial and radial rake angles, lead angle, etc.) is the same and only the number of teeth is different. A worn tool factor is included in the force estimates. The torque and power required by the spindle were calculated as explained in Chapter 6 for the different number of teeth and are given in Table 12.9. The power for the 10- and 12-tooth cutters was greater than 40 KW, so the cutting speed was dropped from 7000 to 6000 rpm in these cases. The feed rate, cycle time, and MRR were calculated using the equations from Chapter 2; the results are given in Table 12.10. It is assumed that part quality and tool life are adequate for any number of teeth. Since the tools with 6, 8, 10, and 12 inserts meet the cycle time requirement of 8 sec, and the part quality is sufficient, the problem is solved. If the production system is a transfer line where the machines are in serial, the feed rate is often decreased (as long as tool life is not affected) to reduce cutting forces while maximizing the cycle time, thus improving part quality and tool

TABLE 12.8
Selected Cutting Speeds for the Corresponding Number of Teeth for the Face Milling Cutter in Example 12.8

Number of Teeth	6	8	10	12
rpm	7000	5200	6000	6000
Angular velocity (rad/sec)	733	544.5	628.3	628.3
Radial velocity (m/min)	2749	2042	2356	2356

TABLE 12.9
Selected Feed and Estimated Torque and Power for the Cutter with Four Different Teeth Designs Defined in Table 12.8 for Example 12.8

Number of Teeth	6	8	10	12
Force/tooth (N)	224	224	224	224
Force/cutter (N)	448	672	896	896
Torque (N m)	28	42	56	56
Power (kW)	20.5	22.9	35.2	35.2

TABLE 12.10
Calculated Feed Rate, Cycle Time, and MRR for the Cutter with Four Different Teeth Designs Defined in Table 12.8 for Example 12.8

Number of Teeth	6	8	10	12
Feed (mm/min)	8,400	8,320	12,000	14,400
Cycle time (sec)	5.7	5.8	4.0	3.3
MRR (mm^3/min)	663,600	440,960	408,000	489,600

life. The feed and rpm can also be reduced together to improve tool life; however, it is important to check stability before altering the rpm.

Example 12.9 This example considers the solution of an intermittent chatter problem observed during a semifinish boring operation. A 90 mm diameter bore was machined with a boring bar with three inserts. The spindle speed was 3600 rpm and the feed rate was 1060 mm/min. The spindle speed of 3600 rpm is equivalent to a 60 Hz exciting disturbance; a three-inserted boring bar operating at this speed excites the structure at a frequency of 180 Hz. Intermittent chatter was observed which in some instances produced chatter marks (deflections) up to 1.27 mm in amplitude, even though the average depth of cut for the operation was roughly 0.380 mm. In many cases the chatter diminished and died out as the tool advanced, but in some cases it grew in amplitude, resulting in spindle failure.

Several process changes were implemented in an attempt to eliminate the chatter. These included: (1) using only one insert, rather than three, in the boring bar, and (2) changing the design of the front spindle bearings. The original spindle had a set of two 70 mm angular contact bearings behind a double row of roller bearings. In the modified design, a set of two 25° angular contact bearings with larger balls was substituted for the original double row of angular bearings with smaller balls. After these modifications were made, the system never chattered or failed. However, the modifications resulted in an unacceptably low production rate. A study was undertaken to identify the underlying cause of chatter so that further modifications could be made to increase the production rate without inducing chatter.

The stiffness of each spindle was measured off-line by mounting on a bench and applying static and dynamic loads at the front of the spindle shaft and at the end of the boring bar. In addition, the stiffness of the boring bar was measured when mounted on the production machine. All the measurements were made while the spindle was not rotating. Both the static and dynamic measurements were made using a noncontact gage and a hammer. An impulse force was applied either at the front of the spindle shaft or at the end of the boring bar; a capacitance probe located opposite the excitation location was used to measure the resulting deflection. The frequency response of the system was analyzed to characterize the system dynamics and identify the source of the chatter.

Typical dynamic responses of the modified spindle on the bench and on the machine are shown in Figure 12.47 and Figure 12.48. The results of the static spindle stiffness test are shown in Figure 12.49. The stiffness of the spindle with the tool is reduced by roughly 60 to 75% compared to the stiffness of the spindle itself. The natural frequencies for both spindles are shown in Figure 12.50. The natural frequency of the spindle with the tool on the bench was slightly higher than the frequency measured in the machine. As expected, the chatter

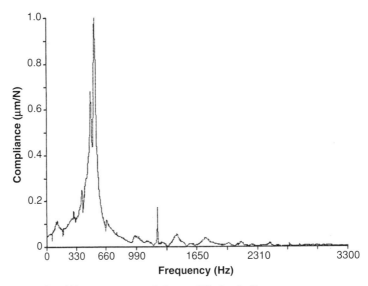

FIGURE 12.47 Dynamic stiffness response of the modified spindle.

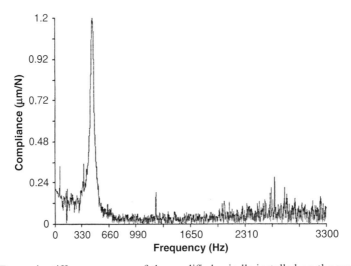

FIGURE 12.48 Dynamic stiffness response of the modified spindle installed on the machine.

frequency (about 480 Hz) coincides with the natural frequency of the spindle with the tool in the machine station, which was between 470 and 500 Hz.

These results show that the original spindle is significantly stiffer than the modified spindle. The dynamic stiffness results shown in Figure 12.51 indicate that the original spindle had minimum natural frequencies of 1000 and 1570 Hz, compared to 800 Hz for the modified spindle. This result might have been anticipated from the bearing designs of the two spindles. The bearings in the original spindle had smaller balls than the bearings in the modified spindle; the smaller balls should have provided higher stiffness and increased heat generation. The static stiffness results indicated that changing the spindle design did not address the root cause of the chatter.

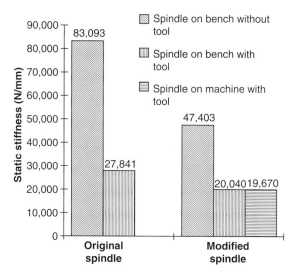

FIGURE 12.49 Comparison of the static stiffnesses of the original and modified spindles with boring tools on the bench and the modified spindle with tooling installed on the machine.

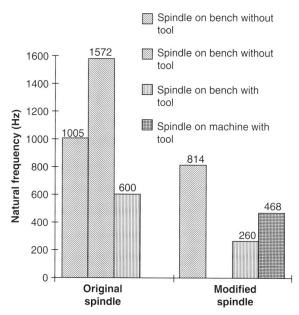

FIGURE 12.50 Comparison of the natural frequencies of the original and modified spindles with boring tools on the bench and the modified spindle with tooling installed on the machine.

Dynamic stiffness was characterized by constructing a stability lobe diagram as shown in Figure 12.52. The stability lobes for the original and modified spindles with tooling are shown in Figure 12.53 and Figure 12.54, respectively, for the bench tests and in Figure 12.55 for the modified spindle on the machine. Chatter vibration can be avoided by selecting the width of cut to be below the lobes or well between lobes farther to the right on the graph. The axial cumulative width of cut for one insert is located below the lobes in both figures, while that for three inserts is much closer to the lobes. The process may cross the limiting lobe curves (i.e.,

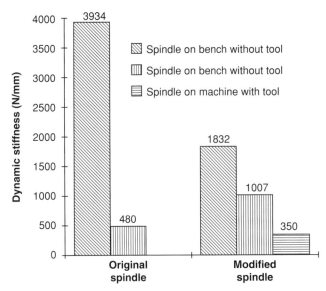

FIGURE 12.51 Comparison of the dynamic stiffness of the original and modified spindles with boring tools on the bench and the modified spindle with tooling installed on the machine.

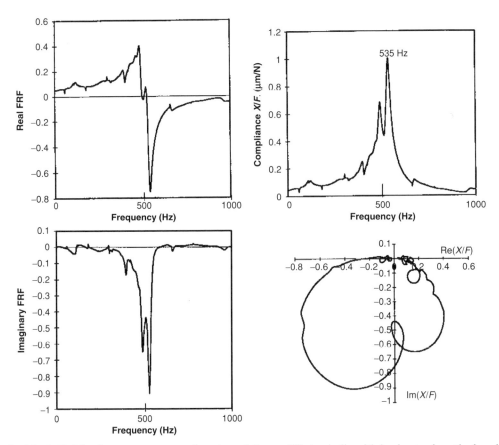

FIGURE 12.52 The frequency response function of the modified spindle with boring tool on the bench.

Machining Dynamics

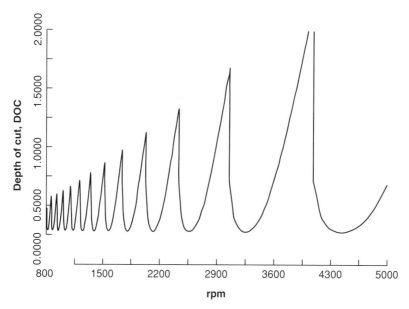

FIGURE 12.53 Stability lobe diagram for the original spindle with a boring tool on the bench.

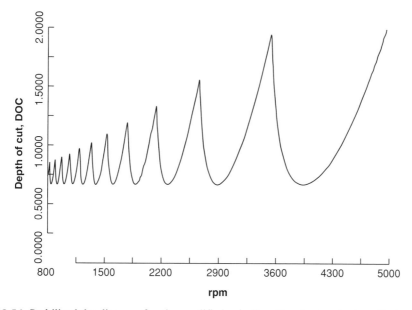

FIGURE 12.54 Stability lobe diagram for the modified spindle with a boring tool on the bench.

become unstable) due to variations in the spindle speed during operation when the three-insert boring bar is used. This explains why the chatter occurred when three inserts were used but did not occur when a single insert was used.

The following conclusions can be drawn from the above results:

- The static and especially the dynamic stiffness of the tool on the machine station is insufficient for the operation to be performed with a three-insert boring bar.

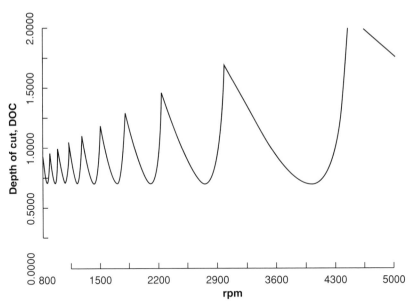

FIGURE 12.55 Stability lobe diagram for the modified spindle with a boring tool installed on the machine.

- The chatter frequency is the same as the lowest natural frequency of the boring bar.
- The system stability was improved using one insert instead of three because the cumulative width of cut was reduced significantly.
- It would be worthwhile to investigate using two inserts instead of one, since this might permit an increase in the production rate without resulting in chatter.

Example 12.10 As an example of the identification of modal parameters, we consider the characterization of the dynamic properties of a single-spindle drill press.

An electrodynamic exciter was mounted between the spindle and the table. The drilling machine structure was excited in the Z-direction at point 1 as shown in Figure 12.56, using a pink noise random signal with bandwidth of 200 Hz generated by a function generator. The random vibration signal was measured with accelerometers at points 1, 2, 3, and 4 in Figure 12.56. The direct and cross transfer functions between the input force signal and the output displacement signals were estimated using a dual channel FFT real-time analyzer (RTA). Each set of data was analyzed using a transform size of 1024 points, a frequency range of 100 Hz (with a sampling rate of 2000 Hz), 32 sample averaging, and a Hanning window. A 32 signal average was used to reduce the variance of the spectrum since the 3 dB bandwidth of the Hanning window is 0.36 Hz. The direct transfer function (TF) at point 1 is shown in Figure 12.57. The auto spectrum of the input and output signals are shown in Figure 12.58. Although the spectrum of the input signal to the exciter from the function generator is almost flat over the frequency range of interest as shown in Figure 12.59, the spectrum of the actual input force signal to the structure is not quite flat. The coherency spectrum at point 1 (Figure 12.60) indicates that the TF is accurate over a relatively wide range of frequencies, that is, that noise effects are acceptably small except at very low frequencies.

To identify the dynamic characteristics of the structure, the real part of the TF was fit using a nonlinear least-squares curve fitting routine based on the model in Equation (12.37) or (12.68). Initial values were estimated from visual inspection of the shape of the TF. Assuming

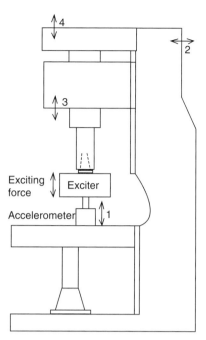

FIGURE 12.56 Schematic diagram of the experimental setup for vibration test on the drilling machine structure.

a 3DOF system, 30 points of the TF from 21 to 50 Hz were fit. The curve fit TF and the original TF are shown in Figure 12.61. The decomposition of the three modes is shown in Figure 12.62, from which the effect of each mode on the vibration at point 1 can be identified. The dynamic characteristics of the structure in the normal coordinate system were determined using the analysis described in Section 12.3 and are summarized in Table 12.11. The mode shapes of the structure are found by curve-fitting the real parts of all the cross TFs using the natural frequencies, stiffness coefficients, and damping factors in Table 12.11. The eigenvectors or mode shapes were calculated using Equation 12.34 and are summarized in Table 12.12 and illustrated in Figure 12.63.

Example 12.11 Consider a single-point boring bar used for roughing and finishing a deep hole. The body of the bar is cylindrical with a diameter of 50 mm, a length of 350 mm ($L/D = 7$), and a lead angle of 30°. The bar is made of steel. The bar is clamped in a CAT-50 hydraulic toolholder as explained in Chapter 5. Estimate the limit of stability and the corresponding MRR assuming that the bar acts as an SDOF system.

Solution: The stiffness of the boring bar including the tool–toolholder–spindle interface can be calculated as explained in Chapter 5. However, let us assume the bar is a cantilevered beam. In this case, the stiffness of the bar is derived using the material in Chapter 4 and Chapter 5, assuming a damping ratio of 0.04:

$$k_c = \frac{3E\pi D^4}{64L^3} = 4293 \text{ N/mm}$$

and using Equation (12.66)

$$b_{\lim} = \frac{2k_c \zeta (1+\zeta)}{k_d \sin^2 \kappa} = 0.67 \text{ mm}$$

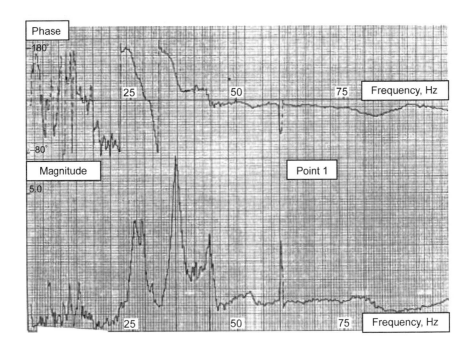

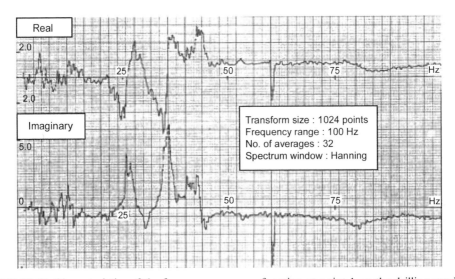

FIGURE 12.57 Characteristics of the frequency response function at point 1 on the drilling machine.

The depth of cut of 0.65 mm is small for a roughing operation and could be increased by increasing the diameter of the boring bar or using a stiffer bar material as explained in Chapter 4. Changing the material of the boring bar to heavy metal with $E = 330{,}000$ MPa instead of 206,700 MPa for steel, the $b_{\lim} = 1.07$ mm. The above minimum depth of cut is sufficient for a finish cut even using the steel bar. Selecting a roughing feed of 0.2 mm/rev and a cutting speed of 100 m/min, the MRR is estimated using Equation (2.7) to be

$$\mathrm{MRR}_{\lim} = Q = Vf\, b_{\lim} = 21{,}430 \text{ mm}^3/\text{min}$$

Machining Dynamics

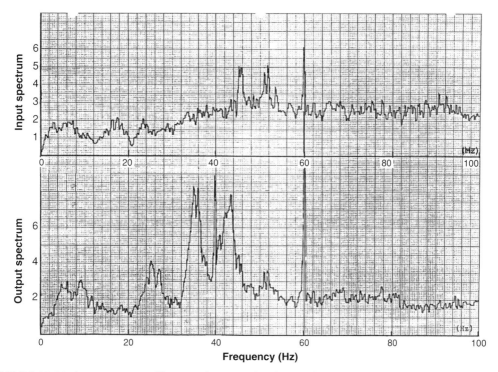

FIGURE 12.58 Autospectrum of input and output signals at point 1 on the drilling machine.

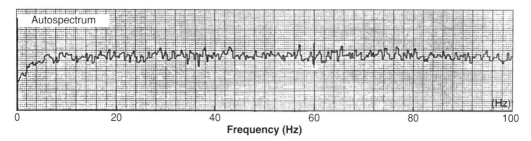

FIGURE 12.59 Autospectrum of the input signal to the exciter generated by the function generator for the identification of the drilling machine structure.

12.11 PROBLEMS

Problem 12.1 Select a spindle speed (rpm) for a four-flute end mill cutter so that the fundamental forcing frequency (FFF) and first two harmonics do not fall within 100 Hz of the main resonance peak in Figure 12.64. The spindle speed must be between 6,000 and 18,000 rpm. Plot the frequencies of the FFF and first two harmonics in Figure 12.64.

Problem 12.2 Dynamic test results for a carbide end mill in a vertical machining center are shown in Figure 12.65. The machine has a 14,000 rpm maximum spindle speed and the spindle power is not a constraint. Estimate the time required to machine a pocket 800 mm long, 400 mm wide, and 20 mm deep from a solid block workpiece using the maximum estimated material removal rate (MRR).

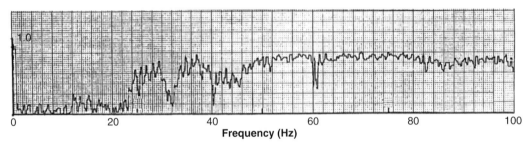

FIGURE 12.60 Coherency of transfer function at point 1 on the drilling machine.

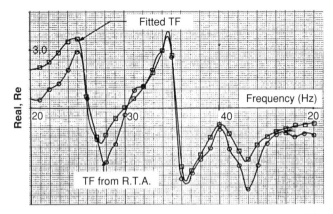

FIGURE 12.61 Curve fitting of transfer function at point 1 on the drilling machine.

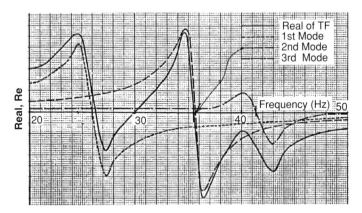

FIGURE 12.62 Decomposition of real part of the transfer function at point 1 on the drilling machine.

Problem 12.3 A two-flute carbide end milling cutter with an overhang of 87 mm out of the toolholder is used in an operation. The feed is 0.15 mm/tooth. The stability lobes and the MRR lobes were estimated and are shown in Figure 12.66 and Figure 12.67, respectively. Determine: (a) the maximum axial depth of cut and MRR in Figure 12.66 and Figure 12.67 if maximum spindle speeds of 18,000, 24,000, and 30,000 rpm are selected; (b) name two physical changes that can be made to the cutter geometry or setup to alter Figure 12.66 to look more like Figure 12.68, to make the system more productive at 30,000 rpm.

Machining Dynamics

TABLE 12.11
Modal Parameters of the Drilling Machine Structure Estimated by Curve Fitting and Decomposition

Mode	Natural Frequency	Damping Ratio	Stiffness
1st	26	0.051	1.85
2nd	35.6	0.026	2.62
3rd	42	0.007	11.93

TABLE 12.12
Mode Shapes of the Drilling Machine Structure Estimated by Curve Fitting

Position	1st Mode	2nd Mode	3rd Mode
1	1.0	1.0	1.0
2	0.013	0.016	−0.304
3	0.407	−0.5	0.493
4	0.409	−0.45	0.574

Problem 12.4 A pocket is to be machined in an aluminum structure ($K_c = 2.8 \times 10^8$ N/m^2). An end milling process will be used on a horizontal machining center capable of a maximum spindle speed of 15,000 rpm. The workpiece is determined to be stiff (2.3×10^9 N/mm) compared to the long end-mill (1.1×10^7 N/m) required to machine the deep pocket. Use the approximate values for the first bending mode of a 2-tooth end mill obtained from experimental modal analysis ($f_n = 315$ Hz, $k = 1.1 \times 10^7$ N/m, $\zeta = 0.02$) to:

(a) create an analytical frequency response function (FRF);
(b) create stability lobes for a 1-, 2-, 3-, and 4-tooth endmill assuming all four tools have the same dynamic values. (Note: for a full slot, 1- and 2-tooth end mills have 1-tooth in cut, and a 3- and 4-tooth end mills have two teeth in cut.)

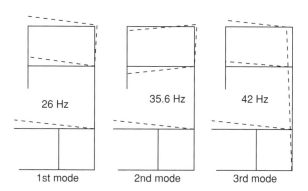

FIGURE 12.63 Mode shapes at point 1 on the drilling machine.

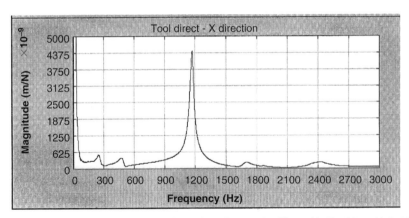

FIGURE 12.64 Frequency response function for a four-flute end mill used in Problem 12.1. (Courtesy of D3 Vibrations, Inc.)

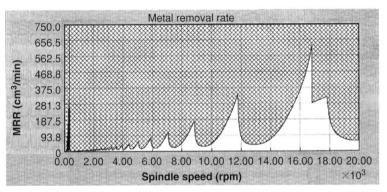

FIGURE 12.65 Frequency response function for an end mill used in Problem 12.2. (Courtesy of D3 Vibrations, Inc.)

Determine:

   (i) What is the maximum negative real value of the FRF? Why is this important?
   (ii) What is b_{lim} for each of the four tools?
   (iii) What is the maximum axial depth of cut (ADOC) for each tool at 4000 and 5000 rpm?
   (iv) The tool you are using is 25 mm in diameter and the feed rate is 0.05 mm/tooth. If it is suggested that you must use a two-flute tool with 0.05 mm/tooth feed rate, remove between 5 and 10 mm ADOC, and rotate between 4,000 and 10,000 rpm, then what rpm and ADOC will you choose to maximize the MRR and stay at least 1000 rpm and 2 mm ADOC away from a stability line? What is the volume of material removed per minute?

Problem 12.5 A 25.4 mm slot is to be machined in a steel structure (cutting stiffness $K_c = 7 \times 10^8$ N/m^2). An end milling operation will be designed for the full slot on a horizontal machining center capable of a maximum spindle speed of 15,000 rpm. A four-flute end mill is suggested with a feed of 0.1 mm/tooth. The workpiece is determined to be much stiffer than the long 25.4 diameter end mill required to machine the slot. Use the approximate values for

Machining Dynamics

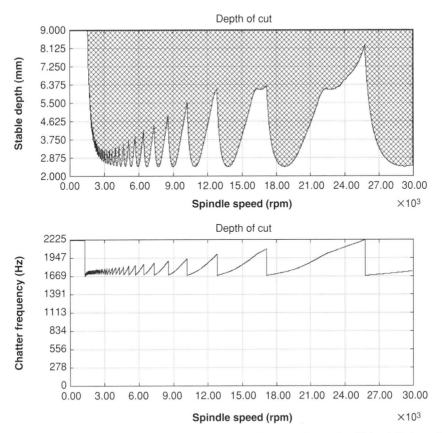

FIGURE 12.66 Stability chart for a 15 mm diameter two-flute carbide end mill for full slot milling at 0.15 mm/tooth in an aluminum workpiece. (Courtesy of D3 Vibrations, Inc.)

the first bending mode of a 2-tooth endmill obtained from experimental modal analysis ($f_n = 5000$ Hz, $k = 2 \times 10^7$ N/m, $\zeta = 0.05$) to:

(a) create an analytical frequency response function (FRF);
(b) create stability lobes for the end mill. (Note: for a full slot, a four-flute end mill has two teeth in cut.)

Questions:
1. What is the maximum negative real value of FRF? Why is this important?
2. What is $b_{\lim}$ for each of the two tools?
3. What is the maximum axial depth of cut (ADOC) for each tool at 15,000 rpm?

Problem 12.6 Derive the solution of Equation (12.71) for regenerative chatter used to plot the stability chart of b versus rpm.

Problem 12.7 Consider a 2DOF milling system with the following characteristics: $k_1 = 4.4 \times 10^6$ N/m, $f_{n1} = 60$ Hz, $\zeta_1 = 0.04$ and $k_2 = 5.6 \times 6$ N/m, $f_{n2} = 66$ Hz, $\zeta_1 = 0.04$. Assume the directions of individual modes of vibration are measured to be 30° and 120° as illustrated in Figure 12.17. Calculate the critical limiting width of chip assuming the specific force for the workpiece material is $k_d = 2400$ N/mm².

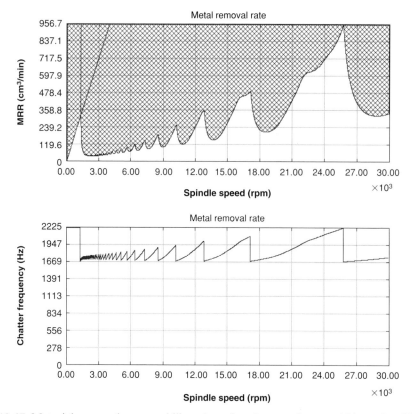

FIGURE 12.67 Material removal rate stability chart for the two-flute carbide end milling of an aluminum workpiece based on the stability chart in Figure 12.66. (Courtesy of D3 Vibrations, Inc.)

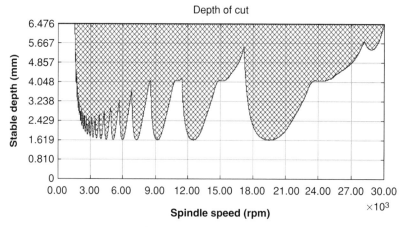

FIGURE 12.68 Stability chart for the same carbide end mill as in Figure 12.66 with long overhang milling of a full slot in an aluminum workpiece. (Courtesy of D3 Vibrations, Inc.)

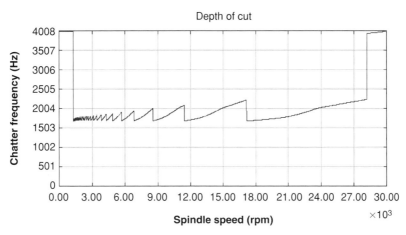

FIGURE 12.68 *Continued*

REFERENCES

1. L. Andren, L. Hakansson, A. Brandt, and I. Claesson, Identification of motion of cutting tool vibration in a continuous operation — correlation to structural properties, *Mech. Systems Signal Process.* **18** (2004) 903–927
2. *Technology of Machine Tools*, Vol. 3: Machine Tool Mechanics, Lawrence Livermore National Laboratory Report UCRL-52960-3; distribution by SME
3. J. Tlusty and T. Moriwaki, Experimental and computational identification of dynamic structural models, *CIRP Ann.* **25** (1976) 497–503
4. J. Tlusty and F. Ismail, Dynamic structural identification tasks and methods, *CIRP Ann.* **29** (1980) 251–255
5. Y. Altintas, *Manufacturing Automation — Metal Cutting Mechanics, Machine Tool Vibrations, and CNC Design*, Cambridge University Press, Cambridge, UK, 2000
6. E.I. Rivin, *Stiffness and Damping in Mechanical Design*, Marcel Dekker, New York, 1999
7. D.J. Seagalman and E.A. Butcher, Suppression of regenerative chatter via impendance modulation, *J. Vibr. Control* **6** (2000) 243–256
8. H. Shinno and Y. Ito, Computer aided concept design for structural configuration of machine tools: variant design using directed graph, *J. Mechanisms Transmiss. Automat. Des.* **109**:3 (1987) 372–376
9. H. Shinno and Y. Ito, Structural description of machine tools, 1st report: description method and some applications, *Bull. Jpn. Soc. Mech. Engr.* **24**:187 (1981) 251–258
10. H. Shinno and Y. Ito, Structural description of machine tools, 2nd report: evaluation of structural similarity, *Bull. Jpn. Soc. Mech. Engr.* **24**:187 (1981) 259–265
11. J.N. Dube, Computer aided design of machine tool structure with model techniques, *Computers Struct.* **28**:3 (1988) 345–352
12. J.G. Bollinger, Computer aided analysis of machine dynamics, *Proceedings of the International Conference on Manufacturing Technology by The American Society of Tool and Manufacturing Engineers*, Dearborn, MI, September 1967, 77–88
13. R.C. Bahl and P.C. Pandey, Computer analysis of machine tool structures using flexibility method, *Mech. Eng. Bull.* **7** (1976) 42–51
14. R.C. Bahl and P.C. Pandey, Comparative study of computer aided methods for the design of machine tool structures, *Mech. Eng. Bull.* **8** (1977) 33–38
15. O.I. Averyanov and A.P. Bobrik, Methodology of forming standardized subassemblies for machine tools, *Soviet Eng. Res.* **3** (1983) 55–57
16. R.C. Dorf, *Modern Control Systems*, 4th Edition, Addison-Wesley, Reading, MA, 1986
17. D.C. Karnopp, D.L. Margolis, and R.C. Rosenberg, *System Dynamics*, 2nd Edition, Wiley, New York, 1990

18. M.P. Norton, *Fundamentals of Noise and Vibration Analysis for Engineers*, Cambridge University Press, New York, 1994
19. R.C. Dorf, *Time-Domain Analysis and Design of Control Systems*, Addison-Wesley, Reading, MA, 1965
20. T.H. Glisson, *Introduction to System Analysis*, McGraw-Hill, New York, 1984
21. H.P. Neff, *Continuous and Discrete Linear Systems*, Harper and Row, New York, 1984
22. S.A. Tobias, *Machine Tool Vibrations*, Blackie and Son, London, 1965
23. J. Tlusty and K.M. Polacek, Experiences with analysing stability of machine-tool against chatter, *Proceedings of the 9th International MTDR Conference*, Pergamon Press, New York, 1968
24. J. Tlusty, A method of analysis of machine tool stability, *Proceedings of the 6th MTDR Conference*, 1965, 5
25. D.B. Welbourne and J.K. Smith, *Machine-Tool Dynamics: An Introduction*, Cambridge University Press, New York, 1970
26. C.W. Bert, Material damping: an introduction review of mathematical models, measures and experimental techniques, *J. Sound Vibr.* **29**:2 (1973) 129–153
27. R.H. Lyon, *Machinery Noise and Diagnostics*, Butterworths, London, 1987
28. S. Timoshenko, D.H. Young, and W. Weaver, Jr., *Vibration Problems in Engineering*, 4th Edition, Wiley, New York, 1974
29. J.W.S. Rayleigh, *Theory of Sound*, 2nd Edition, Vol. 1, Dover Publications, New York, 1945
30. T.K. Caughey, Classical normal modes in damped linear dynamic systems, *ASME J. Appl. Mech.* **27** (1960) 269–271. See also *ASME J. Appl. Mech.* **32** (1965) 583–588
31. R.A. Fraser, W.J. Duncan, and A.R. Collar, *Elementary Matrices*, Cambridge University Press, London, 1946
32. K.A. Foss, Coordinates which uncouple the equations of motion of damped linear dynamic systems, *ASME J. Appl. Mech.* **25** (1958) 361–364
33. J. Peters and M. Mergeay, Dynamic analysis of machine tools using complex modal method, *CIRP Ann.* **25** (1976) 257–261
34. D. Dilley and T. Delio, Machine tool vibration monitoring using audio signal analysis, *1st Annual Manufacturing Technology Summit* by SME, Dearborn, MI, August 2004, 10–11
35. E.I. Rivin, Machine-tool vibration, in: C.M. Harris, Ed., *Shock and Vibration Handbook*, 4th Edition, McGraw-Hill, New York, 1996, Chapter 40.
36. Methods for Performance Evaluation of CNC Machining Centers, ASME Standard B5.54, 1992
37. Weck, M., *Handbook of Machine Tools*, Vols. 1–4, Wiley, New York, 1984
38. J. Tlusty, Analysis of the state of research in cutting dynamics, *CIRP Ann.* **27** (1978) 583–589
39. J. Tlusty, *Manufacturing Processes and Equipment*, Prentice Hall, Englewood Cliffs, NJ, 2000
40. F.C. Moon, *Dynamics and Chaos in Manufacturing Processes*, Wiley, New York, 1998
41. G. Stépán, Modelling nonlinear regenerative effects in metal cutting, *Philos. Trans. Roy. Soc.* **359** (2001) 739–757
42. A.M. Gouskov, S.A. Voronov, H. Paris, and S.A. Batzer, Cylindrical workpiece turning using multiple-cutter tool heads, *Proceedings of the ASME 2001 Design Engineering Technical Conference*, Pittsburgh, PA, paper no. DETC2001/VIB-21431
43. J.E. Halley, A.M. Helvey, K.S. Smith, and W.R. Winfough, The impact of high-speed machining on the design and fabrication of aircraft components, *Proceedings of the 1999 ASME Design Engineering Technical Conference*, Las Vegas, NV, paper no. DETC99/VIB-8057
44. J. Tlusty, Machine dynamics, in: R.I. King, *Handbook of High-Speed Machining Technology*, Chapman and Hall, New York, 1985, Chapter 3, 48–153
45. S.A. Tobias and W. Fishwick, The chatter of lathe tools under orthogonal cutting conditions, *ASME Trans.* **80** (1958) 1079–1088
46. J. Tlusty and M. Polacek, Beispiele der Behandlung der selbsterregten Schwingung der Werkzeugmaschinen, *Proceedings of the 3rd Fo Ko Co*, October, Vogel-Verlag, Wuerzburg, 1957, 131
47. H.E. Merritt, Theory of self-excited machine-tool chatter: contribution to machine-tool chatter research — 1, *ASME J. Eng. Ind.* **87** (1965) 447–454
48. J. Tlusty and S.B. Rao, Verification and analysis of some dynamic cutting force coefficient data, *Proc. NAMRC* **6** (1978) 420–426

49. E.E. Mitchell and E. Harrison, Active machine tool controller requirements for noise attenuation, *ASME J. Eng. Ind.* **96** (1974) 261–267
50. Y. Antilas and E. Budak, Analytical prediction of stability lobes in milling, *CIRP Ann.* **44** (1995) 357–362
51. S. Smith and J. Tlusty, Efficient simulation programs for chatter in milling, *CIRP Ann.* **42** (1993) 463–466
52. J. Tlusty, W. Zaton, and F. Ismail, Stability lobes in milling, *CIRP Ann.* **32** (1983)
53. D. Montgomery and Y. Altintas, Mechanism of cutting force and surface generation in dynamic milling, *ASME J. Eng. Ind.* **113** (1991) 160–168
54. W.R. Winfough and S. Smith, Automatic selection of the optimum metal removal conditions for high-speed milling, *Trans. NAMRI/SME* **23** (1995) 163–168
55. J. Tlusty, S. Smith, and T. Delio, Stiffness, stability, and loss of process damping in high-speed machining, *Fundamental Issues in Machining*, ASME PED Vol. 43, 1990, 171–191
56. The MetalMAX Machine Tool Analyzer and Dynamic Machining Process Optimizer, Manufacturing Laboratories, Inc. (MLI), Las Vegas, NV
57. S.A. Tobias and W. Fishwick, The vibrations of radial drilling machines under test and working conditions, *Proc. Inst. Mech. Engr.* **170** (1956) 232–256
58. M.K. Das and S.A. Tobias, The relation between the static and the dynamic cutting of materials, *Int. J. Mach. Tool Des. Res.* **7** (1967) 63–89
59. M.M. Nigm, M.M. Sadek, and S.A. Tobias, Determination of dynamic cutting coefficients from steady state cutting data, *Int. J. Mach. Tool. Des. Res.* **17** (1977) 19–37
60. J. Peters and P. Vanherck, Machine tool stability tests and the increamental stiffness, *CIRP Ann.* **17** (1969) 225–232
61. J. Tlusty and F. Ismail, Special aspects of chatter in milling, *ASME J. Vibr. Stress Rel. Des.* **105** (1983) 24–32
62. W.J. Endres, Prediction Stability in Machining Using Time-Domain Simulation — A Comparison to Frequency Analysis, Technical Report, MEAM Department, University of Michigan, 1995
63. H.J.J. Kals, On the calculation of stability charts on the basis of the damping and the stiffness of the cutting process, *CIRP Ann.* **19** (1971) 197–303
64. J. Peters, P. Vanherck, and H. Van Brussel, The measurement of the dynamic cutting force coefficient, *CIRP Ann.* **20** (1971) 129–136
65. W.A. Knight and M.M. Sadek, The correction for dynamic errors in machine tool dynamometers, *CIRP Ann.* **19** (1971) 237–245
66. J. Gradisek, E. Govekar, and I. Grabec, Time series analysis in metal cutting: chatter versus chatter-free cutting, *Mech. Systems Signal Process.* **12** (1998) 839–854
67. M.K. Khraisheh, C. Pezeshki, and A.E. Bayoumi, Time series based analysis for primary chatter in metal cutting, *J. Sound Vibr.* **180** (1995) 67–87
68. T. Insperger, G. Stépán, P.V. Bayly, B.P. Mann, Multiple chatter frequencies in milling processes, *J. Sound Vibr.* **262** (2002) 333–345
69. T.L. Schmitz, M.A. Davies, K. Medicus, and J. Snyder, Improving high-speed machining material removal rates by rapid dynamic analysis, *CIRP Ann.* **50**:1 (2001) 263–268
70. A. Halanay, Stability theory of linear periodic systems with delay, *Revue de Mathéematiques Pures et Appliquées* **6**:4 (1961) 633–653 (in Russian)
71. J.K. Hale, *Theory of Functional Differential Equations*, Springer-Verlag, New York, 1977
72. I. Minis and R. Yanushevsky, A new theoretical approach for the prediction of machine tool chatter in milling, *ASME J. Eng. Ind.* **115** (1993) 1–8
73. Y. Altintas and E. Budak, Analytical prediction of stability lobes in milling, *CIRP Ann.* **44**:1 (1995) 357–362
74. E. Budak and Y. Altintas, Analytical prediction of chatter stability in milling. Part I. General formulation, *ASME J. Dynam. Systems Meas. Control* **120** (1998) 22–30
75. E. Budak and Y. Altintas, Analytical prediction of chatter stability in milling. Part II. Application of the general formulation to common milling systems, *ASME J. Dynam. Systems Meas. Control* **120** (1998) 31–36

76. M.A. Davies, J.R. Pratt, B. Dutterer, and T.J. Burns, Stability prediction for low radial immersion milling, *ASME J. Manuf. Sci. Eng.* **124** (2002) 217–225
77. T. Insperger and G. Stépán, Stability of high-speed milling, *Proceedings of the ASME 2000 DETC, Symposium on Nonlinear Dynamics and Stochastic Mechanics*, Orlando, FL, ASME AMD-241, 2000, 119–123
78. P.V. Bayly, J.E. Halley, B.P. Mann, and M.A. Davies, Stability of interrupted cutting by temporal finite element analysis, *Proceedings of the ASME 2001 Design Engineering Technical Conference*, Pittsburgh, PA, paper no. DETC2001/VIB-21581
79. J. Tian and S.G. Hutton, Chatter instability in milling systems with flexible rotating spindles — a new theoretical approach, *ASME J. Manuf. Sci. Eng.* **123** (2001) 1–9
80. S. Smith and J. Tlusty, An overview of modeling and simulation of the milling process, *ASME J. Eng. Ind.* **113** (1991) 169–175
81. T. Insperger, B.P. Mann, G. Stépán, and P.V. Bayly, Stability of up-milling and down-milling. Part 1. Alternative analytical methods, *Int. J. Mach. Tool Manuf.* **43** (2003) 25–34
82. B.P. Mann, T. Insperger, P.V. Bayly, and G. Stépán, Stability of up-milling and down-milling. Part 2. Experimental verification, *Int. J. Mach. Tool Manuf.* **43** (2003) 35–40
83. B. Balachandran, Non-linear dynamics of milling process, *Philos. Trans. Roy. Soc.* **359** (2001) 793–820
84. M.X. Zhao and B. Balachandran, Dynamics and stability of milling process, *Int. J. Solids Struct.* **38** (2001) 2233–2248
85. D.M. Esterling, Y. Ren, and Y.S. Lee, Time-domain chatter prediction for high speed machining, *Trans. NAMRI/SME* **30** (2002)
86. W.T. Corpus and W.J. Endres, A high-order solution for the added stability lobes in intermittent machining, *Proceedings of the ASME 2001 DETC, Symposium on Machining Processes*, Orlando, FL, ASME MED-11, 2001, 871–878
87. J. Gradisek, E. Govekar, and I. Grabec, Using coarse-grained entropy rate to detect chatter in cutting, *J. Sound Vibr.* **214**:5 (1998) 941–952
88. H.M. Shi and S.A. Tobias, Theory of finite amplitude machine tool instability, *Int. J. Mach. Tool. Des. Res.* **24** (1984) 45–69
89. G. Stépán and T. Kalmár-Nagy, Nonlinear regenerative machine tool vibration, *Proceedings of the 1997 ASME Design Engineering Technical Conference*, Sacramento, CA, paper no. DETC97/VIB-4021
90. G. Stépán and R. Szalai, Nonlinear vibrations of highly interrupted machining, *Proceedings of the 2nd Workshop of COST P4 WG2 on Dynamics and Control of Mechanical Processing*, Budapest, Hungary, 2001, 59–64
91. J. Tlusty, Dynamics of high-speed milling, *ASME J. Eng. Ind.* **108** (1986) 59–67
92. M.A. Davies, J.R. Pratt, B. Dutterer, and T.J. Burns, The stability of low radial immersion machining, *CIRP Ann.* **49**:1 (2000) 37–40
93. P.V. Bayly, J.E. Halley, M.A. Davies, and J.R. Pratt, Stability analysis of interrupted cutting with finite time in the cut, *Proceedings of the ASME 2000 DETC*, Orlando, FL, ASME MED-11, 2000, 989–994
94. P.V. Bayly, J.E. Halley, B.P. Mann, and M.A. Davies, Stability of interrupted cutting by temporal finite element analysis, *ASME J. Manuf. Sci. Eng.* **125** (2003) 220–225
95. P.V. Bayly, B.P. Mann, T.L. Schmitz, D.A. Peters, G. Stépán, and T. Insperger, Effects of radial immersion and cutting direction on chatter instability in end-milling, *Proceedings of the ASME IMECE*, New Orleans, LA, No. IMECE2002-34116, ASME, 2002
96. B.P. Mann, P.V. Bayly, M.A. Davies, and J.E. Halley, Limit cycles, bifurcations, and accuracy of the milling process, *J. Sound Vibr.* **277**(2004) 31–48
97. B.P. Mann, K.A. Young, T.L. Schmitz, M.J. Bartow, and P.V. Bayly, Machining accuracy due to tool or workpiece vibrations, *Proceeedings of the ASME IMECE*, Washington, DC, No. IMECE2003-41991, ASME, 2003
98. J. Tlusty, F. Ismail, and W. Zaton, Use of special milling cutters against chatter, *Proc. NAMRC* **11** (1983) 408–415
99. F. Koenigsberger and J. Tlusty, *Machine Tool Structures*, Vol. 1, Pergamon Press, Oxford, 1970

100. J. Tlusty, Basic non-linearity in machining chatter, *CIRP Ann.* **30** (1981) 299–304
101. H.K. Tonshoff and J.P. Wulfsberg, et al., Developments and trends in monitoring and control of machining processes, *CIRP Ann.* **37** (1988) 611
102. H. Ota and T. Kawai, Monitoring of cutting conditions by means of vibration analysis, *Trans. Jpn. Soc. Mech. Eng. Ser. C* **57** (1991) 2752 (in Japanese)
103. M. Rahman, In-process detection of chatter threshold, *ASME J. Eng. Ind.* **110** (1988) 44
104. M. Higuchi and M. Doi, A study on the estimation of the chatter commencing in turning process, *Trans. Jpn. Soc. Mech. Eng. Ser. C* **52** (1986) 1697 (in Japanese)
105. T. Matsubare and H. Yamamoto, Study on regenerative chatter vibration with dynamic cutting force, *J. Jpn. Soc. Precision Eng.* **50** (1984) 1079 (in Japanese)
106. E. Kondo, H. Ota, and T. Kawai, A new method to detect regenerative chatter using spectral analysis. Part I. Basic study on criteria for detection of chatter, *ASME Manuf. Sci. Eng.* MED-Vol. 2–1/MH-Vol. 3–1, 1995, 617–627
107. M. Weck, E. Verhaag, and M. Gather, Adaptive control for face-milling with strategies for avoiding chatter-vibrations and for automatic cut distribution, *CIRP Ann.* **24** (1975) 405
108. S. Smith and J. Tlusty, Update on high-speed milling dynamics, *ASME J. Eng. Ind.* **112** (1990) 142–149
109. R.J. Allemang and D.L. Brown, Machine-tool vibration, *Shock and Vibration Handbook*, 4th Edition, McGraw-Hill, New York, 1996, Chapter 21
110. R.L. Kegg, Cutting dynamics in machine tool chatter-contribution to machine tool chatter. Research-3, *ASME J. Eng. Ind.* **November** (1965) 464–470
111. H. Van Brussel and P. Van Herck, A new method for the determination of the dynamic cutting coefficient, *Proceedings of the 11th MTDR Conference*, 1970, 105–118
112. H. Opitz, Application of a process control computer for measurement of dynamic coefficients, *CIRP Ann.* **21** (1972) 99–100
113. E.J. Goddarol, The Development of Equipment for the Measurement of Dynamic Cutting Force Coefficients, MSc Thesis, UMIST, Manchester, April 1972
114. J. Tlusty and B.S. Goel, Measurement of the dynamic cutting force coefficient, *Proc. NAMRC* **2** (1974)
115. R. Snoey and D. Roesems, Survey of modal analysis applications, *CIRP Ann.* **20** (1979) 497–510
116. L.D. Mitchell, Improved methods for the fast Fourier transfom (FFT) calculation of the frequency response function, *ASME J. Mech. Des.* **104** (1982) 277–279
117. K.B. Elliott and L.D. Mitchell, The improved frequency response function and its effect on modal circle fits, *ASME J. Appl. Mech.* **106** (1984) 657–663
118. R.J. Allemang, Experimental modal analysis bibliography, *Proceedings of the IMAC2 Conference*, February, 1984
119. L.D. Mitchell, Modal analysis bibliography — an update — 1980–1983, *Proceedings of the IMAC2 Conference*, February, 1984
120. D.J. Ewins, *Modal Testing: Theory and Practice*, Wiley, New York, 1991 (reprinted)
121. M.M. Sadek and S.A. Tobias, Comparative dynamic acceptance tests for machine tools applied to horizontal milling machines, *Proc. Inst. Mech. Engr.* **185** (1970/1971) 319–337
122. J.W. Cooley and J.W. Tukey, An algorithm for the machine computation of complex function series, *Math. Comput.* **19** (1965) 297–301
123. D.L. Brown, R.J. Allemang, R. Zimmerman, and M. Mergeay, Parameter Estimation Techniques for Modal Analysis, SAE Paper 790221, 1979
124. G.T. Rocklin, J. Crowley, and H. Vold, A Comparison of H1, H2, Hv frequency response functions, *Proc. IMAC Conf.* **1** (1985) 272–278
125. S.M. Pandit, T.L. Subramanian, and S.M. Wu, Modeling machine tool chatter by time series, *ASME J. Eng. Ind.* **97** (1975)
126. K.F. Eman and S.M. Wu, A comparative study of classical techniques and the dynamic data system (DDS) approach for machine tool structure identification, *Proc. NAMRC* **13** (1985) 401–404
127. F.A. Burney, S.M. Pandit, and S.M. Wu, A new approach to the analysis of machine tool system stability under working conditions, *ASME J. Eng. Ind.* **99** (1977) 585–590
128. T.L. Subramanian, M.F. DeVries, and S.M. Wu, An investigation of computer control of machining chatter, *ASME J. Eng. Ind.* **98** (1976) 1209–1214

129. T. Moriwaki, Measurement of cutting dynamics by time series analysis technique, *CIRP Ann.* **22** (1973) 117–118
130. T. Moriwaki and T. Hoshi, System identification on digital techniques, new tool for dynamic analysis, *CIRP Ann.* **23** (1974) 239–246
131. S.M. Pandit and S.M. Wu, *Time Series and System Analysis with Applications*, Wiley, New York, 1983
132. S.M. Wu, Dynamic data system: a new modeling approach, *ASME J. Eng. Ind.* **99** (1977) 708–714
133. M.P. Norton, *Fundamentals of Noise and Vibration Analysis for Engineers*, Cambridge University Press, New York, 1994
134. A.H. Slocum, E.R. Marsh, and D.H. Smith, A new damper design for machine tool structures: the replicated internal viscous damper, *Precision Eng.* **16** (1994) 539
135. C.C. Ling, J.S. Chen, and I.E. Morse, Jr., An experimental investigation of damping effects on a single point cutting process, *ASME Manuf. Sci. and Eng. MED-Vol. 2–1/MH-Vol. 3–1*, 1995, 149–164
136. E.I. Rivin and W. D'Ambrogio, Enhancement of dynamic quality of a machine tool using a frequency response optimization method, *Mech. Systems Signal Process.* **4** (1990) 495
137. E.I. Rivin and H. Kang, Enhancement of dynamic stability of cantilever tooling structures, *Int. J. Mach. Tools Manuf.* **32** (1992) 539
138. B.W. Wong, B.L. Walcott, and K.E. Routh, Active vibration control via electromagnetic dynamic absorbers, *Proceedings of the 4th IEEE Conference on Control Applications*, Albany, NY, 1995
139. M.A. Marra, B.L. Walcott, K.E. Rough, and S.G. Tewani, H vibration control for machining using dynamic absorber technology, *Proceedings of the 1995 ACC*, Seattle, WA, 1995, 739–743
140. J. Slavicek, The effect of irregular tooth pitch on stability in milling, *Proceedings of the 6th MTDR Conference*, Pergamon Press, London, 1965
141. P. Vanherck, Increasing milling machine productivity by use of cutter with non-constant cutting-edge pitch, *Proceedings of the Advanced MTDR Conference*, Vol. 8, Pergamon Press, London, 1967, 947–960
142. B.J. Stone, Advances in machine tool design and research, *Proceedings of the 11th International MTDR Conference*, Vol. A, Pergamon Press, London, 1970, 169–180
143. W. Chen, Cutting forces and surface finish when machining medium hardness steel using CBN tools, *Int. J. Mach. Tools Manuf.* **40** (2000) 455–466
144. S. Smith and T. Delio, Sensor-based chatter detection and avoidance by spindle speed selection, *ASME J. Dynam. Systems Meas. Control* **114** (1992) 486–492
145. T. Delio, S. Smith, and J. Tlusty, Use of audio signals in chatter detection and control, *ASME J. Eng. Ind.* **114** (1992) 486–492
146. Y. Altintas and P.K. Chan, In-process detection and suppression of chatter in milling, *Int. J. Mach. Tools Manuf.* **32** (1992) 329–347
147. T.R. Comstock, F.S. Tse, and Z.R. Lemon, Application of controlled mechanical impedances for reducing machine tool vibration, *ASME J. Eng. Ind.* **91** (1969)
148. C.L. Nachtigal and N.H. Cook, Active control of machine tool chatter, *ASME J. Eng. Ind.* **92** (1970) 238
149. R.G. Klein and C.L. Nachtigal, The Active Control of a Boring Bar Operation, ASME Paper 72-WA/AUT-19, 1972
150. K.C. Maddux and C.L. Nachtigal, Wideband Active Chatter Control Scheme, ASME Paper 71-WA/AUT-10, 1971
151. C.L. Nachtigal and K.C. Maddux, An Adaptive Active Chatter Control Scheme, ASME Paper 72-WA/AUT-10, 1972
152. E.E. Mitchell and E. Harrison, Design of a hardware observer for active machine tool control, *ASME J. Dynam. Systems Meas. Control* **99** (1977) 227
153. T. Takemura, T. Kitamura, and T. Hoshi, Active suppression of chatter by programmed variation of spindle speed, *CIRP Ann.* **23** (1974) 121–122
154. T. Inamura and T. Sata, Stability analysis of cutting under varying spindle speed, *CIRP Ann.* **23** (1974) 119–120

155. T. Takemura, T. Kitamura, T. Hoshi, and K. Okushima, Active suppression of chatter by programmed variation of spindle speed, *CIRP Ann.* **23** (1974) 121–122
156. J.-Y. Yu, X.-J. Han, and B.-D. Wu, Study on turning with continuously varying spindle speed, *Proc. ASME Manuf. Int.* **1** (1988) 327–331
157. S.C. Lin, R.E. DeVor, and S.G. Kapoor, The effects of variable speed cutting on vibration control in face milling, *ASME J. Eng. Ind.* **112** (1990) 1–11
158. T. Hosi, et al., Study of practical application of fluctuating speed cutting for regenerative chatter control, *CIRP Ann.* **26** (1977) 175–180
159. H. Zhang, J. Ni, and H. Shi, Machining chatter suppression by means of spindle speed variation, *Proceedings of the First S.M. Wu Symposium on Manufacturing Science*, 1994, 161–175
160. K. Jemielniak and A. Widota, Suppression of self-excited vibration by the spindle speed variation method, *Int. J. Mach. Tool Des. Res.* **23** (1984) 207–214
161. T. Delio, Method of Controlling Chatter in a Machine Tool, U.S. Patent No. 5,170,358, December, 1992
162. S. Smith and W.R. Winfough, The effect of runout filtering on the identification of chatter in the audio spectrum of milling, *Trans. NAMRI/SME* **22** (1994) 173–178
163. S. Smith and J. Tlusty, Stabilizing chatter by automatic spindle speed regulation, *CIRP Ann.* **41** (1992) 433–436
164. S. Smith, et al., Automatic chatter avoidance in milling using coincident stable speeds for competing modes, ASME PED Vol. 68, 1994, 737–747
165. S.Y. Tsai, On Line Identification and Control of Machining Chatter in Turning Through the Dynamic Data System Methodology, PhD Dissertation, Mechanical Engineering, University of Wisconsin–Madison, 1983
166. K.F. Eman, Machine Tool System Identification and Forecasting Control of Chatter, PhD Dissertation, Mechanical Engineering, University of Wisconsin–Madison, 1979
167. Y. Miyoshi and H. Nakazawa, Time series forecasts of chatter stability limits (1st report) — establishment of foundational method, *Bull. Jpn. Soc. Precision Eng.* **13** (1979) 89–94

13 Machining Economics and Optimization

13.1 INTRODUCTION

Economic considerations are obviously important in designing a machining process. There is generally more than one approach for machining a particular part; each approach will have an associated cost and level of part quality. The initial method of producing a new part, including the machine tools, cutting tool materials and geometries, speeds and feeds, and coolant, is generally determined from previous experience with similar parts, handbook recommendations, catalog data, or rules of thumb. These sources provide plausible starting points, but rarely yield the most efficient approach. In high-volume operations, changes are continuously made as experience with the specific part is accumulated. This can be a tedious process and often results in comparatively inefficient practices being used over much of the production run. A more efficient methodology to optimize machining practices would be desirable to reduce the time required to identify the best process more systematically.

The manufacturing cost per part is made up of several components, including the machining cost, tool cost, tool change and its setup costs, and handling cost. These costs vary most significantly with the cutting speed as shown in Figure 13.1. At a certain cutting speed the costs reach a minimum which corresponds to the optimum cutting speed. A large

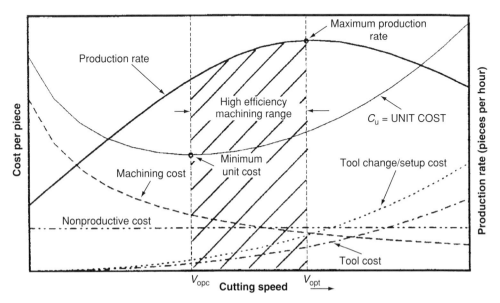

FIGURE 13.1 Machining cost and production rate versus cutting speed.

amount of machining information is constantly generated in a shop or manufacturing plant, which if properly analyzed could clarify cost relations and simplify the identification of an optimum process. It has long been recognized that computers could be used to analyze cutting data and rapidly identify an optimum.

Efforts to use computers to select machining parameters date back to the early 1960s. Many systems have been or are being developed by government agencies, industries, research institutions, and universities to meet specific requirements [1]. Computer-aided selection of limited machining parameters has become possible based on look-up tables or mathematical formulas. Systems which perform such tasks can transform the selection of machinability information from an experience-based activity to a more scientific discipline so that an efficient/optimum machining process can be identified as shown in Figure 13.2. The most common class of such systems, computer-aided process planning (CAPP) systems, are summarized in Section 16.5. The analysis methods reviewed in Chapter 8 are also useful for this purpose.

This chapter discusses the economics of machining operations with particular emphasis on analytical and computer methods. Computerized optimization systems (COSs) are discussed in Section 13.2. Basic economic considerations are discussed in Section 13.3. Based on these considerations, analytical methods for optimizing various classes of machining systems and operations are discussed in Section 13.4–Section 13.7. In many ways this discussion shows how information reviewed in previous chapters, especially Chapters 3-5 and Chapters 9–11, can be used to make the basic economic decisions which are central to successful process design. Finally, four numerical examples are presented in Section 13.8.

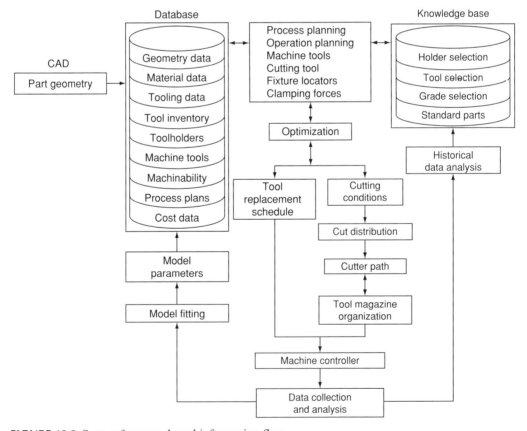

FIGURE 13.2 System framework and information flow.

13.2 ROLE OF A COMPUTERIZED OPTIMIZATION SYSTEM

This section discusses the general features and functions of a COS. In describing the function of such systems, it is helpful to distinguish three levels of optimization corresponding to increasingly sophisticated analysis capabilities.

The common manufacturing database of a computer integratd manufacturing (CIM) system typically consists of many kinds of interacting information sources. One of the important databases for any CIM system is the one that stores machinability data (Figure 13.2). The basic role of a computerized machinability database system is to select the cutting speed, feed rate, and depth of cut (if needed) for each operation in the process given the following characteristics of the machining operation: workpiece material, type of machining operation, machine tool, cutting tool, cut distribution, cutter path, tool replacement schedule, and operating parameters other than speed and feed. In addition to these basic functions, the *first level* of a COS can use computerized machinability systems to select the optimum cutting conditions based on economic criteria, select appropriate cutting tools from the available inventory, determine the type of cut (i.e., single- or multipass operations, number of passes, depth of cut for each pass), and select suitable machine tools from the available inventory. It is not uncommon to consider the surface finish and surface quality/integrity as manufacturing specifications which eventually control the selection of cutting parameters. In principle, the computer selected speeds and feeds can be inserted into the NC program automatically, but this has not been done on a widespread basis in practice.

When machining centers are managed based on experience, the tools are typically set up in the magazine in an order determined by the operator's knowledge and experience. This may result in very significant underutilization of the machine, especially for predetermined (rather than random-select) tool changing systems. This problem can also be avoided by considering optimal tool order loading strategies in an automatic tool-change (ATC) magazine based on the part process sequence. This is particularly important when several operations on different faces of a part are performed, so that a number of tool changes and pallet rotations are necessary.

The *second level* of a COS optimizes the manufacturing process and sequence of operations (in the case of drilling, reaming, tapping, etc.) on a single- or multistage machining system (e.g., dial machines, conventional transfer lines, flexible/convertible transfer lines, multiaxis machining centers with single or multiple spindles, and flexible manufacturing systems) using the specified part shape, size, and dimensional accuracy. Often the features to be machined are extracted from a computer-aided design (CAD) model using a feature extraction algorithm as described in Section 16.5. Such a planning system will reduce dependence on skilled human personnel for the development of an efficient part process sequence and tool magazine organization, and can be designed to take advantage of the machine capabilities and system flexibility. The benefits of such an approach include gains in production time, machine utilization, and the number of parts produced between tool magazine reloadings, which results in lower manufacturing part costs. The second level of a COS is sometimes referred to as CAPP system as discussed in Chapter 16. Several CAPP systems are being developed as a link between design and manufacturing, filling the gap between CAD and computer-aided manufacturing (CAM).

The second level of a COS interacts with the first level. It uses detailed part geometry, machine motion, and tooling data files to calculate individual machine motions and cutting times based on the optimum cutting conditions identified by the first level of the COS. This analysis can provide information on gross productivity, and estimate of the required investment, and detailed data on individual machine stations and station performance. Such computer programs analyze both the whole system and individual machining stations to balance work loads, increase productivity, and decrease investment. Compared to manual methods, such machine and part processing simulations can significantly reduce the total time required to develop and analyze manufacturing systems. Several variations of the machining system, the processing sequence, and tooling designs can be evaluated with minimal effort once the initial analysis of the system is completed.

The *third level* of a COS is used to determine optimum combination of all machining parameters including the machine tool, cutting tool material and geometry, and cutting fluid. The optimization must include all system objectives as outputs and should use the entire range of working conditions as inputs. Approaches such as expert system concepts and neural networks for modeling manufacturing process parameters have been used to develop third-level COSs.

A knowledge-based expert system [2, 3] includes a machinability database (experimental results and theoretical predictions including information on machine tools, cutting tools, coolants, etc.), inference engine, knowledge base, and working memory. A wealth of symbolic (rules of thumb) knowledge gained through experience can be incorporated into the knowledge base. The knowledge is collected, analyzed, and systematically organized. The inference engine interacts sequentially with the databases to find the best combination of process parameters (selection of machine tools, cutting tools, coolant, etc.), from which the first or second level of the COS can be used to determine the optimum process.

Neural networks [4–6] process information through the interactions of a large number of simple processing elements or nodes. Knowledge is represented by the strengths of the connections between elements. Each piece of knowledge is a pattern of activity spread among many processing elements, and each processing element may be involved in the partial representation of many pieces of data. There is an input layer and an output layer to receive data and send information out to users. Neural networks have the ability to learn using an algorithm that maps input vectors to output vectors.

13.3 ECONOMIC CONSIDERATIONS

To reduce manufacturing costs, the direct concern of factory floor management has traditionally been the improvement of machine utilization and the reduction of part lead time. Machine time and part lead time are composed of:

$$\text{machine time} = \text{setup time} + \text{loading/unloading time} + \text{cutting time} + \text{idle time}$$

$$\text{part lead time} = \text{setup time} + \text{loading/unloading time} + \text{cutting time} + \text{waiting time} + \text{moving time}$$

Machine utilization and lead time efficiency are defined, respectively, as:

$$\text{Machine utilization} = \frac{\text{Cutting time}}{\text{Machine time}}$$

$$\text{Lead time efficiency} = \frac{\text{Cutting time}}{\text{Lead time}}$$

Much effort has been put into improving machine utilization and lead time efficiency through the improvement of software and hardware. Software improvements include better scheduling and control strategies, but are difficult to implement in the case of long setup times, batch (job-shop) production environments, and machine flexibility or capability limitations. Hardware improvements including the development of new machines or machining systems with better versatility and flexibility can be used to improve both machine utilization and lead time efficiency. These improvements, which are essentially improvements in processing speed and versatility, generally increase capital costs and decrease operating costs.

Reducing the setup time, loading–unloading time, and cutting time for a workpiece reduces not only the machine time and lead time, but also the waiting time for other workpieces. In fact, the waiting time for job-shop part applications typically accounts for more than 90% of the lead time. Queuing theory indicates the reduction of some fraction of the part processing time requires the reduction of a higher fraction of the part waiting time. Therefore, it is very important to define the relationship between the increased machine investment cost and reduced factory operating cost including the cost of workpiece congestion (work-in-progress or WIP inventory cost). The major aim of a queuing model is to determine the relationship between the throughput rate and average WIP inventory level. Justifying investment requires economic analysis of all operations of the machine(s).

An engineering methodology is necessary to evaluate the value of investment on new or current manufacturing technologies for high-volume or job-shop applications in a new or existing factory environment. This approach can provide a better understanding of factory operations and guidelines for investment decisions for new technologies, workpiece routing methods, scheduling methods, new machine designs, or the determination of proper WIP inventory levels. For the machine tool investment problem, several alternatives can be evaluated based on some evaluation criteria such as functional (technical) capacity, production cost, reliability, product quality, etc. as shown in Figure 13.3 [7–9]. The relative significance of each factor depends on each investment case. However, production cost is the most basic factor; among the sub-factors, capital cost, labor cost, depreciation cost, and inventory cost are the most significant considerations in addition to the material cost. The production cost is given by:

The operating cost of the production
= (capital opportunity cost) + (depreciation cost) + (labor cost)
= (capital investment)(interest rate)(processing time)/utilization
  + (capital investment)(depreciation rate)(processing time)/utilization
  + (labor cost rate)(processing time)

Reliability is important because, as the capability of machines increases, the failure of machines has a bigger effect on the material flow. This requires an analytical evaluation of routing changes for the workpiece during extended breakdowns. The part quality becomes important because a bad workpiece detected at an inspection station is either reprocessed or

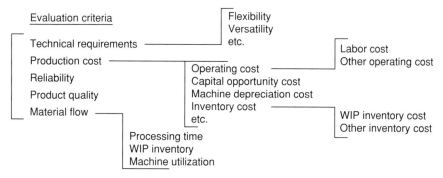

FIGURE 13.3 Some factors to consider for each machine for a machine investment problem.

discarded. In the case of a discard, part processing expenses and raw material expenses are lost. Inspection cost may also increase as part quality decreases. Understanding the cause of poor quality and corrective measures (if they involve modifying machines) contribute additional costs.

Most of the literature on machine investment problems deals with simple machines (or one-step operations) without consideration of material flow, and as a consequence most of the results are not applicable to multimachine operations [10–15]. The expected increase in system productivity with an increase in the machine capacity is often not observed because minimal consideration is given to material flow. Therefore, in introducing a new technology, it is important to consider not only the cutting capacity and setup time, but also the effect on factory material flow. The variables of interest in this case are part flow times, WIP inventory levels, machine utilization, product mix, and workpiece processing rates. The material flow for a job-shop can be most naturally modeled as a queuing network [16–22].

13.4 OPTIMIZATION OF MACHINING SYSTEMS — BASIC FACTORS

The most important factors which influence a process are the workpiece material and finished part requirements, type and structure of machine tool, cutting tool material and geometry, toolholder, cutting fluid, and cutting conditions. A knowledge on these factors, generally based on experience, must be incorporated into the knowledge base.

The specific information for each group of factors necessary to implement a COS is discussed in the remainder of this section.

Workpiece and material: The material type, microstructure, hardness, ductility, machinability rating, heat treatment, and initial surface condition affect all the selection parameters discussed next. In addition, the surface finish and integrity, starting size and shape, final size and shape, tolerances, and lot size affect the selection parameters.

Machine tool selection: For a given part size and geometry, a family of machine tools will be available from the database provided by the user. The operation selection and best tool orientation will screen the available machines further or, vice versa, the available machine tools will determine the best processing and tooling requirements.

The artificial intelligence approach requires that several machine tool performance parameters be available in the database. These parameters should describe the machine type (number of axes, spindle orientation, etc.), table type (rotary, indexing, etc.), table rotational speed, pallet change time, workpiece weight limit, capabilities of each axis (acceleration/deceleration, rapid traverse rate, maximum feed rate, maximum thrust, resolution increment and full travel distance), spindle capabilities (speed versus torque/power response, maximum speed, maximum thrust and radial forces, and acceleration/deceleration rate), tool-change time, tool-spindle interface style, magazine capacity, coolant type(s), and accuracy and repeatability of the axes.

Cutting tool selection: Tool parameters to be selected include the tool type and style, material and grade, and geometry (tool angles). The tooling database should contain the tool description (type and style), diameter, geometry, chip breaker, material (grade), hardness, coating, number of teeth, tool performance with respect to chip ejection, hole location, and maximum allowable feed rate, surface speed or rpm, forces, and torque.

Toolholder selection: The proper selection of the toolholder is critical to tool performance as discussed in Chapter 5. The database should contain the available types of toolholders for the available cutting tools and their performance characteristics. The limitations of each toolholder type with respect to roundness and location as a function of the L/D ratio of the tool should be included. A comparative assessment of the toolholder stiffness and damping as a function of the L/D ratio of the tool should also be available.

Cutting fluid selection: The selection of the cutting fluid depends mainly on the operation, tool type, workpiece material, and cutting conditions. Some applications require coolant (i.e., gun drilling) while others prevent coolant application (i.e., ceramic tools). The coolant application method is also important.

Cutting conditions selection: The selection of cutting conditions depends on all of the factors listed above and can be optimized as will be discussed in the remainder of this chapter. The basic parameters to be selected are the cutting speed, feed rate, depth of cut, and length and width of cut, which determine cutting forces and tool life.

13.5 OPTIMIZATION OF MACHINING CONDITIONS

The proper selection of machining conditions — the cutting speed, feed rate, and depth of cut — has a strong impact on machining performance. The optimization of cutting conditions was initially investigated by Gilbert [23] in 1950 and is still an active area of research. Many new concepts and optimization procedures have been developed to determine optimum conditions, taking into account as many influencing variables as possible [24–26]. However, the use of these advanced techniques in machinability database systems to date has been very limited, even though there is a strong economic need for such techniques.

Many investigations on the optimization of cutting conditions have been confined to single-stage manufacturing, namely, on a single machine tool using a single cutting tool. Single-stage optimizations may involve single- or multipass operations. A single-pass analysis typically addresses the problem of determining optimum values of the cutting speed and feed rate assuming that the depth of cut is given. A multipass analysis typically addresses the problem of determining optimum cutting speeds, feeds, depths of cut, and number of passes simultaneously for a given total depth of cut. Turning is the most commonly analyzed machining operation due to its simplicity. Since a product requiring only one machining operation is seldom found in practice, optimization of multistage manufacturing systems has been attempted by a few researchers.

The steps involved in optimizing cutting conditions include:

- formulating an objective function based on the desired economic criteria;
- defining all the constraints applicable to the machining system; and
- minimizing/maximizing the objective function subject to the constraints.

Formulation of the objective function based on the desired economic criterion has been investigated extensively in the literature. Economic criteria are normally functions of cost elements, metal removal rates, and tool life. Representative objective functions are related to only one criterion, such as the production cost, production time [27–31], profit rate [32, 33], or to a combined criterion based on a weighted sum of these factors [34].

An objective function based on production cost does not consider the influence of production time. Maximum production objectives identify the cutting conditions that best balance the metal removal rate and tool life to produce the highest output. Minimum cost objectives identify the cutting conditions that best balance the metal removal rate and tool life for the lowest cost (see Figure 13.1). Maximum rate of profit criteria (or maximum efficiency criteria) have been proposed to include and eventually achieve a balance between the contributions of both minimum production cost and total time criteria into the objective function [32, 33]. However, this criterion cannot adequately predict the optimum cutting conditions; it has been found that unless the profit margin is very high, the optimum conditions predicted by the maximum profit rate criterion lie close to the minimum production cost conditions [33]. Also, it cannot be readily applied since prior knowledge of the percentage of the product sale price (usually unknown) attributed to the particular operation being optimized is needed to determine the operation's profit rate. A more effective technique, especially for complex production systems, may be multiple criteria optimization [34], which considers interactions between the various criteria. A difficulty with this approach is that the order of importance of the different criteria must be specified, and this introduces a subjective element into the analysis.

13.6 FORMULATION OF THE OPTIMIZATION PROBLEM

The formulation of the optimization problem requires specification of equations to represent the economic and physical parameters of the machining process and the entire machine–tool–workpiece system. The physical parameters are obtained through tests, previous production runs, or existing data from machinability database systems.

13.6.1 FORMULATION OF OBJECTIVE FUNCTION

Both the production cost and total production time are considered in the formulation of the objective function. The equations for determining cutting parameters, metal removal rate, machining cost and time are quite general and may be used for a variety of machining operations such as turning, milling, drilling, etc. The total production time for an operation is given by

$$T_{u_i} = t_{m_i} + \frac{t_{m_i}}{T_i} t_{l_i} + t_{cs} + t_{e_i} + t_{r_i} + t_{p_i} + t_{\alpha_i} + t_{d_i} + t_{x_i} \tag{13.1}$$

where t_{m_i} is the cutting time and t_{l_i}, t_{cs}, t_{e_i}, t_{r_i}, t_{p_i}, t_{α_i}, t_{d_i}, and t_{x_i} are the tool's loading/unloading, interchange, magazine traveling, approach, table index, acceleration, deceleration, and rapid travel location times, respectively, for the ith operation or pass. The tool's loading/unloading time, t_{l_i}, represents the time to replace a worn or broken tool (not usable for further machining) in the machine. The interchange time, t_{cs}, represents the time required for a tool on the work spindle to be replaced from the tool magazine by the tool required for the next operation. All these times are required because the tool-change time between consecutive operations in an NC machine is not constant, but varies depending on the speed and acceleration/deceleration time of the spindle and machine slides, tool switch time, machine travel distance (which allows for tool change between operations), and the tool magazine indexing time. The assumption of constant tool-change time is often erroneous, especially when applied to high-speed machining centers for which the tool-change time is as significant or perhaps more important than the machining time. T_i is the life of the tool used in the ith operation.

For a turning, boring, and drilling operation the cutting time is equal to:

$$t_{m_i} = \frac{\pi D_i L_i}{1000 V_i f_i} \tag{13.2}$$

For a milling operation,

$$t_{m_i} = \frac{\pi D_i (L + \varepsilon)_i}{1000 n_{t_i} V_i f_i} \tag{13.3}$$

For a tapping operation, Equation (13.2) is used with the thread pitch m_i substituted for the feed f_i. D is either the diameter of the machined surface or the rotary tool diameter. L, V, and f are the workpiece length to be machined, cutting surface speed, and feed of cutting tool, respectively. n_t and ε are, respectively, the number of inserts (edges) per cutter body and the overtravel of milling cutter on the workpiece (Chapter 2).

The machine slide's rapid travel time between two machine coordinate locations t_{x_i} is

$$t_{x_i} = \frac{V_s^2 + 7.2\alpha_s \Delta S}{3600\alpha_s V_s} + t_{s_i} \tag{13.4}$$

where $\Delta S = \sqrt{X_i^2 + Y_i^2 + Z_i^2}$
and V_s and α_s are the speed and acceleration of the machine slides, respectively. ΔS is the machine rapid travel distance, and X_i, Y_i, and Z_i are the machine traveling coordinates for the ith operation. The deceleration function of the slides consists of two parts, linear and nonlinear. A two-part rampdown is used to maximize speed during deceleration, yet minimize deceleration discontinuity which causes system oscillation or shock. The time required to accomplish the nonlinear deceleration portion is called settling or stabilization time of the machine slides, t_{s_i}, which is a function of the servo control characteristics.

The time required for the spindle to accelerate or decelerate during the tool change is

$$t_{\alpha_i} = t_{d_i} = \frac{V_i}{1.8\alpha D_i} \tag{13.5}$$

where α is the acceleration of the spindle. Linear accelerations and decelerations are assumed in the previous two equations.

The tool magazine indexing (traveling) time is almost proportional to the number of slots that the magazine has to travel:

$$t_{e_i} = \frac{|j_i - j_{i-1}|S_{cs}}{1000 V_{cs}} + \frac{V_{cs}}{1800\alpha_{cs}}, \quad j = 1, 2, \ldots, M^T \tag{13.6}$$

where S_{cs} is the distance between adjacent slots in the tool magazine and α_{cs} and M^T are the acceleration and the number of slots of the tool magazine, respectively. The tool switch time, t_{cs}, is the time required for a tool in a tool chain slot to be inserted into the spindle (Chapter 3).

A general tool life equation applied to all processes is

$$T_i(\underline{s}) = R_{ci} \prod_{k=1}^{p} s_{ik}^{a_{ki}}, \quad \forall i \in A_c, \forall k \in A_b, s \in \{f, d_c, V, V_B, \ldots\} \tag{13.7}$$

where s_{ik} is the kth cutting parameter for the ith operation and R_c is a tool life constant.

The production cost for an operation is given by

$$\begin{aligned} C_{u_i} &= C_o t_{m_i} + \frac{t_{m_i}}{T_i}(C_o t_{l_i} + C_{t_i}) \\ &+ C_o(t_{cs} + t_{e_i} + t_{r_i} + t_{p_i} + t_{\alpha_i} + t_{d_i} + t_{x_i}) \quad \forall i \in A_c \end{aligned} \tag{13.8}$$

where C_o and C_t are the operating cost and tool cost, respectively. The relationships between the cutting speed and these costs are shown in Figure 13.1. The machining cost decreases with increasing cutting speed, while both the tool cost and the tool-change/setup cost increase with increasing cutting speed. The cutting speed for minimum cost per part is determined by taking the partial derivative of Equation (13.8) with respect to speed and setting it equal to zero. The result is

$$V_{opc} = V_{\min \cos t} = \frac{K_t}{f_h^a d_c^b \left[\left(\frac{1}{n}-1\right)\left(t_l + \frac{C_t}{C_o}\right)\right]^n} = \frac{\left[(-a_3-1)\left(t_l + \frac{C_t}{C_o}\right)\right]^{1/a_3}}{(R_c f_h^{a_1} d_c^{a_2})^{1/a_3}} \quad (13.9)$$

where f_h and d_c are the highest possible feed and depth of cut; the tool life constants a, b, n, a_1, a_2, and a_3 are from the tool life equation, Equation (13.7). When the maximum production rate is required, the time per part per operation, given by Equation (13.1), must be minimized. In this case, the tool-change/replacement times are important. The highest permissible production rate is again found by selecting the highest possible feed and calculating the cutting speed from [35]

$$V_{opt} = V_{\max prod} = \frac{K_t}{f_h^a d_c^b \left[\left(\frac{1}{n}-1\right)t_l\right]^n} = \frac{[(-a_3-1)t_l]^{1/a_3}}{(R_c f_h^{a_1} d_c^{a_2})^{1/a_3}} \quad (13.10)$$

The cutting speed for maximum production rate (V_{opt}) is always higher than that of minimum cost (V_{opc}) as shown in Figure 13.1. The speed range V_{opc} to V_{opt} is referred to as the high efficiency machining range or "HI-E" range, which means that any machining speed in this range is preferable from an economic standpoint. In cases in which the production rate is fixed (as in transfer lines and dial machines), the economic tool life or maximum production rate should not be used. Instead, the tool cost should be kept as low as possible, that is, by selecting the cutting conditions most suitable for the required production rate (cycle time). However, it is desirable to use V_{opt} in bottleneck operations.

There are several cases that require the estimation of the optimum cutting parameters based on tool life or tool wear equations with constraints on the metal removal rate, thrust force, torque, etc [24, 36, 37].

Tool life and especially tool cost are functions of the number of regrinds. The effective number of total regrinds, K_r, before the tool should be scrapped are determined at the optimum cutting conditions, especially for brazed tools, which have a very short regrindable length. A tool life reduction between regrinds after the tool has already been reground q times is usually observed and the tool life is therefore given by an exponential function:

$$T_{irj} = e^{\beta_r \cdot (j-q)} T_{ir1}, \quad j = 1, 2, \ldots, q, q+1, \ldots, K_r+1 \quad (13.11)$$

where r is the number of tool steps for multistep tools, j represents the number of the tool regrinds, and

$$\begin{array}{ll} \beta_r = 0 & j < q \\ -0.1 < \beta_r < 0 & q < j < K_r + 1 \end{array}$$

The appropriate value of K_r therefore can be determined by evaluating the operation's objective function U_r for various number of regrinds and selecting the optimum K_r:

$$\frac{U(K_r) - U(1)}{U(1)} < \varepsilon_x < \frac{U(K_r+1) - U(1)}{U(1)}$$

for a specified ε_x ($0 < \varepsilon_x < 1$).

The expected productive tool life is

$$E[T_{irj}] = T_{pirj} = P_s T_{irj} - P_f t_{m_r} \quad \text{for } \forall r \in A_r \quad (13.12)$$

where P_s and P_f are the probabilities for tool survival and tool failure, respectively, and A_r is the set consisting of all tool steps.

The mean time for tool replacements for a mixed servicing approach is

$$\overline{T}_s = P_f T_{s1} + P_s T_{s2} \tag{13.13}$$

where T_{s1} and T_{s2} are the times required to accomplish a replacement-by-failure and a forced replacement, respectively.

The total production time per part is

$$T_t = \sum_{i=1}^{n} T_{u_i} + t_h \tag{13.14}$$

where t_h is the loading and unloading time for the part and n is the number of operations or passes performed on the part in a given setup.

The profit rate is given by

$$P = \frac{S - C_u}{T_u} \tag{13.15}$$

where S is the sale price of the machined part. Until the maximum profit criterion was developed there had been no way to indicate how the optimum point should be chosen from the HI-E range from V_{opc} to V_{opt}. The cutting speed for maximum profit lies between V_{opc} and V_{opt}, and provides an economic balance between the cost and production rate [38, 39].

The objective function is a weighted sum of the production cost and the production time:

$$U(f, V, d) = w_1 C_u + w_2 \lambda T_u \tag{13.16}$$

where w_1 and w_2 are the weighting coefficients representing the relative importance of the production cost and the total production time criteria, respectively, and λ is a conversion factor (see below). It is usually assumed that the weighting factors should satisfy the condition

$$w_1 + w_2 = 1, \quad 0 \leq w_1 \leq 1, 0 \leq w_2 \leq 1 \tag{13.17}$$

Their specific values should be selected based on the relative importance of cost and production rate for the prevailing business conditions. Usually, the optimization problem is solved for different trial weight values to determine the optimum weight distribution for a particular case. These weight values can be used for optimizing processes for similar parts. Hence, the weighting coefficients are usually determined in part by analytical means and in part by experience.

The objective function is normalized using a constant multiplier, λ:

$$\lambda = \frac{C_{u,\min}}{T_{u,\min}} \tag{13.18}$$

where $C_{u,\min}$ and $T_{u,\min}$ are the minimum production cost and total time, respectively, under the defined process constraints. The minimum cost and total time are determined by using the optimum values of the cutting speeds, V_{opc} and V_{opt}, in Equations (13.8) and (13.1) and the highest possible feed, f_h, in both equations. An analytical equation for λ cannot be developed because the highest feed is a function of depth of cut under the applicable process constraints.

13.6.2 Constraints

Physical limitations on cutting conditions due to the characteristics of the machine–tool–workpiece system should be identified from previous experience and taken into account in the optimization process. A number of deterministic constraints exist, such as allowable maximum cutting force, cutting temperature, depth of cut, speed, feed, machine power, vibration and chatter limits, and constraints on part quality. Some constraints, such as those on speed, feed, etc., are simple boundary constraints, while others must be computed from linear or nonlinear equations. The following general equation can be used:

$$G_{ij}(s) = B_{ij} \prod_{k=1}^{p} s_{ij}^{b_{kij}} \quad \forall i \in A_c, \forall k \in A_b, j \in 1, 2, \ldots, M_i^c \tag{13.19}$$

where b_{kij}, B_{ij}, and E_{ij} are empirical constants for the ith operation and jth constraint. Mathematical models for some of the commonly used constraints as a function of cutting speed, feed, and depth of cut are given in Refs. [34, 40].

In the case of automated production, a chip form constraint is required so that no undesirable chips are produced. However, chip form constraints have not been studied in detail to date. The chip-breaking capability of the tools must be analyzed as a function of machining variables (as discussed in Chapter 11) and this data must be stored in the database. Using this data, a chip disposal constraint can be identified for use in the optimization.

13.6.2.1 Depth of Cut

In general, tool life is less affected by changes in the depth of cut than by changes in either the feed or speed, particularly for roughing cuts. The most favorable compromise between the tool life and the metal removal rate in rough cutting is therefore often obtained when the highest permissible depth of cut is used. However, a heavier depth of cut could result in decreased tool life in semiroughing and finishing cuts. A 50% increase in the depth of cut will typically produce only a 15% reduction in tool life when the depth of cut exceeds ten times the feed rate [37]. The effect of the depth of cut is somewhat greater when the starting depth of cut is less than ten times the feed rate. The selection of the maximum depth of cut is dependent on: (a) tool material and geometry, (b) the cutting force, (c) the available machine horsepower, (d) the stability of the tool–work–machine system, (e) the required dimensional accuracy, and (f) surface finish requirements.

Limits due to material considerations (a) are usually provided by the tool manufacturer; the tool material and geometry determines the strength of the cutting edge and deflection characteristics of the tool under an applied force. Cutting force limits (b) become significant when deflection and chatter are of concern. Chatter is a major factor limiting the maximum depth of cut because it affects the surface finish, dimensional accuracy, tool life, and machine life. The prediction of the maximum chatter-free depth of cut and the speed intervals in which resonant frequencies occur are explained in detail in Chapter 12.

13.6.2.2 Feed

The maximum allowable feed has a pronounced effect on both the optimum spindle speed and the production rate. Feed changes have a more significant impact on tool life than depth of cut changes. In many cases, the largest possible feed consistent with the available machine power and surface finish requirements is desirable in order to increase machine utilization.

Machining Economics and Optimization

It is often possible to obtain much higher metal removal rates without reducing tool life by increasing the feed and decreasing the cutting speed. This technique is particularly useful for roughing cuts, in which the maximum feed is dependent on the maximum force the cutting edge and the machine tool are able to withstand. The maximum permissible feed f_h used to calculate the cutting speeds V_{opc} and V_{opt}, can be determined from

$$f_h = \min\{f_{c,\max}, f_{s,\max}, f_{F,\max}\} \qquad (13.20)$$

where $f_{c,\max}$, $f_{s,\max}$, and $f_{F,\max}$, are the maximum allowable feeds, respectively, due to the chip-breaking and surface finish requirements and to force limitations. In general, the maximum feed in a roughing operation is limited by the maximum force that the cutting tool, machine tool, workpiece, and fixture are able to withstand. The maximum feed in a finish operation is limited by the surface finish requirement (Chapter 10).

13.6.2.3 Cutting Speed

The cutting speed has a greater effect on tool life than either the depth of cut or feed. When the cutting speed is increased by 50%, the tool life based on flank or crater wear typically decreases by 80 to 90% [41]. The influence of cutting speed on tool life is discussed in detail in Chapter 9.

The cutting speed has only a secondary effect on chip breaking when varied within conventional ranges. Depending on the type of tool and the workpiece materials, certain combinations of speed, feed, and depth of cut are optimal from the point of view of chip removal. As discussed in Chapter 11, charts providing the feasible region for chip breaking as a function of feed versus depth of cut are often available from tool manufacturers for a specific insert or tool.

The constants in the deterministic constraint relation, Equation (13.19), must be identified from experimental data and are thus subject to inherent uncertainty. Experimental results contain only partial information because the number of experiments conducted is always limited, the measurements involve errors, and important (often unknown) factors may not be recorded or controlled. Also, physical constraints change during the life of the cutting tool and thus during the machining of each part, since the machining process changes with tool wear. Hence, most of the constraint equations, and especially the surface finish model, should be functions not only of the speed, feed, and depth of cut, but should also include tool wear as an independent variable. The experimental evaluation of tool life, even in particularly simple conditions, may require a great deal of testing due to the inherent variability in tool wear processes. General constraints can be developed for a specific work material and tool geometry family for a specified range of cutting conditions. Determination of constraint equations is a very tedious process but generally does not require testing as extensive as that required for estimating parameters in a tool life equation over a broad range of conditions. Constraint equations, however, are applicable only to the test conditions, and are thus valid only for a limited range of cutting conditions for a particular application. Changes in the size or geometry of the part and the machine, tool, or coolant generally require additional testing to update constraint relations. Some of the parameters in constraint equations can be predicted to an accuracy within 10 or 20% using analytical models which account for the effects of the tool geometry, workpiece and tool material properties, and machine variables, but which use test data as inputs to account for the size or geometry of the part and machine characteristics [34, 42–48]. This approach can be used to advantage for materials used in high-volume applications over long periods of time, since either turning test data or empirical tool life data is often available in this case.

13.6.3 PROBLEM STATEMENT

The machining optimization problem for a single-pass, multipass, and single-stage multifunctional system can be formulated as follows:

Minimize/maximize the objective function $U_i(\underline{s})$

The objective function measures the effectiveness of the decisions about inputs as reflected in quantifiable outputs. A qualitative graph of minimum cost per piece and the maximum production rate (i.e., the minimum time per piece) is shown in Figure 13.1.

Subject to the constraints

$$G_{ij}(\underline{s}) \leq 0 \quad \forall i \in A_c \text{ and } j \in 1, 2, \ldots, M_i^c \tag{13.21}$$

The machining optimization problem for a multistage machining system (Figure 13.4) can be formulated as:

Minimize the objective function

$$U\left(\underline{\underline{s}}\right) = \sum_{i=1}^{N} U_i(\underline{s}) \quad \forall i \tag{13.22}$$

Subject to the constraints

$$F_{ij}(s) = G_{ij}(s) - E_{ij} < 0 \quad \forall i\, j \tag{13.23}$$

$$\Omega_i(s) = 0 \quad \forall i: T_{u_i} < t_x \tag{13.24}$$

The equality constraint, Equation (13.24), is applied only when the production time of the ith operation or station is lower than the system maximum cycle time t_x. E_{ij} are the upper bound values for the inequality constraints. The condition for optimization for Equation (13.22) is

$$U\left(\underline{s}^*\right) = \min_{\underline{s}^* \in S} U\left(\underline{s}\right), \quad U: S \to E_1$$

$$S = \left\{\underline{s}^* \in E_1^n, F\left(\underline{s}^*\right) 0 \leq \text{ and } \Omega\left(\underline{s}^*\right) = 0\right\}, \quad \underline{s}^* \in [V, f, \text{depth of cut}, \ldots]$$

is the vector of decision variables defined in the n-dimensional Euclidean space of variables E_1^n. The optimum cutting parameters are given by the matrix $\underline{\underline{s}}^*$.

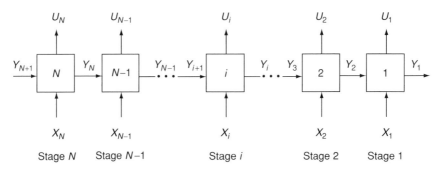

FIGURE 13.4 A multistage flow machining system.

13.7 OPTIMIZATION TECHNIQUES

13.7.1 SINGLE-PASS OPERATION

The optimization problem for a single-pass operation is straightforward because it can be reduced to the two-variable problem of determining the cutting speed and feed. Several methods of solution for this problem have been proposed by various investigators [49]. The simplest method is *linear programming* [50, 51], which requires a linear objective function and linear constraints. In this method, the various constraints are expressed as power transforms and the mathematical relations are linearized by logarithms. Standard linear programming procedures like the Simplex algorithm [52, 53] can be used for optimization. *Nonlinear programming* methods may also be used. The sequential unconstrained minimization technique (SUMT) with the Davidon–Fletcher–Powel algorithm [27] can be used; this is a penalty function method. There are several problems with this approach which can adversely affect results. Another approach is to use a logarithmic transformation [54] and solve the resulting equations using Rosen's gradient method with linear constraints. Some constraints are not readily transformed and must be approximated by piecewise linear functions, which significantly complicates the use of this search procedure [55]. The above methods are all deterministic approaches. A probabilistic approach using *change-constrained programming* has been suggested by Iwata et al. [56]; in this method, normal distributions for the constants in the constraints are used. The probabilistic nature of tool life affects the optimum cutting conditions and can be represented using a tool life probability distribution function [27, 57, 58]. *Polynomial geometric programming* can be used assuming an extended Taylor tool life equation [59–65]. Geometric programming can also be used based on quadratic polylognomials (QPL) [66]. *Probabilistic geometric programming* has been applied to multiobjective functions [67]. *Goal programming*, a special type of linear programming developed to solve problems involving complex and usually conflicting multiple objectives, has been applied with both linear [68–70] and nonlinear [71] goals. *Genetic algorithms* (GAs), based on the principles of natural biological evolution, have been used to optimize not only the cutting conditions but also system modeling and process optimization in automated process planning systems [37, 72–74]. Finally, a *fuzzy optimization* approach can be used for selecting optimal machining conditions based on fuzzy set theory; this method manipulates the uncertainty present in the empirical equations or experimental data [75, 76]. Genetic optimization of fuzzy rules has been carried out with the help of an object-oriented genetic optimization library (GOL) [26]; the GOL can help the development of the fuzzy expert system for machining data selection. *Simulated annealing algorithms*, based on the annealing process, have also been used to search for the optimal process parameters [24]. A computer-aided graphical approach for the optimization of the cutting conditions can be very helpful as illustrated in Example 13.4.

13.7.2 MULTIPASS OPERATION

The optimization problem for a multipass operation is generally more complex than for a single-pass operation [40, 65, 77–83]. A multipass process, particularly multipass turning application, is a four-variable problem in which the number of passes and depth of cut for each pass must be determined through a dynamic programming procedure while the optimum cutting speed and feed for each pass is determined using single-pass optimization methods. Dynamic programming [84] is very useful in handling a multipass problem in which every cutting pass is independent of the previous passes, such as in turning operations. The decision variable is the depth of cut to be taken in the ith pass, which is represented as $d(i,j)$, and the state variable is the diameter of the workpiece at the ith pass, D_i. $d(i,j)$ is interpreted as the depth of cut that starts at diameter D_i and consists of j sections of size d. Referring to

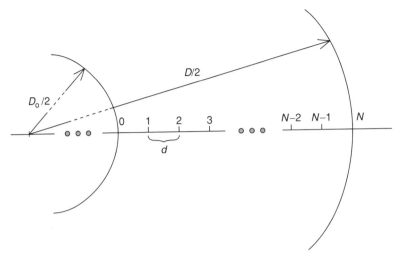

FIGURE 13.5 The division scheme for the total depth of cut to be removed in a multipass operation.

Figure 13.5, DC is divided into N equal sections d, which are the N discrete decision states for dynamic programming. The minimum increment for the depth of cut is $DC/N = d$, where d is defined arbitrarily and must be smaller than the maximum depth of cut allowed for a particular machine–tool–workpiece system. The optimum number of passes M is determined by the dynamic programming approach; each pass consists of a certain number of sections d. The starting point for dynamic programming is the inner section of the workpiece, D_o, which will result after the required material has been removed. From the inner to the outer section of the workpiece, the stages are denoted by first section, second section, and so on up to the Nth section (Figure 13.5). As noted above, the optimization of each stage of the dynamic programming approach is carried out using one of the above methods for a single-pass optimization.

13.7.3 Single-Station Multifunctional System

The productivity of a single-station multifunctional system (SSMS) is dependent not only on the cutting time, but more importantly on the noncutting time; the part process sequence for the various operations, which significantly affects the part cycle time, should be considered in the optimization. There are three stages to planning part processes for such systems: (a) determining the cutter trajectory from geometric models of machining processes [85, 86], (b) developing mechanistic models for the machining operations [87, 88], and (c) applying optimization methodologies for the machining system.

A fundamental analysis of an SSMS was developed by Hitomi et al. [89, 90], in which the optimum cutting speeds were analyzed based on a minimum production cost criterion with due-date constraints by neglecting all necessary constraints at each stage of the system. In a later publication, Hitomi and Ohashi [91] proposed a computational algorithm for determining the optimum sequence of operations for a random-select tool changer type (using the classic traveling salesman problem) by considering the tool-change time as a function of the tool location in the tool magazine and providing the time matrix for extracting tools, so that the noncutting time could be properly accounted for. Their scheme optimized the noncutting time attributed only to the tool magazine indexing and did not account for the cutter trajectory. Constraints between operations, such as precedence relations and contiguous allocations, were not considered in the mathematical algorithm as these constraints are very difficult to describe in an integer programming approach.

Tang and Denardo [92] considered the minimization of the number of tool switches. The job sequence was considered fixed, and tooling decisions were the only decisions to be made. The objective of this problem was to determine the set of tools to be placed on the machine at each instant so that the total number of tool switches was minimized. They considered the case in which the tool capacity of the machine magazine is lower than the number of tools required in order to process a certain number of jobs.

A comprehensive optimization scheme for the part sequence of operations in SSMSs was proposed in Ref. [93]. The mathematical model and methodology for the determination of the optimum sequence of operations were analyzed with an equal emphasis on tool magazine organization for high-speed SSMSs. The tools required for the completion of all the operations by a machine are stored in its limited capacity tool magazine (storage chain). The number of tool magazine loading instants becomes important when large batch or mass production of identical parts is considered. The capability of storing tool information, such as total tool life, accumulated tool life during machining, etc., should be also available. The setup organization of the tools in the tool magazine is important and is part of the design process. Therefore, the tool switch time, machine rapid travel time required for tool change, the tool magazine indexing time for other than random-select ATC types, and the loading/unloading time of the tool magazine, all affect the noncutting machining time and should be considered in the mathematical analysis. The optimization approach in Ref. [93] incorporates machine–tool–workpiece related restrictions (physical constraints) and shows how some of the required assumptions of current schemes can be relaxed when appropriate mathematical schemes are used and sufficient information on the tool and the machines is available from previous tests or experience.

The analytical approach to SSMS optimization can be based on one of the following three schemes:

a. *For mass production of the same part type with a cycle time constraint.* The optimum cutting conditions are defined based on the minimum production time for each operation. The optimum sequence of operations is determined. If the cycle time of the optimum sequence is shorter than the available time, the difference is distributed to the bottleneck operation(s), or uniformly to all operations, in order to increase the tool life and decrease the probability of tool failure for critical operations. This is accomplished by minimizing production cost using a production time equality constraint.
b. *For mass production of the same part type without a cycle time constraint.* The optimum cutting conditions based on the combined criterion are defined. The optimum sequence of operations is determined. For both of the aforementioned cases, the optimum loading of tool types in the tool magazine is determined in order for the SSMS to operate at its maximum continuous time between tool magazine reloadings.
c. *For production of multiple part types of short or medium batch sizes.* The batch order of the production schedule is determined. Independent optimization of each batch is considered. If there is a due-date (cycle time) constraint, Rule (a) is used. Otherwise, Rule (b) is used to solve the problem.

In Rule (c), the optimum loading of tool types in the tool magazine is determined for the completion of the maximum number of part types in different batches. A sufficient stock of tools must be available in the machine to ensure the progress of production for long periods without stoppages for refilling the magazine.

The optimization approach consists of finding the sequence of the machining operations of a part, organizing the tools in the magazine, and determining magazine reloading intervals in order to maximize the machine utilization while reducing the cycle time. The optimization of the part process needs to be reevaluated each time a redundant tool change is encountered.

This problem can be approached in several ways, two of which we will discuss. The first approach consists of determining the shortest time path that the machine requires to complete all the operations in the part by considering the constraints between operations and the tool organization in the ATC magazine. The optimum sequence of operations is determined using a network which involves the transformation of a sequential multifunctional decision process into a series of single operation processes. The problem becomes somewhat more complicated if an index (rotary) table is used to machine multiple faces on the part; if a tool is used in multiple faces, it must be decided whether to keep the tool and perform table indexes to finish all the features requiring the same tool or perform tool changes to finish individual faces sequentially. Hence, the multifunctional machining problem becomes a sequence of single operation problems for which each operation is optimized independently of the others. An algorithm for determining the optimum sequence of the operations is sought, which consists of a two-step procedure:

Step A1 — obtain the optimum cutting parameters, such as cutting speed, feed, and depth of cut, based on the predetermined tool types and tool life equation for each operation.

Step B1 — determine the optimum sequence of operations which encompass the N1 distinct tools relative position in the ATC magazine. The optimum number of the different tools in the magazine should be used in order to increase the number of machined parts between tool magazine reloadings. However, when the tool life or operation constraints are not available, the proposed scheme in Step B1 can be applied by using suggested cutting conditions from previous work. The number of multiple tools required for each tool type in order to fill the tool magazine as densely as possible can also be determined.

The optimization of the dynamic sequence of part operations is carried out by optimizing the tool organization and determining the optimum cutting conditions for the operations from Step A1. It is a combination optimization problem for various operations with precedence, pairwise, and order constraints. The shortest path algorithm is at the core of many problems in network optimization. The machining process of a part can be represented by nodes and arcs, where in some cases the tool rapidly travels from node to node (i.e., when drilling holes at different locations) while in other cases, the tool removes metal from node to node (i.e., during milling, turning, etc.). An arc represents the connection between two nodes. Operations such as drilling, reaming, etc., are represented by a node. On the other hand, operations such as milling, turning, etc., for which the starting and ending locations differ, are represented by two nodes (one at the starting location and one at the ending location).

The distance between any pair of nodes is given by the matrix $[D^*] = [d_{ij}]$. The part process is represented by a network in which an arc corresponds to the time required to rapidly travel the arc distance given by the matrix $[H] = [h_{ij}]$ of dimension $(n_1 \times n_1)$. The noncutting time between nodes is considered in the minimization function and depends on link travel times and on delays at operation switchings, which are attributed to the tool-change time including acceleration/deceleration of spindle. The entries h_{ij} can be positive or zero and are calculated based on Equations (13.4)–(13.6):

$$h_{ij} = t_{d_i} + t_{x_i} + t_{cs_i} + t_{e_i} + t_{p_i} + t_{x_j} + t_{a_j} \tag{13.25}$$

The distance and time matrices are symmetric. Therefore, the shortest path can be defined from a specific origin to a specific destination which can be the same, but this is not necessary.

The applied prerequisites for the processing sequence for certain cases are represented in matrix form. The matrix $[S_n^a]$ of dimension $(n \times n)$ contains the precedence constraints between operations:

$$S_{n_{ij}}^a = \begin{cases} 1 & \text{if operation } j \text{ precedes operation } i \\ 0 & \text{otherwise} \end{cases}$$

Precedence constraints are used especially for holemaking operations; for example, a hole must be drilled before it can be tapped.

Pairwise constraints are encountered during milling, turning, and other operations in which the tool usually starts and finishes at different part locations which constitute pairs. These constraints are represented by a vector $\{S_n^b\}$ of dimension $(1 \times n)$ with elements

$$S_{n_i}^b = \begin{cases} 1 & \text{if operation } i \text{ has a pairwise constraint} \\ 0 & \text{otherwise} \end{cases}$$

Order constraints are also encountered in part processing. This type of restriction occurs: (a) for individual operations with different starting and finishing locations (i.e., milling, etc.) in which the operation should be completed in a given direction only; and (b) between two operations which have to be processed adjacently in a specific order. They are defined by the matrix $[S_n^c]$ of dimension $(n \times n)$:

$$S_{n_{ij}}^c = \begin{cases} 1 & \text{if operations } i \text{ and } j \text{ are performed in order} \\ 0 & \text{otherwise} \end{cases}$$

or

$$S_{n_{ii}}^c = \begin{cases} 1 & \text{if operation } i \text{ has an order constraint} \\ 0 & \text{otherwise} \end{cases}$$

All constraints can be represented by a combination of precedence and pairwise constraints. The constraint $S_{n_{ii}}^c = 1$ or $S_{n_{ij}}^c = 1$ is equivalent to $S_{n_{ij}}^a = 1$ and $S_{n_i}^b = 1$.

The number of nodes is dependent on the number of operations and the number of active pairwise and order constraints (given by $\{S_n^b\}$ and $[S_n^c]$). The total number of nodes is

$$n_1 = n + 1 + \sum_{i=1}^n S_{n_i}^b + \sum_{i=1}^n S_{n_{ij}}^c \tag{13.26}$$

Two sets of constraints, represented by matrices $[C_n^a]$ and $[C_n^b]$ of dimension $(n_1 \times n_1)$, are applied which are analogous to the earlier constraints $[S_n^a]$, $\{S_n^b\}$, and $[S_n^c]$:

$$C_{n_{ij}}^a = \begin{cases} 1 & \text{if node } j \text{ precedes node } i \\ 0 & \text{otherwise} \end{cases}$$

$$C_{n_{ij}}^b = \begin{cases} 1 & \text{if node } i \text{ is adjacent to node } j \\ 0 & \text{otherwise} \end{cases}$$

$$C^a_{n_{ij}} = C^b_{n_{ij}} = \begin{cases} 1 & \text{if node } i \text{ follows node } j \text{ in order} \\ 0 & \text{otherwise} \end{cases}$$

This optimization problem can be treated as a sequencing problem similar to the traveling salesman problem (TSP), in which the shortest Hamiltonian path or cycle passing through each node exactly once is sought [94].

The optimization of the process operations problem is not equivalent to a simple TSP, but rather to a TSP with precedence, order, and pairwise constraints. Lokin [95] solved the general TSP using an exact branch-and-bound procedure with additional constraints. However, it is more convenient to solve this problem using heuristic methods which incorporate the constraints. The heuristic method is based on GAs and specifically using the genetic edge recombination operator (GERO) [96], which searches for the optimum permutation and allows constraints to be incorporated in an efficient manner. This heuristic method has been successfully applied by Whitley et al. to the standard TSP and has achieved best-known solutions for problems with up to 75 cities.

A heuristic algorithm based on the GERO which considers the constraints $[C^a_n]$ and $[C^b_n]$ can be developed based on the minimization function:

$$U^T = h(\text{oper}(n_1), \text{oper}(1)) + \sum_{i=2}^{n_1} h(\text{oper}(i-1), \text{oper}(i)) \qquad (13.27)$$

for a given sequence of operations. The algorithm randomly generates a population of processes (sequences of operations) for which the function U^T is calculated. Then the GERO is applied with a specified bias value; after a given number of generations are formulated, it stops. The final permutation is the solution. The constraints are incorporated using a penalty approach; a number larger by an order of magnitude than the time of a longer arc time is added to U^T when a constraint is violated.

In the second approach to the optimization problem, the optimum tool order in the tool magazine is determined through a two-step procedure:

Step A2 — Determine the optimum cutting parameters for each operation. The method discussed in Step A1 (of the first approach) may be used. Next, find the optimum process sequence by neglecting the tool magazine traveling time, t_{e_i}, and the tool interchange time, t_{cs}, in the time matrix $[H]$. Finally, use the analysis discussed in Step B1 below to determine the required number of tools for each operation.

Step B2 — Determine the optimum tool sequence in the tool magazine. The problem is complicated since there is a large number of possible tool combinations. However, the tools may be organized in the order of processing and multiple tools may be inserted in order of replacement determined based on an algorithm for the optimum tool type number. This need occurs for tools with lower tool life or reliability so that redundant tools are available to avoid unscheduled stoppages caused by premature tool failure. Tool replacement between the tool magazine and the tool room should also be arranged to avoid operating interruption. Simulation may be used to improve the tool sequence.

13.7.4 MULTISTAGE MACHINING SYSTEM

The optimization problem for a multistage machining system is generally very complex compared to those for a single- or multipass operation. The majority of high-volume products have traditionally been produced using conventional transfer lines, although more recently flexible transfer lines (FTLs) or flexible manufacturing systems (FMSs) have been used or considered. The productivity of a conventional transfer line, FTL, or FMS, is affected by the cutting conditions, the rapid machine travel rate, and the speed of the transfer mechanism.

FMSs used for multiproduct applications require the optimization of two functions, process planning and scheduling. However, the configuration of an FMS for a first-time installation or expansion is critical to process planning; an appropriate configuration must be chosen that satisfies both technical (design and dimension) and economic (operating cost, capital expenditure, and production quantity) requirements. The alternatives for the design and configuration of an FMS can be evaluated based on performance evaluation models or optimization models [97–104]. Performance models can be developed using static allocation models, analytical models based on queuing theory, or simulation models.

Scheduling will not be discussed in this section but is very important to the production optimization of an FMS [85, 105–110]. Watson and Egbelu [111–113] integrated both functions. They proposed a heuristic algorithm that allocates jobs to machines while at the same time generating optimum machining parameters for each job in order not to violate any job or machine related restriction. Ham et al. [114] developed a scheduling model based on multiple production stages; both the optimum job sequence for groups and jobs within each group were determined. The optimum sequence of operations in a multistage manufacturing system was also analyzed by Kishinami and Saito [115] using dynamic programming by relating the optimum sequence of operations to the optimum job allocation for each operation.

Balancing of a multistage manufacturing system, so that the total time required at each workstation is approximately the same, is an important aspect of optimization. Unfortunately, a perfect balance is not achievable in most practical situations. In this case the slowest station determines the overall production rate of the line. By reducing the machining time by increasing the cutting speed and feed rates at stations with long process times, and increasing the machining time by reducing the cutting speed and feed combinations at stations with idle time, an improvement in cycle time and line balance can be achieved. Reducing idle time by reducing the cutting speed and feed combinations prolongs tool life, reduces tool failure, and reduces the probability of machine failure. Therefore, the use of all or a substantial portion of the idle time should result in higher system uptime and higher productivity.

A fundamental analysis of a multistage machining system was described by Hitomi [116–120] in which the optimum cutting speeds and the cycle time were analyzed based on one of the following three criteria: minimum production cost, maximum production rate, or optimum cycle time. He determined the optimum cutting speeds for both single- and multi-product production [119]. However, his method does not consider the optimization of other cutting parameters, such as feed or depth of cut, and in addition, neglects all necessary constraints at each stage of the system. In a later publication [121], Hitomi applied a simplifying approximation in order to optimize the cutting speed and feed. The cycle time of the stations was controlled by the largest optimum production time of the individual stations.

The optimum cutting conditions for a job requiring multistage machining operations were determined by Rao and Hati [122] using a constrained mathematical programming approach based on one of the previous two optimization criteria, and by Iwata et al. [123, 124], on the basis of the total minimum production cost while keeping the total production time as short as possible.

Discrete cutting speeds and feeds were used by Sekhon [125] in his optimization procedure, which is simpler since there is a finite number of speed and feed combinations. This procedure is well suited to existing conventional transfer lines. However, during the design analysis of a conventional transfer line, continuous speeds and feeds are used to study the processes in order to select discrete speeds and feeds for each station.

Iwata et al. [123, 124] and Sekhon [125] did not consider the idle time for the various machining stations. Hitomi [117–120] and Rao and Hati [122] attempted to minimize the idle time through a trial-and-error method which did not guarantee optimality. An effective optimization method for the cutting conditions should be based on the consideration of minimum idle time for all machining stations.

In this section, the analysis of a fundamental model and a methodology for determining optimum cutting conditions for each station of a flow-type multistage machining system will be discussed. The optimization minimizes idle time at each station in order to reduce production costs and increase production output. This minimization is achieved using a zero idle time constraint in the optimization calculations.

The transition variables for a multistage system are most conveniently represented as the elements of an $N \times K$ matrix for the objective function. Its rows represent the machining operations and its columns the corresponding machining variables such as the speed, feed, depth of cut, and flank wear for all stations (i.e., s_{ik} corresponds to V_i, f_i, d_{ci}, V_{bi}, respectively, for $k = 1, 2, \ldots, K$ and that $i = 1, 2, \ldots, N$). In addition, it is assumed that more than one operation can be performed in a machining station.

An N multistage system is shown in Figure 13.4; each stage corresponds to a machining station. The decision process is characterized by the input parameters $Y(D, d_c, L, V_b, \text{etc.})$ and the decision variables $X(V, f)$. The objective function is denoted $U(X, Y)$. The aim is to obtain the optimum values of the speed and feed for each machining station based on the input parameters, using as much of the available machining time as possible by minimizing the idle time. This problem can be solved for different initial values of the input parameters (depth of cut, tool flank wear, etc.); a new matrix $[s_{ij}]$ is obtained for each solution.

An analytical solution of the objective function, with equality constraints for the production time and cycle time for each operation, will be analyzed next. The formulation of the problem involves some assumptions which allow the fundamental mathematical models for a single operation to be used for the analysis of the more complicated multistage system. These assumptions include:

- Suppose a conventional transfer line, FTL, or FMS consists of Z stations with a total of N operations ($Z \leq N$) to be performed. It is assumed that the minimum ideal production rate of P_x parts per hour is required; in other words, the maximum cycle time is t_x. t_x is the gross cycle time which includes the machining time, tool-change time, and the transfer time for the part between stations.
- Each part has at most N tasks or operations (work elements) on the machines which are distributed in the same order as the order of the machines.
- The parameters of the initial raw material state and the finished part state (dimensions, mechanical and physical material properties, tolerances etc.) including surface finish specifications are known.
- If the theoretical cycle time is not specified, t_x is assumed to be the optimum cycle time of either the longest or the bottleneck operation.
- The economic parameters for each operation or station, such as operator cost, machine operation cost, and tool cost, are known.
- The tool life equation and physical constraints for each operation and machining station are known from previous experience.
- If a particular part configuration requires multipass operations due to machine or tool limitations (such as face milling, end milling, turning, boring, reaming, or drilling deep holes), the optimization of these operations is accomplished first based on the analysis discussed in Ref. [77]; this allows for multiple passes to be divided into a number of single passes which represent the minimum rational work elements. In addition, the characteristics of each pass are known.

- The precedence constraints between work elements [93] are known.
- A specified time is devoted to manual or automatic tool changing. The establishment of a tool-change interval schedule is preferable especially for manual tool-change machining stations.
- The cutting speed and feed variables are continuous.

The optimization model reduces to a two-variable function of the cutting speed and feed. The tool life equation as a function of speed and feed, Equation (13.7), is substituted into Equations (13.1) and (13.8), resulting in the following general equations for the production time and production cost, respectively:

$$T_{u_i} = \frac{A_{1i}}{V_i f_i} + B_{1i} V_i^{e_{1i}} f_i^{e_{2i}} t_{l_i} + t_{h_i} + t_{r_i} + t_{p_i} \quad \forall i \tag{13.28}$$

$$C_{u_i} = C_0 \frac{A_{1i}}{V_i f_i} + B_{1i} V_i^{e_{1i}} f_i^{e_{2i}} (C_0 t_{l_i} + C_{t_i}) + C_0(t_{h_i} + t_{r_i} + t_{p_i}) \quad \forall i \tag{13.29}$$

where A_{1i}, B_{1i}, e_{1i}, and e_{2i} are constants. The optimization for the objective functions $U_i(s) = T_{u_i}$ and $U_i(s) = C_{u_i}$ considering only inequality constraints given by Equation (13.19) was discussed above for single-pass operations. However, the solution for $U_i(s) = C_{u_i}$ is more difficult when the stations are subject to the production time equality constraint, Equation (13.24):

$$\Omega_i(s) = T_{u_i} - t_x = 0 \tag{13.30}$$

This constraint reduces the idle production time at the machining stations which usually occurs during line balancing. When more than one operation is performed at a station, t_x is a portion of the system cycle time; however, the total production time for all operations at the station should be less than or equal to t_x. This objective function, Equation (13.29), which is nonlinear and subject to an equality constraint of identical nonlinearity in addition to inequality constraints which may be also nonlinear, is the subject of the present analysis. In general, optimization problems subject to equality and inequality constraints have been solved by the Lagrange or substitution methods. The Lagrange method cannot be applied in the present problem due to the shape of the constraint surface $\Omega_i(s) = 0$ and the isometric surfaces $U_i(s) = $ constant (see the first numerical example below); only a singular solution can be obtained using the Lagrange method. However, if the substitution method is used, the representation of one parameter (i.e., V) as a function of the other parameter (f) within Equations (13.28) and (13.29) is generally impossible. Therefore, a closed form solution has been difficult to achieve. A trial-and-error method may be used, but this requires a great deal of knowledge about the feasible region. Therefore, it is recommended that this problem be approached using the following change of variables:

$$x_i = V_i^{e_{1i}} f_i^{e_{2i}} \tag{13.31}$$

$$y_i = V_i f_i \tag{13.32}$$

As shown in Ref. [126], substituting Equations (13.31) and (13.32) into Equations (13.28) and (13.29) yields the solution

$$x_i = \frac{A_{1i} + A_{2i}y_i}{B_{1i}t_{l_i}y_i} \tag{13.33}$$

$$y_i > -\frac{A_{1i}}{A_{2i}} = y_i^{**} \quad \text{for } A_{2i} = t_{h_i} + t_{r_i} + t_{p_i} + t_x \tag{13.34}$$

The inequality constraints, Equation (13.19), are also reduced to two-variable inequalities of the cutting speed and feed and are included in the analytical solution for the objective function. In the case in which the tool life and inequality constraints are not only functions of the speed and feed variables, analytical solutions may be obtained at various levels of the excluded variables. From these solutions, the significance or sensitivity of the excluded variables may be determined, and an appropriate solution may be selected. Analytical solutions for the function F in Equation (13.23) for inequality constraints can be obtained for both one variable (either speed or feed) and two variable (both speed and feed) cases. The one variable solution is a simplified case of the two variable case. Hence, the solution for two variable constraints is considered first.

In the case in which the inequality constraints are functions of two variables, cutting speed and feed, for each operation, the constraints described by Equation (13.23) are given by

$$F(y'_{ij}) = \left[\left(-\frac{A_{1i} + A_{2i}y'_{ij}}{B_{1i}t_{l_i}} \right)^{b_{2ij}-b_{1ij}} y'^{a_{1i}b_{2ij}-a_{2i}b_{1ij}}_{ij} \right]^{1/e_{2i}-e_{1i}} - \frac{E_{ij}}{B_{ij}} < 0 \tag{13.35}$$

where y'_{ij} corresponds to the variable y_i (of the objective function given by Equation 13.32) of the ith operation for its jth inequality constraint. The solution of the previous equation is very complicated since it is influenced by the signs and magnitudes of the exponents a_{1i}, a_{2i}, b_{1ij}, and b_{2ij}. The optimum solution is [126]

$$y'_{ij,\text{opt}} = \min\{y'_{ij} | y'_{ij} \in S^{**}_{ij} \wedge y'_{ij} > y^{**}_i\} \tag{13.36}$$

where y_i^{**} is given by Equation (13.34) and S_{ij}^* is defined in Ref. [126].

In some cases some of the constraints are functions of only one variable, either the feed or the cutting speed. The feed is more commonly the variable of interest, since the cutting force is often observed to be independent of speed within conventional cutting speed ranges for light alloy materials. When a constraint is assumed to be a function of either the feed or the speed only, it can be transformed to an inequality which is a function of y'_{ij} given by

$$F(y'_{ij}) = \left[-\frac{A_{1i} + A_{2i}y'_{ij}}{B_{1i}t_{l_i}} y'^{-(e_{1i}+1)}_{ij} \right]^{b_{2ij}/e_{2i}-e_{1i}} - \frac{E_{ij}}{B_{ij}} \le 0 \tag{13.37}$$

and

$$F(y'_{ij}) = \left[\frac{-B_{1i}t_{l_i}}{A_{1i} + A_{2i}y'_{ij}} y'^{(e_{2i}+1)}_{ij} \right]^{b_{1ij}/e_{2i}-e_{1i}} - \frac{E_{ij}}{B_{ij}} \le 0 \tag{13.38}$$

The solution(s) lies in one of several intervals which are defined in Ref. [126].

The solution for the optimum cutting speed and feed for each machining operation and station with idle time, after a multistage machining system is balanced, is obtained by using the previous mathematical analysis based on minimum production cost (Equation 13.29), equality constraints (Equation 13.32), and inequality constraints (Equation 13.19). The computational scheme is as follows:

Step 1 — Determine the lower bound y_i^{**} of y_i' from Equation (13.34).
Step 2 — Determine the intervals of the sets S_{ij}^*, S_{ij}^{**}, or S_{ij}^{***}, which fulfill the inequality constraints given by Equations (13.35), (13.37), and (13.38), respectively.
Step 3 — Determine the intervals which are the solutions of both the cutting speed and feed constraints given by $V_{i,\min} < V_i < V_{i,\max}$ and $f_{i,\min} < f_i < f_{i,\max}$ which are functions of y (Equations 13.39 and 13.40)
Step 4 — Determine the smallest y_i value (y_i^s) which lies in all the derived y_i intervals from Steps 1, 2, and 3:

$$y_i^s = \min\left\{y_i | y_i \in \bigcap_{j=1}^{M_i} S_{ij}^{\diamond} \wedge y_i > y_i^{**} \; \forall i\right\}$$

where

$$S_{ij}^{\diamond} = S_{ij}^* \cup S_{ij}^{**} \cup S_{ij}^{***} \; \forall i,j$$

Step 5 — Calculate the optimum cutting speed and feed by substituting $y_i = y_i^s$ into

$$V_i = \left[\frac{-B_{1i}t_{l_i}}{A_{1i} + A_{2i}y_i} y_i^{(e_{2i}+1)}\right]^{1/e_{2i}-e_{1i}} \quad (13.39)$$

$$f_i = \left[-\frac{A_{1i} + A_{2i}y_i}{B_{1i}t_{l_i}} y_i^{-(e_{2i}+1)}\right]^{1/e_{2i}-e_{1i}} \quad (13.40)$$

These steps are repeated for each *i*th machining operation belonging to a machine station which is subject to the time equality constraint. A mathematical proof that y_i^s is the optimum solution for the objective cost function subject to an equality time constraint is given in Ref. [126]. This solution results in a reduction of the total production cost and an improvement of tool life as shown in Ref. [127].

A multistage machining system may be optimized based on the following criteria:

(i) Specified cycle time, t_x
(ii) Minimum production time for the bottleneck operation
(iii) Cycle time for minimum production cost for the bottleneck operation

Case (iii) is very rare in a high-productivity and high-efficiency production environment, and will not be discussed.

The following procedure can be used to effectively utilize all stations for case (i):

Step 1: Define all machine and process constraints and the tool life equation in deterministic or probabilistic form for each operation.
Step 2: Determine the depth of cut for all multipass processes and the level(s) of flank wear to be used in the analysis.

Step 3: Determine the operations which can be performed with multispindle heads and multistep or combination tools.

Step 4: Determine the optimum machining parameters (cutting speed and feed) for each operation using the minimum production cost objective function, $U_i(s) = C_{u_i}$ for $i \in A_s$, where $A_s = \{1, 2, \ldots, N\}$.

Step 5: Balance the stations by selecting the operations with respect to their t_{m_i} value and their order with respect to the precedence constraints. Any heuristic algorithm can be applied at this step for balancing the multistage system based on the information from the previous four steps. The line balancing problem reduces to arranging or grouping the individual processing tasks (minimum rational work elements) at the workstations so that the total time required at each workstation is approximately the same and the required cycle time is met. Manual and computerized line balancing methods for assembly systems have been developed based on several of heuristic approaches such as ranked positional weights [128, 129], Comsoal [130], Calb [131], and Alpaca [132] methods, which can be integrated in the present heuristic approach. Unfortunately, a perfect balance may not be achievable in most practical situations due to the precedence constraints between work elements. The balancing process will determine the number of stations N_c required to accomplish the N operations.

Step 6: Calculate the total production time for each station, $T_{u_l}^* \ \forall l \in A_c$ (where $A_c = \{1, 2, \ldots, N_c\}$), based on the cutting parameters determined in Step 4.

Step 7: If there exists a station q such that

$$q \in \{A_c | T_{u_l}^* \leq T_{u_q}^* \leq t_x \ \forall l \in A_c\}$$

then the optimum cycle time is assumed to be $T_{u_q}^*$. Otherwise, it is assumed that N_c–N_z stations satisfy the condition $T_{u_l}^* \leq t_x \ \forall l \in A_{cl}$, where

$$A_{cl} = \{l \in A_c | T_{u_l}^* \leq t_x\}$$

is a subset of A_c.

Step 8: Determine the optimum machining parameters based on the minimum production time $T_{u_i}^{**}$, for the ith operations which belong in the number of stations N_z in the set A_c–A_{cl}. For this purpose, set $U_i(s) = T_{u_i}$.

Step 9: Based on the cutting conditions determined in the previous step, calculate the total production time $T_{u_i}^{**}$ for each of the N_z stations. If there exists a station r such that

$$r \in \left\{A_c - A_{cl} | T_{u_l}^* < T_{u_r}^* < t_x \ \forall l \in A_c - A_{c_l}\right\}$$

the cycle time t_x is satisfied by all stations. If $r\{\notin\}\{A_c - A_{c_l}\}$, proceed to Step 11.

Step 10: The cutting conditions for each station with idle time are optimized based on the mathematical analysis discussed above. The machine idle time is considered for every station $l \in A_{c2}$ where $A_{cl} = \{l \in A_c | \in T_{u_l}^* < (0.95)t_x\}$. Distribute the idle time of each station in the set A_{c2} to its corresponding operations. The optimum cutting conditions are determined for all these operations by minimizing the production cost $U_i(s) = C_{u_i}$ of each operation subject to the equality constraint, Equation (13.30). For the A_{c2} stations, the idle time is used effectively to reduce the total production cost by improving tool life if allowed by the lower bounds on the speed and feed constraints. The optimum cutting conditions for the stations in the set A_c–A_{c2} were determined in either Step 4 or 8. Hence, the optimization of the multistage system is complete.

Step 11: If $r \notin \{A_c - A_{cl}\}$, the specified production time t_x cannot be met. The stations requiring longer production time t_x, lie in the subset $A_{c3} = \{l \in A_c - A_{cl} - A_{c2} | T_{u_l}^* > t_x\}$ and have to be re-evaluated in order to decrease the cutting time by substituting tools suitable for more aggressive cutting conditions. Another option could be to reprocess the part so that the production time is distributed more uniformly within the available stations. The use of multispindle heads, multiple spindles, and multifunctional tools are the final options for reducing the production time and number of stations. After changes are made in the stations in subset A_{c3} and in any other stations, return to Step 1 and repeat the above procedure for all the modified stations.

The algorithm is repeated every time the setup conditions or the part process are changed until the expected production time at an acceptable cost is met.

When it is necessary to establish the production cycle time based on the minimum production time of the bottleneck operation in a multistage system, case (ii), the procedure differs from the aforementioned one and consists of the following steps:

Step 1: Similar to Step 1 of case (i).
Step 2: Similar to Step 2 of case (i).
Step 3: Similar to Step 3 of case (i).
Step 4: Obtain the optimum machining parameters for all the N operations based on the minimum production time using the objective function $U_i(s) = T_{u_i} \ \forall i$.
Step 5: Similar to Step 5 of case (i).
Step 6: Similar to step 6 of case (i).
Step 7: Find the station q such that $q \in \{A_c | T_{u_q}^* \leq T_{u_i}^* \ \forall i\}$. The minimum production time is $x = T_{u_q}^*$ and station q governs the cycle time.
Step 8: Optimize the operations in all the stations with idle time, $l \in \{A_c | T_{u_l}^* \leq (0.95) t_x\}$, by using the maximum available time for machining in order to reduce the production cost and improve the tool life if allowed by the lower bounds of the speed and feed constraints.

Therefore, the optimum machining parameters are determined by minimizing the production cost $U_i(s) = C_{u_i}$, subject to the equality constraint, Equation (13.30).

Whenever the total tool life for any station is not satisfactory under the derived cutting conditions, changing the tool should be considered in order to increase its life. The algorithm for case (ii) should be repeated every time a tool is changed. A detailed example of a multistage optimization based on the above algorithm is given in Ref. [127].

13.7.5 Cutting Tool Replacement Strategies

One major issue in machining systems analysis is tool balancing and tool regulation. The utilization of the machine tool can be improved by ensuring the availability of the required cutting tools when needed, eliminating machine interruptions due to manual tool changes and setup (during machine cycle) and monitoring tool wear and life in order to reduce scrap. This can be achieved by maintaining an adequate supply of usable tools, monitoring tool usage, wear and premature failure, and incorporating mathematical control algorithms for the automatic replacement of tools.

Generally, conservative and deterministic approaches are used in a production environment to define useful tool life since current sensing technologies do not yet provide consistent, reliable tool monitoring approaches. The most common method is forced replacement at specified intervals. These deterministic approaches minimize risk but do not optimize tool and machine. Tool replacement decisions affect the system utilization significantly and have a large economic impact. Therefore, a tool replacement strategy based on a machining cost/production time criterion may be preferable [133–152].

The best tool replacement strategy will minimize the machine idle time due to tool change or tool magazine loading and tool failure while reducing tool redundancy and duplication in the tool magazine. The optimal policy for single tool is characterized by one parameter, the time to replacement. This policy is called periodic replacement. The policy should be based on a model in which the tool is subject to random failures, but in which the state of the tool is always known with certainty (based on sensors). Then the tool can be replaced when it has failed or after a sufficiently long time has passed since the last replacement, whether or not the tool failed (mixed replacement by failure and forced replacement before failure). This policy is of interest if the cost of replacement after failure is greater than the cost of replacement before failure. A second policy which may be considered is to replace a tool when it is still usable, provided that the cost of replacement is less than the likelihood of failure over the next interval between scheduled tool magazine or tool machine changes multiplied by the cost of changing the tool and repairing the partially machined part after failure in operation.

The best methodology for regulating tools depends on the type of manufacturing system used (i.e., conventional transfer line, FTL, FMS, etc.). There should be a minimum number of tool positions required on machine tools to economically process a certain number of parts. Some scheme must be adapted when the number of tools required is greater than the number of tool positions on the machine turret or tool magazine. The tool changing strategy can be optimized by adjusting the optimum tool life and the cutting conditions.

There are two approaches to the solution of the tool magazine system reliability problem: (a) the use of cutting tools of high reliability, and (b) the use of redundant tools. The use of redundant tools in the machine's magazine results in an increase in production tooling costs, but tends to increase productivity and reduce production costs. The mathematical solution for the optimum number of redundant tools in a magazine can be determined when the reliabilities of the individual tools are known [147, 153–157]. The reliability of single or multistep tools, which is estimated at a prespecified flank wear limit based on the tool wear response discussed in Chapter 9, is the most important parameter in determining the optimum redundancy and therefore should be investigated in detail through experiments or current production experience [133, 136]. The strategy for multistep tool replacement depends on the probability of tool survival for each tool step based on experience from multistep or identical single-step tools. In the case of NC machine or transfer line operations, experimental data on tool durabilities can be acquired by testing the tool in production over a period sufficiently long to ensure statistical reliability.

All distinct tools in the magazine can be assumed to act in series; therefore their cumulative reliability is

$$R_s = \prod_{i=1}^{m} R_i(t) \tag{13.41}$$

for m distinct cutting tools. When redundant tools are used, then the redundant tools can be assumed to act as a standby parallel system with reliability

$$R_i(t) = P_0(t) + P_1(t) + \cdots + P_{n_i}(t) \tag{13.42}$$

In this equation, $P_j(t)$ is the probability that exactly j tools fail ($j = 1, 2, \ldots, n_i$) and n_i is the number of redundant tools for the ith cutting tool. The total number of tools in the magazine is

Machining Economics and Optimization

$$n = \sum_{i=1}^{m} n_i$$

The time between failures is usually assumed to follow an exponential distribution:

$$R_i(t) = P_0(t) = e^{-\lambda t} \tag{13.43}$$

In this case, the number of failures occurring over an interval of time follows a Poisson distribution. Failures are assumed to be independent. Physically this means that the system experiences no effects of wearout. In some cases the cutting tool experiences various types of wear as the cutting process or machine tool change, so that the tool failure rate changes over time. In these cases the exponential distribution would not be an appropriate model. However, for system level reliability calculations, the exponential is usually a good model [158]. The time to system failure distribution will approach an exponential distribution as the number of tools is increased even though tool wearout, which may not follow an exponential distribution, is the cause of failure. For n_i standby components for the ith cutting tool in the magazine, the reliability of the system can be calculated from [146, 149, 151]

$$R_s = \prod_{i=1}^{m} R_i(t) = \prod_{i=1}^{m} \left[e^{-\lambda t} \sum_{j=0}^{n_i} \frac{(\lambda t)^j}{j!} \right] \tag{13.44}$$

The value of mean-time-to-failure (MTTF) for each cutting tool type for the above system is

$$\text{MTTF} = (n_i + 1)/\lambda$$

The optimum number of redundant tools should be determined based on reliability in conjunction with cost and production time.

13.7.6 Cutting Tool Strategies for Multifunctional Part Configurations

The majority of tools have a single function and are used for only one operation or are made to machine a single part feature. However, the use of multifunctional cutting tools has become necessary in single-spindle manufacturing systems in order to increase productivity and quality, and to reduce cost.

Multifunctional cutting tools include combination and multistep tools. The term "combination" is used for tools that are used in more than one distinct operation while the term "multistep" is used for tools that are employed for manufacturing a multistep configuration (e.g., a hole with multiple diameters at different depths) in one motion. For example, the valve guide and seat in an engine head may be manufactured using either a generating head tool or a plunge (form) tool. The generating head tool is considered a combination tool because it bores and reams the guide in one machine motion and then bores the seat in another. On the other hand, the plunge tool is a multistep tool since it machines the guide and seat in one motion.

The design of multistep tools was discussed in Chapter 4. Multistep tools are often used to drill or ream multidiameter holes (see Figure 13.6), and to drill and counterbore, drill and chamfer, ream and chamfer, or drill and thread simultaneously. These tools have more than one cutting edge along their axis and can machine a complex part feature in a single tool motion during which two or more tool steps generally cut at the same time. Stepped holes can be bored using either a multistep boring bar or a single-point tool. Single-point boring tools used to bore multistep holes can be treated as multistep tools for analysis purposes.

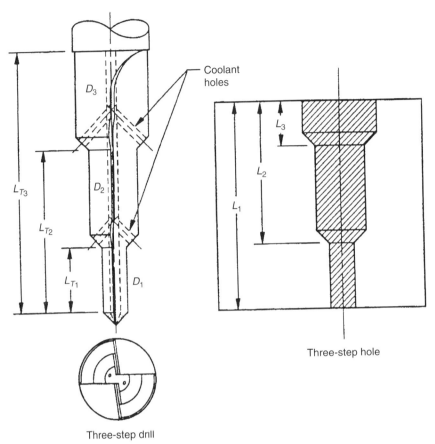

FIGURE 13.6 Configuration of a multistep drill and the corresponding generated hole.

Combination tools are designed to be used for different distinct operations. Examples of such tools include generating head tools used for the machining of the valve seat and the guide in an engine head and drill–thread mill tools used for drilling and thread milling a hole independently. Tools for various operations are being combined on one tool body to eliminate tool changes or improve the geometric relationship between features (e.g., perpendicularity or concentricity between features).

13.7.6.1 Formulation of the Multistep Tool Problem

Multistep tools fitted with indexable inserts or replaceable tool bits require a different mathematical analysis than solid or brazed tools, since the wear of one indexable step does not necessarily require the regrinding or disposal of the other unworn steps on the tool.

The cutting conditions generally vary from step to step on a multistep tool. The tool steps can operate at a constant spindle speed (rpm), constant surface speed, varying surface speed or spindle rpm, and varying feed rates. The best strategy for a particular application depends on the tool (geometry, size, and material), operation economics, and the capability of the machine tool control system. It is especially common to vary the spindle rpm as different parts of the tool cut for the following reasons: (a) to reduce the surface speed for larger diameter steps to prevent excessive cutting temperatures and tool wear, (b) to optimize the speed when steps are made of different tool materials, and (c) to optimize part quality and surface finish for each step.

The simplest method is to use constant spindle rpm and feed, which results in different surface speeds at the various tool steps. On NC machines, it is also easy to operate at a constant surface speed by varying the spindle rpm according to the diameter of the step as each step cuts the workpiece. This approach should be used when there is a significant difference in tool step diameters. Current control systems allow speed or feed changes within 50 to 100 msec in a controllable manner. The least desirable strategy is to operate the various tool steps at different surface speeds and spindle rpms. This approach is sometimes used as a compromise to provide acceptable cutting conditions for steps with a significantly different diameters. In particular, this method is used for single-step tools with a large length to diameter ratios, such as drills and reamers, in high-speed machining applications; it can eliminate the need for a bushing at the entrance by allowing the tool to enter the part at conventional (low) speeds and feeds which are subsequently increased to the optimum values. The feed is usually kept constant for all the tool steps, except in special applications in which surface finish and chip ejection are significant concerns. The mathematical analysis of the variable feed approach, where a feed change is considered for at least one tool step, is similar to that for the variable speed approach.

Analysis is performed to determine the optimum value of the spindle rpm and feed which maximizes the tool life and reliability. Tool life and reliability analyses for single-step tools can be applied to multistep tools of the same family in early design stages. The probability of survival for multistep tools is usually smaller than that for single-step tools and can be analyzed using static or dynamic models in a preliminary stage and verified later by testing prototype tools. This approach can be used to evaluate possible design configurations and to determine the required reliability levels for the tool steps. The problem can be solved for different values of tool reliability as a function of tool flank wear and cutting time.

An analytical solution of the optimization (using production cost, production time, and composite objective functions) can be determined for constant rpm, constant surface speed, or variable speed and feed cases [159]. The regrinding policy for multistep tools, that is, whether (a) to regrind all the tool steps every time a step is worn or (b) to resharpen different steps at different intervals, should be established ahead of time. Policy (a) is generally preferred because it maintains the concentricity of the tool steps and narrow tolerances on the step lengths. Case (b) is usually used for indexable tools since they provide the flexibility to index a tool at one step without necessarily doing so at other tool steps.

Analytical solution for constant rpm: The average production time for solid and brazed multistep tools at a constant rpm is

$$\overline{T}_u = t_{m_x} + \overline{Y}_{kj} t_l + t_y, \quad \text{where } t_{m_x} = \frac{L_{T_{n-1}} + L_n}{fN}$$

and

$$\frac{1}{\overline{Y}_{kj}} = \text{int} \left[\frac{\sum_{j=1}^{K_k+1} T_{P_{kj}}}{(K_k + 1) t_{m_k}} \right], \quad \forall k \in A_c \qquad (13.45)$$

on the basis of the kth tool step which satisfies the condition

$$Y_{kj}^* = \max\{Y_{rj}^* \, \forall r\}, \quad \text{where } Y_{rj}^* = \frac{L_r N_r^{a_3} f_r^{a_1}}{T_{rj}}$$

The production cost for a multistep tool operation using regrinding policy (a) is

$$C_{u_j} = \frac{C_o[L_{T_{n-1}} + L_n]}{Nf} + Y_{kj}(C_o t_l + C_t^*) + C_o t_y \quad (13.46)$$

$$\frac{1}{Y_{kj}} = \text{int}\left[\frac{T_{p_{kj}}}{t_{m_k}}\right] \quad \text{and} \quad C_t^* = \frac{C_t + K_k \sum_{r=1}^{n} C_{gr} t_{sr}}{K_k + 1}, \quad j = 1, 2, \ldots, K_k + 1$$

where C_t and C_t^* are the tool cost and the tool cost between regrinds, respectively, and C_{gr} and t_{sr} are, respectively, the grinding cost and time available for tool regrinds for the rth tool step. The production cost for the regrinding policy (b) is discussed in Ref. [159].

The production time and cost for boring multistep configurations with n steps using a single-point tool at constant rpm are, respectively:

$$T_{u_j} = \frac{L_1}{Nf} + \frac{t_l}{X_j^*} + t_y \quad \forall i, r \quad (13.47)$$

$$C_{u_j} = \frac{C_o L_1}{Nf} + \frac{C_o t_l + C_t^*}{X_j^*} + C_o t_y \quad (13.48)$$

where

$$\frac{1}{X_j^*} = \sum_{r=1}^{n} \frac{t_{m_r}}{T_{Pr_j}} \quad (13.49)$$

X_j^* is the number of parts machined (or machine cycles) over which flank wear accumulates. Tool wear curves for boring a step hole at a constant rpm, with changes in cutting speed from V_1 to V_2, are shown in Figure 13.7. Applications of this type are discussed further in Ref. [160].

Analytical solution for constant surface speed: For NC machines, the spindle speeds are often varied to avoid exceeding a maximum surface speed when steps of different diameters cut. The cutting cost and time for solid and brazed tools in these cases are given by [159]

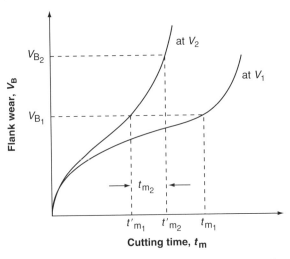

FIGURE 13.7 Flank wear transition for a single-point boring tool at two different speeds.

$$T_{u_j} = \frac{\pi P_c}{1000 Vf} + \frac{\pi t_l Z_{xj}}{1000 V^{a_3+1} f^{a_1+1}} + t_y \qquad (13.50)$$

$$C_{u_j} = \frac{C_o \pi P_c}{1000 Vf} + \frac{\pi Z_{xj}[C_o t_l + C_t^*]}{1000 V^{a_3+1} f^{a_1+1}} + C_o t_y \qquad (13.51)$$

where

$$P_c = \sum_{i=1}^{n} D_i(L_i - L_{i+1}), \quad D_i = D_{i-1} \, \forall i \in A_d$$

where the set A_d consists of the tool steps in which the spindle rpm stays the same and

$$Z_{xj} = \max \left[\frac{\sum_{r=i}^{n} D_r^{a_3+1}(L_r - L_{r+1})}{R' D_i^{a_3} \exp(d V_{B_i}^{a_4}) \exp[\beta_i(j-q)]} \right] \forall i$$

The number of tool cycles between regrinds is calculated using an approach similar to that used for case (a) in the constant rpm case [159]. For the variable speed approach to be effective, the time necessary for a speed transition should be shorter than, and generally a small fraction of, the tool step machining time during which the transition takes place. Therefore, the speed change from N_p to N_{p+1} should satisfy the inequality

$$\frac{\pi f \left| N_{p+1}^2 - N_p^2 \right|}{3600 \alpha} \leq L_r - L_{r-1} \qquad (13.52)$$

where α is the spindle acceleration in rad/sec².

The case of boring multistep configurations in a lathe is similar to that of the constant rpm discussed earlier. The production time and cost are given by

$$T_{u_j} = \frac{\pi \sum_{r=1}^{n} D_r [L_r - L_{r+1}]}{1000 Vf} + \frac{t_l}{X_j^*} + t_y \quad \forall i, r \qquad (13.53)$$

$$C_{u_j} = \frac{C_o \pi \sum_{r=1}^{n} D_r [L_r - L_{r+1}]}{1000 Vf} + \frac{C_o t_l + C_t^*}{X_j^*} + C_o t_y \qquad (13.54)$$

A variable surface speed and rpm strategy is sometimes used for multistep tools, for example, in cases in which the tool steps are made of different materials due to economic cost and design limitations, when the optimum surface speed of one step is significantly different than that of another, or when the surface quality requirement for one step is much more stringent than for another. The equations for analyzing such cases are given in Ref. [159].

It is desirable to use optimum cutting speed(s) and feed(s), which can be determined using the optimization procedures described above. Physical limitations or constraints on the cutting conditions always exist and must be taken into account in the optimization. The solution for the maximum cutting speed(s) based on minimum operation time and cost for a multistep tool using regrinding policy (a) discussed above can be determined from Equations (13.45) and (13.46) respectively, using Equations (13.9) and (13.10), and the highest possible feed. In the same manner, the maximum cutting speed for all the other above cases can be calculated using Equations (13.47), (13.48), (13.50), (13.51), (13.53), and (13.54).

13.7.6.2 Formulation of the Combination Tool Problem

The tool life equation for combination tools can be estimated from the tool life equations for similar conventional (single function) tools. The tool life for the different tool functions can be assumed to be independent. Tool life predictions for a combination tool can be obtained when the reliability of the different tool functions is either known or can be estimated from experience with single function tools as discussed earlier for multistep tools. The best tool design and cutting conditions are those which result in uniform or balanced tool life for all the tool functions.

Mathematical relations between the different tool functions must be defined in order to analyze a combination tool [159]. The useful tool life is dependent on the tool life and reliability of all functions since the failure of one function could result in disposal of the tool. The total production time for a tool with n functions is

$$T_{u_j} = \sum_{i=1}^{n} t_{m_i} + Y_{kj}t_l + \sum_{i=1}^{n} (\Delta t_{\alpha_i} + t_{r_i} + t_{x_i}) + t_{c_s} + t_{d_n} + t_p + t_h \qquad (13.55)$$

The parameter Y_{kj} is defined as in Equation (13.45). $\Delta t_{\alpha i}$ is the acceleration/deceleration time required when changing the spindle rpm between the tool functions. Either regrinding policy (a) or (b) discussed above for multistep tools can be used especially with indexable tooling. The production cost for the tool is

$$C_{u_j} = C_0 \sum_{i=1}^{n} t_{m_i} + Y_{kj}(C_0 t_l + C_t)$$
$$+ C_0 \sum_{i=1}^{n} (\Delta t_{\alpha_i} + t_{r_i} + t_{x_i}) + C_0(t_{c_s} + t_{d_n} + t_p + t_h) \qquad (13.56)$$

Optimizing these production time and cost equations is a multivariable problem. However, assuming the tool functions act independently, a realistic assumption in practice, allows the cutting conditions for each function to be optimized independently based on the following production time and cost functions:

$$T_{u_i}^* = t_{m_i} + \frac{Y_{ij}t_l}{n} + t_{\alpha_i} \qquad (13.57)$$

$$C_{u_i}^* = C_0 t_{m_i} + Y_{ij}\frac{C_0 t_l + C_{tf}}{n} + C_0 t_{\alpha_i} \qquad (13.58)$$

The maximum cutting speeds for the objective functions given by Equations (13.55) and (13.56) can be determined using the procedure discussed above for multistep tools.

13.7.6.3 Solution Scheme for Multistep Tools

The simulation algorithm for justifying the use of a multistep tool can be based on the following heuristic scheme:

Machining Economics and Optimization

Step 1: Define the maximum surface speed of the tool material used for the tool steps.

Step 2: Determine the tool life equation for the tool steps if it is not known from single-step tools. The same tool life equation can be used for all steps made of the same material and used for similar operations.

Step 3: Evaluate the total production cycle time and cost when conventional single-step tools are used to machine the multistep part configuration. Determine the optimum cutting conditions for all the required single-step tools.

Step 4: Specify whether through the tool coolant will be used in order to remove the chips; this is especially important for multistep tools.

Step 5: Determine the maximum or optimum spindle rpm and feed for the multistep tool. Calculate the production time and cost. Use the constant rpm approach solution unless the diameter changes between the steps are significant or different tool materials are used for different steps.

Step 6: Calculate the amount of lifetime sacrificed when the multistep tool is used at a constant rpm [159]. Continue with Step 7 if the percentage of the tool cost lost is significant. Otherwise, proceed to Step 10.

Step 7: Determine the feasibility of using the constant surface speed or variable rpm approaches. This step should be considered when motorized machine spindles are used.

Step 8: Evaluate the advantage of using constant surface speed or variable rpm approaches for different tool steps. Determine the maximum or optimum spindle speed for all the steps, the production time and cost reduction when constant surface speed or variable rpm are used in place of the constant rpm approach for significantly different tool step diameters.

Step 9: Calculate the time required for a speed transition at a tool step if the constant surface speed or variable speed approach is used, in order to ensure that process limitations are met and these approaches are feasible.

Step 10: Determine the most viable approach for a multistep tool. If Step 9 is satisfied and there is an advantage in the approach in Step 8 as opposed to that of Step 5, consider using a constant surface speed or variable rpm approach. Otherwise, accept the constant rpm approach.

Step 11: Compare the viable operation approach in Step 10 for a multistep tool with that for the conventional process in Step 3, and determine if the multistep tool approach is justified.

Step 12: Changes in feed for various steps can be analyzed, if necessary, using the approach used for Step 8, since the variable feed approach is similar to the variable speed approach.

13.7.6.4 Solution Scheme for Combination Tools

Combination tools require a different heuristic scheme than multistep tools in order to optimize cutting conditions for each function while balancing the expected tool life. Since the cutting conditions are usually different for each function, when the tool is utilized in a transfer line or FMS any available idle time can be used to improve the tool life of the tool function with the shortest expected life. In this case, the objective function should be optimized using an equality constraint on the number of parts produced by each tool function. In general, the cutting conditions of a combination tool are considered to be balanced if the numbers of parts produced between regrinds by each tool function are equal or integer multiples, and ideally when all tool functions require regrinding or replacement at approximately the same time.

The following heuristic scheme can be used to optimize combination tools:

Step 1: Define all the machine and process constraints and the tool life equation for each tool function.
Step 2: Determine the optimum speed and feed for all tool functions using minimum production cost as the objective function.
Step 3: Calculate the total production time T_{u_j} for the tool based on the cutting conditions determined in Step 2.
Step 4: Calculate the expected tool life T_{pij} and the Y_{ij} term for each tool function.
Step 5: Determine $Y_{rj} = \max\{y_{ij}\ \forall i\}$, which is the upper bound for Y_{ij}, and evaluate the cutting conditions for all tool functions other than the rth by using the equality constraint

$$Y_{ij} = t_{m_i}, \quad T_{pij} = mY_{rj}, \quad m = 1, \tfrac{1}{2}, \tfrac{1}{3}, \ldots$$

Determine the optimum cutting conditions based on the objective functions discussed above and the equality constraint [160].

Step 6: Determine the cutting time $t_{m_i}\ \forall i \in r$ and calculate the total machining time t_c,

$$t_c = \sum_{i=1}^{n} t_{m_i}$$

If the total machining time is longer than the allowable time t_{cr} for the process, especially when the tool is used in a transfer line or FMS station, proceed with the following steps. Otherwise, go to Step 13.

Step 7: Determine the optimum cutting conditions for all the tool functions based on the minimum production time objective function, and use them to calculate the machining time, $t_{m_i}^*$, and the term Y_{ij} for all functions.

Step 8: Organize the tool functions in a set A_c in decreasing order with respect to the magnitude of Y_{ij}, so that $A_c = \{u, \ldots, q\}$ with $Y_{qj} = \min\{Y_{ij}\ \forall i\}$ and $Y_{uj} = \max\{Y_{ij}\ \forall i\}$.

Step 9: Calculate the total machining time,

$$t_c = \sum_{i=1}^{n} t_{m_i}^*$$

If $t_c > t_{cr}$ go to Step 12. Otherwise, proceed with the next step.

Step 10: If $t_c \geq et_{cr}$ (where $0.95 \leq e \leq 1$), proceed with Step 13. Otherwise, attempt to balance the number of tool cycles available on each tool function during its life span, which means that the Y_{ij} terms must be balanced. Therefore, set

$$t_{m_i}^{**} = t_{m_i}^*, \quad \forall i$$

For $k = 1$ to $n-1$ proceed with Step 11.

Step 11: Iterate for $i = A_{c1}$ to A_{ck} (where A_{ci} are the elements of set A_c) in order to determine the optimum cutting conditions for the ith tool function based on minimum production cycle time, using the equality constraint $Y_{ij} = Y_{k+1,j}$. For each ith iteration calculate the quantities

$$t_{m_i}^* \quad \text{and} \quad t_c = \sum_{l=A_{cl}}^{A_{ci}} t_{m_i}^* + \sum_{l=A_{ci+1}}^{A_{cn}} t_{m_i}^{**}$$

and check the condition $t_c \geq et_{cr}$; continue this iteration on i and go back to the iteration on k in Step 10 if the inequality condition is not satisfied. Otherwise, proceed with Step 13.

Step 12: Change the tool material(s) in order to increase the allowable cutting conditions, if possible, or reduce the number of tool functions to reduce the total machining time for the tool.

Step 13: The optimum conditions for all the tool functions have been determined based on the tool cycle time.

13.8 EXAMPLES

Example 13.1 A gray iron casting with a diameter of 150 mm is rough turned to 144 mm for a length of 200 mm plus a 3 mm approach distance. An indexable carbide-coated insert with eight corners is used in the turning operation with a cost of $20. The tool life equation is $VT^{0.2} = 230$ (metric units) for the above tool in the current workpiece material. The maximum allowable feed for the cutting tool is 0.35 mm at the maximum depth of cut of 4 mm. The time for loading/unloading an insert/tool is about 3 min. The part handling (load/unload) time is 26 sec for the casting mounted on the fixture. The time for the tool to rapid travel across the part is 4 sec. The operating cost is $80 per hour. Estimate the individual cost curves similar to Figure 13.1 and show the optimum cutting speed.

Solution: The production cost is given by Equation (13.8). There are four components in the cost equation: the machining cost, tool cost, tool-change (loading/unloading the tool in the machine) cost, and nonproductive (including the part handling, rapid travel location times, etc.) cost given by the following equations, respectively:

$$C_o t_m, \quad C_o t_l \frac{t_m}{T}, \quad C_t \frac{t_{mc}}{T}, \quad C_o(t_h + t_x + t_{cs})$$

The difference between the machining time t_m and t_{mc} is that the t_m includes the approach time while t_{mc} uses only the time that the part is in contact with the tool. The above four costs and the total cost (the sum of the four costs) are calculated for a range of cutting speeds using a spreadsheet. A plot is shown in Figure 13.8. The optimum cutting speed can be estimated either from the graph in the figure (corresponding to the lowest cost) or calculated from Equation (13.9) to be 127 m/min.

Example 13.2 Estimate the production cost for a rough turning operation on a free-machining steel bar 70 mm in diameter and 200 mm long, when 3 mm is to be removed from the OD of the bar using a carbide tool. Some of the characteristics of the turning lathe and cutting tool are given in the Table 13.1.

Solution: The production cost is given by Equation (13.8):

$$C_u = C_o t_m + \frac{t_m}{T}(C_o t_l + C_t) + C_o(t_{cs} + t_h + t_x)$$

The machining time is given by Equation (2.5) (or Equation 13.2) assuming 2 mm for the tool approach distance:

$$t_m = (L + L_e)/f_r = (L + L_e)/(fN) = (200 + 2)/(0.3 \times 700) = 0.962 \text{ min} = 58 \text{ sec}$$

The cutting speed $V = \pi D N = \pi(70)(700)/1000 = 154$ m/min

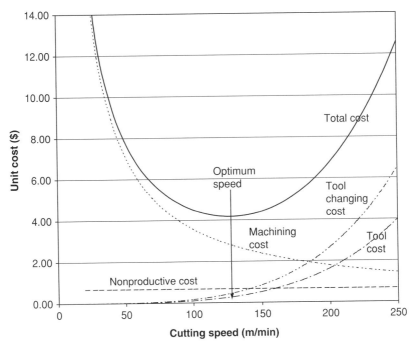

FIGURE 13.8 Various machining costs versus cutting speed.

TABLE 13.1
Tool and Economic Parameters for Example 13.2

Parameter	Value	Parameter	Value
Machine tool		Tool life equation, $VT^n = C_t$	
Spindle speed	700 rpm	Tool life constant, C_t	500
Feed, f	0.3 mm/rev	Tool life exponent, n	0.25
Rapid feed rate, f_{rapid}	6000 mm/min	Cutting speed units	m/min
Operating cost, C_o	$60 per hour	Tool interchange time between operations, t_{cs}	8 sec
Part load/unload time, t_h	20 sec	Number of cutting edges per insert, n_t	3
Tool load/unload time, t_l	1 min		
Tool cost, C_{te}	$9 per insert		

The tool life equation $VT^{0.25} = 500$ or $T = 500^4 \times 154^{-4} = 111$ min

$t_h = 20$ sec $= 0.33$ min
$t_{cs} = 8$ sec $= 0.133$ min
$t_l = 1$ min

The rapid travel time for the tool after the completion of the turning operation is

$t_x = (L + L_t)/f_{rapid} = (200 + 80)/6000 = 0.047$ min $= 2.8$ sec
$C_o = \$60$ per hour $= \$1$/min
$C_{te} = C_t/n_t = 9/3 = \$3$ per cutting edge

$$C_u = 1 \times 0.962 + \frac{0.962}{111}(1 \times 1 + 3) + 1(0.133 + 0.333 + 0.047) = \$1.51$$

Example 13.3 Calculate the minimum production cost for Example 13.1 assuming that the maximum allowable feed for the insert is 0.5 mm/rev. The above tool life equation is modified to include the effect of feed with an exponent $a = 0.5$.

Solution: The extended Taylor equation is

$$VT^n f^a = K_t \quad \text{or} \quad VT^{0.25} f^{0.5} = 500$$

The feed rate for the turning operation has an upper bound of 0.5 mm/rev. The optimum cutting speed for this operation can be determined from Equation (13.90):

$$V_{opc} = \frac{K_t}{f_h^a \left[\left(\frac{1-n}{n}\right)\left(t_l + \frac{C_{te}}{C_o}\right)\right]^n} = \frac{500}{0.5^{0.5}\left[\left(\frac{1-0.25}{0.25}\right)\left(1 + \frac{3}{1}\right)\right]^{0.25}} = 380 \text{ m/min}$$

The minimum production cost will occur at the optimum speed using the maximum allowable feed:

$$t_m = (L + L_e)/f_r = (L + L_e)/(fN) = (200 + 2)/(0.5 \times 1729) = 0.234 \text{ min} = 14 \text{ sec}$$

The spindle rpm $N = V/(\pi D) = (380)(1000)/(\pi 70) = 1729$ rpm
The tool life equation $VT^{0.25} f^{0.5} = 500$ or $T = 500^4 \times 380^{-4} \, 0.5^{-2} = 5.2$ min

$$C_u = 1 \times 0.234 + \frac{0.234}{5.2}(1 \times 1 + 3) + 1(0.133 + 0.333 + 0.047) = \$0.928$$

Therefore, the production cost is reduced from \$1.51 to \$0.928 by using an optimum cutting speed.

Example 13.4 Optimize the cutting conditions for a multitool (multispindle) lathe operation for the part in Figure 13.9. The CNC lathe has two turrets, which allows two tools to cut the part simultaneously. The cutting parameters and cutting tool characteristics are given in Table 13.2. The cutting tools for turning and boring have the same tool life equation. The strategy of changing each tool individually will be considered. Calculate the minimum production cost and production time.

Solution: There are four operations, namely OP1 (turning the 120 mm diameter), OP2 (turning the 70 mm diameter), OP3 (drilling the 25 mm hole), and OP4 (boring the 25.6 mm hole) as shown in Figure 13.9. OP1 and OP3 are performed simultaneously by the two independent turrets, as are OP2 and OP4. The spindle speed for the workpiece for each set of operations is determined based on the cycle times of the individual operations. The optimum cutting speeds for each individual operation can be estimated using Equations (13.9) and (13.10) based on the minimum production cost and time, respectively, using the maximum allowable feed for each tool. This analysis is conceptually similar to multistage machining system analysis (see Section 13.7) for the operations occurring simultaneously by individual turrets. Therefore, operations occurring by different turrets should be balanced to minimize the idle time of either turret.

The optimum speed and the corresponding rpm and machining time for each operation based on minimum cost (using Equations 13.9, 2.1, and 13.2) are:

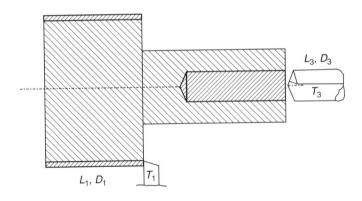

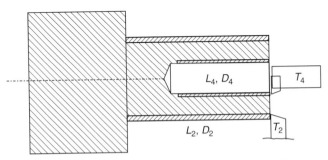

FIGURE 13.9 Manufacturing process for a twin-turret lathe in Example 13.3.

TABLE 13.2
Machining Parameters Used for Example 13.4

Parameter	Value	Parameter	Value	Parameter	Value
D_1	120 mm	f_{tmin}	0.12 mm/rev	Drill a_3	−3.7
D_2	70 mm	f_{tmax}	0.4 m/rev	Drill R_c	3.0×10^8
D_3	25 mm	f_{bmin}	0.075 mm/rev	t_h	20 sec
D_4	25.6 mm	f_{bmax}	0.15 mm/rev	tl_1, tl_2, tl_4	3 min/edge
L_1	75 mm	f_{dmin}	0.15 mm/rev	tl_3	4 min/edge
L_2	100 mm	f_{dmax}	0.3 mm/rev	t_{cs}	2 sec
L_3	63 mm	Turn a_1	−1	T_x	0.05 min/operat
L_4	60 mm	Turn a_3	−3	C_o	\$1/min
N_{min}	100 m/min	Turn R_c	3.4×10^6	$C_{te1}, C_{te2}, C_{te4}$	\$4/edge
N_{max}	5000 m/min	Drill a_1	−1.4	C_{t3}	\$30/drill/regrind

Note: a_1, a_3 = exponents in tool life equation correspond to feed and speed parameters, respectively.

$$V_{opc1} = 85 \text{ m/min}, \quad V_{opc2} = 85 \text{ m/min}, \quad V_{opc3} = 99 \text{ m/min}, \quad V_{opc4} = 117 \text{ m/min}$$

$$N_{opc1} = 225 \text{ rpm}, \quad N_{opc2} = 405 \text{ rpm}, \quad N_{opc3} = 1265 \text{ rpm}, \quad N_{opc4} = 1460 \text{ rpm}$$

$$t_{mc1} = 0.84 \text{ min}, \quad t_{mc2} = 0.62 \text{ min}, \quad t_{mc3} = 0.18 \text{ min}, \quad t_{mc4} = 0.27 \text{ min}$$

Machining Economics and Optimization

The optimum speeds and corresponding machining times for each operation based on the minimum production time (using Equation 13.10) are:

$$V_{opt1} = 112 \text{ m/min}, \quad V_{opt2} = 112 \text{ m/min}, \quad V_{opt3} = 162 \text{ m/min}, \quad V_{opt4} = 156 \text{ m/min}$$

$$N_{opt1} = 298 \text{ rpm}, \quad N_{opt2} = 511 \text{ rpm}, \quad N_{opt3} = 2063 \text{ rpm}, \quad N_{opt4} = 1937 \text{ rpm}$$

$$t_{mt1} = 0.63 \text{ min}, \quad t_{mt2} = 0.49 min, \quad t_{mt3} = 0.11 \text{ min}, \quad t_{mt4} = 0.21 \text{ min}$$

Since OP1 and OP3 occur together and $t_{mc1} > t_{mc3}$ but $N_{opc3} > N_{opt1}$, the spindle rpm $N_{opc1} = 225$ rpm is selected to control these two operations. However, the machining time for drilling the hole at 225 rpm is $t_{m3} = 0.756$ min and $t_{m3} > t_{mt1}$. The tool life for the tool in OP1 can be improved by reducing the feed below the maximum value used to estimate the V_{opt1}. Therefore, equating the machining time for OP1 to that of OP3 ($t_{m1} = t_{m3} = 0.756$ min), a feed of 0.34 mm/rev is selected ($f = 75$ mm/[0.756 min × 298 rpm] = 0.34 mm/rev). The reduction of the feed from 0.4 to 0.34 mm/rev results in a tool life improvement from 6 to 7.3 min. The machining time and cost for these operations are

$$T_{u13} = t_{m3} + \frac{t_{m1}}{T_1} t_{l1} + \frac{t_{m3}}{T_3} t_{l3} + (t_{cs} + t_x)_{1+3}$$

$$T_{u13} = 0.756 + \frac{0.756}{7.3} 3 + \frac{0.756}{13,900} 4 + 2\left(\frac{2}{60} + 0.05\right) + \frac{20}{60} = 1.23 \text{ min}$$

$$C_{u13} = C_o t_{m3} + \frac{t_{m1}}{T_1}(C_o t_{l1} + C_{t1}) + \frac{t_{m3}}{T_3}(C_o t_{l3} + C_{t3}) + C_o(t_{cs} + t_x)_{1+3}$$

$$C_{u13} = 1 \times 0.756 + \frac{0.756}{6.05}(1 \times 3 + 4) + \frac{0.756}{13,900}(1 \times 4 + 30) + 1\left(\frac{2}{60} + 0.05\right) = \$1.65$$

Similarly, OP2 and OP4 occur simultaneously and $t_{mc2} > t_{mc4}$ but $N_{opc4} > N_{opt2}$, so the spindle rpm $N_{opc2} = 511$ rpm is selected to control these two operations. However, the machining time for boring the hole at 511 rpm is $t_{m4} = 0.783$ min and $t_{m4} > t_{mt2}$. The tool life for the tool in OP2 can be improved by reducing the feed below the maximum value used to estimate the V_{opt2}. Therefore, equating the machining time for OP2 to that of OP4 ($t_{m2} = t_{m4} = 0.783$ min), a feed of 0.25 mm/rev is selected, which results in a tool life improvement from 14 to 22 min. The machining time and cost for these operations are

$$T_{u24} = t_{m4} + \frac{t_{m2}}{T_2} t_{l2} + \frac{t_{m4}}{T_4} t_{l4} + (t_{cs} + t_x)_{2+4}$$

$$T_{u24} = 0.783 + \frac{0.783}{22} 3 + \frac{0.783}{327} 3 + 2\left(\frac{2}{60} + 0.05\right) = 1.06 \text{ min}$$

$$C_{u24} = C_o t_{m4} + \frac{t_{m2}}{T_2}(C_o t_{l2} + C_{t2}) + \frac{t_{m4}}{T_4}(C_o t_{l4} + C_{t4}) + C_o(t_{cs} + t_x)_{2+4}$$

$$C_{u24} = 1 \times 0.756 + \frac{0.756}{22}(1 \times 3 + 4) + \frac{0.756}{327}(1 \times 3 + 4) + 1\left(\frac{2}{60} + 0.05\right) = \$1.21$$

Finally, the total production time (Equation 13.14) and cost are:

$$T_u = T_{u13} + T_{u24} + t_h = 1.23 + 1.06 + \frac{20}{60} = 2.63 \text{ min}$$

$$C_u = C_{u24} + C_{u24} + C_o t_h = 1.65 + 1.21 + 1\frac{20}{60} = \$3.20$$

Example 13.5 Optimize the cutting conditions for the single-pass turning operation defined in Table 13.3. The machining parameters and the coefficients and exponents for the tool life and constraint equations used in a single-pass turning operation are given in the table. The maximum and minimum values for feed, speed, power, and cutting force in the table are based on a knowledge of machine limitations and handbook information for the given workpiece and tool materials.

Solution: The contours for the production cost, production rate, and the constraints on the force F (600 and 900 N), surface roughness SR (5 and 8 μm), spindle power HP (4 kW), and temperature θ (400 and 550°C), given by Equation (13.19), are shown in Figure 13.10 for a 2.54 mm depth of cut. The objective function is given by Equation (13.16). The constant multiplier is calculated by Equation (13.18) to be 0.212 based on $f_h = 0.56$ and calculated values of $V_{opc} = 145$ (Equation 13.9), $V_{opt} = 149$ m/min (Equation 13.10), $C_{u,min} = \$0.6754$ (Equation 13.8), and $T_{u,min} = 3.183$ min (Equation 13.1).

The response contours for the profit rate (P) at sale prices of \$0.8 and \$1.6 are shown in Figure 13.11 by the solid and dashed lines, respectively; the profit rate increases with increasing feed and sale price. The horizontal span of the profit rate contours increases to the right with increasing sale price. The highest profit rate contours (achieved at the higher sale price) are located to the right of the lower profit rate contours (achieved at the lower sale price). This is indicated by the contour span for $S = \$1.6$ (denoted by C_2 in Figure 13.11) which is located to the right on the horizontal speed axis for the contour denoted by C_1 corresponding to $S = \$0.8$. The contours for the objective function with weights w_1 and w_2 of 0.8 and 0.2, respectively, are shown as dotted lines in the figure. The optimum profit value for contours with $S = \$0.8$ (solid lines) lies to the left of the 100 m/min speed, while that for the

TABLE 13.3
Machining Parameters Used for the Turning Example 13.5

Parameter	Value	Parameter	Value	Parameter	Value
L	203 mm	SR_{max}	8 m	R_c	1.396×10^9
D	152 mm	HP_{max}	5 kW	T_l	0.5 min/edge
V_{min}	30 m/min	F_{max}	1100 N	t_{cs}	0.2 min/tool
V_{max}	200 m/min	θ_{max}	500°C	t_x	0.13 min/pass
f_{min}	0.254 mm/rev	a_1	−1.16	C_o	\$0.1/min
f_{max}	0.762 mm/rev	a_2	−1.4	C_t	\$0.5/edge
SF_{max}	2 μm	a_3	−4	w_1	0.6
				w_2	0.4

Note: SF = surface finish; SR = surface roughness; HP = spindle power; F = cutting force; θ = average cutting temperature at the tool–chip interface a_1, a_2, a_3 = exponents in tool life equation corresponding to feed, depth of cut, and speed parameters, respectively, given by Equation (13.7).

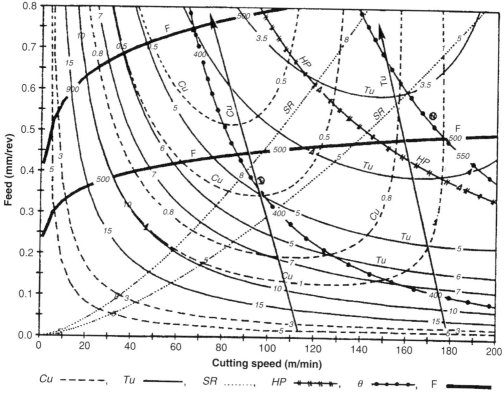

FIGURE 13.10 Contours of production cost, production time, and the constraints for power, force, surface roughness, and temperature using a 2.54 mm depth of cut.

contours with $S = \$1.6$ lies to the right of this speed. Figure 13.11 shows that the optimum conditions are close to the conditions of low profit when emphasis is placed on the production cost by using a large value of the weight w_1. On the other hand, when the w_1 and w_2 values are interchanged so that larger emphasis is placed on the production time, the lower value contours of the objective function will move toward the higher profit rate contours for $S = \$1.6$. When both weight coefficients are set equal to 0.5, the objective function contours are close to the higher profit rate contours. In general, a good practice would be to use weight coefficients $w_1 \geq w_2$ since in this case the optimum value lies in the area of maximum profit at the lower sale price.

Example 13.6 Optimize conditions for a single-stage multifunctional system (CNC machine). The optimum sequence of operations for a simple part made of gray cast iron, shown in Figure 13.12, is considered to demonstrate the analytical approach, the solution results, and the computational algorithms discussed above. The types of operations with the corresponding tool number, the corresponding nodes for the 13 operations, and the part processing dimensions are given in Table 13.4.

The top surface of the part is first milled before other operations are performed. Tool life data, described by Equation (13.7), are provided in Table 13.5. The constraints encountered in all operations and their upper and lower limits are given in Table 13.6. The parameters of the physical constraints, described by Equation (13.19), are given in Table 13.7 for most of the operations. All the constraint models were deterministic. Table 13.8 lists cost data for the

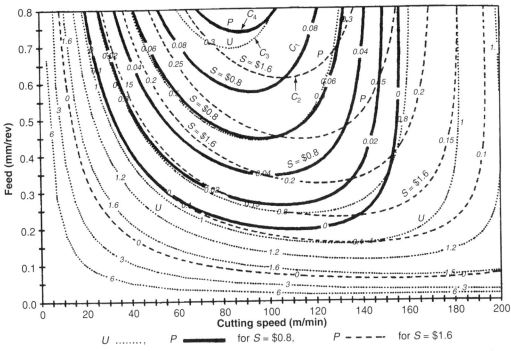

FIGURE 13.11 Contours of the objective function $U = 0.8C_u + 0.2(0.212)T_u$ and the profit rate for a sale price of $0.8 and $1.6 using a 2.54 mm depth of cut.

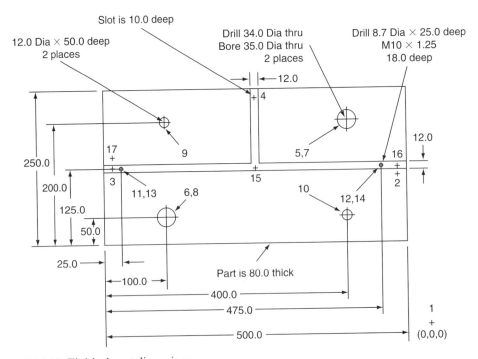

FIGURE 13.12 Finished part dimensions.

Machining Economics and Optimization

TABLE 13.4
Characteristics of Various Processes Used in Example 13.6

Operation, No.	Node No.	Length (mm)	Diameter (mm)	Depth of Cut (mm)	Tool Offset
Face milling, 1	2,17	850	324	4	200
End milling, 2	3,16	510	12	10	50
End milling, 3	4,15	130	12	10	50
Drilling, 3	5,6	92	34	—	140
Boring, 4	7,8	84	35	0.5	140
Drilling, 5	9,10	56	12	—	100
Drilling, 6	11,12	30	8.7	—	50
Tapping, 7	13,14	20	M10–1.25	—	38

Note: Face milling cutter has 14 teeth and end milling cutter has 4 teeth.

TABLE 13.5
Tool Life Parameters for Various Tools Used in Example 13.6

Tool Number	Tool Material	Constant R_i	Exponent a_{1i} (speed)	a_{2i} (feed)	a_{3i} (depth of cut)
1	Coated WC	—	—	—	—
2	Coated WC	5×10^6	-1.05	—	-2.4
3	WC tipped	20.549×10^{12}	-3.5	—	-7.0
4	Uncoated WC	3.772×10^7	-0.91	-1.09	-3.125
5	WC tipped	20.549×10^{12}	-3.5	—	-7.0
6	WC tipped	20.549×10^{12}	-3.5	—	-7.0
7	HSS	753.4	0.0	—	-1.825

Note: a_1, a_2, a_3 = exponents in tool life equation corresponding to feed, depth of cut, and speed parameters, respectively, given by Equation (13.7).

TABLE 13.6
Constraints for the Machine for Various Operations Used in Example 13.6

Tool No.	Speed Limit (m/min) Upper	Speed Limit (m/min) Lower	Feed limit (mm/rev) Upper	Feed limit (mm/rev) Lower	Max. Power (kW)	Max. Force (N)	Max. Temp. (°C)	Max. Roughness (μm)
1	50	250	0.03	0.25	—	—	—	—
2	70	300	0.03	0.15	7	4000	—	—
3	100	350	0.12	0.45	10	8000	—	—
4	60	350	0.1	0.6	5	1100	500	1
5	50	250	0.1	0.45	5	2000	—	—
6	50	250	0.1	0.45	5	2000	—	—
7	4.5	17	—	—	—	—	—	—

TABLE 13.7
Parameters for Physical Constraint Relationships Used in Example 13.6

Constraint Number, j	Tool Number, i	Constant, B_{1ij}	Speed Exponent, b_{1ij}	Feed Exponent, b_{2ij}	Depth of Cut Exponent, b_{3ij}	Diameter Exponent, b_{5ij}
Power (kW)	2	0.277	1.0	1.0	—	—
Force (N)	2	83	1.1	0.75	—	—
Power (kW)	3,5,6	17.09	1.0	0.8	0.0	0.8
Thrust (N)	3,5,6	191	0.0	0.85	0.0	1.3
Power (kW)	4	0.0373	0.91	0.78	0.75	0.0
Force (N)	4	844	−0.1013	0.725	0.75	0.0
Temp. (°C)	4	75	0.4	0.2	0.105	0.0
Surface roughness (μm)	4	14785	−1.52	1.0	0.25	0.0

TABLE 13.8
Economic Parameters for the Operations Used in Example 13.6

Parameter	Tool Number						
	1	2	3	4	5	6	7
C_o ($/min)	0.2	0.2	0.2	0.2	0.2	0.2	0.2
C_{ti} ($/tool)	56.0	6.0	10.0	4.0	6.0	5.0	4.0

cutting tools while the parameters for the machining center are given in Table 13.9. The drill cost includes the regrinding cost (corresponding to the tool cost between regrinds).

Solution: Given these data, the problem is to obtain the optimum cutting conditions for each operation based on the objective function given by Equation (13.17). Assuming the coefficients of the objective function are $w_1 = 0.4$, $w_2 = 0.6$, and $\lambda = 0.163$, the resulting optimum speed, feed, and tool life for each operation are given in matrix s^*:

$$s^* = \begin{bmatrix} 120 & 0.12 & 300 \\ 150 & 0.40 & 78.4 \\ 150 & 0.40 & 78.4 \\ 75 & 0.35 & 60.7 \\ 75 & 0.35 & 60.7 \\ 270 & 0.38 & 4.9 \\ 270 & 0.38 & 4.9 \\ 70 & 0.33 & 120.9 \\ 70 & 0.33 & 120.9 \\ 70 & 0.33 & 120.9 \\ 70 & 0.33 & 120.9 \\ 12 & -- & 8.6 \\ 12 & -- & 8.6 \end{bmatrix}$$

The problem constraints for the 13 operations are established next. The elements (2,3), (4,5), (6,7), (8,9), (10,11), and (12,13) of the tool matrix $[T_n]$ are equal to one. Several operations, such as drilling and boring, drilling and tapping, etc., impose precedence relations. The

TABLE 13.9
Machining Parameters Used for Example 13.6

Parameter	Value	Parameter	Value
t_{li}	2.0 min	S_{cs}	20 mm
t_{cs}	0.033 min	α	60 rad/sec^2
t_h	0.3 min	α_s	1.27 rad/sec^2
V_s	7.62 m/min	α_{cs}	0.762 rad/sec^2
V_{cs}	20.32 m/min		

precedence constraint for the operations are given by the elements (1,2), (1,3), (1,4), (1,5), (1,6), (1,7), (1,8), (1,9), (1,10), (1,11), (1,12), (1,13), (4,6), (6,7), (10,12), and (11,13) of the matrix $[S_n^a]$; these elements are equal to one while all other elements in the matrix are zero. Pairwise constraints are present for operations 1, 2, and 3, and are represented by the value of one for the elements 1, 2, and 3 in vector $\{S_n^b\}$. The end milling operation (number 3) has an order constraint since the operation can be proceeded only in one direction; thus $S_{n33}^c = 1$.

The 13 operations ($n = 13$) are described by a network with 17 nodes ($n_1 = 17$). Node 1 is the origin (0,0,0), while nodes 2, 3, ..., $n+1$ correspond to the operations 1, 2, ..., n. The drilling, boring, and tapping operations correspond to a single node. Nodes $n+2, n+3, \ldots, n_1$ are used to describe the face milling and end milling operations. Hence, the face milling operation is represented by nodes 2 and 17, while the end milling operations by nodes 3,16 and 4,15.

The next step is to determine the optimum sequence of operations without considering t_{cs} and t_{e_i} in Equation (13.25). The settling time for the machine slides, t_{si}, is assumed to be zero since it is a function of the servo control system. For this purpose, we assume that the spindle starts and finishes at node 1, which is the tool switch location. After completion of the operation(s) with one tool, the tool for the next operation(s) is placed on the spindle automatically by going back to node 1. We also assume that the part position with respect to the machine coordinates is such that the lower left corner of the top surface of the part is located at $(-900,-250,300)$.

The operations' constraints are described in the node constraint matrices $[C_n^a]$ and $[C_n^b]$. The elements (2,1), (3,1), (3,2), (4,1), (4,2), (5,1), (5,2), (6,1), (6,2), (7,1), (7,2), (7,5), (8,1), (8,2), (8,6), (9,1), (9,2), (10,1), (10,2), (11,1), (11,2), (11,3), (12,1), (12,2), (12,3), (13,1), (13,2), (13,11), (14,1), (14,2), (14,12), (15,1), (15,2), (15,4), (16,1), (16,2), and (17,1) of matrix $[C_n^a]$ are equal to one. The elements (2,17), (3,16), (4,15), (15,4), (16,3), and (17,2) of matrix $[C_n^b]$ are equal to one.

The distance time in $[H]$ was obtained from Equation (13.25) in which the terms t_{cs} and t_{e_i} are assumed to be zero. Next, using the previously mentioned genetic algorithm for the TSP with a population size of 1000, a maximum number of generations 40,000, and a bias of 2, the optimum sequence, (1–2–17–4–15–3–16–6–5–11–12–14–13–9–10–8–7), was obtained. The objective function had the value of 4.607 min. It is obvious that all operations to be performed with a specific tool will be performed together because tool change is time consuming. The same objective function values were found with sequences:

(1–2–17–4–15–3–16–6–5–11–12–13–14–10–9–8–7),
(1–2–17–4–15–3–16–6–5–12–11–14–13–9–10–8–7),
(1–2–17–4–15–3–16–6–5–11–12–10–9–14–13–8–7),
(1–2–17–6–5–4–15–3–16–12–11–13–14–10–9–8–7),

due to the symmetry between some operations. However, the values of objective function for sequences (1–2–17–4–15–16–3–6–5–12–11–9–10–13–14–8–7) and (1–2–17–6–5–4–15–3–16–11–12–10–9–13–14–8–7) were 4.815 and 4.944, respectively, and require 4.5 and 7.3% longer noncutting time, respectively, when compared to the optimum sequence. On the other hand, t_{cs} and t_{e_i} may be included in h_{ij} and the aforementioned sequence could change depending on the tools' location in the magazine.

The tool organization in a 60-slot magazine ($M^T = 60$) was also analyzed to determine the optimum tool type number. Seven different tools are required based on Figure 13.12. The expected tool life and cutting time per part were estimated and are given in Table 13.10. Estimates of the number of tools for each operation required in the tool magazine were obtained based on the procedure described in Ref. [93]. A total of 60 tools were organized in the order of processing and the order of replacement so that the seven tools are usually close together. Their suggested order in the tool magazine is (1,2,3,6,7,5,4,4,4,4,3,7,4,1,4,3,2,7,4,4, 1,4,3,4,7,4,4,4,1,3,2,4,7,4,4,3,4,1,4,7,4,4,3,4,2,4,1,7,4,4,3,4,7,1,4,5,3,4,4,2).

The problem was simulated for two different instants of the tools' location in the magazine which will occur during production based on the above tool arrangement. An evaluation of the optimum operations sequence, incorporating the tool organization in the magazine, is given in Table 13.11. The slot number in the magazine for the corresponding operation is also given

TABLE 13.10
Estimates of Number of Tools Required for Mass Production Used in Example 13.6

Tools	1	2	3	4	5	6	7
Tool life (min)	300	78.4	60.7	4.9	120.9	120.9	8.9
Cutting time (min)	2.5245	0.3965	0.6514	0.171	0.1632	0.0592	0.08376
No. of tools	7	5	9	28	2	1	8

TABLE 13.11
The Optimum Sequence of Operations Used in Example 13.6

Tool Slot Location	Node Sequence	Tool Slot Location	Node Sequence
0	1	0	1
1	17	29	2
1	2	31	17
2	4	31	4
2	15	36	15
2	3	36	3
2	16	35	16
3	6	35	9
3	5	5	10
6	12	5	12
6	11	6	11
7	13	6	13
7	14	33	14
5	10	33	5
5	9	31	6
4	8	31	8
4	7	29	7
Total time	5.029 min	Total time	5.362 min

Machining Economics and Optimization

in the table. When the tools for all the operations are located next to each other in the order of the optimum sequence found earlier without considering the terms t_{cs} and t_e, the total time U^T to process a part is 5.029 min. However, when the tools are scattered in the magazine at a later production time, the total time U^T becomes 5.632 min. The part production time is 9.841 and 10.444 min, respectively, for these two cases. The tools' location in the former case requires 6.1% shorter production (cycle) time than in the latter case.

Example 13.7 Optimization of a multistep tool. Consider two multistep solid carbide head drills with oil holes having three steps, $A_c = \{1,2,3\}$, with diameters D_1, D_2, and D_3, which are used to drill two different holes with lengths L_1, L_2, and L_3, respectively (see Figure 13.6). The work material is 390 aluminum. The same tool life equation is used for all steps on both tools. The tool life equation and the economic parameters for the process are given in Table 13.12, while the operating constraints are given in Table 13.13. Both the constant rpm and surface

TABLE 13.12
Tool and Economic Parameters for Example 13.7

Parameter	Value for Drill #1	Value for Drill #2
D_1 (mm)	8	8
D_2 (mm)	10	17
D_3 (mm)	12	25
L_1 (mm)	70	20
L_2 (mm)	40	14
L_3 (mm)	15	8
t_{h_i} (min)	0.15	0.15
t_{cs_i} (min)	0.033	0.033
t_{r_i} (min)	0.01	0.01
t_l (min)	2.0	2.0
α (rad/sec^2)	1000	1000
VB$_i$ (mm)	0.3	0.3
R_i	1273.8 × 10^{15}	1273.8 × 10^{15}
R'	42.1 × 10^9	42.1 × 10^9
a_3	−2.989	−2.989
a_1	−0.5433	−0.5433
c	−0.14833	−0.14833
d	−5.069	−5.069
P_1	1.0	1.0
P_2	1.0	1.0
P_3	1.0	1.0
C_o ($)	0.4	0.4
K_i	8	8
C_{gi} ($/min)	1.00	1.00
C_t ($)	130	100
t_{s1} (min)	5	5
w_1	0.4	
f_h (mm/rev)	0.25	0.25
N_{max} (rpm)	6000	6000
F_{max} (N)	500	1300
P_{max} (kW)	7	5

TABLE 13.13
Physical Constraint Relationships for Example 13.7

Constraint	Constant, B_{fi}	Speed Exponent, a_{1i}	Feed Exponent, a_{2i}	Diameter Exponent, a_{3i}	Wear Exponent, a_{5i}
Power (kW)	0.007743	0.7822	0.5961	0.582	0.7643
Thrust (N)	191	0.0	0.5496	1.3	0.3779

speed approaches are evaluated. The first and second tool steps are assumed to operate at the same rpm in all cases, $A_d = \{2\}$. Optimize the cutting conditions.

Solution:

(a) Constant rpm approach for drill #1

The value Y_{kj} or $Y_{kj}^*(Y_k = \text{int}(T_{mk}/T_{pk}))$ is estimated for the three tool steps using Equations (13.45) and (13.46) to determine the maximum number of parts the tool can machine between tool reconditions. This drill has $L_k = 40$ mm and $D_k = 10$ mm because $Y_{2j} > Y_{1j} > Y_{3j}$. The maximum spindle rpms for minimum production time and cost, $N_{\max\,t}$ and $N_{\max\,c}$, respectively, obtained at $f_h = 0.25$ mm/rev using Equations (13.9) and (13.10) ($N = V/\pi D$), are given in Table 13.14. The parameters $T_{u,\min}$, $C_{u,\min}$, and λ were calculated to be 0.273 min, \$0.258, and 0.947, respectively, and resulted in the objective function $U(N,f) = 0.4C_u + 0.568T_u$. T_u and C_u are obtained from Equations (13.1) and (13.8), respectively. Therefore, the optimum cutting conditions are 5700 rpm and 0.263 mm/rev, which result in a tool life of 37.4 min (see Table 13.14).

(b) Constant rpm approach for drill #2

The drill step $L_k = 8$ mm and $D_k = 25$ mm is selected as the critical one because $Y_{3j} > Y_{2j} > Y_{1j}$. The maximum spindle rpm for minimum production time and cost, obtained at 0.25 mm/rev feed, are given in Table 13.12. $T_{u,\min} = 0.234$ min and $C_{u,\min} = \$0.177$, resulting in $\lambda = 0.757$. Therefore, $U(N,f) = 0.4C_u + 0.454T_u$ and the optimum cutting conditions are 2800 rpm and 0.263 mm/rev, which result in a tool life of 20.3 min (see Table 13.14).

TABLE 13.14
The Optimum Cutting Parameters for Example 13.7

	Approaches					
	Constant Spindle rpm			Constant Surface Speed		
Parameters	Drill #1	Drill #2	Parameters	Drill #1	Drill #2
$N_{\max t}$ (rpm)	14,684	6,618	$V_{\max t}$ (m/min)	413	415
$N_{\max c}$ (rpm)	4,272	2,007	$V_{\max c}$ (m/min)	120	126
f_h (mm/rev)	0.25	0.25	f_h (mm/rev)	0.25	0.25
N_{op} (rpm)	5,700	2,800	V_{op} (m/min)	160	134
f_{op} (mm/rev)	0.263	0.263	f_{op} (mm/rev)	0.263	0.263
T_{pj} (min)	37.4	20.3	T_{pj} (min)	52.4	89.0
T_u (min)	0.261	0.222	T_u (min)	0.258	0.229
C_u (\$/hole)	0.124	0.104	C_u (\$/hole)	0.120	0.097

(c) Constant surface speed approach for drill #1

The estimation of the optimum speed or rpm is defined for the tool step with the maximum Z_{xj} (see Equation 13.52). It was found that $Z_1 > Z_2 > Z_3$. Therefore, the maximum cutting surface speeds for minimum production time and cost are 413 and 120 m/min, respectively, based on which $T_{u,min} = 0.279$ min, $C_{u,min} = \$0.26$, and $\lambda = 0.932$. Hence, $U(N,f) = 0.4C_u + 0.559T_u$ and the optimum cutting speed and feed are 160 m/min and 0.263 mm/rev, which correspond to $N_1 = N_2 = 6366$ rpm and $N_3 = 4244$ rpm, respectively, for the first and third tool steps. The speed transition from N_1 to N_3 requires a deceleration time, t_α, of 0.222 sec, which corresponds to a linear length, S_a, of 5.16 mm on the tool. Since $S_a \ll L_{T2} - L_{T1}$, the speed transition is achievable. The tool life is 52.4 min (see Table 13.14).

The production time and cost for drill #1 at the optimum cutting conditions for a constant spindle rpm and a constant surface speed are given in Table 13.14; the difference between the results for the two approaches is insignificant even though they result in significantly different tool lives. The response of the production cost as a function of the cutting speed for both approaches is shown in Figure 13.13. The constant surface speed curve has a response similar to that of the constant spindle rpm curve of the first tool step D_1 when compared with that of the third tool step because $Y_{11j} > Y_{13j}$. The constant surface speed approach resulted in a 40% increase in tool life and 1.5% increase in the number of holes drilled between tool regrinds.

(d) Constant surface speed approach for drill #2

In this case, $Z_3 > Z_2 > Z_1$. Therefore, the maximum surface speeds for minimum production time and cost are 415 and 126 m/min, respectively, which are used to obtain $T_{u,min} = 0.264$, $C_{u,min} = \$0.159$ and $\lambda = 0.603$. Hence, $U(N,f) = 0.4C_u + 0.362T_u$, and the optimum cutting

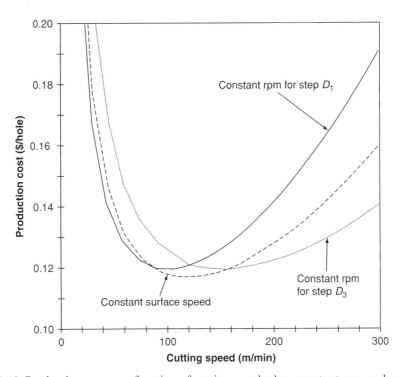

FIGURE 13.13 Production cost as a function of cutting speed when constant rpm and surface speed approaches are used on multistep drill #1 in Example 13.7.

speed and feed are, respectively, 134 m/min and 0.263 mm/rev. The corresponding spindle rpm for the first two steps ($N_1 = N_2$) is 5332 rpm, while for the third step (N_3) it is 1706 rpm. The speed transition from N_1 to N_3 requires an acceleration time t_α of 0.38 sec, which corresponds to a linear length S_a of 5.86 mm. Since $S_a \ll L_{T2} - L_{T1}$, the speed transition is achievable. The tool life under the optimum conditions is 89 min.

The production time and cost for drill #2 at the optimum cutting conditions for the constant spindle rpm and surface speed approaches are given in Table 13.14. A 3 and 7% difference in production cost and time, respectively, were found between the two approaches. The production cost response for both approaches as a function of cutting speed is shown in Figure 13.14, which shows the small difference in the cost for the two approaches at the corresponding optimum speeds. The constant surface speed curve has a similar shape to that of the constant spindle rpm curve based on the third tool step, D_3, because $Z_3 > Z_1$. On the other hand, the tool life is significantly different between the two approaches. The constant surface approach resulted in a 338% increase in tool life and a 167% increase in the number of holes drilled between tool regrinds.

The insignificant difference of the number of holes drilled between regrinds for drill #1 with both approaches is shown in Figure 13.15 for $\beta = 0.07$ and $q = 4$ in Equation (13.11). The production approach did not affect the number of holes produced by the drill with a geometry of $D_1/(L_1 - L_2) < D_2/(L_2 - L_3) < D_3/L_3 < 1$. On the other hand, Figure 13.16 shows a significant difference in the number of holes obtained between regrinds for the two approaches when drill #2 is used. The constant surface speed approach is the best strategy for a drill with a geometry such that $D_3/L_3 > D_2/(L_2 - L_3) > D_1/(L_1 - L_2) > 1$.

The traditional manufacturing approach using three conventional (single-step) tools for drilling the multistep hole produced with the multistep drill #1 is considered next. The drill

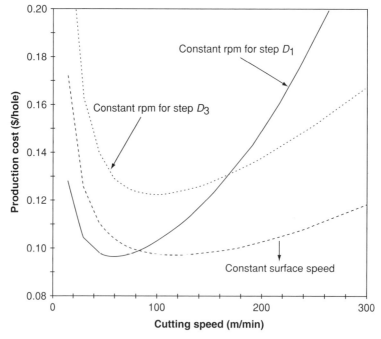

FIGURE 13.14 Production cost as a function of cutting speed when constant rpm and surface speed approaches are used on multistep drill #2 in Example 13.7.

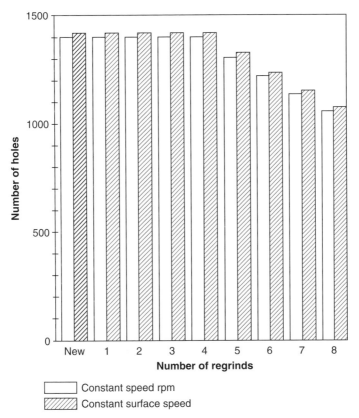

FIGURE 13.15 Number of holes produced between all available regrinds when constant rpm and surface speed approaches are used with multistep drill #1 in Example 13.7.

designations with the corresponding hole diameter sand depths drilled by each tool are given in Table 13.15. The costs for the three drills are $90, $60, and $40, respectively, with eight available regrinds for each drill. Other cost parameters, tool-change time, part loading time, tool life equation, and physical constraints are assumed to be the same as for the multistep drill #1. The optimum spindle rpm and feed for each of the three drills were calculated based on the objective function $U(N,f) = 0.4C_u + 0.559T_u$ and are given in Table 13.15. The number of holes drilled between regrinds by the single-step tools is larger than the 1402 holes obtained with the multistep tool. However, the multistep drill reduced the production cycle time and

TABLE 13.15
Productivity of Single-Step Tools Used for the Hole Drilled with Multistep Drill #1 in Example 13.7

	Drill Diameter (mm)	Drilled Depth (mm)	N_{op} (rpm)	f_{op} (mm/rev)	C_u ($/hole)	T_u (min)	T_p (min)	Number of Holes
A	8	30	5140	0.263	0.114	0.278	99.3	4474
B	10	25	4720	0.263	0.102	0.247	51.0	2532
C	12	15	3680	0.263	0.090	0.222	29.6	1910
				Total	0.306	0.747		

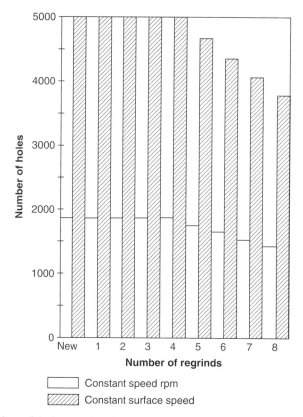

FIGURE 13.16 Number of holes produced between all available regrinds when constant rpm and surface speed approaches are used with multistep drill #2 in Example 13.7.

cost by 65 and 60%, respectively. The breakeven point with respect to production cost is obtained at a production batch of 154 holes. Therefore, the multistep tool is superior at either high production rates or in medium or mass production applications.

13.9 PROBLEMS

Problem 13.1 A 25 mm through hole is made in a medium carbon steel part using a two-flute drill with 140° point angle. The hole is made through a 50 mm thick plate. The tool life equation is $VT^{0.24} = 150$ (metric units) for this tool and workpiece material. The maximum allowable feed for the cutting tool is 0.15 mm per flute. The time for loading/unloading the drill in the machine tool is about 2 min. The part handling (load/unload) time is 20 sec. The rapid feed rate for the machine is 25,000 mm/min. The drill cost is $100 with five maximum regrinds at $30 per regrind. The operating cost is $80 per hour. Estimate the individual cost curves similar to Figure 13.1 and show the optimum cutting speed.

Problem 13.2 Determine the optimum spindle rpm for minimum production cost in boring a 50 mm diameter hole 60 mm deep. The depth of cut is 3 mm and the maximum feed allowed is 0.2 mm. The tool life equation is $VT^{0.3}f^{0.6} = 450$. The load/unload time for the boring tool is 1 min. The tool cost is $4/edge. The operating cost is $80 per hour.

Problem 13.3 Determine the optimum spindle rpm for minimum production time in Problem 13.2.

Problem 13.4 The part in Figure 13.17 is being machined in a lathe from bar steel with $D_1 = 60$ mm. A triangular carbide insert is used for the rough turning operation with zero lead angle and a nose radius of 1.59 mm. The limits of this insert for feed and depth of cut are 0.6 mm/rev and 5 mm, respectively. The feed limits for the drill and reamer per edge are 0.2 and 0.1 mm. A triangular carbide insert with a nose radius of 0.762 mm is also used for finish turning. The depth of cut for the finish operation is 0.4 mm. The surface finish requirement for turning the 44 mm diameter is 2s μm. The cost of the roughing insert is $7 with 6 useful edges, while the finishing insert costs $8 with three useful edges. The cost of the 12 mm diameter two-flute carbide drill is $70, and on average it can be reground six times. The cost of the 12.5 mm diameter four-flute carbide reamer is $100 and on average it can be reground three times. The cost of regrinding is $15 for either the drill or the reamer. The operating cost is $60 per hour. The constants for the tool life equation $VT^n f^a = K_t$ are given in Table 13.16. The tool load/unload time is 2 ;min, and the tool index time is 5 sec. Calculate the minimum production cost and the corresponding production time.

Problem 13.5 A step hole in a block of soft steel, shown in Figure 13.18, is drilled and counterbored with a step carbide drill (single tool). The hole diameters are 10 and 14 mm. The maximum allowable cutting speed for the tool is 100 m/min and the feed is 0.13 mm per tooth. A standard two-flute solid carbide step drill is used with a 120° point angle. The tool life is $VT^{0.35} f^{0.5} = 550$. The load/unload time for the boring tool is 2 min. The tool cost is $150 and

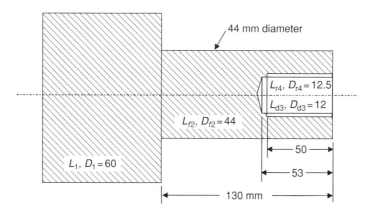

FIGURE 13.17 Configuration and finish dimensions for the part machined in Problem 13.4.

TABLE 13.16
Tool Life Constants for Problem 13.4

Tool Type	a	n	K_t
Rough insert	0.5	0.38	150
Finish insert	0.5	0.4	200
Drill	0.5	0.3	100
Reamer	0.5	0.5	180

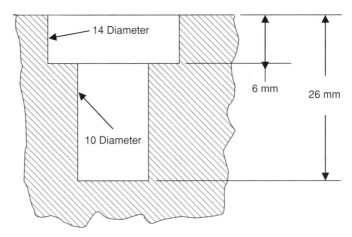

FIGURE 13.18 Configuration of the step hole machined in Problem 13.5.

the regrinding cost is $40 with four regrinds total. The operating cost is $60 per hour. Determine the optimum cutting conditions for machining the stepped hole with a single-step tool.

REFERENCES

1. P. Balakrishnan and M.F. DeVries, A review of computerized machinability data base systems, *Proc. NAMRC* **10** (1982) 348–356
2. B. Gopalakrishnan, Computer integrated machining parameter selection in a job shop using expert systems, *J. Mech. Working Technol.* **20** (1989) 163–170
3. L. Zeng and H.P. Wang, A patchboard-based expert-systems model for manufacturing applications, *Int. J. Adv. Manuf. Technol.* **7** (1992) 38–43
4. G. Sathyanarayanan, I.J. Lin, and M.K. Chen, Neural network modeling and multiobjective optimization of creep feed grinding of superalloys, *Int. J. Prod. Res.* **30**:10 (1992) 2421–2438
5. D.F. Cook and R.E. Shannon, A predictive neural network modeling system for manufacturing process parameters, *Int. J. Prod. Res.* **30**:7 (1992) 1537–1550
6. Y.H. Pao, *Adaptive Pattern Recognition and Neural Networks*, Addison-Wesley, Reading, MA, 1989
7. H.T. Klahorst, How to justify multi-machine systems, *Am. Machinist*, **September** (1983) 67–70
8. K. Mundy, The financial justification of FMS, *Proceedings of the 1st International Machine Tool Conference*, Birmingham, UK, June 1984
9. P.L. Primrose and R. Leonard, The financial evaluation of flexible manufacturing modules (FMM), *Proceedings of the First International Machine Tool Conference*, Birmingham, UK, June 1984
10. E. Porteus, Investing in reduced setups in the EOQ model, *Manage. Sci.* **31**:8 (1985) 998–1010
11. G. Keller and H. Noori, Justifying new technology acquisition through its impact on the cost of running an inventory policy, *IEEE Trans.* **20**:3 (1988) 284–291
12. A.M. Spence and E.L. Porteus, Setup reduction and increased effective capacity, *Manageg. Sci.* **33**:10 (1987) 1291–1301
13. U. Karmarkar and S. Kekre, Manufacturing configuration, capacity and mix decisions considering operational costs, *J. Manuf. Systems* **6** (1987) 315–324
14. D.J. Vee and W.C. Jordan, Analyzing trade-offs between machine investment and utilization, *Manage. Sci.* **35**:10 (1989) 1215–1226
15. C.H. Fine and R.M. Freund, Optimal investment in product-flexible manufacturing capacity, *Manage. Sci.* **36**:4 (1990) 449–466

16. J.P. Buzen and P.J. Denning, Measuring and calculating queue length distributions, *IEEE Spectrum*, **April** (1980) 33–44
17. H.C. Co and R.A. Wysk, The robustness of CAN-Q in modeling automated manufacturing systems, *Int. J. Prod. Res.* **24**:6 (1986) 1485–1503
18. N.M. Bengtson, Using operational analysis in simulation: a queuing network example, *J. Oper. Res.* **39** (1988) 1125–1136
19. P.J. Denning and J.P. Buzen, The operational analysis of queuing network models, *Computing Surveys* **10**:3 (1978) 225–261
20. R. Suri, Robustness of queuing network formulas, *J. ACM* **30**:3 (1983) 564–594
21. R. Suri and R.R. Hildebrant, Modeling flexible manufacturing systems using mean-value analysis, *J. Manuf. Systems* **3**:1 (1984) 27–38
22. R. Suri and M. Tomsicek, Rapid modeling tools for manufacturing simulation and analysis, *Proceedings of the 1988 Winter Simulation Conference*, 1988, 25–32
23. W.W. Gilbert, Economics of machining, *Machining-Theory and Practice*, ASM, Metals Park, OH, 1950, 465–485
24. B.Y. Lee, H.S. Liu, and Y.S. Tarng, Modeling and optimization of drilling process, *J. Mater. Process. Technol.* **74** (1998) 149–157
25. M. Alauddin, M.A. El Baradie, and M.S.J. Hashmi, Optimization of surface finish in end milling Inconel 718, *J. Mater. Process. Technol.* **56** (1996) 54–65
26. S.V. Wong and A.M. S. Hamouda, Development of genetic algorithm-based fuzzy rules design for metal cutting data selection, *Robot. Computer Integr. Manuf.* **18** (2002) 1–12
27. S.K. Hati and S.S. Rao, Determination of optimum machining conditions — deterministic and probabilistic approaches, *ASME J. Eng. Ind.* **98** (1976) 354–359
28. D. Ermer and S. Kromodihardjo, Optimization of Multipass Turning with Constraints, ASME Paper 80-WA/Prod-22, 1980
29. M.A. Shalaby and M.S. Riad, A linear optimization model for single pass turning operations, *Proceedings of the 27th International MATADOR Conference*, 1988, 231–235
30. D. Ermer, Optimization of the constrained machining economics problem by geometric programming, *ASME J. Eng. Ind.* **93** (1971) 1067–1072
31. K. Hitomi, Analysis of optimal machining speeds for automatic manufacturing, *Int. J. Prod. Res.* **27** (1989) 1685–1691
32. S.M. Wu and D. Ermer, Maximum profit as the criterion in the determination of the optimum cutting conditions, *ASME J. Eng. Ind.* **88** (1966) 435–442
33. G. Boothroyd and P. Rusek, Maximum rate of profit criteria in machining, *ASME J. Eng. Ind.* **98** (1976) 217–220
34. J.S. Agapiou, The optimization of machining operations based on a combined criterion. Part I. The use of combined objectives in single-pass operations, *ASME J. Eng. Ind.* **114** (1992) 500–507
35. E.J.A. Armarego and R.H. Brown, *The Machining of Metals*, Prentice-Hall, Englewood Cliffs, NJ, 1969, Chapter 9
36. S.K. Choudhury and I.V.K. Appa Rao, Optimization of cutting parameters for maximizing tool life, *Int. J. Mach. Tools Manuf.* **39** (1999) 343–353
37. W.T. Chien and C.S. Tsai, The investigation on the prediction of tool wear and the determination of optimum cutting conditions in machining 17–4PH stainless steel, *J. Mater. Process. Technol.* **140** (2003) 340–345
38. M.F. DeVries, Basic Machining Economics, SME Technical Paper MR70-538, 1970
39. E.J.A. Armerego and J.K. Russel, Maximum profit rate as a criterion for the selection of machining conditions, *Int. J. Mach. Tool Des. Res.* **6** (1966) 15–23
40. D.Y. Jang, A unified optimization model of a machining process for specified conditions of machined surface and process performance, *Int. J. Prod. Res.* **30** (1992) 647–663
41. Carboloy, Inc., HI-E (HI Efficiency Machining) Technical Information Catalog, 1993
42. J.W. Sutherland, G. Subramani, M.J. Kuhl, R.E. DeVor, and S.G. Kapoor, An investigation into the effect of tool and cut geometry on cutting force system prediction models, *Proc. NAMRC* **16** (1988) 264–272

43. J.S. Agapiou and M.F. DeVries, On the determination of thermal phenomena during a drilling process. Part I. Analytical models of twist drill temperature distributions, *Int. J. Mach. Tools Manuf.* **30** (1990) 203–215
44. S. Hinduja, D.J. Petty, M. Tester, and G. Barrow, Calculation of optimum cutting conditions for turning operations, *Proc. Inst. Mech. Engrs.* **199B** (1985) 81–92
45. D.A. Stephenson and S.M. Wu, Computer models for the mechanics of three-dimensional cutting processes. Parts 1 and 2, *ASME J. Eng. Ind.* **110** (1988) 203–215.
46. D.A. Stephenson, Material characterization for metal cutting force modeling, *ASME J. Eng. Mater. Technol.* **111** (1989) 210–219.
47. D.A. Stephenson and J.S. Agapiou, Calculation of main cutting edge forces and torque for drills with arbitrary point geometries, *Int. J. Mach. Tools Manuf.* **32** (1992) 521–538
48. J.S. Agapiou and D.A. Stephenson, Analytical and experimental studies of drill temperatures, *ASME J. Eng. Ind.* **116** (1994) 54–60
49. A.M. Abuelnaga and M.A. El-Dardiry, Optimization methods for metal cutting, *Int. J. Mach. Tool Des. Res.* **24** (1984) 11–18
50. D.A. Milner, Use of Linear Programming for Machinability Data Optimization, ASME Paper No. 76-WA/Prod-5, 1976
51. D.S. Ermer and D.C. Patel, Maximization of the production rate with constraints by linear programming and sensitivity analysis, *Proc. NAMRC* **2** (1974) 436–449
52. J.A. Nelder and R. Mead, A simplex method for function minimization, *Comput. J.* **7** (1965) 308–313
53. D.M. Olsson and L.S. Nelson, The Nelder–Mead simplex procedure for function minimization, *Technometrics* **17** (1975) 45–51
54. K. Challa and P.B. Berra, Automated planning and optimization of machining processes: a systems approach, *Comput. Ind. Eng.* **1** (1976) 35–45
55. D.L. Kimbler, R.A. Wysk, and R.P. Davis, Alternative approaches to the machining parameter optimization problem, *Comput. Ind. Eng.* **2** (1978) 195–202
56. K. Iwata, et al., A probabilistic approach to the determination of the optimum cutting conditions, *ASME J. Eng. Ind.* **94** (1972) 1099–1107
57. C. Giardini, A. Bugini, and R. Pagagnella, The optimal cutting conditions as a function of probability distribution function of tool life and experimental test numbers, *Int. J. Mach. Tools Manuf.* **28** (1988) 453–459
58. A.K. Sheikh, L.A. Kendall, and S.M. Pandit, Probabilistic optimization of multitool machining operations, *ASME J. Eng. Ind.* **102** (1980) 239–246
59. B. Gopalakrishnan and F. Al-Khayyal, Machine parameter selection for turning with constraints: an analytical approach based on geometric programming, *Int. J. Prod. Res.* **29** (1991) 1897–1908
60. P.G. Petropoulos, Optimal selection of machining rate variables by geometric programming, *Int. J. Prod. Res.* **13** (1975) 390–395
61. C.S. Beightler and D.T. Phillips, Optimization in tool engineering using geometric programming, *AIIE Trans.* **2** (1970) 355–360
62. A.G. Walvekar and B.K. Lambert, An application of geometric programming to machining variable selection, *Int. J. Prod. Res.* **8** (1970) 241–245
63. D.S. Ermer, Optimization of constrained machining economics problem by geometric programming, *ASME J. Eng. Ind.* **93** (1971) 1067–1072
64. H. Eskicioglu, M.S. Nisli, and S.E. Kilic, An application of geometric programming to single-pass turning operations, *Proceedings of the International. MTDR Conference*, Birmingham, 1985, 149–157
65. B.K. Lambert and A.G. Walvekar, Optimization of multipass machining operations, *Int. J. Prod. Res.* **9** (1978) 247–259
66. C.L. Hough, Jr. and R.E. Goforth, Optimization of the second-order logarithmic machining economics problem by extended geometric programming, *AIIE Trans.* **13** (1981) 234–242
67. V.I. Vitanov, D.K. Harrison, N.H. Mincoff, and T.V. Vladimirova, An expert system for the selection of metal-cutting parameters, *J. Mater. Process. Technol.* **55** (1995) 111–116
68. R.H. Philipson and A. Ravindran, Application of goal programming to machinability data optimization, *ASME J. Mech. Des.* **100** (1978) 286–291

69. R.M. Sundaram, An application of coal programming technique in metal cutting, *Int. J. Prod. Res* **16** (1978) 375–382
70. G.W. Fischer, Y. Wei, and S. Dontamsetti, Process-controlled machining of gray cast iron, *J. Mech. Working Technol.* **20** (1989) 47–57
71. P.C. Subbarao and C.H. Jacobs, Application of nonlinear goal programming to machining variable optimization, *Proc. NAMRC* **6** (1978) 298–303
72. F. Cus and J. Balic, Optimization of cutting process by GA approach, *Robot. Computer Integr. Manuf.* **19** (2003) 113–121
73. E.E. Goldberg, *Genetic Algorithm in Searching, Optimization, and Machine Learning*, Addison-Wesley, Reading, MA, 1989
74. W.J. Hui and Y.G. Xi, Operation mechanism analysis of genetic algorithm, *Control Theory Appl.* **13** (1996) 297–303
75. D. Dubois, A fuzzy-set-based method for the optimization of machining operations, *Proceedings of the International Conference on Cybernetics and Society*, IEEE Systems, Man and Cybernetics Society, October 1981, 331–334
76. X.D. Fang and I.S. Jawahir, Predicting total machining performance in finish turning using integrated fuzzy-set models of the machinability parameters, *Int. J. Prod. Res.* **32** (1994) 833–849
77. J.S. Agapiou, The optimization of machining operations based on a combined criterion. Part II. Multipass operations, *ASME J. Eng. Ind.* **114** (1992) 508–513
78. A.N. Shuaib, S.O. Duffuaa, and M. Alam, Optimal process plans for multipass turning operations, *Trans. NAMRI/SME* **20** (1992) 305–310
79. I. Yellowley and E.A. Gunn, The optimal subdivision of cut in multi-pass machining operations, *Int. J. Prod. Res.* **27** (1989) 1573–1588
80. J.R. Crookall and N. Venkataramani, Computer optimization of multipass turning, *Int. J. Prod. Res.* **9** (1971) 247–259
81. H.J.J. Kals, J.A.W. Hijink, and A.C.H. Van der Wolf, Computer aid in the optimization of turning conditions in multi-cut operations, *CIRP Ann.* (1977) 465–471
82. F.P. Tan and R.C. Creese, A generalized multi-pass machining model for machining parameter selection in turning, *Int. J. Prod. Res.* **33** (1995) 1467–1487
83. K.P. Rajurkar, Optimization of multistage operations in EDM, *Proc. NAMRC* **16** (1986) 251–256
84. G.L. Nemhauser, *Dynamic Programming*, Wiley, New York, 1966
85. R. Srinivasan and C.R. Liu, On some important geometric issues in generative process planning, *Intelligent and Integrated Manufacturing Analysis and Synthesis*, ASME PED Vol. 25, ASME, New York, 1987, 229–243
86. N. Tounsi and M.A. Elbestawi, Optimized feed scheduling in three axes machining. Part I. Fundamentals of the optimized feed scheduling strategy, *Int. J. Mach. Tools Manuf.* **43** (2003) 253–267
87. R.E. DeVor, W.J. Zdeblick, V.A. Tipnis, and S. Buescher, Development of mathematical models for process planning of machining operations, *Proc. NAMRC* **6** (1978) 395–401
88. G. Chryssolouris and M. Guillot, An AI approach to the selection of process parameters in intelligent machining, ASME PED Vol. 33, ASME, New York, 1988, 199–206
89. K. Hitomi and I. Ham, Group scheduling technique for multiproduct, multistage manufacturing systems, *ASME J. Eng. Ind.* **99** (1977) 759–765
90. K. Hitomi, N. Nakamura, and H. Tanaka, An Optimization Analysis of Machining Center, ASME Paper 76-WA/P–34, 1976
91. K. Hitomi and K. Ohashi, Optimum process design for a single-stage multifunctional system, *ASME J. Eng. Ind.* **103** (1981) 218–223
92. C.S. Tang and E.V. Denardo, Models arising from a flexible manufacturing machine. Part I. Minimization of the number of tool switches. Part II. Minimization of the number of switches instants, *Oper. Res.* **36** (1988) 767–784
93. J.S. Agapiou, Sequence of operations optimization in single-stage multifunctional systems, *J. Manuf. Systems* **10** (1992) 194–208
94. C.H. Papadimitriou and K. Steiglitz, *Combinatorial Optimization Algorithms and Complexity*, Prentice-Hall, Englewood Cliffs, NJ, 1982

95. F. Lokin, Procedures for traveling salesman problems with additional constraints, *Eur. J. Oper. Res.* **3** (1979) 135–141
96. D. Whitley, T. Starkweather, and D. Fuquay, Scheduling problems and traveling salesmen: the genetic edge recombination operator, *Proceedings of the Genetic Algorithms Conference*, June 1989, 133–140
97. H. Tempelmeier and H. Kuhn, *Flexible Manufacturing Systems — Decision Support for Design and Operation*, Wiley, New York, 1993
98. R.A. Maleki, *Flexible Manufacturing Systems — The Technology and Management*, Prentice-Hall, Englewood Cliffs, NJ, 1991
99. H.J. Warnecke and R. Stenhilper, *Flexible Manufacturing Systems*, Springer Verlag, New York, 1985
100. U. Nandkeolyar and D.P. Christy, Evaluating the design of flexible manufacturing systems, *Proceedings of the Third ORSA/TIMS Conference on Flexible Manufacturing Systems: Operations Research Models and Applications*, 1989, 15–22
101. C.H. Chung and I.J. Chen, A systematic assessment of the value of flexibility for an FMS, *Proceedings of the Third ORSA/TIMS Conference on Flexible Manufacturing Systems: Operations Research Models and Applications*, 1989, 27–37
102. N. Alberti, U. La Mare, and S.N. La Diega, Cost efficiency: an index of operational performance of flexible automated production environments, *Proceedings of the Third ORSA/TIMS Conference on Flexible Manufacturing Systems: Operations Research Models and Applications*, 1989, 67–73
103. I.Y. Alqattan, Systematic approach to cellular manufacturing systems design, *J. Mech. Working Technol.* **20** (1989) 415–424
104. C. Doiteaux, B. Rochotte, E. Bajic, and J. Richard, FMS architecture for an optimum quality and process reactivity, *Proceedings of the 29th International MATADOR Conference*, 1992, 225–232
105. B.K. Modi and K. Shanker, Models and solution approaches for part movement minimization and load balancing in FMS with machine, tool and process plan flexibilities, *Int. J. Prod. Res.* **33** (1994) 1791–1816
106. K.E. Stecke, Formulation and solution of nonlinear integer production planning problems for flexible manufacturing systems, *Manage. Sci.* **29**:3 (1983) 273–288
107. P. Afentakis, Maximum throughput in flexible manufacturing systems, *Proceedings of the Second ORSA/TINS Conference on Flexible Manufacturing Systems: Operations Research Models and Applications*, Amsterdam, 1986, 509–520
108. A. Villa and M. Arcostanzo, DOPP — dynamically optimized production planning, *Int. J. Prod. Res.* **26** (1988) 1637–1650
109. R.L. Graham, E.L. Lawler, J.K. Lenstra, and A.H.G. Rinnooykan, Optimization and approximation in deterministic sequencing and scheduling: a survey, *Ann. Discrete Math.* **5** (1979) 287–326
110. A.S. Kiran and S. Alptekin, A tardiness heuristic for scheduling flexible manufacturing systems, *Proceedings of the 15th Conference on Production Research and Technology, Advances in Manufacturing Systems Integration and Processes*, 1989, 559–564
111. P.J. Egbelu, Planning for machining in a multijob, multimachine manufacturing environment, *J. Manuf. Systems* **5** (1985) 1–13
112. E.F. Watson and P.J. Egbelu, Scheduling and machining of jobs through parallel nonidentical machine cell, *J. Manuf. Systems* **8** (1988) 59–68
113. P.J. Egbelu, Machining and material flow system design for minimum cost production, *Int. J. Prod. Res.* **28** (1990) 353–368
114. I. Ham, K. Hitomi, N. Nakamura, and T. Yoshida, Optimal group scheduling and machining-speed decision under due-date constraints, *ASME J. Eng. Ind.* **101** (1979) 128–134
115. T. Kishinami and K. Saito, The optimum sequence of operations in the multistage manufacturing system, *Proceedings of the 16th International MTDR Conference*, 1975, 57–61
116. K. Hitomi, Optimization of multistage machining system: analysis of optimal machining conditions for the flow-type machining system, *ASME J. Eng. Ind.* **93** 1971 498–506
117. K. Hitomi, Analysis of production models. Part II. Optimization of a multistage production system, *AIIE Trans.* **8** (1976) 106–112

118. K. Hitomi, Optimization of multistage production systems with variable production times and costs, *Int. J. Prod. Res.* **15** (1977) 583–597
119. K. Hitomi, Analysis of optimal machining conditions for flow-type automated manufacturing systems: maximum efficiency for multi-product production, *Int. J. Prod. Res.* **28** (1990) 1153–1162
120. K. Hitomi, Analysis of optimal machining conditions for flow-type automated manufacturing systems, *Int. J. Prod. Res.* **29** (1991) 2423–2432
121. K. Hitomi, *Manufacturing Systems Engineering*, Taylor and Francis, London, 1979, 184–193
122. S.S. Rao and S.K. Hati, Computerized selection of optimum machining conditions for a job requiring multiple operations, *ASME J. Eng. Ind.* **100** (1978) 356–362
123. K. Iwata, Y. Murotsu, and F. Oba, Optimum machining conditions for flow-type multistage machining systems, *CIRP Ann.* **23** (1974) 175–176
124. K. Iwata, Y. Murotsu, F. Oba, and Y. Kawabe, Analysis of optimum loading sequence of parts and optimum machining conditions for flow-type multistage machining system, *CIRP Ann.* **24** (1975) 465–470
125. G.S. Sekhon, Application of dynamic programming to multi-stage batch machining, *Computer-Aided Des.* **14** (1982) 157–159
126. J.S. Agapiou, Optimization of multistage machining systems. Part I. Mathematical solution, *ASME J. Eng. Ind.* **114** (1992) 524–531
127. J.S. Agapiou, Optimization of multistage machining systems. Part II. The algorithm and applications, *ASME J. Eng. Ind.* **114** (1992) 532–538
128. M.D. Kilbridge and L. Wester, A heuristic method of assembly line balancing, *J. Ind. Eng.* **12** (1961) 292–298
129. T.O. Prenting and N.T. Thomopoulos, *Humanism and Technology in Assembly Line Systems*, Spartan Books, Hayden Book Co., Hasbrouk Heights, NJ, 1974
130. A.L. Arcus, COMSOAL — a computer method of sequencing operations for assembly lines, *Int. J. Prod. Res.* **4** (1966) 259–277
131. E.L. Magad, Cooperative manufacturing research, *Ind. Eng.* **4** (1972) 36–40
132. W.I. Sharp, Jr., Assembly Line Balancing Techniques, SME Technical Paper MS77-313, 1977
133. Y.S. Kim and W.J. Kolarik, Real-time conditional reliability prediction from on-line tool performance data, *Int. J. Prod. Res.* **30** (1992) 1831–1844
134. A. Jeang and K. Yang, Optimal tool replacement with nondecreasing tool wear, *Int. J. Prod. Res.* **30** (1992) 299–314
135. K.S. Park, Optimal wear-limit replacement wear-dependent failures, *IEEE Trans. Reliabil.* **37** (1988) 293–294
136. K.S. Park, Optimal wear-limit replacement under periodic inspections, *IEEE Trans. Reliabil.* **37** (1988) 97–102
138. J. Sharit and S. Elhence, Computerization of tool-replacement decision making in flexible manufacturing systems: a human–systems perspective, *Int. J. Prod. Res.* **27** (1989) 2027–2039
137. G. Black and F. Proschan, On optimal redundancy, *Oper. Res.* **7** (1958) 581–588
139. J.D. Kettelle, Jr., Least-cost allocations of reliability investment, *Oper. Res.* **10** (1961) 249–267
140. R. Subramanian, K.S. Venkatakrishnan, and K.P. Kistner, Reliability of a repairable system with standby failure, *Oper. Res.* **24** (1976) 169–176
141. M. Messinger and M.L. Shooman, Techniques for optimum spares allocation: a tutorial review, *IEEE Trans. Reliabil.* **19** (1970) 156–166
142. P. Maropoulos and S. Hinduja, A tool-regulation and balancing system for turning centers, *Int. J. Adv. Manuf. Technol.* **4** (1989) 207–226
143. S. Ramalingam, N. Balasubramanian, and Y. Peng, Tool life scatter, tooling cost and replacement schedules, *Proc. NAMRC* **5** (1977) 212–218
144. U. La Commare, S.N. La Diega, and A. Passanati, Tool replacement strategies by computer simulation, *Proceedings of the Fourth International Conference on Production Control in the Metalworking Industry*, 1986
145. U. La Commare, S.N. La Diega, and A. Passanati, Optimum tool replacement policies with penalty costs for unforeseen tool failures, *Int. J. Mach. Tool Des. Res.* **23** (1983) 237–243

146. R.D. Yearout, P. Reddy, and D.L. Grosh, Standby redundancy in reliability — a review, *IEEE Trans. Reliabil.* **35** (1986) 285–292
147. P.D. T. O'Connor, *Reliability Engineering*, Hemisphere Publishing Corporation, New York, 1988.
148. R.A. Dovich, *Reliability Statistics*, ASQC Quality Press, WI, 1990
149. R.E. Barlow and F. Proschan, *Mathematical Theory of Reliability*, Wiley, New York, 1965
150. S.M. Ross, *Introduction to Probability Models*, Academic Press, New York, 1985
151. S.O. Schall and M.J. Chandra, Optimal tool inspection intervals using a process control approach, *Int. J. Prod. Res.* **28** (1990) 841–851
152. A.I. Daschenko and V.N. Redin, Control of cutting tool replacement by durability distributions, *Int. J. Adv. Manuf. Technol.* **3** (1988) 39–60
153. C.P. Koulamas, Simultaneous determination of optimal machining conditions and tool replacement policies in constrained machining economics problems by geometric programming, *Int. J. Prod. Res.* **29** (1991) 2407–2421
154. C.P. Koulamas, Total tool requirements in multi-level machining systems, *Int. J. Prod. Res.* **29** (1991) 417–437
155. L. Zavanella, G.C. Maccarini, and A. Bugini, FMS Tool supply in a stochastic environment: strategies and related reliabilities, *Int. J. Mach. Tools Manuf.* **30** (1990) 389–402
156. G.C. Maccarini, L. Zavanella, and A. Bugini, Production cost and tool reliabilities: the machining cycle influence in flexible plants, *Int. J. Mach. Tools Manuf.* **30** (1990) 415–424
157. B. Mursec, F. Cus, and J. Balic, Organization of tool supply and determination of cutting conditions, *J. Mater. Process. Technol.* **100** (2000) 241–249
158. J.N. Pan, W.J. Kolarik, and B.K. Lambert, Mathematical model to predict the system reliability of tooling for automated machining systems, *Int. J. Prod. Res.* **24** (1986) 493–501
159. J.S. Agapiou, Cutting tool strategies for multifunctional part configurations. Part I. Analytical economic models for cutting tools, *Int. J. Adv. Manuf. Technol.* **7** (1992) 59–69
160. J.S. Agapiou, Cutting tool strategies for multifunctional part configurations. Part II. Discussion of experimental and analytical results, *Int. J. Adv. Manuf. Technol.* **7** (1992) 70–80

14 Cutting Fluids

14.1 INTRODUCTION

The cutting fluid is an important component of the machining system in many applications. A large proportion of metal cutting operations are performed wet, and in these applications fluid consumption, maintenance, and disposal costs account for a significant fraction of the total machining cost.

Cutting fluids provide lubrication between the tool, chip, and workpiece at low cutting speeds, and cool the part and machine tool and clear chips at higher cutting speeds and in holemaking operations such as drilling and reaming. They also help prevent edge build-up and part rust in most circumstances. When properly applied they permit the use of increased cutting speeds and feed rates, and improve chip formation, tool life, surface finish, and dimensional accuracy. A cutting fluid's cooling ability depends largely on the base fluid and the coolant volume. Chip flushing capabilities are determined by the operation geometry and the coolant application method. Lubrication is controlled by the chemical composition of the coolant and the application method. In recirculating systems, in which a large volume of coolant is continuously collected in a sump and reused, coolant performance over time is strongly influenced by coolant maintenance practices such as concentration monitoring, stabilization, tramp oil removal, rancidity control, and filtering.

In general, cutting fluids should be safe to the worker (no dermatitis, not a carcinogen or a suspected carcinogen, no eye, nose or throat irritation, little odor, mist, or foam, nontoxic), have a long useful life, be waste-treatable (easy to dispose of and recyclable), and be chemically inert with respect to the workpiece material (to control rust, corrosion, and undesirable residues on machined surfaces). The types of coolants which are effective for broad classes of work materials and operations, for example, for turning aluminum alloys or milling steels, are broadly understood. The selection and maintenance of a cutting fluid in a specific application, however, is often determined by experience and limited performance testing, and typically represents one of the more arbitrary decisions in process design. This is unfortunate, since improper or suboptimal coolant application can greatly increase machining costs and process variation.

Recent studies have shown that occupational exposure to cutting fluids can lead to adverse health effects. This has led to increased interest in minimizing coolant usage and other measures to limit workforce exposure. The Material Safety Data Sheet (MSDS) describes the characteristics and safe usage practices for a given fluid and should be consulted prior to use. In the United States, The National Institute for Occupational Safety and Health (NIOSH) is charged with recommending occupational safety and health standards and provides documents on the prevalence of hazards, the existence of safety and health risks, and the adequacy of control methods [1].

This chapter provides a broad overview of common cutting fluids, application methods, filtering, maintenance and waste treatment practices, and health and safety issues. It also provides an overview of recent research on the development of dry and near-dry machining technologies.

14.2 TYPES OF CUTTING FLUIDS

14.2.1 Cutting Oils

Cutting oils are mineral, animal, vegetable, or synthetic oils used without dilution with water [2–7]. Petroleum-based mineral oils, including light solvents, neutral oils, and heavy bright and refined oils, are most common due to their low cost [2]. Fatty animal oils are used largely as compounding oils as discussed below. Vegetable oils used include palm oil, rapeseed (canola) oil, and coconut oils. They are three to five times as expensive as mineral oils but are sometimes required in environmentally sensitive applications, especially in Europe [6], for example, in machining titanium and stainless steel and in applications requiring the German WGK-0 or NWG classification [7]. Synthetic esters made from renewable sources are also used as substitutes for mineral oils.

Mineral oils may be classified as straight or compounded, with compounded being the more common. Straight oils are base oils without active additives; compounded oils consist of the base oil mixed with polar and chemically active additives. Common polar additives include fatty animal oils such as lard oil or tallow, and vegetable oils such as palm oil or caster oil derivatives. These additives, which are used in concentrations between 10 and 40%, increase the cutting oil's wetting ability and penetrating properties and thus improve lubricity. Common active additives include chlorine, sulfur, and phosphorous compounds; they are surface reactive and form a metallic film at the tool surface, which acts as a solid lubricant. Additives must be chosen to be chemically compatible with the work material; improper additives can produce staining or corrosion of the part.

Cutting oils are more effective as lubricants than coolants. They are used extensively in grinding and honing operations, where they permit higher metal removal rates with better finish and less surface damage than water-based fluids [8]. They are also used in relatively low speed operations such as broaching, tapping, gear hobbing, and gun drilling, since in these operations cooling capability is less critical.

Cutting oils are stable and provide excellent rust protection. They are relatively costly, however, have limited applicability to high-speed applications, present a potential smoke and fire risk, and are associated with contact dermatitis and other operator health issues [9]. Due to their high cost and fire risk, they are generally used only on individual machines, and not in large recirculating systems.

Vegetable oils have high flash point temperatures compared to high-viscosity mineral oils (typically 230 versus 165°C), so that smoke formation is less of a concern. They also provide better lubricity (because of the dipolar nature of their molecules) and are often used as a polar additive to enhance the lubricity of mineral oils. Even though vegetable oils are significantly more expensive than mineral-oil based formulations, they may be cost effective in some applications due to reduced consumption less dragout than mineral oils and improved tool life and productivity [10].

14.2.2 Water-Based Fluids

Water-based fluids are dilute emulsions or solutions of oils in water, which provide less lubrication but better cooling and chip clearing abilities than cutting oils. Water cools two to three times faster than mineral oils and can retain more than twice the amount of heat. They are used extensively in higher speed operations and large recirculating systems. There are three basic types: soluble oils, semisynthetics, and synthetics [2–6, 11].

Soluble oils are special types of mineral oils emulsified in water at concentrations typically between 5 and 20%, with lower concentrations (less than 10%) being most common in general-purpose machining. (The term "soluble oil" is in fact incorrect, since oils are not soluble in water, but is regrettably in common use.) Soluble oil concentrates contain severely refined base oils (30 to 85%), emulsifiers, and performance additives such as extreme pressure (EP) additives, stabilizers, rust inhibitors, defoamers, and bactericides. Base oils are usually naphthenic or pariffinic mineral oils, with naphthenic oils being more common due to their lower cost. The oil viscosity is typically 100 Saybolt Universal Seconds (SUS) at 100°F (100/100 oils); higher viscosity oils provide better lubricity but are more difficult to emulsify. Emulsifiers are added to form stable dispersions of oil and water; emulsifier particles are located around the oil droplets to give them a negative charge that will bond them with the water molecules. The size of the emulsified oil droplets is very critical to fluid performance; it is easier for the smaller emulsion sizes to penetrate the interface of the cutting zone. Most emulsions have droplet sizes between 2 and 50 μm and give the fluid a translucent white or blue-white appearance. "Pearlescent" microemulsions have droplet sizes between 0.1 and 2 μm; they provide better lubricity but have a tendency to foam and entail greater waste-treatment difficulties. EP additives have the same function as chemically active additives in cutting oils. Stabilizers inhibit the breakdown of the emulsion which occurs over time as metal particles in the fluid combine with the emulsifiers. Bacteriocides are added to control bacterial growth and rancidity.

The emulsifier may be anionic (negative chemical charge; e.g., carboxylates, sulfonates, phosphates, and phosphonates), cationic (positive chemical charge; e.g., amine salts, phosphonium compounds), or nonionic (neutral chemical charge; e.g., polymeric ethers, esters, and amides). Light-duty soluble oils use soap-sulfonate emulsifier systems. Moderate- to heavy-duty soluble oils contain chlorinated paraffins and sulfurized fat emulsifiers. The percentage of the base oil, emulsifier, EP additives, and other additives by weight in general-purpose light-duty soluble oils is 83, 15, 0, and 2%, respectively; for medium-duty soluble oils the corresponding percentages are 70, 15, 10, and 5%, respectively, while for heavy-duty soluble oils the percentages are 30, 20, 40, and 10%, respectively.

Soluble oils provide excellent cooling and reasonable lubricity performance at low concentrations, have inherent rust preventive properties, and can incorporate additional performance additives. Their disadvantages include their hard water sensitivity [12], susceptibility to microbial attack, tendency to foam, potential to cause contact dermatitis, and disposal difficulty.

Semisynthetic cutting fluids are fine emulsions in which the base concentrate is a mixture of mineral oil and additional chemicals, which may include emulsifiers, couplers, corrosion inhibitors, EP additives, and bacteriocides and fungicides. They combine features of both soluble oils and synthetics. The base concentrate usually contains between 5 and 30% mineral oil and 30 to 50% water. The fluid itself typically contains 50 to 70% water. Semisynthetics require higher emulsifier concentrations than soluble oils, especially when oil-soluble chemical additives are used. They produce smaller emulsion particle sizes than soluble oils (from 0.01 to 0.1 μm), which gives them improved lubricity and makes them transparent or transluscent. They can be formulated either to emulsify or reject tramp oils. The advantages of semisynthetics include rapid heat dissipation, excellent wettability, cleanliness, and resistance to rancidity and bacteria; bacterial resistance is enhanced by the small emulsion size and relatively low concentration of mineral oil. Compared to synthetic fluids, they provide better rust prevention, improved flexibility in incorporating additives (since both water- and oil-soluble additives can be used), and fewer waste treatment concerns. Their disadvantages include foaming and residue concerns and susceptibility to contamination by tramp oils.

Synthetic cutting fluids are water-based fluids consisting of chemical lubricating agents, wetting additives, disinfectants, and EP additives. They are true solutions in water and contain no oil. They are used in relatively low concentration (typically 5 to 10%) and produce particle sizes below 0.005 μm. Dilution affects the extreme-pressure lubricity of the fluid, so heavier cuts require a higher concentration. They are generally clear, but are often dyed to indicate their presence in water. Additives are chosen to be low-foaming and stable in water.

Synthetic fluids typically contain rust or corrosion inhibitors, lubricants, and fungicides. Rust inhibitors are necessary since no oil is present. Common inhibitors include borate esters and amine carboxylate derivatives. Both boundary and EP lubricants are added. Typical boundary lubricants include soaps, amides, esters, and glycols. Common EP additives include chlorinated and sulfurized fatty acid soaps and esters. Fungicides are necessary since synthetic fluids are susceptible to yeast and mold growth; bacteria are not a concern since the fluids contain no mineral oils and have a relatively high pH.

Synthetic fluids provide excellent cooling and produce fine surface finishes. Compared to emulsifiable oils, they provide better resistance to bacterial degradation and improved tramp oil rejection, stability, and workpiece visibility. They do not usually cause dermatitis. Their disadvantages include reduced lubricity due to absence of petroleum oils, a tendency to leave hard crystalline residues, high alkalinity, a tendency to foam, and disposal problems.

Figure 14.1 shows historical trends in the usage of cutting oils and water-based fluids [2]. Only straight oils were used prior to 1910. Soluble oils were introduced between 1910 and 1920 to improve cooling properties and fire resistance. Semisynthetic fluids were introduced in the 1950s to further improve cooling and rust inhibition. Synthetics were first widely used in the 1970s, when the price of petroleum increased dramatically. Over the long term, straight oils have been used less as cutting speeds have increased; of the water-based fluids, soluble oils are still most common due to their low cost and comparatively good waste-treatability.

Water-based fluids commonly fail by loss of emulsion stability and wetting ability (which are affected by the metal machined, the material removal rate in relation to the volume of coolant, the type of water used in the system, the water evaporation rate, tramp oils, the pH level, and the loss of stabilizers such as biocides and odor control agents). In general, the most common causes of failure include heat, hard water, tramp oil, oxidative and reactive agents, and microbial growth [13]. The ASTM Standard E1497–94 [14] should be consulted for safe-use guidelines.

14.2.3 Gaseous Fluids and Gaseous-Liquid Mixtures

In some operations gases rather than liquids are used as cutting fluids. This is true in particular in applications in which no fluid residue on the workpiece can be tolerated, as is the case in some aerospace applications. Gaseous fluids used include air, helium, CO_2, argon, and nitrogen [5], with air being the most common due to its low cost. (Freons were used historically but are no longer permitted in the United States.) Air can be compressed to provide better cooling by forced convection. High-pressure air streams can be used to eject chips (though not as effectively as a liquid) assuming the proper shrouding is used. Noise generation is a significant concern with high velocity and pressure air systems. CO_2 provides evaporative cooling when it is compressed and sprayed into the cutting zone. Cryogenic coolants (e.g., liquid nitrogen combined in a stream of air) have also been successfully used on steel, stainless steel, and inconel parts [15]. Liquid argon or nitrogen permits cooling to subzero temperatures.

Air–oil mists, consisting of small droplets of water-based oil mixed with air, have been successfully applied as cutting fluids in many applications. Historically, they were used in high-speed applications with small areas of cut, for example, end milling applications [2].

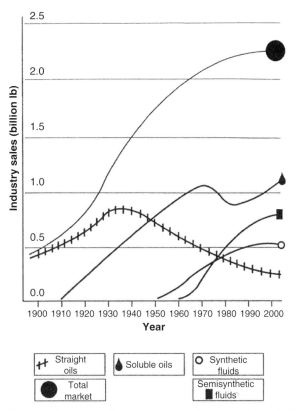

FIGURE 14.1 Historical usage of straight oil and water-based cutting fluids. (After J.C. Childers [2].)

With the development of through-spindle coolant systems, they have been increasingly restricted to low-speed cutting applications such as drilling and gear shaping operations. There are two methods of producing mists: aspirator methods and direct-pressure methods [3]. In aspirator systems a stream of air is directed past an open tube containing oil, creating a partial vacuum which draws oil droplets into the air stream. In direct-pressure systems compressed air is directed through the oil to create a mist. Oil mists are best suited to applications in which flood coolant is impractical, for example, in applications in which the cutting zone is relatively inaccessible. Their major disadvantages are a tendency for the nozzles to clog and exposure of the operator to mist inhalation, which has adverse health consequences as discussed in Section 14.6. Air–oil mists are also used in minimum quantity lubrication (MQL) systems described in Section 14.7.

14.3 COOLANT APPLICATION

The effectiveness of cutting fluids depends to a large extent upon the method of their delivery into the cutting zone. There are four basic methods of applying coolant: low-pressure flood application, high-pressure flood application, through-tool application, and mist application (see Figure 3.52).

In *low-pressure flood application* systems, coolant is delivered through nozzles over work zone at line water pressures. Coolant may be applied through either fixed or fle piping, with fixed systems being common on dedicated systems and flexible tubing being more typical of general-purpose machines. In either case, coolant nozzles should be directed ahead of the cut, with sufficient additional coolant being supplied to cover the workpiece and back of the cutter, especially in milling operations. Insufficient or intermittent coolant application in milling operations can lead to thermal fatigue of the cutter. If the coolant volume is sufficient, low-pressure flood application is effective in clearing chips and cooling the part to maintain dimensional tolerance, but has limited lubricating effectiveness.

In *high-pressure flood application*, coolant is directed through nozzles to impinge ahead of the cutter at higher pressures. The inlet nozzle pressure varies widely with the orifice diameter but is typically between 5 and 50 bar (75 to 750 psig). Considerably higher pressures are used in some grinding operations, and in impingement chip breaking jet systems. Coolant is applied through rigid piping, with nozzles typically being mounted on a ring around the spindle nose in boring, milling, and drilling (Figure 3.52), on the toolholder behind the insert in turning, and on a rigid pipe in front of the wheel in grinding [16]. High-pressure application provides more effective lubricating and chip clearing capabilities, but generates mist which may present a hazard to the operator if not properly collected and filtered. There is also an increased tendency for the coolant to become aerated and foam.

In *through-tool coolant systems*, coolant is supplied through the spindle to coolant passages in the tool under high pressure (see Figure 3.52). Pressures of 13 bar are standard with most machine tools and 35 to 100 bar (500 to 1500 psig) are becoming typical in drilling applications. Through-tool coolant is required to clear chips in many high throughput drilling and deep hole drilling operations, and is also used in grinding (through porous wheels). It requires special seals and pumps which generally must be built into the machine tool. It is very effective in cooling the cutting edge and clearing chips in holemaking operations, and in cooling the wheel in grinding. Effective seals must be installed and maintained to prevent coolant from leaking into the spindle bearings. In drilling, the coolant must be effectively filtered to prevent debris from clogging the tool coolant passages in recirculating systems. Disadvantages of a high-pressure coolant include increased maintenance to ensure that seals, pumps, and rotary unions do not fail, a tendency to generate mists, and increased foaming.

As discussed in the previous section, mist application is applied using either aspirator or direct-pressure systems, and is used especially in drilling and gear shaping operations.

Generally, the coolant pressure should be sufficient to penetrate the vapor barrier pocket generated around the cutting edge. The force with which the fluid penetrates the cutting zone (through the vapor barrier) is somewhat proportional to the coolant velocity and a critical factor for the design of the process. However, the velocity is proportional to the square root of the pressure. Therefore, doubling the coolant pressure results in a roughly 40% increase of the coolant force as explained in Chapter 4. Hence, it is preferable to increase the coolant volume in preference to the pressure to optimize coolant performance.

During machining, airborne mist is generated from cutting fluids. Many factors including the machining conditions, cutting tool design, and fluid application method influence mist generation. Mist generation is usually controlled through mechanical means (i.e., enclosures, ventilation), chemical means (i.e., antimist polymers additive, mist suppression at the source, foam reduction, formulations with low oil concentrations, avoiding contamination with tramp oil, etc.) [17, 18], minimizing fluid delivery pressure and flow rate, the proper design and operation of the cutting fluid delivery system [1], or the use of dry or near-dry machining methods. Once generated, mists are normally controlled or abated through ventilation and filtering as discussed in the next section.

Regardless of the method used to apply coolant, sufficient volume must be supplied to provide adequate cooling and chip clearing capability. Coolant volume is measured by the flow rate in gallons or liters per minute. As a reasonable rule of thumb, the coolant volume should be 1 to 2 gal/min for each HP of cutting energy (or 5 to 10 l/min for each kW) for general machining [19], and 2 to 4 gal/min per HP (or 10 to 20 l/min per kW) for grinding [3]. As will be discussed in the next section, an adequate sump capacity is required to ensure proper filtering and minimize foaming; as a general rule of thumb, the sump capacity should be 3 to 10 times the flow rate per minute [19]. More detailed recommendations of the sump capacity are 5 times the flow rate for steel machining, 7 times the flow rate for cast iron and aluminum machining, 10 times the flow rate for grinding, and 10 to 20 times the flow rate for high stock removal machining and grinding [3].

14.4 FILTERING

Coolants in reciruclating systems entrain and transport a number of impurities, including chips, airborne contaminants, hydraulic and machine way oils, residues left on the part from previous operations, etc. [20]. These impurities must be removed to maintain coolant performance. A variety of methods and systems are used to remove contaminants from cutting fluids. They can be broadly classified as separation methods and filtration methods; generally two or more methods are used in sequence in what is referred to by convention as a filtration system.

Common separation methods include settling tanks, centrifuges, cyclones, and magnetic separators [3, 20]. A *settling tank* is a large tank with two or more baffles (Figure 14.2); as the fluid moves under and over successive baffles, tramp oils and lighter impurities rise to the surface, where they can be skimmed off, and chips and other heavy debris settle on the bottom, where they can be similarly removed. The effectiveness of the tank depends on the settling time, or the volume of the tank divided by the inlet flow rate; if the settling time is too short, not all impurities will settle out, there will be an increased tendency to foam since coolant bubbles will not have time to burst, and coolant temperature will be more difficult to control [12, 16, 21]. Settling times should be 5 min for small systems [3], 7 to 10 min for general-purpose systems [22], and 10 to 15 min for large systems [3, 20]. Centrifuges and cyclones separate debris from the coolant through centrifugal action [3, 20]. In a *centrifuge* dirty coolant is introduced between rotating bowls or cone-shaped disks. Chips and other

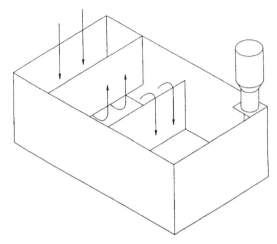

FIGURE 14.2 Settling tank. (After R.H. Brandt [20].)

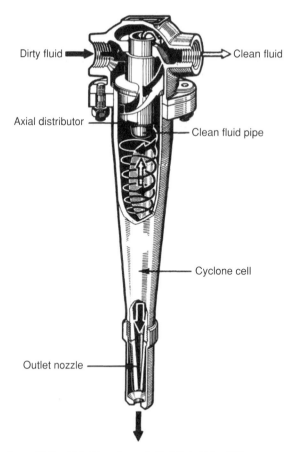

FIGURE 14.3 Chip cyclone. (After T.J. Drozda and C. Wick Eds. [3].)

heavy debris are forced toward the center of the bowls or disks, and tramp oils with low specific gravity are forced outward. The bowls or disks eventually fill with debris and require periodic cleaning. In a *cyclone* (Figure 14.3), dirty coolant is directed into a cone-shaped vessel; the resulting rotary motion of the fluid forces chips and debris outward, so that relatively clean fluid emerges from the center of the device [3, 20, 21]. In a *magnetic separator*, the coolant is directed past a rotating magnetic drum. Chips and fines stick to the drum and are removed by a blade on the opposite side of the drum, generally out of the coolant bath [20]. Magnetic cyclones and belt systems are also used [23]. These systems are obviously effective mainly for ferrous chips, although they will also remove abrasive grains adhering to ferrous fines in grinding systems.

Since separators are not effective in removing impurities with specific gravities near that of the coolant, they are normally used in conjunction with filtering systems. These systems may use either disposable or permanent porous media; the degree of filtering achieved is determined by the pore size of the medium. Common *disposable medium filtering systems* include bag, cartridge, and roll systems (Figure 14.4) [3, 20]. The filtering medium in these systems may be made of paper, cotton, wool, synthetic fibers, or felted materials; they are often precoated with fine particles such as cellulose fibers or diatomaceous earth. In a bag system, coolant is passed through two or more fabric bags with successively smaller mesh sizes. The bags fill with chips and must be changed periodically. Cartridge systems are similar but use

FIGURE 14.4 Disposable filtering media. (After R.H. Brandt [20].)

cylindrical cartridges, similar to an automotive oil filter, rather than bags [24]. Flat-bed roll filters are common in large recirculating systems. In this approach, the coolant is passed through a sheet of filter media. As filtering progresses and the medium becomes clogged, the fluid level in the tank rises, eventually triggering a float which indexes the roll to expose fresh media. Disposable media filters may be driven by gravity, vacuum, or pressure. In a gravity system fluid flow is maintained by gravity. In a vacuum system (Figure 14.5), a negative pressure is applied to the back of the medium to accelerate flow. In a pressurized system, positive pressure is applied to the fluid to force it through the medium. Pressurized systems operate at higher pressures and can produce the largest flow rates. Both gravity and pressurized systems are used with flat-bed roll filters; pressurized systems are especially common when cartridge filters are used.

In *permanent media filtering systems*, a wire mesh or permanent fabric is used as the filtering medium [25]. As in disposable media systems, a porous precoating material is often used to increase effectiveness. Chips and other debris are periodically removed by a cleaning unit; cleaning may be controlled by a timer or activated by a float trigger. Common system configurations include those employing circular meshes, flat belts or screens, and stacked disks separated by spacers. Figure 14.6 shows one especially common approach, in which a wedge-wire mesh belt is used in a flat-bed configuration. Permanent media filters require higher initial investment but may be more economical over time, since disposable medium costs are avoided; they also generate waste in cake form which may be more economically disposed of than the contaminated fabrics produced by disposable media systems.

Apart from cost, factors to be considered in choosing a filtering strategy are the level of filtration necessary, required system capacity, and the effect of the filter medium on the coolant [3]. The level of filtration is generally expressed as the largest dimension of contaminant which can be tolerated. Generally, standard chip removal and cutting fluid filtration systems remove debris over 50 μm in size fairly easily. Greater care and additional filtering stages are required to remove finer debris [26]. A common rule of thumb for general-purpose machining is that coolant should be filtered to one tenth the tolerance band [16]. This can lead to very stringent requirements for precision operations. In large systems employing through-tool coolant, it is common filter between 5 and 10 μm. The required system capacity

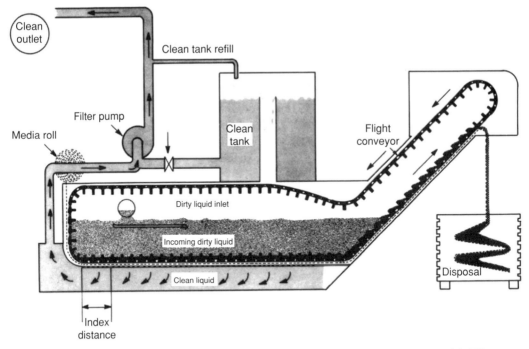

FIGURE 14.5 Vacuum media flat-bed roll filtering system. (After T.J. Drozda and C. Wick [3].)

depends on the coolant flow rate in gallons per minute. Minimum capacities can also be computed from the sump volume; large systems typically pump at least three times the sump capacity in 24 h [21]. Bag filters generally have capacities between 25 and 50 gal/min per square foot of filter area (1000 to 2000 l/min/m^2) [20]. Cartridge filters typically have a capacity of approximated 1 gal/min per foot of cartridge length (1 l/min for 6 cm) [24]. Flat-bed systems have much higher capacities and are most common for large systems. Permanent media systems have the same flow capacity as disposable media systems of comparable pore size. Some types of filters can reduce fluid effectiveness by removing oxidization inhibitors, detergents, and other additives as well as contaminants [3]. Filter system manufacturers provide information on flow rates, filtering levels, and application ranges for specific systems.

Another frequently important component of the fluid filtration system is the *ventilation system* to control mist and airborne contaminants [27–29]. Of major concern are atomized mists with particle diameters of tenths of a micrometer, which are most common in high-speed machining and grinding operations [25] with high-pressure coolant. When mist generation cannot be reduced, effective building ventilation and source isolation through machine enclosures ventilated to appropriate mist collectors must be implemented. The ANSI B-11 Technical Report 2 and the American Conference of Governmental Industrial Hygienists (ACGIH) provide guidelines for proper ventilation systems [1, 30] consisting of physical barriers and ductwork. The key points discussed in the ANSI B-11 Report are summarized in Figure 14.7. Effective systems for removing submicrometer mists include high-efficiency particulate air (HEPA) filters, electrostatic precipitators (ESPs), and filter-bed systems [25].

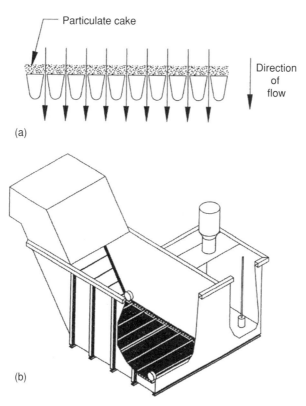

FIGURE 14.6 (a) Wedge-wire permanent media filtering principle. (b) Flat-bed wedge wire filtering system. (After R.H. Brandt [20].)

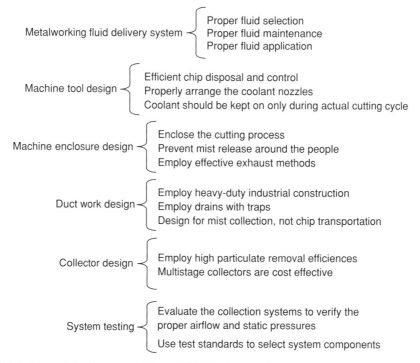

FIGURE 14.7 Key points discussed in the ANSI B–11 Report [1]

14.5 CONDITION MONITORING AND WASTE TREATMENT

The chemical composition and other properties of the cutting fluid should be monitored regularly to maintain acceptable performance. Common parameters monitored include concentration, percent solid and tramp oil, bacteria and fungi count, water quality, and temperature [3, 31].

In water-based systems, the *concentration* of the base oil or synthetic concentrate should be monitored to ensure that it is within acceptable limits. If the concentration is too low, the fluid will not perform properly; if it is too high, the system is being run uneconomically, and the fluid will have an increased tendency to foam. In mineral-oil based fluids, excessive concentration may also promote bacterial growth. Concentration is most often measured using a refractometer, a device which correlates the diffraction of a beam of light by a drop of sample fluid to concentration [22, 32]. Laboratory tests such as oil split methods and chemical titration are also used. Concentration can be difficult to measure accurately if tramp oil contamination is not controlled, since refractometers do not distinguish tramp oils from base oils well.

The *percent solid and tramp oil* are indications of the effectiveness of filtering and of maintenance issues. Excessive solid entrainment, which often produces poor surface finish, indicates that filtering is not effective and should be corrected. Similarly, excessive tramp oil may indicate the presence of a broken hydraulic line, bad seal, malfunctioning lubrication unit, or other component failure. Excessive tramp oil contamination results in pumping problems, poor filter performance and life, and destabilization of the cutting fluid emulsion. Regular machine maintenance minimizes contamination of the cutting fluid with free debris and tramp oils leaking from spindles, slides, and gearboxes. Both percent solid and tramp oil are measured by allowing a fluid sample to stand in a graduated cylinder; over a period of hours, the tramp oil will rise to the top, and solids will settle in the bottom [22]. Centrifuges and filters can also be used for this purpose.

Bacteria and fungi in the coolant should be monitored and controlled. Standard tests for microbial content include plate count and dipslide tests; other methods, including enzyme catalase and dissolved oxygen monitoring, are less commonly used [33]. Excessive microbial content generates objectionable odors, and may cause corrosion, changes in fluid chemistry, and filter blockage. In rare instances there may also be adverse health effects as discussed in the next section. In addition, bacteria consume cutting fluid components, and the by-products of this activity can lower the mix pH. Typically, well-maintained coolant systems have bacterial counts below one million per milliliter. Methods of controlling microbial growth include filtration, the use of chemical biocides, and a variety of less common pasteurization, radiation, and microwave treatments. The best method to control bacteria and fungi is to maintain a clean system and limit oil and water contamination. Excessive use of biocides should be avoided to limit occupational exposure to toxic concentrates.

Important factors in *water quality* are hardness, pH, and chemical content [3, 34]. Fluid effectiveness can be compromised if its hardness exceeds 200 ppm dissolved minerals and organics, or if it contains excessive levels of chloride, sulfate, or phosphate ions [3, 19]. Hard water increases concentrate usage and may lead to corrosion and residue problems. Hardness is a particular concern if wetting agent additives are used [3]. On the other hand, when the water is too soft (total hardness less than 80 ppm) the fluid will have an increased tendency to foam [22, 35]. The ideal water hardness range for metalworking fluid mixes is between 80 and 125 ppm. Treatment of incoming water or an alternative water source is required for excessively hard or soft water [3]. The fluid's pH can be measured using a pH meter or pH paper. New coolants generally have a pH between 8 and 9 [3, 22]. In this range, the fluid is alkaline, which discourages microbial growth. The spoilage range of most coolants is between pH 7 and 8. Low pH may result in increased corrosion, objectionable odors, and destabilization of the coolant. If the pH is above 9.5, it may irritate the skin of operators [3].

Corrosion is a problem not only of water-diluted cutting fluids but also in dry machining [36]. Corrosion results from chemical reactions and increases with increasing temperature in the presence of moisture and oxygen in the atmosphere. Moisture condenses on a workpiece and acts as an electrolyte to form a galvanic cell. The concentration and type of additives used to provide protection against corrosion depend upon the type of metal(s) involved (including ferrous and nonferrous), the cutting fluid, and the anticipated chemical reactions. Some of the factors affecting corrosion are the pH of cutting fluid (>9 protect ferrous metals but will adversely affect nonferrous metals such as aluminum and brass), impurities in the water (e.g., high concentration of ions, >100 ppm chloride, >100 ppm sulfate, or >50 ppm nitrate and conductivities >4 mS/cm are considered aggressive waters), high bacteria counts, and unstable emulsions and fluid concentrations.

The *cutting fluid temperature* should be controlled in operations which generate large amounts of heat, and in precision operations in which thermal expansion may produce significant dimensional errors. Temperature control within 1 to 2°C is normally sufficient [22]. Temperature control can be accomplished using natural convection currents or forced air circulation, or circulating the coolant through a chiller integrated into the tank or sump. Fluids cooled down to 10°C are very effective in difficult operations.

Waste treatment is an important consideration in selecting a coolant [37]. The manner and degree to which used coolants must be treated are governed by local regulations. Generally, straight oils are reprocessed by distillation or vacuum distillation prior to reuse or burning. Soluble oils are treated by adding chemicals such as sulfuric acid, polyelectrolytes, or salts to deactivate the emulsifier to separate the concentrate [6, 32]. Chemical separation is not effective for semisynthetics and synthetics [6], which must generally be treated using special filtration methods [32]. A more thorough discussion of this subject is given in Refs. [32, 37].

14.6 HEALTH AND SAFETY CONCERNS

Occupational exposure to cutting fluids in liquid or mist form can have a number of adverse health effects [1, 11, 38–40]. The common exposure mechanisms are dermal (skin) contact and inhalation [1, 4, 11, 40]; less common mechanisms include ingestion either orally or through an open cut [1, 11, 32, 40–42]. The resulting health effects may include toxicity, dermatitis, respiratory disorders, microbial infections, and cancer.

14.6.1 TOXICITY

Short- or long-term exposure to cutting fluids and fluid additives, especially biocides and fungicides, may induce toxic reactions [3]. Acute toxicity may occur through accidental ingestion through splashes or handling food with dirty hands, absorption through cuts, or inhalation of mists; chronic toxicity may result from long-term exposure to mists. Coolants and additives are regulated in all jurisdictions; in the United States, coolant suppliers must provide a Material Safety Data Sheet (MSDS), and biocides are regulated as pesticides by the Environmental Protection Agency (EPA). The documentation required by regulation provides information on permissible exposure levels, necessary protective equipment, and countermeasures in case of ingestion. Coolants used in recirculating systems may leach heavy metals or lead from the work material [32]; if this is a possibility, coolants should be monitored for metal content and changed as needed to avoid unacceptable exposure levels.

14.6.2 Dermatitis

Dermal contact with cutting fluids can cause contact or allergic dermatitis [3–5, 43]. Dermatitis is the most common health problem associated with cutting fluids; estimates are that it afflicts between 0.3 and 1% of machinists in the United States [44]. There are two common mechanisms for skin irritation: blocking of hair follicles by fines or other debris in used fluid, and removal of protective oils from the skin, particularly when the coolant pH is high (over 9) [3]. The first mechanism leads to condition called oil acne, an old ailment first reported in the United States in 1861 [6]. The second mechanism has become common in recent years with the increased use of water-based fluids, and is also associated with some types of additives, notable biocides [3]. Dermatitis is more easily prevented than treated; common prevention methods include ensuring that fluid concentration levels are not too high, providing gloves, hand creams, and low pH soaps to operating personnel, transferring operators who show particular sensitivity to dermatitis, substituting alternative additives for those suspected of causing a problem, improved filtering, and education on the workforce on proper hygiene and methods of minimizing exposure [3, 35].

14.6.3 Respiratory Disorders

Exposure to cutting fluid mists has often been reported to cause acute respiratory difficulties including coughs, increased airway secretions, asthma, bronchitis, and airway constriction which can result in shortness of breath [35, 40, 45–50]. When these symptoms persist or get worse over time, the condition is sometimes called occupational asthma [45]. Since these difficulties result from mist exposure, they occur primarily in water-based fluid applications [48]. As discussed in detail in Ref. [45], a number of cutting fluid components or additives may act as sensitizing agents. In the United States, greatly reduced permissible fluid exposure levels have been proposed, and as a result most new production installations have machine enclosures and mist collectors to minimize exposure.

Another respiratory disorder associated with cutting fluid mists is hypersensitive pneumonitis (HP) [42, 45, 51]. Although the causes of HP are not well understood, it seems that a small percentage of operators become hypersensitive upon exposure to some component of the mist, and develop an adverse reaction upon further exposure. It may ultimately result in pulmonary fibrosis or similar build-up in the lungs [45, 51]. A sensitizing agent may be mycobacteria or other Gram-positive bacteria, but other causes have also been suggested. Switching biocides to alter the bacterial makeup of the fluid may help, but the safest course appears to be to transfer affected operators to a job which does not entail mist exposure. The Occupational Safety and Health Administration (OSHA) recommends an 8-h, time-weighted permissible exposure limit of 0.5 mg/m^3 total particulate [26].

14.6.4 Microbial Infections

Coolant sumps often sustain populations of bacteria, yeasts, and fungus [33]. Most microbes feed off mineral oils, so microbial growth is a particular problem for coolants with high mineral oil content, either through high concentration or free oil contamination, and less of a problem for synthetic fluids which contain no mineral oil. Other factors contributing to bacterial growth are a bacteria-rich water supply, poor housekeeping, and low fluid pH.

Excessive microbial growth in the sump is regarded as a problem mainly because it produces objectionable odors [35, 52] and leads to fluid breakdown. It has traditionally been felt that microbial infection is unlikely [3]. Most coolant systems are comparatively nutrient-poor and unlikely to support true human pathogens, which generally require a nutrient-rich environment. Isolated exceptions are reported, however. As noted above, HP may be caused by exposure to mycobacteria or other microbial agent. Microbial infections may also be a problem for operators whose immune systems are compromised [33]. There have also been reports of flu-like outbreaks associated with the bacteria that causes Legionnaire's disease [33]. Control measures recommended in these cases included diligent microbial monitoring and avoiding allowing sumps to stagnate for long periods of time.

14.6.5 CANCER

Long-term exposure to cutting fluids has been associated with increased incidence of several types of cancer [41, 53–55], including skin, scrotal, laryngeal, rectal, pancreatic, bladder, and digestive cancers. Exposure routes include both dermal contact and inhalation. Details of specific correlations are given in Ref. [41]. Broadly, risks are increased when cutting with straight mineral oils rather than water-based fluids, and in grinding operations, which generate more mists.

There is some indication that greater risk is associated with exposure to older oils. Prior to the 1950s, the mineral oils used were primarily raw or mildly refined petroleum and shale oils, which contained significant concentrations of substances now known to be carcinogenic; in addition, work practices at that time resulted in more severe dermal exposure [56]. Carbon tetrachloride was also used as a cutting fluid prior to 1980 [57]; it is now classified as a potential occupational carcinogen by NIOSH and a probable human carcinogen by the EPA [40, 58, 59]. Since the mid-1970s, substantial changes have been made in fluid composition which reduce concentrations of harmful substances [41], and work practices have been modified to reduce exposure. These changes, coupled with substantially lower permissible exposure levels issued by regulatory agencies in recent years, will hopefully reduce operator risk in the future.

14.7 DRY AND NEAR-DRY MACHINING METHODS

There has been increasing interest recently in reducing or eliminating the use of cutting fluids in machining. This would be of benefit for three reasons. First, it would reduce or eliminate exposure of operators to health risks as discussed in the previous section. Second, it would reduce machining costs. One study by an automotive company indicates that 16% of the cost of machined parts is directly attributable to cutting fluids (including fluid management, disposal, and equipment) [60]. While this percentage varies for different applications, there is no question that the costs associated with the purchase, maintenance, and disposal of cutting fluids are invariably significant. Third, large central coolant systems require in-ground trenches and sumps which greatly limit equipment flexibility.

Some processes are readily carried out without a cutting fluid for some workpiece materials; for example, cast iron parts can be machined dry under conventional cutting conditions if the tool is oriented properly for chip ejection, and aluminum parts are routinely milled and turned without coolants using PCD tooling. Dry machining is more difficult to perform effectively in many other operations, especially at higher cutting speeds. Without coolant, it becomes difficult to effectively clear chips (especially in holemaking operations), control dimensional distortion due to part heating, and prevent build-up on the tool. Special machine architectures which shed chips passively and low-friction tool coatings help address some of these problems [61]. It is unlikely, however, that coolants will be completely eliminated in the near future, especially in holemaking and grinding processes.

Significant reduction in coolant use, however, is achievable in many operations [62]. Strategies include reducing cycle times to reduce the amount of coolant used per part, extending coolant life to reduce waste disposal volumes per unit time, and reducing coolant flow rates [63]. One promising method of reducing coolant flow rates is to use MQL or microdispensing coolant systems [64–67]. In these systems, a low-volume vegetable oil or water-based fluid mist is applied through the tool. Vegetable oils are used in MQL systems; water-based fluids are used primarily for cooling applications because water evaporation removes heat. The required amount varies by application and fluid type, but is typically between 1 oz and 2 gal of fluid every 8 h, with lower amounts being used in MQL systems. The lubricant is virtually consumed in the machining operation without airborne mists. There are two basic approaches for applying the mist in MQL systems. In the first (Figure 14.8), the mist is generated by externally mixing the oil and shop air and supplying it through the spindle. In the second (Figure 14.9), oil and air are supplied through the spindle and mixed internally near the tool holder. While the second method would seem to be more robust, both have been found effective in production machining of aluminum and steel parts. In non-MQL systems, fluid mixed with air is applied externally. These systems are simple and less expensive, but are only effective in drilling shallow holes and in low-speed processes such as gear cutting and broaching. A particularly promising area of application for MQL is in crankshaft oil hole drilling; in this application, an MQL system consuming approximately 50 ml/h of vegetable oil permits the use of significantly higher penetration rates than conventional through-tool coolant. The chips generated are nearly dry and easily recycled. Many machine tool builders now offer MQL systems as standard equipment, and further applications are being widely investigated. Lubricant supply, tool materials, tool coatings, tool geometries,

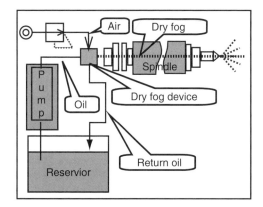

FIGURE 14.8 A minimum quantity lubrication (MQL) system with external mixing. (*Source:* Yasunaga, Inc.)

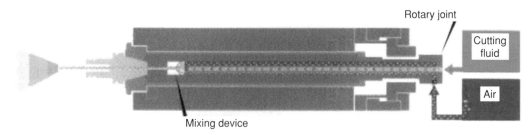

FIGURE 14.9 A minimum quantity lubrication (MQL) system with internal mixing. (*Source:* Horkos, Inc.)

cutting conditions for specific workpiece materials, and machine tool characteristics including peripheral equipment for MQL systems are discussed in Ref. [68].

14.8 TEST PROCEDURE FOR CUTTING FLUID EVALUATION

The growing environmental concerns over the use of cutting fluids has increased the need for a common procedure for evaluating or comparing cutting fluids. The current practice has been to evaluate two or more cutting fluids in the same machine, under the same operating and cutting conditions, and to measure cutting forces, spindle power, part quality (surface finish, dimensional control, etc.), and/or tool wear as the main criteria for performance comparison. Recommended procedures for cutting fluid performance testing or Machinability Test Guidelines are available for turning (plunging), drilling, milling, and grinding processes [11]. Such tests require the specification of the cutting tool, cutting conditions, cutting fluid, and criteria for ending the test (e.g., tool wear considering uniform flank wear along the cutting edge or grinding ratio G).

Some of the critical areas to be evaluated for longer-term tests are tool life, odor, foam, rust, residue, stability, filterability, dermatitis, biological control, pH, machine cleanliness, and overall usage rate as a function of the condition of the coolant. Information about solving cutting fluid problems and extending the life of the cutting fluids is provided in Ref. [11].

As noted above, cutting fluid costs often account for 10 to 15% of the total machining cost, while cutting tool costs typically account for a smaller component (on the order of 5% in high-volume machining). In some operations, it is often better when possible to eliminate the cutting fluid and accept somewhat lower tool life and higher tooling costs.

REFERENCES

1. American National Standards Institute, *American National Standard Technical Report: Mist Control Considerations for the Design, Installation, and Use of Machine Tools Using Metalworking Fluids*, ANSI, New York, NY, 1997, B11 Ventilation Subcommittee, ANSI B-11 TR 2–1997
2. J.C. Childers, The chemistry of metalworking fluids, in: J.P. Beyers, Ed., *Metalworking Fluids*, Marcel Dekker, New York, 1994, 165–189
3. T.J. Drozda and C. Wick, Eds., *Tool and Manufacturing Engineers Handbook*, Vol. I: Machining, 4th Edition, SME, Dearborn, MI, 1983, 4.1–4.34
4. J.K. Howel, W.E. Lucke, and J.C. Steigerwald, Metalworking fluids: composition and use, *Proceedings of the AAMA Metalworking Fluids Symposium*, Dearborn, MI, March 1996, 13–22
5. Sandvik Coromat, Inc., *Modern Metal Cutting*, 1996, Chapter XII, 22–41
6. J.V. Owen, Picking a coolant, *Manuf. Eng.* **120**:5 (1998) 92–100
7. G. Littlefair, Trends in cutting fluid, *Metalworking Production*, December 11, 2000, 26–27
8. G.S. Cholakov, T.L. Guest, and G.W. Rowe, Lubricating properties of grinding fluids: 1. Comparison of fluids in surface grinding experiments, *Lubric. Eng.* **February** (1992) 155–163
9. ASTM, *ASTM Standard E1687–95: Determining Carcinogenic Potential of Virgin Base Oils in Metalworking Fluids*, American Society for Testing and Materials, Philadelphia, PA, 1997
10. S.R. Steigerwald, Vegetable oil improves machining of medical devices, *Prod. Mach.* November/December (2003) 44–46
11. *Pollution Prevention Guide to Using Metal Removal Fluids in Machining Operations*, Developed for the United States Environmental Protection Agency at Cincinnati by IAMS and its Int. National Industrial Working Group
12. R.B. Aronson, Fluid management basics, *Manuf. Eng.* **126**:6 (2001) 90–98
13. J. Burke (Eaton Corporation), Metalworking fluid failure, *Machining & Metalworking Conference*, Detroit, MI, September 4, 1999

14. ASTM, *ASTM Standard E1497-94: Safe Use of Water-Miscible Metalworking Fluids*, American Society for Testing and Materials, Philadelpia, PA, 1997
15. A.B. Chattopadhyay, A. Bose, and A.K. Chattopadhyay, Improvements in grinding steels by cryogenic cooling, *Precision Eng.* **7**:2 (1985) 93-98
16. S. Malkin, *Grinding Technology*, Society of Manufacturing Engineers, Dearborn, MI, 1989, 215-216
17. S. Kalhan (Lubrizol Corporation), Mist control by using shear stable antimist polymers, *Machining & Metalworking Conference*, Detroit, MI, September 4, 1999
18. E. Gulari, C.W. Manke, S. Yurgelevic, and J.M. Smolinski, Suppression and management of mist with polymeric additives, *In the Industrial Metalworking Environment: Assessment and Control of Metal Removal Fluids*, American Automotive Manufacturing Association, Washington, DC, 1998, 291-307
19. M. Hoff, Critical coolant questions, *Manuf. Eng.* **124**:5 (2000) 142-148
20. R.H. Brandt, Filtration systems for metalworking fluids, in: J.P. Beyers, Ed., *Metalworking Fluids*, Marcel Dekker, New York, 1994, 273-303
21. H. Urdanoff and R.J. McKinley, Why do coolants fail? *Manuf. Eng.* **130**:5 (2003) 122-130
22. G.J. Foltz, Cooling fluids: forgotten key to quality, *Manuf. Eng.* **130**:1 (2003) 65-69
23. J. Mackowski, Magnetic filters keep coolant clean, *Manuf. Eng.* **124**:4 (2000) 74-79
24. C. Likens and A. Venner, Keep good coolant from going bad, *Am. Machinist* **April** (2000) 72-75
25. A. Shanley, Mist collection: the pressure is on, *Cutting Tool Eng.* **55**:6 (2003) 46-51
26. A. Richter, Recovery processes — options and considerations for processing and removing metal chips, *Cutting Tool Eng.* **55**:7 (2003)
27. W.J. Johnston, Metal removal fluid mist control: system considerations, *In the Industrial Metalworking Environment: Assessment and Control of Metal Removal Fluids*, American Automotive Manufacturing Association, Washington, DC, 1998, 213-221
28. J.B. D'Arcy, D. Hands, and J.J. Hartwig, Comparison of machining fluid aerosol concentrations from three different particulate sampling and analysis methods, *In the Industrial Metalworking Environment: Assessment and Control*, American Automotive Manufacturing Association, Washington, DC, 1998, 196-199
29. S.J. Cooper, D. Leith, and D. Hands, Vapor generation in laboratory and industrial mist collectors, *In the Industrial Metalworking Environment: Assessment and Control of Metal Removal Fluids*, American Automotive Manufacturing Association, Washington, DC, 1998, 233-238
30. ACGIH, *Industrial Ventilation: A Manual of Recommended Practice*, 22nd Edition, American Conference of Governmental Industrial Hygienists, Committee on Industrial Ventilation, Cincinnati, OH, 1995
31. J.J. Werner, A practical analysis of cutting fluids, *Lubric. Eng.* **April** (1984) 234-238
32. Optimising the Use of Metalworking Fluids, Guide GG199, Environmental Technology Best Practice Programme, United Kingdom, September 1999
33. L.A. Rossmoore and H.W. Rossmoore, Metalworking fluid microbiology, in: J.P. Beyers, Ed., *Metalworking Fluids*, Marcel Dekker, New York, 1994, 247-271
34. E.E. Heidenreich, Metalworking Fluids — A Practical Approach to Recycling, SME Technical Paper MR84-918, 1984
35. G.J. Foltz, Metalworking fluid management and troubleshooting, in: J.P. Beyers, Ed., *Metalworking Fluids*, Marcel Dekker, New York, 1994, 305-337
36. G.J. Foltz, Corrosion and metalworking fluids, *Prod. Mach.* **May/June** (2003) 40-43
37. P.M. Sutton and P.N. Mishra, Waste treatment, in: J.P. Beyers, Ed., *Metalworking Fluids*, Marcel Dekker, New York, 1994, 367-394
38. C.R. Mackerer, Health effects of oil mists: a brief review, *Toxicol. Ind. Health* **5** (1989) 429-440
39. P.J. Beattie and B.H. Strohm, Health and safety aspects in the use of metalworking fluids, in: J.P. Beyers, Ed., *Metalworking Fluids*, Marcel Dekker, New York, 1994, 411-422
40. NIOSH, *What You Need to Know About Occupational Exposure to Metalworking Fluids*, U.S. Department of Health and Human Services, March 1998, DHHS (NIOSH) Publication No. 98-116
41. G.M. Calvert, E. Ward, T.M. Schnorr, and L.J. Fine, Cancer risks among workers exposed to metalworking fluids: a systematic review, *Am. J. Ind. Med.* **33** (1998) 282-292
42. R.B. Aronson, What's happening with coolants, *Manuf. Eng.* **130**:6 (2003) 81-88

43. C.G.T. Mathias, Contact dermatitis and metalworking fluids, in: J.P. Beyers, Ed., *Metalworking Fluids*, Marcel Dekker, New York, 1994, 395–410
44. E.O. Bennet, *Dermatitis in the Metalworking Industry*, Pamplet SP-ll, STLE, Park Ridge, IL, 1983
45. D.J.P. Basett, Review of acute respiratory health effects, *Proceedings of the AAMA Metalworking Fluids Symposium*, Dearborn, MI, March 1996, 147–152
46. S.M. Kennedy, I.A. Graves, D. Kriebel, E.A. Eisen, T.J. Smith, and S.R. Woskie, Acute pulmonary responses among automobile workers exposed to aerosols of machining fluids, *Am. J. Ind. Med.* **15** (1989) 627–641
47. K.D. Rosenman, M.J. Reilly, D. Kalinowski, and F. Watt, Occupational asthma and respiratory symptoms among workers exposed to machining fluids, *Proceedings of the AAMA Metalworking Fluids Symposium*, Dearborn, MI, March 1996, 143–146
48. M.S. Hendy, B.E. Beattie, and P.S. Burges, Occupational asthma due to an emulsified oil mist, *Br. J. Ind. Med.* **42** (1985) 51–54
49. H.J. Cohen, A Study of formaldehyde exposures from metalworking fluid operations using hexahydro–1,3,4-tris(2-hydroxyethyl)-*S*-triazine, *Proceedings of the AAMA Metalworking Fluids Symposium*, Dearborn, MI, March 1996, 178–183
50. E. Brandy and D. Bennet, Occupational airway diseases in the metal-working industry, *Tribol. Int.* **18** (1985) 169–176
51. B.G. Shelton, W.D. Flanders, and G.K. Morris, *Mycobacterium* sp. as a possible cause of hypersensitivity pneumonitis in machine workers, *Emerg. Infect. Dis.* **5**:2 (1999)
52. E.O. Bennett and D.L. Bennett, Cutting fluids and odors, *Clinic on Metalworking Coolants*, Dearborn, MI, 1987
53. E.A. Eisen, P.E. Tolbert, M.F. Hallock, R.R. Monson, T.J. Smith, and S.R. Woskie, Mortality studies of machining fluid exposure in the automotive industry: III. A case–control study of larynx cancer, *Am. J. Ind. Med.* **25** (1994) 185–202
54. E.A. Eisen, P.E. Tolbert, R.R. Monson, and T.J. Smith, Mortality studies of machining fluid exposure in the automotive industry: I. A standardized mortality ratio analysis, *Am. J. Ind. Med.* **10** (1992) 803–815
55. E.A. Eisen, Case–control studies of five digestive cancers and exposure to metalworking fluids, *Proceedings of the AAMA Metalworking Fluids Symposium*, Dearborn, MI, March 1996, 25–28
56. C.M. Skisak, Metal working fluids, base oil safety, *Proceedings of the AAMA Metalworking Fluids Symposium*, Dearborn, MI, March 1996, 44–49
57. E. Usui, A. Gujral, and M.C. Shaw, An experimental study of the action of CCl_4 in cutting and other processes, *Int. J. Mach. Tool. Des. Res.* **1** (1961) 187–197
58. E. Ward, Overview of preventable industrial causes of occupational cancer, *Environ. Health Perspect. Suppl.* **103**:S8 (1995) 197–204
59. Agency for Toxic Substances and Disease Registry, Fact Sheet for Carbon Tetrachloride, CAS# 56-23-5, September 2003
60. D. Graham, Going dry, *Manuf. Eng.* **124**:1 (2000) 72–78
61. F.M. Kustas, L.L. Fehrehnbacher, and R. Komanduri, Nanocoatings on cutting tool for dry machining, *CIRP Ann.* **46** (1997) 39–42
62. H.K. Toenschoff, F. Kroos, W. Sprintig, and D. Brandt, Reducing use of coolants in cutting processes, *Prod. Eng.* **1**:2 (1994) 5–8
63. T. Kondo, Environmentally clean machining in Toyota, *Int. J. Jpn. Soc. Precision Eng.* **31** (1997) 249–252
64. J.F. Kelly and M.G. Cotterell, Minimal lubrication machining of aluminum alloys, *J. Mater. Process. Technol.* **120** (2002) 327–334
65. M. Rahman, A.S. Kumar, and M.U. Salam, Experimental evaluation on the effect of minimal quantities of lubricant in milling, *Int. J. Mach. Tools Manuf.* **42** (2002) 539–547
66. J. McCabe, Dry holes, dry drilling study, *Cutting Tool Eng.* **54**:2 (2002) 44–51
67. G. Landgraf, Dry goods — factors to consider when dry or near-dry machining, *Cutting Tool Eng.* **56**:1 (2004) 28–35
68. K. Weinert, I. Inasaki, J.W. Sutherland, and T. Wakabayashi, Dry machining and minimum quantity lubrication, *CIRP Ann.* **53**:2 (2004) 511–518

15 High Throughput and Agile Machining

15.1 INTRODUCTION

Transfer lines and similar dedicated manufacturing systems capable of producing a single product have been used in high-volume production for many years. This practice was acceptable in the past because the product could be expected to change little over the 15- or 20-year service life of the system. In recent years, however, increasingly frequent product redesigns and new product introductions has created a need for more flexible high-volume production systems. A variety of such systems, based on CNC technology, have emerged to meet this new competitive requirement. These include flexible systems based on high-speed horizontal machining centers, described in Chapter 3, and systems based on multitasking machine tools which combine a turning center (with one or two turning spindles) and a machining center (a separate vertical machining spindle) [1, 2].

Although fully flexible systems have been investigated, the most successful implementations in high-volume manufacturing to date have been systems capable of producing parts within a given family with minimal changeover [3, 4]. These systems include convertible and flexible transfer lines, agile systems, and cellular systems, some of which were discussed briefly in Chapter 3. Successful operation of these systems requires proper programming strategies as well as changes in operating philosophy. Metal removal rates must be increased and noncutting time minimized to achieve acceptable levels of investment. Moreover, tooling, fixturing, and materials handling methods quite different from those used on traditional transfer lines must be employed.

This chapter summarizes high throughput machining, agile machining systems, and tooling, fixturing, and material handling concerns for high-volume flexible or agile production.

15.2 HIGH THROUGHPUT MACHINING

CNC-based production systems typically require more investment for a given volume than single-purpose systems. The base hardware (stations) are more expensive. Moreover, since only a single spindle is usually cutting as compared to multiple spindles in many transfer machine stations, cutting times are typically longer and proportionately more time is consumed in noncutting tasks such as tool changes and pallet rotations. Increased investment is also often required for multiple fixtures, special tooling, and more complex materials handling equipment. CNC systems offer a significant advantage in changeover and lead times; nonetheless, to make them economical it is essential to minimize cutting and noncutting times. The strategies used to accomplish this are referred to as high throughput machining (HTM).

HTM, sometimes called high efficiency machining, should not be confused with high-speed machining (HSM), which refers to the use of high spindle speeds. As the name implies, the object of HTM is to achieve high throughput measured in parts per hour or shift. High spindle speeds contribute to achieving high throughput, but are not the sole or even primary enabler; in fact, in single-spindle applications requiring frequent tool changes, increasing spindle speeds beyond a given level (roughly 8000 rpm in many automotive applications) is counterproductive since the extra time required for spindle acceleration and deceleration exceeds the cutting time saved. Moreover, if the rapid travel distance is small, the axis never reaches the upper speed. For example, a machine tool with a 60 m/min rapid feed rate reaches its top speed after 100 mm of travel, while an HSM with a 120 m/min feed rate will reach the top speed after 200 mm of travel. HTM, rather, relies on several strategies, which include:

1. *Proper selection of CNC architecture.* The CNC architecture (three-, four-, or five-axis) should be selected based on the part geometry and machined content. Parts requiring complex contour machining require five-axis capability. Parts requiring primarily flat surfaces and threaded holes can often be machined with a three-axis machine with a B-axis on the work table, although five-axis machines should be considered, especially for parts requiring holes at multiple angles.
2. *Design of the system for minimum investment.* For simple parts, all machining may be done on a single machine. For more complex parts, more than one fixturing is required to complete all operations, and machining is usually performed sequentially on a number of machines. The machines may be arranged in a long serial line or in smaller cells [5]. Both alternatives should be investigated to determine which requires the least investment for the target volume. It is also important to consider the value of flexibility, changeover flexibility, and the uncertainty in product demand.
3. *Proper tooling selection/design.* Tooling for agile systems differs significantly from dedicated systems. Special tools (e.g., multistep and multifunctional tools) are used to drill multidiameter holes in one operation to eliminate time-consuming tool changes, and large milling cutters are often not used due to dimensional limits in tool magazines.
4. *Selection of optimum cutting speeds and feeds and machining strategy.* Cutting speeds and feed rates should be selected based on system constraints and volumes. Agile systems are typically operated at higher spindles speeds than dedicated systems to improve productivity. Higher spindle speeds may be used because the number of parts produced per shift is lower than in a dedicated system, so that one-shift tool life can be maintained at a higher speed. However, lower depths of cut are often necessary due to comparatively low system rigidity. The cutting strategy (e.g., up-milling versus down milling) has a significant influence on tool life and part quality. Safety is also an important consideration due to the high kinetic energy of the tooling at high spindles speeds; a high standard of passive and active safety technology must be built into the system.
5. *Proper fixture design.* Fixtures in agile systems must be designed to provide adequate support while maintaining access from multiple planes to minimize pallet rotations and refixturings.
6. *Selection of an optimum materials handling strategy.* Materials handling is an important concern in agile machining, especially as daily volumes increase. Lack of careful attention to materials handling can result in increased cost and reduced flexibility and reliability.

Tooling, fixturing, and materials handling issues are discussed in more detail in Section 15.4 and Section 15.5.

15.3 AGILE MACHINING SYSTEMS

Agile manufacturing is similar to flexible manufacturing but implies speed in responding to market changes in addition to flexibility. In the broadest sense, agile manufacturing strategies are enterprise-wide, covering design, marketing, etc. in addition to production [6, 7]. Early work on this subject was carried out at Lehigh University under a U.S. Government Advance Research Programs Agency (ARPA) grant beginning in 1991. Later other universities and national laboratories conducted further research [8, 9]. A number of CNC-based machining systems referred to as agile systems have been installed, primarily in automotive production [10–14], but also in munitions [15] and other sectors.

Typically cited characteristics of agile systems include ease and speed of installation, scalability, reusability, and reconfigurability [16]. Scalability means that additional increments of production capacity (e.g., cells) can be added without disrupting ongoing production. Speed of installation (on the order of weeks, rather than months) is required to respond to market changes rapidly. Reconfigurability implies that machines can be rearranged to make a different product without excessive changeover time. Reusability is required to minimize investment over time. Speed of installation, reconfigurabilty, and reusability are all facilitated through the use of standard machining centers as system building blocks.

The components of an agile system are CNC machine tools, an automated materials handling system, and control software. Most agile systems to date have used three-axis machining centers with horizontal spindles and B-axis table rotation capability that allows multiside machining as shown in Figure 15.1. This architecture is a good compromise between flexibility and cost. Horizontal spindles are preferred because they are comparatively easy to repair and maintain, and because they simplify chip management. Machines are arranged either serially or in three- or four-machine cells, depending on the materials handling method chosen (Figure 15.2–Figure 15.6). The best arrangement in a given application depends on the manufacturing task to be undertaken and available floor space. Many, but not all, agile systems include relatively inflexible auxiliary equipment (such as parts washers) or dedicated stations to carry out specialized operations (e.g., cylinder hones).

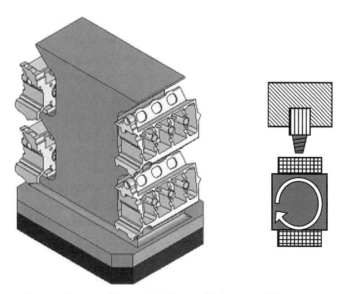

FIGURE 15.1 A tombstone fixture offers multiside machining capability when used in a horizontal CNC machine with B-axis (table rotation).

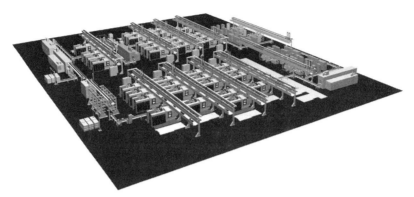

FIGURE 15.2 An agile system consisting of serial cells with overhead gantry loading. (*Source:* Halle.)

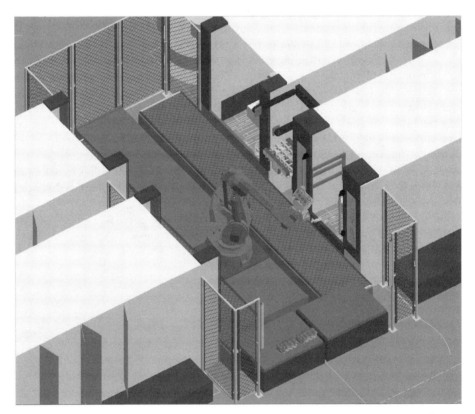

FIGURE 15.3 An agile machining cell with four machining centers and robotic part loading. (*Source:* General Motors Corporation.)

The use of three-axis machines provides flexibility to accomodate two-dimensional feature changes. *B*-axis capability is required for more complex parts; for example, cylinder head machining requires *B*-axis capability to process angular intake and exhaust features without excessive refixturing. To progress beyond this level of flexibility requires three-dimensional machining capability as shown in Figure 15.1. In this case, cylinder heads are located horizontally for multiple part fixturing to reduce tool change times and improve productivity, and parallel kinematic or hybrid machine architectures, providing five-axis machining

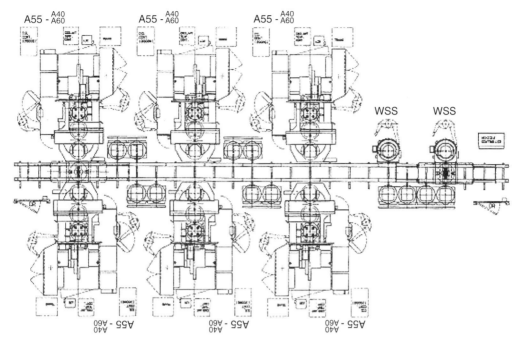

FIGURE 15.4 An agile machining cell with six machining centers, palletized part transfer, and a staging station for pallet loading. (*Source:* Makino.)

capability, are used. A second example is shown in Figure 15.6, in which three- and five-axis machines are used to provide the required agility at minimum cost. The same process using parallel processing with five-axis machines would require more machines but would be linearly scalable.

Flexibility, reconfigurability, and reuse of equipment have been demonstrated in existing systems. Systems which make more than one product with minimal changeover time have been in operation for many years, and machinery from some older systems has been redeployed to create new systems for new products. Production experience has, however, brought to light several difficulties not envisioned by early proponents of agile manufacturing, mainly related to high-volume production. Materials handling in these applications is often accomplished using fixed automation, such as gantries or rail-guided vehicles. The cost and complexity of extending these systems, as well as floor space limitations, has hindered scalability in some cases. Moreover, some early systems required extended ramp-up times and never met planned production volumes due to lack of familiarity with fixturing, tooling, gaging, and control strategies very different from those used in dedicated systems. Subsequent systems built by organizations which have been through this initial learning curve, however, have ramped up more rapidly and achieved full planned volume.

15.4 TOOLING AND FIXTURING

As was noted in Section 15.2 tooling in agile systems is often run at higher cutting speeds than in transfer systems. For complex components, cycle times are longer in agile than in dedicated systems; typical cycle times are 5 to 6 min for agile systems versus 30 sec for a transfer machine. (This is because agile machines perform a larger number of operations in a single setup than dedicated systems.) Also, the percentage of cutting time is larger in transfer

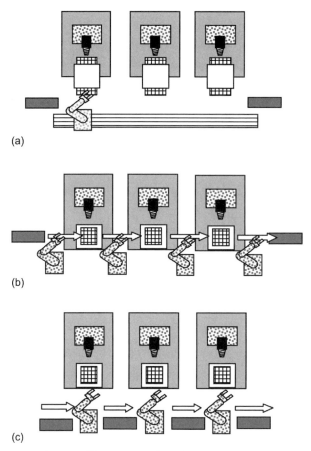

FIGURE 15.5 Various materials handling systems for agile/flexible machining. (a) A single rail-guided robot system for supporting one to three processes in a parallel machining system. (b) Serial machining system interlinked by either power roller conveyors or overhead gantries to create a cell. (c) Serial machining system interlinked with individual systems (robot, gantry, guided vehicle AGV, rail-guided transporter).

than in agile systems. Higher spindle speeds, and the resulting higher tool wear rates, are therefore acceptable because fewer parts per shift are machined. Depending on tool changer capacity and access time, it is also possible in some cases to include redundant tools in the tool magazine in an agile system, further reducing the number of parts which must be machined per shift by a given tool. Higher spindle speeds are also desirable to maintain productivity because lower depths of cut are also often used due to the comparatively high fixture and machine compliance in agile systems.

Special step and combination (multistep and multifunctional) tools are also used more widely used on agile systems than has traditionally been the case in transfer systems. This is particularly true for multidiameter holes. For example, the hole shown in Figure 15.7, with two diameters and two chamfers, would normally be drilled using standard tools in two to four steps on a transfer machine. On an agile system, it would more likely be drilled in one pass using a combination tool to minimize tool changes and increase accuracy. If frequent regrinding is required, the use of combination tools can significantly increase per-part tooling costs; this is more likely to be an issue for ferrous than aluminum parts. In other cases, an end mill may be used to generate different diameter holes using high-speed contouring or circular

High Throughput and Agile Machining 793

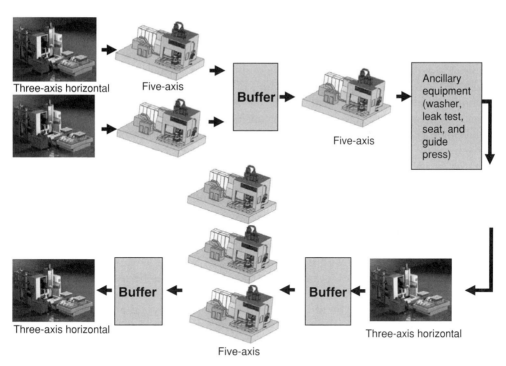

FIGURE 15.6 Agile cylinder head machining layout using three- and five-axis CNC machines.

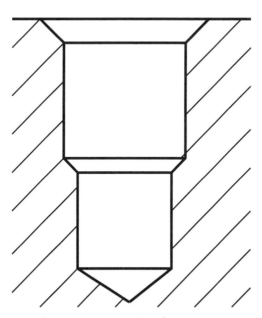

FIGURE 15.7 A hole with two diameters and two chamfers, which would likely be made in multiple passes on dedicated equipment but in a single pass with a combination tool in an agile system.

interpolation algorithms instead of using different diameter drills. Similarly, multifunctional or multitasking tools such as drill-reamers or thread mills may also be used as discussed in Chapter 4. Rules for designing multiple diameter holes for combination tooling are discussed in Section 16.4. It is also common to standardize hole diameters for parts made on agile systems to minimize tool changes. This is less of an issue for larger diameter holes, since in many cases they may be finished by circular interpolation using a smaller diameter tool.

Fixtures for agile systems are generally more compliant than those for dedicated systems. For complex parts, the fixture must provide multiple access planes so that many features can be machined in a single setup. Figure 15.8 and Figure 15.9 show dedicated and agile fixtures for cylinder head machining operations. In the dedicated case, machining loads come from only one direction, and the fixture provides considerable support in that direction. In the agile case, cutting forces will generally act on multiple planes, and the structure is more compliant to accommodate tool access. The fixture shown in Figure 15.9 is a tombstone fixture, shown schematically in Figure15.10. This type of fixture is very common for horizontal spindle machining centers in general, and in agile system applications in particular. They permit multiple loading of small- or medium-sized parts and multiside machining capability when the machine has B-axis table capability (as shown in Figure 15.1). Tombstones are often used with a two-position pallet changer or multiple pallets on rotary or rail-guided vehicle material handling systems as shown in Figure 15.11. One drawback of a tombstone is that the horizontal stiffness decreases with increasing distance from the base. As a result, lower feed rates are often used on features located away from the fixture base to avoid chatter, especially in roughing operations.

A second, less common type of fixture used in agile systems is the window frame fixture (Figure 15.12). In this fixture, the part is located on pins in the base and clamped from the

FIGURE 15.8 Typical fixture for machining an engine head on a transfer machine. The part is mounted on a pallet with locators and clamped from the top. The transfer mechanism is at the bottom of the photo. Horizontal loads are supported by a heavy machine structure. (*Source:* General Motors Corporation.)

FIGURE 15.9 Typical tombstone-type agile fixture for an engine head. (*Source:* General Motors Corporation.)

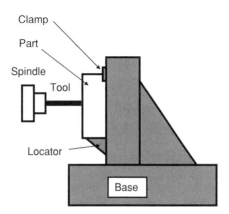

FIGURE 15.10 Schematic illustration of a tombstone fixture.

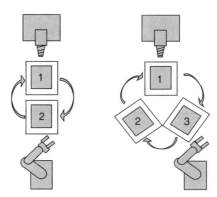

FIGURE 15.11 Agile/flexible system using a single machining center with index table for parts requiring multiple fixturing setups or machining various parts set up on different pallets.

top. It provides maximum access, included access on opposite faces perpendicular to the spindle, if the machine has *B*-axis capability. As a trade-off, part distortion may be increased because clamping forces are supported through the part. (Through clamping is also sometimes used on tombstone and dedicated fixtures.) Also, for aluminum parts, the locating holes may deform if the part is clamped and unclamped three or four times, leading to a serious loss in accuracy.

Adaptor plates like those shown in Figure 15.13 have also been used in agile systems to provide flexibility for a limited family of parts. In this case the part is bolted to a plate with locating elements. The plate is then mounted on a fixture base or tombstone. The hole pattern on the adapter plate can be designed to fit different part variants, which can be easily mounted by an overhead gantry robot. Since the bolts secure the part through threaded holes, hydraulic or mechanical clamps are not needed, and there is significantly less distortion than in through-clamping methods. This method, however, requires either a plate designed to be adaptable to different parts, or an inventory of multiple plates for different parts.

Parts and fixtures are often mounted on pallets which travel to different stations in agile systems. This arrangement reduces part variation especially when parts are machined in several stations due to the elimination of the hole-pin tolerance. A pallet may accept one or more parts depending on their size and shape as shown in Figure 15.1 and Figure 15.9. In a common arrangement, pallets are loaded in a central staging area and transported to machines by a materials handling system (Figure 15.4). In this case parts do not have to be located and probed in each machine, saving load time and wear on the locators [17]. Using large numbers of pallets limits flexibility; machines may be fully programmable, but an inventory of dedicated tooling and fixtures must be maintained to manufacture multiple parts with short changeover times. In an alternative arrangement, two pallets are mounted at each machine; while one part is being machined, the second pallet is loaded robotically or manually, the parts having been transferred to the machine through a conveyor loop or cart. If the cutting cycle time exceeds the pallet load time, this greatly reduces machine wait time.

As the preceding discussion indicates, one ironic aspect of flexible and agile machining implementations, at least for high volumes, is that they employ flexible machines but dedicated tooling, fixturing, and (frequently) materials handling. This is rightly recognized as a barrier to efficient inventory, reconfigurability, and rapid changeover capability. Considerable research has been conducted on flexible and reconfigurable fixtures, and there appears to be reasonable hope that flexibility in this area will be increased.

Currently several hydraulic or mechanical modular fixture systems which can be reconfigured are commercially available; almost all, however, are reconfigured manually and

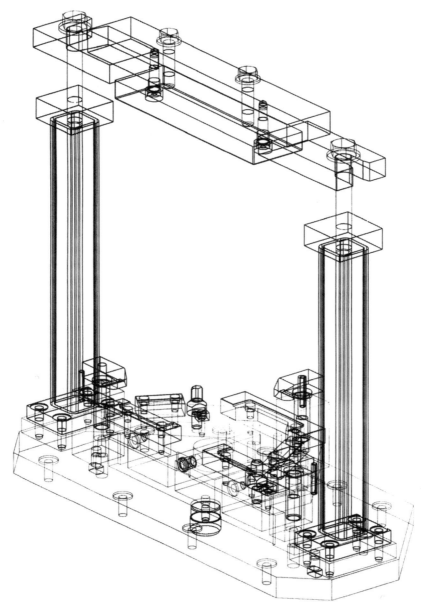

FIGURE 15.12 Exploded view of a windowframe fixture. In this type of fixture the part is located by pins on the base and clamped from the top.

require long setup times, ranging from several hours to several days. (Some of these are described in Chapter 5.) Much research on automated reconfigurable fixtures has focused on bed-of-nails (BON) methods, in which the part is located and clamped using protruding pins. Figure 15.14 shows one such fixture, Mazak's Form-Lok chuck, clamping an irregular workpiece. A similar concept was explored by Lamb Technicon through the Intelligent Fixture System (IFS) program, funded through a U.S. federal research grant (NIST/ATP) [18, 19]. This system, shown in Figure 15.15, is intended to provide automated reconfigurability for high-volume automotive powertrain applications. In the figure, it is shown set up to support and clamp a family of similar engine cylinder heads on a pallet-like structure. It

FIGURE 15.13 Examples of engine cylinder heads fixtured using adapter plates.

FIGURE 15.14 Flexible fixture using the bed-of-nails conformable surface method: Form-Lok chuck holding an irregularly shaped workpiece. (*Source:* Mazak.)

requires three additional stations to: (a) identify the specific incoming part design by vision, (b) use CMM(Co-ordinate measuring machine) probing to inspect the part to determine its exact three-dimensional location after clamping, and (c) fine-adjust the spatial orientation of the part by a micropositioner before the part is presented for machining. Consequently, the total system investment is comparatively high and the operational procedures are complex and require significant programming, coordination, and synchronization among the four stations. There are also

High Throughput and Agile Machining

FIGURE 15.15 IFS "bed-of-nails" fixture shown holding an engine head. (*Source:* Lamb Technicon.)

serious concerns about the short stroke range, indentation marks, and possible cutting tool path interference from the protruding nails.

An alternate approach for high-volume production is being investigated at the General Motors Research & Development Center [20–23]. This system, the Automated Reconfigurable Machining Fixture (ARMF), addresses the shortcomings inherent in the various earlier studies. It is capable of rearranging the locating, supporting, and clamping elements in a three-dimensional space quickly and automatically using appropriate modular fixture elements. Many types of workpieces of significantly different dimensions and geometries can be set up for machining on this fixture with minimum labor, material cost, and production delay. Figure 15.16 shows this system holding two different aluminum engine cylinder heads with significant size and shape differences. The ARMF system consists of a powerful electro-permanent magnetic chuck and several specially designed autonomous modular hydraulic elements and the system is to be operated directly by any multiaxis NC machining center.

FIGURE 15.16 Two different engine heads clamped on the ARMA magnetic agile fixture. (*Source:* General Motors Corporation.)

15.5 MATERIALS HANDLING SYSTEMS

Materials handling is a significant challenge in designing and operating agile machining systems, especially as required volumes increase. Materials handling is a comparatively straightforward process in dedicated serial systems; in this case, parts are loaded at the beginning of the system and advance through a fixed series of stations at fixed times. In contrast, in an agile system, cycle times may vary more and parts may be transferred to a number of machines from a given point, depending on which stations are busy or inoperative. Thus, more complex control logic and mechanisms are required for materials handling in the agile case, and there is more potential for delay and uncertainty. In many ways the materials handling system is the backbone of an agile system, and should be tailored to be compatible with the machine tools, support equipment, and size, shape, and weight of the parts [3, 4].

Three approaches to materials handling are commonly used in large agile systems: conveyor loops, overhead gantries, and vehicular or cart transport.

When a *conveyor loop* is used, parts are loaded at a central loading or staging area and conveyed to machines on rollers or tracks (Figure 15.4 and Figure 15.6). Part loading at the staging area may be done manually or by a robot, and parts may be loaded onto a pallet or directly onto the conveyor. Similarly, a pallet storage carousel with room for several pallets and an integrated workpiece transportation mechanism can be used with one or more CNC machines. Upon arrival at a machine, palletized parts are loaded directly, while loose parts are loaded either manually or by robot into the machine or onto an unused fixture in dual fixture setups (Figure 15.5 and Figure 15.11). Conveyor loop systems are comparatively flexible, especially if manual loading is used, but conveyors must still be rebuilt if the machine configuration is changed. For systems that operate when some machines are down, a control system is required to route parts to operative but unused machines. Depending on the distances between machines, this can introduce uncertainty or excessive waiting in the machine. The part routing control system should also store information on the sequence of machines visited by a particular part to facilitate troubleshooting when out-of-tolerance parts are detected. In palletized systems, this is facilitated through encoding chips on the pallets; in free transfer systems a part marking and tracking system is required. When robotic loading is used there is increased potential for parts to be dropped or misloaded; this is normally a significant concern only for relatively heavy parts.

When a *gantry system* is used (Figure 15.2 and Figure 15.17), parts are conveyed between machines using overhead gantries. Parts are loaded into each machine through a safety door that separates the gantry from the machine tool area. This system allows for parts to be removed (e.g., for gaging) and reintroduced to the system. Part buffering stages can also be included in gantry systems. Gantry systems are fixed automation systems which limit flexibility in part routing and machine layout changes. Complex gantry systems are expensive, often costing more than the machining centers themselves, and may also be subject to significant downtime. When operating properly, however, they limit misloading due to improper orientation and similar errors. Many early agile systems used gantry transport.

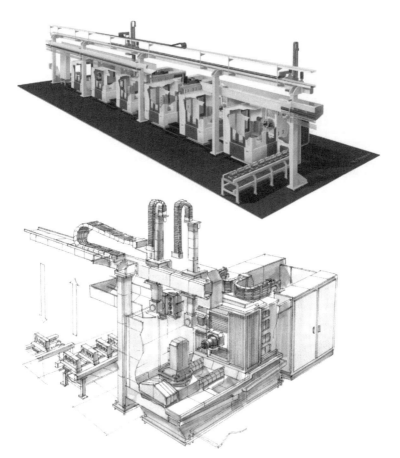

FIGURE 15.17 Agile cylinder head system based on six parallel machining centers, interlinked by an overhead gantry and power roller conveyors to form a cell. (Courtesy of Cross Huller.)

When *vehicular transport* is used, parts are conveyed in batches from a receiving or storage areas by vehicle to machines, where they are loaded manually or by robot. Vehicles used for transport include carts, fork trucks, and wire- or track-guided vehicles. Manual carts are flexible but require proper workforce training to limit overcyling and uncertainty. Fork trucks are also flexible but are often perceived as safety risks and avoided in North American plants. Guided vehicles may also be perceived as safety risks, and have traditionally had lower reliability as vehicles tend to get off track and stop. Whatever transport method is chosen, the number of parts transported in each batch must be carefully chosen, often using discrete event simulation programs, to minimize machine starving and in-process inventory. Vehicular transport is the most flexible materials handling option but to date has seen relatively limited use due to reliability concerns. Recently developed automated vehicles which navigate via glabal positioning systems (GPS) rather than tracks or wires on the floor increase flexibility and show promise in increasing system reliability [24].

REFERENCES

1. Mazak Inc., Mazak Integrex III Series, Multitasking Machine Tool Technical Manual
2. S. Krar and A. Gill, *Exploring Advanced Manufacturing Technology*, Industrial Press, New York, 2003
3. R.A. Maleki, *Flexible Manufacturing Systems — The Technology and Management*, Prentice-Hall, Englewood Cliffs, NJ, 1991
4. H. Tempelmeier and H. Kuhn, *Flexible Manufacturing Systems — Decision Support for Design and Operation*, Wiley, New York, 1993
5. A.W. Hallmann, Flexible manufacturing for auto parts, *Tooling and Production*, January 2003
6. P.T. Kidd, *Agile Manufacturing: Forging New Frontiers*, Addison-Wesley, Reading, MA, 1994
7. S. Goldman, R. Nagel, and K. Preiss, *Agile Competitors and Virtual Organizations*, Van Nostrand Reinhold, New York, 1995
8. D. Whitney, et al., Agile pathfinders in the aircraft and automobile industries — a progress report, *1995 Agile Conference*, Atlanta, GA, 1995
9. M.P. Groover, *Automation, Production Systems, and Computer-Integrated Manufacturing*, 2nd Edition, Prentice-Hall, Englewood Cliffs, NJ, 2000, Chapter 27
10. R. Dove, On cells at Kelsey-Hayes, *Production*, February 1995
11. T. Beard, High speed machining: general production. Fast and flexible, *Modern Machine Shop*, August 1999
12. G.S. Walsh, Dedicated automation to give way to agility, *Automotive Design and Production*, February 2001
13. Ford furthers flexible manufacturing effort, *Manuf. Eng.* **133**:1 (2004)
14. R.R. Burleson and S.J. Rosenberg, Totally Integrated Munitions Production: Affordable Munitions for the 21st Century, Lawrence Livermore National Laboratory Report UCRL-ID-139738 rev2, September 13, 2000
15. R.B. Aronson, High volume production, *Manuf. Eng.* **131**:1 (2003)
16. R. Dove, Agile cells and agile production, *Production*, October 1995
17. C. Koepfer, Can palletizing reduce your setup? *Modern Machine Shop*, May 1998
18. D. Chakraborty, E.C. De Meter, and P.S. Szuba, Part location algorithms for an intelligent fixturing system — Parts 1 & 2, *J. Manuf. Systems* **20**:2 (2001) 124–134 and 135–148
19. Lamb Technicon Corporation, Intelligent Fixturing System. Technical brochures
20. R. Wilson, Magnetic manufacturing technology, *Automotive Ind.* **184**:8 (2004) 39
21. C.H. Shen, Y.T. Lin, J.S. Agapiou, G.L. Jones, M.A. Kramarczyk, and P. Bandyopadhyay, An innovative reconfigurable and totally automated fixture system for agile machining applications, *Trans. NAMRI/SME* **31** (2003) 395–402
22. C.H. Shen, Y.T. Lin, J.S. Agapiou, P.A. Bojda, G.L. Jones, and J.P. Spicer, Reconfigurable Workholding Fixture, U.S. Patent 6644637, 2003

23. C.H. Shen, Y.T. Lin, J.S. Agapiou, and P. Bandyopadhyay, Reconfigurable fixtures for automotive engine machining and assembly applications, in: O. Dashchenko, Ed., *Transformable Factories and Reconfigurable Manufacturing in Machine Building*, Springer Verlag, Berlin, 2005, Chapter 42
24. G.S. Vasilash, Chrysler Group's approaches to advanced manufacturing engineering, *Automotive Des. Prod.* **116**:2 (2004)

16 Design for Machining

16.1 INTRODUCTION

Machining operations are costly compared to most other manufacturing processes because they are relatively slow to perform, consume a significant amount of energy, and require substantial capital investment and skilled labor. It has long been known that machining costs can be reduced through proper part design [1]. Many recent books on mechanical design list simple rules which can be used to avoid designing parts with unnecessary machined content or features which are difficult to machine accurately [2, 3]. Similar rules are also sometimes found in machining reference works [4–6]. Although in some cases the recommended rules are overly simplified because they do not take into account part volume and equipment concerns, they indicate a growing awareness of the importance of considering machining issues in the design stage.

This chapter summarizes rules for designing parts to improve machined part quality and reduce machining costs. The primary emphasis is on high-volume applications, although the more general rules are also applicable to batch production. In addition, only conventional and abrasive processes such as turning, boring, milling, drilling, and grinding are considered. More detailed discussions of batch production concerns and rules for designing parts for nontraditional machining operations are discussed in Refs. [4, 5].

Section 16.2 discusses machining cost components. General design for machinability rules are discussed in Section 16.3. Special considerations for transfer machines, CNC machining systems, and holemaking operations are discussed in Section 16.4. Methods of applying design for machining rules and design for manufacturability (DFM) programs are briefly considered in Section 16.5. Examples are reviewed in Section 16.6 and Section 16.7.

16.2 MACHINING COSTS

Very few purchasers of machined products are overly concerned with how a particular part is fabricated. All are quite interested, however, in what it costs. The goal of design for machining is therefore to minimize machining costs while still meeting functional requirements. It is therefore useful to review the components of machining cost before considering rules for minimizing them. These components include:

1. *Raw material costs:* The cost of unmachined stock, which may be in the form of a standard bar or slab, casting, or forged blank.
2. *Labor costs:* The wages for the machine operator, usually measured in units of standard hours.
3. *Setup costs:* The cost of special fixtures or tool setups and the wages paid to setup personnel.

4. *Tooling costs:* The cost of perishable tooling, including inventory, and any special tooling required for the operation.
5. *Equipment costs:* The cost of the machine tools, including required capital expenditures, facilities costs, maintenance costs, and machine depreciation.
6. *Scrap and rework costs:* The cost of repairing or disposing of finished or partially finished parts of unacceptable quality.
7. *Programming costs:* The cost of writing NC programs to generate the required toolpaths.
8. *Engineering costs:* Salaries paid to engineers for process design, validation, and other overhead functions.

Design for machining rules may address any of the first seven cost categories. The most significant cost component in a given application depends on several factors. Material costs are most significant for parts made of expensive materials or machined from complex castings or forgings. In high-volume production, material, labor, tooling, and scrap costs are generally the most significant. In small or medium lot production, setup and programming costs are proportionally more significant. Equipment and tooling costs are of greatest significance when producing complex or precision parts which require special tooling or investment in precision machines. Specific design rules generally have a strong impact on one or a few cost components; therefore the rules which should be emphasized in a given application vary depending on the complexity of the part, the cost and form of the raw material, and the required production volume.

Engineering costs are generally not reduced through use of design rules for machining; in fact, the application of these rules usually increases overhead as engineering effort is invested up front to realize benefits in later stages of manufacture.

16.3 GENERAL DESIGN FOR MACHINING RULES

Strictly speaking, to optimally design a part for machining, it is essential to know what kind of equipment will be used to manufacture it. A knowledge of the type of equipment currently available in the intended manufacturing facility, and the capacities and tolerance capabilities of this equipment, is necessary to identify features which would require investment in new or precision equipment, and to avoid these features to the extent possible. There are, however, some general rules for designing parts for machining which apply regardless of the type of equipment available. These rules are discussed in the following subsections.

16.3.1 CHOOSE MATERIALS FOR OPTIMUM MACHINABILITY

Nothing has a greater impact on machining costs and quality than the nature of the work material itself. The nature of the work material determines machining system characteristics such as motor power and bearing sizes, the tool materials and geometries which can be used, the range of cutting speeds and other cutting conditions, the perishable tooling costs, and the tolerances and surface finishes which can be achieved. The material choice is determined largely by material property and other functional requirements independent of machining, but in many cases at least a narrow spectrum of materials is available which satisfy these requirements. When this is the case, the relative machinability of the candidate materials and the standard forms in which each material can be obtained should be considered.

As discussed in detail in Chapter 11, machinability is a loosely defined concept reflecting the ease with which a material can be machined. Machinability generally decreases with material strength and ductility, and varies most significantly between different material classes or base chemistries. The common classes of metallic work materials, in rough order of decreasing machinability, include

1. Magnesium alloys
2. Aluminum alloys
3. Copper alloys
4. Cast irons
5. Ductile irons
6. Carbon steels
7. Low alloy steels
8. Stainless steels
9. Hardened and high alloy steels
10. Nickel-based superalloys
11. Titanium alloys
12. Uranium alloys

Although functional requirements rarely permit much latitude in this area, materials from earlier classes on this list should be substituted for those lower on the list whenever possible.

There is also a large range of machinability for materials with the same general base chemistry, based on the primary processing, heat treatment, and alloy content. Cast materials are often more difficult to machine than wrought materials of the same chemistry, due to greater variability and the presence of cast layers; this is most notably the case for cast versus wrought steel. Chilled irons are more difficult to machine than standard cast irons. Specific heat treatments can improve the machinability of low-carbon steels and aluminum alloys as discussed in Chapter 11. Metal-matrix composites are generally less machinable than the base matrix metal. Similarly, powder metals are usually less machinable than cast or wrought alloys of equivalent chemistry, but are more nearly net shape and thus require less machining. Most material classes include "free machining" alloys which contain additives that improve tool life or chip and burr characteristics, generally at the cost of reduced hardness, strength, and ductility. All of these factors should be taken into account when selecting the best material for a given application.

16.3.2 MINIMIZE THE NUMBER OF MACHINED FEATURES

Features should be machined only when they require dimensional or surface finish tolerances which cannot be produced by a primary operation. This is most often the case when the feature in question is used for bearing, locking, locating, press fitting, or dynamic balance, or when subsequent assembly considerations dictate a close dimensional tolerance. When tolerances permit, alternative methods of producing features such as casting in holes, chamfers, and undercuts should be considered to minimize the number of machining operations required.

Minimizing the number of machined features reduces machining cycle times, tool costs, and capital equipment requirements. To date, elimination of unnecessary machining operations has had a greater practical impact than any of the other designs for machining rules [2].

16.3.3 MINIMIZE THE MACHINED STOCK ALLOWANCE

The amount of material which must be machined away to produce the final part should be minimized. Excessive stock to be removed increases material costs, equipment costs, cycle times, and fixed tooling costs (since tool wear per part increases with the stock allowance).

For parts produced from cast or forged blanks, the stock allowance can be controlled by controlling the blank dimensions. Most machined features require roughly 1.5 mm of stock for roughing and finishing in two passes. A greater amount of stock should be provided only when primary process tolerances are of this magnitude or when the primary process leaves a thick surface layer of material with unacceptable mechanical properties. An excessively thick surface layer is more likely to occur with cast than forged blanks. For surfaces with relatively open tolerances requiring only a single machining pass, the stock allowance should be reduced to 1 mm when primary process tolerances permit.

For parts made from bar stock or other standard cold-worked shapes, machined part dimensions should be chosen when possible to be slightly smaller than the dimension of a standard available size. Rotational parts made from cold-rolled steel bars, for example, should have finished diameters roughly 1.25 mm smaller than the diameter of a standard bar size. The stock allowance should be increased for larger diameter bars to account for increased out-of-roundness tolerances in the raw bar. Similarly, if hot-rolled raw stock is used, the machined stock allowance should be increased (to roughly 2 mm) to ensure that surface scale is removed. Powder metal (P/M) parts should be used when possible since they are near-net shaped and require little or no machining.

16.3.4 Optimize Dimensional and Surface Finish Tolerances

The most open dimensional and surface finish tolerances compatible with the part function should be specified for all machined features. Excessively stringent dimensional or surface finish tolerances increase machining costs by requiring the use of additional finishing passes and reduced feed rates, which increase machining time, and by dictating more frequent tool changes to avoid the degradation of surface finish and increased tendency to burr which accompanies tool wear.

Figure 16.1 and Figure 16.2 [4] show the range of dimensional and surface finish tolerances which can be achieved using various machining processes under general machining conditions. It is particularly desirable to avoid the use of final grinding and honing operations, as these operations are relatively slow and capital intensive and may produce environmentally hazardous byproducts. In general, dimensional tolerances less than 25 μm and surface finish tolerances less than 0.4 μm often require the use of grinding or honing operations. (As discussed below, dimensional tolerances less than 50 μm may be difficult to achieve consistently without grinding in transfer machine applications.) Aluminum and magnesium workpieces are generally less likely to require grinding and polishing operations because very fine finishes can be produced through turning and milling with polycrystalline diamond (PCD) tooling. In high-volume applications, holes with a diametrical tolerance less than 50 μm generally require an additional reaming process to ensure statistical capability.

It should be noted that achievable tolerances in specific applications may depend on the part geometry (compliance), gaging and compensation capability of the tooling, tool change schedules (often between shifts in mass production), supports built into the fixture, and tool guiding jigs such as bushing plates for drills. These factors are often especially important in high-volume production; previous production experience with similar parts should be used when available to estimate realistic tolerances in these applications.

Tolerance allocation is discussed from an analytical viewpoint in Section 16.6.

16.3.5 Standardize Features

Machined features should be standardized to the extent possible. For example, hole diameters should be selected from a limited range of sizes, and the minimum number of different diameters (ideally, one diameter) should be used on a given part. Thread forms should also

Design for Machining

Size range, mm	Achievable tolerance, ± mm							
0–15	0.003	0.005	0.008	0.013	0.02	0.031	0.051	0.08
15–25	0.004	0.0065	0.01	0.015	0.025	0.038	0.064	0.10
25–38	0.005	0.008	0.013	0.02	0.031	0.051	0.08	0.13
38–70	0.0065	0.01	0.015	0.025	0.038	0.064	0.10	0.15
70–115	0.008	0.013	0.02	0.031	0.051	0.08	0.13	0.20
115–200	0.01	0.015	0.025	0.038	0.064	0.10	0.15	0.25
200–350	0.013	0.02	0.031	0.051	0.08	0.13	0.20	0.30
350–500	0.015	0.025	0.038	0.064	0.10	0.15	0.25	0.38

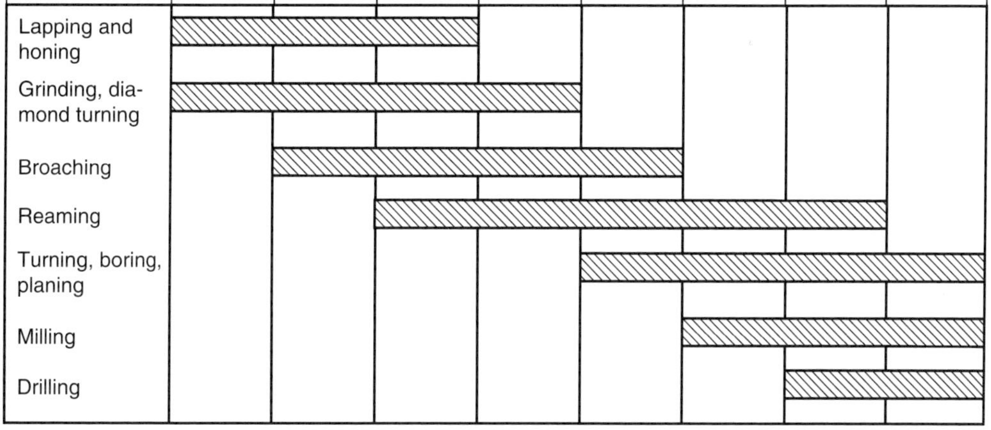

Lapping and honing
Grinding, diamond turning
Broaching
Reaming
Turning, boring, planing
Milling
Drilling

FIGURE 16.1 The dimensional tolerance achievable through various machining operations under general machining conditions as a function of feature size. (After Bakerjian [4].)

be standardized based on the hole diameter and work material, and consideration should also be given to standardizing the depth of blind holes based on their intended function. Standardization simplifies and reduces the cost of maintaining a tool inventory. It also promotes interchangeability of tools between operations, reducing the likelihood of tool shortages. As discussed below, it also reduces the number of tool changes required when CNC equipment is used, reducing machining cycle times.

In standardizing features, dimensions which can be produced from standard rather than special tools should be chosen. This is easily done for hole diameters and tap styles. Internal corner radii used on rotational parts should be chosen to match the nose radii available on standard inserts; this simplifies the CNC programming required to generate radii and reduces the need to stock special inserts.

16.3.6 Minimize the Number of Machined Orientations

Parts should be designed so that the number of machined orientations is minimized. A machined orientation is defined as spatial or directional orientation of a feature with respect to a reference plane or orientation on the part. For example, planes at distinct angles to a reference plane, parallel planes at different depths from a reference position, or holes with axes at different angles to a reference direction (often a normal to a reference plane) may all define distinct machined orientations (Figure 16.3).

Features with different machined orientations generally cannot be machined with a single tool in a single fixturing or pallet orientation. Therefore, minimizing the number of machined

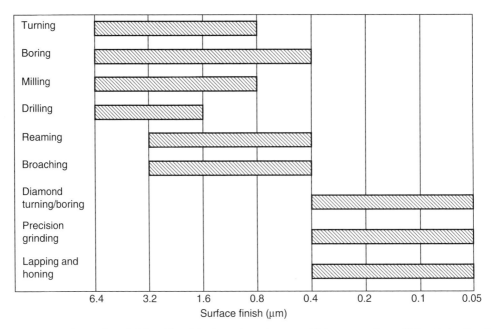

FIGURE 16.2 The surface finish achievable through various machining operations. (After Bakerjian [4].)

orientations minimizes the number of fixturings, pallet rotations, tool changes, and machining passes required to finish a part. This in turn minimizes processing time and increases machined accuracy.

16.3.7 Provide Adequate Accessibility

In general, a feature becomes more difficult to machine when it becomes more difficult to access with standard tools. Accessibility is most often a problem for features located in internal cavities or on remote faces of the part. Examples of features which should be avoided are shown in Figure 16.4 and include increased diameters in an internal bore, slots and sharp corners at the bottom of a bore, countersinks or counterbores on the exit surfaces of holes (especially when they must be highly concentric), holes on internal surfaces, and surfaces or corners blocked by overhanging features.

Inaccessible features often require the use of special tooling, tools with long overhangs (length/diameter ratios >5) which may be prone to deflection or unstable vibration, specialized machine attachments such as right angle drives, or excessive fixturings or pallet rotations. The machining cost is increased and the achievable tolerance is often limited by these requirements.

16.3.8 Provide Adequate Strength and Stiffness

The cutting forces generated during machining act between the tool and the part and may cause breakage, deflection, or unstable vibrations if the strength and stiffness of the system is inadequate. The part may be the weakest or most compliant element of the system, particularly if it is made of a material such as aluminum which is relatively weak, a material with a high yield strength and comparatively low elastic modulus such as titanium, or if its geometry is structurally weak.

Care should be taken to design the part so that it has adequate strength and stiffness in the expected directions of loading. This can be done by thickening sections over which heavy

Design for Machining   811

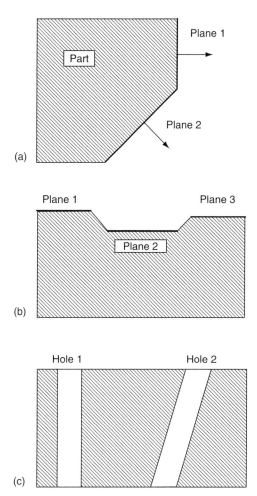

FIGURE 16.3 Examples of features with different machined orientations. (a) Planes with normals in different directions, which may require refixturing or pallet rotations to complete. (b) Parallel planes at different depths, which may require multiple passes, spindle extensions, or tool changes to complete. (c) Holes with axes in different directions, which may require refixturing or pallet rotations to complete.

loading is expected or by adding ribs or other structurally stiffening features to support thin sections. Since cutting forces increase with the metal removal rate, particular attention should be paid to surfaces which will be subjected to roughing cuts. Thin-walled sections and areas where large diameter holes are to be drilled should also be examined and stiffened with ribs or other structural features when possible. Analysis using simple models or finite element analysis (FEA) can be used to estimate deflection at the tool point and determine whether stiffening is needed for specific features as explained in Chapter 5 and Chapter 8.

If it is not possible to stiffen the part significantly, cutting forces can be reduced by proper design of the tool and by removing large amounts of stock in multiple passes, although this will increase cutting time. Inadequate part stiffness can also be compensated for by designing supporting elements into the fixture, although this significantly increases fixture costs and setup times and often generates maintenance problems. In extreme cases stress sensitive parts may be impossible to machine using conventional operations, and may have to be processed using a nontraditional operation such as electrodischarge machining (EDM).

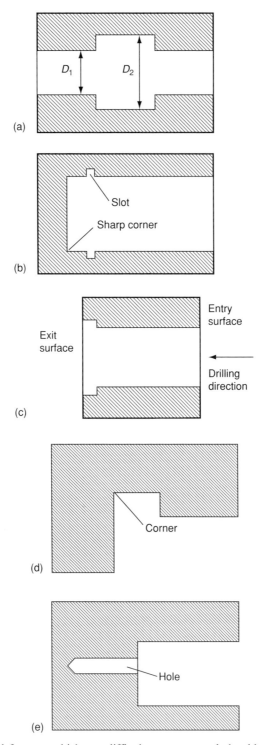

FIGURE 16.4 Machined features which are difficult to access and should be avoided. (a) Increased diameter on an internal bore. (b) A slot and a sharp corner at the bottom of an internal bore. (c) A counterbore on the exit surface of a through hole. (d) A corner blocked by an overhanging surface. (e) A hole on an internal surface.

16.3.9 Provide Surfaces for Clamping and Fixturing

Parts must be clamped to a chuck or fixtured securely before they can be machined. In designing a part, it is important to consider the possible ways it may be held, and to determine if the most likely workholding methods present access problems or part deflection concerns.

Rotational parts are held in lathes between centers or in chucks or collets. Parts finished on both ends and held in chucks or collets must be reversed at some point to complete all required operations. In this case, a clear section of the part with a constant diameter and without a tight surface finish tolerance should be provided for clamping when possible. If this is not possible, an additional grinding operation may be required to produce the part.

Prismatic parts with irregular or curved surfaces may present fixturing difficulties if clamps must be applied to curved surfaces. This can result in point loading and surface deformation or damage, particularly when fixturing for roughing cuts. When possible, clamping pads with flat surfaces should be designed into such parts. If swing clamps or similar devices are used in automatic fixtures, adequate clearance for the clamp motion must be ensured.

Clamping and fixturing concerns are particularly critical for structurally weak parts, especially when clamping stresses are transmitted through the part (e.g., when the part is held in a vise or window-frame fixture). In many applications, clamping forces exceed machining forces and can contribute significantly to deflections and form errors. Once the principal clamping force directions are determined, the structural stiffness of the part in these directions should be examined, and if necessary ribs or other stiffening elements should be added when possible. As noted above, FEA can be very useful for this purpose [7–9].

16.4 SPECIAL CONSIDERATIONS FOR SPECIFIC TYPES OF EQUIPMENT AND OPERATIONS

16.4.1 Special Considerations for Transfer Machines

The fixed cycle times and specialized nature of individual mechanisms characteristic of transfer machines result in a number of special design rules for parts manufactured using such systems. In addition, some of the general design for machining rules, such as using open tolerances when possible, are more critical for parts made using transfer machines, while other rules, such as maintaining adequate accessibility, are less critical.

Stations on transfer machines can be equipped with multiple spindles, so that patterns of holes can often be drilled simultaneously at a single station. To facilitate this, a minimum separation between holes should be maintained to leave room for distinct spindles. The magnitude of the separation required varies with the hole diameter and the particular type of equipment; for 12 mm holes, a minimum separation of 50 mm is typically adequate. Since a single feed slide is normally used on multispindle heads, it is also desirable to make all holes in a given pattern of roughly equal depth; a great variation in depth complicates tool setup and may require an excessive feed stroke length, which can increase cycle time, and also leads to uneven tool life, which may dictate more frequent tool changes.

Features on parts designed for transfer machine manufacture should be grouped so that multiple features can be machined simultaneously at a single station. Features which require close dimensional tolerances relative to each other should be machined at a single station. When possible, flat surfaces on a given part face should be at equal depths so that all can be machined with a single milling cutter. The amount of stock removed at each station should also be balanced to the extent possible so that all tools wear out at roughly the same time. Light cuts at odd depths which would require additional unbalanced stations should be avoided.

It is more critical to avoid tight tolerances on transfer machine parts than on parts to be manufactured on CNC equipment. The use of tight tolerances which increase the frequency of required tool changes has an exaggerated impact in transfer machine operations, since all or part of a line may need to be shut down to perform a tool change on a single station. Ideally tools on all stations should be changed on schedule between production shifts to maximize machine utilization. Tolerancing becomes a more critical issue as the machinability of the part material decreases; fairly tight tolerances can be consistently achieved on aluminum and magnesium parts since tool wear rates are low, but it may be difficult to achieve dimensional tolerances less than 0.05 mm consistently when machining iron or steel parts in high volumes.

Maintaining adequate access to features such as interior holes is not as critical for parts manufactured on transfer machines since special fixtures and jigs (fixtures incorporating guiding components such as bushings) can be more easily incorporated; the additional cost of such jigs and fixtures is more easily justified for the high-production runs typical of transfer machines.

16.4.2 Special Considerations for CNC Machining Systems

In operations on CNC machining and turning centers, tools are typically cutting for a smaller fraction of the processing time than in production machining systems; more of the processing time is taken up by noncutting functions such as tool changes, pallet rotations, and axis moves. Most actions which consume noncutting time also have a negative impact on part quality because they introduce additional tolerance components due to the finite accuracy and repeatability of the system. In designing parts to be manufactured on CNC machinery, therefore, the primary concern from a machining viewpoint should be to minimize noncutting functions by appropriate standardization and grouping of features.

Standardizing feature dimensions can reduce the number of required tool changes and thus the processing time. A minimum number of hole diameters should be used on a given part, and holes used for functions such as clearance or mounting should have the same diameter. Care should also be taken to avoid designing in pockets of widely varying dimensions, so that tool changes between end mills of different diameters can be avoided. Standardizing features to eliminate tool changes also simplifies tool magazine management as discussed below.

Pallet rotations can be minimized by reducing the number of machining orientations as discussed in the previous section. Standardizing features reduces axis motion time by eliminating excess tool changes; tool changes consume axis motion time because the spindle or table must generally be moved to a set position for a tool change. In addition to standardizing features, axis motion time can be reduced by reducing the distance between features which must be machined using the same tool. For example, when drilling a pattern of holes with equal diameter, the holes should be placed as close together as possible to minimize transit time between holes. (Note that in this respect hole patterns should be designed differently for parts to be manufactured on NC machines than for those manufactured on transfer machines.)

To further minimize tool changes for medium to long production runs, multidiameter or stepped holes should be designed so that they can be machined using stepped or combination holemaking tools rather than discrete tools as discussed below. The added cost for these special tools is generally not justified for small batch production.

Finally, tool magazine management can be of concern when producing complex parts on CNC machines. Generally, the tool magazine on a CNC machine will have a finite capacity (i.e., can hold only a specified number of tools). Some magazine types, such as belt or chain magazines, have a minimum access time to position tools for tool change, and this time often

Design for Machining

depends on the number of slots between tools used in succeeding operations. Operations should be grouped so that each tool retrieved from the magazine is used for some minimum period greater than the tool access time, so that the machine does not have to wait during the tool access cycle. This becomes a more difficult constraint to satisfy as the access time increases. Parts should be designed so that the number of tools required for the operations scheduled for a given machine does not exceed the number of slots in the tool magazine; if this cannot be arranged, additional time will be required to reload the tool magazine. If the number of slots available exceeds the number of tools required, the surplus slots can be loaded with redundant tools of the type which wear most rapidly, and the NC program can be written to access these additional tools after a specified number of parts have been produced, reducing the number of magazine reloads required. Tool magazine management can be simplified through standardization of features to reduce the number of tools required and through the use of stepped or combination tools to produce multidiameter holes.

16.4.3 SPECIAL CONSIDERATIONS FOR HOLEMAKING OPERATIONS

Holemaking operations such as drilling, reaming, and tapping are time consuming operations often used to produce critical features such as locating holes. As a result, quality issues and machining time constraints are often particularly critical, so that special care should be taken in the design phase to simplify required holemaking operations.

When possible, holes with increased internal diameters, interrupted holes, and holes intersecting inclined entry and exit surfaces should be avoided (Figure 16.4a and Figure 16.5). Holes with *increased internal diameters* require additional boring operations to produce

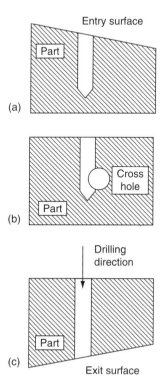

FIGURE 16.5 Examples of hole geometries which should be avoided when possible. (a) A hole drilled into an inclined entry surface. (b) Intersecting holes. (c) A through hole with an inclined exit surface.

and cannot be produced using step or compound tools. Holes drilled into *inclined entry surfaces* or at *compound angles* often represent unique machined orientations which may require additional fixturings or pallet rotations. It is also more difficult to maintain location accuracy and adequate tool life for these features; the drill has a tendency to "walk" at the entry of such holes, and unbalanced loading which may result in cutting edge chipping or margin rubbing is also present. In many cases a preliminary spotfacing operation will be required to avoid these difficulties. *Interrupted or intersecting holes* also generate unbalanced loads on the drill and may result in straightness errors, excessive vibration, or drill chipping, especially when drilling iron or steel with solid carbide drills. Burrs may also form at intersections, requiring an additional deburring operation. Intersecting holes are often used to produce lubrication passages; when they are unavoidable, an on-center rather than scalloped design should be chosen (Figure 16.6) [10] to minimize load unbalance, burr formation, and the likelihood that straightness errors will cause the drill to miss the target hole in deep drilling applications. Drilling through *inclined exit surfaces* also results in unbalanced loading which may cause excessive vibration, drill chipping, burr formation, and straightness errors. If such holes are unavoidable, it is advisable to drill them through relatively thick sections of the part (i.e., at depths greater than two drill diameters), so that the initial hole can act as a bushing and support the drill during exit.

For large batch or mass production, multidiameter holes should be designed to be manufactured with step or combination drills rather than discrete drills and counterbores. Specifically, the diameter of such holes should decrease in a stepwise fashion with hole depth, all steps in the hole should have a minimum axial length generally greater than the step diameter, and the difference in the diameters of adjacent steps should not exceed 50% of the larger diameter. As noted above, interior steps with increased diameter (Figure 16.3a) require use of a boring bar after drilling and should be avoided, especially when they are greater than three times the drill diameter below the surface. Stepped and combination tools are particularly attractive for CNC equipment because they eliminate tool changes. Cost analyses can be

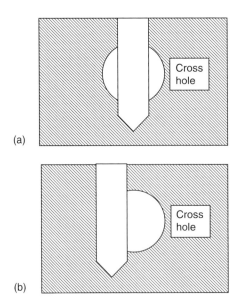

FIGURE 16.6 (a) On-center and (b) scalloped intersecting holes. When intersecting holes are unavoidable, the on-center configuration should be used to minimize drill chipping and burr formation.

used to determine the conditions under which the additional cost of a stepped or combination tool over discrete standard tools is justified (see Chapter 13).

Lists of simple design rules often state that blind holes should be avoided [3]. It is often difficult to remove chips from blind holes in parts manufactured on vertical spindle machines, and deep blind holes should be avoided when using such equipment. Blind holes should also be avoided when drilling magnesium parts, since fines which can result in a fire hazard may be generated during spindle reversal at the bottom of such holes. When drilling materials other than magnesium on horizontal spindle equipment, however, the preference for through versus blind holes is not as easily justified. In these applications chip removal does not present as serious a problem, and the burr formation and additional tool wear generated by vibration and feed surging at exit make through hole drilling less attractive.

16.5 CAPP AND DFM PROGRAMS

The application of design for machining rules requires cooperation between part designers and manufacturing engineers. This may occur in a variety of settings depending on the complexity of the part and organizational considerations [11]. For simple parts, only one designer and a shop foreman or process engineer may be responsible for the part design and manufacturing plan, and communication issues generally do not arise. In larger organizations or for complex parts, however, additional personnel may be involved on both sides, and design review meetings or workshops may be scheduled to ensure manufacturing input into the design. In this case, computer programs may be useful in quantifying machining concerns and structuring the discussion. The two types of programs most commonly used are computer-aided process planning (CAPP) and design for machining (DFM) codes.

CAPP programs [10–22] work with computer-aided design (CAD) systems to extract the relevant geometric features of a part and produce a process sequence or set of tool paths which is optimal in some sense. Two steps are involved in this approach: feature extraction and application of best practices. In the feature extraction stage, the CAD data for the part are analyzed to identify and classify the features which require machining. The extraction algorithm is specific to the CAD system used, and particularly to how data are stored and whether they are parameterized or not. Many systems can be used to classify features. Once features are extracted and basic rules such as precedence rules are applied, a best practices database is used to generate an initial process plan. The best practice database is specific to the organization and contains information on the preferred tool types and speeds and feeds for a given class of features. The program often performs analysis at this stage to balance stock between operations, to optimize for minimum cycle time or minimum cost (as discussed in Chapter 13), and to apply other rules identified by the organization. The process sequence may in theory also be optimized, but in practice is usually specified based on experience as part of the best practice database or Bill of Process (BOP). Design for machining rules and achievable tolerance constraints may also be considered in the analysis. Based on this information, and especially on breakdowns of processing times or cost, features which appear particularly difficult or expensive to machine can be identified and considered for redesign. A great deal of manufacturing data is required in process planning, such as selection of machines, fixtures, operations, cutting tools and holders, parameter selections for machining, etc. as shown in Figure 16.7.

CAPP programs are most easily applied to production with CNC equipment and medium to large production volumes. It is an optimization approach which is normally used at a relatively late stage of the design process when a complete initial design is available.

DFM programs [11, 23–25] are artificial intelligence programs written specifically to apply DFM rules. General purpose DFM programs include modules for assembly, stamping, and

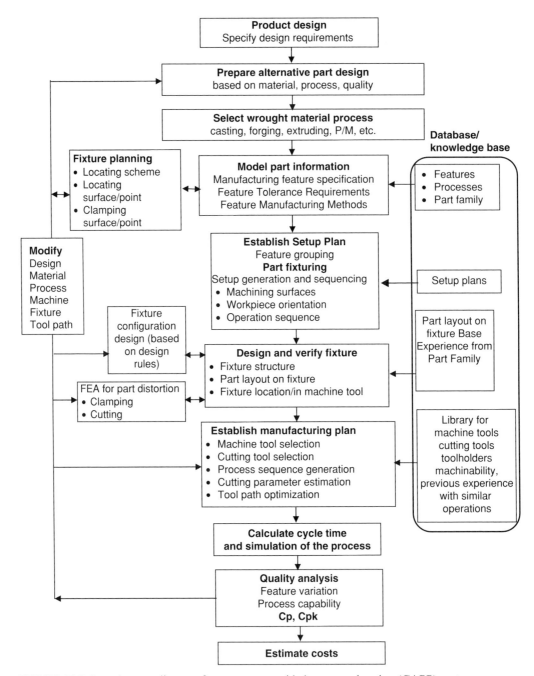

FIGURE 16.7 Input/output diagram for a computer-aided process planning (CAPP) system.

other processes as well as machining. Since desirable machining practices vary depending on the volume of production and the machine tools available, it is difficult to write a widely applicable general purpose design for machining module. Some large companies have proprietary in-house codes used to apply design for machining rules in a manner tailored to their business operations.

A DFM program typically has an input module and an analysis module. Data input is not as automated as in CAPP programs; rather than reading required geometric information from a CAD file, part features and dimensions must generally be input manually according to some format and classification scheme. This is partly because DFM programs are intended to be applied at an earlier stage of the design process (when no complete CAD model of the part may be available), and partly because additional subjective information, such as the perceived relative machinability of various materials or the relative penalty associated with given undesirable features, is often required.

Once data are input, the analysis module is used to compute a relative machinability score for the design as entered. The algorithm used to compute the score varies from program to program, but in general the score depends on the complexity of the design and the penalties associated with difficult to machine materials or features. In DFM workshops, some rough estimate of the machining cost may also be computed (e.g., using a spreadsheet) for the given design. The output of the program is a detailed breakdown of components of the score due to individual features, which often clearly identifies the feature(s) most responsible for complexity or excessive cost.

Unlike CAPP programs, DFM programs are used for comparison rather than formal optimization. Usually several design alternatives are compared benchmark designs, which typically include both successful existing designs and competitor's products which have been torn down and analyzed. Based on the DFM score the best design is chosen and refined, and particularly troublesome features are identified for possible redesign. For complex parts, the process may be repeated at various stages of the design (e.g., at an early stage and before fabrication of the first prototype). DFM programs can be used for parts manufactured on either CNC or dedicated production equipment. They are well suited for designing complex parts for mass production, and are currently more widely used than CAPP programs in these applications.

16.6 PART QUALITY MODELING

As discussed in Section 16.3, tolerance allocation has an important impact on manufacturing cost. Recent research on part quality modeling can provide insight into impact of tolerances on specific part features on cost, and can aid in designing both the part and the manufacturing process and system to minimize the adverse consequences of poor tolerance decisions.

During the early 1970s, articles were published on how to assign tolerances to minimize costs [26, 27]. In the late 1980s, some researchers emphasized the application of statistical tolerancing in order to reduce manufacturing costs [28, 29]. Variation simulation techniques have also been developed for tolerancing and dimensioning in process planning [30]. Such models take into account fixturing tolerances as well as machining tolerances. Another approach is error budgeting, which has been used to estimate overall workpiece accuracy [31]. Linear state space modeling of dimensional machining errors has also recently been used to transform and accumulate machining errors as the workpiece is machined in a multistation process [32, 33]. These variation simulation approaches tend to neglect the influence of fixture positional errors on the geometric accuracy of a part. Another group of researchers concentrated on the analysis of a part's geometric error due to fixture positional errors [9, 34–36]. In one study, the static errors due to clamping and cutting forces were estimated in a very simple part using finite element analysis on the part and fixture [37]. Software companies have done significant work in variation simulation and tolerance analysis [38, 39]. Commercially available packages use GD&T information and Monte Carlo sampling techniques to estimate the conformance of the assembly to part print tolerances.

A complete quality model should include all of the types of machining errors and specifically geometric (machine and fixture errors or GD&T tolerances), static (clamping, cutting, spindle, and workpiece errors), and dynamic (wear and thermal) errors and their interactions. Models for static and dynamic errors are more complicated and less widely available. A modeling methodology described in Ref. [7] will be reviewed to illustrate this approach. This methodology takes into account geometric and static errors and has the potential to include dynamic errors as well as interactions among errors. It propagates the various errors in machining the part from station to station as shown in Figure 16.8. The error at the final station is affected by the errors from all previous stations.

The station level variability (or error) E_{ik} is a sum of the geometric, static, and dynamic variations/errors (E_g, E_s, E_d). Typical variation/error components are shown in Figure 16.9. Each variation/error is modeled separately. If there are N operations (stations) in the system (i.e., $k = 1, 2, \ldots, K-1, K, \ldots, N$ as shown in Figure 16.8), the error for the jth feature and the ith quality characteristic Q_{ij} is given by

$$Q_{ij} = \sum_{k=1}^{N} Q_{ijk} \qquad (16.1)$$

where the error at each station

$$Q_{ijk} = \sum_{X=1}^{m} C_X \cdot E_{X_{ijk}} \qquad (16.2)$$

where, $C_X = 1$ if the E_X type of error contributes to the error at the kth station, otherwise $C_X = 0$; Q_{ijk} is the ith quality characteristic of the jth feature at the kth machining station; $E_{X_{ijk}}$ is the type of Xth error at the kth station contributing to the jth feature and the ith quality characteristic. The ith quality characteristic includes various types of tolerances such

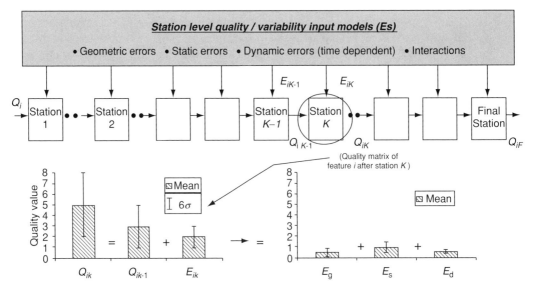

FIGURE 16.8 Quality modeling methodology for propagation of various machining errors within and between stations.

Design for Machining

Station sources of errors

Part variation	E_1: Casting
	E_2: Temperature
	E_3: Hardness

Machine variation	E_4: Positioning accuracy
	E_5: Positioning repeatability
	E_6: Thermal growth
	E_7: Spindle accuracy
	E_8: Machine/spindle stiffness
	E_9: Machine/spindle dynamics

Fixture variation	E_{10}: Locators accuracy
	E_{11}: Workpiece displacement within the fixture
	E_{12}: Clamping force
	E_{13}: Workpiece locatable surfaces/features
	E_{14}: Machining chips

Tool and holder variation	E_{15}: Positioning
	E_{16}: Face & axial runout
	E_{17}: Machining chips
	E_{18}: Tool wear

| Number of parallel stations per operation | E_{19}: Machine accuracy within stations |
| | E_{20}: Fixture accuracy within stations |

| Gage variation | E_{21}: Gage accuracy & repeatability |

FIGURE 16.9 Sources of variations and influence diagram for a machining station.

as form (flatness, straightness, circularity, cylindricity), profile, orientation (perpendicularity, angularity, parallelism), location (position and concentricity), and runout (circular and total).

Any tolerance stack-up analysis can be used in combination with Equation (16.1) to define the *i*th quality type error at the *k*th station for each feature *j*. The various variations for each feature are summed at each station. Then the mean (η_{ijk}) and variance (σ_{ijk}^2) of the distribution for Q_{ijk} can be determined at any station.

Any standard variation simulation program can be used to simulate error propagation through the system. Most packages accept multiple statistical distributions (e.g., Gaussian, Weibull, Pearson, plant data, etc.) for errors [39]. The basic approach is to follow the manufacturing process sheets in constructing the model. Each station in the line is modeled. The model of the fixture is positioned in the machine, the model of the in-process part is placed on the fixture, and then the machining operation is performed by placing the machined feature on the in-process part. The modeler must know or select an appropriate station level variation/error model to specify the variation (or tolerance) which occurs for each type of error, which can contribute to machining error at this station. If more than one machining operation is performed on a single fixture, this procedure is repeated for each operation.

Figure 16.10 shows the structure of the propagation model for part quality. The key product input variables (KPIV) are used as inputs to describe the variability of system setup parameters (shown on the top level). The group shown at the next level includes the geometric,

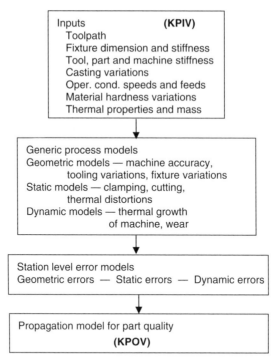

FIGURE 16.10 Static level propagation model for machining part quality using the 3DCS software package.

static, and dynamic models required at each station. These comprise the generic process models, which are used to model individual machining processes. The simplest process models can be used directly as the station level error model, for example, a machine accuracy which might be specified by a mean, η, and a variance, σ. In complicated cases, such as for clamping distortions, the process model may be an FEA run for a number of boundary conditions to generate an intermediate database. The station level error model is an algorithm (which the user must create) that determines the part distortion as a function of the station level process parameters such as variations in clamping force or part geometry as it affects fixturing. In the final or fourth level, the key product output variables (KPOV) or errors are tracked and accumulated on a station-by-station, process-by-process basis.

An example of the use of this methodology is given in the next section (Example 16.3).

16.7 EXAMPLES

Example 16.1 *Shaft Support Bracket:* Figure 16.11 shows an initial design of a shaft support bracket [6]. This part was designed to be bolted to a mating housing wall to provide support and lubrication for a long shaft. The features to be machined include the shaft bore, an oil hole for lubrication, and holes for locating dowel pins and bolts. To prevent binding of the shaft, the diameter of the bore must be machined accurately, and the location of the bore center with respect to the dowel pin hole centers must be held to a close tolerance.

The part, made of nodular cast iron, was to be machined in high volumes on a horizontal spindle CNC machining center. The critical tolerances dictated that the dowel holes be machined first using short drills, and that the bore be produced by a single-point boring bar without refixturing the part.

Design for Machining

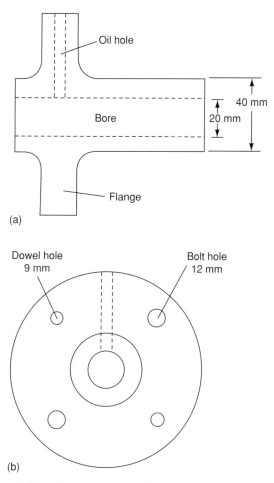

FIGURE 16.11 (a) Side and (b) end views of the initial design for a shaft support bracket.

The initial design presented a number of difficulties from a machining viewpoint. Different diameters were used for the dowel and bolt holes, so that a tool change would be required to produce these features. This would increase processing time and reduce the location accuracy of the holes relative to one another. The bore was long (roughly 100 mm) and would require a boring step of significant length. The oil hole was also long (roughly 50 mm) relative to its diameter and would require a long processing step. Finally, there is no obvious way to fixture the part automatically; presumably it would be held in a vise on the outer surfaces of the flange, which are contoured and would not present flat clamping surfaces. To prevent rotation of the part due to the torques involved in the boring operations, a high clamping pressure would be required; since the clamping stresses are transmitted through the part, this could result in distortion of the bore and an out-of-roundness error upon unclamping.

The part was redesigned to address these difficulties as shown in Figure 16.12. Three significant changes were made. First, the diameters of the dowel and bolt holes were standardized to eliminate the tool change. Second, the casting was changed so that the center of the bore had a larger diameter than the ends. This increased the mass of the part, but reduced the length of the bore which would have to be machined to roughly 40 mm. In addition, the depth of the oil hole was reduced by 5 mm, and the hole no longer exited on a machined portion of the bore,

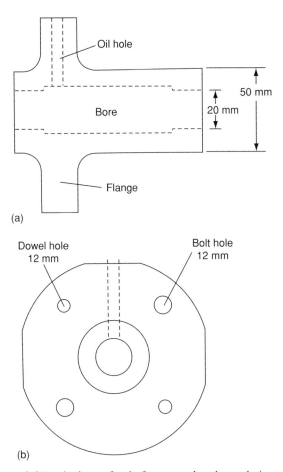

FIGURE 16.12 (a) Side and (b) end views of a shaft support bracket redesigned to simplify machining.

eliminating a possible exit burr on the machined bore. Finally, flat surfaces were cast on three faces of the flange to provide clamping pads and to eliminate a contoured entry surface when drilling the oil hole. The revised part could easily be fixtured by clamping in a vise on the flat surfaces parallel to the oil hole; by mounting the vise on an angle plate on the machine table, the hole could be drilled horizontally from the side of the vise. A lower clamping pressure could be used because the flats provide a positive stop which would resist rotation. The flat surface provided for the oil hole entry also reduced the hole depth by an additional 5 mm.

With these changes, the processing time required to machine these features was reduced from 173 to 119 sec, a reduction of over 33%; in addition, the design changes should improve part quality by simplifying achievement of the critical tolerances and by permitting the use of reduced clamping pressures.

Example 16.2 *Rotor Housing:* Figure 16.13 shows a cross section of the initial design of a diecast aluminum rotor housing to be machined in several steps on CNC turning centers [6]. In the operation shown, internal surfaces are to be cut using boring and grooving tools. To minimize axial tolerances between normal surfaces, two long boring passes (labeled "first cut" and "second cut" in the figure) are planned. In addition, two grooves for retaining rings are also required.

Design for Machining

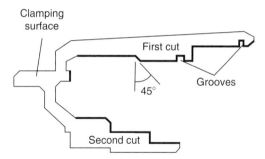

FIGURE 16.13 Cross section of the initial design for a rotor housing.

The boring passes are difficult to plan for in the initial design. There are several internal sharp corners; in addition, there is an internal angled cut at 45° to the part axis which cannot be accessed by a standard tool that will clear other internal surfaces. The grooves also have different axial widths. For the initial design, therefore, two grooving tools, two standard boring tools, and additional special tools to reach the internal sharp corners would be required to make the desired cuts.

Figure 16.14a shows a revised design which simplifies the machining. Upon consultation, it was determined that the dimensions of the grooves could be standardized, eliminating the need for one of the grooving tools. The internal sharp corners were replaced by radiused corners to permit machining with a standard boring insert. Finally, the initial 45° angled cut was replaced with a 60° angled cut which could be produced with a standard 55° boring insert mounted in a standard –5° lead boring bar as shown in Figure 16.14b. In the revised design, the required cuts can be made with two standard boring tools and a single grooving tool, saving at least three tool changes. Because the special tools required to machine the sharp internal corners would wear rapidly, the revised design also results in increased tool life and improved part quality.

Example 16.3 *Cylinder Head:* The modeling methodology described in Section 16.6 can be illustrated by evaluating the quality of an engine cylinder head, shown in Figure 16.15, through its machining process [7]. The various functions performed are shown in Figure 16.16. Station level error models predicted from FEA were used to provide the variation of the static errors. Monte Carlo variation simulation was used to propagate errors between stations.

The manufacturing processes modeled were: (1) mill the locating pads, (2) drill and ream locator holes, (3) rough mill deckface, (4) drill, (5) counterbore and (6) finish bore the exhaust seats and guides pockets, (7) drill, (8) counterbore, (9) finish bore the intake seats and guides pockets, (10) press valve seats and guides, (11) finish mill deckface, (12a) (12b) (12c) finish bore exhaust seats and guides, and (13a) (13b) (13c) finish bore intake seats and guides. The deckface of the cylinder head was affected by five of the manufacturing stations (1, 2, 3, 10, 11), while the seats and guides were affected by all of the above stations.

A number of software routines are used to track accumulating errors or tolerances for the factors shown in Figure 16.16. The kinematic and static errors in the machining system are shown in Figure 16.17 and Figure 16.18. After a model of the process has been built, the tolerances are applied and the quality is simulated. A Monte Carlo method was applied to stack-up all anticipated and predicted deviations/errors during manufacturing of the selected part features. One advantage of the Monte Carlo method is that it lets the user enter a probability distribution for each KPIV or estimated KPOV. The Monte Carlo simulation will run a specified number of parts M. Each part travels from station to station through the machining modeled process, and a random error (within its statistical distribution) for each

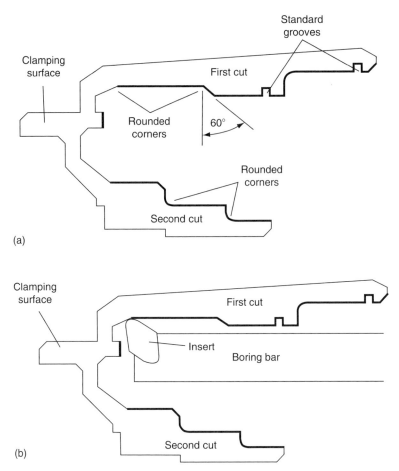

FIGURE 16.14 (a) Cross section of a rotor housing redesigned to simplify machining. (b) Access to internal features using a boring bar with a standard 55° boring insert.

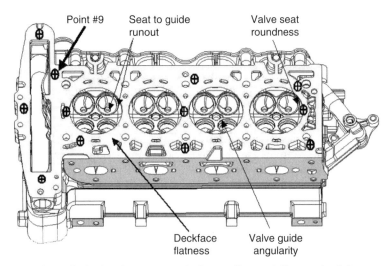

FIGURE 16.15 Engine cylinder head — case study for quality modeling methodology.

Design for Machining 827

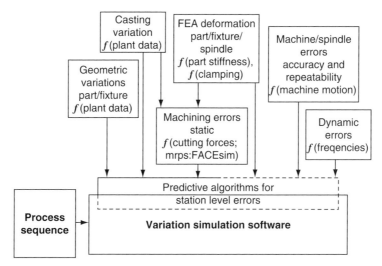

FIGURE 16.16 Quality modeling information/data flow.

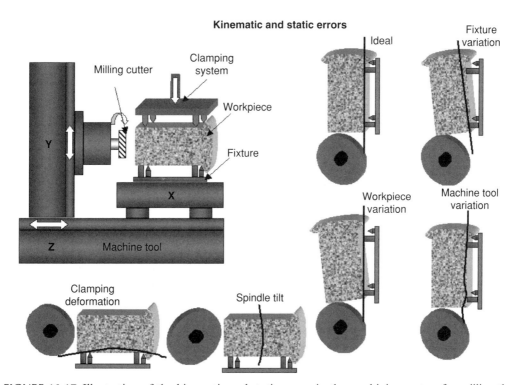

FIGURE 16.17 Illustration of the kinematic and static errors in the machining system for milling the deckface of a cylinder head.

KPOV can be assigned for each machined feature. After the desired number of iterations are run, standard statistics are used to determine the nominal and the variance (tolerance) values for the features. In this example 3000 samples were used, which gives accuracy of better than 2%.

The deckface height and flatness were calculated using a measured pad height distribution (CMM data) for 219 parts. The height of 12 points on the deckface (shown in

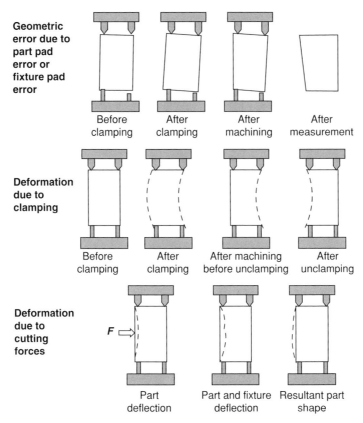

FIGURE 16.18 Illustration of the procedure for estimating the distortion of the deckface.

Figure 16.15) was measured by a CMM and also predicted by the model. The height variation of the primary machining locating pads and the fixture pad height deviation affects significantly the machined deckface surface height through the geometric and static errors (Figure 16.18).

The flatness error is defined for each part as the range (maximum minus the minimum value) of the height of the surface at twelve points. A comparison of the predicted flatness versus measurements for 219 parts is shown in Figure 16.19 after manufacturing station 11. The two distributions have about the same shape, but the model prediction is approximately 0.010 mm higher than the measurements. Considering the limitations of the input data (values which had to be estimated), the model provides fairly good agreement with the measured SPC data.

A sensitivity analysis was performed for the finish milling of deckface at station 11 to evaluate the contributions of the pad height error, clamping, cutting, and spindle tilt on the deckface height for nine points. The results are shown in Figure 16.20. The sensitivities are different at each different point, but some general trends can be identified: (a) the clamping contribution is very significant for most of the points, (b) the spindle tilt is significant at the points far from the centerline of the cutter (by the intake and exhaust sides) as expected, and (c) the contribution from cutting deflection tends to be the smallest of the four factors. From a part design viewpoint, this indicates that particular emphasis should be given to minimizing clamping distortions by stiffening the part in the expected clamping direction.

A number of different parameters have the potential to contribute to part-to-part variation. These parameters are shown in Figure 16.9, Figure 16.10, and Figure 16.15–Figure 16.18.

Design for Machining   829

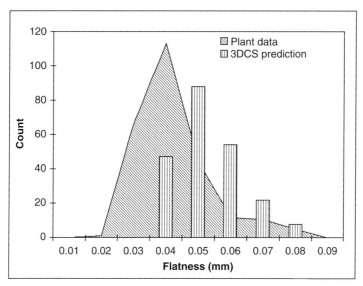

FIGURE 16.19 Comparison of flatness for 219 heads after the finish milling operation.

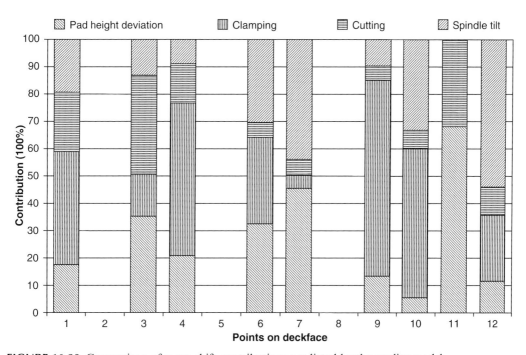

FIGURE 16.20 Comparison of mean shift contributions predicted by the quality model.

The variation is defined in terms of the square of the standard deviation. However, some errors, that is, pad height, have a compounding effect on the measurement, clamping, cutting, and tilting effect. In this case, a better approach is to examine the contributions from each effect by turning off each of the four variables in turn. The results of this approach are shown in Figure 16.21. This figure shows that most of the contributions to the variation come from two sources, both of which are related to the variations in machining the locating pads. Figure 16.21 also shows that clamping distortion also can make a significant contribution to the

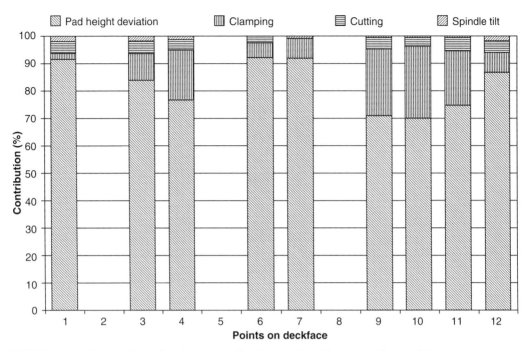

FIGURE 16.21 Comparison of variance contribution predicted by the quality model.

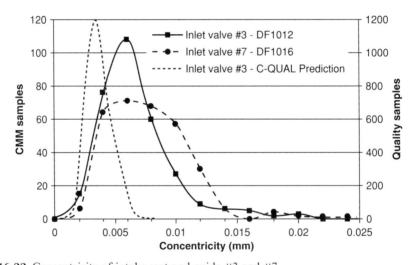

FIGURE 16.22 Concentricity of intake seat and guide #3 and #7.

part-to-part variations in the height of the deckface. At nearly all the points on the deckface, the clamping distortion is much larger than the cutting distortion. The other variations such as the machine accuracy tend to be relatively small, on the order of a few micrometers (assuming the machine is properly maintained), and do not contribute significantly to the variation. In terms of part design, in addition to providing stiffness as discussed above, these results indicated that tight tolerances must be specified on the locating pads.

The concentricity of the seats and guides was evaluated through the model based on 219 parts. Figure 16.22 shows the concentricity for two valves machined in a single station with twin spindles. The model prediction is shown as well which indicates how much the concentricity can vary from station to station. A detailed inspection of the results for all the 16 valves and guides on a station-by-station basis shows that each station has its own unique distribution of concentricity because some of the errors, that is, runout between the tool and spindle, vary from station to station. The model predicts the spread of the variation based on the statistical distribution for the errors. However, some of the errors can be reduced (or adjusted) during the assembly of the tool in the spindle. The reduction of the error by adjustment was not considered in the model.

REFERENCES

1. O.W. Boston, *Metal Processing*, 2nd Edition, Wiley, New York, 1951, 1–8
2. J.G. Bralla, *Design for Excellence*, McGraw Hill, New York, 1996, 46–47
3. C.V. Starkey, *Engineering Design Decisions*, Edward Arnold, London, 1992, 178–179
4. R. Bakerjian, Ed., *Tool and Manufacturing Engineer's Handbook, Vol. VI: Design for Manufacturability*, 4th Edition, Society of Manufacturing Engineers, Dearborn, MI, 1992, Chapter 11
5. G. Boothroyd and W.A. Knight, *Fundamentals of Metal Cutting and Machine Tools*, Marcel Dekker, New York, 1989, 399–438
6. D.A. Stephenson, Design for machining, *Metals Handbook, Vol. 20: Materials Selection and Design*, ASM, Materials Park, OH, 1997, 754–761
7. J. Agapiou, E. Steinhilper, F. Gu, and P. Bandyopadhyay, A predictive modeling methodology for part quality from machining lines, *Trans. NAMRI/SME* **31** (2003) 629–636
8. J. Xie, J.S. Agapiou, D.A. Stephenson, and P. Hilber, Machining quality analysis of an engine cylinder head using finite element methods, *SME J. Manuf. Processes* **5** (2003) 170–184
9. Y. Rong and Y. Zhu, *Computer-Aided Fixture Design*, Marcel Dekker, New York, 1999
10. J.S. Agapiou, An evaluation of advanced drill body and point geometries in drilling cast iron, *Trans. NAMRI/SME* **19** (1991) 79–89
11. H.W. Stoll, Tech report: design for manufacture, *Manuf. Eng.* **100**:1 (1988) 67–73
12. B.W. Nieble, Mechanized Process Selection for Planning New Designs, ASTME Paper 737, 1965
13. E.J. A.M. van Houten and A.H. van't Erve, PART, a parallel aproach to computer aided process planning, *Proceedings of CAPE 4*, Edinburgh, 1988
14. J. Nolen, *Computer-Automated Process Planning for World-Class Manufacturing*, Marcel Dekker, New York, 1989
15. H.P. Wang, *Computer Aided Process Planning*, Elsevier Science, London, 1991
16. H.A. El Maraghy, Evolution and perspectives of CAPP, *CIRP Ann.* **42**:2 (1993) 1–11
17. H.-C. Zhang and L. Altig, *Computerized Manufacturing Process Planning Systems*, Chapman and Hall, London, 1994
18. T.C. Chang and R.A. Wysk, *An Introduction to Automated Process Planning Systems*, Prentice-Hall, Englewood Cliffs, NJ, 1995
19. Z. Huang and D. Yip-Hoi, High-level feature recognition using feature relationship graphs, *Computer-Aided Des.* **34** (2002) 361–382
20. S. Yao, X. Han, Y. Rong, S. Huang, D.W. Yen, and G. Zhang, Feature-based computer aided manufacturing planning for mass customization of non-rotational parts, *Proceedings of the 23rd Computers and Information in Engineering (CIE) Conference*, Chicago, IL, September 2–6, 2003, DETC2003/CIE-48195
21. S. Yao, X. Han, and Y. Rong, Automated setup planning for part families, *Proceedings of the 2004 Japan–USA Symposium on Flexible Automation*, Denver, CO, July 19–21, 2004
22. Y. Rong, X. Han, S. Yao, and W. Hu, A Computer Aided Production Planning System for Mass Customization of Non-Rotational Parts, SAE Technical Paper 2004-01-1248, 2004
23. S.K. Gupta and D.S. Nau, Systematic approach to analysing the manufacturability of machined parts, *Computer-Aided Des.* **27**:5 (1995) 323–342
24. F.G. Mill, J.C. Naish, and C.J. Salmon, Design for Machining with a Simultaneous-Engineering Workstation, *Computer-Aided Des.* **26**:7 (1994) 521–527

25. G. Boothroyd, Product design for manufacture and assembly, *Computer-Aided Des.* **26**:7 (1994) 505–552
26. F.H. Speckhart, Calculation of tolerance based on minimum cost approach, *ASME J. Eng. Ind.* **94** (1972) 447–453
27. M.F. Spotts, Allocation of tolerances to minimize cost of assembly, *ASME J. Eng. Ind.* **92** (1973) 762–764
28. K.W. Chase, W.H. Greenwood, B.G. Loosi, and L.F. Hauglund, Least cost tolerance allocation for mechanical assemblies with automated process selection, *Failure Prevention and Reliability*, ASME DE-Vol. 16 (1989) 165–171
29. W.J. Lee and T.C. Woo, Tolerances: their analysis and synthesis, *ASME J. Eng. Ind.* **112** (1990) 113–121
30. D. Fainguelernt, R. Weill, and P. Bourdet, Computer aided tolerancing and dimensioning in process planning, *CIRP Ann.* **35** (1986) 381–386
31. D.D. Frey, K.N. Otto, and W. Pflager, Swept envelopes of cutting tools in integrated machine and workpiece error budgeting, *CIRP Ann.* **46** (1997) 475–480
32. D. Djurdjanovic and J. Ni, Linear space modeling of dimensional machining errors, *Trans. NAMRI/SME* **29** (2001) 541–547
33. Maier-Speredelozzi and J.S. Hu, Selecting manufacturing system configurations based on performance using AHP, *Trans. NAMRI/SME* **30** (2002) 637–644
34. R. Weill, I. Darel, and M. Laloum, The influence of fixture positional errors on the geometric accuracy of mechanical parts, *Proceedings of the CIRP Conference on PE & MS*, September 1991
35. S.A. Choudhuri and E.C. De Meter, Tolerance analysis of machining fixture locators, *ASME J. Manuf. Sci. Eng.* **121** (1999) 273–281
36. P. Chandra, S.M. Athavale, S.G. Kapoor, and R.E. DeVor, Finite Element Based Fixture Analysis Model for Surface Error Predictions Due to Clamping and Machining Forces, ASME MED-Vol. 2, 1997, 245–252
37. Y. Zhang, W. Hu, Y. Rong, and D.W. Yen, Graph-Based Setup Planning and Tolerance Decomposition for Computer-Aided Fixture Design, *Intl J. of Prod Research* 39 (14), 2001, 3109–3126
38. VSA-GDT/UG & VSA–3D/UG Variation Systems Analysis, Inc. Training Manual, Southfield, MI, 1998
39. Dimensional Control Systems — DCS Training Manual, Troy, MI, 2000

Index

Abrasives, for grinding, 237–239
Abrasive wear, tool, 512, 513, 517–520 (*see also* Tool wear)
    coated carbide, 517
    ceramics, 519, 520
    cermet, 518, 519
    high speed steel, 516
    PCBN, 520
    silicon nitride, 520
    tungsten carbide, 517
Acme automotive threaded connection, 272
Active vibration control, 662–669
Adhesive wear, tool, 512 (*see also* Tool wear)
Advant Edge, 406
Agile machining systems, 789–802
    fixtures, 794–799
    materials handling, 800–802
    tooling, 791–794
Aluminum alloys, machinability of, 593–596
    cast, 594
    eutectic, 594
    hypereutectic, 594
    metal-matrix composites, 596
    wrought, 595
Aluminum oxide (alumina) tools, 151, 519 (*see also* Ceramic tool materials)
    wear of, 519
Asymptotic borderline of stability, 645, 647
Attritional wear, tool, 512 (*see also* Tool wear)
Automatic tool changing systems (*see* Tool changing systems)
Autoregressive moving average (ARMA) model, 658, 668
Average roughness, 554
Average uncut chip thickness, 461, 479
Axial rake angle (*see* Milling cutters, Rake angle, Tool angles)
Axis drives, 106–111
    controllers, 110, 111
    direct, 107, 108
    hydraulic, 106
    motors, 106–108
    transmissions, 108–110

Back rake angle (*see* Rake angle)
Balancing, of toolholders, 341–347
Ball end milling, analysis of, 470, 471 (*see also* Milling, Milling cutters)
Ballscrews, 108, 109 (*see also* Parallel Kinematic Machines)
Bar chuckers, 76
Bearing ratio, 554
Bearings, 104–5, 114–123
    aerodynamic, 114, 115
    aerostatic, 105, 114, 115
    air, 117
    arrangement, 120, 121
    ball, 104, 118
    catch, 116
    comparison of, 115
    guideway, 104–5
    hydrodynamic, 114–116
    hydrostatic, 105, 114–116
    lubrication, 121, 122
    magnetic, 114–117
    preload, 122, 123
    roller, 104, 118, 119
    rolling element, 114, 115, 118
    rotary, 114, 115
    selection of, 115, 119
    service life, 119, 120
    sliding contact, 104
    spindle, 114–123
Bed-of-nails fixtures, 797–799
Belt grinding, 52
Bevel angle, 184, 186
Bickford point drill, 205
Bode plot, 623
Boring, 20, 21, 173–180, 462–465, 558–562, 661, 683–687 (*see also* Boring bars)
    analysis of, 462–465
    chatter in, 175, 176, 683–687
    machines for, 20
    surface finish in, 558–562
    thermal expansion in, 463
    tooling for, 173–180, 661
Boring bars, 173–180, 661
    adjustable, 176–180
    anitvibration, 661
    materials, 176
    multiple step, 179, 180
    rigidity of, 173–5

Boundary element method (BEM), 439, 440, 447
  cutting temperatures, 439, 440, 447
  thermal expansion, 447
Brass, machinability of, 597
Brazed tools, 162, 163
Breakout, 588, 589
Broaching, 33–35
  machines for, 34
Bronze, machinability of, 597
Brush honing, 51, 245, 247
BTA drilling, 23–25, 212, 213 (*see also* Deep hole drilling)
  tooling for, 212, 213
Built-up edge (BUE), 382, 411–413, 505, 506, 515, 551, 561, 566, 595
  aluminum alloys, 595
  mechanics of, 411–413
  methods of avoiding, general, 413, 515
Burning, surface, grinding, 570–572
Burnishing, roller, 52–54, 245, 247
  tooling for, 245, 247
Burn limit, grinding, 570
Burrs, 588–592, 597
  breakout, 588, 589
  compression, 588
  drilling, 590–592
  exit, milling, 589, 590
  formation, analysis of, 588, 589
  milling, 589, 590
  Poisson, 588
  rollover, 588, 589
  types, 588, 589

CAD/CAM systems, 459, 707, 817–819
  analysis, integration with, 459
CAPP systems, 706, 707, 817–819
Capto (Sandvik) connection, 301, 302 (*see also* Toolholder/spindle connections)
Carbide (*see* Tungsten carbide)
Carbon steel, machinability of, 601–603
Cast iron (*see* Iron)
CAT-V toolholders, CAT-V taper, 280–294, 306–313 (*see also* Toolholder/spindle connections, Tool retention systems)
  advantages of, 280
  face contact type, 289–294
  shortcomings of, 281, 284
  speed limitation of, 281
  stiffness of, 284
CBN (see Cubic boron nitride, Polycrystalline cubic boron nitride)
Cellular manufacturing systems, 88
Center column system (*see* Rotary transfer machines)

Centerless grinding, 45, 46
Ceramic tool materials, 150–152, 519, 520
  alumina based, 151, 519
  carbide composite based, 151, 519
  properties, 152
  silicon nitride based, 151, 520
  types, 151, 152
  wear of, 59, 520
  whisker reinforced, 152
Cermets, 149, 150, 518, 519
  compositions, 149
  grades, 150
  wear of, 518, 519
Chamfered inserts, milling, 191, 563, 564
Change constrained programming, 719
Chase control, 662–664
Chatter, 175, 176, 631, 632, 636–669 (*see also* Stability analysis)
  control of, 658–669
  mechanisms, 636–653
  prediction, 653–658
  nonregenerative, 652, 653
  regenerative, 636–651
Chemical vapor deposition, 156, 157
Chemical wear, tool, 514, 517–520 (*see also* Tool wear)
  ceramics, 519
  cermets, 518
  PCBN, 520
  tungsten carbide, 517
Chip breakers, 167, 187, 188, 582–588
  analysis of, 582, 583
  groove type, 167, 584, 585
  obstruction type, 584
  pattern type, 584–586
Chip charts, 586, 587
Chip control, 187, 188, 582–588
  drilling, 586–588
  turning, 582–586
Chip flow angle, 375, 376, 398
Chip formation, 382–389
  mechanics of, 385–389
Chip hammering, 507, 508
Chips, 381–389, 408–411 (*see also* Chip control)
  continuous, 381
  discontinuous, 381, 408–411
  shear localized, 381
  types of, 381
Chip thickness, 373, 461, 479
  average uncut, 461, 479
  measurement of, 373
Chisel edge, drill, 197, 198, 203, 204, 471, 472, 476
  thrust force estimation, 476
  web thinning, 203, 204

Chisel edge wear, drill, 530, 533
Chucks, 315, 316, 322–326, 349 (*see also* Tool clamping systems)
    hydraulic, 322, 323
    milling, 315, 323–325
    shrink-fit, 325, 326
    side lock, 315, 316
    turning, 349
    weldon, 315
    whistle-notch, 316
Chuckers, 76
CNC automatics, 78
CNC lathes, 77, 78
CNC machine tools, 77–91
Coated carbides, wear of, 517, 518 (*see also* Tool coatings)
Colding tool life equation, 523, 524
Collets, 316–322, 330–338 (*see also* Tool clamping systems)
    accuracy of, 319, 320, 330–338
    coolant fed, 320, 321
    sizes, 317, 318
    types of, 318, 320
Combination tools, 661, 738–741, 792–794, 816, 817
Compacted graphite iron (*see* Iron)
Complex plane method, 647, 648
Compliance matrix, 488
Composite (two-part) tools, 390, 391
Compression burrs, 588
Computer-aided process planning, 706, 707, 817–819
Computerized optimization systems, 707, 708
Computer simulation (*see* Machining process analysis)
Constitutive equations, 404, 406
Constraints, optimization, 716–718, 723
    cutting speed, 717
    depth of cut, 716
    feed, 716, 717
    order, 723
    pairwise, 723
    precedence, 723
Controlled contact tools, 391
Coolants, 126, 127, 217–219, 767–785 (*see also* Cutting fluids)
    drilling, requirements for, 217–219
Coolant systems, 126, 767–785
    filtering, 126, 773–777
Copper alloys, machinability of, 597
Corrosion (*see* Chemical wear)
Cost, total production, 709, 713
Costs, machining, 708–710, 805, 806
Counterboring, 24, 25

CRAC system, 666
Crater wear, tool, 504–506, 508, 510, 511, 531 (*see also* Tool wear)
    drills, 531
    measurement of, 508, 510, 511
Creep feed grinding, 47
Critical spindle speed, 635
Critical width of cut, 642–645 (*see also* Limiting width of cut)
Cross frequency response function, 630
Cryogenic chip breaking, 588
Cryogenic cutting fluids, 770
Cryogenic tool treatments, 162
Cubic boron nitride, 239 (*see also* Polycrystalline cubic boron nitride)
Curvic couplings, 272, 275, 278 (*see also* Toolholder/spindle connections)
Cutoff tools, 172, 173
Cutting, mechanics of, 371–424
Cutting edge engagement, 31, 166, 167, 638
Cutting fluids, 126, 127, 767–785 (*see also* Coolants)
    application of, 126, 127, 771–773
    elimination of, 781
    emulsifiable oils, 769
    filtering, 773–777
    gaseous, 770
    health and safety concerns, 779–781
    liquid, 768–770
    maintenance of, 773–779
    mist application, 770, 771
    neat oils, 768
    semisynthetics, 769
    soluble oils, 769
    straight oils, 768
    synthetics, 770
Cutting force coefficients, 460, 461, 465–467, 478–482, 639, 641, 646, 648
    dynamic, 639, 646
Cutting forces, 371–380, 394–402, 404–409, 462, 478–482 (*see also* Finite element analysis, Oblique cutting, Orthogonal cutting, Machining Process Analysis)
    analytical models for, 394–399
    components, 374–378
    correction for tool wear, 462
    empirical models for, 378–380, 478–482
    measurement of, 371–373
    numerical models for, 399–402, 404–409
Cutting pressures (*see* Cutting force coefficients)
Cutting process simulation (*see* Machining Process Analysis)
Cutting ratio, 385
Cutting speed, influence on temperatures, 433

Cutting stiffness, 639, 641
Cutting temperatures, 387, 388, 425–457
   drilling, 444–446
   factors affecting, 432, 433
   interrupted cutting, 441–444
   measurement of, 425–432
   numerical models for, 437–442
   process analysis models of, 462
   steady-state, analytical models for, 434–439
Cutting tool replacement strategies (see Tool replacement strategies)
Cutting tools, 141–263, 661, 791–794
   agile systems, 791–794
   basic types of, 162, 163
   boring, 173–180, 661
   coatings, 155–162
   deep hole drilling, 212–215
   drilling, 192–223
   end milling, 182, 183, 187, 188
   face milling, 181
   materials, 141–155
   microdrilling, 205, 216, 217
   milling, 179, 181–192, 315, 323–325
   reaming, 223–228
   threading, 172, 228–237
   turning, 163–173
Cutoff tools, 172, 173
CVD (see Chemical vapor deposition)
Cylindrical grinding, 45, 46

Damped natural frequency, 620
Damping of mechanical systems, 620, 621, 629, 630
   dry frictional, 621
   hysteretic, 621
   modal, 629, 630
   proportional, 629
   viscous, 620, 621, 630
Damping ratio, 620
Database systems, machinability, 577
Da Vinci, Leonardo, 2, 4
Deburring, 54, 221, 222
   hole, tooling for, 221, 222
Deep hole drilling, 23–25, 212–215
   machines for, 25
   tooling for, 212–215
Deflection, cantilevered tools, 174, 175
Deformation zone, primary, 383–389, 434, 435
   experimental study of, 383–385
   heat generation in, 434, 435
   shear flow stress in, 387
   strain in, 385, 386
   strain rate in, 386, 387
   temperature in, 387, 388

Deformation zone, secondary (see Tool-chip contact)
Depleted uranium alloys, machinability of, 610, 611
Design for machining, 805–832
   CNC machines, 814–815
   general rules, 806–813
   holemaking operations, 815–817
   programs, 817–819
   transfer machines, 813–814
Dial systems (see Rotary transfer machines)
Diamond, 154, 155, 161, 239 (see also Polycrystalline diamond)
   coatings, 161
   grinding wheels, 239
Die threading, 42
Diffusion wear, tool, 513, 514, 517–519, 521 (see also Tool wear)
   ceramics, 519
   cermets, 518
   PCD, 521
   tungsten carbide, 517
Direct drives (see Linear motors)
Direct transfer function, 640
Discontinuous chip formation, mechanics of, 408–411
Dish, surface, face milling, 564, 565
Disk grinding, 52
DN (speed factor or index), 119
Drawbars, 124, 125, 287, 290, 295, 296
   CAT-V connections, 287, 290
   HSK connections, 295, 296
Dressing, grinding wheels, 242
Drilling, 21–23, 192–223, 377, 378, 444–446, 470–478, 570, 571, 586–588, 590–592, 815–817 (see also Drills)
   analysis of, 470–478
   accuracy, 221
   basic equations, 22, 23
   burr formation in, 590–592
   chip breaking, 586–588
   chip removal, 217–220
   DFM rules for, 815–817
   force components in, 377, 378
   machines for, 21, 22
   surface finish in, 565, 566
   temperatures in, 444–446
   tooling, 192–223
   white layer formation, 570, 571
Drills, 192–223, 530–534, 586–588, 816, 817 (see also Chisel edge)
   body geometry, 193–198
   breakage of, 532–534
   chip breaking, 586–588

chip removal, 217–220
coolant requirements, 217–219
deep hole, 212–215
double margin, 208
four-flute, 197, 198
fracture of, 532–534
G, 207, 208
half-round, 215, 216
indexable, 192, 209, 210
life, 219, 530–534
materials, 196, 197
micro, 205, 216, 217
multimargin, 208
multitip, 212–215
point geometries, 197–208
racon point, 202
regrindable, 192
replaceable head, 209, 210
spade, 192, 208, 209
spiral point, 203, 205
split point, 203, 204
step, 210–212, 816, 817
structural properties, 193–197
subland, 210–212
three-flute, 197, 198
trepanning, 216
wear of, 530–534
web thinning, 203, 204
Dry machining, 447, 450, 451, 781
Ductile iron (*see* Iron)
Dynamic analysis, machine tool structures, 618
Dynamic compliance, 622
Dynamic cutting force coefficients, 639, 646
Dynamics of machining, 617–703
Dynamic stiffness index, 633, 635
Dynamometers, 371–373

Economics of machining, 705–766 (*see also* Optimization)
Edge build-up (*see* Built-up edge)
Edge chipping, of cutting tools, 507, 518–521, 528–530, 535–537
 ceramics, 519, 520
 cermets, 518
 milling, 535–537
 PCBN, 520
 PCD, 521
Edge frittering, of cutting tools, 507 (*see also* Edge chipping)
Effective friction coefficient, 389 (*see also* Friction, tool-chip)
Effective lead angle, 460, 461, 479
Eigenvalues, MDOF system, 628
Eigenvectors, MDOF system, 628

Ejector drilling, 23–25, 212, 213 (*see also* Deep hole drilling)
 tooling, 212, 213
End cutting edge angle, 170, 171, 173 (*see also* Tool angles)
End milling, 28–30, 182, 183, 187, 188, 467–470, 562–565 (*see also* Milling, Milling cutters)
 analysis of, 467–470
 chip breaking, 187, 188
 chucks for, 315, 316, 322–326
 contact (entry/exit) angles, 187, 189
 cutters, 182, 183, 187, 188
 surface finish in, 562–565
End turning, 395, 480
Engagement, cutting edge, 31, 166, 167, 638
Exit burr, milling, 589, 590
Experimental modal analysis, 655–657
Extended Taylor tool life equation, 523, 540–542
External honing, 51, 52

Face milling, 28–30, 181, 465–467, 562–565 (*see also* Milling, Milling cutters)
 analysis of, 465–467
 cutters, 181
 surface dish in, 564, 565
 surface finish in, 562–565
Feed, 18, 433
 definition of, 18
 influence on cutting temperatures, 433
 per tooth, definition of, 18
 rate, definition of, 18
Filters, cutting fluid, 773–777
Finite difference models, for cutting temperatures, 439
Finite element models, 350, 351, 404–409, 437–442, 484–493, 569, 617, 618
 cutting forces, 404–409
 cutting temperatures, 437–442
 Eulerian formulations, 404, 405, 407
 fixtures, 350, 351
 Lagrangian formulations, 404, 406
 machine tool structures, 617, 618
 process analysis, use in, 484–493
 residual stresses, 569
Fixtures, 347–352, 794–799
 adapter plates, 796, 798
 agile, 794–799
 analysis of, 350–352
 bed-of-nails, 797–799
 clamps, 348
 flexible, 349
 locators, 348
 magnetic, 350
 modular, 349, 350

Fixtures (*cont'd*)
  operation efficiency, 349
  phase change, 350
  supporting structure of, 347, 348
  tombstones, 349, 794, 795
  tooling cubes, 349
  types of, 349, 350
  windowframe, 794, 797
Flank wear, tool, 504, 506–512, 530, 533 (*see also* Tool wear)
  drills, 530, 533
  measurement of, 508–512
Flexible manufacturing systems, 88, 89, 725
Foot formation, in milling, 536, 537, 589, 590
Forced vibration, 630–635
Forces (*see* Cutting forces)
Ford Plant, 11
Forecasting control, 668, 669
Form tools, 173, 174
Fourier transform, 621
Four-flute drills, 197, 198
Fracture, tool, 505, 507, 508, 520–530, 532, 533, 535–537 (*see also* Edge chipping)
Free machining steel, 604
Free vibration, 630
Frequency plane method, dynamic analysis, 619, 620
Frequency response function (FRF), 622, 633 (*see also* Transfer function)
Fretting, of CAT-V toolholders, 284
FRF (*see* Frequency response function)
Friction angle, 389
Friction, tool-chip, 389–394 (*see also* Tool-chip contact)
  coefficient of, 389
  experimental study of, 390–392
  stress distribution, 392–394
Fundamental forcing frequency, 633, 635, 669, 670

Genetic algorithms, 719
Geometric programming, 719
Goal programming, 719
G-drill, 207, 208, 565
Grain size, effect on machinability, 591
G-ratio, definition of, 48
Gray iron (*see* Iron)
Grinding, 44–48, 236–242, 447, 448, 566–568, 570–572, 768, 772 (*see also* Grinding wheels)
  basic equations, 47, 48
  belt, 52
  burning, surface, 570–572
  burn limit, 570
  centerless, 45, 46

  coolant application in, 768, 772
  creep feed, 47
  cylindrical, 45, 46
  disk, 52
  machines for, 45–47
  surface finish in, 566–568
  surface integrity in, 570–572
  thermal expansion in, 447, 448
  wheels, 236–242
Grinding wheels, 236–242
  abrasives, 237–239
  bonds, 239, 240
  dressing of, 242
  grades, 240–242
  truing of, 242
  types of, 237, 238
Groove-type chip breakers, 167, 584, 585
Grooving tools, 172, 173
Guideways, 102–5
  bearings for, 104–5
Gundrilling, 22–25, 213–215, 565 (*see also* Deep hole drilling)
  tooling for, 213–215

Hard turning, 19–20, 447, 561
  surface finish in, 561
  thermal expansion in, 447
Harmonics, 633
Harmonizer system, 666, 667
Hartig, E., 7
Heat generation, in cutting, 434–436
Health effects, of cutting fluids, 779–781
Hexapod machines, 93, 96–99
High throughput machining, 787, 788
High Speed Steel (HSS), 9, 145, 146, 516
  cobalt enriched, 146
  development of, 9
  grades, 145
  sintered, 146
  wear of, 516
Hole deburring tools, 221, 222
Honing, 49–52, 243–247
  brush, 51, 245, 247
  external, 51, 52
  tooling for, 243–247
HP (Hypersensitive pneumonitis), 780
HSK interfaces, 295–299, 306–313 (*see also* Toolholder/spindle connections)
  accuracy of, 298
  advantages of, 295
  drawbars for, 296, 297
  forms, 296
  operational factors, 299
  torque capacity of, 297

Hydraulic chucks, 322, 323
Hypersensitive pneumonitis, 780

Impedance, 623, 624
Impingement chip breaking, 588
Inclination angle, 184, 374–377, 398, 399, 461, 462, 473, 480 (*see also* Tool angles)
  chip flow angle, relation to, 375–377
  cutting forces, effect on, 376, 398, 399
  drilling, 473
  end turning, 480
  milling, 184
  turning, 461, 462
Incremental stiffness method, 648
Indexable drills, 192, 209, 210
Inertance, 622, 624
Infrared temperature measurements, 430–432
Inner modulation, 638
Inserted blade tools, 162, 163
Inserts, indexable, 164–169, 172, 173, 191, 192, 209, 210, 563, 564
  chamfered, milling, 191, 563, 564
  clamping methods, 169, 170
  cutoff, 172, 173
  dimensions, 166
  drilling, 192, 209, 210
  edge preparations, 168, 169
  grooving, 172, 173
  milling, 190–192, 563, 564
  shapes, 165, 166
  thread turning, 172
  turning, 164–167
  wiper, 169, 599, 560, 563, 564
Internal grinding, 47
Interrupted cutting, 441–444 (*see also* Milling)
  temperatures in, 441–444
Iron, machinability of, 596–601
  cast, 596–601
  compacted graphite, 600, 601
  ductile, 600
  gray, 598, 599
  malleable, 599
  nodular, 600
  white, 601

Jigs, 347 (*see also* Fixtures)

KM system, Kennametal, 299–300, 306–313 (*see also* Toolholder/spindle connections)
K_n, K_f, K_t, K_r, 460, 461, 465–467, 478, 478 (*see also* Cutting force coefficients)
Krossgrinding, 244

Lathes, 2, 3, 17, 18, 77, 78, 96
  CNC, 77, 78
  turret, 78, 96
Lead angle, 170, 182, 186, 460, 461, 479 (*see also* Tool angles)
  effective, 460, 461, 479
  milling, 182, 186
  turning, 170
Lead time efficiency, 708
Lean manufacturing systems, 89
Length of contact, cutting edge, 166, 167, 638
Length of contact, tool-chip (*see* Tool-chip contact)
Limit chip analysis, 653
Limiting width of cut, 632, 642–645, 653
Linear motors, 107, 108
Linear programming, 719
Locating, 3-2-1 principle, 348
Low alloy steel, machinability of, 603, 604
Low radial immersion milling, 650, 651
Lumped mass dynamics analysis, 618
Lumped mass systems, vibration of, 618–630

Machinability, 577–616, 806, 807 (*see also* Tool life testing)
  aluminum alloys, 593–596
  brass, 597
  bronze, 597
  copper alloys, 597
  criteria, 578–580
  databases, 577
  definitions of, 577
  depleted uranium alloys, 610, 611
  factors affecting, 578–580, 591, 592
  indices, 581, 582
  iron, cast, 596–601
  iron, compacted graphite, 600, 601
  iron, ductile, 600
  iron, gray, 598, 599
  iron, malleable, 599
  iron, nodular, 600
  iron, white, 601
  magnesium alloys, 592, 593, 596
  metal matrix composites, 596, 597
  nickel alloys, 609, 610
  powder metals, 606, 607
  ratings, 581, 582
  steels, carbon, 601–603
  steels, free machining, 604
  steels, low alloy, 603, 604
  steels, metal matrix composite, 596, 597
  steels, microalloyed, 604
  steels, powder metal, 606, 607
  steels, stainless, 604–606

Machinability (cont'd)
  testing, 580, 581
  titanium alloys, 607, 608
  uranium alloys, 610, 611
Machinability ratings, 581, 582
Machine tools, 1–14, 71–139
  CNC, 77–91
  foundations, 102
  historical development of, 1–14
  production, 72–76
  spindles, 111–126
  structures, 91–102
  types, 71
Machine tool vibration, 175, 176, 630–669
  chatter, 175, 176, 631, 636–669
  free, 630
  forced, 630–635
  self-excited, 636–669
  types of, 630–632
Machine utilization, 708
Machining centers, 79–87
  high speed, 82–87
Machining costs, 708–710, 805, 806
Machining dynamics, 617–703
Machining economics, 705–766 (see also Optimization)
Machining Process Analysis, 459–502
  ball end milling, 470, 471
  baseline material testing for, 478–482
  boring, 462–465
  definition of, 459
  deflections, 484–490
  drilling, 470–478
  end milling, 467–470
  face milling, 465–467
  examples, 482–487, 489–493
  force equations for, 478–482
  temperatures, 462
  turning, 460–462
Magnesium alloys, machinability of, 592, 593, 596
  metal-matrix composites, 596
Magnetic tool treatments, 162
Malleable iron (see Iron)
Manganese sulfide, 607
Margin wear, drill, 530, 533
Material considerations, tool wear, 514–521
Material properties, influence of on temperatures, 433
Material properties, thermal, 433, 435, 436
Materials handling, agile systems, 800–802
Maudslay, Henry, 5, 7
MDOF systems, vibration of, 626–630
Mechanical cracking, tool, 505
Mechanical equivalent of heat, 3

Mechanics of cutting, 371–424
Mechanistic models, 459 (see also Machining Process Analysis)
Metallurgical methods, for temperature measurement, 429, 430
Metal-matrix composites, machinability of, 596, 597
  aluminum, 596
  magnesium, 596
  steel, 596, 596
  titanium, 596
Microalloyed steels, 604
Microdrilling, 25–27, 205, 216, 217
  tooling for, 205, 216, 217
Microsizing, 48, 49, 243
  tooling for, 243
Microstructure, effect on machinability, 591, 592
Milling, 28–33, 179, 181–192, 377, 378, 465–471, 536, 537, 562–565, 589, 590, 643, 650, 651
  (see also Milling cutters)
  analysis of, 465–471
  basic equations, 29–33
  burr formation in, 536, 537, 589, 590
  critical width of cut in, 643
  end, 28–30, 182, 183, 187, 188, 467–470, 562–565
  face, 28–30, 181, 465–467, 562–565
  force components in, 377, 378
  inserts, 190–192, 563, 564
  low radial immersion, 650, 651
  machines, 29
  peripheral, 28–30, 562–565
  spindle tilt in, 564
  surface finish in, 562–565
  tooling, 179, 181–192
  types of, 28
Milling cutters, 179, 181–192, 315, 323–325, 534–537, 563, 564
  chucks for, 315, 323–325
  contact (entry/exit) angles, 187, 189
  density, 186
  design, 183–190
  double negative, 184, 185
  double positive, 184, 185
  edge clamping methods, 191, 192
  fracture of, 534–537
  inserts for, 190–192, 563, 564
  pitch, 186
  rotary, 182, 183
  shear angle, 184, 185
  thermal cracking of, 534, 535
  tool angles, 183–186
  types, 181–183
  unequal tooth spaced, 186, 661
  white noise, 186, 661

Minimum Quantity Lubrication, 782, 783
Minimum work assumption, 396, 397, 403
Mists, cutting fluid, 779–781
Mobility, 622–624
Modal analysis (*see* Lumped mass systems)
Modal analysis, experimental, 655–657
Modal damping, 629, 630
Modal matrix, MDOF system, 627
Mode coupling, 652, 653
Modular toolholders, 267–270
   accuracy of, 268, 269
   structure of, 267.268
Motors, 106–108, 111–113
   axis drive, 106–108
   spindle, 111–113
MQL (Minimum Quantity Lubrication), 782, 783
Multipass operations, optimization of, 719, 720
Multiple degree of freedom systems (*see* MDOF systems)
Multistage machining systems, optimization of, 724–731
Multistep tools, 734–738, 792–794, 816, 817
Mushet, Robert, 8, 9

Natural frequency, 620, 628
Newcomen, Thomas, 3
Nickel alloys, machinability of, 609, 610
Nodular iron (*see* Iron)
Nonlinear programming, 719
Nonregenerative chatter, 636, 652, 653
Nonuniqueness, of cutting process, 403
Normal rake angle (*see* Rake angle)
Nose radius, 166, 167, 433, 558, 559
   cutting temperatures, effect on, 433
   surface finish, effect on, 558, 559
Nose radius wear, 505, 507 (*see also* Tool wear)
Notch wear, tool, 505, 509 (*see also* Tool wear)
Numerical control, 13, 77
   historical development of, 13
Nyquist plot, 624, 625

Objective functions, 711–715 (*see also* Optimization)
Oblique cutting, 1, 2, 374–376, 398, 399, 437
   force components in, 374–376
   shear plane theory of, 398.399
   temperatures in, 437
Obstruction-type chip breakers, 584
Optical measurements, surface finish, 556, 557
Optimization, 705–766 (*see also* Constraints, Tool replacement strategies)
   combination tools, 738–741
   computerized systems, 707, 708
   constraints, 716–718

   machining conditions, 711, 712
   machining systems, 710, 711
   multipass operations, 719, 720
   multistage machining systems, 724–731
   multistep tools, 734–738
   numerical examples, 741–758
   objective functions, 711–715
   problem formulation, 712–718
   single pass operations, 719
   single station multifunctional system, 720–724
   techniques, 719–741
   tool replacement strategies, 731–741
Optimum cutting speeds, 714
Order constraints, 723
Orthogonal cutting, 1, 2, 394–397, 399–402, 404–409, 434–442 (*see also* Finite element analysis, Shear plane model, Shear zone theories, Slip line theory)
   finite element models of, 404–409, 437–442
   shear plane theory of, 394–396
   shear zone theories of, 399–402
   slip line theories of, 396, 397
   temperatures in, 434–442
Outer modulation, 638
Overlap factor, 641
Oxidation wear, tool, 514, 517, 518 (*see also* Tool wear)
   cermets, 518
   tungsten carbide, 517
Oxide coatings, 158
Oxley's model, 399–402, 437

Pairwise constraints, 723
Parallel kinematic machine (PKM), 93, 96–99, 129, 130
   stiffness of, 99, 129, 130
Parsons, John, 13
Pattern-type chip breakers, 584–586
PCBN (*see* Polycrystalline cubic boron nitride)
PCD (*see* Polycrystalline diamond)
Peak height, maximum, 554
Peak-to-valley height, maximum, 554
Peclet number, 433
Peripheral milling, 28–30, 562–565 (*see also* End milling)
   surface finish in, 562–565
Photoelastic methods, 391
Physical vapor deposition, 156, 157
Planing, 33
Plastic deformation, in chip formation, 382–389
Plastic deformation, of cutting edge, 505, 507, 513, 515–517
Plunge grinding, 47

Poisson burrs, 588
Polycrystalline cubic boron nitride (PCBN), 153, 154, 161, 162, 196, 520, 599, 600
  coatings, 161
  drills, 196
  grades, 153
  iron, machining with, 599, 600
  sintered tools, 162
  wear of, 520
Polycrystalline diamond (PCD), 154, 155, 162, 196, 197, 520, 521, 592–596
  aluminum alloys, machining with, 593–595
  drills, 196, 197
  grades, 154, 155
  magnesium alloys, machining with, 592, 593
  metal-matrix composites, machining with, 596
  sintered tools, 162
  wear of, 520, 521
Powder metal tools, 146
Powder metals, machinability of, 606, 607
Power, cutting (*see* Specific cutting power)
Power, deformation (*see* Specific cutting power)
Power, frictional (*see* Specific cutting power)
Precedence constraints, 723
Predictive control, 663, 664
Principal coordinates, MDOF system, 628
Prism systems (*see* Rotary transfer machines)
Profilometers, 552–556
Proportional damping, 629
Pulsed magnetic tool treatments, 162
PVD (*see* Physical vapor deposition)
Pyrophoricity, 610

Quick-change toolholders, 269, 270, 302–305 (*see also* Toolholder/spindle connections)
Quick-stop devices, 383, 384

Racon point drill, 202
Radial rake angle (*see* Milling cutters, Rake angle, Tool angles)
Rake angle, 171, 172, 182–185, 201, 203, 227, 231, 379, 479, 504 (*see also* Tool angles)
  axial, 183–185
  back, 171
  cutting forces, effect on, 379, 479
  drilling, 201, 203
  effective, due to crater wear, 504
  milling, 182–185
  radial, 183–185
  reaming, 227
  side, 171
  tapping, 231
  turning, 171, 172
Reamers, 223–228
  geometry of, 226–228
  types of, 224–226
Reaming, 27, 223–228, 330, 566 (*see also* Reamers)
  attachments for, 329, 330
  surface finish in, 566
  tooling for, 223–228
Receptance, 624
Reconfigurable manufacturing systems, 89
Regenerative chatter, 636–651
Reliability analysis, tool life, 524, 525
Residual stresses, 568, 569
Resonant natural frequency, 620
Roller burnishing, 52–54
Rollover burrs, 588, 589
Rotary cutters, milling, 182, 183
Rotary index systems (*see* Rotary transfer machines)
Rotary transfer machines, 72, 73
Roughness, 554 (*see also* Surface finish)
Rumford, Count, 3, 6

Salomon, C., 443
Screw machines, 76
SDOF system, vibration of, 620–625
Seals, spindle bearing, 124
Self-excited vibration (*see* Chatter)
Shape memory, 610, 611
Shaping, 33
Shear angle, 385, 386, 394–396
Shear angle cutters, milling, 184, 185
Shear angle formulas, 396
Shear flow angle, 399
Shear flow stress, in primary deformation zone, 387–389
Shear localized chip formation, 381, 607, 609
Shear plane method, stability analysis, 646, 647
Shear plane theory, 394–399, 434–439
  agreement with data, 396, 397
  oblique cutting, 398, 399
  orthogonal cutting, 394–396
  temperature models based on, 434–439
Shear zone theories, 399–402
Shrink-fit toolholders, 276–280, 325, 326
Sialon, 151 (*see also* Ceramic tool materials)
Side cutting edge angle, 170, 171, 173 (*see also* Tool angles)
Side rake angle (*see* Rake angle, Tool angles)
Signal magnitude analysis, 654
Silicon nitride tools, 151, 520, 599, 600 (*see also* Ceramic tool materials)
  iron, machining with, 599, 600
  wear of, 520

Simulation of cutting processes (*see* Machining Process Analysis)
Single degree of freedom system (*see* SDOF system)
Single pass operations, optimization of, 719
Single stage machining systems, optimization of, 720–724
Sintered tools, 146, 162
Skiving, 35
Slides, 102–105
Slip line theories, orthogonal cutting, 396, 397
   agreement with data, 397
Slot milling cutters, 181, 182
Solution wear, tool (*see* Diffusion wear)
Spalling, 514, 518, 534, 535 (*see also* Tool coatings, Tool wear)
Specific cutting power, 19, 380–382
   boring, 380
   deformation, 381, 382
   drilling, 381
   frictional, 381
   milling, 381
   turning, 19, 381
Spindles, 111–126
   balance, 124, 125
   bearings, 114–123
   box, 112, 113
   motorized, 113, 114
   motors, 111–113
   preload, 122, 123
   rigidity, 124
   seals, 124
   speed factor (DN), 119
   tool retention, 124, 125
Spindle tilt, milling, 564
S-plane method, dynamic analysis, 619
Split point drill, 203, 204
Stability analysis, 642–650 (*see also* Chatter)
   asymptotic borderline of stability, 645, 647
   complex plane method, 647, 648
   incremental stiffness method, 648
   shear plane method, 646, 647
   stability conditions, 642, 643, 647
   stability lobes, 645
   time domain method, 648–650
   Tlusty's method, 639–646
Stability chart method, 653
Stability charts, 645
Stability limit, 640, 641, 642, 647
Stabler's rule, 376
Stainless steel, machinability of, 604–606
   austenitic, 605
   duplex, 605
   ferritic, 605
   free machining, 605
Steel, machinability of, 596, 597, 601–606
   carbon, 601–603
   free machining, 604
   low alloy, 603, 604
   metal-matrix composite, 596, 597
   microalloyed, 604
   powder metal, 606, 607
   stainless, 604–606
Stellite, 146
Step tools, 210–212, 734–738, 792–794, 816, 817 (*see also* Drills)
Stewart platform, 96
Strain, in primary deformation zone, 385, 386, 399
   oblique cutting, 399
Strain rate, in primary deformation zone, 387
Stress distribution, tool-chip contact, 392–394
STS drilling, 212, 213 (*see also* BTA Drilling, Deep hole drilling)
Stylus measurements, surface finish, 552–556
Superabrasive grinding wheels, 237
Surface finish, 551–568, 571, 573 (*see also* Surface parameters)
   boring, 558–562
   characterization of, 552–557
   drilling, 565, 566
   geometric, 551, 558, 559, 562
   grinding, 566–568
   measurement of, 552–557
   milling, 562–565
   natural, 551, 561
   reaming, 566
   turning, 558–562
Surface integrity, 551, 568–572
   residual stresses, 568, 569
   surface burning, grinding, 570–572
   white layer formation, 569–571
Surface parameters, 552–557
   average roughness, 554
   bearing ratio, 554
   peak height, 554
   peak-to-valley height, 554
   ten point height, 554
   valley depth, 554
Swarf (see Chips)
Swiss automatics, 76
System identification, 655–658

Tapered toolholders (*see* CAT-V toolholder, Toolholder/spindle connections)
Tapping, 35–38, 228–235, 328, 329 (*see also* Taps)
   attachments for, 328, 329
   basic equations, 36–38

Tapping (*cont'd*)
  machines for, 36
  tooling for, 228–235
Taps, 228–235
  cut, 228–234
  geometry of, 228–232
  materials, 235
  roll form, 234, 235
Taylor, Frederick W., 9, 10
Taylor tool life equation, 523, 537–539
Temperature, in primary deformation zone, 387, 388, 434, 435, 438
Temperatures, cutting (*see* Cutting temperatures)
Temperature sensitive paints, 432
Ten point height, 554
TF (*see* Transfer function)
Thermal cracking, tool, 443, 504, 505, 534, 535 (*see also* Tool wear)
  milling, 534, 535
Thermal expansion, 446–448
Thermal number, 433
Thermal properties, of tool and work materials, 435, 436
Thermal softening, tool, 513, 515–517 (*see also* Tool wear)
  high speed steel, 516
  tungsten carbide, 517
Thermocouples, 425–429
  conventional, 428, 429
  tool-work, 425–428
Thermoplastic shear (*see* Shear localized chip formation)
Threading tools, 172, 228–237
Thread milling, 38, 40–42, 235–237
  basic equations, 41, 42
  machines for, 41
  tooling for, 235–237
Thread rolling, 43, 44
Thread turning, 38, 39, 172
  inserts for, 172
Thread whirling, 42, 43
Three flute drills, 197, 198
3-2-1 principle, locating, 348
Thrilling, 40, 41, 236, 237
Time domain method, dynamic analysis, 620, 648–650
Time series analysis, vibration, 648–650, 668, 669
Time, total production, 712
TiN coatings, 158, 159
Titanium alloys, machinability of, 597, 607, 608
  metal-matrix composites, 597
Titanium diboride coatings, 160
Tool angles, 170–173, 183–186, 201, 203 (*see also* Rake angle, Inclination angle)
  boring, 173
  drilling, 201, 203
  milling, 183–186
  turning, 170–172
Tool breakage, 505, 507, 508, 520–530, 532, 533, 535–537 (*see also* Edge chipping)
Tool changing systems, 126, 127
  cycle time, 127
  double arm, 127
  single arm, 127
  tool placement, 127
  turret, 127
Tool-chip contact, 389–394, 434–436
  experimental study of, 390–392
  heat generation along, 434–436
  length of, 392
  stress distribution along, 392–394
Tool clamping systems, 313–343 (*see also* Collets, Chucks, Toolholding systems)
  collet chucks, 316–322
  comparison of, 330–343
  corogrip, 328
  hydraulic chucks, 322, 323
  milling chucks, 323–325
  milling cutter drives, 315
  powRgrip, 327, 328
  reaming attachments, 329, 330
  shrink-fit chucks, 325, 326
  side-lock chucks, 315, 316
  SINO-T, 328
  tapping attachments, 328, 329
  tribos, 327
Tool coatings, 155–161, 514, 517, 518, 534, 535
  coating methods, 156, 157
  diamond, 161
  materials, conventional, 157–161
  spalling, 514, 518, 534, 535
  wear of, 517, 518
Tool fracture, 528–530, 532, 533, 535–537 (*see also* Edge chipping, Tool life)
  drilling, 532, 533
  milling, 535–537
Toolholder/spindle connections, 270–313 (*see also* CAT-V toolholder, HSK interfaces)
  3-lock, 291, 292
  ABS (Komet), 303, 304
  Beta, 303, 304
  Big-Plus, 289, 290
  Capto (Sandvik), 301, 302
  CAT-V taper, 280–294
  CAT-V DFC, 292–294
  comparison of, 306–313
  curvic coupling, 272, 275, 278
  curvic flange, 272

Index 845

cylindrical shank, 271, 271, 277, 303, 304
flange mount, 315
HSK, 295–299
hydraulic expansion sleeve, 275, 276
KM (Kennametal), 299–300
mono-flex, 292
Morse taper, 272
quick-change, 302–305
shrink-fit, 276–280
threaded (Acme automotive), 272
WSU-1, 292
Toolholding systems, 265–347 (see also Modular toolholders, Quick-change toolholders, Tool clamping systems)
   balancing of, 341–447
   modular, 267–270
   quick-change, 269, 270, 302–305
   turning, 300–302, 305–309
Tool life, 522–525, 528–542(see also Tool life equations, Tool life testing, Tool wear)
   drilling, 530–534
   equations, 522–525, 537–542
   fracture, 528–530, 532, 533, 535–537
   milling, 534–537
   testing, 521, 522, 537–542
Tool life equations, 522–525, 537–542
   Colding, 523, 524
   extended Taylor, 523, 540–542
   probability-based, 524, 525
   Taylor, 523, 537–539
Tool life testing, 521, 522
   accelerated tests, 522
   specialized tests, 522
   standard tests, 521
Tool replacement strategies, 731–741
   combination tools, 738–741
   multifunctional part configurations, 733, 734
   multistep tools, 734–738
Tool retention systems, 124, 125, 287, 290, 295, 296
Tools (see Cutting tools)
Tool wear, 462, 503–558 (see also Tool life, Tool life testing)
   abrasive, 512, 513, 517–520
   adhesive, 512
   aluminum oxide ceramics, 519
   attritional, 512
   ceramics, 519, 520
   cermets, 518, 519
   chemical, 514, 517–520
   chip hammering, 507, 508
   coated carbides, 517, 518
   correction of cutting forces for, 462
   corrosion, 514
   crater, 504–506, 508
   diffusion, 513, 514, 517–519, 521
   drilling, 530–533
   edge chipping, 507
   flank, 504, 506, 507
   high speed steels, 516
   material considerations, 514–521
   measurement of, 508–512
   mechanical cracking, 505
   mechanisms of, 512–516
   monitoring, 537
   nose radius, 505, 507
   notch, 505, 509
   oxidation, 514, 517, 518
   PCBN, 520
   PCD, 520, 521
   prediction of, 525–58
   silicon nitride ceramics, 520
   solution, 513, 514, 517–519, 521
   spalling, 514, 518, 534, 535
   thermal cracking, 505, 510, 534
   thermal softening, 513
   troubleshooting, 515, 516
   tungsten carbide, 517
   types, 504–511
Tool-work thermocouple method, 425–428
Tooth passing frequency, 633
Top rake angle (see Rake angle)
Total production cost, 713
Total production time, 712
Transfer function (TF), 622, 640, 655–658 (see also Frequency response function)
   measurement of, 655–658
Transfer machines, 10–12, 72–76, 724, 725
   conventional, 72, 74–6, 724, 725
   convertible, 76
   flexible, 76, 724, 725
   part transfer in, 72, 74
   rotary, 72, 73
Traveling salesman problem, 720, 724
Trepanning, 23, 24, 216
   tooling for, 216
Tribos chuck, 327
Tungsten carbide (WC), 12, 146–149, 517, 525–58
   development of, 12
   grades, 147, 148
   grade selection, 148, 149
   wear of, 517
   wear, prediction of, 525–58
Turnbroaching (see Skiving)
Turning, 17–19, 163–173, 300–302, 305–309, 349, 377, 460–462, 558–562, 582–586
   analysis of, 460–462
   basic equations, 17–19
   chip breaking, 167, 582–586

Turning (*cont'd*)
  chucks, 349
  force components in, 377
  inserts for, 164–167
  surface finish in, 558–562
  tool angles, 170–172
  toolholders for, 300–302, 305–309
  tooling, 163–173
Turning centers, 77, 78
Two-part tools, 390, 391

Ultrasonic vibration, 26
Unbalance, 343, 344
Uncut chip area, 461
Uniqueness, of cutting process, 403
Unit cutting power, 19, 380–382 (*see also* Specific cutting power)
Uranium alloys, machinability of, 610, 611

Valley depth, maximum, 554
Variational principles, 403, 405
Velocity modified temperature, 400
Vibration analysis methods, 617, 618
Vibration control, 658–669
  active, 662–669
  chase, 662–664
  CRAC system, 666
  damping, 660
  dynamic absorption, 660, 661
  forecasting, 668, 669
  isolation, 659
  predictive, 663, 664
  stiffness improvement, 659
  tool design, 661, 662
Vibration, machine tool (*see* Machine tool vibration)

Waterjet chip breaking, 588
Watt, James, 4
Waviness, 553, 556
  cutoff, 556
Ways, 102–5
WC (*see* tungsten carbide)
Web thinning, drill, 203, 204
Weldon chucks, 315
Whistle-notch chucks, 316
White layer formation, 569–571
  drilling, 570, 571
White iron (*see* Iron)
White, Mansuel, 9
White noise cutters, milling, 186, 661
Wilkinson, John, 4–6
Wiper inserts, 169, 599, 560, 563, 564
Work hardening, 505, 592
  machinability, effect on, 592
  notch formation, effect on, 505